Converting Power into Chemicals and Fuels

Converting Power into Chemicals and Fuels

Power-to-X Technology for a Sustainable Future

Martin Bajus

WILEY

This edition first published 2023

Registered Office(s)
John Wiley & Sons, Inc., 111 River Street, Hoboken, NJ 07030, USA
John Wiley & Sons Ltd, The Atrium, Southern Gate, Chichester, West Sussex, PO19 8SQ, UK

Editorial Office
9600 Garsington Road, Oxford, OX4 2DQ, UK

For details of our global editorial offices, customer services, and more information about Wiley products visit us at www.wiley.com.

Wiley also publishes its books in a variety of electronic formats and by print-on-demand. Some content that appears in standard print versions of this book may not be available in other formats.

Library of Congress Cataloging-in-Publication Data
Names: Bajus, Martin, 1943- author. | John Wiley & Sons, publisher.
Title: Converting power into chemicals and fuels : power-to-X technology for a sustainable future / Martin Bajus.
Description: [Hokoben, NJ] : Wiley, [2023] | Publication place and date from CIP data view.
Identifiers: LCCN 2023017550 | ISBN 9781394184293 (hardback) | ISBN 9781394184262 (pdf) | ISBN 9781394185764 (epub) | ISBN 9781394185771 (ebook)
Subjects: LCSH: Energy storage. | Energy conversion. | Renewable energy sources.
Classification: LCC TK2980 .B35 2023 | DDC 621.31/26–dc23/eng/20230501
LC record available at https://lccn.loc.gov/2023017550

Cover Image: © Kittikorn Nimitpara/Moment/Getty
Cover Design: Wiley

Set in 9.5/12.5pt STIXTwoText by Integra Software Services Pvt. Ltd., Pondicherry, India

To my loving wife Mária

and grandchildren,

Rebeka, Kristina, Sorayah,

Jakub and David

Contents

About the Book

The *Converting Power into Chemicals and Fuels: Power-to-X Technology for a Sustainable Future* concept covers the activities involved in taking surplus renewable electricity from wind, solar, water or nuclear energy and converting it into other energy carriers (the "X") to be able to store the energy for later use and absorb energy fluctuations.

The first step in the process is to convert the renewable power into hydrogen by electrolysis (Power-to-Hydrogen). Hydrogen, the smallest molecule we know, does not emit carbon dioxide when burnt. It can be used immediately, or it can be stored in pressurised tanks and retrieved when supply is low.

There are several different utilisation pathways: feeding hydrogen into the gas network; displacing some of the carbon dioxide containing natural gas (Power-to-Gas); or through a methanation process with carbon dioxide converting the hydrogen into methane. The methane can be injected into the natural gas network replacing the fossil natural gas (also Power-to-Gas). The carbon dioxide source for the methanation process could therefore be biogas produced from biowaste in biogas plants or wastewater plants.

Other concepts include the production of methanol or ammonia to be used in fuel cells in cars and ships, or synthetic fuels to be used in conventional car and jet engines (Power-to-Liquids). This is all achieved through synthesis that involves hydrogen and a carbon dioxide source that could come from the process of converting waste into biogas.

The generated "green hydrogen" from renewable energies can also be used in fuel refining (hydrogenation) in conventional refineries as well as a basic chemical in many different industries (Power-to-Chemicals, Power-to-Plastics).

Finally, the stored hydrogen can also be converted back into electricity when required via fuel cells (Power-to-Power).

Preface

I have written this book at a time when the global oil consumption averages about 100 million barrels per day or two litres per person. At a price of $50–$130 per barrel, petroleum is one of the most affordable commercial liquid products ($0.3–$0.6/litre). Technological advances and efficiency improvements over the last century have enabled this level of scalability and affordability. However, largely because of the prevalent use of petroleum for energy, global carbon dioxide emissions have reached 100 million metric tons per day, averaging 13 kg per person in the world or 43 kg per person in the US The transition to alternative energy sources suggests that global oil consumption will peak soon, even though proven world oil and coal reserves are sufficient for another 50 and 100 years, respectively.

By 2050, the world population is projected to increase by more than 20% from today's 8 billion to 9.7 billion, and the global gross domestic product (GDP) is expected to more than double. Not only will energy demand grow, but the demand for infrastructure, housing, and consumer goods will also grow. All this demand growth will undoubtedly increase the consumption of raw materials and eventually lead to a material challenge for natural resources and environmental sustainability.

The good news is that the petroleum, gas and petrochemical industries have the technology and assets needed for offshore wind turbines, blue and green hydrogen production, and carbon dioxide capture and storage. They also have the refinery units and technology to produce renewable fuels. These industries are prepared for the journey to complete this crucial energy transition to a lower-carbon world. Lummus Technology introduced the industry's first net-zero ethane cracker. They announced the launch of a major enhancement to their leading ethane feed steam cracker that can achieve zero carbon dioxide emissions from an ethylene plant.

Strengthening the development of science, chemical processes, and chemical technology in the field of electrochemicals or/and electrofuels also means strengthening the economy and energy independence. We will convert more parts of light hydrocarbons from crude oils and natural gas into petrochemicals with the rise and increased use of electric vehicles. Different analysis predict different changes to gasoline usage due to the electric vehicle evolution. To respond to this trend, national oil companies are adapting their product from hydrocarbons into petrochemicals. It is also possible to consider a zero-gasoline refinery where a refinery is dedicated to producing olefins, aromatics, and synthesis gas production for petrochemicals. Some refineries in Europe will reduce gasoline production and increase production of olefins and aromatics for petrochemicals. Petrochemical processes, hydrocarbon technologies and green engineering have paved the way for incorporation of electrochemical technologies into modern chemical industry.

Nowadays, it is difficult to imagine the global energetic matrix free of fossil transportation fuels, especially in developing economies. Despite this, recent forecasts and growing demand for petrochemicals, as well as the pressure to minimize the environmental impact produced by fossil fuels, creates a positive scenario and acts as a driving force for closer integration between refining and petrochemical assets. In some scenarios the zero fuels refineries grow in the middle term, especially in developed economies.

The focus of the closer integration between refining and petrochemical industries is to promote and take advantage of the opportunities existing between both downstream sectors to generate value to the whole crude oil production chain. The synergy between refining and petrochemical processes raises the availability of raw material for petrochemical plants and makes the supply of energy for these processes more reliable whilst at the same time ensuring a better refining margin to refiners due to the high added value of petrochemical intermediates when compared with transportation fuels. The development of crude-to-chemicals technologies reinforces the necessity of closer integration of refining and petrochemical assets by brownfield refineries aiming to face the new market that tends to be focused on petrochemicals against transportation fuels. It's important to note the competitive advantage of the refiners from the Middle East who have easy access to light crude oils that can be easily applied in crude-to-chemicals refineries. Crude oil-to-chemicals refineries are based on deep conversion processes that require high capital spending, and this fact can put pressure on the refiners with restricted access to capital, again reinforcing the necessity to look for close integration with the petrochemical sector aiming to achieve competitiveness.

At the extreme end of the petrochemical integration trend there are the zero fuels refineries.It is still difficult to imagine the downstream market without transportation fuels, but it seems a serious trend and the players in the downstream sector need to consider the focus change in their strategic plans as opportunity or threat, mainly considering the pressure over the transportation fuels due to the decarbonization necessity and new technologies.

Due to large production of biodiesel and green diesel, there is the possibility of having an oversupply of diesel. If that occurs, diesel can be converted to chemicals. This is a strategy of research aiming to anticipate the oversupply of diesel, where steam cracking of green diesel created olefins and benzene, toluene, and xylenes. Waste materials are targeted as raw feedstocks for biodiesel production. Solid waste from agriculture mining waste are among the most studied materials. With this concept, there are possibilities to synergize a bio-based economy and circular economy. Hence, the adoption of 5R principles (reduce, reprocess, reuse, recycle, and recover) and the use of renewable resources has been consolidated in the daily life of citizens and regulates the actuation of every industrial activity according to the circular economy.

Decarbonization has left numerous challenges for C_1-technology. After carbon dioxide capture, the next challenge is carbon dioxide utilisation. The most prospective carbon dioxide utilization will be carbon dioxide hydrogenation to methanol to produce methane (methanation) or methanol. Dry reforming of methane is an interesting application of carbon dioxide as a sustainable C_1 source in current commercial processes. We showed in the first edition of *Petrochemistry* (2020: Wiley) the latest references on kinetics and thermodynamics of carbon dioxide reforming of methane to understand the mechanism of coke formation. The conversion of carbon dioxide to methane and methanol is a strategic topic as methane and methanol are applied as hydrogen storage.

Hydrogen is an ideal electrochemical and electrofuel of the future. One of the main challenges of hydrogen fuelled vehicles is the appropriate technology to produce hydrogen on board the vehicle. There is clear trend to produce hydrogen from carbon dioxide and methane, both of which are greenhouse gases. Numerous governments are promoting green hydrogen from water electrolysis. However, the production cost of green hydrogen is still significantly higher than hydrogen from natural gas. Currently, the cheapest production of hydrogen is still from catalytic reforming and steam pyrolysis of naphtha, producing hydrogen as a by-product.

Storage of hydrogen in porous nanomaterials has stimulated crowded research activity in metal-organic frameworks, hydrides, composites. Storage of hydrogen in liquid form such as blue ammonia has recently been commercialized. This will stimulate more research activities in the transformation of petrochemicals into fuel additive production over novel catalysts. The United States has successfully introduced gasoline blended with 10% ethanol (E10) and is developing 2-methyltetrahydrofuran as a fuel additive. Dicyclopentadiene is used as a feedstock to produce endo- or exotetrahydrodicyclopentadiene. A fuel with such properties as high energy density, lower viscosity, and lower freezing point is desirable to be used in missile-bearing jets at higher altitudes. High molecular weight alcohol and ether fuels with their advanced autoignition propensities and oxygenated molecular structures are promising future fuel candidates for compression-ignition engine application, because they can provide improved combustion efficiencies and reduced pollutant emissions.

Converting Power into Chemicals and Fuels seeks to elucidate the pivotal role of petrochemical processes in actively pursuing the transition from fossil fuel scenarios to more sustainable energy supply systems. The transformation of energy systems into a sustainable future will be impossible without chemical energy conversion. As energy cannot be created, we always deal with conversion processes. Many of them involve molecular or solid energy carriers, thus it is evident that chemical technology is at the centre of the energy challenge. Chemical science can control the energetic costof the conversion of energy carriers.

The global demand for hydrocarbons – as petrochemical feedstock, as fuels for transport and for other uses – is expected to increase until at least 2040. These products have an unrivalled energy density and are easy to transport, making them an ideal means to carry and store energy. While alternatives are being developed for some of their current uses (e.g., in passenger cars, where electrification is expected to play a major role), hydrocarbons remain difficult to replace in heavy-duty and marine transport, in aviation and as a feedstock for petrochemical technology.

It is therefore of great importance for the energy and technological value chain that carbon emissions of hydrocarbon technologies be progressively reduced. The petrochemical and refining technology is well placed to evolve its business model with this objective, by increasingly using combinations of new feedstocks – such as captured carbon dioxide, waste, and biomass, in very efficient manufacturing. It can also expand its use of surplus renewable electricity and hydrogen on-site and further exploit synergies with other industries in integrated clusters. The flexibility and resilience of the hydrocarbon technology infrastructures, including those for the distribution of products, will allow this transformation to occur at a comparatively low cost and provide immediate benefits in term of carbon dioxide reduction. COP26 concluded in November 2021 in Glasgow (with nearly 200 countries agreeing the Glasgow Climate Pact to keep the rise in average global temperature at 1.5°C alive and finalise the outstanding elements of the Paris Agreement.

The petrochemical and refining technologies are already engaged in low-carbon transition, through investment in R&D projects and the early deployment of new technologies. These technologies, which have already been proven at different technology readiness levels (High-TRL 9), need to be implemented at scale. Innovative solutions will allow the use of new feedstocks and will cut greenhouse gas (GHG) emissions from refineries and from the use of their products. In the process, the European Union will develop and reinforce its global leadership in low-carbon technologies, which will be exported around the world where they are needed. Examples include the conversion of refineries to bio-refineries, the development of sustainable hydrogen and biofuels produced from surplus renewable electricity. These are just the tip of the iceberg of the chemical industry's extensive R&D.

In addition to a reduction in carbon dioxide emissions, the European Union energy strategy addresses air quality and the transition to a more to a circular economy. This implies maintaining the value of products, materials and resources in the economy for as long as possible and minimising the generation of waste. The petrochemical and refining technologies are deeply embedded in these important areas, with innovations and initiatives that aim to improve air quality and minimise waste – or, when possible, re-use it.

Power-to-X denotes methods for converting renewable energy into liquids or gases, which can be stored, distributed, or converted to valuable products. Furthermore, Power-to-X can provide grid stability in connection with fluctuating electricity from renewable sources. One of the essential steps in determining the feasibility of a Power-to-X system for the market is the thermodynamic, techno-economic, and environmental assessment through mathematical modelling and simulation. In this book we present different Power-to-X system configurations with their performance, environmental impact, and cost. We provide an introduction to various Power-to-X processes including all the stages from the power generation to the upgrading of the final product (X), followed by several key system-level Power-to-X studies, which consist of thermodynamic, techno-economic, and life cycle assessment analyses.

The Power-to-X chemical technology hydrogen pillar is as follows:

- Hydrogen Pillar–Electrochemicals → Power-to-e-Chemicals and Electrofuels → Power-to-e-Fuels.
  - These include hydrogen → Power-to-Hydrogen production through water electrolysis as a means of storing surplus renewable electricity in chemical bonds.
  - Hydrogen can be used for transportation in fuel cell vehicles (FCV), but can also be react with carbon dioxide to form other fuels → Power-to-Fuels
  - We present a technical, environmental, and economic comparison of direct hydrogen use in fuel cells, and production of methane → Power-to-Methane methanol → Power-to-Methanol; and dimethyl ether → Power-to-DME for use in internal combustion engines for light-duty vehicle applications.
  - With respect to their suitability as diesel fuels for the transport sector, the Power-to-Fuels products: dimethyl ether → Power-to-DME; oxymethylene dimethyl ether → Power-to-$OME_{3\text{-}5}$; and n-alkanes → Power-to-FTdiesel.

Power-to-Chemical-technologies and Power-to-Fuel-processes thus pave the way for the integration of surplus renewable energies in the petrochemical and transport sector. Electrofuel technologies could first be introduced to enhance the carbon today's fuels derive from biomass and wastes. Approximately 7% of global oil demand will be replaced by electric vehicle (EV) even though the real growth of electric vehicle s depends on numerous factors such as the price of a battery, subsidy from government, and the availability of rare-earth and lithium elements.

The integration of carbon dioxide via methanol and methane would require comparably low research development effort and would allow use of large parts of the existing petrochemical infrastructure and hydrocarbon technology. The designed chemical models include carbon capture and utilization (CCU) technologies for the direct conversion of carbon dioxide into olefins, BTX aromatics, carbon monoxide and hydrogen, ethylene oxide, and styrene. These electrochemical technologies are currently at early research and development stages with TRLs below seven.

The bottom-up model of the chemical technology yields future production pathways to produce the 20 large-volume chemicals: acrylonitrile, ammonia, benzene, caprolactam, cumene, ethylene, ethylene glycol, ethylene oxide, methanol, mixed xylenes, phenol, polyethylene, polypropylene, propylene, propylene oxide, p-xylene, styrene, terephthalic acid, toluene, and vinyl chloride. Production pathways are represented by more than 160 processes based on engineering-level data. Thereby, flows of energy and materials are determined in detail throughout entire supply.

Acknowledgments

Many thanks to my wife Mária for supporting my efforts in bringing together these concepts in the form of a book and also for her direct participation in the generation of graphics.

General Literature

Aadnoy, B. and Looye, R. (2019). *Petroleum Rock Mechanics Drilling Operations and Well Design*. Gulf Professional Publishing.

Abdel-Aal, H.K. and Alsahlawi, M.A. (2013). *Petroleum Economics and Engineering*, 3e. CRC Press.

Aguado, J. and Serrano, D. (1999). *Feedstock Recycling of Plastic Wastes* (ed. J.H. Clark). RSC Clean Technology Monographs, Royal Society of Chemistry, Cambridge, UK.

Albright, L.F., Crynes, B.L., and Corcoran, W.H. (1983). *Pyrolysis: Theory and Industrial Practice*. New York: Academic Press.

Arpe, H. (2010). *Industrial Organic Chemicals*, 5e. New York: Wiley.

Arpentinier, P., Cavani, F., and Trifiro, F. (2001). *The Technology of Catalytic Oxidation, 1. Chemical, Catalytic & Engineering Aspects*. Paris: Editions TECHNIP.

Arpentinier, P., Cavani, F., and Trifiro, F. (2001). *The Technology of Catalytic Oxidation, 2. Safety Aspects*. Paris: Editions TECHNIP.

Austin, G.T. (1984). *Shreve's Chemical Process Industries*, 5e. New York: McGraw Hill Book Company.

Bajus, M. (1987). Thermal conversion of petroleum hydrocarbons into petrochemicals. D.Sc Thesis, Slovak University of Technology in Bratislava, Faculty of Chemical and Food Technology, Bratislava, Slovak Republic.

Bajus, M. (1989). Sulfur Reports, Volume 9 (1), *Sulfur Compounds in Hydrocarbon Pyrolysis*. Glasgow, UK: Harwood Academic Publishers, 25–71.

Bajus, M. (1996). *Alternatívne palivá, interné skripá Katedry technológie ropy a petrochémie*. Bratislava: FCHPT-STU.

Bajus, M. (2002a). Organická technológia a petrochémia, Uhľovodíkové technológie. Vydavateľstvo STU, 178 s., Bratislava.

Bajus, M. (2002b). Organická technológia a petrochémia, Uhľovodíkové technológie. Elektronická CD učebnica, Vydavateľstvo DIVYD, Bratislava.

Bajus, M. (2017). *Petrochemistry: Hydrocarbon Technology*. Editions, Slovak University of Technology, Faculty of Chemical & Food Technology, Slovak Chemical Library.

Bajus, M. (2019). Chemické procesy v organickej technológii a petrochémii, e-Učebnica, Vydavateľsto Spektrum STU, Bratislava.

Bajus, M. (2020). *Petrochemistry: Petrochemical Processing, Hydrocarbon Technology, and Green Engineering*. Chichester: Wiley.

Bajus, M., Buláková, Ľ. et al. (2006). 150 rokov plynárenstva na Slovensku. Vydal Slovenský plynárenský a naftový zväz, Bratislava.

Bajus, M., Lederer, J., and Bělohlav, Z. (2016). Plenary lecture: hydrocarbon technology, trends and outlook in petrochemistry. *4th International Conference on Chemical Technology*, 87. Mikulov, Czech Republic.

Banerjee, D. (2020). *Thermal Processing of Hydrocarbons to Petrochemicals*. PennWell Books.

Bazzanella, A.M. and Ausfelder, F. (2017). Technology study: low carbon energy and feedstock for the European chemical industry. Dechema Gesellschaft fur Chemische Technik und Biotechnology e.V., Cefic.

Bisio, A. and Boots, S. (eds.) (1995). *Encyclopaedia of Energy Technology and the Environment, Volume 1–4*. Chichester: Wiley.

Boodhoo, K. and Harvey, A. (eds.) (2013). *Process Intensification Technologies for Green Chemistry: Engineering Solutions for Sustainable Chemical Processing*. Wiley.

BP statistical review of world energy. (2022). 1922–2018. bp.com/statisticalreview.

Braun, R.C. (ed.) (2011). *Thermochemical Processing of Biomass, Conversion into Fuels, Chemicals and Power*. Chichester: Wiley.

Burnham, A.K. (2017). *Global Chemical Kinetics of Fossil Fuels*. Springer.

Cenka, M. et al. (2001). *Obnovitelné zdroje energie*. Praha: FCC Public.

Centi, G., Cavani, F., and Trifiró, F. (2001). Fundamental and applied catalysis. In: *Selective Oxidation by Heterogeneous Catalysis* (ed. M.V. Twigg and M. Spencer). New York: Kluwer Academic/Plenum Publishers.

Červený, L., a koklektív. (1995). *Zakladní pochody prumyslové organickej syntézy*. Bratislava: STU.

Comyns, A.E. (1999). *Encyclopedic Dictionary of Named Processes in Chemical Technology*, 2e. Boca Raton, London: CRC Press.

Corma, A. and Martinez, A. (2002). Catalysis on Porous Solids. In: *Handbook of Porous Solids* (ed. F. Schutz, K.S.W. Sing, and J. Wetkamp). Weinheim: Wiley-VCH.

Cutler, J.C. (Editor-in-Chief) (2004). *Encyclopaedia of Energy, Volume 1-6*. San Diego, CA: Elsevier Inc., Academic Press.

Dawe, R.A. (ed.) (2000). *Modern Petroleum Technology, Vol. 1, Upstream*, 6e. Chichester: Wiley.

Devins, D. (1982). *Energy, Its Physical Impact on the Environment*. New York: Wiley.

Ehrfeld, W., Hessel, V., and Loewe, H. (2000). *Microreactors-New Technology for Modern Chemistry*. Weinheim: Wiley-VCH, 9.

El-Gendy, L.S. (2018). *Biodesulfurization in Petroleum Refining*. Scrivener Publishing/Wiley.

Favennec, J.P. (ed.) (2001). *Petroleum Refining, 5. Refinery Operation and Management*. Paris: Editions TECHNIP.

Fink, J. (2015). *Petroleum Engineer's Guide to Oil and Fluids*, 2e. Oxford: Gulf Professional Publishing, Elsevier.

Furimsky, E. (2020). CO_2 hydrogenation to methanol and methane over carbon-supported catalysts. *Ind. Eng. Chem. Res.* 59 (35): 15393–15423.

Germain, J.E. (1969). *Catalytic Conversion of Hydrocarbons*. London and New York: Academic Press.

Goyal, M.R., Kumar, A., and Gupta, A.K. (2018). *Novel Dairy Processing Technologies: Techniques, Management, and Energy Conservation*. Oakville, Canada: Apple Academic Press, Inc.

Harker, J.H. and Backhurst, J.R. (1981). *Fuel and Energy*. London: Academic Press.

Hoogers, G. (2003). *Fuel Cell Technology Handbook*. New York: CRC Press.

Hougen, O.A. and Watson, K.M. (1947). *Chemical Process Principles*, vol. 3. New York: Wiley.

Hsu, Ch. and Robinson, P.R. (eds.) (2017). *Springer Handbook of Petroleum Technology*. Cham: Springer International Publishing AG.

Hsu, C.S. and Robinson, P.R. (2019). *Petroleum Science and Technology*. Cham: Springer.

Humphrey, J.L. and Keller, G.L. (1997). *Separation Process Technology*. New York: McGraw-Hill.

Iglesia, E., Spivey, J.J., and Fleisch, T.H. (2001). *Natural Gas Conversion VI*, vol. 136, Amsterdam: Elsevier.

Ince, A.C., Colpan, C.O., Hagen, A., and Serincan, M.F. (2021). Modelling and Simulation of Power-to-X systems: a review. *Fuel* 304: 121354.

Jones, D.S.J. and Pujado, P. (eds.) (2006). *Handbook of Chemical Processing*. Dordrecht: Springer.
Ke, L., Chunshan, S., and Velu, S. (2010). *Hydrogen and Syngas Production and Purification Technologies*. Hoboken, New York: AIChe, Wiley.
Kelland, M.A. (2014). *Production Chemicals from the Oil and Gas Industry*. Boca Raton: CRC Press.
Kiss, A.A. (2014). *Process Intensification Technologies for Biodiesel Production: Reactive Separation Processes*. Springer Briefs in Applied Sciences and Technology, Cham: Springer.
Kizlink, J. (2011). *Technologie chemických látek a jejich použití*, 4. přepracované a doplneněné vydání, vydavatelství: VUTIUM.
Lee, S. (2012). *Biofuels and Bioenergy*. Boca Raton: CRC Press.
Leprince, P. (ed.) (1995). *Petroleum Refining, 3. Conversion Process*. Paris: Editions TECHNIP. (1995).
Lucas, A.G. (ed.) (2000). *Modern Petroleum Technology, Vol. 2, Downstream*, 6e. Chichester: Wiley.
Luque, R. and Balu, A.M. (eds.) (2013). *Producing Fuels and Fine Chemicals from Biomass Using Nanomaterials*. Boca Raton: CRC Press.
Luque, R. and Leung-Yuk Lam, F. (2018). *Substainable Catalysis: Energy-Efficient Reactions and Applications*. Weinheim: Wiley-VCH.
Maxwell, I.E. and Stork, W.H.J. (2001). Hydrocarbon processing with zeolites. In: *Introduction to Zeolite Science and Practise*, vol. 137, 2e (eds. H. van Bekkum, E.M. Flamigen, P.A. Jacobs, and J.C. Jansen). Amsterdam: Elsevier.
Marko, Š., a ďalší. (1988). *Energetické zdroje a premeny*. Bratislava: Alfa.
Meyers, R.A. (2005). *Handbook of Petrochemicals Production Processes*. New York: McGraw-Hill.
Meyers, R.A. (2018). *Handbook of Petrochemicals Production*, 2e. New York: McGraw Hill.
Minteer, S. (2006). *Alcoholic Fuels*. Boca Raton: CRC Press.
Moulijn, J.A., Makkee, M., and Van Diepen, A. (2005). *Chemical Process Technology*. Chichester: Wiley.
Noronha, F.B., Schmal, M., and Sousa-Aguiar, E.F. (2007). *Natural Gas Conversion VIII*, vol. 167. Amsterdam: Elsevier.
Olah, G.A. and Molnar, A. (1995). *Hydrocarbons Chemistry*. New York: Wiley.
Olivé, G.H. and Olivé, S. (1984). *The Chemistry of the Catalyzed Hydrogenation of Carbon Monoxide*. Berlin, Heidelberg: Springer-Verlag.
Parmaliana, A., Sanfilippo, D., Frusteri, F., Vaccari, A. and Arena, F. (eds.) (1998). *Natural Gas Conversion V*, vol. 119, Amsterdam: Elsevier.
Pašek, J. (1996). *Uhlíkaté suroviny*. Praha: VŠCHT.
Pastorek, Z., Kára, J., and Jevič, P. (2004). *Biomasa obnovitelný zdroj energie*. Praha: FCC Public.
Pasupathi, S., Gomez, J.C.C., Su, H. et al. (2016). *Recent Advances in High-temperature PEM Fuel Cells*. London: Academic Press.
Pines, H. (1981). *The Chemistry of Catalytic Hydrocarbon Reactios*. London: Academic Press.
Roberts, P. (2019). *A Dictionary of Oil & Gas Industry Terms*. Oxford: Oxford University Press.
Rojey, A., Jaffret, C., Cornot-Gandolphe, S., and Durand, B. (1997). *Natural Gas Production Processing Transport*. Paris: Editions TECHNIP.
Rossillo-Calle, F., de Groot, P., and Hemstock, S.L. (eds.) (2007). *The Biomass Assessment Handbook, Bioenergy for a Sustainable Environment*. London: Earthscan.
Sabu, T. (2014). *Biomaterial Applications: Micro to Nanoscales*. Boca Raton: CRC Press.
Sánchez-Bastardo, N., Schlögl, R., and Holger Ruland, H. (2021). Methane pyrolysis for zero-emission hydrogen production: a potential bridge technology from fossil fuels to a renewable and sustainable hydrogen economy. *Ind. Eng. Chem. Res.* 60 (32): 11855–11881.
Shah, Y.T. (2014). *Water for Energy and Fuel Production*. Boca Raton: CRC Press.
Sheldon, R.A. (1982). *Chemicals from Synthesis Gas*. Dordrecht: D. Reidel Publishing Company.
Shi, F. (ed.) (2014). *Reactor and Process Design in Sustainable Energy Technology*. Amsterdam: Elsevier.

Shukla, A.K. and Marchesini, F.H. (2017). *Analytical Characterization Methods for Crude and Relative Products*. Weinheim: Wiley-VCH.
Skrzypek, J., Sloczynski, J., and Ledakowicz, S. (1994). *Methanol Synthesis*. Waszawa: Polish Scientific Publishers PWN.
Smil, V. (2003). *Energy at the Crossroads*. Cambridge (MA): The MIT Press.
Soni, V.K., Singh, G., Vijayan, B.K. et al. (2021). Thermochemical recycling of waste plastics by pyrolysis: a review. *Energy Fuels* 35: 12763–12808.
Sorensen, B. (2000). *Renewable Energy*. Boston: Academic Press.
Sorensen, B. (2005). *Hydrogen and Fuel Cells, Emerging Technologies and Applications*. London: Academic Press.
Spath, P.L. and Dayton, D.C. (2003). Prelimanary screening – technical and economic assessment of synthesis gas to fuels and chemicals with emphasis on the potential for biomas – derived syngas. ADA 436529/NREL/tp-510 – 34929.
Speight, J.G. (2011). *Handbook of Industrial Hydrocarbon Processes*, 1e. Burlington: Elsevier.
Speight, J.G. (2014). *The Chemistry and Technology of Petroleum*, 5e. Boca Raton: CRC Press.
Speight, J.G. (2017). *Rules of Thumb for Petroleum Engineers*. Hoboken: Scrivener Publishing/Wiley.
Spitz, H.P. (1988). *Petrochemicals, The Rise of an Industry*. Hoboken: Wiley.
Stocchi, E. (1999). *Industrial Chemistry*, vol. 1. New York: Ellis Horwood Limited.
Stolten, D. and Emonts, B. (2016). *Hydrogen Science and Engineering: Materials, Processes, Systems and Technology*. Weinheim: Wiley-VCH.
Surový, J. (1991). *Chemicko-inžinierska termodynamika*. Bratislava: Alfa.
Syad, A.M. (2015). *Fundamentals of Heat Exchanger Design*. London: LAP Lambert Academic Publishing.
Sylvester, E. (2010). Course: chemical technology (organic). Module VII., Lecture 2. https://www.academia.edu/33302746/Course_Chemical_Technology_Organic_Module_VII_Lecture_7_Aromatic_Production.
Trambouze, P. (ed.) (2000). *Petroleum Refining, 4. Materials and Equipment*. Paris: Editions TECHNIP.
Treese, S.A., Pujado, P.R., and Janes, D.S. (eds.) (2015). *Handbook of Petroleum Processing*, 2e. Berlin: Springer-Verlag.
Uner, D. (2017). *Advances in Refining Catalysis, (Chemical Industries)*, 1e. Boca Raton: CRC Press.
van Santen, R.A., van Leeuwen, P.W.N.M., Mouljin, J.A. and Averill, B.A. (eds.) (1999). *Catalysis: An Integrated Approach Vol. 123*, 2e. Elsevier.
Varfolomeev, S.D. (2014). *Chemical and Biochemical Technology: Material and Processing*. Boca Raton: CRC Press.
Vasconcelos, B.R. and Lavoie, J.-M. (2019). Recent advances in power-to-X technology for the production of fuels and chemicals. *Front. Chem.* 7.
Vertes, A.A., Qureshi, N., Blaschek, H.P., and Yukawa, H. (eds.) (2010). *Biomass to Biofuels, Strategies for Global Industries*. Chichester: Wiley.
Veselý, V., Mostecký, J., Bajus, M. et al. (1989). Petrochémia, Alfa Bratislava, 678.
Vision 2050 (2018). A pathway for the evolution of the refining industry and liquid fuels. Fuels Europe.
Wauquier, J.-P. (ed.) (1995). *Petroleum Refining, 1. Crude Oil, Petroleum Process Flowsheets*. Paris: Editions TECHNIP.
Wauquier, J.-P. (ed.) (2000). *Petroleum Refining, 2. Separation Processes*. Paris: Editions TECHNIP.
Wikipedia, The Free Encyclopedia. https//en.wikipedia.org/wiki/Petrochemical.
Winter, F., Avinash, A., Hrdlička, J., and Varjani, S. (eds.) (2019). *CO_2 Separation, Purification to Chemicals and Fuels*. Singapore: Springer Nature.
Wittcoff, H.A. and Reuben, B.G. (1980). *Industrial Organic Chemicals in Perspective, Part One: Raw Materials and Manufacture*. New York: Wiley.

Wittcoff, H.A. and Reuben, B.G. (1996). *Industrial Organic Chemicals*. New York: Wiley.

Wittcoff, H.A., Reuben, B.G., and Plotkin, J.S. (2013). *Industrial Organic Chemicals*, 3e. New York: Wiley.

Wohlleben, W. (2015). *Safety of Nanomaterials along Their Lifecycle*. Boca Raton: CRC Press.

Zangeneh, F.T., Sahebdelfar, S., and Ravanchi, M.R. (2011). Conversion of carbon dioxide to valuable petrochemicals: an approach to clean development mechanism. *J. Nat. Gas Chem.* 20: 219–231.

Zhu, F. (Xin X.). (2019). *Efficient Petrochemical Processes, Technology, Design and Operation*. Hoboken: Wiley.

Nomenclature

Bo	Bodenstein number	
D_{ax}	Axial dispersion coefficient	$[m^2\,s^{-1}]$
D_i	Diffusion coefficient	$[m^2\,s^{-1}]$
$\Delta_R H$	Heat of reaction	$[kJ\,mol^{-1}]$
H_i	Henry coefficients	[MPa]
L	Length	[m]
$\dot{m}$	Mass flow	$[kg\,s^{-1}]$
T	Temperature	[K]
ΔT	Temperature difference	[K]
t_D	Time constant of diffusion	[s]
u	Velocity	$[m\,s^{-1}]$
y	Molar concentration	

Greek Letters

β_i	Mass transfer coefficient
η	Process efficiency
τ	Time constant of convection/residence time

Abbreviations and Acronyms

ABR	Acrylonitrile butadiene rubber
ABS	Acrylonitrile butadiene styrene
AC	Activated carbon
ADIP	Diisopropylamine
AEL	Alkaline electrolysis
AFC	Alkaline fuel cell
AI	Artificial intelligence
AMR	Asset mechanical reliability
APC	Advanced process control
API	American Petroleum Institute
APM	Asset performance management
ARDS	Atmospheric residue desulphurization
ASTM	American Society for Testing and Materials
ASU	Air separation unit
ATR	Auto-thermal reformer
BAC	Biomass-derived activated carbon
BACT	Best available control technology
BAT	Best available technology
BDE	Bond dissociation energy
BECCS	Bioenergy with carbon capture and storage
BDD	Boron-doped diamond
BET	Brunauer–Emmett–Teller
BFG	Blast furnace gas
BTL	Biomass-to-Liquid
BIO-route	Use of biomass
BOE	Barrel oil equivalent
BOFG	Basic oxygen furnace gas
BP	British Petroleum
BPT	Best practice technology
BSEC	Brake specific energy consumption
BSFC	Brake specific fuel (energy) consumption
BTE	Brake thermal efficiency
BtG	Biomass-to-Gas
2-BTHF	2-butyltetrahydrofuran
BTX	Benzene–toluene–xylene
BTXE	Benzene-toluene-xylene-ethylbenzene
BTDC	Before top dead centre (in a spark ignition)

BUE	Biomass utilization efficiency
C2CNT	Carbon dioxide to carbon nanotube
Ca–Br	Calcium–Bromine cycle
CAGR	Compound annual growth rate
CAMERE	Process produced methanol from carbon dioxide
CANDU	Canada deuterium-uranium; Canadian pressured heavy-water reactor
CAP	Operating capacity
CAP	Chinese atomic plant (CAP-1400)
CAT	Catalyst
CAPEX	Capital expenditure
CAPEX; OPEX	Network capture systems for carbon dioxide
CB	Carbon black
CBP	Pyrolytic carbon black
CCS	Carbon capture and storage
CCR	Capital charges
CCU	Use of captured carbon dioxide
CCUS	Carbon capture use and storage
CDM	Clean development mechanism
CE	Circular economy
CERI	Canadian Energy Research Institute
CFPCI	Chinese Federation of Petroleum and Chemical Industry
CFPP	Cold filter plugging point
CG-FAW	Glycerol-fatty acids waste
CGN	China General Nuclear
CHP	Combined heat and power
CI	Compression ignition
CI	Carbon intensity
CIZO	Cu-In-Zr-O catalyst from mixed oxide sites
CIS	Commonwealth of Independent States
CLC	Chemical looping combustion
CMK	Ordered mesoporous carbon
CMR	Continuous microwave reactor
CN	Cetane number
CNFs	Carbon nanofibers
CNG	Compressed natural gas
CNNC	China National Nuclear Corporation
CNL	Canadian National Laboratories
CNT	Carbon nanotubes
CNY	Chinese Yuan Renminbi
CNSC	Canadian Nuclear Safety Commission
COC	Crude oil to chemicals
COMPER	Refinery of Petrobras
COP-21	The Paris Agreement; The Conference of The Paris (Paris Climate Conference); Conference of Parties
COP-26	The Pact of Glasgow (The 2021 United Nations Climate Change Conference, held at Centre in Glasgow, Scotland)
CORDEL	Cooperation in reactor design evaluation and licensing
CO_2 to CNTs	Carbon dioxide to carbon nanotubes
CO_2-FTS	Carbon dioxide to Fischer–Tropsch synthesis

CO_2RR	Carbon dioxide reduction reaction
CTL	Coal-to-Liquid
COG	Coke ove gas
COP	Coefficient of performance
COTC	Crude oil to chemicals
COTS	Commercial off-the-shelf
CPmax	Maximum in-cylinder pressure
CPO	Cyclopentadiene polymer composite (which is also called MOF)
CPSI	Channels per square inch
CRDI	Common rail direct injection
CRI	Carbon recycling international
CRT	Creditors reliability test
CRTs	Cathode ray tubes
CSBR	Conical spouted bed reactor
Cu/Zn-H	Zn deposited from high concentration (6M) on Cu substrate
Cu/Zn-L	Zn deposited from low concentration (0.6M) on Cu substrate
CVD	Chemical vapour deposition
DAC	Direct activated carbon
DAFC	Direct ammonia fuel cell
DA-SOFC	Direct ammonia solid oxide fuel cell
DBT	Dibenzyl toluene
DC	Direct cost
DCN	Derived cetane number
DCU	Delayed coking unit
DDR	Polymer composite material
DEA	Diethanolamine
DFAFCs	Direct formic acid fuel cells
DFT	Density functional theory
DHR	Deep-pool low-temperature heating reactor
DI	Direct injection
DIPA	Diisopropylamine
DS	Desilication stabilization
DME	Dimethyl ether
DMF	Dimethylformamide
DMC	Dimethyl carbonate
DMSR	Denatured molten salt reactor
DNBE	Di-n-butyl ether
DOE	United States Department of Energy
DPC	Diphenyl carbonate
DRIFT	Diffuse reflectance infrared Fourier transform
DTG	Derivative thermogravimetry analysis
ECR	Electrocatalytic reduction
ECR	Electrochemical carbon oxide reduction technology
EEE	Energy, economy, and environmental index
EIS	Electrochemical impedance spectroscopy
EJ	Exajoule ($1x10^{18}$); quintillion joules
EMAR	Electrochemical mediated amine regeneration
EMIPS	Exergetic material input per unit of service
EN590	European diesel fuel standards

ENI	Refining and marketing a chemical in Europe
EOR	Enhanced oil recovery
EPA	Environmental Protection Agency
EPS	Expanded polystyrene
EPSW	Expanded polystyrene waste
EPZ	Emergency planning zone
ESA	Electric swing adsorption
ETBE	Ethyl tertiary butyl ether
ETEM	Environmental transmission electron microscopy
EU-RED	Renewable energy directive of EU (2009)
EURO1-5	European Standard for Fuel Quality
EV	Electrical vehicles
EWE	Energy–water–environment
FA	Furfuryl aldehyde
FAME	Fatty acid methyl ester
FCC	Fluid catalytic cracking
FCC	Food chemical codex
FCEV	Fuel cell electrical vehicle
FCV	Fuel cell vehicle
FE	Faradaic efficiency
FFA	Free fatty acid
FLG	Few layers graphene
FNR	Fast-neutron reactor
FOAK	First of a kind
FT	Fischer–Tropsch
FT-ICR	Fourier transform ion cyclotron resonance
FTIR	Fourier transform infrared
FTO	Fischer–Tropsch to olefins
G	Gaseous
GC	Gas chromatography
GCC	Gulf Cooperation Council
GC/MS	Gas chromatography/mass spectrometry
GCP	Glasgow Climate Pact
GC-TOF	Gas chromatography time-to-flight mass spectrometry
GDE	Gas diffusion electrode
GDP	Gross Domestic Product
GE	Generated Energy
Gen IV	Generation IV of nuclear reactors
GHG	Greenhouse gases
GHSV	Gas hourly space velocity
GIF	Generation IV International Forum
GO	Graphene oxide
GPIC	Gulf Petrochemical Industries Company
GTL	Gas-to-Liquid
GWe	Gigawatt electric power
HALEU	High-assay low-enriched uranium
HAZOP	Hazard and operability study
HCl	Hydrochloric acid
HCO	Heavy cycle oil

HCR	Honeycomb reactor
HCU	Hydrocracking unit
HDPE	High-density polyethylene
HEEP	Hydrogen Economic Evaluation Program
HEO	Hydrogen evolution overvoltage
HER	Hydrogen evolution reaction
HES	High energy sites
HEU	High enriched uranium
HEV	Hybrid electric vehicle
HEO	Hydrogen evolution overvoltage
HER	Hydrogen evolution reaction
HIPS	High-impact polystyrene
HKUST-1	Hong Kong University of Science and Technology
HMF	5-Hydroxymethylfurfural
HNZ	Natural clinoptilolite zeolite
HPA	3-Hydroxypropionaldehyde
HPI	Hydrocarbon processing industry
HPLC	High-performance liquid chromatography
HPR	Hydroprocessing
HPU	Hydrogen production unit
HREELS	High-resolution electron energy loss spectroscopy
HRRmax	Maximum heat release rate
HSER	Health and safety environmental regulation
HTR	High-temperature gas-cooled reactor
HTSC	High-temperature steam electrolysis
HTTR	High-temperature engineering test institute
HUSY	Hierarchical H-style ultrastable Y-zeolite
HVC	High value chemicals
HyS	Hybrid sulfur cycle
IAEA	International Atomic Energy Agency
ICE	Internal combustion engine
IEA	International Energy Agency
IFP	French Institute of Petroleum
IIoT	Industrial Internet of Things
KiOR	Biofuel company burns
IL	Ionic liquid
IMEP	Indicated mean effective pressure
IMH	Ionic microhabit
IMSR	Integral molten salt reactor
INL	Idaho National Laboratory (United States)
IP	Injection pressure
IPCC	Intergovernmental Panel on Climate Change
IPGC	Integrated plasma gasification combined
IPW	Industrial plastic waste
IRPC	International Refining and Petroleum Conference
IRENA	International Renewable Energy Agency
IT	Injection timing
JAEA	Japan Atomic Energy Agency

JAERI	Japan Atomic Energy Research Institute
JSR	Jet-stirred reactor
KPI	Key performance indicator
L	Liquid
LAER	Lowest achievable emissions rate
LCA	Life cycle assessment
LCFS	Low carbon fuel standards
LCO	Light cycle oil
LCOE	Levelized cost of electricity
LDAR	Leak detection and repair
LDPE	Low-density polyethylene
LHV	Low heating value
LLDPE	Linear low-density polyethylene
LMOF	Light-responsive metal–organic framework
LOCA	Loss-of-coolant accident
LOHC	Liquid organic hydrogen carrier
LPG	Liquefied petroleum gas
LRL	Licencing readiness level
LSM	Lanthanum strontium manganite
LWR	Light water reactor
MACT	Maximum achievable control technology
MAS	Magic angle spinning
MDEP	Multinational design evaluation programme
MDR	Methane dry reforming
MEA	Membrane electrode assembly
MEA	Monoethanolamine
MECS	Microencapsulated carbon sorbent
MECs	Microbial electrolysis cells
$MefCO_2$	Methanol from carbon dioxide and water
MEG	Monoethylene glycol
MeOH	Methanol
MFC	Microbial fuel cell
MIPS	material input per unit of service
MMA	Methyl methacrylate
MMbpd	Billions of barrels per day; $1x10^9$ barrels per day
MMR	Micro modular reactor
MOF	Metal organic framework
MONG	Non glycerol organic matter
MoU	Memorandum of understanding
MOX	Mixed oxide fuel
MPC	Mesoporous carbon
MPCs	Mesoporous carbons
MSR	Molten salt reactor
MTA	Methanol to aromatics
MTG	Methanol to gasoline
MTH	Methanol to hydrocarbons
MTO	Methanol to olefins
Mtoe	Million-ton oil equivalent
MTPA	Million tons per annum

MWe	Megawatt electric
MWCNT	Multiwalled carbon nanotube
MZ	Meso-ZSM-5
NADH	Nicotinamide adenine dinucleotide enzymes
NC2I	Nuclear Cogeneration Industrial Initiative
NC-AFM	Non-contact atomic force microscopy
NCBJ	Polish National Centre for Nuclear Research
NCF	Nanoporous copper film
NEA	Nuclear Energy Agency
NGNP	Next generation nuclear plant
NHDD	Nuclear Hydrogen Development & Demonstration
NHE	Normal hydrogen electrode
NHR	Nuclear heating reactor (NHR-200)
NLC	Nuclear Law Committee (NEA)
NMP	No micropores
NNL	National Nuclear Laboratory (United Kingdom)
NO_x	Nitrogen oxides
NPs	Nanoparticles
NRC	Nuclear Regulatory Commision (United States)
NRS-SnO	Nanorod sheet
NSSS	Nuclear steam supply system
NTC	Negative temperature coefficient
NTs	Nanotubes
NZ	Natural zeolite
NZ	Nano-ZSM-5
O&M	Operation and maintenance
OC	Operating cost
OCM	Oxidative coupling of methane
ODH	Oxidative dehydrogenation
OE	Oil equivalent
OECD	Organisation for Economic Co-operation and Development
OGI	Optical gas imaging
OME	oxyethylene ether
OT&E	Operational test and evaluation
OTC	Oil to chemicals
OTS	Operator training simulator
OTSG	Once-through steam generator
P2X	Power-2-X technology
PA-6	Polyamide 6
PAH	Polyaromatic hydrocarbons
PBMR	Pebble bed modular reactor
PC	Poly (bisphenol A carbonate)
PCB	Printed circuit boards
PCFC	Proton-conducting fuel cell
PCPW	Postconsumer plastic film waste
PC-SOFC	Proton-conducting solid oxide fuel cell
PDH	Propane dehydrogenation
PE	Polyethylene
PEI	Polyethyleneimine

PEM	Proton exchange membrane
PEM	Polymer electrolyte membrane electrolysis
PEMFC	Proton exchange membrane fuel cell
PEMFC	Polymer electrolyte membrane fuel cell
PES	Polyether sulfone
PET	Polyethylene terephthalate
PEW	Polyethylene wax
PHP	Process heat plan
PHS	Pumped hydro storage
PHWR	Pressurised heavy water reactor
PITOFMS	Photoionization time-of-flight mass spectrometry (After pyrolysis)
PM	Particulate matter
PMCC	Propylene maximization catalytic cracking
PMI	Process mass intensity
PMMA	Polymethylmethacrylate
PNP	Pincer complexes of osmium
PO	Propylene oxide
POX	Partial oxidation
PP	Polypropylene
PPE	Personal protective equipment
PPM	Preventitive and predictive maintenance
PPS	Polyphenylene sulfide
PS	Polystyrene
PSA	Pressure swing adsorption
PT	Photoluminescence
PtC	Power-to-Chemicals
PtF	Power-to-Fuels
PTFE	Polytetrafluoroethylene
PtG	Power-to-Gas
PtH	Power-to-Hydrogen
PtL	Power-to-Liquids
PtM	Power-to-Methane
PtX	Power-to-X
PU	Polyurethane
PURs	Polyurethane resins
PV	Photovoltaic panels
PVC	Polyvinyl chloride
PVDC	Polyvinyl dichloride
PWR	Pressure water reactor
RAM	Reliability, availability and maintainability
RED-2	Renewable energy directive
re-X	Recycling, remanufacturing, reuse, etc.
RFCCU	Residue fluid catalytic cracking unit
RFS2	Renewable fuel standard
rGO	Reduced graphene oxide
RHCU	Residue hydrocracking
RHE	Reversible hydrogen electrode

RLS	Rate limiting step
ROI	Return on investment
RON	Research octane number
RPET	Recycled polyester
RWGS	Reverse water gas shift
S	Solid
SADARA	Joint venture between Saudi Aramco and Dow Chemical
SAF	Sustainable aviation fuel
SAPO-34	Zeolite
SBCR	Slurry bubble column reactor
SCE	Saturated calomel electrode
SDS	Sustainable devclopment scenario
SEC	Specific energy consumption
SEM	Scanning electron microscopy
SFC	Specific fuel consumption
SI	Spark ignition
S-I	Sulfur-iodine
SMR	Steam methane reforming
SMR	Small modular reactor
SMR	Small medium reactor
SmAHTR	Small, modular, advanced high temperature reactor
SNETP	Sustainable nuclear energy technology platform
SNG	Substitute natural gas
SNG	Synthetic natural gas
SNPA	Societe Nationale des Petroles d'Aquitaine (Process)
SOE	Solid oxide electrolysis
SPARG	Sulfur-passivated reforming
SPC	Single pass conversion
SPIC	State power investment corporation
SST	Sea surface temperature
STR	Stirred tank reactor
SWCNT	Single-walled carbon nanotubes
TAN	Total acid number
TCI	Total cost of investment
TCO	Total cost of ownership
TCR	Thermocatalytic reforming
TEM	Transmission electron spectroscopy
TCGR	The Catalyst Group
TG	Thermogravimetry
TGA	Thermogravimetric analyzer/analysis
THC	Total hydrocarbons
THFA	4-(2-terahydrofuryl)-2-butanol
TOF	Turnover frequency
TOS	Time on stream
TON	Turnover number
TPD	Temperature programmed desorption
TPL	Tire pyrolysis oil
TPSR	Temperature-programmed desorption
TPY	Tons per year

TRISO	Tri-structural ISOtropic nuclear fuel
TRM	Tri-reforming of methane
TRL	Technology readiness level
TSA	Polyaniline composite
TSA	p-Toluene sulfonic acid
TSA	Temperature swing adsorption
UCAP	(TEA) Rectisol
ULS	Ultra-low sulfur
ULSD	Ultralow sulfur diesel
UNCCC	United Nations Framework Convention on Climate Change
USP	United States Pharmacopeia
US-Y	Ultra stable zeolite
UV	Ultraviolet
VGO	Vacuum gas oil
VHTR	Very high temperature reactor
VLS	Vapor–liquid–solid
VPSA	Vacuum pressure swing adsorption
VR	Virtual reality
VRE	Variable renewable energy
VSMR	Very small medium reactor
WEEE	Waste electrical and electronic equipment
WGS	Water gas shift
WHSV	Weight hourly space velocity
WLTPA	Worldwide harmonised light vehicle test procedure
WNA	World Nuclear Association
WPCBs	Waste printed circuit boards
WRAP	charity working with governments, businesses and citizens around the globe to create a world in which resources are used sustainably
WT	Wind turbines
WTW	Well-to-wheel
XRD	X-ray diffraction
XPS	X-ray photoelectron spectroscopy
XRD	X-ray diffraction
YPF	Desulfurization project in Argentina
YSZ	Yttria-stabilized zirconia
ZSM-5(23)	Zeolite Socony Mobil ($SiO_2/Al_2O_3 = 23$)

1

Power-to-Chemical Technology

CONTENTS

Converting Power into Chemicals and Fuels: Power-to-X Technology for a Sustainable Future, First Edition. Martin Bajus.

1.1 Introduction

Chemical technology faces significant challenges relating to the transition towards cleaner energy and more sustainable feedstocks. The key drivers of this transformation are to avoid producing greenhouse gases and the formation of a more circular economy to lower the waste burden on the environment. Both transitions should support the needs of global population growth without affecting the planet's ability to also supply future generations. This is not the first transition that chemical technology has faced, but, where previous industrial transitions were driven by technology developments, the current transitions are primarily driven by a strong societal pull. This means that the goals are clear, but the technical path to achieve these goals is not.

A traditional petrochemical value chain takes in carbon from a fossil origin (oil or gas) and converts it into a broad variety of products that have been optimized for cost efficiency and product quality. This typically comes with a significant energy need that is traditionally delivered via combustion of fossil carbon, thus resulting in significant carbon dioxide emissions and a post-consumer waste disposal problem. Therefore, the challenge will be to develop technologies and partnerships to deliver circular processes that utilize renewable energy and feedstocks.

Carbon capture and utilization (CCU) represents a step forward from carbon capture and storage (CCS), as it allows the captured carbon dioxide to be valorised, repurposed, or reutilised. Various carbon capture and storage methods have recently received a great deal of attention as potential solutions to reduce carbon dioxide emissions from major industrial emitters. In this context, the research into utilizing carbon dioxide as a resource to synthesise chemical feedstock, fuels or commodity products has piqued the interest of academia and industry alike.

Carbon dioxide is one of the most stable molecules in the atmosphere due to the strong C=O (O=C=O) bonds and has the properties of double and triple bonds. Because of the strong C–O bond, the reduction of carbon monoxide requires greater activation energy. Generally, the conversion technologies of carbon dioxide mainly include biological, chemical, thermal conversion, photocatalysis, electrocatalysis and photoelectric cooperative transformation. Among these technologies, electrochemical carbon dioxide reduction (ECR) has attracted the most attention. From a simplistic chemistry standpoint, carbon dioxide conversion typically involves nucleophilic attack to the carbon atom, followed by reaction with electron-donating reagents. Levels of carbon dioxide in the atmosphere were reported to have reached 401.3 ppm in July 2015, well above the safe limit of 350 ppm.

Thermodynamically, carbon dioxide conversion processes consist of multiple uphill steps with many proton/electron transfer intermediates. Additionally, each of these steps has its own individual kinetic reaction barriers that need to be surmounted with extra energy input, making carbon dioxide conversion reaction extremely endothermic. (Figure 1.1a and b) illustrate the overall carbon capture and utilization scheme and standard Gibbs free energy of formation of various C_1 chemicals with different oxidation states of carbon.

Among various carbon dioxide valorisation approaches, electrochemical carbon dioxide reduction reaction (CO_2RR) stands out for its scalability and flexibility of using a wide array of energy (electricity) input, including the ease of renewables integration. The focus of the study of CO_2RR herein is on the conversion of carbon dioxide to a diatomic carbon olefin product (ethylene), which is more valuable than C_1 products due to its longer carbon chain as well as being the most valuable and widely produced olefin products (about 146 Mt/year globally, in 2014). Global ethylene capacity is expected to see considerable growth of 27% from 2008 in 2019 to 264 MTPA in 2023, led by Asia and North America, with petrochemical consumption driven by growth in economic activity and population. Ethylene products are reported to have a market price of ~\$1100/ton, which is much higher than formic acid (~\$400/ton), methanol (~\$400/ton), carbon monoxide (~\$300/ton) and methane (~\$200/ton). However, as the implementation of CO_2RR is still in its infancy, there is very limited understanding of its carbon dioxide emission balance and other types of environmental impacts such as acidification, photochemical oxidation, human toxicity etc., especially at a much larger scale.

Key industrial chemicals such as ethylene and carbon monoxide can already be made by adding electrons to abundant starting materials such as carbon dioxide and water if efficiency is no object. The trick is to do so economically. The process requires only cheap sources of renewable electricity or non-carbogenic nuclear energy. The electrosynthesis would be competitive for producing chemical staples such as hydrogen, carbon monoxide, ethylene and ethanol if electricity cost four cents per kilowatt hour or less – and if the conversion of electrical energy to energy stored in chemical bonds was at least 60% efficient.

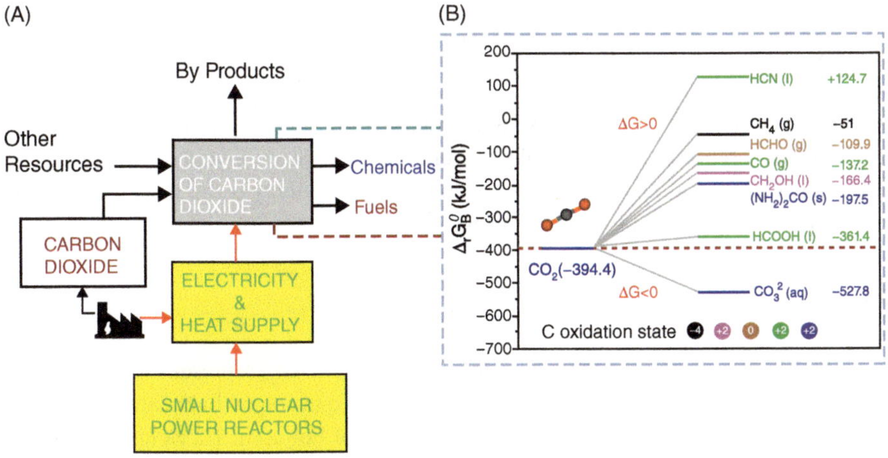

Figure 1.1 (A) The overall scheme of the carbon capture and utilization, showing the resources required and (B) the formation Gibbs free energy of the products. (ΔfG_B^0, B represents products in the reaction; $\Delta G^0 > 0$ indicates non-spontaneous reactions).

If electricity's costs fell further, more compounds would be within reach. In a May 2018 analysis it was found that under stricter market assumptions, including an electricity price of 2 cents / kWh, synthesizing formic acid, ethylene glycol and propanol would all be feasible.

In this regard, Life Cycle Assessment (LCA) is an appropriate tool to evaluate the net environmental impact of a certain product or processes of interest. LCA can consider the different emissions arising from different energy sources, which can provide important complementary information by evaluating the energy options in a broader context. This is important in evaluating the CO_2RR scenarios, as the environmental impacts of multiple energy sources will be considered, including existing electricity grid and other potential renewables. Here I selected renewable hydrogen and biomass derived energy source to represent commonly available alternative energy sources in the region.

1.2 Power-to-Chemical Engineering

1.2.1 Carbon Dioxide Thermodynamics

The two thermodynamic parameters to be considered when analysing the thermodynamics of chemical reaction are enthalpy H and Gibbs energy G. The variations in enthalpy and in Gibbs energy associated with a chemical change will have to be assessed to draw the relevant conclusions for the heat balance and for the equilibrium advancement.

Carbon dioxide and water are located at the bottom of the energetic ladder in thermodynamics (Figure 1.1, Equation 1.1). The reaction of carbon dioxide requires considerable Gibbs energy input:

$$\Delta G^0_{298.15K}\left(CO_2\right) = -394.4\ \text{kJmol}^{-1} \tag{1.1}$$

Thermodynamic consistency was enforced using the DETCHEM^{ADJUST} tool (Deutschmann et al.). It ensures that the chemical equilibrium is represented accurately for any initial composition with the limit of infinite time. To achieve this, all included reactions are required to be microkinetically reversible, i.e., every pair of forward and backward reaction rate coefficients must be linked by an equilibrium constant. With the commonly known relation between equilibrium constant and Gibbs free energy, the rate constants of a pair of forward and backward reactions k_f and k_r must fulfil the equation

$$\frac{k_i(T)}{k_r(T)} = \prod_i \left(c_i^{\ominus}\right)^{\nu_i} \exp\left(-\frac{\Delta_R G}{RT}\right) = F_{c,p} \exp\left(-\frac{\Delta_R G}{RT}\right) \tag{1.2}$$

with $c_i^{\ominus}$ signifying the concentration of species i at standard conditions, ν_i denoting the stoichiometric coefficient of i, $\Delta_R G$ representing the Gibbs free energy of the reaction and $F_{c,p}$ denoting the conversion factor between K_p and K_c. Since the reaction Gibbs free energy is the sum of the Gibbs free energies of the partaking species, it can be expressed as

$$\Delta_R G(T) = \sum_i \nu_i G_i(T) = \sum_i \nu_i \left[H_{0i} + c_{pi}(T - T_0) - TS_{0i} + c_{pi} T \cdot \ln\frac{T}{T_0}\right] \tag{1.3}$$

under the assumption of constant heat capacities. Here, G_i, H_{0i}, S_{0i} and $\bar{c}_{pi}$ represent the Gibbs free enthalpy, the standard enthalpy and entropy and the temperature-averaged heat capacity of species i, respectively. Combining Equation 1.3 with the logarithm of Equation 1.2 yields the following relation:

$$\ln k_f - \ln k_r = \ln F_{c,p} - \sum_m \nu_m \frac{G_i(T)}{RT} - \sum_n \frac{\nu_n}{R}\left[\frac{H_{0n} - c_{pn}T_0}{T} + c_{pn}(1 - \ln T_0) - S_{0n} + c_{pn}\ln T\right] \tag{1.4}$$

Where m denotes species with known thermodynamic properties (i.e., gas-phase species) and n indicates those without (i.e., surface species). Consolidating the unknown thermodynamic functions into one, $y(T)$, all known quantities into $w(T)$ and introducing the adjustments to the rate coefficients $x(T)$ yield a system of equations for the pairs of reversible reactions q:

$$x_{\text{iq}}(T) - x_{\text{rq}}(T) = w_{\text{q}}(T) - \sum_{n} v_{\text{nq}} \frac{G_n(T)}{RT} \cdot y_n(T) \tag{1.5}$$

With

$$w_{\text{q}}(T) = \ln F_{\text{c,p}} - \sum_{m} \nu_m \frac{G_{\text{i}}(T)}{RT} - \ln k_{\text{f}} + \ln k_{\text{r}} \tag{1.6}$$

and $x(T)$ and $y(T)$ in the form

$$x_q(T) = y_n(T) = a + b \ln T + \frac{c}{T} \tag{1.7}$$

The objective is to find thermodynamic functions $x_{\text{q}}(T)$ and $y_{\text{n}}(T)$ that fulfil Equation 1.5 while minimizing the correction terms $x_{\text{m}}(T)$. The unknown thermodynamic properties of surface species can be obtained from $y_{\text{n}}(T)$. The use of DETCHEM^{ADJUST} (Deutschmann et al.) ensures that thermodynamic consistency is reinstated after modifications to the kinetic parameters. The adjustments to the kinetic parameters are minimized to reduce impact on the model performance. The procedure is explained in more detail in a publication by Stotz et al.

Hydrogenation of carbon dioxide makes the reaction possible in thermodynamics and practically significant when hydrogen is supplied from renewable energy. However, the activation energy barriers of reactions involved with carbon dioxide are always very high. Active catalysts are prerequisites for carbon dioxide hydrogenation, which must enable the activation of both hydrogen and carbon dioxide.

Different routes to produce low-carbon hydrogen are available and water is the usual e-technique, but the net reaction can be summarized as water splitting:

$$2H_2O \rightleftarrows 2H_2 + O_2 \qquad \Delta H^0_{298} = 571.8\ \text{kJ/mol} \tag{1.8}$$

The high ΔH^0 value indicates the intense energy requirement of the water splitting, which makes hydrogen production usually the most energy demanding step in the production of chemicals from hydrogen and carbon dioxide.

Carbon dioxide is an abundant, nontoxic, inflammable carbonaceous raw material. It a source in C_1 chemistry for the production of chemicals and already has several important industrial applications. However, carbon dioxide is rather inert and its reactions are energetically highly unfavourable. Despite its reputation as an inert material, many exothermic reactions of carbon dioxide are known (Table 1.1). The Gibbs free energy of reaction (Table 1.1) also shows that favourable reactions typically involve formation of C–O bond and/or water.

Figure 1.2 illustrates Gibbs free energy change of typical carbon dioxide reactions with temperature. It shows that the breakage of carbon-oxygen bond is energetically highly unfavourable. Except for dry reforming, which is favourable only at high temperatures, reactions are typically unfavourable. The free energy change of reaction can be made more favourable using appropriate operating conditions (Figure 1.2). Therefore, developing effective catalysts remains as a useful but difficult endeavour. In bulk chemical technology relatively few favourable reactions exist, but here the volumes involved are very high. In fine chemistry several reactions with a negative Gibbs free energy exist. However, the volumes involved are too modest.

Table 1.1 Enthalpy and Gibbs free energy changes of some exothermic reactions involving carbon dioxide.

Reaction	ΔH^0 (kJ/mol)	ΔG^0 (kJ/mol)
$CO_{2(g)} + H_{2(g)} \rightleftarrows HCOOH_{(l)}$	−37.6	+34.3
$CO_{2(g)} + 3H_{2(g)} \rightleftarrows CH_3OH_{(l)} + H_2O_{(l)}$	−131.0	−10.7
$CO_{2(g)} + 4H_{2(g)} \rightleftarrows CH_{4(g)} + 2H_2O_{(l)}$	−253.1	−132.4
$CO_{2(g)} + CH_{4(g)} \rightleftarrows CH_3COOH_{(l)}$	−15.9	+58.1
$CO_{2(g)} + CH_{4(g)} + 2H_{2(g)} \rightleftarrows (CH_3)_2CO_{(l)} + H_2O_{(l)}$	−65.7	+51.2
$CO_{2(g)} + H_2C = CH_{2(g)} \rightleftarrows H_2C = CHCOOH_{(l)}$	−43.1	+26.2

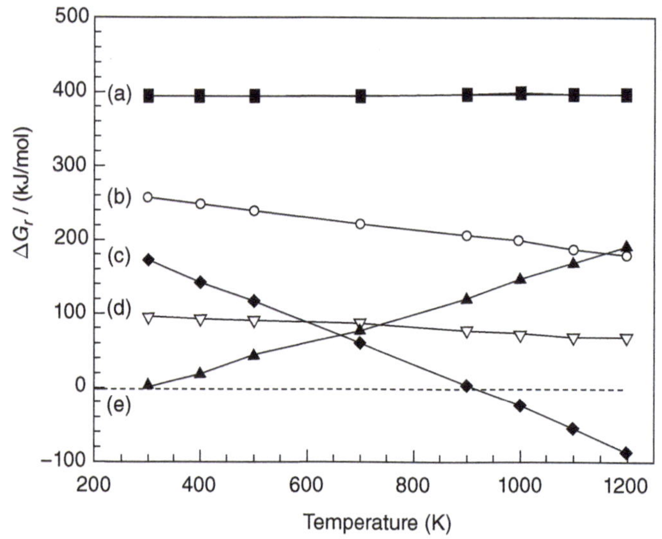

Figure 1.2 Thermodynamics of some carbon dioxide reactions.

a) $CO_2 \rightleftarrows C + O_2$
b) $CO_2 \rightleftarrows CO + ½\, O_2$
c) $CO_2 + 3H_2 \rightleftarrows CH_3OH + H_2O$
d) $CO_2 + 2CH_4 \rightleftarrows C_2H_6 + CO + H_2O$
e) $CO_2 + CH_4 \rightleftarrows 2CO + 2H_2$

An immediate solution for carbon dioxide utilization is its conversion to the more reactive carbon monoxide via a reverse water-gas shift (RWGS) reaction (Equation 1.9)

$$CO_2 + H_2 \rightleftarrows CO + H_2O \tag{1.9}$$

and subsequent use of the synthesis gas technology to create the desired product.

Development of new catalytic processes for chemical fixation of carbon dioxide to produce alcohols, ethers and hydrocarbons supplies new feedstock for the synthesis of petrochemicals and fuels. Completion of the carbon cycle by the synthesis of ethanol, methanol and ethers may contribute to the solution of the two major problems of the twenty-first century, namely global warming and the depletion of oil reserves.

Thermodynamically, carbon dioxide is a highly stable compound that necessitates highly active catalysts and energy for conversion into targeted products. Also, carbonates, carbamates and carboxylates can be formed by catalytically reacting carbon dioxide into organic molecules. However, they do not find large-scale applications for fuels or petrochemicals production.

The enthalpy term (ΔH^0) helps in evaluating thermodynamic stability as well as the feasibility of carbon dioxide transformations into useful chemicals and is represented by the Gibbs–Helmholtz correlation (Equation. 1.3).

ΔG^0 represents the Gibbs free energy between the reactants and the products of a chemical reaction and $T\Delta S^0$ is the entropy term which does not help much in accessing thermodynamic driving force for the chemical reactions involving carbon dioxide. A strong indication of a carbon dioxide molecule's stability can thermodynamically be identified with a high negative standard Gibbs free energy of formation value ($\Delta G^0f = -349.67\,\text{kJ mol}^{-1}$) when compared to other molecules. The structure changes from linear to bent on transferring one electron to the carbon dioxide molecule. In the presence of water, electrochemical reduction of carbon dioxide is completed by hydrogen evolution reaction. Hence, the suppression of hydrogen evolution is important because the applied energy is consumed by hydrogen evolution rather than for carbon dioxide reduction.

The formation of carbon monoxide at pH less than 4.1 from carbon dioxide is carried out at much more positive potential than required for one-electron reduction. Moreover, protonating the reduction product will help in reducing the thermodynamic barrier for carbon dioxide reduction. Thus, understanding the thermodynamics of the carbon dioxide reduction is necessary.

Among the many carbon dioxide conversion technologies, electrochemical carbon dioxide reduction is one of the most promising methods. In recent years, ionic liquids (ILs) have provided opportunities for electroreduction of carbon dioxide due to their unique physical and chemical properties. Ionic liquids are kinds of organic salts composed of an organic cation and an organic anion. In the electrolyte, different compounds can be obtained by electrocatalytic reduction of carbon dioxide at different normal electrode potentials, as shown in Table 1.2.

Carbon dioxide catalytic reduction products are not a single species but a mixture and their selectivity and Faradaic efficiency are related to electrode materials, catalyst, electrolyte solution, system applied potential, temperature, pressure and pH. In addition to these factors, the degree of difficulty in hydrogen evolution reaction (HER) also has a great influence on the conversion efficiency of carbon dioxide. The participation of ionic liquids can effectively inhibit the occurrence of HER and provide an efficient ionic microhabit (IMH) for carbon dioxide dissolution, activation and reduction.

Table 1.2 Thermodynamic equilibrium potential of electrochemical carbon dioxide reduction.

Half-electrochemical reaction	Potential/V (vs. NHE pH = 7)
$CO_2(g) + 8H^+ + 8e^- \rightarrow CH_4(g) + 2H_2O(l)$	−0.24
$2CO_2(g) + 9H^+ + 12e^- \rightarrow C_2H_5OH(l) + 3OH^-$	−0.33
$CO_2(g) + 6H^+ + 6e^- \rightarrow CH_3OH(aq) + H_2O(l)$	−0.38
$CO_2(g) + 4H^+ + 4e^- \rightarrow HCHO(aq) + H_2O(l)$	−0.48
$CO_2(g) + 2H^+ + 2e^- \rightarrow CO(g) + H_2O(l)$	−0.53
$CO_2(g) + 2H^+ + 2e^- \rightarrow HCOOH(aq)$	−0.61
$2H_2O(l) + 2e^- \rightarrow 2OH^- + H_2(g)$	−0.41

In general, reactions involving a low number of electrons transferred are more likely to occur. The electrochemical carbon dioxide reduction (ECR) reaction produces two electron transfer products, for example, carbon monoxide and formic acid, which usually need a low reduction potential. For the multielectron transfer products, for example, methane, methanol, ethanol, ethylene, etc., a higher overpotential is required. One reason is that as the number of electron transfers in a reaction increases, the energy barrier to be overcome gradually increases, so the reactions are less likely to occur. Because the reaction in ECR is a one-way and changing dynamic process, it is difficult to capture its equilibrium state in experiments. To find more suitable reaction conditions for the system to improve the conversion efficiency of carbon dioxide, it is necessary to carry out experimental investigations, one by one, on the factors affecting product selectivity and conversion efficiency. From Table 1.2, it can be seen that the required overpotential for hydrogen is relatively low (E = −0.41 V vs. normal hydrogen electrode (NHE)), which is close to the catalytic reduction potential of carbon dioxide. Therefore, the cathode electrode in an electrocatalytic system is often accompanied by severe HER and forms a competitive reaction with the carbon dioxide reduction reaction. To improve the efficiency of carbon dioxide conversion and energy utilization, the HER should be effectively reduced. Carbon dioxide is a stable molecule: Gibbs energy of formation: $\Delta G^0{}_{298\,K}(CO_2) = -394.4$ kJ mol^{-1}. The action of a suitable catalyst, energy input and a reductant may form methane and methanol. Currently, carbon dioxide is being used in several processes. The concept of producing methanol using carbon dioxide was conceived in the 1920s. The reduction of carbon dioxide to hydrocarbons, alcohols and ethers using renewable energy sources or nuclear energy may deliver a future energy distribution system based on gaseous or liquid fuels without a net upsurge in atmospheric carbon dioxide. The various chemical reactions associated with carbon dioxide are listed in Table 1.3.

Table 1.4 summarizes the reactions that may be part of the overall transformation during the hydrogenation of carbon dioxide towards methanol and methane. The free energies of formation

Table 1.3 Reactions involving carbon dioxide conversion into chemicals/fuels.

Name of the reaction	Reaction	Reaction No.	ΔH^0 at 298.15 K (kJ mol^{-1})
Dry reforming	$CO_{2(g)} + CH_{4(g)} \rightleftarrows 2H_{2(g)} + 2CO_{(g)}$ (Simultaneous consumption of two major greenhouse gases)	1	247.3
Methane decomposition	$CH_{4(g)} \rightleftarrows C_{(s)} + 2H_{2(g)}$	2	74.8
Boudouard equilibrium	$2CO_{(g)} \rightleftarrows C_{(s)} + CO_{2(g)}$	3	−172.5
Reverse water gas shift	$CO_{2(g)} + H_{2(g)} \rightleftarrows CO_{(g)} + H_2O_{(g)}$	4	41.2
CO_2 hydrogenation	$CO_{2(g)} + 4H_{2(g)} \rightleftarrows CH_{4(g)} + 2H_2O_{(g)}$	5	−113.6
Steam reforming	$CH_{4(g)} + H_2O_{(g)} \rightleftarrows CO_{2(g)} + 3H_{2(g)}$	6	206.1
Hydrogenation of CO_2 to methanol	$CO_{2(g)} + 3H_{2(g)} \rightleftarrows CH_3OH_{(g)} + H_2O_{(g)}$	7	−49.5
Direct synthesis of dimethyl ether using CO_2/H_2	$2CO_{2(g)} + H_{2(g)} \rightleftarrows CH_3OCH_{3(g)} + 3H_2O_{(g)}$	8	-
CO_2 hydrogenation to formic acid	$CO_{2(g)} + H_{2(g)} \rightleftarrows HCOOH_{(l)}$	9	32.9
CO_2 hydrogenation to higher alcohols	$2CO_{(g)} + 4H_{2(g)} \rightleftarrows C_2H_5OH_{(g)} + H_2O_{(g)}$	10	−255.5

Table 1.3 (Continued)

Name of the reaction	Reaction	Reaction No.	ΔH^0 at 298.15 K ($kJ\ mol^{-1}$)
CO_2 hydrogenation to higher alcohols	$2CO_{2(g)} + 6H_{2(g)} \rightleftarrows C_2H_5OH_{(g)} + 3H_2O$	11	−173.1
CO_2 reforming of ethanol and higher alcohols	$C_2H_5OH_{(g)} + CO_{2(g)} \rightleftarrows 3CO_{(g)} + 3H_{2(g)}$	12	
Oxidative dehydrogenation in the presence of CO_2	$C_6H_5 - C_2H_{5(g)} \rightleftarrows C_6H_5\text{-}CH{=}CH_{2(g)} + H_{2(g)}$	13	
Oxidative dehydrogenation (ODH) for C_2H_4 production	$C_2H_6 + CO_2 \rightleftarrows C_2H_4 + CO + H_2O$	14	$\Delta H^0_{983} = +135$

estimated at 300, 400, 600 and 800 K indicate driving forces and/or probabilities for the reactions to occur, i.e., the more negative and/or lower the free energy value, the higher the driving force for the corresponding reactions. It has been generally observed that such information is reported usually under standard conditions (298 K). For some reactions, such data indicated favourable driving forces while the estimates at higher temperatures confirmed the opposite trend. The values estimated for temperatures that approach those applied during the hydrogenation of carbon dioxide then are more meaningful.

Trends established based on results in Table 1.4 do not take into consideration the involvement of surface intermediates and/or transition states formed between catalyst surfaces, as well as carbon dioxide and hydrogen reactants. The products that are not favored by the direct hydrogenation of carbon dioxide (e.g., reactions 1, 3, 6 and 8 in Table 1.4) cannot be excluded as intermediates during the transformation from carbon dioxide to methanol and/or methane. Such intermediates can arise from some transition states comprising carbon dioxide and hydrogen adsorbed on active sites. Once formed, their subsequent conversion to methanol via reactions 2, 4 and 5 in Table 1.4 is favourable. Thus, despite a very low probability for the direct formation of formaldehyde and formic acid during the hydrogenation of carbon dioxide, i.e., reactions 1 and 3 in Table 1.4, respectively, these compounds are intermediates during the overall carbon dioxide conversion.

In experimental studies, the involvement of formaldehyde during the hydrogenation of carbon dioxide to methanol was reported in the isotope-label study. In this case, the isotope labelled ^{13}CO was used in the mixture with carbon dioxide and hydrogen over CuO-ZnO based catalysts, in comparison to Re/ZrO_2 and Re/CeC_2 catalysts. Methanol was exclusively formed during the hydrogenation of carbon dioxide over CuO-ZnO catalysts. In this case, the surface format and adsorbed formaldehyde were intermediates during the overall hydrogenation to methanol. In fact, the hydrogenation of format to formaldehyde was the rate-determining step. In the same study conducted over Re/ZrO_2 and Re/CeC_2 catalysts, the transformation of carbon dioxide involved two reaction pathways, i.e., one via the direct hydrogenation of carbon dioxide and the other a partial hydrogenation of carbon dioxide to carbon monoxide, i.e., the RWGS (Reaction 7 in Table 1.4), followed by the formation of methanol. Similar confirmation of the carbon dioxide to methanol pathway was made in the study of during the kinetic measurements involving the isotope labelled $^{14}CO_2$ in the mixture with carbon monoxide over the $Cu\text{-}ZnO/Al_2O_3$ catalyst.

These observations were consistent with the results where it was observed that the rate of Fischer-Tropsch Synthesis reaction over Co/CNF catalyst decreased upon the addition of carbon dioxide to synthesis gas. The transformation of carbon dioxide via surface format intermediate to formaldehyde and methoxy species ending with methanol were considered in several other studies.

Table 1.4 Effect of temperature on the free energy of formation for reaction occurring during hydrogenation of carbon dioxide.

Reaction	Reaction no.	Free Energy (kJ/mol)			
		@300 K	@400 K	@600 K	@800 K
$CO_2 + 2H_2 \rightleftarrows CH_2O + H_2O$	1	56	62	77	94
$CH_2O + H_2 \rightleftarrows CH_3OH$	2	−56	−40	−32	−9
$CO_2 + H_2 \rightleftarrows HCOOH$	3	36	55	83	114
$HCOOH + H_2 \rightleftarrows CH2O + H2O$	4	20	8	−7	−20
$HCOOH + 2H_2 \rightleftarrows CH_3OH + H_2O$	5	−36	−32	−23	−11
$CO_2 + 3H_2 \rightleftarrows CH_3OH + H_2O$	6	0	22	62	104
$CO_2 + H_2 \rightleftarrows CO + H_2O$	7	29	24	16	8
$CO + H_2 \rightleftarrows CH_2O$	8	27	39	62	87
$CO + 2H_2 \rightleftarrows CH_3OH$	9	−28	−2	30	98
$CO_2 + 4H_2 \rightleftarrows CH_4 + 2H_2O$	10	−119	−96	−58	−16
$CO + 3H_2 \rightleftarrows CH_4 + H_2O$	11	−143	−119	−73	−23
$HCOOH + 3H_2 \rightleftarrows CH_4 + 2H_2O$	12		−151	−141	−129
$CH_2O + 2H_2 \rightleftarrows CH_4 + H_2O$	13	−169	−158	−136	−111
$CH_3OH + H_2 \rightleftarrows CH_4 + H_2O$	14	−114	−116	−120	−120
$CH_4 + 0.5O_2 \longrightarrow CH_3OH$	15	−115	−106	−94	−83

Reaction 10 in Table 1.4, involving a direct hydrogenation of carbon dioxide to methane, is thermodynamically favourable. In addition, all potential intermediates between carbon dioxide and methane, i.e., carbon monoxide, formic acid, formaldehyde and methanol can be readily converted to methane (reactions 11–14 in Table 1.4, respectively) providing that sufficient active hydrogen is available at an active site. However, it is believed that, under typical methanation conditions, methanol used as an intermediate to methane may play only a limited role, because all preceding intermediates (if present) are rapidly converted to methane (reactions 11–13 in Table 1.4) before any methanol can be formed. Therefore, reaction 10 in Table 1.4 will dominate the conversion of carbon dioxide to methane, while the contribution from the intermediates may be limited. In fact, such conditions are unfavourable for RWGS (reaction 7 in Table 1.4). With respect to temperature effects, the direct methanation becomes more favourable with decreasing temperature. Yet, all experimental evidence suggests that carbon dioxide activation increases as the temperature increases. This dilemma may be reconciled by selecting a suitable catalyst. In this regard, the ability of a catalyst to activate hydrogen and supply active surface hydrogen to a carbon dioxide molecule in an activated state is critical.

For the hydrogenation of carbon dioxide to methane, the above rationale assumes that at least a stoichiometric amount of hydrogen is available for hydrogenation reactions. Thus, for the complete hydrogenation of carbon dioxide to methane, a H_2/CO_2 ratio of at least 4.0 is necessary. With a ratio of <4.0, the direct hydrogenation of carbon dioxide to methane would be most affected. In this regard, it is essential to distinguish the local H_2/CO_2 ratio on the active site from that of the corresponding hydrogen plus carbon dioxide mixture entering the catalyst bed. The former plays a decisive role during carbon dioxide conversion. Active metals, which can facilitate activated adsorption of both carbon dioxide and hydrogen required for the local H_2/CO_2 ratio of 4.0, are suitable for methanation. The equilibrium aspects of the hydrogen of carbon dioxide to methanol were investigated in detail using the Gibbs free-energy minimization method. Despite decades of experimental research, this topic only attracted wider attention in 2018 when the effect of the H_2/CO_2 ratio on carbon dioxide conversion and methanol selectivity was shown. The conversion increased with the ratio increase. Similarly, the methanol selectivity increased up to a stoichiometric H_2/CO_2 ratio of 3.0. The results were estimated for a total pressure of 5 MPa. According to Le Chatelier's principle, increased total pressure is favourable for methanol formation, because the ratio of the number of reacting moles to the number of product moles was 2.0. This was indeed confirmed by increased conversion and selectivity estimated for 10 MPa in a similar manner as those for 5 MPa. The role of water during the hydrogenation of carbon dioxide to methanol is an important finding in the study. It was evident that, in the presence of water, lower conversions and methanol yields during the hydrogenation of carbon dioxide were observed.

The direct hydrogenation of carbon dioxide to methanol, as well as formaldehyde and formic acid, is not favourable. This is confirmed by the free-energy values for reactions 6, 1 and 3, respectively, in Table 1.4. Then, formic acid- and formaldehyde-like intermediates may be involved during the formation of methanol if a suitable catalyst ensuring the activation of both carbon dioxide and hydrogen at active site is present. Thus, when the targeted product is methanol, the local H_2/CO_2 ratio of ~3.0 is necessary. In this regard, the type of catalyst may play a decisive role.

It is well known that methanol and dimethyl ether (DME) can be catalytically synthesized from carbon dioxide and hydrogen over Cu-ZnO based catalyst, but these reactions exhibit low carbon dioxide conversions because of the thermodynamical limitation. Large hydrocarbon molecules such as gasoline fractions can be selectively produced from methanol and DME over zeolite catalysts with a relatively high conversion. In addition, the equilibrium calculation also confirms that production of large hydrocarbon molecules is favoured to greatly improve carbon dioxide conversion compared to that of methanol and DME. For example, in Figure 1.3, the carbon dioxide conversion for methanol (Equation 1.10), DME (Equation 1.11) and *n*-hexane (Equation 1.12) at 553 K under 3.0 MPa are 13.2%, 20.2% and 90.7%, respectively. Therefore, it is a promising choice to use a hybrid catalyst for directly producing C_5+ hydrocarbons from carbon dioxide, which is composed of methanol synthesis and zeolite catalysts. Employment of a hybrid catalyst having complementary properties can not only directly produce C_5+ hydrocarbons from carbon dioxide via methanol synthesis, but also realize a higher conversion efficiency of carbon dioxide.

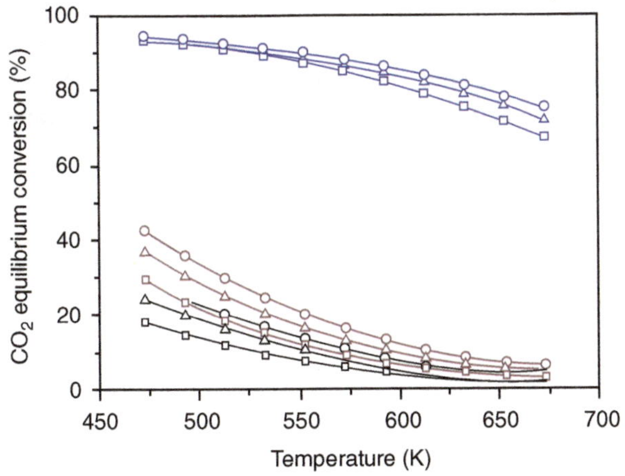

Figure 1.3 Equilibrium conversion of carbon dioxide.

Methanol synthesis, $CO_2 + 3H_2 \rightleftarrows CH_3OH + H_2O$ (black),
DME synthesis, $2CO_2 + 6H_2 \rightleftarrows CH_3OCH_3 + 3H_2O$ (red),
C_6H_{14} synthesis, $6CO_2 + 19H_2 \rightleftarrows C_6H_{14} + 12H_2O$ (blue).

Pressure of 2.0 MPa (square);
3.0 MPa (triangle);
4.0 MPa (circle).

$$CO_{2+}3H_2 \rightleftarrows CH_3OH + H_2O \ \Delta H^0_{298} = -49.8\,kJ/mol \tag{1.10}$$

$$2CO_2 + 6H_2 \rightleftarrows CH_3OCH_3 + 3H_2O \quad \Delta H^0_{298} = -122.5\,kJ/mol \tag{1.11}$$

$$6CO_2 + 19H_2 \rightleftarrows C_6H_{14} + 12H_2O \ \ \Delta H^0_{298} = -738.9\,kJ/mol \tag{1.12}$$

1.2.2 Carbon Dioxide Aromatization Thermodynamics

Due to the nature of the carbon dioxide aromatization reaction(s) that convert a C_1 chemical to C_6-C_{10+} chemical mixture, several parallel and series reactions and products are involved in its mechanism. Moreover, the multi-functionality of the catalysts provides the opportunity for multiple side-reactions in the system. Therefore, a thermodynamic analysis of the whole system as a single process provides less insight than the analysis of the main reactions separately. The following reactions (Reactions 1–8 in Table 1.5) have been considered for the thermodynamic analysis of the system. Ethylene, ethane and toluene are considered as representative products for olefins, paraffins and aromatics, respectively.

A closer look at the Gibbs free energy changes of the above reactions provides some insight towards the limits for the carbon dioxide aromatization reaction via the two reaction pathways. The coupling of the two catalyst functionalities (hydrogenation via metal oxides or Fe-based catalysts and aromatization via H-ZSM-5) enhances the carbon dioxide conversion due to the thermodynamic driving force imposed by the conversion of the intermediate species to more thermodynamically stable oligomers and aromatic products. As shown in Figure 1.4 (A), as temperature increases, unlike the methanol aromatization reaction (reaction (2) in Table 1.5), the overall exothermic methanol synthesis reaction (reaction 1) becomes more thermodynamically unfavourable. Similarly, the overall change in Gibbs free energy with temperature increases for the exothermic ethylene synthesis reaction (reaction 3); however, the ΔG for this reaction is relatively lower than that of methanol synthesis reaction and closer to the ΔG of the ethylene aromatization reaction (reaction 4). This indicates that if carbon dioxide conversion to the intermediate compound is thermodynamically limited and its aromatization reaction is kinetically limited, but still fast, then thermodynamics will favour the olefin pathway, since the olefin production is less endothermic and fast kinetics of olefin aromatization drive carbon dioxide conversion to the right due to Le Chatelier's principle. Also, based on the Gibbs free energies obtained, the formation of methane, ethylene, ethane and toluene are the only reactions occurring spontaneously at relevant

Table 1.5 Aromatization of carbon dioxide (all species in gas phase).

Reaction	Reaction No.
$CO_2 + 3H_2 \rightleftarrows CH_3OH + H_2O$	1
$CH_3OH \rightleftarrows {}^1/_7\, C_7H_8 + H_2O + {}^3/_7\, H_2$	2
$CO_2 + 3H_2 \rightleftarrows {}^1/_2\, C_2H_4 + 2\, H_2O$	3
${}^1/_2\, C_2H_4 \rightleftarrows {}^1/_7\, C_7H_8 + {}^3/_7\, H_2O$	4
$CO_2 + {}^{18}/_7\, H_2 \rightleftarrows {}^1/_7\, C_7H_8 + 2H_2O$	5
$CO_2 + H_2 \rightleftarrows CO + H_2O$	6
$C_2H_4 + H_2 \rightleftarrows C_2H_6$	7
$CO_2 + 4H_2 \rightleftarrows CH_4 + 2H_2O$	8

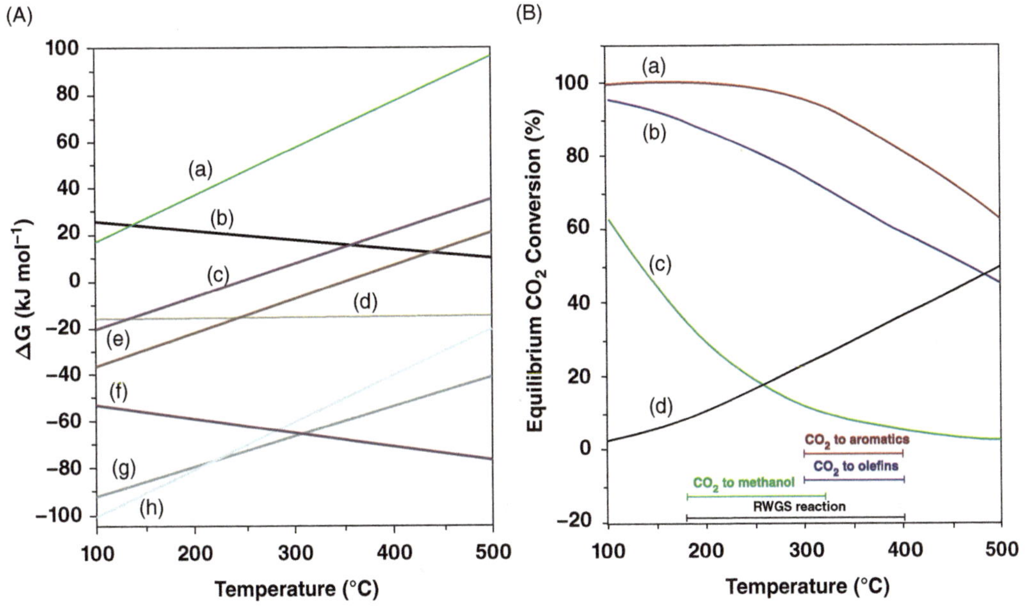

Figure 1.4 A) Gibbs free energy changes.

a) $CO_2 + 3H_2 \rightleftarrows CH_3OH + H_2O$
b) $CO_2 + H_2 \rightleftarrows CO + H_2O$
c) $CO_2 + 3H_2 \rightleftarrows 1/2C_2H_4 + 2H_2O$
d) $1/2C_2H_4 \rightleftarrows 1/7C_7H_8 + 3/7H_2$
e) $CO_2\ 18/7H_2 \rightleftarrows 1/7C_7H_8 + 2H_2O$
f) $CH_3OH \rightleftarrows 1/7C_7H_8 + H_2O + 3/7H_2$
g) $C_2H_4 + H_2 \rightleftarrows C_2H_6$
h) $CO_2 + 4H_2 \rightleftarrows CH_4 + 2H_2O$

B) Equilibrium carbon dioxide conversion at different temperatures for:

a) Methanol synthesis from carbon dioxide hydrogenation,
$CO_2 + 3H_2 \rightleftarrows CH_3OH + H_2O$
b) Carbon dioxide to lower olefins (ethylene),
$CO_2 + 3H_2 \rightleftarrows 1/2C_2H_4 + 2H_2O$
c) Carbon dioxide to aromatics (toluene),
$CO_2\ 18/7H_2 \rightleftarrows 1/7C_7H_8 + 2H_2O$
d) The reverse water-gas shift reaction,
$CO_2 + H_2 \rightleftarrows CO + H_2O$

reaction temperatures, a decreasing trend in relative energies, leading to more spontaneous reactions, is expected as the number of carbon atoms in the alkanes, alkenes and aromatics increases. Methane formation has a considerably lower ΔG value compared to all the other reactions involving carbon dioxide. However, the formation of methane in aromatization studies is limited due to kinetic barriers and the absence of stoichiometric values of CO_2/H_2 for this reaction, i.e., 1/4, as opposed to a typical ratio of 1/3 used in carbon dioxide aromatization studies.

Simulated conditions for equilibrium carbon dioxide conversion: $H_2/CO_2 = 3$, total pressure 4 MPa. The calculations were based on HSC5 chemistry software. Coloured range lines indicate the relevant temperature ranges for each reaction.

Among the main reactions that involve carbon dioxide as the reactant at 4 MPa and H_2/CO_2 ratio of 3.0, the RWGS is the only reaction that becomes more favourable upon a temperature increase;

other reactions have a decreasing equilibrium carbon dioxide conversion as the reaction temperature increases (Figure 1.4(B)). The carbon dioxide equilibrium conversion at 300–360°C is more than 90% for toluene formation. The fact that this value is far from experimentally observed carbon dioxide conversion values indicates that the kinetics is the limiting factor for this reaction at the studied conditions. To better understand this effect, the thermodynamic equilibrium carbon dioxide conversion and product selectivity's were calculated considering a mix of chemicals presented in reactions (a)–(h) (Figure 1.4) at different temperature ranges (250–400°C), 4 MPa pressure and H_2/CO_2 ratio of 3.0. Methane was the most thermodynamically dominant product due to its high stability, with its carbon selectivity being more than 99% in all cases. To further analyse the thermodynamic behaviour of the reaction, a similar analysis was conducted in the absence of methane. This change resulted in having ethane as the thermodynamically favoured chemical with >99% carbon selectivity at all temperatures. Removing the ethane from the system resulted in having >98% toluene as the thermodynamically dominant chemical under similar conditions. In all cases studied, the carbon dioxide conversion stayed above 70%. The comparison of all these results with the experimental catalytic data below indicates that the reaction is far from thermodynamic equilibrium under the typically studied conditions. Moreover, the results indicate that selective aromatic formation requires kinetic limitations to avoid formation of light alkanes.

1.2.3 Reaction Mechanism of Carbon Dioxide Methanation

This section covers the development of carbon dioxide and carbon monoxide over nickel-based catalysts using the mean-field approximation. The microkinetic model is based on theoretical investigations as well as several experimental studies, conducted both in-house and from the literature (Schmider et al.). This dataset encompasses experiments in both fixed-bed and monolithic reactors under various conditions (Section 1.2.11.3).

The catalytic methanation of carbon dioxide and/or carbon monoxide is a key step in Power-to-Gas technologies. The production of synthetic natural gas from hydrogen coming from electrolysers is of special interest for the storage of renewable electrical energy in the form of hydrocarbons, especially as many regions already possess an extensive natural gas grid. Since water is the only significant side product of methanation, the product stream is rather easily introduced into the natural gas grid after dehydration. Carbon dioxide and combined carbon monoxide and carbon dioxide methanation also is of interest for carbon dioxide point sources such as a typical steel plant, where large portions of the exhaust gases consist of carbon monoxide and carbon dioxide.

The catalytic methanation of carbon dioxide and carbon monoxide has been studied since its discovery by Sabatier and Senderens in 1902. The primary application of this reaction has been the purification of synthesis gases via the removal of carbon monoxide. The commercial conversion of carbon monoxide to methane is primarily realized over a nickel catalyst.

$$CO + 3H_2 \rightleftarrows CH_4 + H_2O, \quad \Delta_R H = -206.2\ \text{kJ.mol}^{-1} \tag{1.13}$$

While carbon dioxide methanation research was largely a by-product of work on carbon monoxide methanation, it has attracted more attention recently. Generally, nickel is also primarily used as the catalyst, with ruthenium also drawing some interest.

$$CO_2 + 4H_2 \rightleftarrows CH_4 + 2H_2O, \quad \Delta_R H = -165.0\ \text{kJ.mol}^{-1} \tag{1.14}$$

Both reactions (Equations 1.13 and 1.14) are highly exothermic; thus, high temperatures are unfavourable to the conversion of the carbon oxides. Additionally, high pressures are very conducive to high methane yields. Due to the species involved in these reactions, the water-gas shift (WGS) reaction (Equation 1.15) needs to be taken into account when dealing with methanation systems.

$$CO + H_2O \rightleftarrows CO_2 + H_{2,} \qquad \Delta_R H = -41.2\ \mathrm{kJ.mol^{-1}} \tag{1.15}$$

The reverse water-gas shift (RWGS) reaction utilizes the same reactants as carbon dioxide methanation; therefore, in experiments of this kind, there will likely be carbon monoxide evolution alongside methane production. At the same time, interconversion between carbon monoxide and carbon dioxide as well as reforming reactions of the produced methane may also take place. A typical issue for carbon-containing reaction processes over nickel catalysts is coke formation on the active component. These many chemical interactions call for a microkinetic model, i.e., a detailed, multistep surface reaction mechanism with associated kinetic and thermodynamic data, rather than a microkinetic description. The microkinetic model should include all relevant species and possible reaction routes from/to carbon monoxide and carbon dioxide to/from methane as well as WGS and RWGS and be tested for a wide range of conditions.

The elementary steps of carbon monoxide and carbon dioxide methanation over nickel catalysts have been extensively studied over time. It is generally accepted that the activation of carbon monoxide is achieved via associative adsorption. This is supported by experimental findings supporting the argument that associative carbon monoxide adsorption competes with dissociative hydrogen adsorption on the surface. Subsequently, carbon monoxide(s) (s denotes surface species) dissociates to form a surface carbide species, the hydrogenation of which is thought to be the rate-limiting step (RLS), possibly alongside the dissociation itself. Evidence for this conclusion comes from using dynamic response studies. Carbon formation on the nickel surface may also be the result of the Boudouard reaction, the disproportionation of carbon monoxide(s) to carbon dioxide(s) and surface carbide. While it was originally believed that carbon monoxide methanation on nickel proceeds via oxygenated intermediates such as methanol or formaldehyde, surface studies have not confirmed the presence of such species.

The mechanism of carbon dioxide methanation is a topic of current discussion and its exact route is not generally agreed upon, with experiments at different conditions leading to varying suggestions about the exact pathway. The adsorbed carbon dioxide could react in one or possibly two ways: it might dissociate and form carbon monoxide(s), from where it follows the carbon monoxide methanation mechanism via a surface carbide species (RWGS path). This mechanism was suggested following carbon dioxide methanation investigations and after carbon dioxide pulse adsorption studies. Alternatively, carbon dioxide(s) might react with hydrogen directly and form oxygenated species such as carboxyl, COOH(s), or format, HCOO(s), which then dissociate and form CO(s) or are further hydrogenated towards methane (direct hydrogenation path). In situ diffuse reflectance spectroscopy studies of carbon dioxide methanation have shown the formation of format and carbonate species above 383 K. Density functional theory (DFT) calculations have shown that the direct dissociation into carbon monoxide(s) and oxygen(s) is favourable energetically compared to the formation of format (bonded to the surface via one or two oxygen atoms). Other density functional theory results indicate that the formation of carboxyl (bonded to the surface at the carbon atom) is more favourable than dissociation into carbon monoxide(s) and oxygen(s). On the basis of density functional theory calculations and reaction flow analysis, the carboxyl intermediate COOH(s) was also determined as the most abundant species in the WGS reaction.

There is also evidence for the variation of the reaction path based on the support material of nickel catalysts. This might be a consequence of the catalyst structure, i.e., particle size and exposed crystal faces, adsorption and desorption characteristics of the support material and a difference in the dominant reaction pathways taken. While on more inert supports, the direct dissociation of carbon dioxide and the participation of format are expected, the adsorption of carbon dioxide and the formation of carbonate species on the support itself have been proposed for more basic materials.

The proposed rate-determining steps in the carbon dioxide methanation mechanism are the dissociation of carbon monoxide(s) and the following hydrogenation of CH_x(s), depending on reaction conditions. However, it is disputed which step is rate-limiting under which conditions. For example, carbon monoxide(s) dissociation has been suggested as the rate-limiting step (RLS) from 270 to 400°C, while it is ruled out elsewhere, at least for temperatures below 284°C.

Alternative pathways have also been proposed: a dioxymethylene ($C(OH)_2$(s)) species was originally proposed by Medsforth in 1923 but does not represent a likely intermediate from an energetic standpoint. A possible pathway to format in an Eley–Rideal-type mechanism was presented in another publication based on density functional theory; it is however ruled out as an intermediate on the way to methane as the authors assume its hydrogenation to unstable formic acid HCOOH(s), which decomposes back into format, forming a dead end to the reaction path. Other studies have argued in favour of an Eley–Rideal mechanism based on gravimetric analysis of a nickel catalyst. Deduced from energetic calculations, the presence of additional hydrogenated intermediates such as H_2COH(s) is also considered.

So far, the carbon monoxide and carbon dioxide methanation reactions over nickel have not been described using an elementary, thermodynamically consistent mechanism that includes both reactants. However, there is a multitude of global kinetics available, some of which include both carbon monoxide and carbon dioxide conversion terms. Such models are in general constrained to the systems from which they were developed and their application outside these conditions is risky. Additionally, a significant number of these models neglect kinetic reversibility and are therefore not suitable to describe the equilibrium composition adequately.

The microkinetic model (Schmider et al.) is based on a model formerly developed for methane oxidation and steam reforming over nickel and later extended to also include carbon dioxide reforming reactions. This model was not suited to emulate both carbon monoxide and carbon dioxide methanation reactions as well as co-methanation systems. The goal of their work was to adapt the mechanism to enable the description of all methanation reactions (Schmider et al.). The source kinetic parameters of the reversible elementary steps were derived from transition state theory and semiempirical UBI-QEP calculations for a nickel (111) surface in the limit of zero coverage. Considering the importance of carbide formation in oxygen-free methanation conditions, the selected steps with surface carbon species were corrected for carbon(s)-coverage-dependent activation energies.

A potential energy diagram for a possible path of carbon monoxide methanation on a nickel (111) surface based on heat of formation of surface intermediates and activation energies is shown in Figure 1.5. For the first carbon hydrogenation step, calculations for surfaces coverage $\theta_C = 0$ and $\theta_C = 1$ were performed and display a significant disparity in the activation barrier.

In the model development procedure, the preexponential factors of reactions were altered to improve the predictive quality of the model in comparison to experimental data. The changes in activation energy and the temperature parameter β (Equation 1.13) are rooted in the enforcement of thermodynamic consistency by the DETCHEM^{ADJUST} tool (Deutschmann et al., Gossler et al.).

The surface kinetics developed in the work of Schmider et al. feature 42 elementary (forward and backward) reactions including five gas-phase and 14 surface species. All reactions are reversible. Thermodynamic consistency is ensured from 300 to 2000 K by linking the reaction rate parameters of forward and backward reactions with generated equilibrium constants and thermodynamic functions.

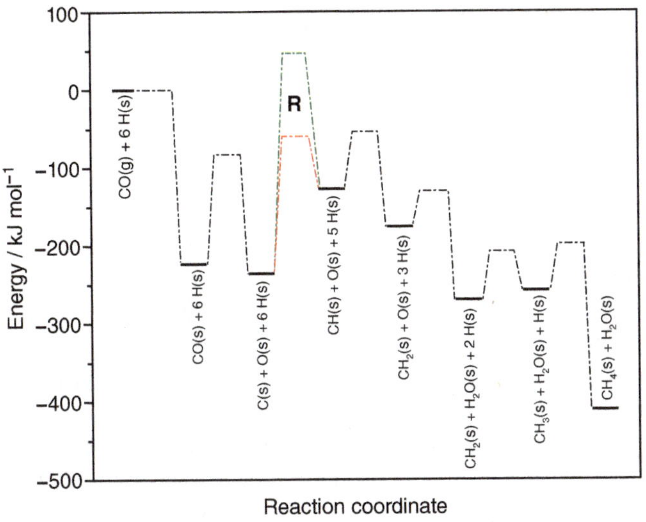

Figure 1.5 Potential energy diagram of one possible methane formation path from carbon monoxide. Potential energy levels of the named species, solid lines; transition states, dashed lines. For reaction R, the transition state with an assumed: $\theta_C = 0$ in green, $\theta_C = 1$ in red.

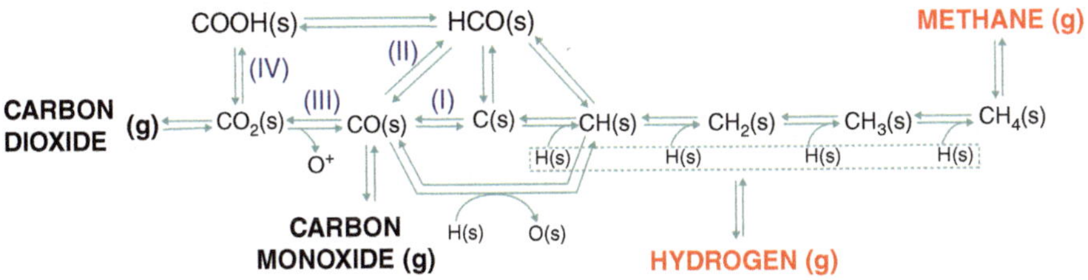

Figure 1.6 Reaction scheme of the kinetic model developed by Schmider et al. Some reactions are omitted for clarity. Featured pathways: (I) carbide pathway; (II) H-assisted carbon monoxide dissociation; (III) direct carbon dioxide dissociation; (IV) H-assisted carbon dioxide dissociation. Schmider et al. 2021/American Chemical Society.

A scheme of the kinetic model is shown in Figure 1.6. It includes several pathways of methane formation from both carbon monoxide and carbon dioxide. Carbon monoxide activation is represented by a direct dissociation of adsorbed carbon monoxide(s) to a surface carbide species (I) and a hydrogen-assisted dissociation, both in a single reaction step and via a formyl intermediate (II). The conversion of carbon dioxide also features multiple pathways. The direct dissociation of carbon dioxide(s) to carbon monoxide(s) (III) is included in addition to the formation of a format/carboxyl species COOH(s) (IV), which can itself further react to carbon monoxide(s) or HCO(s). The formation of methane is included because of stepwise addition of adsorbed hydrogen to CH_x(s) ($0 \le x \le 3$). Water formation proceeds via a hydroxyl (OH(s)) intermediate. It is important to note that in this model, carbon(s) is an active intermediate species and does not block the nickel surface through coke formation.

The mechanism was developed by comparing its performance in the simulations to the experimental data and adjusting the kinetic parameters manually to improve the fit. This process was aided by reaction flow analysis and a process determining the effect of particular parameters on the predicted conversions. All major adjustments of the kinetic model were performed manually, there was no algorithmic optimization procedure. Minor changes to the model were caused by the method used to enforce thermodynamic consistency, which was described. The performance of the mechanism is analysed by comparing its predicted gas composition over a range of temperatures against experimental data from both literature and in-house measurements in either species axial profiles or conversion data by end-of-pipe measurements, if the former was not made available.

To judge the performance of the kinetic model in the thermodynamic equilibrium, the composition of a stoichiometric feed at equilibrium as a function of temperature was evaluated using DETCHEM^{EQUIL} (Deutschmann et al). The results are displayed Figure 1.7, reflecting the fact that methane formation, both from carbon monoxide and carbon dioxide, is thermodynamically suppressed by high temperatures and promoted by high pressures. At atmospheric pressure, the methane mole fraction in the equilibrium is very low above 800 K, while a significant amount is still present at even 1000 K for a pressure of 2 MPa. Additionally, the concentration of carbon is highest at this temperature (800 K), which is noted for the understanding of the effect of carbon deposition on reaction kinetics. Most of the collected experiments reach equilibrium at the upper end of their temperature range; therefore, a thermodynamically consistent kinetic model is required to accurately describe the performance at all temperature ranges.

1.2.4 Water Electrolysis Thermodynamics

Water electrolysis involves the decomposition of water into hydrogen and oxygen (Equation 1.8). The reaction has specific energy requirements based on the temperature and pressure of the water. The Gibbs free energy change of water formation, ΔG, enthalpy change of water formation ΔH and entropy change ΔS (multiplied by temperature T) determine the minimum theoretical required voltage to decompose water as a function of temperature and pressure:

$$\Delta G = \Delta H - T\Delta S \tag{1.16}$$

Note that the reactions taking place at the anode and cathode are dependent on the electrolysis type (materials based), but the overall reaction leads to the same overall voltage requirement. Also

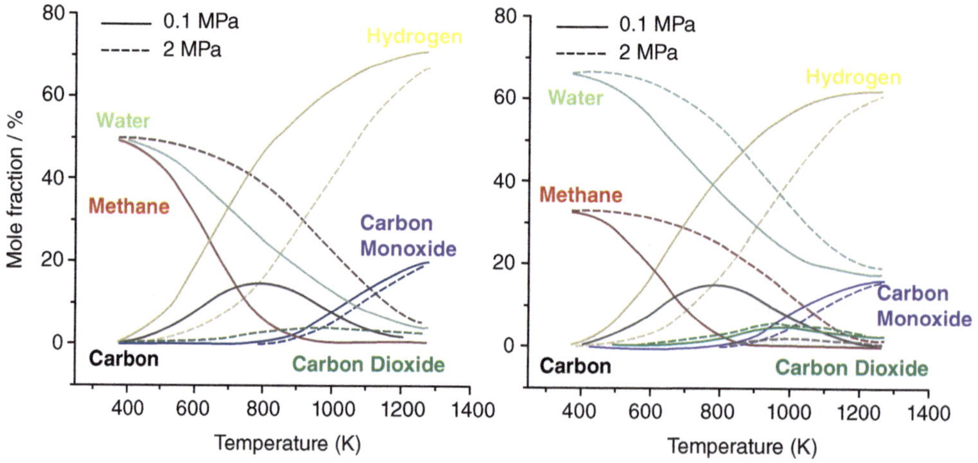

Figure 1.7 Equilibrium positions of stoichiometric feeds for CO (left, $H_2/CO = 3{:}1$) and CO_2 (right, $H_2/CO_2 = 4{:}1$) as a function of temperature for pressures of 0.1 (solid line) and 2 (dashed line) MPa.

note that, for simplicity of the electrolysis model, pressures are assumed to be 0.1 MPa. It is known that efficiencies decrease slightly (due to increased voltage requirements) for higher pressure systems but can lead to power savings because the hydrogen is produced at a higher pressure. However, this analysis does not take into account energy required to compress and store hydrogen, only to produce it through various methods.

The Gibbs free energy determines the equilibrium cell voltage $E_{Equilibrium}$, or the absolute minimum voltage required for electrolysis to occur, as shown in Equation 1.17:

$$\Delta G = nFE_{\text{Equilibrium}} \tag{1.17}$$

For water, n, the number of moles of electrons transferred in the reaction is two and F is the Faraday constant, at 96485 C/mol. At 25°C, 0.1 MPa (standard temperature and pressure), $\Delta G = 237.1$ kJ/mol and $E_{Equilibrium} = 1.23$ V. The equilibrium voltage describes the minimum electrical energy demand for electrolysis. Figure 1.8 shows how temperature affects Gibbs free energy, enthalpy and entropy. Note that there is a drop in energy demand once water becomes a vapour. Also note that, at the equilibrium voltage, the reaction is endothermic and will not proceed without an additional heat source.

Another important voltage to note is the thermoneutral voltage, $E_{Thermoneutral}$. The thermoneutral voltage is the voltage required for no heat to be added (it is the transition point between endothermic and exothermic reactions). It remains relatively constant with temperature, as it depends mostly on the heat of formation of water (285.8 kJ/mol for liquid and 241.8 kJ/mol for gas at atmospheric pressure). It is related to the enthalpy change in Equation 1.18:

$$\Delta \text{H} = \text{nFE}_{\text{Thermoneutral}} \tag{1.18}$$

The thermoneutral and equilibrium voltages are good for describing theoretical minimum energy requirements, but they are lower than the expected required voltage, E_{cell}. Real electrolysis cells

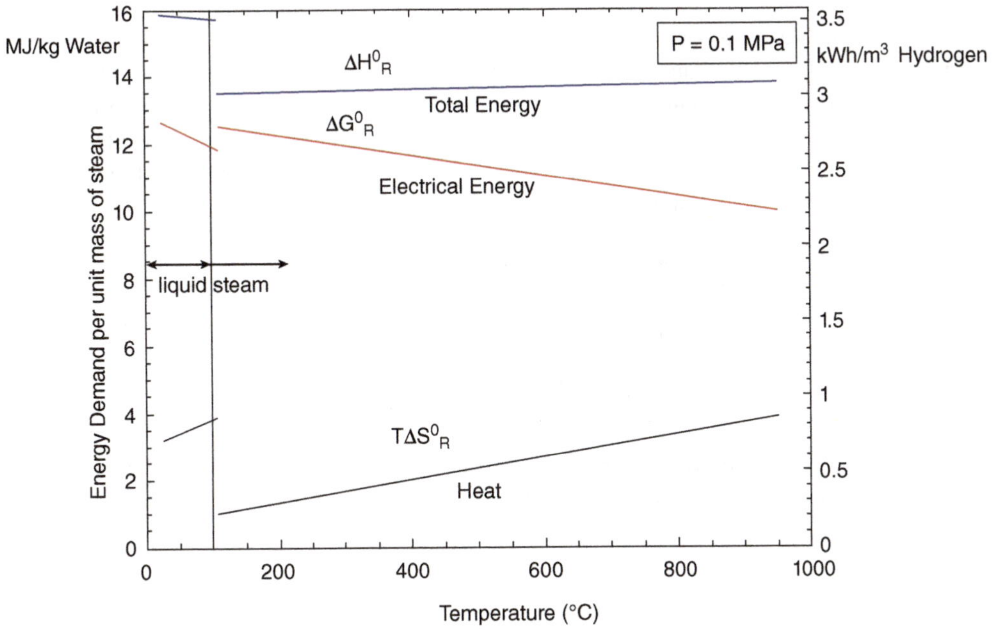

Figure 1.8 Energy demand for water electrolysis at various temperatures and 0.1 MPa.

have problems such as activation overvoltage or overpotentials on the cells and resistances throughout that must be accounted for. These loss terms may increase energy consumption by around 1.5 to 2.2 times more than theoretical. The cell voltage required is described in Equation 1.19.

$$E_{cell} = E_{Equilibrium} + \eta_A + \eta_K + iR_{cell} \tag{1.19}$$

Where η_A is the anode overvoltage, η_K is the cathode overvoltage and *i* is the current density. R_{cell} of all areas specific resistances in the cell: electric, anode, cathode, bubbles (resistance due to partial coverage of electrodes by bubbles) and membranes. It is estimated to be somewhere in the range 0.01–2 cm^2 for simplicity, but the resistance is dependent on materials (electrolysis type) and temperature. The term R_{cell} (V) consists of the ohmic losses in the cell. Resistance can be minimized by minimizing the interelectrode gap and optimizing the current density based on electrode surface morphology and electrolyte concentrations. There may also be a diffusion overvoltage that exists in the case of mass transport limitations. This typically occurs at high current densities. However, in this model, it is assumed that there is a continuous supply of water and the current density stays low enough that mass transport limitations are not reached.

1.2.5 Methane Pyrolysis Reaction Thermodynamic Consideration

1.2.5.1 The Carbon-Hydrogen System

The standard free energy of hydrocarbon formation from their elements [C (graphite) and H_2 (gas)] is a measure of their thermodynamic stabilities. But equilibria of the type:

$$nC + m/2H_2 \rightleftarrows C_nH_m \qquad 0 < _{m/n} \leq 4 \tag{1.20}$$

are not experimentally feasible, with the notable exception of:

$$C + 2H_2 \rightleftarrows CH_4 \quad \Delta H^0_{298^\circ K} = -75.7 \text{ kJ}, \tag{1.21}$$

This equilibrium was approach from both sides, in the presence of metallic catalysts (iron, cobalt, nickel) (see Figure 1.9); above inversion temperature [t_i = 570°C ($K_p = 1$, $\Delta G^0 = 0$)] methane should spontaneously decompose. Indirect thermochemical calculations show that methane is the most stable compound of the paraffins series C_nH_{2n+2} and Park's formula indicates clearly that above room temperature all paraffins (C_nH_{2n+2}) should dissociate to carbon and hydrogen if $n > 2$; the free energy of formation is:

$$\Delta G_f^0 = -14.700 + 6.000n + T + 25 \ ln \ T; \ (\text{cal/mol}; \times 4.18 \text{ J/mol}) \tag{1.22}$$

A similar condition is reached with olefins (C_nH_{2n}):

$$\Delta G_f^0 = 5.700 - 6.500(n-2) + 21.1T24(n-2); \quad (\text{cal/mol}; \times 4.18 \text{ J/mol}) \tag{1.23}$$

with acetylenic and aromatic hydrocarbons. More recent data do not alter these predictions.

Nevertheless, it is interesting to compare the relative stabilities of different hydrocarbons; in the Park's diagram (Figure 1.10), the formation equation is reduced to:

$$C + xH_2 \rightleftarrows CH_{2x} \qquad 0 < x < 2 \tag{1.24}$$

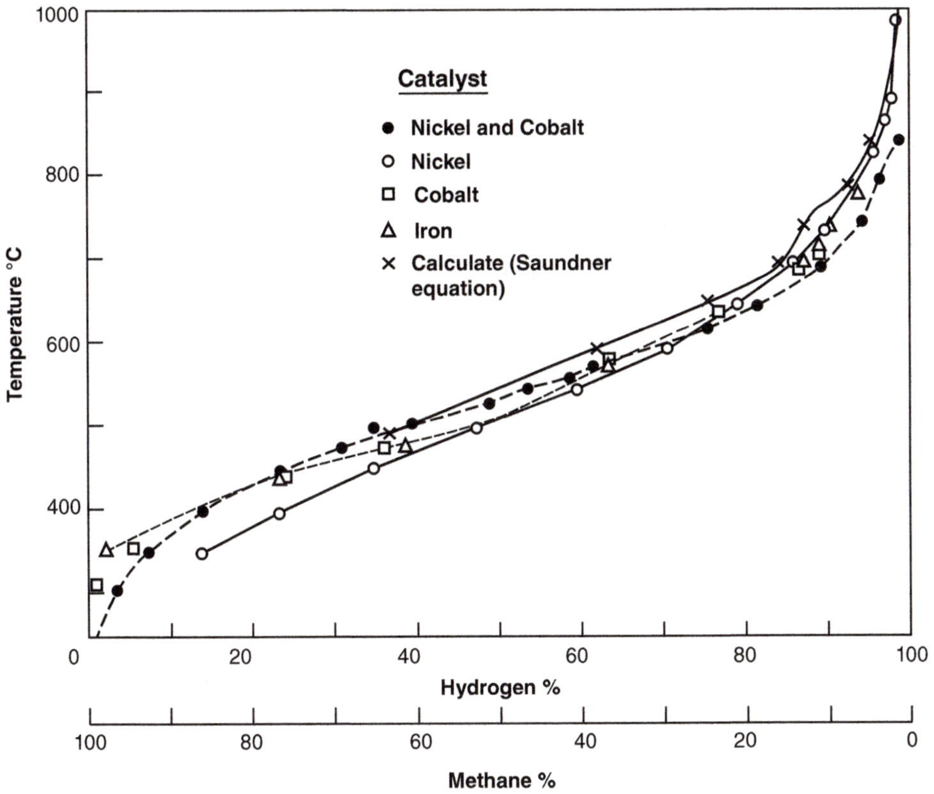

Figure 1.9 The methane equilibrium: $CH_4 \rightleftarrows C + 2\ H_2$. Comparison of observed and calculated values. (From Egloff et al., p. 1632).

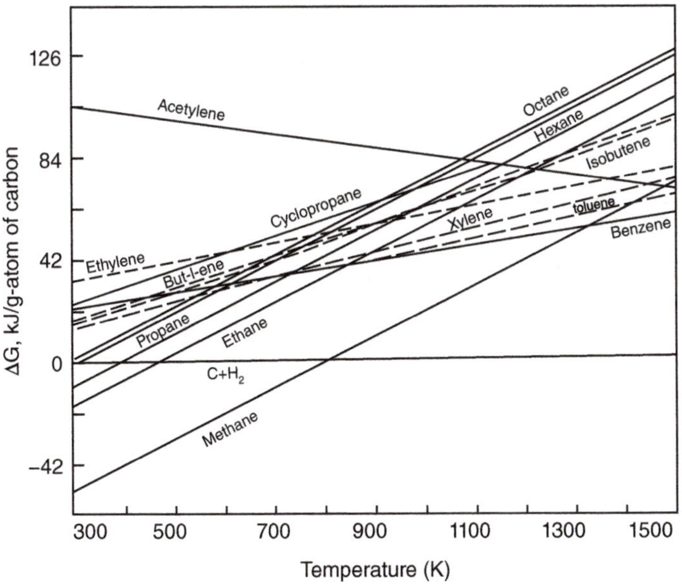

Figure 1.10 Free energy of formation of hydrocarbons, see left side of figure. (From Parks, G.S., Huffman, H.M., (1932), Free Energy of Some Organic Compounds, Reinhold, New York).

to simplify stochiometric calculations and the possible conversions of hydrocarbons are thus predicted in a qualitative way.

At low temperatures, paraffins are more stable than corresponding olefins, allowing the *hydrogenation*:

$$C_nH_{2n+}H_2 \rightleftarrows C_nH_{2n+2} \tag{1.25}$$

whereas the reverse reaction of *dehydrogenation* is favoured at high temperatures. Hydrogenation of acetylenic and aromatic hydrocarbons are exothermic low-temperature reactions.

Typical inversion temperatures are:

$$C_2H_4 + H_2 \rightleftarrows C_2H_6 \quad t_i = 820^\circ C \text{ (decreases with } n\text{);} \tag{1.26}$$

$$C_2H_4 + H_2 \rightleftarrows C_2H_6 \quad t_i = 1150^\circ C \tag{1.27}$$

$$C_6H_6 + 3H_2 \rightleftarrows C_6H_{12} \quad t_i = 280^\circ C \tag{1.28}$$

Free-energy differences between cycloparaffins and paraffins are much smaller (except for low carbon numbers, e.g., 3 or 4), but ring opening by hydrogenolysis is also a low-temperature exothermic reaction:

$$C_nH_{2n} + H_2 \rightleftarrows C_nH_{2n+2} \tag{1.29}$$

At high temperatures, if $n > 6$, dehydrogenation to benzenic or aromatic hydrocarbons is favored over dehydrogenation to olefins:

$$n - C_7H_{16} \rightleftarrows C_6H_5 - CH_3 + 4H_2 \quad \text{ti} = 190^\circ C \tag{1.30}$$

All these hydrogen-containing reactions are hydrogen pressure dependent.

Hydrogenolysis of higher paraffins to lower ones is also a result of the free-energy increase with carbon number:

$$C_nH_{2n+2} + H_2 \rightleftarrows C_mH_{2m+2} + C_pH_{2p+2} \quad n = m + p \tag{1.31}$$

Even in the absence of hydrogen, *cracking* reactions reflect the same tendency:

$$C_nH_{2n+2} \rightleftarrows C_mH_{2m+2} + C_pH_{2p} \quad n = m + p \tag{1.32}$$

The cracking of aromatic side chains is also expected:

$$C_6H_5 - C_nH_{2n+1} \rightleftarrows C_6H_5 + C_nH_{2n} \tag{1.33}$$

in the same temperature range. But, at low temperatures, the reverse exothermic *alkylation* reactions become possible. Inversion temperatures are in the neighbourhood of 400°C. Olefin *polymerization* is also a consequence of the relationship between ΔG^0 and n:

$$2C_2H_4 \rightleftarrows C_4H_8 \quad t_i = 527^\circ C \tag{1.34}$$

Its reversal at high temperatures is olefin cracking, or depolymerization. In the gas phase, these reactions may be displaced by pressure.

Multiple conversion paths are open for each system under a given set of conditions. Thus, a normal paraffin at high temperature may crack to a mixture of lower paraffins and olefins, or dehydrogenate to olefins of same carbon number, or undergo dehydrocyclization to aromatic hydrocarbons, or decompose to carbon and hydrogen, to mention only a few of thermodynamically possible reactions. Actual products of the conversion are controlled by the relative rates of these reactions. Purely thermal processes have high activation energies associated with the high energies of the C-C, C-H and H-H bonds that must be broken and re-arranged (250–418 kJ/mole). The rates are negligibly small at low temperatures when exothermic reactions (hydrogenation, alkylation, polymerization) are possible.

For these reasons, specific hydrocarbon conversion processes are tied up with discovery of highly active and selective catalysts strongly favouring one (or a few) of the numerous possible reaction paths. It is purpose of this book to show how well this goal has been reached in different cases. Of course, introduction of oxygen on the hydrogen-carbon system vastly increases the complexity and the number of possible compounds. The extremely high stability of carbon monoxide, carbon dioxide and water dominate the picture and the complete combustion of the hydrocarbon is overwhelmingly favoured by thermodynamics at all temperatures. Nevertheless, specific catalysts have been found to direct the oxidation towards metastable intermediates (such as carbonyl or carboxyl compounds). The presence of carbon monoxide, carbon dioxide and water in all these processes led us finally to examine reactions of hydrocarbons with these simple molecules.

Hydrocarbons conversion reactions depend on catalysis and it seems necessary first to examine the most likely mechanisms of interaction between hydrocarbons and catalysts of different types. Methane, which is the main component of natural gas, is a suitable raw material in terms of availability due to the existence of huge natural gas reserves. Since steam methane reforming leads to significant carbon dioxide emissions, cleaner processes must be investigated. The thermal decomposition of methane, also known as methane pyrolysis, is an adequate alternative because hydrogen and solid carbon are the only reaction products and thus, the formation of carbon dioxide is prevented during the reaction. The carbon dioxide footprint of methane pyrolysis corresponds to the emissions derived from the required electricity and those generated during the extraction and transportation of natural gas. In any case, the carbon dioxide emissions corresponding to methane pyrolysis are significantly lower than those derived from the well-established fossil fuel-based technologies. Methane pyrolysis is a one-step process, unlike steam methane pyrolysis in which the water–gas shift reaction must also be carried out. Via the WGS reaction the carbon monoxide produced in the reaction between methane and water is converted into carbon dioxide and additional hydrogen. Regarding the energy efficiency, if the sequestration of carbon dioxide is not considered, steam methane reforming is significantly more efficient than methane pyrolysis (75% vs 58%). However, when the implementation of carbon capture and storage systems is considered, the net energy efficiency of both processes becomes very similar (60% for steam methane reforming and 58% for methane pyrolysis). Methane pyrolysis is more advantageous with regard to the energy input requirement (Figure 1.11). According to the standard reaction enthalpies, 37.7 kJ are needed in methane pyrolysis to obtain one mole of hydrogen, whereas in steam methane reforming coupled with the water–gas shift reaction and without taking into account the heat for water evaporation this value amounts to 41.4 kJ per mole of hydrogen. Nevertheless, if the heat required to evaporate liquid water is considered, then 63.4 kJ must be applied to produce one mole of hydrogen in the steam reforming process. The decomposition of methane is also energetically much more favourable than water electrolysis, where 285.8 kJ are required to produce one mole of hydrogen. Despite the advantages of methane pyrolysis, the production of hydrogen from this process is not yet competitive with the mature steam reforming technology. The estimated product costs from the decomposition of methane range from €2600 to €3200 per ton of hydrogen depending on the expected carbon credit. In contrast, one ton of hydrogen generated by steam reforming costs €2000. However, this value could increase in the future if higher penalties for carbon dioxide

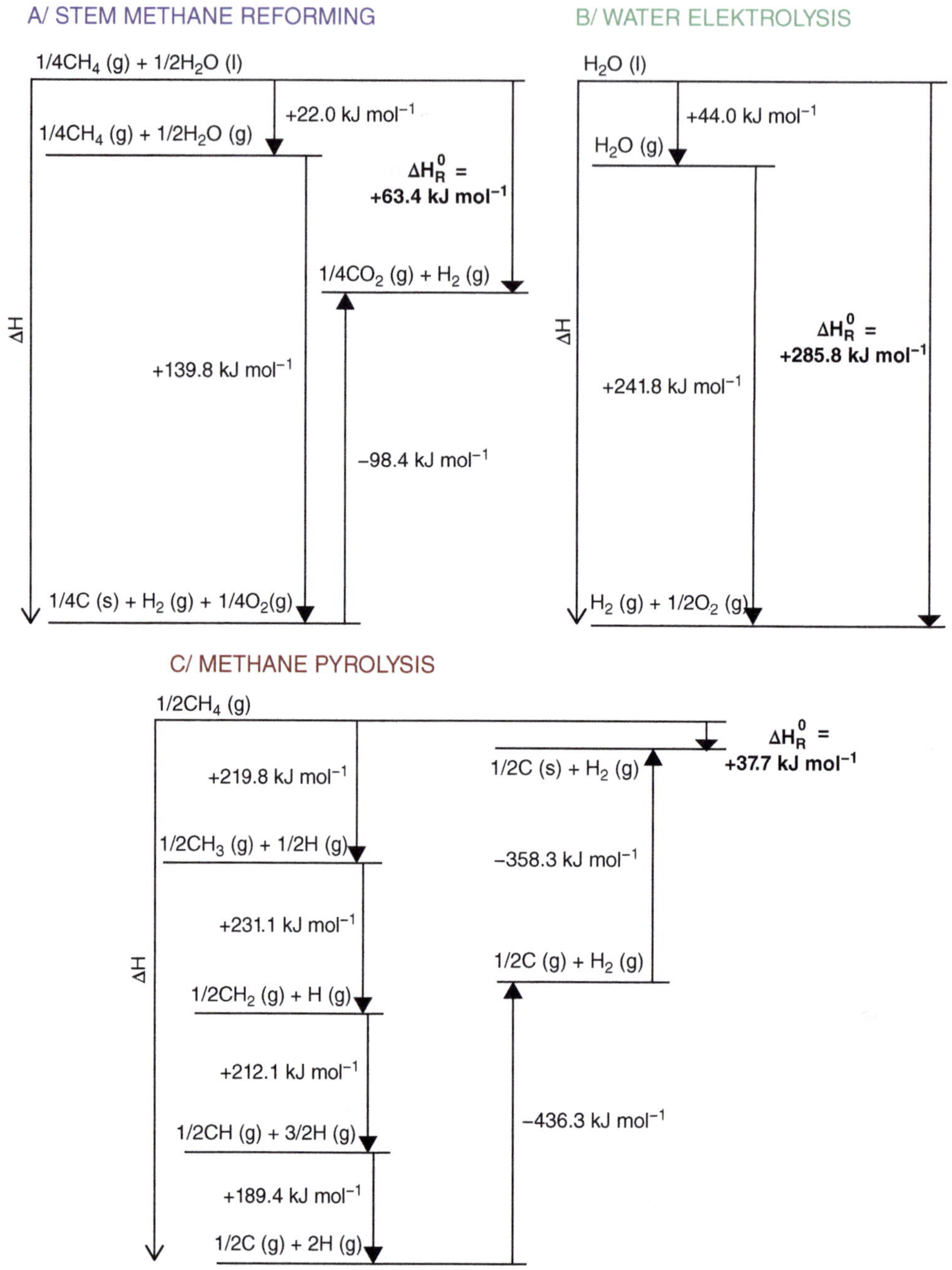

Figure 1.11 Enthalpy diagrams of (A) steam methane reforming, (B) water electrolysis and (C) methane pyrolysis.

Water and carbon dioxide enthalpies

$\Delta_f H^0_{gas}$ (H_2O) = 241.8 kJ/mol.

$\Delta_f H^0_{liquid}$ (H_2O) = 285.8 kJ/mol.

ΔH^0_{vap} (H_2O) = $\Delta_f H^0_{liquid}$ (H_2O) − $\Delta_f H^0_{gas}$ (H_2O).

$\Delta_f H^0_{gas}$ (CO_2) = −393.5 kJ/mol.

Bond energies of methane

ΔH^0_{dis} (CH_3−H) = 439.6 kJ/mol.

ΔH^0_{dis} (CH_2−H) = 462.2 kJ/mol.

ΔH^0_{dis} (CH−H) = 424.1 kJ/mol.

ΔH^0_{dis} (C−H) = 338.7 kJ/mol.

ΔH^0_{dis} (H−H) = 436.3 kJ/mol.

ΔH^0_{sub} (C) = 716.7 kJ/mol.

emissions are imposed. Although methane pyrolysis cannot compete economically today with the traditional process and even though natural gas is a fossil raw material, this technology seems to be an appropriate temporary alternative for carbon dioxide-free hydrogen production that can serve as a bridge in the transition period towards renewable energies.

The decomposition of methane involves an endothermic reaction and hence, the conversion of methane and the production of hydrogen are favoured by high temperature. However, an increase in methane conversion also leads to a higher carbon production rate (Figure 1.12). In the catalytic decomposition of methane, this results in an imbalance between the carbon production and the carbon migration rates through the catalyst particles, which consequently accelerates the deactivation of the catalyst. Any factor that increases the rate of methane decomposition without an equivalent improvement in the rate of carbon transfer promotes the rapid loss of catalytic activity. In the case of metal catalysts, the particles can sinter at high temperatures, which also favours the catalyst deactivation.

Carbon materials are less active than metal catalysts and require higher operating temperatures, usually between 800 and 1000°C. There are also differences in the optimum operating temperature range depending on the type of metal catalyst. Nickel catalysts are more active but deactivate more rapidly than iron catalysts at high temperatures. For this reason, experiments over nickel materials are carried out in a lower temperature range (500–700°C) than iron catalysts (700–900°C) (Figure 1.13).

According to Le Chatelier's Principle, lower reaction pressures shift the equilibrium towards the formation of hydrogen, giving rise to a gas product with a higher hydrogen mole concentration (Figure 1.14A). Since keeping pressures lower than atmospheric is costly and complex, an inert gas, such as nitrogen or argon, is usually incorporated into the feed gas (Figure 1.14B). The addition of an inert gas decreases the partial pressure of methane, while the total pressure is maintained at 0.1 MPa. At constant temperature and for the same amount of catalyst, the mass of carbon deposited per mass of catalyst decreases with a reduction in the partial pressure of methane due to a dilution effect. Therefore, the incorporation of an inert gas has a positive effect on hydrogen production and catalyst stability. Nevertheless, the main disadvantage is the requirement of

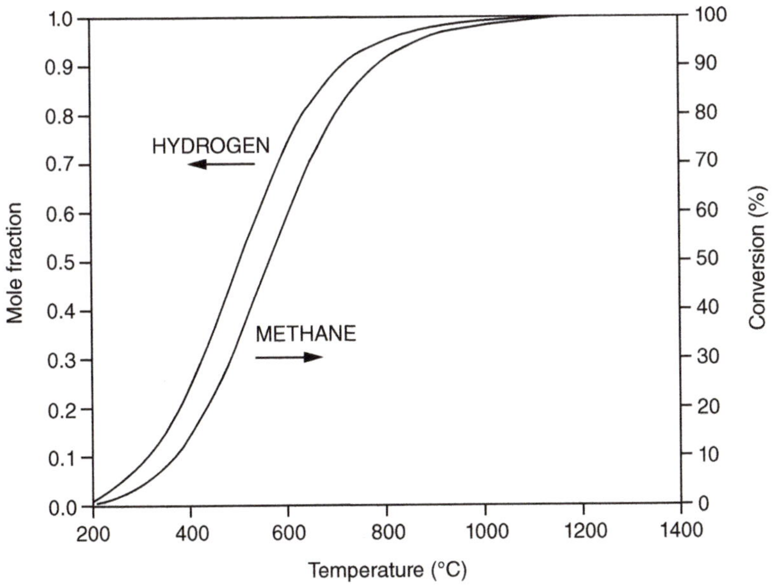

Figure 1.12 Hydrogen mole fraction (in the gas phase excluding carbon) and methane conversion in the thermodynamic equilibrium of methane pyrolysis at 0.1 MPa and different temperatures.

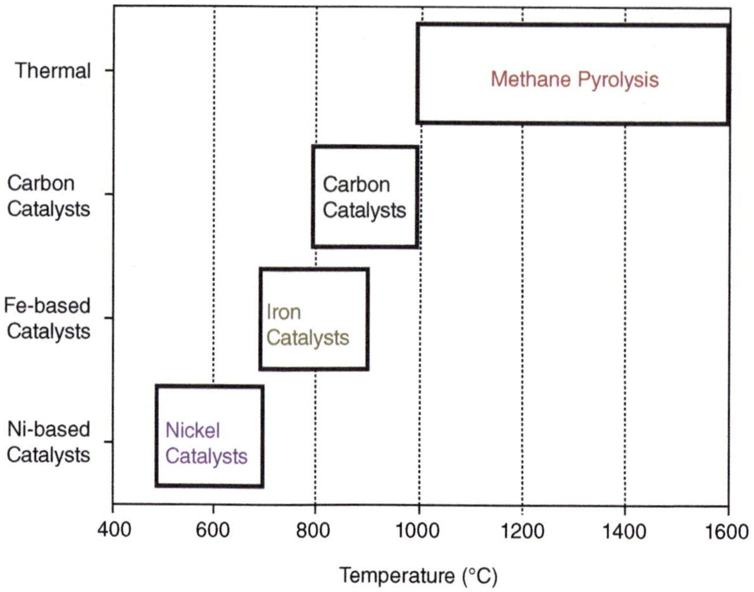

Figure 1.13 Temperature range of applicability of different catalysts for methane pyrolysis.

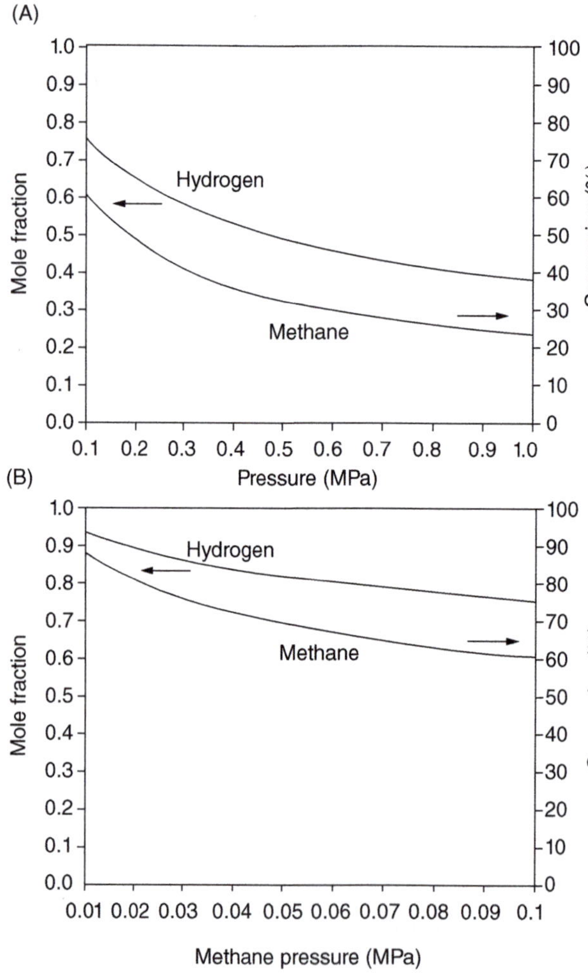

Figure 1.14 Hydrogen mole fraction (in the gas phase excluding carbon and nitrogen) and methane conversion in the thermodynamic equilibrium of methane pyrolysis at 600°C at (A) different total pressures and (B) different methane partial pressures (total pressure = 0.1 MPa), pressure balanced with nitrogen (calculated using Aspen Plus software).

additional separation and purification processes and thus dilution with an inert gas may not be feasible for the industrialization of methane pyrolysis.

1.2.6 Reaction Kinetics and Mechanism

The reaction kinetics are modelled using Arrhenius-type rate expressions of the form

$$k_j = A_j T^{\beta_j} exp\left(\frac{-E_{a,j}}{RT}\right) exp\left(\frac{\epsilon_{ij}\theta_{ij}}{RT}\right)_0^0 \tag{1.35}$$

where k_j is the reaction rate coefficient, A_j is the preexponential factor, β_j is a temperature dependency parameter, θ_{ij} is the surface coverage of species i in reaction j, $E_{a,j}$ is the activation energy of reaction j and R is the ideal gas constant. Equation (1.35) accounts for coverage-dependent changes in the heat of formation of surface intermediate i, resulting in additional coverage-dependent contributions to the activation barrier $E_{a,j}$. The corresponding contributions ϵ_{ij} are incorporated in the calculation of the activation energy according to the repulsive ($\epsilon_i > 0$) or attractive ($\epsilon_i < 0$) self-interactions of adsorbed species i on the surface. However, hereafter, only self-interactions of the most abundant surface intermediates are considered, as those contribute the most according to the degree of rate control theory (Campbell et al.). The production rates of a species $\dot{s}_i$ are then determined by

$$\dot{s}_i = \sum_j \nu_{ji} k_j \prod_i c_i^{\nu_{ji}} \tag{1.36}$$

Here, ν_{ji} represents the stoichiometric coefficient of species i in reaction j, k_j is the reaction rate coefficient and c_i is the concentration of species i in reaction j. The surface coverage of adsorbed species θ_i is evaluated by

$$\frac{\partial \theta_i}{\partial t} = \frac{\sigma_i \dot{s}_i}{\Gamma} \tag{1.37}$$

With t representing time and σ_i indicating the number of catalyst sites occupied by the adsorbate.

The absolute number of actives sites and the surface area in a computational cell are linked by the surface site density Γ (2.6×10^{-5} mol m^{-2} for nickel) (Rostrup-Nielsen). This number is constant; thus, the exposed surface area of nickel, calculated via experimental data presented in the references such as chemisorption measurements or nanoparticle size, determines the number of catalytic sites.

The software tool CaRMeN was employed as an interface between the user and the DETCHEM program package to enhance the workflow when dealing with large amounts of experiments/simulations.(Gossler et al.) It automates sets of calculations, resulting in an accelerated workflow and a lower potential for error.

High gas hourly space velocities (GHSVs) lead to short residence times and low contact efficiencies between the gas molecules and the catalyst. Consequently, the amount of methane adsorbed on the catalytic active sites as well as methane conversion rate decreases. High GHSVs also cause an imbalance between the rates of carbon formation and diffusion through the catalyst particles, which favours the carbon accumulation and accelerates the catalyst deactivation. For instance, in a fluidized-bed reactor an increase in GHSV promotes the mixing between methane and the catalyst particles due to the better gas–solid contact. Nevertheless, excessive values give rise to the formation of more and bigger bubbles, which negatively affects the interaction between the reactants and the catalyst. Therefore, a compromise between a good mixing and the formation of bubbles must be found.

Despite the numerous investigations on the effect of different operating conditions, they all focus on research at the academic level. As there are no industrially developed processes, the operating conditions for industrial application are still unknown. Further studies are required to find suitable experimental conditions for the implementation of an economic process on a large scale.

1.2.7 Thermal Mechanism of Methane Pyrolysis into a Sustainable Hydrogen

Since the 1960s, different reaction mechanisms for the noncatalytic decomposition of methane have been postulated. Pyrolysis tests in shock-tube experiments have allowed measurement of the initial rate of methane dissociation. Most authors agree that the reaction mechanism involves a free-radical scheme with the initiating reaction step corresponding to the dissociation of methane into a methyl radical and a hydrogen atom. A detailed reaction mechanism resulting from experiments at low temperatures (<830°C) was proposed in 1976 (Figure 1.15). This mechanism is based on the cleavage of C–H bonds and the consequent formation of methyl radicals. C_{2+} hydrocarbons are generated by the reaction between CH_3 radicals and other intermediate hydrocarbon species. In the first step, methane splits into a methyl radical and a hydrogen atom to subsequently form ethane and hydrogen molecules. In the second step, the rate of ethane formation falls gradually towards a plateau and ethylene is obtained as a secondary product via the radical chain dehydrogenation of ethane. In addition, under certain conditions ethane can be dissociated into two methyl radicals. In the third step, acetylene and propylene are formed from ethylene via radical chain dehydrogenation and radical

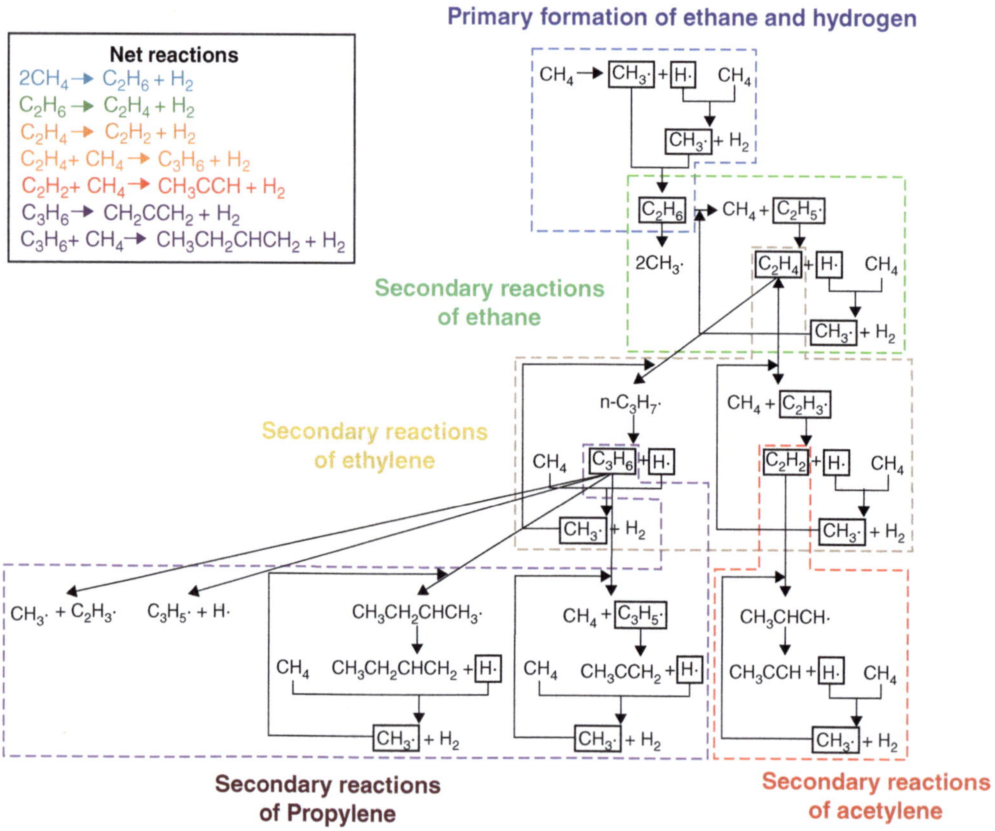

Figure 1.15 Reaction mechanism of the noncatalytic methane pyrolysis.

chain methylation reactions, respectively. At the same time, a sharp increase in the formation rate of ethane is observed. This fact is not explained by the described reaction mechanism and autocatalysis is apparently involved. Unlike other works, where autocatalysis is attributed to the formation of catalytically active carbon, in the present studies the appearance of new radical sources in the autocatalytic region, probably coming from acetylene and propylene, may be responsible for the increased formation rates of ethane and the corresponding products derived from it.

This mechanism was extended in 1985 by adding reverse, isomerization and abstraction and addition reactions for radicals up to C_3 reacting with primary and secondary products up to C_3. The reaction between methane and a methyl radical to produce ethane was evaluated at higher temperatures (>1000°C) (Equation 1.38). Nevertheless, the formation of ethane is unlikely above 1000°C and the direct production of ethylene from methane is more probable (Equations 1.39 and 1.40). According to the same study, the formation of benzene occurs from acetylene (Equation 1.41) and ethylene (Equations 1.42–1.44) and even if both hydrocarbons can form carbon directly, benzene and probably higher condensed aromatics are the main species for carbon growth.

$$CH_3\cdot + CH_4 \rightarrow C_2H_6 + H\cdot \tag{1.38}$$

$$CH_3\cdot + CH_4 \rightarrow C_2H_5\cdot + H_2 \tag{1.39}$$

$$C_2H_5\cdot \rightarrow C_2H_4 + H\cdot \tag{1.40}$$

$$3C_2H_2 \rightarrow C_6H_6 \tag{1.41}$$

$$2C_2H_4 \rightarrow C_4H_8 \tag{1.42}$$

$$C_4H_8 + C_2H_4 \rightarrow C_6H_{12} \tag{1.43}$$

$$C_6H_{12} \rightarrow C_6H_6 + 3H_2 \tag{1.44}$$

Although it is generally accepted that the rate-limiting step is the splitting of methane into a methyl radical and a hydrogen atom, another reaction mechanism with an alternative rate-determining step has also been proposed. In this case, the controlling stage corresponds to the dissociation of methane into methylene and a hydrogen molecule (Figure 1.16):

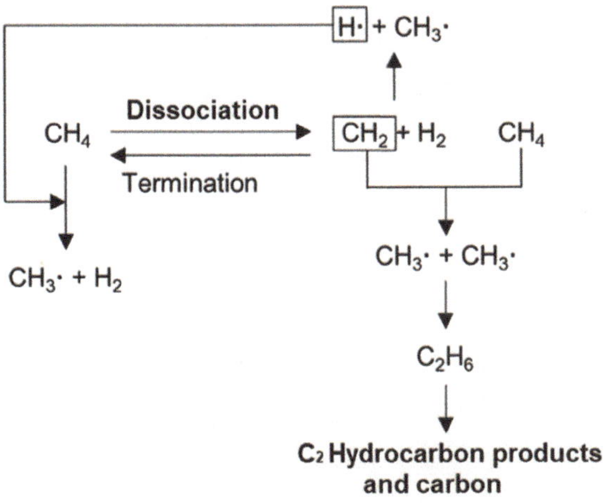

Figure 1.16 Reaction mechanism of noncatalytic methane pyrolysis.

These controversial results in the initiating and rate-controlling step may be related to the different temperature ranges used in methane pyrolysis experiments. The decomposition of methane into a methyl radical and a hydrogen atom is observed in experiments at lower temperatures (<1400°C), whereas the dissociation into methylene and a hydrogen molecule derives from reactions at higher temperatures (>1400°C).

1.2.8 Catalytic Mechanism Splitting of Methane into a Sustainable Hydrogen

Several reaction mechanisms have been postulated to explain the catalytic conversion reaction of methane. Some works have proposed a molecular adsorption mechanism (Figure 1.14A), whereas a dissociative adsorption model has been described in other studies (Figure 1.14B). In the molecular adsorption mechanism, methane is first adsorbed on the catalyst surface and then dissociates following a series of stepwise surface dehydrogenation reactions. Nevertheless, according to the dissociative adsorption model, methane dissociates upon adsorption on the catalytic active sites generating chemisorbed CH_3 and H fragments. This step is followed by the same surface dissociation reactions described by the molecular adsorption mechanism. The dissociative adsorption mechanism emerges from studies on the interaction of methane with metal surfaces at pressures below atmospheric using molecular beam techniques with high-resolution electron energy loss spectroscopy (HREELS) and Auger electron spectroscopy. However, this model has also been applied to methane pyrolysis, which is usually carried out at atmospheric pressure (Figure 1.17).

Different rate-limiting steps have been proposed in the molecular adsorption mechanism. Some works consider the abstraction of the first hydrogen atom from molecularly adsorbed methane to form an adsorbed methyl group as the initiation and rate-limiting step of the decomposition of methane. Nevertheless, the removal of the second hydrogen from the adsorbed methyl fragment or the adsorption of methane on the catalyst surface have also been suggested as rate-determining steps. In the dissociative adsorption mechanism, there is also no agreement on the rate-controlling step of the reaction. Some authors confirm that the dissociation of methane giving rise to a methyl group and a hydrogen atom controls the overall mechanism. On the other hand, according to various works on the kinetics of carbon nanotube formation, the dissociative adsorption of methane followed by the removal of hydrogen from the adsorbed methyl group limits the catalytic decomposition of methane. The kinetic models presented in the later works differ in the number of active

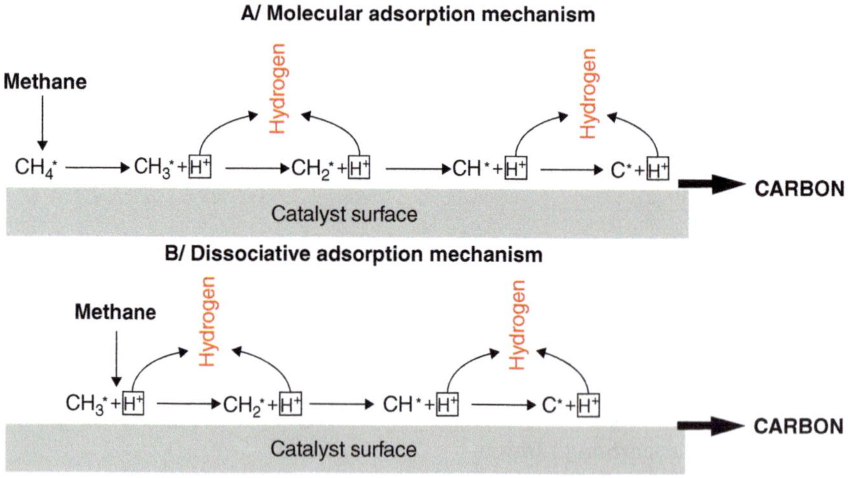

Figure 1.17 Reaction mechanisms proposed for catalytic methane pyrolysis.

site types. The kinetic models for double- and single-walled carbon nanotube synthesis are based on the presence of only one type of active site, whereas that developed for the formation of multi-walled carbon nanotubes considers two different types of active site. Here, CH_x* and H* species are adsorbed on different kinds of active sites.

Other possible mechanisms arise from studies on the decomposition of hydrocarbons focused on the synthesis of carbon nanostructures via chemical vapour deposition. The vapour–liquid–solid (VLS) model was initially developed in 1964 to explain the crystal growth of silicon whiskers. Some years later this mechanism was applied to the growth of filamentous carbon, which includes carbon nanotubes and nanofibers, over a nickel catalyst using acetylene as a carbon precursor. The general VLS mechanism for the decomposition of hydrocarbons comprises several steps (Figure 1.18). First, the hydrocarbon is adsorbed on the catalyst particle surface and dissociates into elemental carbon (Figure 1.18A). Then, carbon is taken into solution, diffuses through the bulk of the metal particle and precipitates at the backside of the catalyst particle (Figure 1.18B). Excess carbon accumulates at the exposed catalyst faces and is transported by surface diffusion around the catalyst particle to form the outer part of the filament. The accumulated carbon deposits force the metal particle away from the support, giving rise to the formation of the carbon filament (Figure 1.18C). Finally, the particle is completely encapsulated by carbon and the filament growth ceases because there is no longer contact between the hydrocarbon and the active catalyst particle (Figure 1.18D).

During the third step (Figure 1.18C) the metal particle is distorted and elongated and the metal is assumed to have the properties of a liquid. When the metal particle is detached from the support, an initial hollow channel appears because the carbon does not have enough time to deposit in this

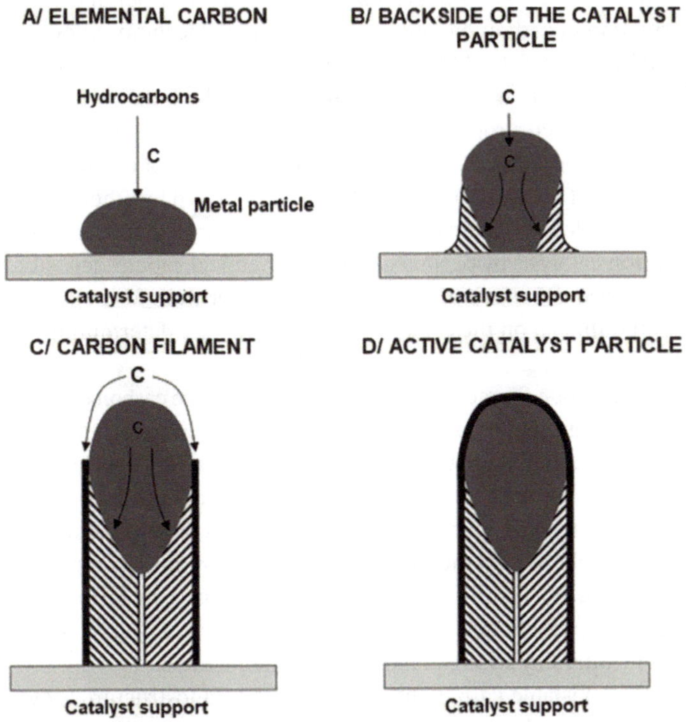

Figure 1.18 Vapour–liquid–solid mechanism for the growth of filamentous carbon from hydrocarbon decomposition over metal catalysts.

region. The formation of filamentous carbon from the decomposition of acetylene was also studied over iron, cobalt and chromium catalysts. The apparent activation energies for carbon growth over iron, cobalt, chromium and nickel agree with those for the diffusion of carbon through the bulk of the corresponding metal. This suggests that carbon diffusion through the metal particle is the rate-determining step in the filament growth process. A thermal gradient may be the driving force for the bulk carbon diffusion. Carbon diffuses from the hotter front face, on which the carbon precursor decomposition occurs, to the cooler backside of the catalyst particle, on which the carbon precipitates. This theory is valid when the decomposition of the carbon precursor is exothermic, as in the case of acetylene. The thermal gradient is maintained by the exothermic decomposition of the hydrocarbon on the front face of the catalyst particle and the endothermic precipitation of the carbon at the backside. However, this hypothesis cannot explain the diffusion of carbon when the decomposition of the hydrocarbon is endothermic, as happens in methane decomposition. To overcome this limitation a different driving force is suggested. In this case, the diffusion of carbon through the metal particle derives from a carbon concentration gradient due to the different carbon solubility at the metal–gas and the metal–carbon interfaces.

Studies using high-resolution electron microscopy techniques have led to a new carbon growth mechanism. This mechanism does not involve the bulk diffusion of carbon species but the diffusion on the surface of the catalyst particle. The carbon species start to dissociate at the contact angle between the metal particle and the catalyst support. The accumulation of carbon at the rear part of the metal particle and the subsequent formation of lateral layers following the contour of the metal surface, cause the metal particle to move away from the support, resulting in the formation of a carbon filament. The surface diffusion of carbon would also explain the hollow channel in the centre of the carbon filaments. Since these studies were performed with ex situ analytical methods, the microscopy images correspond to the "cold" or even deactivated catalyst. The use of in situ analytical techniques becomes thus crucial to understand the growth of carbon nanotubes and nanofibers. Images of the formation of carbon nanofibers from methane decomposition over nickel supported catalysts provided by time-resolved high-resolution in situ transmission electron microscopy demonstrate that carbon atoms do not necessarily diffuse into the bulk of the metal particles. Unlike the VLS mechanism, which proposes the carbon diffusion through the metal particles as the rate-limiting stage, this model suggests that the surface transport of carbon atoms controls the rate of the nanofiber growth. This mechanism was later corroborated by other authors for the growth of nanocarbons and extended to the growth of graphene. The growth rate of carbon nanofibers over nickel, cobalt and iron catalysts by plasma-enhanced chemical vapour deposition shows that the activation energy for carbon surface diffusion is much lower than for bulk diffusion. This would confirm the carbon diffusion on the catalyst surface as the rate-determining step for plasma-enhanced carbon growth.

These mechanisms for the growth of carbon filaments involving bulk or surface carbon diffusion correspond to a tip-growth model. In this growth model the metal particles are located at the tip of the carbon filaments since the carbon accumulated at the rear part of the catalyst particle causes its detachment from the support. The formation of carbon filaments can also follow a base-growth mechanism. Here, carbon precipitates on the apex part of the metal particle as far as possible from the support and crystallizes as a hemispherical dome. Subsequent hydrocarbon decomposition occurs on the lower surface of the particle and carbon diffuses upward, leading to the formation of carbon filaments above the metal particle, which remains attached to the support. The base-growth mechanism of carbon nanotubes is favoured by strong metal–support interactions.

The formation of an intermediate metastable carbide phase during the nanostructure growth may also take place. Here, metal particles undergo partial carburization. The metal is first transformed into a metal carbide, which is subsequently decomposed during the synthesis of carbon filaments. The carbide cycle mechanism, that is, the decomposition of hydrocarbons through intermediate carbide-like compounds, has been reported for the growth of carbon filaments over iron and nickel catalysts. According to this mechanism, a metastable carbide-like intermediate compound is formed at the surface of an active catalyst particle as a result of the decomposition of the hydrocarbon. Then, the dissociation of the intermediate carbide results in the formation of carbon atoms that enter the bulk metal and lead to the supersaturation of metal by carbon. When a critical supersaturation is attained, a graphite phase is formed at the surface of the metal particle and carbon filaments start to grow. The diffusion of carbon atoms from the surface through the bulk of the metal particles to the sites of crystallization into a graphite phase takes place by a carbon concentration gradient. The decomposed intermediate carbide is restored as a consequence of the continuous dissociation of the hydrocarbon. Therefore, this cyclic process is maintained as long as there is a carbon gas source and available catalytic active sites. In different studies, the presence of carbide species for nickel catalysts has been revealed. The characterization of the samples in these works was carried out using ex situ analytical techniques, such as transmission electron microscopy (TEM) or X-ray diffraction (XRD), after catalyst cooling. Therefore, the observed metal carbide phase could be a result of the precipitation of excess carbon on the nickel particle upon catalyst cooling. Thus, it cannot be assumed that nickel carbide is formed as an intermediate compound during the carbon growth. In addition to this, the formation of nickel carbide phases is not confirmed by in situ time- and depth-resolved X-ray photoelectron spectroscopy (XPS) and XRD measurements. This would support the fact that an intermediate carbide is not involved in the formation of carbon nanostructures over nickel catalysts.

In the case of iron catalysts, the formation of intermediate iron carbide phases has been extensively reported. Many of the works on iron catalysts were performed using in situ measurements. In situ electron microscopy images reveal that the carbide phase, which decomposes into metal and carbon, is an intermediate phase in the formation of graphite layers constituting multiwalled carbon nanotubes. Therefore, iron carbide (Fe_3C) would be involved in the growth of carbon nanotubes. Equations. 1.45–1.48 describe the carbon nanotube growth using methane and iron catalysts. The decomposition of metal carbides into surface graphitic carbon (Equation. 1.47) is regarded as the crucial step.

a) Methane decomposition on the catalyst surface leading to surface carbon and hydrogen:

$$\mathrm{CH_4} \rightarrow \text{surface- C} + 2H_2 \tag{1.45}$$

b) Carbon diffusion through the catalyst particles leading to carbide formation:

$$\text{surface-C} \rightarrow \text{carbide-C} \tag{1.46}$$

c) Formation of graphitic carbon on the catalyst surface from the metal carbide:

$$\text{carbide-C} \rightarrow \text{surface graphitic-C} \tag{1.47}$$

d) Carbon nanotubes formation from surface graphite:

$$\text{surface graphitic-C} \rightarrow \text{carbon nanotubes} \tag{1.48}$$

Contrary to the previous results, studies on iron and nickel catalysts based on environmental transmission electron microscopy (ETEM) and in situ time resolved XPS demonstrate that a carbide phase would not be necessary for the formation of carbon nanotubes. The carbide formation only occurs during the carbon nanotube growth over iron catalysts, while nickel carbide is not detected. Therefore, the growth of carbon nanotubes would not require the presence of a carbide phase.

Iron carbide may serve as an active catalyst for carbon growth and methane dissociation. A study on carbon deposition on polycrystalline iron catalysts describes different mechanisms for fresh samples and after the deposition of some carbon, as well as depending on the reaction temperature. Below 600°C the rate-determining step on fresh samples is the diffusion of carbon in iron, whereas after the deposition of some carbon the rate is determined by the carbide-catalysed hydrocarbon decomposition. Above 600°C the surface decomposition of the hydrocarbon is assumed to be the limiting step for the carbon formation. In this case, the geometry of the catalyst surface affects the carbon formation rate on fresh samples. However, after the deposition of some carbon the rate is influenced by the iron carbide active sites instead of the geometry. The formation of γ- and α-Fe phase mixtures takes place in oxide supported iron catalysts during the reduction step prior to the chemical vapour deposition of carbon nanotubes using acetylene as a carbon source. This variation in the iron crystal structures is due to carbon contamination prior to hydrocarbon exposure. Two different growth mechanisms occur depending on the phase composition. For γ-rich iron catalyst mixtures, metallic Fe is the active phase for the growth of carbon nanotubes, whereas the formation of an iron carbide is not a prerequisite. However, for α-rich iron catalyst mixtures, the formation of Fe_3C is dominant and takes part in the carbon growth process. An investigation on nanocarbon growth over iron-based catalysts by plasma-enhanced chemical vapour deposition identifies α-Fe and Fe_3C phases depending on the growth temperature. Although at low temperatures (600°C) only Fe_3C acts as a catalyst for the growth of carbon nanotubes both, α-Fe and Fe_3C, are active at higher temperatures. A recent study on methane pyrolysis over iron catalysts confirms that not only metallic iron but also an iron carbide phase (Fe_3C) is active for the decomposition of methane. The proposed reaction mechanism is summarized by Equations 1.49–1.52:

a) Decomposition of methane on the Fe^0 surface into amorphous carbon and hydrogen:

$$CH_4 \overset{Fe^0}{\leftrightarrow} C_{amorphous} + 2H_2 \tag{1.49}$$

b) Reaction between the amorphous carbon and Fe^0 to form Fe_3C or a mixture of Fe^0 and Fe_3C, which acts as a catalyst of the reaction itself:

$$Fe^0 + C_{amorphous} \leftrightarrow Fe_3C \tag{1.50}$$

$$CH_4 \overset{Fe_3C/Fe^0}{\leftrightarrow} C_{amorphous} + 2H_2 \tag{1.51}$$

c) Diffusion of the formed carbon into Fe_3C to form supersaturated Fe_3C.
d) Decomposition of supersaturated Fe_3C to stoichiometric Fe_3C and transformation of amorphous carbon to graphitic carbon:

$$C_{amorphous} \overset{Fe_3C}{\leftrightarrow} C_{graphite} \tag{1.52}$$

From these results, it can be concluded that different controlling steps have been proposed for the decomposition of methane. There is no general agreement and the reaction mechanism involved in methane conversion as well as the overall rate-limiting step is still unclear. Additionally, the formation of metal carbides and their role in the decomposition of hydrocarbons and nanocarbon growth is still a matter of debate today. Most research on the reaction mechanism focuses on the decomposition of hydrocarbons over metal catalysts and no in-depth studies over carbon catalysts have been found. For this reason, the reaction mechanism over carbon catalysts is far from clear. Further studies are needed to fully unravel the reaction mechanism and develop enhanced catalysts that accelerate the kinetics under optimized experimental conditions. It is essential to elucidate the reaction mechanism of methane pyrolysis and identify the determining step so that future improvements in the catalytic activity can be achieved.

1.2.9 Conversion of Methane over Metal Catalysts into a Sustainable Hydrogen

In the absence of a catalyst the thermal decomposition of methane requires temperatures above 1000–1200°C to achieve relevant reaction rates and methane conversions. Such high operating temperatures are necessary because the symmetrical molecular structure and the strong C–H bonds of methane give this molecule great stability. To reduce the reaction temperature and improve the hydrogen yield, different metal and nonmetal catalysts (generally carbon materials) have been developed for this process over the years. Moreover, the use of different molten metals and salts has recently attracted particular attention.

Transition metals, mainly nickel, iron and cobalt, have been widely investigated as active species for methane pyrolysis. Their partially filled 3d orbitals can accept electrons from the C–H bonds of methane, which facilitates its decomposition. In addition, transition metals offer relatively high solubility and capacity for carbon diffusion through their crystalline structure. Another advantage of metal catalysts is the possibility of obtaining valuable carbon nanotubes as coproduct. Nickel, iron and cobalt are very active under moderate operating temperatures. Their activity exhibits the following trend: Ni > Co > Fe (Table 1.6). Compared to nickel and iron, cobalt catalysts have not received much attention lately. Reasons for this are the lower activity and higher price of cobalt compared with nickel, as well as toxicity problems. However, cobalt is commonly used in small amounts as a promoter of different metal-based catalysts.

1.2.9.1 Nickel Catalysts

Nickel catalysts show the highest initial activity among metal catalysts, although above 600°C they deactivate rapidly due to carbon coking and poisoning, so that the active metal sites are encapsulated within the carbon formed during the reaction. The deactivation of the catalyst occurs when the carbon production rate, i.e., the conversion of methane, is faster than the carbon diffusion rate through the metal particles. This imbalance between the carbon production and carbon diffusion

Table 1.6 Initial activity of nickel, iron and cobalt catalysts in the decomposition of methane for hydrogen production.

Catalyst	t (°C)	P (MPa)	CH_4/N_2(Vol.)	Flow Rate (ml/min)	Space Velocity (ml/ hg_{cat})	Hydrogen Yields (%)
Ni/CeO_2	700	0.1	1/0	150	4500	53
Ni/La_2O_3	700	0.1	1/0	150	4500	60
Ni/SiO_2	700	0.1	1/0	60	7200	73
Fe/CeO_2	700	0.1	1/0	150	4500	51
Fe/La_2O_3	700	0.1	1/0	150	4500	40
Fe/SiO_2	700	0.1	3/7	70	4200	20
Ni/SiO_2	800	0.1	1/0	250	5000	74
Fe/SiO_2	800	0.1	1/0	250	5000	39
Co/SiO_2	800	0.1	1/0	250	5000	48

rates results in the accumulation of carbon over the metal sites, which prevents the contact of the methane molecules with the active particles and consequently deactivates the catalyst. To improve the stability of nickel materials, the use of suitable supports and the incorporation of different dopants have been extensively investigated.

Nickel particles are susceptible to thermal sintering in unsupported catalysts and thus, many efforts have been made to improve their stability using appropriate supports. The metal–support interaction affects the reducibility and dispersion of metal particles. Although a strong metal–support interaction hinders the reduction of nickel oxide species, it also decreases the possibility of sintering and agglomeration of nickel particles, improving their fine dispersion on the support and enabling the formation of small crystallite sizes. Consequently, the stability of the catalysts is improved. In some cases, hardly reducible nickel solid solutions ($Ni_xMg_{1-x}O$) or spinel structures ($NiAl_2O_4$) are formed between the nickel particles and the support because of strong metal–support interactions. The d-orbitals of nickel in these species are filled and cannot accept electrons from the C–H bonds, inhibiting the adsorption and dissociation of methane. Additionally, the difficult reducibility of nickel species prevents the formation of active metal particles. The introduction of a second oxide to the catalyst support prevents the formation of nickel structures and improves activity and stability reducing coking. For instance, the addition of TiO_2 or CeO_2 to the support of Ni/Al_2O_3 catalysts inhibits the formation of $NiAl_2O_4$ and increases the reducibility and dispersion of nickel species, which leads to improved activity and stability. Contrary to these findings, $Ni/Al_2O_3 \cdot MgO$ shows worse catalytic performance than Ni/Al_2O_3, associated with the formation of an inactive Ni–Mg solid solution. Therefore, the performance of nickel catalysts is a compromise between the metal–support interaction and the reducibility and dispersion of the metal particles. The interaction between the metal and the support must be strong enough to avoid the aggregation of the particles and allow their fine dispersion on the support. Nevertheless, metal–support interactions that are too strong impede the reduction of nickel species either by increasing the reduction temperature of the nickel oxide precursors or by the formation of hardly reducible species between the metal and the support that prevent the formation of active metallic nickel. In addition to this, a strong metal–support interaction can inhibit carbon diffusion, leading to a faster catalyst deactivation. Therefore, a suitable metal–support interaction has to result in well-dispersed and easily reducible nickel particles and, at the same time, allow an appropriate carbon diffusion through them.

The role of promoters in metal catalysts is to create a balance between the rates of methane dissociation and carbon diffusion, that is, to modulate the dissociation rate of methane and increase that of carbon diffusion. The addition of a second metal as a promoter, such as palladium or copper, allows working at higher temperatures without rapid catalyst deactivation. Since the decomposition of methane is an endothermic reaction, the possibility of operating at higher temperatures results in better methane conversions and hydrogen yields. Palladium and especially copper are the most common promoters of nickel catalysts. These metals are not active for methane dissociation because of their filled 3d orbitals, but they can significantly affect the electronic properties of nickel. Ni–Cu and Ni–Pd catalysts deactivate above 700°C, although their stability and the deactivation temperature can be increased with increasing the promoter loading. The good stability of promoted catalysts is due to the formation of metal alloys with high lattice constants, which can accumulate larger carbon amounts without deactivation. Additionally, the higher carbon diffusion rate through the corresponding alloys rather than through the pure nickel particles prevents the formation of encapsulating carbon on the active sites. As a result, the generation of filamentous carbon is favoured over the formation of encapsulating carbon, which consequently extends the catalyst lifetime. In addition to this, promoters improve the fine dispersion of nickel particles on the catalyst support and the reducibility of nickel oxide species. The better reducibility is associated with the hydrogen spillover effect induced by the promoter. Copper and palladium are active sites for the dissociation of hydrogen molecules and thus, facilitate the conversion of nickel oxide species into metallic nickel during the reduction step prior to the reaction. The presence of a larger number of weakly interacted nickel species on the support after doping may also benefit the reducibility and prevent the formation of hardly reducible nickel structures.

The dopant loading in nickel catalysts is a crucial parameter. Catalysts with higher promoter loadings are stable at higher temperatures and thus, the deactivation temperature grows with the content of the dopant. The higher lattice constants of highly doped catalysts might explain this fact. Nevertheless, there is an optimized promoter loading for a given reaction temperature. The addition of small dopant amounts results in stable materials with a good metal dispersion and small crystallites that inhibit nickel sintering. However, excessive promoter loadings lead to faster deactivation and decreased thermal stability of the catalyst due to the dramatic reduction in surface area and the formation of large, poorly dispersed nickel crystals.

1.2.9.2 Iron Catalysts

Although iron is less active than nickel for the decomposition of methane, iron catalysts are more resistant to carbon coking and poisoning at high temperatures. At low reaction temperatures (<600°C) nickel-based catalysts exhibit superior catalytic performance, whereas iron materials are still active at higher temperatures at which nickel catalysts deactivate very quickly. Unlike nickel, iron materials are stable up to 700–1000°C. The higher stability is related to the carbon diffusion rate, which is three orders of magnitude higher through iron compared to nickel. Thus, iron catalysts keep a better balance between the rates of carbon production and diffusion, leading to a longer catalyst lifetime. Compared to cobalt- and nickel-based materials, iron catalysts are inexpensive and nontoxic. Furthermore, the carbon coproduct in iron-catalysed reactions is free of harmful metals and thus has the potential to be traded or safely stored. For these reasons, iron catalysts are the preferred option to industrialize the pyrolysis process of methane.

Metal promoters influence different catalyst properties (reducibility, surface area, metal dispersion, carbon capacity) and lead to the formation of bimetallic materials that are catalytically more active and stable compared to the corresponding monometallic iron catalysts. An adequate promoter enhances the reducibility of iron oxide species into metallic iron by decreasing the reduction temperature, probably due to the hydrogen spillover effect created by the promoter. A better reducibility of iron catalysts is observed after the addition of cobalt, copper, palladium, molybdenum and nickel. Another advantage of catalyst doping is the increase in the surface area, which takes place over Fe–Mo and Fe–Co catalysts. Furthermore, the greater metal dispersion in different bimetallic materials (Fe–Co, Fe–Ni) affects positively the catalytic performance. The incorporation of a second metal (Ni, Co) also leads to higher carbon capacities and decreases the carbon deposition rate over the active sites due to the balance between the carbon atom formation, diffusion and precipitation. Consequently, the promoted iron catalysts are more stable and exhibit longer catalyst lifetimes.

Supported and non-supported iron catalysts as well as different iron organometallic precursors have been investigated in the decomposition reaction of methane. No supported and highly iron-loaded materials have a low catalytic activity and deactivate quickly due to the small iron surface area. Bulk iron catalysts are highly susceptible to deactivation because of their large particle size. Additionally, the poor dispersion of metal particles and the formation of a solid solution between the metal and the support are responsible for the low methane conversion over highly metal-loaded catalysts. Thus, the support plays a crucial role in the suitable dispersion of the iron particles and the maintenance of an effective metal surface during the reaction. The incorporation of a support reduces the sintering effect, so that supported catalysts present a remarkably higher stability. For these reasons, iron catalysts prepared on a suitable support are preferred over non-supported materials. Al_2O_3 is the most common support of iron catalysts. The combination of Fe^0 and

Al_2O_3 keeps a balance between the rate of methane decomposition and thus, the carbon formation and the diffusion rate of carbon through the catalyst particles, avoiding the fast formation of encapsulating carbon. The formation of some spinel structures ($FeAl_2O_4$, $MgFe_2O_4$, Fe_2SiO_4) between the metal and the support may also take place. Nevertheless, their role in the catalytic activity is still unclear. $FeAl_2O_4$ species are detected in some Fe/Al_2O_3 catalysts. Although $FeAl_2O_4$ is inactive for methane decomposition, the interaction between the metallic iron particles and $FeAl_2O_4$ may play a positive role as the $FeAl_2O_4$ species can enhance the catalytic activity by preventing the agglomeration of metal particles through the strong bonding between Fe and $FeAl_2O_4$. Different iron species are found in several Fe/Al_2O_3 catalysts depending on the reduction temperature. Here, the samples with a combination of Fe^0 and $FeAl_2O_4$ are more active than those containing Fe^0 and Fe_3O_4. The formation of $FeAl_2O_4$ spinel structures is not observed in other processes, but the strong interaction between iron and the alumina support is also beneficial because it prevents metal particles from agglomeration. Contrary to these results, the strong metal–support interaction and the spinel formation are reported to be unfavourable. Different iron species and hence different catalytic activities, may be observed depending on the catalyst preparation method. For example, Fe/Al_2O_3 catalysts synthesized by impregnation are more active and stable than those prepared by coprecipitation. Catalysts synthesized by coprecipitation contain Fe_2O_3 and Fe species, whereas the catalysts prepared by impregnation present spinel species ($FeAl_2O_4$) and Fe_3O_4 in addition to some Fe_2O_3 and Fe species. The activity is related to the presence of Fe^{2+} in $FeAl_2O_4$ and Fe_3O_4 phases but not to the formation of a spinel structure itself. The appearance of Fe^{2+} in the samples synthesized by impregnation may be positive for the formation of catalytic active sites via in situ reduction during methane decomposition. The impregnation method may also be suitable due to the low interaction between iron and Al_2O_3 and the easier reduction of the resulting catalysts. The appearance of the $MgFe_2O_4$ spinel phase is detected in some Fe/MgO catalysts as well. The existence of $MgFe_2O_4$ indicates a strong metal–support interaction, which hinders the reduction of the iron oxide precursor. The low surface area of the catalyst and the difficult reduction of the metal particles may be the reasons for the poor activity of Fe/MgO.

Different organometallic compounds, such as iron pentacarbonyl ($Fe(CO)_5$) and ferrocene (iron dicyclopentadienyl, $Fe(C_5H_5)_2$), have mainly been intended for the production of carbon nanotubes. $Fe(CO)_5$ decomposes at temperatures higher than 300°C (Equation 1.53), whereas the decomposition of ferrocene takes place above 500°C (Equation 1.54)

$$Fe(CO)_5 \rightarrow Fe + 5CO \tag{1.53}$$

$$Fe(C_5H_5)_2 \rightarrow Fe + H_2 + CH_4 + C_5H_6 + \ldots \tag{1.54}$$

Ferrocene has been more widely used than $Fe(CO)_5$. This is probably due to the low cost, stability and nontoxicity of ferrocene in contrast to $Fe(CO)_5$. Ferrocene is a suitable organometallic compound for carbon nanotube growth since it not only gives rise to small iron metal particles but also acts as a carbon source upon its thermal decomposition. The formed iron particles agglomerate into clusters that serve as a catalyst for the decomposition of the reactive carbon species produced in the gas phase (Equation 1.54). The carbon resulting from these species constitutes the source for the subsequent formation of carbon nanotubes that nucleate and grow on the iron clusters. A hydrocarbon is often injected as an additional carbon source. The hydrocarbon decomposes on the iron particles and produces extra carbon for the formation of larger nanotube amounts. The experimental setup for this process usually consists of a low and a high temperature furnace. Ferrocene sublimes in the first oven at low temperature (>150°C). Then, ferrocene as a vapour is carried by a gas stream (Ar, H_2, N_2 and/or hydrocarbons such as acetylene, methane, benzene) into the second furnace at a higher temperature (800–1100°C). In the second oven ferrocene and the additional hydrocarbon decompose, giving rise to the growth of carbon nanotubes. Ferrocene and $Fe(CO)_5$ have also been employed as iron catalyst precursors for the pyrolysis of methane with the aim of producing hydrogen, although this application has been rarely reported. In this case, the iron clusters derived from the decomposition of the organometallic compound act as an in situ generated catalyst. Since different gaseous products result from the breakdown of $Fe(CO)_5$ (Equation 1.53) and ferrocene (Equation 1.54), the outlet gas must be cleaned to remove the undesirable impurities (CO, C_5H_6) and obtain high quality hydrogen. The poisoning of the outlet gas with unwanted compounds may explain the scarce application of these organometallic catalyst precursors in the decomposition of methane for hydrogen production.

1.2.9.3 Regeneration of Metal Catalysts

Different regeneration methods can be employed to remove the carbon deposits from metal catalysts and restore their activity. The reactivation techniques include combustion with oxygen or air of the carbon by-product and gasification with steam or carbon dioxide. During oxygen/air regeneration the carbon deposits are burned with oxygen, giving rise to carbon dioxide in a complete combustion and carbon monoxide if the oxidation is incomplete. This technique has been used to restore the activity of nickel catalysts. All the carbon on the catalyst surface is eliminated after combustion in air at 550–600°C. The initial activity for hydrogen production is restored after regeneration but the deactivation rate of the regenerated catalyst is much faster compared to the fresh catalyst. This is attributed to the increase in the crystallite size due to particle sintering, the disintegration of the catalyst into fine powder and the change in the face planes of the metal atoms occurring during the regeneration. The disintegration of the catalyst may also be related to the destruction of the porous support during the filament growth. The combustion of carbon involves an exothermic reaction so that the release of heat can give rise to high temperatures in the reactor and harm the catalyst. To avoid damaging the catalyst the regeneration with air should be accomplished in a fluidized-bed reactor since in a fixed-bed reactor some hot spots may be formed. Using a low oxygen concentration can also help to avoid high temperatures in the reactor. The heat released during the oxidation of the carbon can be used to thermally sustain the endothermic reaction of methane decomposition. The regeneration with air is much faster than with steam or carbon dioxide, but unlike these techniques, the initial metallic nickel is converted to nickel oxide during air combustion and the catalyst must be reduced again before the next reaction cycle. In the gasification process with steam, carbon reacts with water steam and a gaseous mixture composed of carbon oxides (CO_x) and hydrogen is obtained. One advantage of this procedure is the avoidance of a new reduction step because the metallic nickel form is preserved. Furthermore, additional hydrogen can be produced by steam gasification, which leads to higher global hydrogen yields. However, regeneration with steam requires a long process time and not all carbon species can be removed. Although a small amount of carbon deposit is not eliminated with steam, neither structural changes in the nickel particles nor a significant loss of catalytic activity occur after several successive decomposition–regeneration cycles. The reactivation process by carbon dioxide gasification results in the formation of carbon monoxide. This method preserves the reduced state of the metal but also requires long regeneration times. The application of carbon dioxide regeneration is limited by the low carbon removal rate and the high endothermicity of the reaction.

All the regeneration methods described here lead to the formation of CO_x products, which is an important drawback considering the clean nature of methane pyrolysis. In addition to this, the carbon by-product is destroyed and the carbon nanotubes cannot be recovered. An additional technique to overcome these problems is catalyst regeneration by using an acid or a base. This procedure enables not only the separation, purification and generation of highly pure and crystalline carbon products, but also the reactivation of metal catalysts. The formation of base-grown instead of tip-grown carbon nanotubes is required to avoid catalyst damage. If the metal particles were located at the tip of the carbon nanotubes, they would be dissolved in the acid or base and the catalyst would be destroyed. Contrary to tip-grown carbon nanotubes, base-grown carbon nanotubes, where the metal particles remain attached to the support, can be easily harvested without sacrificing the catalyst. Nevertheless, the use of homogeneous acids to remove the carbon deposits is contraindicated at the industrial level and should be avoided.

The regeneration of the spent catalyst may improve the economics of the process but constitutes a real challenge for scaling-up. The reactivation technique should be energy efficient and environmentally friendly, with short regeneration times and generate a catalyst with good catalytic performance. Nevertheless, none of the state-of-the-art methods meet these requirements and more in-depth studies are essential to advance the development of the pyrolysis process on a large-scale.

1.2.10 Conversion of Methane over Carbon Catalysts into Clean Hydrogen

Carbon materials are usually less active than metal catalysts and require higher reaction temperatures, normally in the range of 800 to 1000°C, depending on the type of carbon. However, carbon catalysts are more stable and exhibit longer catalyst lifetimes. The application of carbon materials in methane pyrolysis has been widely investigated in recent years due to their significant advantages over metal catalysts for the industrialization of the process. Compared to metal catalysts, carbon materials are significantly cheaper. In addition, the resulting carbon product may also have catalytic effects so that the decomposition of methane could be sustained for longer times without a significant decrease in activity. In this case, the carbon catalyst would be required only for the initiation of the reaction and the separation of the carbon product from the carbon catalyst may not be essential. Consequently, carbon dioxide emissions resulting from the regeneration process could be prevented. Carbons are also resistant to sulfur and other impurities contained in natural gas and hence, it would not be necessary to purify the feed gas before entering the reactor. The nontoxicity of the resulting carbon after reaction and the possibility of its subsequent use or secure storage are additional determining factors for the industrial implementation of methane pyrolysis based on carbon catalysts.

Activated carbons and carbon blacks are the most common carbon materials, but some others, such as graphite, diamond powder, carbon nanotubes, glassy carbon, fullerene soot, fullerenes $C_{60/70}$, acetylene black, coal char and ordered mesoporous carbons (CMK materials), have also been investigated. Amorphous carbons (activated carbon, carbon black, acetylene black, coal char) have a disordered structure with many high-energy sites (HES) on their surface. HES include dislocations, low-coordination sites, vacancies, atoms with free valences, discontinuities, edges, defects and other energetic abnormalities. It is generally accepted that HES constitute the main fraction of active sites in carbon catalysts so that the number of HES determines their catalytic activity. For this reason, amorphous carbons, which have a high defect concentration, are usually more active than well-ordered materials. The carbon atoms in HES react with methane molecules to balance their charge and stabilize themselves energetically, giving rise to the decomposition of methane.

Among the amorphous structures, activated carbons and carbon blacks are the most used materials due to their high activity. Although activated carbons are initially more active than carbon blacks, carbon blacks are more stable and show longer catalyst lifetimes. Different catalyst properties determine the activity and stability of carbon materials, as shown in Table 1.7.

1.2.10.1 Activity of Carbon Catalysts

The threshold temperature, which is the temperature at which hydrogen starts to be produced, has been used as a measure of the initial activity of carbon catalysts. Low threshold temperatures are equivalent to high catalytic activities. Activated carbons (mesoporous and microporous), carbon

Table 1.7 Determining factors of the activity and stability of carbon catalysts.

Determining factors of activity	Determining factors of stability
Defect concentration	Total area
Surface area	External surface area
Concentration of surface oxygenated groups released as CO and CO_2	Pore volume
Concentration of surface oxygenated groups desorbed as CO	Structure (interconnected mesoporosity)

blacks (black pearls 2000 and Vulcan XC72) and CMK catalysts (CMK-3 and CMK-5) exhibit the lowest threshold temperatures and hence, the highest initial activity. Their high initial activity is linked to the large density of graphene defects, which are preferential sites for methane adsorption and dissociation. A direct linear relationship exists between the number of defects on the graphene layers and the threshold temperature as well as the initial reaction rate. This fact strongly supports the idea that the surface defects are the main active sites of carbon catalysts. Additionally, the defect concentration and the degree of order, that is, the crystallinity of the carbons, are correlated parameters. Disordered structures, such as activated carbons and carbon blacks, usually have a high defect concentration and low crystallinity. However, CMK materials present a special behaviour since, despite having a high density of carbon defects, they show an intermediate crystallinity. The initial activity has not only been related to the threshold temperature but also to the initial methane decomposition rate at constant temperature. Despite this difference in the measure of initial activity, there is general agreement that carbons with a greater number of surface defects are catalytically more active.

Although the defect concentration of the carbon structure seems to be the most important parameter affecting the catalytic activity, other factors, such as the specific surface area and the concentration of oxygenated groups, can also influence the catalyst performance. For instance, carbons with higher surface areas usually exhibit superior catalytic activities compared with poor surface area materials. An approximately linear relationship in logarithmic scale has been established between the initial activity of different carbon samples and their surface areas. In other cases, despite the greater catalytic activity of carbons with larger surface areas, the relationship between both parameters is not linear. Hence, the surface area cannot be the only determining factor and the number of defects also plays a crucial role. The increase in surface area results in an increment of the number of active sites (defect concentration). However, the catalytic activity correlates quantitatively better with the defect concentration in the graphene layers rather than with the surface area. Contrary to these results, different activated carbons exhibit similar initial activities regardless of the surface area. This suggests that only a part of the surface area is involved in the decomposition of methane. In other studies, carbons with comparable surface areas show very different catalytic activities, which also indicates an apparent non relationship between the surface area and the activity. This was observed when comparing carbons of a different nature. For instance, activated carbon from hardwood displays a higher activity than carbon black (black pearls 2000) despite their similar surface areas. The same tendency was detected with structurally close carbons with the same surface area, such as carbon black and acetylene black. The higher activity of carbon black is attributed in this case to the larger amount of oxygenated surface groups. Different trends were observed when analysing the activity of several coal chars and activated carbons. When only coal chars are compared, an increase in the surface area leads to an increase in the initial activity for methane decomposition, although the relation is not linear. However, coal chars and activated carbons with very different specific surface areas can show similar activities. Here, the nature of the carbon plays a decisive role.

The concentration of oxygenated groups on the surface of carbon catalysts may also have an effect on the initial activity. Two different mechanisms explain their influence. Oxygenated groups can react directly with methane or can be released as carbon monoxide and/or carbon dioxide, which are active reaction sites for methane decomposition. An approximately linear correlation between the initial methane conversion rate and the concentration of oxygenated groups desorbed as carbon monoxide and carbon dioxide has been reported. Also, a good correlation has been established between the initial reaction rate and the concentration of oxygenated groups desorbed only as carbon monoxide, whereas those groups released as carbon dioxide do not show any influence. The exponential decay of the reaction rate during the initial period of the reaction may be due to the decrease of the surface oxygenated groups but also to the partial coverage of defects (active sites) by the carbon formed during the first stages of methane decomposition. Most of the oxygenated groups should be removed during the heating process before the reaction begins, so that the initial activity cannot be attributed exclusively to them. Although oxygen groups may have an impact on the initial activity, surface defects constitute the main part of active sites.

1.2.10.2 Stability and Deactivation of Carbon Catalysts

Although carbon materials usually display longer catalyst lifetimes and higher resistance to carbon coking and poisoning than metal catalysts, they also gradually become deactivated. The long-term efficiency and stability of carbon catalysts are often evaluated from their capacity for carbon accumulation before deactivation. Catalysts able to accumulate larger carbon amounts provide a more stable and sustained hydrogen production for longer times. The stability of carbon catalysts may be determined by a combination of pore size distribution and specific surface area. Mesoporous carbons with high surface areas often lead to a more sustainable hydrogen production because of their larger capacities for carbon deposition. To the contrary, the catalytic activity over microporous carbons decays more rapidly due to the lower carbon capacity and the greater mass transport limitations occurring in micropores. Even if the pores are not filled with carbon deposits, the narrowing of the pore mouth can also prevent the diffusion of methane molecules into the pores leading to the decrease of the catalytic activity. Additionally, a linear relationship exists between the surface area of the fresh catalyst and the stability, that is, catalysts with higher surface areas can accommodate higher carbon amounts before deactivation and thus provide long-term efficiency and sustainability. According to the evolution of the conversion of methane and the surface area over time, which show the same trend (initial drop followed by shallow decline), the catalyst deactivation may occur because of the loss in surface area. Contrary to these results, no relationship was found between the surface area and the long-term sustainability. Therefore, catalysts with similar surface areas would not necessarily accumulate the same amount of carbon deposits. In this case, the pore size distribution plays a decisive role. Carbon catalysts may also lose their activity by the progressive filling and blockage of the pores. For this reason, the pore volume is an important parameter that affects the catalyst lifetime since a bigger pore volume offers a larger space to accommodate carbon deposits. The pore volume defines the maximum amount of carbon deposits before deactivation and determines the maximum hydrogen production per mass of catalyst. There is actually a linear correlation between the catalyst pore volume and the mass of carbon accumulated until deactivation as well as the total hydrogen production. Furthermore, the deactivation of carbon catalysts may be partly explained by the loss of oxygenated groups on the surface over reaction time. Before catalyst deactivation the carbon product derived from methane decomposition may also have some catalytic effects. This fact is revealed by several kinetic studies, in which the carbon deposition rates are calculated throughout the reaction. Three different zones are identified along the reaction rate curves from the start of the reaction to catalyst deactivation. The first zone corresponds to an exponential decay of the catalytic activity, attributed to the removal of oxygenated groups from the catalyst surface, or to the partial coverage of defect sites by the carbon product. The second zone is characterized by an increase in the reaction rate. This fact denotes an autocatalytic effect, which indicates that the carbon produced from methane is also catalytically active. Nevertheless, the activity of these new active sites is clearly lower than the active sites in the fresh sample. In the last part of the kinetic curve, the reaction rate decreases until the catalyst completely deactivates. This zone may correspond to the deactivation of the new active sites or the drop in the effective surface area and pore volume.

CMK materials, which are ordered mesoporous catalysts and carbon blacks, show the highest stability among carbon catalysts. Carbon blacks have well-defined concentric graphene layers that generate large interparticle spaces. In addition to this, although some carbon blacks have a significant contribution of micropores to the overall surface area, they possess a high external surface area. Depending on the catalyst nature and pore structure, the carbon product remains within the pores, causing their blockage and leading to catalyst deactivation, or leaves the pores and grows on the outer part of the catalyst particles. The latter mechanism is expected to occur in catalysts with a high proportion of external surface area (carbon blacks) or an ordered and interconnected mesoporosity (CMKs). The ability of carbon deposits to move and grow towards the outside part of the particles avoids pore blockage and improves the accessibility of methane molecules even after the deposition of significant amounts of carbon. All this explains the higher resistance of CMK materials and carbon blacks to carbon deactivation.

Activated carbons have been widely investigated for methane pyrolysis due to their high initial activity. However, they show a poor long-term stability. Microporous activated carbons are quickly deactivated due to the micropore blockage caused by carbon deposits. Mesoporous activated carbons exhibit longer lifetimes than microporous materials, but the activity decay is relatively faster in comparison to carbon blacks. The reasons for the low stability of mesoporous activated carbons despite the high contribution of mesopores and the high share of external surface area remain unclear.

1.2.10.3 Regeneration of Carbon Catalysts

Different regeneration methods can restore the original activity of carbon catalysts. The regeneration techniques include combustion with oxygen or air and gasification with carbon dioxide or steam. The activity of carbon catalysts can be partially recovered by burning the carbon deposits with highly diluted oxygen in nitrogen. Nevertheless, the catalyst itself can also react with oxygen because it is usually more reactive than the carbon by-product, resulting in the loss of a part of the original catalyst. Via a carbon dioxide gasification regeneration process, the initial catalytic activity and the carbon accumulation decrease after each reactivation cycle. The reduction of the surface area and the concentration of surface oxygenated groups occur after each reaction–carbon dioxide regeneration cycle. This may be due to the removal of a part of the initial catalyst, which is less resistant to carbon dioxide gasification than the carbon deposits. In fact, after several reaction–regeneration cycles the carbon catalyst consists mainly of carbon derived from the reaction itself, whereas the initial catalyst has been gasified. The decrease of surface oxygenated groups would reduce the initial methane decomposition rate and the lower surface area would decrease the capacity for carbon accumulation and shorten the catalyst lifetime. Concerning the steam gasification process, this method significantly increases the surface area of the deactivated carbon catalyst, which allows almost complete restoration of the original activity. Even after several reaction–regeneration cycles the initial activity is completely recovered by means of the steam activation procedure. In this case, the disordered and highly reactive pyrolytic carbon deposits obtained during methane decomposition are more easily oxidized than the catalyst itself. Therefore steam gasification seems to be the most suitable regeneration technique to recover the initial catalytic activity of carbon catalysts. Furthermore, additional hydrogen is produced during the reactivation with steam and thus, the overall hydrogen yield is enhanced. The activity of carbon catalysts is partially or completely recovered by the previous regeneration procedures. Nevertheless, CO_x emissions are generated in all cases. Therefore, the commercialization or storage of the resulting mixture of catalyst and carbon coproduct are the most suitable options to prevent any carbon dioxide emissions. Unlike nickel and cobalt catalysts, carbon materials are cheaper and nontoxic, both of which are important advantages over metal catalysts when aiming to industrialize the process.

1.2.10.4 Co-Feeding to Extend the Lifetime of Carbon Catalysts

Co-feeding of methane with minor amounts of other hydrocarbons can improve the catalytic activity of carbon materials and partially overcome deactivation problems. The addition of a second compound to the methane feed gas aims to generate a catalytically active carbon product to keep a good activity for longer periods of time. Methane has been co-fed with saturated (propane), aromatic (benzene) and unsaturated (acetylene, ethylene) hydrocarbons. An accelerating effect on the decomposition rate of methane takes place when aromatic and unsaturated hydrocarbons are introduced. Moreover, a steady-state hydrogen production is achieved. Carbons produced from ethylene and especially from acetylene and benzene, are catalytically more active for the decomposition of methane than the carbon derived from methane itself. There is a good correlation between the activity and the crystallite size of the carbon product. The smaller crystallites produced from benzene, acetylene and ethylene possess a higher surface concentration of HES and consequently lead to greater catalytic activities. Nevertheless, carbon deposits from propane display a similar activity to that of methane-derived carbon and its incorporation does not result in a significant stability improvement. This may be due to the comparable size and structure of the carbon products derived from the same family of saturated hydrocarbons (methane and propane). In contrast, a beneficial effect of adding alkanes, such as ethane, has been observed in the noncatalytic methane pyrolysis. Here, the activation of methane and the subsequent formation of methyl radicals occur by the attack of radical species generated from the pyrolysis of ethane. These methyl radicals are successively incorporated into the pyrolysis products via radical reactions. Thus, methane can be activated by radicals generated from coexisting molecules without the use of a catalyst or operation at extremely high temperatures. The incorporation of an inert gas causes an analogous effect. In this case, the activation of methane molecules occurs upon collision with the molecules of the inert gas, which accelerates the overall reaction. The co-feeding with propylene, ethylene and ethanol can also reduce the deactivation of carbon catalysts and stabilize the catalytic activity for longer times. Carbon dioxide as a co-fed gas allows the partial regeneration of the deactivated catalyst by in situ carbon gasification. Although carbon dioxide leads to the stabilization of carbon catalysts, its incorporation is not favourable from the energetic and environmental point of view.

The co-feeding of methane is questionable for the industrialization of the process. The viability may be determined by the cost and the required amount of the co-fed compounds. Depending on the final hydrogen application, the feasibility of the industrial implementation may be dependent on whether additional purification steps are necessary to remove the unconverted co-fed compounds from the final gaseous product. Therefore, further research is needed to elucidate whether the addition of a second compound to the methane feed gas is worthwhile on an industrial level.

1.2.11 Reactors

1.2.11.1 Conversion, Selectivity and Yields

In a closed system, the conversion of a reactant A is designated by:

$$X = \frac{n_{A^{\circ}} - n_A}{n_{A^{\circ}}} \tag{1.55}$$

Conversion is dimensionless and is frequently expressed in %.

For selectivity

$$S = \frac{n_C}{n_{A^\circ} - n_A} \tag{1.56}$$

For yield

$$Y = \frac{n_C}{n_{A^\circ}} \tag{1.57}$$

This gives the basic expression:

$$Y = S.X \tag{1.58}$$

where $n_A{}^\circ$ = initial amount of species A_a (mol) and n_c = amount of species C_c (mol).

The case where the only values accessible are element mass concentrations is frequent in hydrocarbon technology. For differential reactor volume (with plug flow) the balance is written as show below:

$$N\,\mathrm{d}n_A = r_A \mathrm{dV} \tag{1.59}$$

where:
N = feed of reactant A in mol/time (reactor output), $N_A{}^0$ = amount of converted moles on mol of feed, r_A = total rate conversion of feed (mol/volume$^{/}$time) owing to order of reaction and V = volume of a reactor.

This allows the calculation of volume of a reactor:

$$\frac{V}{N} = \int_0^{n_A} \frac{dn_\mathrm{A}}{r_\mathrm{A}} \tag{1.60}$$

Reactors involving a single fluid phase present only a little difficulty in scaling up. Homogeneous reactions may be carried out using one of the following reactor systems:

- (semi)-bath reactor (fine chemistry, biotechnology);
- continuous tubular reactor (steam cracking);
- continuous stirred-tank reactor/cascade of stirred-tank reactors (homogeneous catalysis).

1.2.11.2 Modelling Approach of the Structured Catalytic Reactors

The numerical simulations of the reactor configurations were performed using the DETCHEM^{CHANNEL} code, part of the DETCHEM program package (Deutschmann et al.). For experiments conducted in a monolithic reactor, a single channel is simulated using the conditions listed in the corresponding reference.

For fixed-bed experiments, the reactor is replicated by simulating an imaginary path through the fixed bed as a channel. The dimensions of this cylindrical reactor are calculated based on the properties of the packed bed using an approximation that estimates the channel diameter to be equal to that of the powder bed particles (Giehr). The velocity of the reactive flow is corrected for the gain in open surface area in a channel compared to a fixed bed. The volumetric flow rate and, consequently, the linear velocity are recalculated using the bed porosity to obtain a value for the open-faced area fraction. The porosity, if not stated explicitly, is calculated using the approximation by

Pushnov. An available simulation code for fixed-bed reactors was not compatible with all parameters from the dataset. For the remaining experiments, the results of channel and fixed-bed simulations agreed very well. The code resolves the path of the reacting flow through the equivalent channel in the steady state two-dimensionally using the boundary layer approximation. This leads to the following set of governing equations

$$\frac{\partial(r\rho u)}{\partial z}+\frac{\partial(r\rho v)}{\partial r}=0 \tag{1.61}$$

$$\frac{\partial(r\rho uY_i)}{\partial z}+\frac{\partial(r\rho vY_i)}{\partial r}=-\frac{\partial}{\partial r}(rJ_i)+r\dot{\omega}_i W_i \tag{1.62}$$

$$\frac{\partial(r\rho u^2)}{\partial z}+\frac{\partial(r\rho uv)}{\partial r}=-r\frac{\partial p}{\partial z}+\frac{\partial}{\partial r}\left(\mu r\frac{\partial u}{\partial r}\right) \tag{1.63}$$

$$\frac{\partial p}{\partial r}=0 \tag{1.64}$$

$$\rho=\frac{p\bar{M}}{RT} \tag{1.65}$$

where r is the radial coordinate, ρ is the density, z is the axial coordinate, u is the axial component of velocity, v is the radial component of velocity, p is the pressure, Y_i is the mass fraction of species i, μ is the viscosity, T is the temperature, J_i is the radial diffusion flux of species i, $\dot{\omega}_i$ is the gas-phase production rate of species i, W_i is the molecular mass of species k and $\bar{M}$ is the mean molar mass. As the simulations are carried out using isothermal conditions and the solid and gas phases consequently have the same temperature, no enthalpy balance is required. The simulations resolve the channel two-dimensionally to account for a velocity gradient due to wall effects, utilize the mean-field approximation and treat the experiments as isothermal processes. In the mean-field approximation, the state of the reactive surface is represented by the assumed average of the states present in the evaluated computational unit, characterized by the temperature and coverages with the various species Θ_i (Deutschmann et al.).

1.2.11.3 Reactor Concept for Catalytic Carbon Dioxide Methanation

The catalytic methanation of carbon dioxide (Section 1.2.3; Equation 1.14) is a highly exothermic reaction firstly described by Sabatier in 1902, which is recently gaining attention with regards to carbon dioxide-consuming processes producing energy carriers. Due to the reaction equilibrium, high methane yields are obtained for carbon dioxide methanation at low temperature and high pressure.

In technical applications, temperatures above 200°C are used with pressures ranging from 0.5 to 10 MPa, depending on the specific application. There are various catalysts available to accelerate the methanation reaction; in commercial applications nickel-based systems are dominant because of their high catalytic activity and low price.

The vast majority of reactor concepts discussed for catalytic carbon dioxide methanation are:

- fixed-bed reactors;
- structured reactors;
- fluidized-bed reactors; and
- slurry bubble column reactors.

Fixed-bed reactors present the state of the art for large-scale methanation applications. In adiabatic operation mode, multiple fixed-bed reactors are used and reaction heat is removed by intercooling between the reactors and/or staged addition of feed gas. This operation mode causes high heat stress on catalyst and reactor material; typical reactor temperatures range between 250 and 700°C.

Using cooled fixed-bed reactors usually leads to significantly lower temperatures in the reactor. Hence, the overall process is simplified and fewer reactor steps are needed. However, one of the main drawbacks of cooled fixed-bed reactors is the formation of temperature hotspots due to the restricted heat transfer through the catalyst bed and reactor walls to the cooling medium. In tube bundle reactors limited heat transfer restricts the individual tube diameter and consequently more reactor tubes are needed.

Efficient heat management is vital for all reactor concepts to reach the usual requirements on process efficiency ($\eta > 75\,\%$), load flexibility and methane concentrations ($y_{CH4} > 95\,vol\,\%$) in catalytic methanation. There are several reactor concepts available that show improved heat transfer to fulfil these requirements.

In fluidized-bed reactors gas flow introduced into the reactor fluidizes the catalyst particles and causes a high degree of mixing. This effect and the high heat capacity of the catalyst particles allows for nearly isothermal operation and the avoidance of temperature hotspots. Regarding process design, a single reactor is most often sufficient to reach the desired conversion for injection into the natural gas grid. However, attrition processes reduce catalyst particle size and particles are involuntarily discharged from the reactor.

In structured reactors metallic structures are often part of the reactor interior or are used as catalyst carrier significantly enhancing the heat transfer from the catalyst to the cooling medium on the outer shell of the reactor tube. As these structures show significantly higher radial heat conductivity compared to fixed-bed reactors, lower hotspot temperatures are achieved. Structured reactors are characterized by high GHSV and main drawback is the complex procedure to immobilize the catalyst on the structures.

Slurry bubble column reactors include a liquid phase in the reactor, directly present on the catalyst surface where the heat of reaction is produced. Due to its high heat capacity and heat conductivity this liquid facilitates heat management. The reactor design allows for highly dynamic operation modes due to its thermal indolence dampening hotspots or cold spots caused by times of high or low load. However, slurry bubble column reactors show low values for GHSV due to additional mass transfer resistances in the liquid phase that are not present in most other methanation reactors.

For both reactor systems research at the Engler-Bunte-Institut (EBI) focuses on reaction kinetics as well as heat and mass transfer phenomena. Furthermore, hydrodynamic behaviour in the slurry bubble column reactor (SBCR) is investigated (Mörs et al.). Götz identified dibenzyl toluene (DBT) as suitable fluid, which is recently gaining attention due to its application as liquid organic hydrogen carrier (Preuster et al.). Lefebvre determined reaction kinetics in the three-phase reactor. With regard to hydrodynamics, Götz developed a novel gas holdup correlation for the SBCR that can be used to describe the homogeneous flow regime.

Schollenberger determined reaction kinetics for the honeycomb reactor (HCR). Recent research focusses on heat transfer phenomena and experimental determination of the effective axial and radial heat conductivity in commercial honeycombs.

Catalytic methanation has been extensively studied in the context of Power-to-Gas applications (Held et al.). Generally, the reactor types used can be divided into two-phase and three-phase systems. In commercial applications mostly adiabatic and polytropic fixed-bed reactors are applied, which therefore form the state of the art for catalytic methanation. Other two-phase reactor concepts are structured reactors such as microchannel or honeycomb reactors. Fluidized-bed reactors can be operated with two or three phases, whereas slurry reactors are exclusively operated with three phases. Structured reactors, fluidized-bed reactors and slurry reactors are subject to ongoing research.

At the Engler-Bunte-Institut two innovative methanation reactor concepts were developed and are subject to ongoing research activities with focus on dynamic operation and optimized heat transfer characteristics for Power-to-Gas applications: a two-phase honeycomb reactor (HCR) and a three-phase slurry bubble column reactor.

1.2.11.4 Monolithic Reactors

The monolithic (honeycomb) reactor is a two-phase structured fixed-bed reactor that contains coated catalyst carriers made of stainless steel. It is designed as a multitube reactor, in which the metallic catalyst carriers are placed in parallel tubes. A schematic drawing of the honeycomb-like bodies is shown in Figure 1.19(A).

They are made of a combination of corrugated and plain metal sheets, which are jointly coiled up. The layers are form-fit pressed in a cladding tube. Typically, honeycomb structures are characterized by the number of parallel channels per square inch (CPSI). For the discussed application, honeycomb structures of 100–600 CPSI are used corresponding to channel diameters of 0.1–2.8 mm (Chapter 12). This fits the typical dimensions of microreactor channels, but the honeycombs are by definition not microreactors (Ernig et al.). The feed gas flow enters the channels coated with the catalyst and if the reactor temperature is high enough (above 200°C) the catalytic methanation reaction starts. Carbon dioxide and hydrogen are converted to methane in the porous catalyst layer and reaction heat is released mostly at the channel inlets. As a result, characteristic temperature

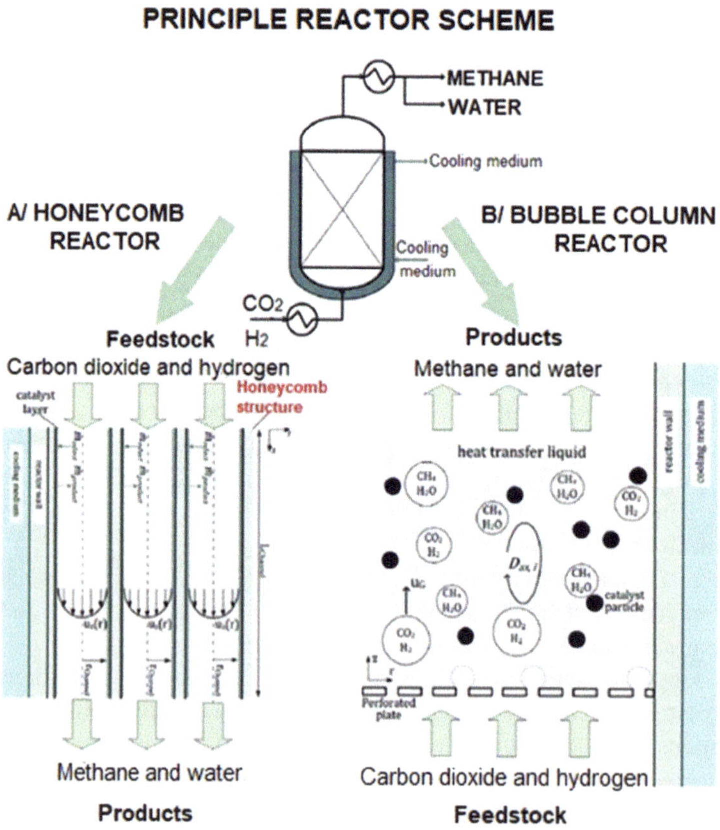

Figure 1.19 Schematic drawing of the honeycomb reactor (A) and the three-phase slurry bubble column reactor (B).

profiles with a defined peak at the inlet and an outlet temperature close to the cooling medium temperature ($\Delta T < 10$ K) are obtained (Schollenberger).

For typical conditions of the honeycomb reactor the Reynolds number in the channels is in the range of 10–500. Based on a channel diameter of 1 mm and low Reynolds numbers the channels can be described as ideal plug flow reactor ($Bo > 100$) with nearly complete cross mixing in a radial direction (Ernig et al.). This illustrates the intensification of transfer phenomena in the channels.

Reaction conditions are set to allow for a maximum hotspot temperature of 550°C in the reactor. To achieve this, the temperature inside the reactor is measured using temperature sensors and the temperature of the cooling medium is adjusted accordingly. For the described reactor system, the cooling medium temperature is set to 220°C. Pressures ranging from 0.6 to 2.1 MPa (absolute) are chosen since higher pressures influence the thermodynamic equilibrium to only a small extent. The catalyst coating on the metallic honeycombs is a commercially available product.

The slurry bubble column reactor presented has three distinctive phases: the commercially available solid catalyst (particle size of 50–100 μm) is suspended in a heat transfer liquid and is fluidized by the educt gases. A schematic drawing of all involved components is shown in Figure 1.19 (B).

Heat management in the SBCR is implemented by the heat transfer fluid, which shows high heat capacity and, thus, enables efficient heat transfer from the catalyst particles to the cooling medium in the cooling jacket. The educt gases entering the bubble column through a perforated plate at the bottom enable back mixing resulting in isothermal operation (Mors et al.). Requirements on the heat transfer fluid are high educt and product gas solubility, high heat capacity and high thermal stability.

Reaction conditions are set to a temperature of approx. 320°C and pressure of 2.0 MPa (absolute). Kinetically a high temperature is favourable, but its maximum is limited by the thermal stability of the liquid phase present in the reactor and the thermodynamic equilibrium presenting the maximum conversion and corresponding methane yield.

1.2.11.5 Mass Transfer in the Honeycomb and Slurry Bubble Column Reactor

An overview of mass transfer phenomena is presented in Figure 1.20. For both systems the reaction takes place in the catalyst particle or in the catalyst layer, so transfer of gaseous educts and products to and from the catalyst must be considered.

In the honeycomb reactor, educts are transferred from the bulk phase to the catalyst layer (thickness of about 100 μm) where the reaction takes place resulting in a concentration gradient in the gas phase. This process can be described by mass transfer coefficients β_i (1). From the outer surface

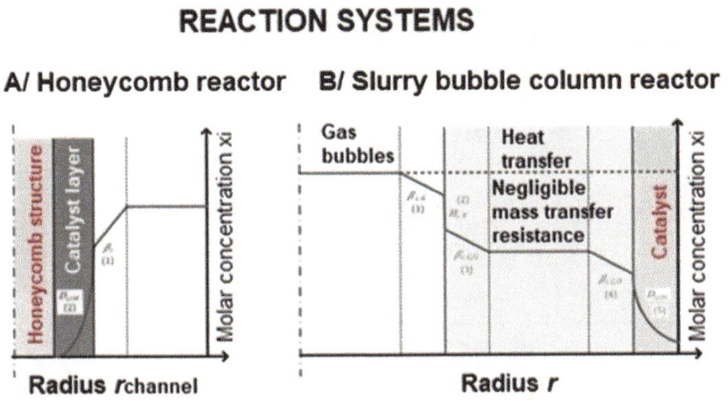

Figure 1.20 Mass transfer phenomena in the honeycomb reactor (A) and the three-phase reaction system (B) based on theoretical considerations.

of the catalyst layer, the gas molecules diffuse to the active sites in the catalyst. This process can be described using molecular diffusion coefficients D_i (2) (Figure 1.20 (A)).

In the SBCR introducing a heat transfer liquid in the reactor offers benefits regarding heat management and dynamic operation, but it also introduces additional mass transfer resistances. A schematic concentration profile is presented in Figure 1.20 (B).

Transfer of the educt gas components from the bulk phase to the gas-liquid layer is described by the mass transfer coefficients $\beta_{i,G}$ (1). The Henry coefficients $H_{i,X}$ of the components describe solubility in DBT (2); this is a rate-limiting step especially for hydrogen as it shows low solubility in most liquids. Solved in the liquid phase, mass transfer from the gas-liquid layer (3) is considered using the mass transfer coefficients $\beta_{i,G/L}$. The bulk liquid phase can be assumed as ideally mixed.

The mass transfer resistance from the liquid phase to the solid catalyst particles is described using the mass transfer coefficients $\beta_{i,L/S}$ (4). At the catalyst, particle gas molecules diffuse from the surface to the active catalytic sites in the pores of the particle (5). This process can be described using the diffusion coefficients $D_{i,cat}$.

Mass transfer in the SBCR plays a crucial role determining the effective reaction rate. In the HCR, mass transfer resistances are minimized. Since the channel diameter is in the range of 1 mm, radial diffusion plays a significant role and convective transfer is not dominant. The intensification of transfer phenomena leads to high conversion rates and high specific methane yields compared to established fixed-bed reactors.

In the SBCR, mass transfer resistances are limiting the effective reaction rate leading to low GHSV values compared to two-phase reactor systems. To reach high conversion, larger reactor volumes are needed, which is a slight drawback compared to established fixed-bed reactor systems.

1.2.11.6 Heat Transfer in Honeycomb and Slurry Bubble Column Reactors

Both reactor concepts introduced show advanced heat management characteristics compared to conventional methanation reactor concepts. An overview of temperature profiles in the HCR and the SBCR is shown in Figure 1.21.

In general, in honeycomb reactors, the approaches to describe heat transfer in the honeycomb channels are analogous to mass transfer mechanisms. A characteristic radial temperature profile is shown in Figure 1.21 (A). Reaction heat is released in the catalyst layer and is transferred to the

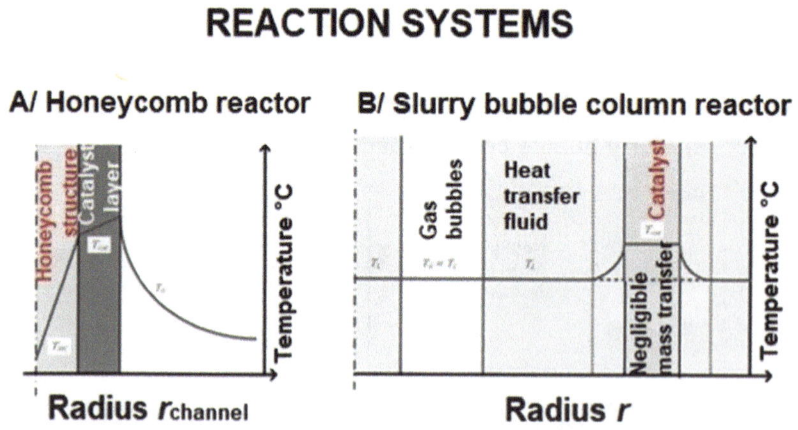

Figure 1.21 Heat transfer phenomena in the honeycomb reactor (A) and the three-phase reaction system (B) based on theoretical considerations.

cooling medium by conduction through the catalyst carrier. Depending on the feed volume flow and the outer heat transfer coefficient an axially and radially pronounced temperature gradient is obtained. Consequently, the HCR is operated in polytropic mode. Radial conduction through the catalyst carrier dominates heat transfer, whereas convective heat transfer in the channels is almost negligible. Due to radial heat transfer resistances in the catalyst carrier higher temperatures are obtained with increasing throughput as more reaction heat is produced. Temperature restrictions of the catalyst (hotspot temperature $T_{HS} < 550°C$) are limiting the throughput of the honeycomb reactor.

To describe heat transfer in metallic honeycombs, established approaches in the literature are based on monoliths and do not consider additional resistances between the coiled metal layers (Schollenberger et al., Visconti et al.). For the metallic honeycombs, heat transfer is described by a model developed at the EBI for cubic cells based on the interconnection of the heat transfer resistances.

In the SBCR, reaction heat is removed using a heat transfer liquid. A temperature profile of the reaction zone is presented in Figure 1.21 (B). The mixing effect in the bulk liquid phase causes a constant temperature in the fluid phases. At the catalyst particle, temperature is higher because of the reaction heat released. However, the temperature increase is not critical since convection is high. The combination of high heat capacity of the liquid phase and high degree of mixing enables isothermal reactor operation.

The liquid phase also enables dynamic operation of the reactor, because of its heat management capabilities. Due to inevitable fluctuations in renewable energy supply, educt gas load of the methanation reactor can fluctuate. The high heat capacity of the system prevents the reactor from cooling down or heating up (too) fast, so periods of high or low gas load can be buffered and the necessary temperatures of operation are assured. The transient operation of the SBCR has been investigated by Lefebvre et al.

In the honeycomb reactor the throughput is limited by heat transfer. However, all relevant steps have been identified and can be quantified for reactor scale-up. The slurry bubble column reactor is characterized by isothermal operation and heat transfer is technically not an issue. This presents a major advantage compared to two-phase reactors, in which characteristic temperature hotspots are unavoidable. Hence, no limiting heat transfer resistances are present, enabling dynamic operation of the reactor system.

1.2.11.7 Process Design

The two reactor systems have been experimentally investigated at various scales with the aim of obtaining reliable data for scale-up to commercial applications. Lab-scale plants at the Engler-Bunte-Institute provide first insights into the systems and are used in ongoing research projects for academic research accompanying the experiments at pilot- and demo-scale (Held et al.). For methanation in the honeycomb reactor a lab-scale plant was built in 2011 for basic research. Within the scope of the KIC DemoSNG project (Bajohr et al.), in 2014 a mobile demo plant was designed with an installed methanation output of 60 kW. During the project the demo plant was transported to Sweden to the project partner CORTUS SE and operated with a syngas from a biomass gasification. During the operation of the plant a total carbon conversion (carbon monoxide and carbon dioxide) of more than 99 % was achieved.

In 2017 the up to now biggest plant using honeycomb reactor technology was built in Falkenhagen. This Power-to-Gas plant with an installed electric power of 1 MW and the accompanying research activities are part of the European Union-funded project STORE&GO. Commissioning of the plant was completed in 2018 and the plant has been on stream for more than 1200 hours. The produced substituted natural gas is injected in a natural gas transportation grid, which means that all relevant German requirements ($y_{H2} < 2$ vol% and $y_{CH4} > 95$ vol%) are met. The pilot plant demonstrates the fully integrated operation of a Power-to-Gas plant with high efficiency. Ongoing activities at the Engler-Bunte-Institute focus on refining the technology in cooperation with industrial partners for a plant size of approx. 20 MW.

For methanation in the slurry bubble column reactor a first lab-scale plant was built in 2009 after preceding bench-scale tests. The plant is equipped with both a SBCR and an autoclave reactor to investigate different aspects of the reaction system.

In 2019, a unique slurry bubble column reactor plant was commissioned as part of the KIT Energy Lab 2.0 with a methane output equivalent to 100 kW. The reactor has a diameter of about 250 mm and a total height of roughly 3 m. The dimensions allow a realistic evaluation of the reaction system for technical application as influences of the reactor diameter on gas holdup can be neglected for diameters above 150 mm. Basic research activities will be performed to give reliable information on long-term operation under technical conditions, e.g., with synthesis gas from the adjacent bioliq® gasification plant or with highly dynamic load changes that can be expected in a technical Power-to-Gas process.

Commissioning of the pilot plant and first tests on dynamic methanation were completed successfully in June 2019. In August 2019, further tests were performed on stationary and dynamic operation. Different educt gas ratios and gas loads were set to investigate influences on carbon dioxide conversion with the prospect of using the reactor in combination with a downstream liquefaction unit.

Further scale-up of honeycomb reactor and slurry bubble column reactor designs will use different approaches. The honeycomb reactor consists of multiple pipes containing metallic honeycomb structures; so far scale-up has been performed by numbering-up. Recent research activities aim at alternative reactor designs like radial flow reactors as numbering-up is no longer beneficial for methanation reactors significantly above the 10-MW scale.

For the slurry bubble column reactor, hydrodynamics and its impact on reactor performance has to be observed very closely during the scale-up procedure. At pilot scale, hydrodynamic data should be acquired under process conditions as close as possible to those of the full-scale unit to avoid false estimation of important design parameters such as the gas holdup or gas velocities of regime transition.

1.2.11.8 Comparison and Outlook

Methane is a highly versatile chemical energy carrier, which offers huge storage capacities and a well-established network for distribution and utilization. Catalytic methanation requires high temperatures to obtain the reaction rates necessary for high product yields per pass. However, temperature is limited by the thermodynamic equilibrium of the highly exothermic reaction. Suitable reactor concepts should address this issue and still allow for high throughput and methane yield.

Future methanation applications need plants of different sizes and with various abilities regarding (load) flexibility. In large-scale plants of more than 10 MW, substituted natural gas output could be used to continuously convert a minimal fluctuating base load of electric energy and/or could be combined with, for example, (biomass) gasification giving a constant base load with a Power-to-Gas option in times of surplus electricity. Plants with a substituted natural gas output of less than 10 MW could help to stabilize decentralized Power-to-Gas systems by offering a highly dynamic synthesis as storage technology.

The honeycomb methanation system shows good load flexibility and high specific methane output, which is especially favourable in Power-to-Gas plants of high base load. In comparison, the slurry bubble column reactor shows lower specific methane production due to mass transfer limitations. However, the possibility of operating the slurry bubble column reactor in a highly dynamic way presents a main advantage for its aimed application in Power-to-Gas plants with low or even no base load.

For both reactor concepts, scale-up from laboratory to pilot scale has been successfully performed. Ongoing projects focus on system optimization and the move to commercial scale. A promising approach is the implementation of a honeycomb reactor as a second reactor step after the SBCR to achieve high specific methane output combined with highly dynamic operation.

1.3 Potential Steps Towards Sustainable Hydrocarbon Technology: Vision and Trends

The petrochemical/chemical and allied products technology has certain well-defined characteristics. This comprehensive section is focused on the vision, business trends, digital transformation and RAM modelling. The refinery of 2030 will be very different from the refinery of 2020. The oil and gas industry is at a turning point and there is no doubt that COVID-19 is disrupting the oil and gas industry. The global chemical technology sector continues to expand exponentially as developing nations' demand for chemical/petrochemical products continues to increase. The refining business needs to change dramatically over the next decade. Refinery executives worldwide are tuned to the signals of change, but some are not sure of the best course of action. Major process units and equipment in the refinery of 2030 will be smart plant building blocks. The production of key plastics will more than double between 2010 and 2050.

In last two decades micro-technologies and micro-equipment have been of great interest in chemical engineering. Small series of high-performance micro heat expanders, reactors, mixers, extractors, pumps and valves has been already developed and produced. The uptake of hybrid, electric and high-efficiency vehicles, along with the pace of renewable power, are key variables in forecasting the demand for hydrocarbon fuels. 2019 was a year of successes within the recycled polyester (RPET) industry. The digital transformation offers new tools that enable downstream producers to improve the yield of valuable products, while also reducing energy consumption and increasing throughput.

The chemical/petrochemical and allied products technology has certain well-defined characteristics, that have been discussed in *Petrochemistry* (Bajus) (Table 1.8). Characteristics 11 and 12 will be discussed in the next section.

An oil refinery or petroleum refinery is an industrial process plant where crude oil is transformed and refined into more useful products such as petroleum naphtha, gasoline, diesel fuel, asphalt base, heating oil, kerosene, liquefied petroleum gas, jet fuel and fuel oils. Petrochemicals feedstock like ethylene and propylene can also be produced directly by cracking crude oil without the need of using refined products of crude oil such as naphtha.

Table 1.8 Characteristics of chemical/petrochemical technology.

1)	Maturity and its consequences
2)	Participation in international trade
3)	Competition from the developing countries
4)	Capital intensity and economics of scale
5)	Criticality and pervasiveness
6)	Freedom of market entry
7)	Strong regulation
8)	High research and development expenses
9)	Dislocations
10)	Feedstock recycling
11)	Business trends
12)	RAM modelling

Oil refineries are typically large, sprawling industrial complexes with extensive piping running throughout, carrying streams of fluids between large chemical processing units, such as distillation columns. In many ways, oil refineries use much of the technology and can be thought of as types of chemical plants.

The crude oil feedstock has typically been processed by an oil production plant. There is usually an oil depot at or near an oil refinery for the storage of incoming crude oil feedstock as well as bulk liquid products.

Petroleum refineries are very large industrial complexes that involve many different processing units and auxiliary facilities such as utility units and storage tanks. Each refinery has its own unique arrangement and combination of refining processes largely determined by the refinery location, desired products and economic considerations.

An oil refinery is considered an essential part of the downstream side of the petroleum industry. Some modern petroleum refineries process as much as 127,000 to 143,000 cubic meters (800,000 to 900,000 barrels) per day. According to the *Oil and Gas Journal*, across the world a total of 700 refineries were in operation on the 31 December 2020 for a total capacity of 13,951,000 million cubic meters (87.75 million barrels) per day. Jamnagar Refinery has been the largest oil refinery since 25 December 2008, with a processing capacity of 197,000 million cubic meters (1.24 million barrels) per day. Located in Gujarat, India, it is owned by Reliance Inc.

1.3.1 Technology Readiness Levels

Another important aspect in fuel cost calculation is the technological maturity of the various technologies since costs need to be predicted. To estimate technological maturity, the technology readiness levels (TRL) are an appropriate method.

The technology readiness levels are defined in the same way by the United States Department of Defense as well as by the European Commission. The scale of TRL is from 1 to 9. TRL 1 as the lowest level means that basic principles of a technology are known. Technology readiness level 9 is the highest level and it describes a technology as proven in an operational environment. Where a topic description refers to a technology readiness level, the following definitions apply, unless otherwise specified (Table 1.9).

Table 1.9 Technology readiness levels, definitions, descriptions and supporting information.

TRL	Definition	Description	Supporting Information
1	Basic principles observed and reported	Lowest level of technology readiness. Scientific research begins to be translated into applied research and development (R&D). Examples might include paper studies of a technology's basic properties.	Published research that identifies the principles that underlie this technology. References to who, where, when.
2	Technology concept and/or application formulated.	Invention begins. Once basic principles are observed, practical applications can be invented. Applications are speculative and there may be no proof or detailed analysis to support the assumptions. Examples are limited to analytic studies.	Publications or other references that outline the application being considered and that provide analysis to support the concept.

Table 1.9 (Continued)

TRL	Definition	Description	Supporting Information
3	Analytical and experimental critical function and/or characteristic proof of concept	Active R&D is initiated. This includes analytical studies to physically validate the analytical predictions of separate elements of the technology. Examples include components that are not yet integrated or representative.	Results of laboratory tests performed to measure parameters of interest and comparison to analytical predictions for critical subsystems. References to who, where and when these tests and comparisons were performed.
4	Component and/ or breadboard validation in a laboratory environment	Basic technological components are integrated to establish that they will work together. This is relatively "low fidelity" compared with the eventual system. Examples include integration of "ad hoc" hardware in the laboratory.	System concepts that have been considered and results from testing laboratory-scale breadboard(s). References to who did the work and when. Provide an estimate of how breadboard hardware and test results differ from expected system goals.
5	Component and/ or breadboard validation in a relevant environment	Fidelity of breadboard technology increases significantly. The basic technological components are integrated with reasonably realistic supporting elements so they can be tested in a simulated environment. Examples include "high-fidelity" laboratory interaction of components.	Results from testing laboratory breadboard system are integrated with other supporting elements in a simulated operational environment. How does the "relevant environment" differ from the expected operational environment? How do the test results compare to expectations? What problems, if any, were encountered? Was the breadboard system refined to more closely match the expected system goals?
6	System/ subsystem model or prototype demonstration in a revelant environment.	Representative model or prototype system, which is well beyond that of TRL 5, is tested in a relevant environment. Represents a major step up in a technology's demonstrated readiness. Examples include testing a prototype in a high-fidelity laboratory environment or in a simulated operational environment.	Results from laboratory testing of a prototype system that is near the desired configuration in terms of performance, weight and volume. How did the test environment differ from the operational environment? Who performed the tests? How did test compare with expectations? What problems, if any, were encountered? What are/were the plans, options, or actions to resolve problems before moving to the next level?
7	System prototype demonstration in an operational environment	Prototype near or at planned operational system. Represents a major step up from TRL 6 by requiring demonstration of an actual system prototype in an operational environment (e.g., in an aircraft, in a vehicle, or in space).	Results from testing a prototype system in an operational environment. Who performed the tests? How did the test compare with expectations? What problems, if any, were encountered? What are/were the plans, options, or actions to resolve problems before moving to the next level?
8	Actual system completed and qualified through test and demonstration	Technology has been proven to work in its final form and under expected conditions. In almost all cases, this TRL represents the end of true system development test and evolution (DT&E) of the system. System is "flight qualified" through test and demonstration.	Results of testing the system in its final configuration under the expected range of environmental conditions in which it will be expected to operate. Assessment of whether it will meet its operational requirements. What problems, if any, were encountered? What are/were the plans, options, or actions to resolve problems before finalizing the design?

(Continued)

Table 1.9 (Continued)

TRL	Definition	Description	Supporting Information
9	Actual system proven through test and demonstration.	Actual application of the technology in its final form and under mission conditions, such as those encountered in operational test and evaluation (OT&E). Examples include using the system under operational mission conditions.	OT&E reports.

On the one hand, the comparison of different productions costs calculated by different groups shows the need of standardized calculations, especially regarding the hydrogen production. Sensitivity analyses are important. On the other hand, the key role of the hydrogen supply was highlighted. Mostly, Hydrogen-to-Fuel technologies are already available at high TRL. However, there is a need for research in the field of electrolysis to achieve the cost targets.

Technology readiness levels of different Power-to-Fuel technologies including electrolysis and carbon dioxide supply technologies are given in a study published in 2016 by the German Environmental Agency (Table 1.10). Depending on the technology, the carbon dioxide extraction from industrial exhaust gas has a TRL of 6–9, whereas the carbon dioxide absorption from air has a technology readiness level of 6. To give an example, the energy required for the chemical absorption-based (MEA) post combustion capture technology for cement as well as black coal power plants has a technology readiness level of 9 and requires 1.2–1.5 MJ electrical energy per kg of carbon dioxide.

Table 1.10 Technology readiness level (TRL) of carbon capture and utilization in the chemical technologies.

Production processes	TRL
1. **Hydrogen production via low carbon emission routes**	
• Alkaline electrolysis	7–9
• PEM-electrolysis	7–8
• High-temperature solid-oxide electrolysis	5–7
• Methane pyrolysis	4–5
• Thermochemical processes	4
• Photocatalytic processes	2–3
2. **Methanol production from hydrogen and carbon dioxide**	
• Conventional methanol production	9
• Low-carbon methanol production	7
• DME synthesis from methanol	9
3. **Ethylene and propylene via hydrogen-based methanol**	8–9
• Conventional ethylene and propylene	9
• Low-carbon production of ethylene and propylene via MTO	8–9

Table 1.10 (Continued)

Production processes		TRL
4. **Benzene, toluene and xylenes via hydrogen-based methanol**		7
Conventional production of BTX		9
• Low-carbon production of BTX via hydrogen-based methanol		7
5. **Production of synthetic fuels via hydrogen-based syngas**		
1. Fischer-Tropsch synthesis		5–7
2. Low-carbon methane/SNG production from hydrogen and CO_2		6–7
6. **Low-carbon chemical production based on biomass as feedstock**		
• Biomethanol production		6–7
• Bioethanol production		7–9
• Bioethylene production		8–9
• Biopropylene production		6–7
• BTX production from biomass		6–7
7. **Ammonia production**		
• Conventional ammonia production		9
• Low-carbon ammonia production		7
8. **Chlorine production**		9
9. **Direct use low-carbon electricity**		
• Electricity based steam production		7
• Advanced heat management via steam recompression		6–7
• From industrial exhaust gas		6–9
• CO_2 absorption from air		6
• Chemical absorption-based (MEA)		9
11. **Hydrogen-to-fuel technologies**		
• Fischer-Tropsch process		9
• Methanol via H_2 and CO_2		9
• DME synthesis from methanol		9
• Methanol-to-gasoline		9
By Perez-Fortes and Tzimas for methanol process		6–7
By Buddenberg et al. for methanol process		8
By Buddenberg et al. for methanol-to-gasoline process		6
• Synthesis gas for Fischer-Tropsch technology		6
• Synthesis of OME_{3-5}		4
12. **CCU implementation**	**TRL > 7**	**TRL < 7**
• Benzene ←	Methanol-to-aromatics	$CO_2 + H_2$
• Ethylene ←	MTO	$CO_2 + H_2$

(Continued)

Table 1.10 (Continued)

Production processes		TRL
		$CO_2 + CH_4$
• Ethylene oxide ←		$C_2H_4 + CO_2$
• Methanol ←		From CO_2 and H_2 (direct hydrogenation)
• Propylene ←	MTO	$CO_2 + H_2$
• Styrene ←		$C_8H_{10} + CO_2$
• Toluene ←	Methanol-to-aromatics	$CO_2 + H_2$
• p-Xylene ←	Methanol-to-aromatics	$CO_2 + H_2$
• Xylene (mixed)	Methanol-to- aromatics	$CO_2 + H_2$
• Hydrogen ←		
		From water electrolysis
• Methane ←		From CO_2 (Sabatier reaction)
• Carbon dioxide ←		Capture from highly concentrated
• Carbon monoxide ←		$CO_2 + H_2$

According to this study, the technology readiness levels of alkaline electrolysis, polymer-electrolyte membrane electrolysis (PEM) and high temperature electrolysis (SOC) are 9, 8 and 5, respectively. In contrast to that, according to Sterner, in 2014, the technology readiness levels of PEM and SOC were 7 and 6, respectively. In a study published in 2016, a full market penetration of the PEM is expected in 2040.

Nowdays, due to relatively low technology readiness levels, it is difficult to assume and forecast electrolysis costs. For 2030, Stolten and Scherer predict investment costs of installed capacity of alkaline and PEM electrolysers of 585 €/kW. According to Mergel, in 2013, the costs for alkaline electrolysers were 1000 €/kW and for PEM higher than 2000 €/kW. These costs will drop to 500 €/kW and below 600 €/kW, respectively. According to the PlanDeLyKad study, today's research goal is 500 €/kW. After 2030, the study forecasts a significant cost advantage for the PEM electrolysis (305 €/kW) compared to the alkaline electrolysis (520 €/kW).

As listed in the study of the German Environmental Agency, the technology readiness levels of various Hydrogen-to-Fuel technologies is relatively high: the Fischer-Tropsch process, the methanol direct synthesis via hydrogen and carbon dioxide, DME synthesis from methanol and Methanol-to-Gasoline process have technology readiness levels of 9. In contrast to that, Perez-Fortes and Tzimas estimate a technology readiness level of 6–7 and Buddenberg et al. a technology readiness levels of 8 for the methanol process from hydrogen and carbon dioxide. The estimations of Perez-Fortes and Tzimas as well as of Buddenberg et al. based on the presence of a pilot plant created by the company CRI in Iceland that has been operating since 2007. Furthermore, Buddenberg et al. estimate the technology readiness levels of 6 for the Methanol-to-Gasoline process. To provide the synthesis gas for the Fischer-Tropsch process, a reactor for the reverse water gas shift reaction is needed. This technology has a TRL of 6 and thus the technology readiness level for the whole process drops to 6. For the direct synthesis of OME_{3-5} from methanol and formaldehyde, Burger et al. estimate a TRL of 4 in 2017 (Table 1.10).

In chapter 2, where the production of the 20 chemicals relies on carbon dioxide-based methanol and methane, feedstock mass flows increase substantially by 287% due to the carbon dioxide used and water coproduced (Figure 2.2): 3.72 Gt carbon dioxide and 0.59 Gt hydrogen from water electrolysis are annually converted into methanol and methane. Methanol is further processed into olefins and BTX, which are the basis of the other large-volume chemicals. Methane substitutes natural gas to produce both heat and ammonia. Carbon dioxide can be captured from two different sources representing the upper and lower bounds for the climate impact of carbon dioxide supply: a highly concentrated industrial point source (~100% carbon dioxide) and ambient air with about 400 ppm carbon dioxide. This technology leads to the methanol economy envisioned by Asinger and promoted by Olah. Since all processes in these technologies are at technology readiness levels of 7 and higher, these processes are denoted TRL > 7 (Table 1.10). The integration of carbon dioxide via methane and methanol would require comparably little research and low development efforts and would allow use of wide parts of the existing petrochemical infrastructure and technologies.

Further production processes include carbon capture and utilization technologies for the direct conversion of carbon dioxide into olefins, aromatics BTX, carbon monoxide, ethylene oxide and styrene (Table 1.10). These technologies are currently at early research and development stages with TRLs below 7. Therefore, these processes are denoted TRL < 7. The direct synthesis of chemicals from carbon dioxide reduces the amount of carbon dioxide and hydrogen needed to 2.77 and 0.38 Gt, respectively, but mass flows are still about 182% larger compared with the conventional technologies. Due to the early development stages of these technologies, the realization of these technologies would require substantial research and development efforts and novel production facilities. Since these processes are based on low-TRL technologies, all calculations are subject to increased data uncertainty.

1.3.2 A Vision for the Oil Refinery of 2030

The refinery of 2030 will be very different from the refinery of 2022. At its core, the multibillion-dollar physical asset-conversion, storage and separation units will still be there. People will still drive refining, but in different jobs. However, it will become an intelligent, flexible and adaptive plant. The refinery of 2030 thrives by being fully demand-driven, with decisions made quickly at the appropriate level. Across the management chain, knowledgeable workers in the 2030 refinery will take actions based on their impact on satisfying demand better than competitors. Integrated systems and visual interfaces show them how their individual and collective actions impact customer delivery and satisfaction, as well as how to optimize the asset to meet demand profitably.

No doubt exists that COVID-19 is disrupting the oil and gas industry. Oil prices are falling, demand is down, supply is up and storage capacity is limited. In these difficult times, the industry must swiftly act to keep the business viable to come back strong when conditions improve. None of this should be news to you. What are you doing about it? Amid the tactical decisions to keep your company afloat, it is crucial to take a step back and think about your long-term strategy. The oil and gas industry has faced similar downturns in the past and the current one is unlikely to be the last. Therefore, refineries should take this opportunity during the slowdown to future-proof their operations.

The refining business needs to change dramatically over the next decade. Refinery executives worldwide are tuned to the signals of change, but some are not sure of the best course of action. What are the challenges? How will technology disrupt the hydrocarbon processing business? What is the path forward for future success?

Different analysis predicted different gasoline displacements due to the electric vehicle evolution. Total (a French oil company) predicted that 30% of personal vehicles in 2030 will be personal electric vehicles (PEV). ExxonMobil (an American oil company) recently released a forecast, Energy Outlook 2040, where they said that 40% of light vehicles in 2040 will be electric vehicles (EVs).

To respond to this trend, national oil companies are adapting their products from crude oil and fuels to chemicals. A noteworthy innovation is the direct crude oil to chemicals (COTC) or oil to chemicals (OTC) which was started by ExxonMobil (EM). In 2014, EM built a COTC refinery in Singapore to convert sweet crude oil to chemicals. In 2018, Saudi Aramco announced grand projects on COTC to convert sour crude oil to chemicals. Due to the abundant availability of crude oil, SABIC also announced its intent to contribute in COTC.

1.3.3 The Transition from Fuels to Chemicals

Why the future of oil is in chemicals, not fuels? With gasoline consumption expected to wane. The Crude-to-Chemicals complexes could dominate the petrochemical industry by the 2020s. Are hydrogen and ammonia the fuel of the future?

Two Saudi state-owned firms, the oil company Saudi Aramco and the petrochemical maker Sabic, are planning a new complex (the Yanbu project), that could prove to be a bellwether for the next decade in petrochemicals. By 2025, the partners expect to have a facility that will make petrochemicals – 9 million metric tons of them – directly from 400,000 barrels (bbl) per day of Arabian light crude oil.

Whereas most refineries convert just 12–15% of incoming oil into petrochemicals, some 45% of the output of the Yanbu facility will be chemicals, including olefins, aromatics, glycols and polymers. The partners are hardly alone in diverting their refinery product slates away from gasoline, diesel and other fuels and towards petrochemicals. ExxonMobil has practiced direct crude cracking technology in Singapore since 2014 and may build another such unit in China. Several facilities under construction in China will transform 40% of their oil into *p*-xylene and other petrochemicals. Aramco is considering another chemical-laden refining project in India. Over the next decade, oil may be the next big thing in petrochemicals. This is a change from the 2010s, when billions of dollars flowed into the US to build crackers and downstream petrochemical plants to process low-cost ethane from shale gas into ethylene and its derivatives.

The driver this time is the market more than it is cheap raw material supply. By 2030, demand for gasoline and other fuels will be on the decline. The petrochemical sector, in contrast, still has room to grow and oil companies and engineering firms are aware of this. They are installing new equipment and even designing new processes to seize on the trend. The trend is so strong that Honeywell UOP uses the term "refinery of the future" to refer to flexible technologies that enable refining complexes to make a wide variety of chemicals. But the shift will have consequences. The larger scale of a refinery compared with a chemical plant means existing petrochemical markets could be swamped with large amounts of product.

In the refinery of the future oil refining is fully integrated with a petrochemical facility. Renewable energy integration within the refining process in a nearly carbon dioxide-neutral operation. I focus on the most important challenges from the process and catalyst development points of view and highlight research opportunities and undergoing efforts. The refinery of the future will have carbon dioxide neutrality, including greening of the footprint, maximal production of chemicals, zero-waste generation and renewable energy integration, all of which should, in my opinion, play a key role.

I believe, in the future, refineries will be built around Oil-to-Chemicals processes, in contrast to the current state of the art where oil is first fractionated and the resulting streams are treated separately to manufacture an extensive portfolio of products. The first challenge at hand for

Oil-to-Chemical technologies is the nature of crude itself: the presence of heavy hydrocarbons leads to very rapid coke formation, not only in the conversion processes but also especially in process lines and support equipment. The main strategies to avoid these issues consist of different pre-treatment units designed to avoid downstream fouling. These include the use of molten salts to reduce carbon and tar deposition, hydrotreating and/or integrated solvent deasphalting units and partial vaporization of the feed. Once the crude has been conditioned, two main oil-to-chemicals approaches are being considered: (i) a steam cracking centred process and (ii) fluid catalytic cracking centred processes.

Oil companies seem willing to take that risk. They are looking into their crystal balls and seeing a future in which the world's appetite for fossil fuels isn't insatiable. Auto sales will taper off in the developing world. Cars everywhere are becoming more fuel efficient and increasingly popular electric vehicles consume no gasoline or diesel at all. The US Energy Information Administration predicts that US fuel consumption for light-duty vehicles will decline by 1% annually through to 2050. According to BP's 2018 *Energy Outlook*, the share of the average oil barrel dedicated to transportation fuel will peak at 58% in 2025 and then begin to decline. Oil consumed by industry, buildings and power will also slump. Chemicals, however, will continue to grow, from 16% of oil demand in 2020 to 20% by 2040 (Figure 1.22). Petrochemical demand will increase with gross domestic product, while motor fuel demand will only show modest growth and may even decline in certain regions, given environmental/legislative pressure and the introduction of battery-powered electric vehicles.

The Yanbu project is significant because it is an acknowledgment by Saudi Aramco, the world's largest oil producer, that the future of oil is chemicals. Aramco pledged $100 billion in petrochemicals investment over the next decade. The tremendous growth in the demand for chemicals provides Aramco with a fantastic window of opportunity. The relationship between refining and petrochemicals was once largely an arm's-length one. Refineries focused on fuels and those that made the refining cut called naphtha would sell it to operators of ethylene steam crackers. These plants crack naphtha at high temperatures into ethylene, propylene and other basic chemical building blocks (Pillar A of Petrochemistry, see Figure 1.23). With the new focus on chemicals, refineries are being reconfigured (Chapter 5). The typical refinery begins with fractional

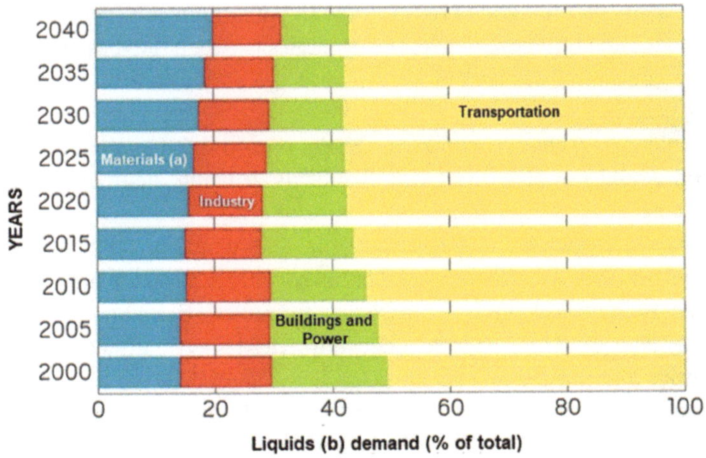

Figure 1.22 The decline of the oil consumed by industry, buildings and power and growing of chemicals. Materials are non combustion uses of liquids, such as petrochemicals, lubricants and bitumen. Liquids include crude oil, natural gas liquids, biofuels, coal-derived liquids, gas-derived liquids and refinery processing gains.

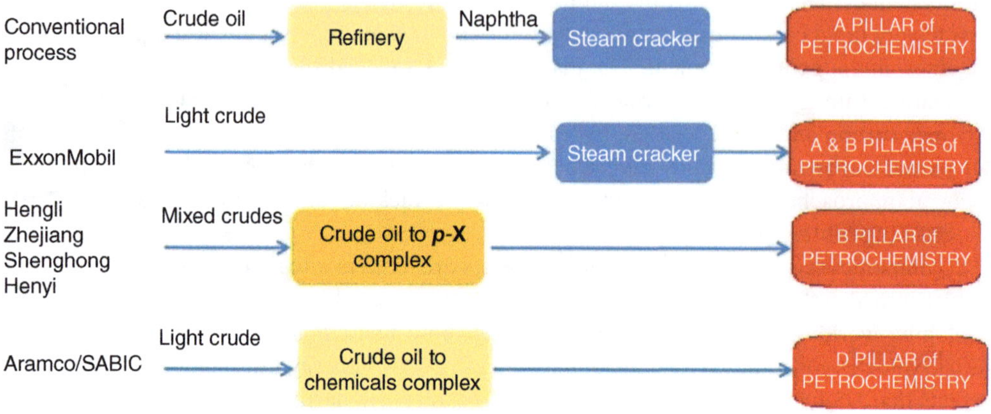

Figure 1.23 Crude-to-Chemicals concepts (IHS Markit, 2019).

distillation, which breaks crude oil into its constituent parts. These products range from light, short-carbon-chain molecules, such as gasoline and naphtha, to heavy diesel and vacuum gas oil. Refiners further process vacuum gas oil in either a hydrocracker or a fluid catalytic cracker (FCC), both of which break long molecules into shorter ones (Chapter 6).

Hydrocracking excels at making diesel. The process produces no olefins because all double bonds are quickly hydrogenated. The naphtha from the hydrocracker is sent either to an ethylene steam cracker to produce olefins or to a reformer to be refashioned into aromatics. To increase chemical output, more and more hydrocrackers are targeting naphtha rather than diesel production.

Alternatively, FCC units break down vacuum gas oil into gasoline. The process also produces olefins, namely, propylene, which is 5–20% of the output of most such units. Because of the strong incentives to make chemicals, refiners are running their FCCs at high severity – at high temperatures and long residence times – to increase propylene yield to well above 20%.

The linkage between oil refining and petrochemicals is as old as these industries themselves. And certainly, examples abound of companies – such as ExxonMobil, Reliance Industries and Sinopec – with a reputation for integration between refining and petrochemicals.

Historically, there have been players that integrated refineries with petrochemicals, not mainly to target maximum chemicals production but more to reap the synergies between the refinery and the petrochemical site. These synergies include shared utilities and the ability to feed naphtha directly into an adjacent steam cracker.

The evolution of refiners into intentional chemical makers has occurred in small increments. In the past, the basic configuration would not have produced chemicals. It would have produced gasoline, diesel, jet fuel and some heavy oil for maritime fuel.

The gateway chemical was typically propylene from the FCC unit. That would be a first step if someone was interested in petrochemicals. After that, refineries either sold naphtha on the merchant market or constructed their own naphtha-based ethylene crackers or aromatics-extraction units.

Now chemicals aren't an afterthought. Anytime we look at new refineries today or any significant refinery upgrades, we almost always find that a major goal is petrochemical production. New-build refineries are integrating petrochemical production at a scale that we really haven't seen before.

A prominent example of this trend is Zhejiang Petrochemical, which is building a Crude-to-Chemicals complex in two phases in China's Zhejiang Province. The scale of the project is staggering. The company is building two refineries, each capable of processing 400,000 bbl of oil per day. Overall, nearly 50% of the output – some 20 million tons – will be petrochemical, twice as much as the complex planned for Yanbu, Saudi Arabia.

UOP is providing a host of technologies for both phases of the Chinese complex, including hydrocracking to break down vacuum gas oil into naphtha and reforming and extraction to make aromatics from the naphtha. All told, the complex will produce 8.8 million tons of *p*-xylene, the main precursor for polyester fibre and polyethylene terephthalate, a polymer commonly used in bottles. Propane dehydrogenation units licensed from UOP will produce more than 500,000 tons of propylene.

The massive Zhejiang Petrochemical project is not the only one in China. According to the consulting firm IHS Markit, Hengli Petrochemical is building a 400,000 bbl. per day refinery that will yield 40% chemicals, largely *p*-xylene. Another project with a hefty output of *p*-xylene is Shenghong Petrochemical, which will process 300,000 bbl. per day.

In China, such privately owned refineries are driven by incentives beyond the attractive growth outlook for petrochemicals. Taxes on chemicals are lower than on fuels. Large, state-owned firms are hard to compete with in the domestic fuel market, yet Chinese refiners are prohibited from exporting fuels.

Refinery scale brought to bear on petrochemical production could forever change petrochemical markets, cautions are a lot bigger than chemical plants. A 400,000 bbl. per day refinery cranking out 50% chemicals will make about 10 million tons of chemicals per year. The largest naphtha-based ethylene crackers make only 2 million tons of products per year.

When these companies start to reconfigure the whole refinery with the target of producing chemicals, they effectively elevate the petrochemical production to a refinery scale, which could easily mean 5–10 times higher.

The scale is already proving to be large compared with the markets themselves. For example, the Chinese *p*-xylene projects will put out a combined 11.8 million tons annually. China imports 11 million tons of *p*-xylene per year, meaning today's South Korean, Taiwanese and Japanese suppliers will need to find new markets.

Aramco's Yanbu project will make about 3 million tons of ethylene per year. That alone is a quarter of the 12 million tons of capacity that chemical companies have been building on the US Gulf Coast. It will only take a few of these projects to significantly affect global supply and demand.

However, as the Crude-to-Chemicals trend continues, refineries might shrink in size. Because the profits for fuels are slim, refineries have to be big to justify the expense. A Crude-to-Chemicals facility can be smaller in scale. That is one of the numerous benefits.

Many of the plants under construction today are using off-the-shelf refining and petrochemical processes to make chemicals. Increasingly, engineering and chemical firms are tailoring processes for turning crude into chemicals. There is now a pipeline of technologies with different catalysts or variations of existing processes that can improve this conversion of oil to chemicals. Examples include petrochemical FCC units and slurry hydrocracking (Chapter 6).

At its Singapore complex, ExxonMobil developed a system to process crude oil, rather than naphtha, directly in a steam cracker. The main challenge of steam cracking of crude oil is the heavy part of the crude. Heavy crude can coke the cracker, leading to expensive downtime for maintenance.

ExxonMobil patents describe a system for evaporating crude in a "flash pot." The setup, which it likely uses in Singapore and might use in China, feeds the vapours into the steam cracker, leaving behind the fouling residues.

At Chevron Lummus Global (CLG), R&D is focused on turning refinery equipment into chemical manufacturing machinery. There is a lot of research that CLG put into understanding which molecules are worth producing – and maximize the production of those molecules – and which molecules CLG try to avoid.

Last year, Saudi Aramco signed a joint development agreement with McDermott and Chevron Lummus Global to further develop a thermal Crude-to-Chemicals process. McDermott is lending its expertise in steam cracking and other technologies and CLG, its hydrocracking and heavy oil conversion know-how. The Saudi firm also signed an agreement last month with Axens and TechnipFMC to work on a high-severity FCC process for chemicals.

The Saudi Aramco partnership has been making progress. All three companies have significant intellectual property contributions and all three work together to develop a scheme to capture all contributions in an integrated way. The future will largely be about tearing down the old walls between refining and chemicals. Aramco is going into the next wave of more intimately integrated sites, where there is not necessarily a clear demarcation of what is the refinery and what is the petrochemical site.

Momentum towards efficiency is driving integration from oil production to finished goods, while a recognition of the need for a positive environmental image for the energy industry is driving innovation in carbon capture and new uses for performance chemicals to reach a carbon-neutral balance.

Sustainability initiatives are steering refiners to implement ambitious programs to reduce both energy and water use in refining and chemicals. Global cybersecurity concerns are driving innovation to enable digital transformation, while simultaneously protecting the safety and integrity of strategic assets.

The uptake of hybrid, electric and high-efficiency vehicles, along with the pace of renewable power, are key variables in forecasting hydrocarbon fuels demand. The market for petrochemical feedstocks depends both on economic growth in Asia and, conversely, a strong push to avoid non-recyclable plastics.

Advancing ideas can be facilitated and expedited during industry events. Any time that the industry can come together and share knowledge on best practices and technological advancements provides a significant catalyst for increasing plant efficiency, profitability and safety. The senior-level hydrocarbon processing industry (HPI) leaders met in Helsinki, Finland for Hydrocarbon Processing's International Refining and Petrochemical Conference (IRPC in June 2019). IRPC EurAsia's focus is on the evolution of the downstream processes.

How do you position a refinery in the next 10 years to make the right capital investments for change but be increasingly flexible? Will digital technologies be transformative in providing refiners a strategic cost and margin advantage? Can digital transformation make flattened refining organizations exciting destinations for the "best and the brightest" graduates – the future key people resources? How can organizations reduce cyber vulnerability and, simultaneously, increase agility and innovation?

The connected supply chain of the 2030 refinery ties together its supply chain. The refinery of 2030 thrives by being fully demand driven, with decisions made quickly at the appropriate level. Across the management chain, knowledgeable workers in the 2030 refinery will take actions based on their impact on satisfying demand better than competitors. Integrated systems and visual interfaces show them how their individual and collective actions impact customer delivery and satisfaction, as well as how to optimize with its operations, planning and trading and sources solutions and applies powerful artificial intelligence (AI)-assisted analytics to understand the business opportunities presented by demand changes.

The refinery of 2030 will operate with knowledge worker "pods" comprising planners, optimization experts, operators and reliability experts. Planning and scheduling will be fully integrated, with the planner supported by an AI expert dashboard displaying economic, sales, sustainability and operations trade-offs. Creation of the best schedule will be fully autonomous.

Prescriptive maintenance systems will flag to the planner and reliability expert situations that could interrupt achieving the production plan, along with alternatives for traders to purchase optimum crudes and operators to maintain operations at the best possible asset utilization, safety and profitability.

Using AI-assisted advanced process control (APC), the integrated plan and schedule will autonomously provide closed-loop optimization. By 2030, data sensors will be pervasive, but new real-time analytics deployed at the edge will provide process stream compositions and catalyst nano sensors will drive better control over process performance. These technologies will drive the APC and underlying control to operate through dynamically adjusted setpoints to achieve and exceed operations.

Major process units and equipment in the refinery of 2030 will be smart plant building blocks. This built-in intelligence will greatly improve the ability of the refinery to operate at process operating limits. It will also be a critical element of asset flexibility. The asset owner will use this intelligence as an AI-based decision tool to model plant reconfiguration to shift product suites based on demand, to execute the reconfiguration rapidly and to reconfigure the monitoring and control models for operating optimization.

The miniaturization of devices achieves small process volume and hold-up, large specific transport rates and high surface-to-volume ratios. The combination of these features results in extremely short response times considerably simplifying process control and eases the handling of large heat and mass fluxes thus making the synthesis of dangerous compounds safer by reducing hazardous waste. For chemical technologies, the applications of miniaturized reaction systems or micro reactors, is centred on gas and liquid phase reactions covering simple microscale mixing of different fluids' volumetric titrations, heterogeneous and homogeneous catalysis, catalytic oxidation, pyrolysis, heterocyclic synthesis, photochemical reactions and micro fuel cell applications. Due to higher yields and selectivities, the application of environmentally more favourable reaction routes and the possibility of replacing large plants with small plants for distributed production according to actual demand, it is to be expected that micro technologies will eventually contribute to their sustainable development.

In last two decades micro technologies and micro equipment have been of great interest in chemical engineering. Small series of high-performance micro heat expanders, reactors, mixers, extractors, pumps and valves have been developed and produced. The next-generation prescriptive maintenance system will not only flag future failure modes for equipment and units, but also will link to the planning and process models used to propose operating changes that will delay or eliminate the future failure risk.

The further development of sensors, prescriptive maintenance AI solutions, unit analytics, etc., will provide the intelligence to understand physical asset conditions, trends, risks and emergencies, obviating most of the need to send people into the plant to inspect equipment and units for issues. AI-enabled virtual solutions will be able to not only plan but also manage start up and shutdown procedures largely remotely. Advances in plant robotics solutions will enable performance of many tasks without sending people into danger. This concept is far beyond and an alternative view to, the concept of "3D augmented reality." Conversely, it is the process and intelligent reliability, planning and execution systems that will drastically reduce the need for people to enter dangerous areas at dangerous times.

The refinery of 2030 will exist in an even more stringent "social license to operate" environment. Every decision-maker and worker in the refinery will have immediate visibility into the impacts of their tactical and strategic decisions on enterprise sustainability metrics and key performance indicators. Online optimization models will constantly optimize both energy and water use in the asset and throughout the value chain. Personnel will keep a running scoreboard of sustainability "points" the workers can earn, recognition for when their actions help the enterprise improve its sustainability investment standing.

Enterprise-wide, next-generation risk models provide a dashboard in the executive suite to assess an increase or reduction of risk being undertaken by different businesses and capital and operational expenditures budgetary decisions. Cyber defensive capabilities and warnings will be built into every layer of the future digital refinery. These concepts will help the refinery of 2030 play a key role in the sustainable, highly diversified and integrated energy enterprise of the future.

1.3.3.1 Crude Oil to Chemicals Investments

As mentioned, big players such as Saudi Aramco have made great investments in crude oil to chemicals (COC) technologies aiming to achieve even better integrated refineries and petrochemical plants, raising considerably its competitiveness in the downstream market. The major technology licensors as Axens, UOP, Lummus, Shell, ExxonMobil, etc. have applied resources to develop technologies capable to allow a closer integration in the downstream sector, aiming to allow refiners extract the maximum added value from the processed crude oil, an increasing necessity in a scenario where the refining margins are under pressure. In 2019, The Catalyst Group (TCGR) gathered data on capital investments in Crude-to-Chemicals projects (Table 1.11).

It is expected that some of these capital investments were postponed due to the economic crisis provoked by the COVID-19 pandemic, but these data reinforce the trend in the market. A typical concern related to the Crude-to-Chemicals enterprises is related to the operation costs in comparison with the traditional routes.

The current trend of reduction in transportation fuels demand at the global level and the capacity of maximum adding value to crude oil can be a competitive differential to refiners. Due to the high capital investment needed for the implementation that will allow the conventional refinery to achieve the maximization of chemicals, capital efficiency becomes also an extremely important factor in the current competitive scenario, as well as the operational flexibility related to the processed crude oil slate.

Table 1.11 Crude oil to chemicals investments (The Catalyst Group, 2019).

Announced crude oil-to-chemicals investments 2019	$ billion
China	
Zhejiang Petroleum and Chemicals, Zhoushan, Greenfield, 2019	26
Hengli Petrochemicals, Changxin Island, Greenfield, 2019	11
Shenghong Petrochemicals, Lianyuangang, Greenfield, 2019	11.84
Ningbo Zhongjin Petrochemical (subs Rongsheng Petrochemical), Ningbo, Revamp, 2018	5(est)
Saudi Aramco/NORINCO/Panjun Sincen (Huajin Aramco Petrochemical), Liaoning Province, Greenfield, 2024	10+
SABIC/Fuhaichuang Petrochemical, Zhangzhou, Greenfield, NA	NA
SINOPEC/SABIC (Tianjin Petrochemical), Tijanjin, Revamp, Operating pre-2017	45
PetroChina, Dalian, Revamp, Operating pre-2017	combined
PetroChina, Yunnan, Revamp, Operating pre-2017	(est)
CNOOC, Huizhou, Revamp, Operating pre-2017	
SINOPEC, Lianyungang, Greenfield, NA	2.8
SINOPEC, Caofeidian, Greenfield, NA	4.2
SINOPEC, Gulei, Greenfield, 2020	4.26
Total China	**$ 120.1**
Other Asia	
Hengyi Group, Pulau Muara Besar, Greenfield, Brunei, 2020	20
Saudi Aramco/ADNOC/India Consortium, Raigad, Greenfield, India, 2025	44

Table 1.11 (Continued)

Announced crude oil-to-chemicals investments 2019	$ billion
Petronas/Saudi Aramco (RAPID), Pengerang, Greenfield, Malaysia, 2019	2.7
ExxonMobil (Singapore Chemical Plant), Jurong Islang, Revamp, Singapore, 2023	<1
Petramina/Rosneft, Tuban, East Java, Greenfield, Indonesia, 2025	15
Total other Asia	**$ 82.7**
Middle East	
ADNOC, Al Ruwais, Revamp, UAE, 2025	45
Saudi Aramco/SABIC, Yanbu, Greenfield, Saudi Arabia, 2025	30
Saudi Aramco/Total, Jubail, Greenfield, Saudi Arabia, 2024	5
KNPC/KIPIC (Al-Zour Refinery), Al Ahmadi, Greenfield, Kuwait, 2019	13
Oman Oil Company/Kuwait Petroleum International (Duqm Refinery), Greenfield, Oman, NA	15
Total Middle East	**$ 108**
Europe	
MOL Group, Croatia, Revamp, Hungary, 2030	4.5
Total Europe	**$ 4.5**
Total Greenfield $ 215 Total Revamps $ 100 Total Global	**$315billion**

1.3.3.2 Available Crude-to-Chemicals Routes

Nowadays, there are three technically available routes that are being considered for capital investments to Crude-to-Chemicals refining complexes. Figure 1.23 presents the Crude-to-Chemicals concepts.

The conventional routes consider the processing of crude oil in a conventional crude oil refinery, producing petrochemical intermediates like naphtha, which is supplied to a petrochemical asset like a steam cracking unit. The ExxonMobil route is based on the direct feed of selected crude oils, normally light and low contaminants crudes, to petrochemical assets, while the Chinese enterprise Hengli Zhejiang Shenghong Henyi project considers the feed of mixed crude oil slate to a crude to *para*-xylene (*p*-X) complex to support the Chinese domestic market that has a high demand for light aromatics (BTX). A conventional highly integrated refining hardware is capable to achieve 15 to 20% of petrochemicals yield while a Crude-to-Chemicals refinery can reach up to 40%. The Aramco/Sabic concept is based on a high complexity refining hardware to convert selected crude oil (light) to maximize the yield of petrochemical intermediates, mainly light olefins.

Although the advantages presented by closer integration between refining and petrochemical assets are clear, it is important to realise that the players of downstream industry are facing a transitive period where, as presented in Figure 1.23, the transportation fuels are responsible for a great part of the revenues. In this business scenario, it is necessary to define a transition strategy where the economic sustainability achieved by the status (transportation fuels) needs to be invested to build the future (maximize petrochemicals). Keeping eyes only on the future or only in the present can be a competitive mistake.

1.3.4 Business Trends: Petrochemicals 2025

The global petrochemical sector continues to expand exponentially as the demand from developing nations for petrochemical/chemical products continues to increase. In its "Oil 2018" report, the International Energy Agency (IEA) forecasts that approximately 25% of the increase in oil consumption to 2023 – nearly 1.7 MMbpd – will be from demand for petrochemical feedstocks. The IEA's long-term petrochemical forecast shows an even greater demand for oil from the petrochemical sector. In the organization's "The future of petrochemicals" report, the IEA forecast that the petrochemical sector will account for one-third of oil demand growth to 2030, increasing to nearly half to 2050. According to the report, production of key plastics will more than double between 2010 and 2050.

Growing demand centres in the Asia-Pacific region will be met by billions of dollars of new petrochemical production capacity in Asia, the Middle East and the US. These three regions are investing heavily to boost petrochemical processing capacity to satisfy demand.

Many non-OECD countries are witnessing petrochemical demand rates increasing faster than GDP growth. Several nations will remain dependent on petrochemical imports, while others are investing to satisfy domestic chemicals demand. Multiple factors will determine the future of the petrochemical industry: supply/demand factors, plastics recycling, regulations, feedstock costs, partnerships/mergers/acquisitions, digitalization, etc. Many nations are investing heavily to satisfy domestic petrochemicals demand and/or export petrochemical products to increase revenues.

More than 280 new petrochemical projects were announced from 2016 to 2018. New petrochemical project announcements increased from 68 in 2016 to 88 in 2017 to 125 in 2018. This trend represents a year-on-year increase of 29% and 42%, respectively. Most new petrochemical project announcements have been in the Asia-Pacific region. The region continues to invest a substantial amount of capital to increase petrochemical capacity to mitigate imports and satisfy increasing domestic demand. Over the three years 2016–2018, the petrochemical sector has maintained 36–41% of new project market share and has been the leader in new project announcements since 2016. Over the past year, Asia-Pacfic and the US have been the leaders in new petrochemical project announcements, accounting for nearly 70% of all new petrochemical project announcements in 2018 (Figure 1.24).

Last year the Boxscore Database tracked nearly 470 active petrochemical projects around the world. A breakdown on active refining project market share is provided in Table 1.12.

In total, the Boxscore Database is tracking nearly $510 billion in active petrochemical projects around the world. Asia-Pacific is the leader in capital expenditures (CAPEX) in the near term. The Boxscore Database tracks more than $200 billion in active petrochemical projects in Asia. The US has more than $100 billion in active petrochemical projects, followed by the Middle East with nearly $90 billion in active projects. These three regions represent not only 76% of total active projects, but also nearly 79% of announced CAPEX in petrochemical capacity additions globally. A breakdown of CAPEX in the petrochemical sector, by region, is provided in Table 1.12.

According to the IEA petrochemicals report, nearly all regions, except for Europe, will increase production of primary chemicals up to 2050. The largest capacity growth is seen in the Middle East

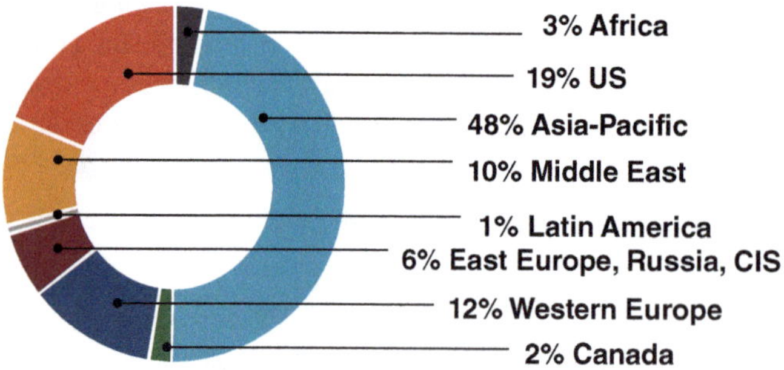

Figure 1.24 Market share of new petrochemical project announcements by region.

Table 1.12 Market share of new petrochemical project announcements by world region, December 2017–December 2018, March 2019.

Year	2019		2017–2018	
World petrochemical projects	**Number**	**%**	**%**	**CAPEX (in $ billion)**
World region				
Africa	29	6	3	32
Asia-Pacific	158	34	47	205
Canada	7	1	2	10
Western Europe	22	5	12	48
Eastern Europe, Russia and CIS	55	12	6	
Latin America	28	6	1	19
Middle East	63	13	10	88
USA	106	23	19	106
Total world	468	100.0	100.0	508

and Asia. The Middle East is forecast to increase production of high-value chemicals, ammonia and methanol from approximately 70 MMtpy to more than 150 MMtpy by 2050. Nearly all Middle Eastern nations are investing in new petrochemical production capacity to mitigate the reliance on crude oil export revenues. The region has already boosted production capacity, led by Saudi Arabia and more investments have been announced.

The Asia-Pacific region's petrochemical buildout is forecast to increase by nearly 200 MMtpy to nearly 500 MMtpy by 2050. China and India will be the leaders in new petrochemical capacity additions; however, multiple Asian nations have announced capital-intensive projects. Many of these projects will help mitigate growing supply and demand gaps and will be built alongside refining operations. The move towards refining and petrochemical integration is a primary focus of many operators around the world, especially in Asia. The ability to share feedstocks allows producers to produce petrochemical products more efficiently.

1.3.4.1 Asia-Pacific

To help mitigate imports and to satisfy robust demand, many Asia-Pacific nations are investing in the expansion and debottlenecking of petrochemical units, as well as the construction of grassroots petrochemical complexes. These capital-intensive investments include increased integration between refining and petrochemical facilities to increase efficiency and value.

China aims to grow its domestic petrochemical production during the next 5–7 years, driven by the ongoing modernization of major production facilities and an increase in foreign investments. According to a report by the Chinese Federation of Petroleum and Chemical Industry (CFPCI), the Chinese petrochemical industry's aggregate profit in 2018 grew by more than 30% on the year, to 839.39 billion CNY ($125 billion). In addition, the level of profitability of the industry increased by 6.8% to the highest figure since 2012. CFPCI analysts expect the industry to achieve at least the same growth rates this year, due to continued demand for Chinese petrochemicals, both in the domestic market and abroad.

The Construction Boxscore Database tracked nearly 160 active petrochemical projects in the region (2019). China accounts for nearly half of all petrochemical projects in the region with India coming second with 15% market share in active petrochemical projects. When broken down by activity level, more than 70% of the region's petrochemical projects are in preconstruction phases, with approximately 43% in the planning/proposal phase.

China's Belt Road Initiative calls for the massive development of domestic petrochemical production capacity to mitigate imports. According to an ICIS report, China plans to add 21 new steam crackers and ten new refineries in petrochemical industry parks. This buildout could equate to nearly 27 MMtpy of new ethylene production by the mid-2020s, reaching more than 42 MMtpy.

Due to increasing demand for petrochemical products, India will continue to heavily invest in additional petrochemical production capacity. According to IHS, India's domestic ethylene capacity increased from 4 MMtpy in 2014 to 7.2 MMtpy in 2018. However, with petrochemicals demand forecast to increase substantially, additional capacity will be needed to satisfy demand and mitigate imports. According to a report by McKinsey and Co., India's petrochemicals shortfall could increase to 25 MMtpy by 2025 if investments in new production capacity are not made. However, the nation's operators have announced plans to invest more than $30 billion to boost petrochemical capacity.

1.3.4.2 Middle East

In response to the decrease in oil prices that occurred from 2014 to 2016 and to mitigate reliance on oil export revenues and diversify their product portfolios, nearly all nations in the Middle East have announced capital investments in new downstream processing capacity, especially in the production of petrochemicals. The region has the largest cost advantage in ethylene production and several countries are building, or planning to build, mixed-feed crackers, as well as ethylene derivatives, ammonia/urea and other petrochemical capacity.

Saudi Arabia is the leader in new petrochemical capacity investments. The Kingdom plans to nearly triple petrochemical production capacity from 12 MMtpy in 2016 to 34 MMtpy by 2030. The country will accomplish this goal by adding petrochemical capacity to existing refineries, as well as building grassroots facilities. Several other nations are investing heavily in domestic output.

1.3.4.3 United States

Due to the shale gas boom, the US has witnessed a renaissance in its domestic petrochemical industry. Cheap, readily available shale gas feedstock has allowed the country to become one of the world's lowest-cost ethylene producers. In turn, the US is building millions of tons of additional ethylene and ethylene derivatives production capacity.

The Construction Boxscore Database tracks 115 active petrochemical projects in the US. More than 70% of the country's petrochemical projects are in Louisiana and Texas. These two states will be instrumental in boosting ethylene and ethylene derivatives, methanol and specialty chemicals capacity into the early 2020s.

By the end of the decade, the US is also forecast to add more than 10 MMtpy of new ethylene capacity. This boom in ethylene capacity includes more than 9 MMtpy of grassroots facilities, as well as more than 1.2 MMtpy in ethylene capacity expansion projects. Most of the country's grassroots facilities will include derivative units. In total, US petrochemical producers will invest nearly $20 billion in new ethylene capacity by 2020. A second wave of new ethane crackers could add more than 5 MMtpy of new capacity after 2020. In total, capital expenditures for both ethane cracking project waves could top $50 billion by the mid-2020s.

The largest ethylene derivative capacity expansion will occur in the production of polyethylene (PE). By 2020, the US will add approximately 8 MMtpy of new PE capacity, most of which will be integrated into new ethane cracking operations; however, grassroots PE plants are also being built. According to Platts, PE surplus in the US reached more than 4 MMtpy in 2017 and could increase to 7.5 MMtpy in 2020. This trend will make exporting US PE a must for producers.

1.4 Digital Transformation

The oil and gas industry is at a turning point. Oil prices have witnessed new levels of volatility, causing producers to accept that this environment may be the new norm. While put in place for good reason, heightened regulatory compliance requirements are complicating the situation. Simultaneously, the industry is determining how to create and retain knowledge as its workforce undergoes one of the largest generational transitions in its history, risking business continuity.

2019 was a year of successes within the recycled polyester (RPET) industry. The continued war against plastics and the heightened awareness of plastic pollution have driven an unprecedented level of interest in recycled polyester. As such, significant momentum has been building behind collection and recycling commitments. Beyond regulatory commitments, such as the EU's single-use plastics directive that was announced earlier in the year, ambitious recycled content sustainability goals have been publicized by many brand owners. Some brand owner targets exceed the EU single-use plastics directive, further highlighting the momentum behind the shift towards achieving a circular economy.

In the midst of this environment, downstream producers are striving to improve performance by reducing energy usage and costs and to reduce safety-related incidents, while optimizing yields and improving operator performance. As resources continue to be cut, how can downstream operators achieve these goals?

The answer lies in digital transformation. Technology is changing industries and downstream oil and gas is no different. Pivotal technologies such as cloud computing, the Industrial Internet of Things (IIoT), digital twins and augmented reality (AR) and virtual reality (VR) are changing the way companies do business for the better. However, the trick is ensuring that refineries make the right technology investments to meet these goals and use these technologies to enhance the customer experience.

1.4.1 Benefits of Digital Transformation

In the present economic environment, capital budgets and overhead are being cut. Simultaneously, energy costs are increasing and feedstock prices are shifting constantly, forcing manufacturers to monitor economic conditions and forecasts daily. It is imperative that manufacturers identify new ways to reduce energy costs across all facets of the business, including process design, implementation and production.

The digital transformation offers new tools that enable downstream producers to improve the yield of valuable products, while also reducing energy consumption and increasing throughput. Think of it as the analogy to digital conversion. Taking what is going on in the physical world – a company's physical assets – and converting it to digital information. Then taking that information and applying digital toolsets to the existing data that has been collected, creating the ability to determine what value can be extracted from this information. An example is the use of artificial intelligence and machine learning to spot anomalies in the operational behaviour of critical assets that may indicate a future breakdown.

Using digital technology, manufacturers can create a complete digital twin of the plant. The digital twin pertains to the asset and the operations life cycle and how the two connect together to create the goods and services produced. It allows operators to take real-time process data and combine it with current economic conditions, which, in turn, enable them to make informed decisions and to evaluate what-if scenarios in batch processing and manufacturing at an expedited rate. Information sharing increases, enabling stakeholders to increase their ability to visualize results and key performance indicator (KPI) data across processes and overall plant production.

For example, among the digital tools, cloud computing specifically offers several opportunities to reduce costs in process design. Due to the nearly infinite processing power and storage available through cloud-based architecture, process design can accelerate, while also reducing capital investment costs for process modelling and training. This eliminates the need for data centres and the costs that go into operating them. Utilizing cloud computing, downstream producers can set up cloud-based servers and computing resources as needed, whether that be three servers one day and 15 the next. Through a cloud-based architecture for process design, information accessibility is increased, availability is enhanced and total cost of ownership (TCO) is significantly reduced. As process design information is digitally stored in the cloud, it becomes easier to combine this information with other sources of data and to drive the enterprise towards a completely digital planning and operations model.

Through digital transformation, business units that have historically existed in a vacuum can now connect to each other in real time to create a unified supply chain model. With the ability to view all the data in one place, such as a combination of real-time economic and market data sources and plant and production status data, companies are able to maximize profitability. Rather than acting as disparate sources, internal teams become armed with the same information, making them better able to understand what drives their process and how to improve the customer's experience.

In a unified supply chain model, planning and operations are fused together to make up the digital value chain. A complete 360° view of the digital value chain emerges, allowing all aspects of the enterprise to be visualized, analysed and optimized – from demand to fulfilment and follow-up. Inputs to the enterprise, such as feedstock and raw materials, are then analysed in real time against planning, operations, scheduling and distribution. Full plant models can also be managed simultaneously within a supply and distribution network. Fast optimization, combined with user-configurable visualizations and reporting, allows the impact of uncertainties and data changes to be evaluated and understood as they occur. Feasible and robust schedules are then brought to the forefront to guide business decisions that are not only easier to understand, but also easier to explain across the entire enterprise.

1.4.2 A New Workforce and Workplace

At present, there is a new generation of mobile-enabled workers, a new regulatory environment and new technology. All are shaping the way industry does business. As a generation of highly experienced refinery operators begins to retire, it is imperative that oil and gas companies determine the safest and best way to train their replacements as fast as possible. At the same time, training budgets are being cut, increasing the need to deliver safe training through a more economic vehicle. Digital transformation offers several options to address these challenges. For example, VR and/or AR are digital toolsets capable of connecting control room operators and maintenance and field personnel in a single, realistic learning environment. In life-like, 3D environments and high-fidelity, dynamic simulations, operator training is accelerated across several categories: equipment understanding; hazard and operability study (HAZOP) design evaluation; operator training simulator (OTS) programs, including procedure training, safety scenarios and crew training; maintenance planning and execution; and real-time enterprise asset management.

These AR and VR training solutions offer minimized project risk by preventing delays during plant commissioning and start-up, while maximizing return on investment (ROI). Mobile technology enables workers to collect data throughout their daily rounds, rather than collecting information on paper and reporting later. It also enables a remote workforce that can connect virtually. People can perform their duties from wherever they are (directly from their handheld device), which enables them to access a digital twin of the plant – thus increasing uptime and reducing operating and maintenance costs.

Mobile technology can also help companies comply with safety, health and environmental regulations (HSER), which is a paramount focus in the downstream processing industry. Many plants operate under a "zero accidents" culture – the idea that all accidents can be prevented through policy and procedure adherence. Many factors contribute to a good safety culture, including skilled employees, continuous safety risk assessment (HAZOP) and good practices and standards, alongside primary and secondary process safety.

Mobilizing the operator workforce helps ensure that regulatory compliance is consistently met. Anything that was previously recorded on paper (such as repair procedures and audit logs) becomes digital as workers log their maintenance and operations records daily via their mobile devices. These records are automatically stored in a central location and backed up to the cloud and, additionally, regulatory audit trails are automatically generated.

Technology is also changing the way plants and refineries operate as a whole, offering new ways to capture, share and store information across the enterprise. Since capital expense constraints limit investment in new equipment, companies are turning to asset performance management (APM) solutions. APM combines enterprise data capture with asset management, advanced workflow, mobility, predictive analytics and risk-based management to drive maximum return on asset investment through reduced downtime, lowered TCO, decreased maintenance costs and improved overall equipment effectiveness. This allows operators and maintenance engineers to automatically generate work orders based on data reported from predictive models, allowing potential equipment failures to be addressed before they occur. Advanced data capture and analytic capabilities provide detailed insight into the health of production assets, avoiding unplanned downtime.

An ARC Advisory Group market study on common failure patterns found that 82% of failure types are random. Only 18% are predictable and preventable with traditional maintenance methods. Machine learning helps identify inefficiencies and abnormalities in equipment operation long before regular inspection. Engineers can reference operational models and digital twins for recent abnormalities in design vs. operational performance. As data is collected at an operational level and fed into other digital toolsets, it is imperative that oil and gas enterprises make the right digital technology investments.

1.4.3 Technology Investment

Choosing the right technology investment requires analysis of four key technology pillars to ensure successful digital transformations and optimal ROIs: comprehensive value chain; open and system agnostic; digital ecosystems; and flexible and agile implementation.

Digital transformation is a data-driven approach to continuous process improvement. Although it harnesses digital technology, it still requires the buy-in of workers, as well as the collaboration of processes and assets. Eventually, processes and assets will collide to bridge the gap between operations technology and information technology.

However, the larger piece of this puzzle is determining what customers are gaining by partnering with your company. With the array of technology disruptions that have already taken place, downstream companies need to ensure that they are shaping the customer experience in a different (better) way than their current or future competitors. A good example of this disruption at work is the story of Netflix and Blockbuster. Netflix came out of nowhere, leveraging new technology and delivering a customer experience that was miles ahead of anyone else in the industry. Ultimately, Netflix was the reason Blockbuster went out of business. However, could Blockbuster see this coming? In such a turnkey changing time, there are many industrial businesses that are concerned about this type of disruption catching them on the back foot.

Digital transformation does not happen all at once. The key is getting started now. Remember to start small in your strategy and adoption, but to give yourself a competitive edge. Start now to maintain or improve your competitive market and industry position.

1.4.4 The Greening of the Downstream Industry

The global push to neutralize carbon-based greenhouse gas emissions by 2050 not only holds great promise for the future of the planet, but it also stands to create millions of jobs, spur rapid growth and drive technological innovation. Businesses in every industry around the world have rolled out environmental and social responsibility initiatives, often with bold plans to decarbonize their operations in line with the global 2050 net-zero goal, although these plans differ in scope and strategy. Since the transportation industry accounts for roughly 30% of all greenhouse gas in the US, new kinds of renewable fuels made from organic sources (such as plant oils or animal waste) are expected to play a leading role in the transition to an environmentally sustainable economy.

Meridian Energy Group Inc. was building two grassroots refineries in the US. The company has announced that these facilities will be among the greenest-operating refineries in the world.

The Davis refinery will be the:

- first full-conversion, grassroots refinery in the US in more than 40 years;
- the cleanest refinery ever built anywhere in the world, producing the cleanest fuels in the US;
- the first major refinery built by a development-stage firm; and
- a model for similar refineries in other shale basins.

Meridian's goal is to redefine the standards of emissions controls of the refining industry. The physical and operational design of Davis refinery includes implementation of state-of-the-art, lowest achievable emissions rate (LAER) technology and air pollution control equipment – considered best available control technology (BACT) – to minimize emissions throughout the plant. The design even includes BACT requirements and controls that would be considered maximum achievable control technology (MACT).

Meridian will ensure that the required strict air quality standards are met. For example, a comprehensive and enhanced smart leak detection and repair (LDAR) program will be used to monitor fugitive emissions throughout the plant. This innovative solution greatly exceeds the present minimum regulatory requirement, which mandates periodic (e.g., monthly, or quarterly) walkthroughs of the facility to monitor each potential source of unintended leakage, utilizing a hand-held screening device and for repairs to be completed within a specified timeframe. Unintended leaks usually account for a large percentage of hydrocarbon emissions at refineries.

For this reason and to minimize the potential for unintended leaks at the Davis refinery, Meridian's LDAR program will utilize optical gas imaging (OGI) technology to continuously scan the equipment. Because fugitive gases are normally invisible to the human eye, OGI cameras use specialized infrared filters that allow them to "see" gas, should it escape. The smart LDAR proposed by Meridian will allow automated early detection and notification of unintended leaks as they occur and the visual confirmation provided by OGI will be extremely useful in pinpointing the exact source to immediately begin the repair process.

Since 2015, OGI has been identified by the US Environmental Protection Agency (EPA) as the "best system of emissions reduction" for detecting fugitive emissions from new equipment installations, upgrades and modified sources.

Gas processing, liquefaction, oil refining, bulk petrochemicals and chemicals plants represent a large population of frontline engineers and technical managers. Their professional lives will become increasingly impacted by the growing adoption of disruptive digital technologies and digital platforms. Why focus on frontline engineers? Frontline engineers and managers represent a large cohort of people who are and will continue to be, impacted the most by Industry 4.0 shifts and trends. They constitute a large proportion of the overall labour pool and, in many countries, they are big influencers of competitiveness of domestic manufacturing.

1.4.4.1 Sustainable Alkylation Technology

Traditional acid-based alkylation processes use either hydrogen fluoride or sulfuric acid to catalyse the reaction of mixed C_4 olefins feedstocks into high-octane C_8 alkylates for gasoline blending. Both hydrogen fluoride and sulfuric acid alkylation processes are inherently unsafe due to the highly corrosive nature of the strong acid catalysts. Refiners using acid-based alkylation technologies require exotic metallurgies for process equipment and may require costly safety systems to protect refinery personnel and the public. The disposal and regeneration of spent acid catalysts are also problematic from an environmental perspective.

Since 2015, a series of high-profile refinery explosions in California, Wisconsin and Pennsylvania have raised alarm over using hydrogen fluoride in refinery operations near densely populated urban centres in the continental US. If accidentally released, hydrogen fluoride has a propensity to become airborne as a dense, toxic vapour cloud. North American regulators and local stakeholder groups have called on refiners to examine their options for mitigating the risks of aging hydrogen fluoride alkylation units.

In China, the Environmental Protection Ministry has imposed heavy penalties on alkylation operators with inadequate safe practices and is closely monitoring practices relating to the treatment of spent sulfuric acid. The highly emissions-intensive process of regenerating large quantities of spent sulfuric acid from alkylation operations may present a future risk to refiners if regulations are further tightened for industrial emitters.

1.4.4.2 Ecofriendly Catalyst

The catalyst can play a significant role in the production of higher quality fuel chemicals as required by standards which, step-by-step, are going to be introduced all over the world due to the growing awareness of the damage to human health and the environment caused by existing products. Fuel reformulation has been seeded by the growing awareness of the potential damage humankind was causing to the ecosystem and to itself. Fuel reformulation means that fuels are defined on the basis of their chemical composition with additional engine-technology related standards rather than on pure performance bases. These standards, which are getting more and more stringent, can be met in different ways, mainly via improving catalyst and process operating conditions.

Market and regulatory factors are pressuring refiners to adopt safe and sustainable processes to produce clean-burning and environmentally friendly fuels. Global, long-term demand for high-quality gasoline is being driven by macro factors, such as a steadily growing consumer base and the adoption of high-compression engines, which have low vapour pressure requirements. Concurrently, the implementation of stringent fuel and emissions standards, including low-sulfur or ultra-low-sulfur content specifications, are increasing the reliance on octane-boosting blending stocks, namely alkylate.

Alkylation technology is an inherently safe, commercial process that uses a proprietary composite ionic liquid catalyst – a non-volatile, non-aqueous liquid salt – to facilitate the alkylation reaction. The alkylation catalyst is non-hazardous and non-corrosive, allowing all process equipment to be manufactured using low-cost carbon steel. The catalyst is regenerated onsite under moderate operating conditions, which provides the added benefits of reduced emissions compared to alternative technologies. A small quantity of chemically benign solid waste by product is removed from the unit every few days to two weeks.

The technology has been under development for 20 years. Significant process improvements have been made in recent years, concurrent with widespread commercial adoption in the Asia-Pacific area. In 2005, the first commercial field demonstration was successfully performed at the PetroChina Lanzhou refinery by retrofitting an existing, 65,000-metric-tpy sulfuric acid alkylation unit with the proprietary catalyst.

Table 1.13 Catalyst consumption.

Item	Designbasis	Commercial
Catalyst active reagent, kg/t alkylate	6.2	2.5
Organic chloride activator, kg/t alkylate	0.5	1.1
Total consumables, kg/t alkylate	6.7	3.6

Table 1.13 compares the commercial and design catalyst consumption for the process. During operation, the majority of the catalyst is regenerated onsite via the composite ionic liquid regeneration unit and catalyst activity is maintained through the addition of an organic chloride activator compound. A small amount of spent catalyst is removed from the process as a chemically benign waste stream and the loss is made up through the addition of catalyst active reagents. The design catalyst consumption was based on performance of an earlier commercial unit commissioned in 2013. However, due to recent process and catalyst manufacturing improvements, total catalyst consumables for the Harbin unit decreased by 46%, leading to significant ongoing cost savings for the operator.

EU member states were among the first to start talking about the circular economy, developing ideas and implementing and adopting them in the industries in their own states. The uptake to reuse within the oil and gas industry is very poor, the opportunity to reuse is never more evident, or possible, as when the equipment is being decommissioned. The novel nature of this technology allows for continuous improvement to identify and realize optimization and cost-saving opportunities, which are beneficial for operators and their owners. The robustness of the technology is demonstrated through its ability to produce alkylate that meets or exceeds design specifications, even during deviation from prescribed design feed specifications. Despite already exceeding alkylate RON expectations, if required, the Harbin Petrochemical unit is capable of producing even higher quality alkylate products by operating at lower temperatures.

1.5 RAM Modelling

RAM refers to reliability, availability and maintainability. RAM analysis is a well-known method of estimating the production availability of a system by assessing failure modes, frequencies and consequences, all the while paying attention to the effect on production. The main objectives of RAM analysis are to increase system productivity, increase the overall profit and reduce the total life cycle cost, which includes lost production cost, maintenance cost, operating cost, etc.

RAM modelling is a Monte Carlo mathematical methodology that allows a user to evaluate different design scenarios, assuming a given process/equipment configuration, pipe flow, tank management and equipment failure/repair characteristics. Model outputs are asset utilization, availability and the costs to achieve that asset utilization over a given time period. It is important to note that, since Monte Carlo modelling is a probabilistic mathematical technique, model outputs are also probabilistic (e.g., outputs are probability distributions, rather than specific discrete values).

RAM models can be built to any degree of detail. However, in practice, these models should be built only to the detail required to answer the specific business/technical questions under consideration. The company Sadara uses three different levels of RAM models: RAM1, RAM2 and RAM3.

Although most engineering and construction companies are more interested in minimizing the initial investment, operating companies such as Sadara have a different objective: they want to balance the initial capital investment against the production and the conversion costs necessary to make their products over the life of the assets.

During the design phase, it is important to factor in or discern how different equipment and process units will affect availability and asset utilization, so that the plant can meet business production targets within cost constraints. This action is more difficult than it sounds, since even an experienced engineer has difficulty predicting actual production in a large, highly integrated site, given different unit availabilities, complex process unit configurations, equipment failure and repair, seasonal effects, fluctuations in supply and demand and intermediate storage capacity. This is a classic optimization problem that requires the right methodology and tools that reason, Sadara used RAM modelling as a key part of its approach.

1.5.1 RAM1 Site Model

RAM1 is the highest-level model and is a single model built for the entire site. Constructing the initial model required incorporating process unit configurations, tankage sizes, import/export capabilities and turnaround schedule estimates, as well as power, utilities and infrastructure information. As opposed to the RAM2 models used at the plant level, the RAM1 model uses simple uptime/downtime estimates for each major process unit.

Initial RAM1 model performance data was based on historical data from Sadara's parent companies and from worldwide refining and petrochemical best practices. Over time, these initial data estimates will be replaced with actual operating data.

Process units, tanks and other major assets are modelled as simple blocks with little underlying supporting detail. What is important in these models are the interconnections and dependencies between the units.

In the initial phases of the Sadara program, RAM1 models were crucial for determining which process units would be built and the ultimate storage configuration. Production revenue determined from the RAM1 model was then balanced against the costs to build, operate and maintain these units, as determined from engineering estimates. However, the RAM1 model remains evergreen and is updated as new business input becomes available.

Primary outputs of the RAM1 model are "utilized hours," along with storage and import/export capabilities over a 20-year time frame. The term "utilized hours" is best understood as the percentage of time that the unit must be available at full production. Simply put, it becomes the yearly asset mechanical reliability (AMR) percentage over that 20-year time horizon. The output of the RAM1 model sets the target availabilities for each of the underlying 26 process unit models.

1.5.2 RAM2 Plant Models

Whereas there is only one RAM1 model, there are RAM2 models built for each of the 26 process units and utilities. RAM2 models are much more detailed than the RAM1 model. Blocks are added for each major equipment item, using typical failure and repair characteristics for that equipment. The model is then run and adjustments are made that add or subtract redundancy, using more/less reliable equipment (obviously with cost differences) until the RAM1 and RAM2 model outputs converge. Outputs from the RAM2 models include the asset utilization of the plant, the total number of failures for all equipment and the percentage of production losses that each equipment caused, in addition to tank full/empty events.

Of importance is the Pareto chart of equipment vs. asset utilization loss for every piece of equipment in the unit, often called the "sensitivity" level. These charts can often be used to determine what equipment performance could be improved to increase overall asset utilization. They are frequently used by reliability engineers to identify potential projects that would increase asset utilization for that unit. The empty/full events of the tanks suggest whether the tanks are properly sized, or if they need to be increased to improve asset utilization or decreased to save capital cost.

1.5.3 RAM3 Models

RAM3 models are smaller and usually more specific, models that are often built to answer a specific question. They have frequently been used for spare parts optimization or case studies. The information needed to conduct a spare parts optimization analysis includes failure and repair characteristics, replenishment time, the cost of spare parts and spare parts production sensitivity. Model outputs are the minimum/maximum stocking levels that guarantee production is never lost because of spare parts unavailability and that spare parts are not overstocked.

1.5.4 RAM Modelling Benefit

Sadara has received benefits from RAM modelling from numerous sources. The following are some of the more common benefit areas, such as the ability to determine:

- when redundancy is needed or when the system is over-designed;
- how reliable a given piece of equipment should be;
- when, where and how much tankage is needed;
- which equipment improvement projects make the most sense economically;
- optimal quantities of spare parts for given equipment;
- effectiveness of preventive and predictive maintenance (PPM) and turnaround frequency/duration; and
- load-shedding strategies when unplanned events occur.

One benefit often achieved is the determination of whether redundancy is needed or not. Determining the reliability of a piece of equipment or a component within a given piece of equipment is not a trivial matter. RAM modelling makes it possible to accurately determine the cost/benefit of improved reliability or to select between vendors. This was used extensively during the RAM modelling effort. This aspect of RAM modelling remains an important part of Sadara's reliability program on an ongoing basis.

By utilizing RAM modelling, Sadara saved more than $1 billion in capital spending, while simultaneously increasing production capabilities and achieving business objectives. These benefits accrued from the elimination of process trains, decreased capital equipment spend, a more robust electrical grid and supporting components, the addition or elimination of redundancy where appropriate, adjustment of tank sizes and spare parts optimization. In 2018, Sadara successfully completed its performance test, the creditor's reliability test (CRT). There is no doubt that RAM modelling played an important role in achieving that important milestone.

At the very least, asset utilization modelling should be an essential part of every capital project. Models created during the design phase can then be leveraged to answer future operational and maintenance issues based on a rigorous analysis rather than on a gut feeling. Used this way, asset utilization models complement the process simulation models that process engineers already maintain for each unit.

1.6 Conclusions

Over the next few decades, the energy and chemicals markets will face a very important reshaping. Chemicals represent one of the fastest-growing crude-oil-demand sectors. The use of oil in the petrochemicals sector is likely to become a key source of oil demand after the 2020s. In this scenario, the role of oil as a feedstock rather than a fuel will be greatly emphasized. From an economic and practical point of view, liquid hydrocarbons are the ideal feedstock to produce carbon-based goods, from polymers to new lightweight materials to pharmaceuticals, solvents and thousands of products that have become part of our daily lives. Indeed, considering an average price of fuel at ~$600/ton and petrochemicals at ~$1200/ton, refineries focusing in the production of petrochemicals are expected to become much more profitable and sustainable.

The refinery of 2030 will thrive by being fully demand-driven, with decisions made quickly at the appropriate level. Across the management chain, knowledgeable workers in the 2030 refinery will take actions based on their impact on satisfying demand better than competitors. For chemical technologies, the applications of miniaturized reaction systems or micro reactors, is centred on gas and liquid phase reactions covering simple microscale mixing of different fluids' volumetric titrations, heterogeneous and homogeneous catalysis, catalytic oxidation, pyrolysis, heterocyclic synthesis, photochemical reactions and micro fuel cell applications. Using digital technology, manufacturers can create a complete digital twin of the plant. The digital twin illustrates the asset and the operations life cycle and how the two connect together to create the goods and services produced. By utilizing RAM modelling, Sadara saved more than $1 billion in capital spending, while simultaneously increasing production capabilities and achieving business objectives.

The refinery of the future will be achieved through breakthroughs in three intertwined areas:

- maximization of chemicals production with integrated carbon capture and utilization, leading to reduce emissions along with the use of renewable energy in the process;
- process intensification by miniaturization of physical footprint where applicable and overall maximum reduction in the number of unit operations, thereby significantly reducing energy and capital intensity. Integration of new intelligent process control systems will also support this by implementing rapidly corrective and preventive actions in line with market demands; and
- discovery of scalable multifunctional catalysts able to upgrade (removal of impurities) and crack crude to light olefins and aromatics in one single reaction vessel.

Indeed, the success of this refinery of the future will rest on the development of economically efficient processes based on minimum steps processes. The combination of new multifunctional catalysts, new reactor concepts, process intensification and smart integration of renewable energy should result in the development of new processes surpassing an 80% wt yield to chemicals from crude oil on an economical and environmentally responsible manner.

The oil and gas industry is rebranding itself as an energy provider and is transitioning to selling low-carbon-intensity energy products that include renewable wind and solar power, green and blue hydrogen, low-carbon LNG, biofuels, renewable diesel, e-fuels, e-gasoline and sustainable aviation fuel.

Improving energy efficiency and stopping methane leakage are the cheapest ways to reduce carbon dioxide emissions. For new facilities, or major plant expansions, project developers should consider incorporating new technologies that have better energy efficiency and lower carbon intensity. For major turnarounds requiring catalyst changeouts, operators should consider newer formulated catalysts for higher yields and lower energy intensity.

The pathways to reduce carbon dioxide emissions include:

- improving efficiency in existing and new facilities by maintaining energy recovery equipment, stopping or minimizing routine flaring (such as minimizing flaring on startups and shutdowns) and identifying and stopping methane leakage;
- using new process technologies and catalysts to improve yields and reduce energy intensity;
- electrifying process equipment using renewable electric power (e.g., wind, solar, hydro and nuclear power); and investing in battery storage and in using and selling renewable power;
- incorporating cost-effective CCUS technology, joining regional CCUS networks and using captured carbon dioxide in e-chemistry to produce e-fuels; and

- producing, utilizing and selling low-carbon hydrogen, as well as using hydrogen as an energy carrier and to produce e-fuels.

Pathways to reduce carbon emissions for the oil and gas industry include producing and selling energy products with low-carbon intensities, such as:

- biofuels, renewable fuels and e-fuels. This includes producing renewable diesel, sustainable aviation fuels, synthetic fuels, e-gasoline, low-carbon LNG and renewable electricity, as well as green and blue hydrogen.

The seventh pathway will help reduce carbon dioxide emissions from the production of petrochemicals:

- circular carbon pathway. This pathway includes recycling plastics by either mechanical reprocessing or chemical means (pyrolysis, gasification); using renewable feedstocks like bio-naphtha, hydrogenated vegetable gasoil (diesel) and plastic pyrolysis oil to produce the base chemicals ethylene and propylene; and producing synthetic chemicals, such as ethanol and methanol, by using renewable hydrogen and captured carbon dioxide.

The good news is that the oil and gas and petrochemical industries have the technology and assets needed for offshore wind turbines, blue/green hydrogen production and carbon dioxide capture and storage. They also have the refinery units and technology to produce renewable fuels. These industries are prepared for the journey to complete this crucial energy transition to a lower-carbon world.

Further Reading

Alabdullah, M.A. et al. (2020). A viewpoint on the refinery of future: catalyst and processes challenges. *ACS Catal.* 10: 8131–8140.

Albrecht, F.G., König, D.H., Baucks, N. and Dietrich, R.-U. (2017). A standardized methodology for the techno-economic evaluation of alternative fuels – a case study. *Fuel* 194: 511–526.

Aresta, M., Dibenedetto, D. and Angelini, A. (2014). Catalysis for the valorisation of exhaust carbon: from CO2 to chemicals, materials and fuels. Technological use of CO_2. *Chem. Rev* 114: 1709–1742. Doi: 10.1021/cr4002758.

Asinger, F. (1986). *Methanol–Chemie- und Energierohstoff: Die Mobilisation der Kohle.* Berlin, Germany: Springer.

Bajohr, S., Schollenberger, D., Buchholz, D. et al. (2014). Kopplung der PtG-Technologie mit thermochemischer Biomassevergasung: Das KIC-Projekt "Demo SNG". *GWF, Gas/Erdgas* 155 (7): 470–475.

Bajus, M. (2020). *Petrochemistry: Petrochemical Processing, Hydrocarbon Technology and Green Engineering.* Chichester: Wiley.

Beck, R. (2019). Executive viewpoint: a vision for the refinery of 2030. *Hydrocarbon Processing*, May 2019.

BP energy oulook. (2023). https://www.bp.com/en/global/corporate/energy-economics/energy-outlook/energy-outlook-downloads.html (Accessed March 2023).

Buddenberg, T., Bergins, C. and Schmidt, S. (2016). Power to fuel as a sustainable business model for cross-sectorial energy storage in industry and power plants. In: *5th Conference on Carbon Dioxide as Feedstock for Fuels, Chemistry and Polymers*, Cologne, Germany.

Buehler, J. (2022). Overview of decarbonization pathways for the oil and gas and petrochemical industries – Part 2. *Hydrocarbon Processing*, June 2022.

Burger, J. (2017). Large-scale production of OME – state of the art. In: *OME Infotag*. DECHEMA, Frankfurt.

Campbell, C.T. (2017). The degree of rate control: a powerful tool for catalysis research. *ACS Catal.* 7: 2770–2779. doi: 10.1021/acscatal.7b00115.

Carugo, M. and Valentine, J. (2022) Accelerating the transition to renewable fuels. *Hydrocarbon Processing*, October 2022.

Cui, Y., He, B., Liu, X. and Sun, J. (2020). Ionic liquids-promoted electrocatalytic reduction of carbon dioxide. *Ind. Eng. Chem. Res.* 59 (46): 20235–20252.

Delgado, K.H. (2014). *Surface Reaction Kinetics for Oxidation and Reforming of H_2, CO and CH_4 over Nickel-based Catalysts.* Dissertation, Karlsruher Institut fur Technologie, Fakultät fur Chemie und Biowissenschafen.

Department of Defense, Assistant Secretary of Defense for Research and Engineering (2011). *(ASD (R&E)) Technology Readiness Assessment (TRA) Guidance.*

Deutschmann, O. (2015). Modeling of the interactions between catalytic surfaces and gas-phase. *Catal. Lett.* 145: 272–289. doi: 10.1007/s10562-014-1431-1.

Deutschmann, O., Tischer, S., Correa, C. et al. (2018). *DETCHEM software package v2.7.* www.detchem.com.

DIN EN ISO 10991:2010-03 (2009). *Mikroverfahrenstechnik – Begriffe.* Berlin: Beuth Verlag.

Egloff, G., Shaad, R.E. and Lowry, C.D. (1930). The Decomposition of the Paraffin Hydrocarbons. *J. Phys. Chem.* 34: 1617–1740.

Ernig, G. and Klemm, E. (2017). *Chemische Reaktionstechnik*, 6e. Berlin: Springer.

EU (2017). Horizon 2020 – work programe 2016–2017, 20. *General Annexes: G. Technology readinness level (TRL).*

Furimsky, E. (2020). CO_2 hydrogenation to methanol and methane over carbon-supported catalysts. *Ind. Eng. Chem. Res.* 59 (35): 15393.

Gentilucci, A., Spagnuolo, L., Hughes, A. et al. (2022). Sustainable steel/chemicals production: removal of impurities for valorization of steel mill gases. *Hydrocarbon Processing*, October 2022.

Gerden, E. (2019). Asia: China eyes significant growth in petrochemicals. *Hydrocarbon Processing*, September 2019.

Giehr, A. (2019). *Dry reforming of CH_4: experimental investigations and microkinetic modeling studies on a co-based catalyst.* Doctoral Dissertation, Karlsruher Institut für Technologie: Karlsruhe.

Gossler, H., Maier, L., Angeli, S. et al. (2019). CaRMeN: an improved computer-aided method for developing catalytic reaction mechanisms. *Catalysts* 9: 227. doi: 10.3390/catal9030227.

Götz, M. (2014). *Methanisierung im Dreiphasen-Reaktor.* Dissertation, Karlsruher Institut für Technologie (KIT).

Götz, M., Lefebvre, J., Mörs, F. et al. (2017). Novel gas holdup correlation for slurry bubble column reactors operated in the homogeneous regime. *Chem. Eng. J.* 308: 1209–1224. doi: 10.1016/j.cej.2016.09.101.

Hayek, F. and Moran, M. (2019). How did asset utilization modeling save Sadara $1B in capital cost. *Hydrocarbon Processing*, March 2019.

Held, M., Schollenberger, D., Sauerschell, S. et al. (2020). Power-to-Gas: CO_2 methanation concepts for SNG production at the Engler-Bunte-Institut. *Chemie Ingenieur Technik* 92 (5): 595–602. doi: 10.1002/cite.201900181.

Ho MT, Wiley DE (2016). 28 – liquid absorbent-based post-combustion CO_2 capture in industrial processes A2 – feron. In: *Absorption-based Postcombustion Capture of Carbon Dioxide* (ed. H.M. Paul), 711–715. Woodhead Publishing.

Lefebvre, J., Trudel, N., Bajohr, S. and Kolb, T. (2018). A study on three-phase CO_2 methanation reaction kinetics in a continuous stirred-tank slurry reactor. *Fuel* 217: 151–159. doi: 10.1016/j.fuel.2017.12.082.

Lefebvre, J., Bajohr, S. and Kolb, T. (2019). A comparison of two-phase and tree-phase CO_2 methanation reaction kinetics. *Fuel* 239: 896–904. doi: 10.1016/j.fuel.2018.11.051.

Lefebvre, J., Götz, M., Bajohr, S. et al. (2015). Improvement of three-phase methanation reactor performance for steady-state and transient operation. *Fuel Process. Technol.* 132: 83–90. doi: 10.1016/j.fuproc.2014.10.040.

McMullen, J. (2020). Business trends: managing risk and uncertainty: the importance optimizing your value chain. *Hydrocarbon Processing*, June 2020.

Mergel, J., Carmo, M. and Fritz, D. (2013). Status on technologies for hydrogen production by water electrolysis. In: *Transition to Renewable Energy Systems* (ed. D. Stolten and V. Scherer), 423–450. Wiley-VCH Verlag GmbH & Co. KGaA.

Mörs, F., Graf, F. and Kolb, T. (2019). Reaktoren für Dreiphasen-Reaktionen: Suspensionsreaktoren. In: *Handbuch Chemische Reaktoren: Grundlagen und Anwendungen der Chemischen Reaktionstechnik* (ed. W. Reschetilowski). Berlin: Springer.

Mörs, F., Ortloff, F., Graf, F. and Kolb, T. (2019). Hydrodynamik in Blasensäulen – Messung von relativem Gasgehalt und Blasengröße. *Chem. Ing. Tech.* 91 (7): 1059–1065. doi: 10.1002/cite.201800107.

Muraza, O. (2021). Highlighting the greener shift in transportation energy and fuels based on novel catalytic materials. *Energy Fuels* 35 (1): 25–44.

Newton, M. (2018). Business trends. *Hydrocarbon Processing*, November 2018.

Nezam, I., Zhou, W., Gusmaäo, G.S. et al. (2021 March). Direct aromatization of CO_2 via combined CO_2 hydrogenation and zeolite-based acid catalysis. *J. CO2 Util.* 45: 101405.

Nichols, L. (2017–2018). Construction boxscore database. *Hydrocarbon Processing*, December 2017–December 2018.

Nichols, L. (2019a). Industry perspectives: the evolution of the downstream. *Hydrocarbon Processing*, May 2019.

Nichols, L. (2019b). Business trends: petrochemicals in 2025: three regions to dominate the surge in petrochemical capacity growth. *Hydrocarbon Processing*, March 2019.

O'Brien, C., Stoots, J., Herring, M. et al. (2010). High temperature electrolysis for hydrogen production from nuclear energy technology summary. *Technical Report, Idaho National Laboratory (INL)*. No. INL/EXT-09-16140.

Olah, G.A., Goeppert, A. and Prakash, G.K.S. (2011). *Beyond Oil and Gas: The Methanol Economy*. Weinheim, Germany: Wiley-VCH.

Parks, G.S. and Huffman, H.M. (1932). *Free Energy of Some Organic Compounds*. New York: Reinhold.

Pérez-Fortes, M., Schöneberger, J.C., Boulamanti, A. and Tzimas, E. (2016). Methanol synthesis using captured CO_2 as raw material: techno-economic and environmental assessment. *Appl. Energy* 161: 718–732.

Pinsky, R., Sabharwal, P., Hartvigsen, J. and O'Brien, J. (2020). Comparative review of hydrogen production technologies for nuclear hybrid energy systems. *Prog. Nuc. Eng.* 123: 103317.

Press release (2019 March 26). *Methanisierungsanlage in Falkenhagen geht in Betrieb und liefert synthetisches Methan – weiterer Schritt für eine erfolgreiche Energiewende*. Karlsruher Institut für Technologie, Uniper SE, ThyssenKrupp Industrial Solutions AG, DVGW.

Press release (2019). *Energy Lab 2.0–Dreiphasen-Methanisierung geht in Betrieb*. Karlsruher Institut für Technologie, Karlsruhe. www.elab2.kit.edu.

Preuster, P., Papp, C. and Wasserscheid, P. (2017). Liquid organic hydrogen carriers (LOHCs): toward a hydrogen-free hydrogen economy. *Acc. Chem. Res.* 50 (1): 74–85. doi: 10.1021/acs.accounts.6b00474.

Pushnov, A.S. (2006). Calculation of average bed porosity. *Chem. Pet. Eng.* 42: 14–17. doi: 10.1007/s10556-006-0045-x.

Rio, R. *Proactive asset management with IIoT and analytics*. https://www.arcweb.com/blog/proactive-asset-management-iiot-analytics.

Rossini, F.D. et al. (1947). *Selected Values of Properties of Hydrocarbons*. American Petroleum Institute, Research Project 44.

Rossini, F.D. (1950). *Chemical Thermodynamics*. New York: Wiley.

Rostrup-Nielsen, J.R. (1984). Catalytic steam reforming. In: *Catalysis: Science and Technology* (ed. J.R. Anderson and M. Boudart). New York: Springer.

Sánchez-Bastardo, N., Schlögl, R. and Holger Ruland, H. (2021). Methane pyrolysis for zero-emission hydrogen production: a potential bridge technology from fossil fuels to a renewable and sustainable hydrogen economy. *Ind. Eng. Chem. Res.* 60 (32): 11855–11881.

Schemme, S., Samsun, R.C., Peters, R. and Solten, D. (2017). Power-to-Fuel as a key to sustainable transport systems – an analysis of diesel fuels produced from CO_2 and renewable electricity. *Fuel* 205: 198–221.

Schmider, D., Maier, L. and Deutschmann, O. (2021). Reaction kinetics of CO and CO_2 methanation over nickel. *Ind. Eng. Chem. Res.* 60 (16): 5792–5805.

Schmidt, P., Weindorf, W., Roth, A., et al. (2016). *Power-to-Liquids – potentials and perspectives for the future supply of renewable aviation fuel*. German Environment Agency.

Schollenberger, D., Bajohr, S., Gruber, M. et al. (2018). Scale-up of innovative honeycomb reactors for Power-to-Gas applications – The project store & go. *Chem. Ing. Tech.* 90 (5): 696–702. doi: 10.1002/cite.201700139.

Schollenberger, D., Bajohr, S., Reimert, R. and Kolb, T. (2018). Einsatz von Wabenreaktoren zur Methanisierung bei Power-to-Gas-Prozessen. *GWF, Gas + Energie* 10: 108–115.

Service, R.F. (2019). *Can the world make the chemicals it needs without oil?* https://www.sciencemag.org/news/2019/09/can-world-make-chemicals-it-needs-without-oil.

Sterner, M. and Stadler, I. (2014). *Energiespeicher: Bedarf, Technologien, Integration*. Berlin: Springer Vieweg.

Stolten, D. and Scherer, V. (2013). *Transition to Renewable Energy Systems*. Weinheim, Germany: Wiley-VCH Verlag GmbH & Co. KGaA, 969.

Stotz, H., Maier, L., Boubnov, A. et al. (2019). Surface reaction kinetics of methane oxidation over PdO. *J. Catal.* 370: 15–175. doi: 10.1016/j.jcat.2018.12.007.

Tullo, A.H. (2019). Why the future of oil is in chemicals, not fuels. *Chem. Eng. News* 97 (8).

Van Utyvanck, P. (2020). Petrochemicals: was 2019 the year of recycled PET? *Hydrocarbon Processing*, February 2020.

Visconti, C.G., Groppi, G. and Tronconi, E. (2013). Accurate prediction of the effective radial conductivity of highly conductive honeycomb monoliths with square channels. *Chem. Eng. J.* 223: 224–230. doi: 10.1016/j.cej.2013.02.095.

Wikipedia (2020). *Circular economy*. https://en.wikipedia.org/wiki/Circular_economy.

Yang, X., Su, X., Chen, D. et al. (2020). Direct conversion of syngas to aromatics: a review of recent studies. *Chin. J. Catal.* 41: 561–573. doi: 10.1016/S1872-2067(19)63346-2.

Zangeneh, F.T., Sahebdelfar, S. and Ravanchi, M.R. (2011). Conversion of carbon dioxide to valuable petrochemicals: an approach to clean development mechanism. *J. Nat. Gas Chem.* 20: 219–231.

2

The Green Shift in Power-to-Chemical Technology and Power-to-Chemical Engineering: A Framework for a Sustainable Future

CONTENTS

Converting Power into Chemicals and Fuels: Power-to-X Technology for a Sustainable Future, First Edition. Martin Bajus.

2.1 Introduction

The drive for efficiency in chemical technology merges the best of the ideals of the Enlightenment and the Renaissance. Certainly, there is a premium on rationalism, with an aspiration of mechanistically sound reaction design and process development. But at the same time, the aesthetic appeal of the new ideas that culminate in the advances we now see routinely is unmistakable. There is no constraint on curiosity when we consider the boundary conditions of efficient, environmentally friendly processes. On the contrary, these considerations spawn new concepts and approaches, ranging from postmodern expansions of photochemistry, reconsideration of seminal thinking about solvation, importation of physical and mechanical phenomena, to reaction development – the creativity born of efficiency considerations now drives major technology innovation in chemistry.

For more than three decades, green chemical technology and green engineering have provided a framework for chemists and chemical engineers to do their part in contributing to the broad scope of global sustainability. The research results highlight just some of the most impactful science in this arena, across a very wide swath of chemical space. For example, in organometallic chemistry, catalysis and materials, there appears scientific data that best highlights the diverse nature of this type of organometallic chemistry, kinetics and catalysis and that are likely to impact development of more sustainable chemical processes and the transition to a circular economy. Key coverage includes catalysis with emerging feedstocks and synthetic methods for preparing materials and chemicals in a sustainable way to help bring critical innovations from a research setting to commercialization. Many results in the green chemical technology and sustainability space are core considerations in applied chemistry and chemical engineering. Data coming from environmental science and technology clearly illustrate the interdisciplinary systems approach that is required to address key environmental challenges that impede sustainability efforts, from remediation of pollution to design of next-generation safer and functional chemicals. Interdisciplinary results are connecting across industry and academia.

With a similar goal there is a tradition of bridging industrial and academic research. Its focus on chemical processes as the science that enables the safe, environmentally friendly and economical manufacturing of chemicals is evident. The highlight aspects of catalysis, synthetic methodology development and synthetic strategy exploration that are needed to enable circular economies.

The environmentally sustainable recovery of resources from wastes is also a burgeoning area of research. The benchmarks for degradation rates of common plastics will help prioritize future research efforts on chemical degradation to close the loop. Parker et al. explore the depolymerization of polyethylene with palladium catalysts to understand approaches to the remediation of polyolefins

The ternary Pd complexes react with hexane in a linear ion trap mass spectrometer, forming the C–H activation product and releasing hydrogen, methane, ethylene and other alkenes respectively. Density functional theory (DFT) calculations agree well with the experiments in predicting low barriers for these reactions proceeding via a metathesis mechanism.

Today's world faces the grand dual challenges of providing affordable and reliable energy whilst reducing emissions. The demand for energy, provided in large part, historically, by the combustion of fossil fuels, continues to grow in tandem with the world's population and economy. Meeting that dual challenge requires a multifaceted approach, including research on clean or renewable energy sources (wind, solar, geothermal, hydro and nuclear energies) to avoid carbon dioxide emissions and the storage and conversion of energy carriers (electricity, hydrogen, or ammonia) for the energy mix. Additionally, carbon dioxide capture and sequestration will be required to facilitate the transition toward this future. The role of crude oil and the disposition of hydrocarbon resources in this future energy landscape are yet to be defined and it will evolve as the energy transition evolves. This highlights a new perspective on using petroleum to address the materials challenge and the scientific research and technologies required to achieve it.

An emerging new opportunity is the conversion of those abundant heavy aromatic hydrocarbons into non-combustible durable materials. Unlocking this opportunity requires mastering reactivities and a thorough understanding of petroleum structures at a higher level than is required just to extract the energy content. Petroleum is one of the most complex mixtures on the planet and unravelling that complexity has long stood as a challenge for analytical and molecular characterization science. The 150 years of research, from physical property measurements in the nineteenth century and simple elution chromatography, gas and liquid chromatography in the early twentieth century, to Fourier transform ion cyclotron resonance (FT-ICR) mass spectrometry and petrol omics in recent decades, have tremendously advanced our mastery of this complexity and helped us produce today's fuels and chemicals. As we move forward into advanced materials and new products, we need to broaden our analyses to a more accurate understanding of molecular structures and their chemical reactivities.

Recent studies in the Energy & Fuels (2022) of Yunlong Zhang with non-contact atomic force microscopy (NC-AFM) provide a glimpse of exact molecules and their polynuclear aromatic motifs in the heaviest petroleum fractions. Coupling these structural insights with molecular modelling and aromatic theories is shedding light on their reactivities. These new structure–reactivity perspectives are opening the doors to non-combustible uses of petroleum and enabling us to explore pathways for controlled molecular weight growth from small molecules in making materials.

This represents a paradigm shift from the historical focus on molecular weight reduction that was required to convert petroleum to clean fuels. Oil-derived carbon will then be sequestered in materials rather than being emitted as carbon dioxide in fuel combustion. Carbon-based materials may also help reduce emissions through the displacement of steel or cement in infrastructure and energy consumption through reducing weight of transportation. This research has garnered significant attention, especially for the disposition of lower value carbon-intensive heavy oils, bitumen, oil sands and coal-derived liquids.

Petroleum evolved from an Age of Illumination in the nineteenth century, with kerosene for lighting as a primary product, to the Age of Energy in the twentieth century, with the growth in hydrocarbon-fuelled transportation. An emerging new Age of Materials is upon us by mastering the structures and reactivities required to grow structures from small molecules to large molecules, with petroleum providing a source of building blocks to meet society's materials demands.

2.2 Eco-Friendly Catalyst

Heterogeneous catalysts will play important roles at different steps of solar refinery, starting in:

- water splitting with photocatalysis materials such as nitride photoelectrodes;
- production of chemicals and fuels from renewable hydrogen;
- available green hydrogen used to produce chemicals, via reverse water gas shift (RWGS);
- utilization of carbon oxides (carbon monoxide and carbon dioxide) to produce methane;
- production of methanol as energy storage; and
- carbon dioxide utilization in methane reforming to produce synthetic gas (dry reforming in the CALCOR and SPARG processes).

Electrification trends will also stimulate the application of catalytic materials in harvesting solar energy. The solar refinery is a conceptual process to utilize solar energy to generate electricity, which can be used to produce renewable hydrogen, the hydrogen then being used to produce chemicals and fuels together with carbon from biomass or waste (carbon dioxide).

2.2.1 Development of Catalysts Supported on Carbons for Carbon Dioxide Hydrogenation

Developments of catalysts for the hydrogenation of carbon dioxide to either methane or methanol have been dominated by the catalysts comprising transition metals supported on oxidic supports. In recent years, growing interests in catalysts supported on carbon supports has been seen. Attention has been paid to carbon nanomaterials such as carbon nanotubes (CNTs), carbon nanofibers (CNFs) and graphene-derived supports, i.e., graphene oxide (GO), reduced graphene oxide (rGO) and few layers graphene (FLG). Among amorphous carbons, so far, only activated carbon (AC) has been tested. Among active metals, transition metals (e.g., Cu, Fe, Ni, Co, Mn, Mo and Zn) and noble metals (e.g., Pd, Pt, Rh and Ru) have been evaluated as part of monometallic and bimetallic catalytic phases. However, the traditional Cu-ZnO phase, in combination with various supports, has been attracting most of the attention as a catalyst for the conversion of carbon dioxide to methanol. Other transition metals have been evaluated during the hydrogenation of carbon dioxide to methane. Significantly higher hydrogenation activity of noble metals indicates the difference in catalyst performance and mechanism, compared with that of catalysts containing transition metals.

It has been established that some catalytic phases can activate carbon dioxide to enable its conversion under hydrogenation conditions. Most recently, Dang et al. gave detailed accounts of catalyst structures developed for the hydrogenation of carbon dioxide. Both transition metals- and noble metals-containing catalysts supported on oxidic supports have been used for the hydrogenation of carbon dioxide to methanol using copper and zinc oxide as the metallic components. On the other hand, the catalysts containing transition metals (e.g., Ni, Co, Fe, etc.) and noble metals (e.g., Pd, Ru and Rh) have been evaluated for the hydrogenation of carbon dioxide to methane. Among supports, the conventional γ-Al_2O_3 has been used most frequently, in addition to silica, titanium and zirconium dioxide and other oxides. The activity and selectivity of these catalysts could be further improved by the addition of various dopants, i.e., Ga, B, Ce, V, etc. The operating parameters (e.g., temperature, pressure, H_2/CO_2 ratio, plant design, etc.), which influence the economy of industrial production of methanol over the conventional Cu-ZnO catalyst were discussed in detail by Alsuhaibani et al. The studies on the other types of catalysts for the hydrogenation of carbon dioxide also have been studied.

Carbon materials have been attracting attention as supports for catalysts used in various applications. Amorphous carbons such as activated carbon and carbon black (CB) have been used extensively as supports of hydroprocessing (HPR) catalysts used for upgrading conventional and nonconventional feeds to fuels. Carbon nanomaterials such as carbon nanotubes, carbon nanofiber and graphene-derived supports such as graphene oxide, reduced graphene oxide and few layers graphene have been evaluated as supports for catalysts used for the hydrogenation of various biomass-derived feeds to fine chemicals and HPR to fuels. There are numerous cases which show that carbon-supported catalysts outperformed the corresponding catalysts supported on traditional oxidic supports. This was attributed to suitable textural properties such as surface area, porosity and particle size. The surface of the carbons can be functionalized to enhance the anchoring of active metals. The hydrophobic nature of carbons ensured high stability, even under aqueous conditions. Similarly, as in the case of other hydrogenation reactions, carbon supports such as AC, CB, CNTs, CNFs, GO, rGO and FLG have been included in studies on the conversion of carbon dioxide to hydrocarbons and alcohols. The performance of the traditional Cu-ZnO phase supported on carbon supports was compared with other active phases consisting of transition metals and noble metals. Both monometallic and bimetallic catalytic phases have been evaluated.

In industrial operations, one-pass conversion of carbon dioxide to methanol varies from ~5% to 15%. The recycling of the unconverted carbon dioxide has been practiced, improving the efficiency of the process. Relatively low conversions could be attributed to the adverse effect of water by-products on catalyst performance. Thus, for every mole of methanol and methane produced, 1 and 2 mol of water, respectively, are produced. The conversion could be increased to ~30% by conducting the hydrogenation of carbon dioxide in two steps. In the first step, the partial hydrogenation of carbon dioxide to carbon monoxide, i.e., reverse water–gas shift (RWGS) reaction, followed by in situ water removal, was performed. The second step involves the hydrogenation of carbon monoxide to methanol. The reactor configurations, such as once-through reactor with recycling of unconverted gases after separation of methanol and water by condensation, as well as by the membrane permeable to water, were compared in the study of Leonzio et al. Therefore, the schemes during which the contact of water with hydrogenation catalyst can be minimized could be identified. Apparently, the catalysts supported on oxidic supports are much more sensitive to water than those supported on carbon supports. In various hydrogenation applications under aqueous conditions, the latter exhibited higher activity and stability. This suggests that studies on the development of novel catalysts supported on carbon supports for the hydrogenation of carbon dioxide applications is an attractive area of research. Enhanced performance of catalysts during the conversion of carbon dioxide to either methane or methanol can be achieved via alternative routes such as electrocatalytic and photocatalytic hydrogenation of carbon dioxide. Similar catalysts formulations have been evaluated for these applications. Generally, the conversions of carbon dioxide and selectivity to the product of interest are higher, compared with catalytic routes.

2.2.2 Properties of Carbon Supports

The database on the properties and preparation of various carbon solids is rather extensive. This included both amorphous and highly crystalline carbon solids, which have been evaluated as potential supports for the catalyst used for the hydrogenation of carbon dioxide to methanol and/or methane. Structurally, highly ordered nanomaterials in their pristine form such as graphene, CNTs and CNFs, as well as the corresponding materials modified either by functionalization or doping, have been evaluated with a primary focus on their applications during the hydrogenation of carbon dioxide to either methane or methanol. Pristine graphene is a single two-dimensional sheet of polycyclic, hexagonally arranged, sp_2-bonded aromatic carbon which can be separated from graphite by exfoliation. CNTs have similar hexagonal geometry with sp_2 hybridization. Both single-walled carbon nanotubes (SWCNTs) and multiwalled carbon nanotubes (MWCNTs) have been used. The SWCNTs appear as a graphene sheet wrapped into a seamless cylinder with a wall thickness approaching the diameter of a carbon atom. The MWCNTs comprise several graphene materials; extensive information on various structural aspects of CNTs can be accessed in the literature.

Three types of CNFs have been investigated, platelet, ribbon and spiral. The platelets and ribbon CNFs consist of graphene layers aligned either perpendicularly or in parallel, respectively, while the spiral form of CNFs consists of platelets oriented parallel to the fibre axes. This allows some bending along the CNF axes. An extensive database on various aspects of CNFs can be accessed in the literature. Rather than repeat this information here, the following studies provide a solid basis for readers to acquire a deeper understanding of the topic of carbon nanomaterials.

Among amorphous carbons, activated carbon and carbon black have been used as the supports of catalysts used for the hydrogenation and hydro processing of various oxygenates and petroleum-derived feeds. Attempts have been made to use AC as the support of catalysts used for the hydrogenation of carbon dioxide, in comparison with carbon nano supports and oxidic supports, while little information on similar applications of CB has been found. Because of low crystallinity, the surface area of ACs frequently exceeds 1000 m^2/g. For most ACs, the pore size varies over a range of 1–10 nm, although pores >10 nm are present and serve as channels for reactants diffusing into mesopores and micropores. Both granular and powder forms of ACs are available commercially.

Several studies highlight advances in catalysis and reaction engineering that have led to waste and energy minimization for processes that still pose significant challenges. On the catalysis front, we see an increase in the scope of synthetic transformations facilitated by supported catalysts. In one example, Wang et al. described the preparation of a reusable magnetic Ag–Fe_3O_4 catalyst supported on cellulose microspheres for reduction of nitrophenols. We also see increased exploration into the utility of base metals in catalysis and organocatalytic processes being performed under mild conditions and with low catalyst loading: for highly enantioselective epoxidation of α,β-unsaturated ketones, Majdecki et al. employed an amide-based *Cinchona* alkaloid at loadings as low as 0.5 mol% as a hybrid phase-transfer organocatalytic.

On the application front, the quest for reducing traditionally high "E-factors" in the pharmaceutical sector continues. The E-factor is a metric of how environmentally friendly, or harmful, a chemical process is. Specifically, the E-factor is the ratio of kilograms of waste generated per kilogram of product synthesized. Thus, the lower the E-factor, the more environmentally friendly a process should be. Of course, this is only a single metric and other factors, like the toxicity of the waste produced, should also be considered. In pharmaceutical companies, the E-factor for the synthesis of drug products is typically in the range of about 25 to 100.

In two critical perspectives, colleagues from major pharmaceutical companies highlight the need to develop less wasteful and toxic methods for peptide and oligonucleotide synthesis and purification methods, in light of the growing utility of biological peptide-based pharmaceuticals. Rossen highlights insights from process chemistry that can reduce the impact of synthetic chemistry at smaller scales. Hayler and colleagues provide a perspective from the pharmaceutical industry on the role of catalysis with abundant metals.

Catalysis with abundant rather than precious metals continues to be an active area of research and a key component of sustainable chemistry research. Publications focus on cobalt and iron catalysts for hydrosilylation and nickel precursors and heterogeneous catalysts for Suzuki–Miyaura coupling, one of the most widely used methods for C–C bond formation. Putten et al. also report the role of manganese in mediating C–C bond formation from organoboranes. The Suzuki reaction is an organic reaction, classified as a cross-coupling reaction, where the coupling partners are a boronic acid and an organohalide and the catalyst is a palladium(0) complex. It was first published in 1979 by Akira Suzuki and he shared the 2010 Nobel Prize in Chemistry with Richard F. Heck and Ei-ichi Negishi for their contribution to the discovery and development of palladium-catalysed cross-couplings in organic synthesis. This reaction is also known as the Suzuki–Miyaura reaction or simply as the Suzuki coupling. It is widely used to synthesize polyolefins, styrene's and substituted biphenyls. Several reviews have been published describing advancements and the development of the Suzuki reaction. The general scheme for the Suzuki reaction is when, a carbon–carbon single bond is formed by coupling an organoboron species (R_1-BY_2) with a halide (R_2-X) using a palladium catalyst and a base.

Waste reduction also continues to be tackled in the development of alternative solvents, such as less hazardous task-specific ionic liquids deep eutectic solvents and increasing the scope and efficiency of reactions in water, such as amidation amine synthesis and transfer hydrogenation. We also see the development of new biomass-derived solvents with greener profiles, such as glycerol-derived 1,2,3-triethoxypropane and systematic solvent selection protocols for "greener" solvents.

2.3 Hydrogen

Environmental issues related to greenhouse gas emissions are progressively pushing the transition toward a fossil-free energy scenario, in which renewable energies such as solar and wind power will inevitably play a key role. However, for this transition to succeed, significant issues related to renewable energy storage have to be addressed. Power-to-Technologies have gained increased attention since they actually convert renewable electricity to chemicals and fuels that can be more easily stored and transported. Hydrogen production through water electrolysis is a promising approach since it leads to the production of a sustainable fuel that can be used directly in hydrogen fuel cells or to reduce carbon dioxide in chemicals and fuels compatible with the existing infrastructure for production and transportation. Carbon dioxide electrochemical reduction is also an interesting approach, allowing the direct conversion of carbon dioxide into value-added products using renewable electricity.

The chemical technology is based on conversion processes, where hydrogen and carbon are essential elements. Reactions with mixtures of hydrogen, carbon monoxide and carbon dioxide are well known in the chemical technology as synthesis gas-based routes and they can be used to build-up all major platform chemicals for the chemical industry's value chain. In contrast to the usual gasification processes to transform fossil carbon feedstocks into synthesis gas, subject to the corresponding high carbon dioxide emissions, alternative pathways can be designed using hydrogen from low-carbon electricity as a reactant. From nine target products, ammonia can be directly synthesized using hydrogen and nitrogen. For the other petrochemicals products from pillar A (see Chapter 1), olefins ethylene and propylene; from pillar B, benzene, toluene, xylenes; from pillar C, methanol; but also from pillar D, synthetic fuels, a carbon source is required and carbon dioxide can serve as such a source, currently available as end-of pipe waste from more or less concentrated point sources such as flue gas of fossil-fired power plants, industrial processes such as lime/cement production, steel manufacturing and also within the chemical industry. In effect, these routes offer an opportunity for the chemical technology to reduce its dependence on fossil fuels, to reduce industrial CO_2 emissions as well as recycle and valorise emitted carbon dioxide.

Hydrogen acts as key high-energy containing reaction partner in the transformations of carbon dioxide to the aforementioned target products, this energy being necessary to activate and convert both kinetically inactive and thermodynamically stable carbon dioxide molecules (Chapter 1). Realizing a low-carbon chemical technology therefore requires a shift in the production of hydrogen from the classic steam reforming of methane to new, carbon dioxide lean, technologies. Different routes to produce low-carbon hydrogen are available and we will start looking at these technologies in Chapter 6.

In Chapter 6, attention will be given to technologies for sustainable hydrogen production, focusing on water electrolysis using renewable energy as well as on its remaining challenges for large scale production and integration with other technologies. Furthermore, recent advances on Power-to-Technologies to produce key C_1 chemicals (formic acid, formaldehyde, methanol and methane) and fuels (gasoline, diesel and jet fuel) will also be discussed with focus on two main pathways: carbon dioxide hydrogenation and carbon dioxide electrochemical reduction.

Other opportunities do exist for chemical storage of renewable energy and are now referred as Power-to-X Technologies. As the name implies, the concept evolves around converting power (electricity) to chemicals (X), which can be very diverse hence the utilization of the "X." The most well-known technology to produce chemicals out of electricity is water electrolysis producing hydrogen and oxygen. Hydrogen produced through electrolysis has been acclaimed as one of the optimally sustainable fuels since its combustion leads only to the production of water, which involves minimal impact on GHG emissions. Production of hydrogen through electrolysis is still limited in many cases by issues related to cost and storage.

Hydrogen is one of the cleanest energies with potential to have a zero carbon emission. However, hydrogen storage is a challenge for hydrogen energy applications. The safety, cost and transportation of compressed and liquified hydrogen hinder the widespread application of hydrogen energy. Chemical absorption of hydrogen in solid hydrogen storage materials is a promising hydrogen storage method due to its high storage and transportation performance. Hydrogen storage density, dehydrogenation temperature and dehydrogenation dynamics are the main challenges for the hydrogen storage materials. The goal of the system gravimetric capacity was set by the Department of Energy to be 6.5 wt % with a working temperature from −40 to +60°C. The theoretical densities of most present hydrogen storage materials are even lower than 6.5 wt %, which make it impossible to reach the system gravimetric capacity goal. Only hydrogen storage materials with high theoretical density (≥10 wt %) with further modification have the possibility of reaching the goal. However, most of the reviews focus on the research progress of general hydrogen storage materials already investigated, many of which have low density. Hydrogen storage materials with high theoretical density including metal borohydrides, metal alanates, ammonia borane, metal amides and amine metal borohydrides have been reviewed by Abdalla et al. The pyrolysis and hydrogen absorption conditions of the hydrogen storage materials have been summarized, especially the improvements of the hydrogen storage materials. Furthermore, the challenges of storing hydrogen materials have been pointed out. Potential hydrogen storage materials and possible modification methods have also been presented and discussed.

Large-scale, affordable, "blue" hydrogen production from natural gas, along with carbon capture, utilization and storage (CCUS), is necessary to bridge the gap until large-scale hydrogen production using renewable energy becomes economically viable. The cost of carbon dioxide already makes blue hydrogen via steam methane reforming (SMR) competitive against grey hydrogen (without CCUS) and a newly available process based on gas partial oxidation (POX) technology and pre-combustion capture dioxide capture solvent technology further increases the affordability of blue hydrogen for greenfield projects.

2.3.1 Different Colours and Costs of Hydrogen

The basic understanding of hydrogen as an energy carrier and the semantics behind its various names, requires answers to two essential questions: what is it and how is it produced?

I could discuss at length about what hydrogen is, from its atomic weight to its final use, but to understand the basics we need to make the distinction of hydrogen as a carrier and not an energy source. What this means is that it needs a primary source of energy to be produced – solar, electricity, hydro, nuclear power, or gas, to name a few. And it is the specifics of the production process, including the energy source it utilises, that determine whether hydrogen will be dubbed green or blue. Or grey, pink, or yellow in fact.

2.3.1.1 Blue Hydrogen

Blue hydrogen is when natural gas is split into hydrogen, carbon monoxide and carbon dioxide either by steam methane reforming or auto thermal reforming, but the carbon dioxide is captured and then stored. As the greenhouse gasses are captured, this mitigates the environmental impacts on the planet. The capturing is done through a process called carbon capture usage and storage.

2.3.1.2 Green Hydrogen

Green hydrogen is hydrogen produced by splitting water by electrolysis. This produces only hydrogen and oxygen. We can use the hydrogen and vent the oxygen to the atmosphere with no negative impact.

To achieve the electrolysis we need electricity, we need power. This process to make green hydrogen is powered by renewable energy sources, such as wind or solar. That makes green hydrogen the cleanest option – hydrogen from renewable energy sources without carbon dioxide as a by-product.

2.3.1.3 Grey Hydrogen

Grey hydrogen has been produced for many years. It is a similar process to blue hydrogen – steam methane reforming or autotherming reforming are used to split natural gas into hydrogen carbon monoxide and carbon dioxide. But the carbon dioxide is not being captured and is released into the atmosphere.

2.3.1.4 Pink Hydrogen

Similar to green hydrogen, pink hydrogen is made via electrolysis, but using nuclear energy as its source of power.

2.3.1.5 Yellow Hydrogen

Another type of hydrogen made by electrolysis is yellow, where electrolysis is achieved solely through solar power (unlike green which could use a combination of renewable energy sources such as wind or solar).

2.3.1.6 Multi-Coloured Hydrogen

The future is a transition from grey, through blue, to green hydrogen. One thing that is clear is the important role hydrogen will play in energy transition.

2.3.1.7 Hydrogen Cost

Cost estimates in 2019–2021 (without costing CO_2 emissions).

- ING: about €1.50/kg for grey hydrogen, €2.50/kg for blue hydrogen and €5–6/kg for green hydrogen.
- Bloomberg New Energy Finance: green hydrogen ranging from \$2.50/kg to \$6.80/kg, but potentially reducing to \$1.40/kg by 2030.
- Aurora Energy Research in 2020 put levelized costs of production (higher heating value, real, 2018) at around €40/MWh (€1.6/kg) for conventional grey hydrogen, €50/MWh (€2/kg) for blue hydrogen and €80/MWh (€3.2/kg) for green hydrogen.
- S&P Global Platts in August 2020: conventional grey hydrogen at €1.24/kg, blue hydrogen at €1.31/kg and green hydrogen (polymer electrolyte membrane electrolysis including capex) at €3.43/kg.
- International Atomic Energy Agency (IAEA) figures in 2019 were: \$4.15/kg for normal electrolysis, \$3.23/kg for dedicated nuclear high-temperature steam electrolysis (HTSE) and \$2.50/kg for off-peak HTSE, competitive with large-scale steam reforming of methane (grey hydrogen).
- LucidCatalyst in 2020: \$2/kg yellow hydrogen from conventional nuclear power, potentially dropping to half that by 2030. In basic energy terms, \$2/kg is \$16.7/GJ (lower heating value) – well above most natural gas prices. LucidCatalyst suggests that \$1.50/kg for clean hydrogen would be competitive generally with high oil prices, but under \$1/kg would be needed with cheap oil.
- IEA: \$0.7–1.6/kg for grey hydrogen from steam reforming of natural gas (unabated), but \$3.2–7.7/kg for green hydrogen from electrolysis.
- Platts: green hydrogen €4.05/kg March 2021, compared with grey hydrogen €1.64/kg.

As well as the financial cost, the energy cost is significant. Input energy exceeds the energy content of the hydrogen output by a factor of at least 1.5.

Let's take the United Kingdom, one of Petrofac's key markets, as an example. The country's national energy system is changing rapidly as the United Kingdom makes plans to reach the legal net zero target by 2050. Although the United Kingdom is unlikely to produce sufficient quantities of green hydrogen to meet the targets and the demand, we are likely to see a transition period of 20–30 years where blue hydrogen plays a key role. And this is where the oil and gas industry, including Petrofac, can help by using existing, transferable knowledge and capability to produce blue hydrogen via CCUS. We can use the existing United Kingdom offshore infrastructure to pipe the carbon dioxide back to offshore reservoirs. Either as enhanced oil recovery (EOR) to help keep up reservoir pressure, or into depleted reservoirs for safe storage.

To meet the global energy demand, while achieving the national and global energy efficiency targets, the industry is looking at every potential technology. There is great potential in both the blue and green hydrogen and both will play an important role in energy transition. With our knowledge and capability, we are there to support this transition.

A growing number of national governments and energy companies have announced net-zero-emissions ambitions. Although renewable electricity is expanding rapidly, without low-carbon hydrogen as a clean-burning, long-term-storable, energy-dense fuel, a net-zero goal is difficult to achieve, especially when it comes to decarbonizing fertilizer production and hard-to-abate heavy industries such as steel manufacturing and power generation. Hydrogen also has potential as a transport and heating fuel that could repurpose existing gas distribution infrastructure or be introduced into existing natural gas supplies.

Consequently, hydrogen plays an important part in many green strategies. The European Union's hydrogen strategy, published in July 2020, describes it as "...essential to support the European Union's commitment to reach carbon neutrality by 2050 and for the global effort to implement the Paris Agreement while working towards zero pollution."

Momentum is building with a succession of commitments to hydrogen by various companies and governments. For example, in June 2020, Germany announced a €9B hydrogen strategy and the International Energy Agency stated, "Now is the time to scale up technologies and bring down costs to allow hydrogen to become widely used." Over the past three years the number of companies with membership in the international Hydrogen Council – which predicts a tenfold increase in hydrogen demand by 2050 – has jumped from 13 to 81 and includes oil and gas companies, automobile manufacturers, trading companies and banks.

In 2018, global hydrogen production was 70 MMtpy. Today's demand is split between use for upgrading refined hydrocarbon products and as a feedstock for ammonia production for nitrogen fertilizers. Nearly all hydrogen production comes from fossil fuels: it accounts for 6% of natural gas and 2% of coal consumption, as well as 830 MMtpy of carbon dioxide emissions – more than double the United Kingdom's emissions. Grey hydrogen is a major source of carbon dioxide emissions. If hydrogen is to contribute to carbon neutrality, it must be produced on a much larger scale and with far lower emission levels.

Over the long term, the answer is likely to be "green" hydrogen, which is produced from the electrolysis of water powered by renewable energy. This supports the integration of renewable electricity generation by decoupling production from use. Hydrogen becomes a convertible currency, enabling electrical energy to be stored and used as an emissions-free fuel and chemical feedstock.

Green hydrogen projects are starting. For example, a Shell-led consortium is at the feasibility stage of the NortH2 wind-to-hydrogen project in the North Sea and a Shell–Eneco consortium secured the right to build the 759-MW Hollandse Kust Noord project at a subsidy-free Dutch offshore wind auction in July 2020; this project will include a green hydrogen technology demonstration.

However, electrolysis alone will not meet the forecast demand. It is expensive at present and there is insufficient renewable energy available to support large-scale green hydrogen production. To put the scale of the task into perspective, meeting today's hydrogen demand through electrolysis would require 3,600 TWh of electricity, more than the European Union's annual use. Moreover, using the current European Union electricity mix would produce grey hydrogen from electrolysis with 2.2 times the greenhouse gas emissions of producing grey hydrogen from natural gas. This is because nearly half (45.5%) of the net electricity generated in the European Union comes from burning natural gas, coal and oil and generating electricity from natural gas, for example, has a 44% efficiency.

An alternative is blue hydrogen produced from natural gas, coupled with CCUS. Hydrogen production via electrolysis has a similar efficiency to blue hydrogen production, but the levelized cost of production is significantly higher for electrolysis at €66/MWh, compared with €47/MWh for SMR–CCUS.

In addition, it is widely acknowledged that scaling up blue hydrogen production will be easier than delivering green hydrogen. For example, the European Union strategy states, "Other forms of low-carbon hydrogen (i.e., blue) are needed, primarily to rapidly reduce emissions ... and support the parallel and future uptake of renewable (green) hydrogen."

However, the strategy goes on to claim that neither green nor blue hydrogen production is cost-competitive against grey; the hydrogen costs estimated for the European Union are €1.5/kg for grey, €2/kg for blue and up to €5.5/kg for green. These costs are based on an assumed natural gas price for the EU of €22/MWh, an electricity price of €35/MWh–€87/MWh and a capacity cost of €600/kW.

With the cost of carbon dioxide at \$25/t–\$35/t, blue hydrogen becomes competitive against grey even with higher capital costs and green still may be more than double the price of blue hydrogen by 2030. Some forecasts indicate that cost parity will occur around 2045. This competitiveness between blue and grey hydrogen (when considering carbon dioxide costs) is based on SMR technology, but other technologies are available to further increase blue hydrogen affordability for greenfield projects.

2.4 Alternative Feedstocks

2.4.1 Carbon Dioxide-Derived Chemicals

Among the many useful chemicals that can be synthesized directly from carbon dioxide, methane, methanol, higher alkanes, lower alkenes, formic acid and formaldehyde have a strategic use in the chemical technology, either as platform chemicals and/or for the production/storage of energy. The following topics investigate the recent advances in the synthesis of the chemicals using both carbon dioxide hydrogenation and carbon dioxide electrocatalytic reduction.

Power-to-Technologies can also involve carbon-based structures which would make them more compatible with the existing infrastructure both for transportation and for large-scale energy production. One especially abundant carbon-based feedstock suitable for Power-to-Technologies is carbon dioxide, which in turn is also the focus of the worldwide climate change efforts. Using electrocatalytic systems or through the utilization of hydrogen as reducing agent, carbon dioxide can now be used to produce simple C_1 molecules such as methanol and methane. Reports from open literature shows that Power-to-Technologies could eventually lead to the production of gasoline, diesel and even jet fuel. Hence, these technologies would at the same time allow storage of renewable energy, while reducing carbon dioxide emissions at the source and producing liquid transportation fuels.

The previously described reaction pathways using carbon dioxide as carbon source are based on hydrogen as the reaction partner. Other transformations of carbon dioxide utilise alternative reactants providing the necessary energy to enable carbon dioxide conversion. Of particular importance is the use of carbon dioxide as co-monomer for different polymers, as this opens the pathway to relatively large products with high added value. These include transformations with alternative high-energy containing reactants such as epoxides and unsaturated compounds or the conversion of carbon dioxide into target molecules with even lower energy content than carbon dioxide, e.g., organic carbonates and polymers. Figure 2.1 provides an overview of different CO_2-based products and current status of deployment. The hydrogen-based processes are included.

Apart from urea, which is by far the largest current CO_2-based product, other commercial processes include the synthesis of cyclic carbonates and salicylic acids, both produced in amounts below 0.1 million tons/year (TRL9). Two groups of polymers can be produced using carbon dioxide as building block: poly(propylene)carbonate and polycarbonate etherols. Novomer Inc. in the US and Covestro (former Bayer Material Science) are active in this field with plants in operation (TRL7–9).

For formic acid a 1kg/day pilot plant for electrochemical carbon dioxide reduction has been erected by DNV, Norway and Mantra Energy Alternatives Inc. is building a 100 kg/day plant in Vancouver, British Columbia (TRL7). Mineral carbonation (mineralization) is used in semi-commercial plants for the treatment of industrial waste, contaminated soil, slags from metallurgy and production of cement-like construction materials (7–9).

Many other processes is under investigation at lab scale, with substantially lower TRL. Because of its large potential, one approach to mention specifically is the direct synthesis potential of 0.125 t CO_2/t DME compared to the current state-of-the-art process with an intermediate methanol stage, corresponding to a 30% reduction. Further synthesis pathways investigated include the direct synthesis of sodium acrylate from ethylene and carbon dioxide or electrocatalytic processes to convert carbon dioxide into ethylene and propylene.

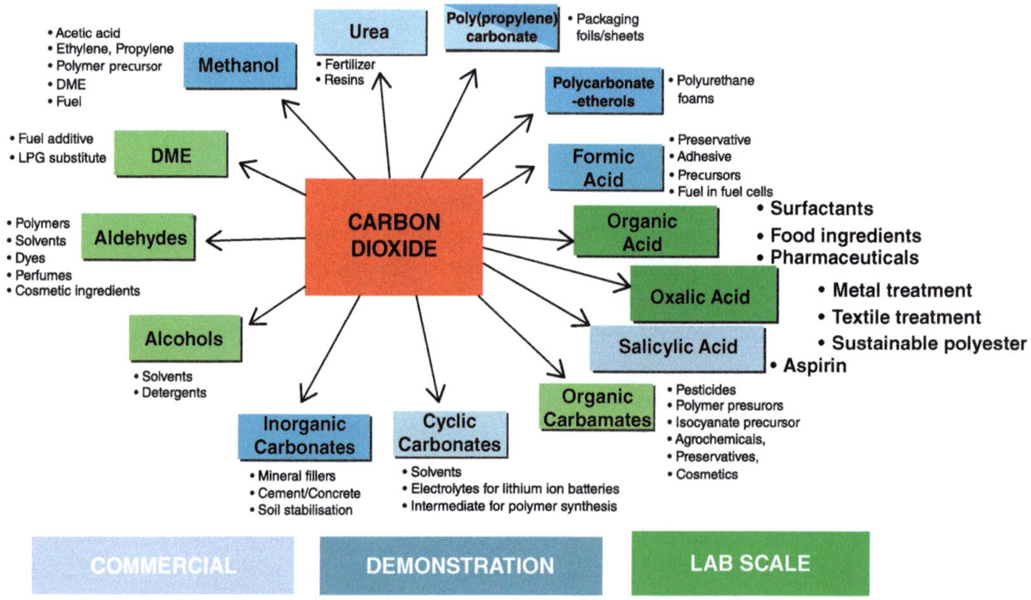

Figure 2.1 Target products of chemical carbon dioxide utilisation routes and status of deployment.

2.5 Alternative Power-to-X-Technology

2.5.1 Power-to-X-Technology to Produce Electrochemicals and Electrofuels

Power-to-X-technology has gained increased attention since it tackles issues related to the production of carbon neutral chemicals and fuels from carbon dioxide and to the storage of renewable energy. The Power-to-X technology includes two main steps: hydrogen production through water electrolysis using renewable energy and carbon dioxide hydrogenation to chemicals and fuels. Alternatively, carbon dioxide can be directly reduced into useful products via electrochemical reduction also using renewable energy.

Electrochemical carbon dioxide reduction offers numerous benefits in comparison to other technologies like (i) directing the intermediates towards reduction, (ii) modular and quick industrial upgrading, (iii) reduced formation of by-products, (iv) operation at ambient conditions without any supplementary chemical requirements, (v) avoiding complex start-up processes, (vi) offering a simpler and flexible process, (vii) avoiding costly high pressure/temperature reactors, (viii) enabling the process production rate to be quickly varied with easily switch on/off and (ix) the products of electrocatalytic carbon dioxide reduction rarely need separation, as the mixture of hydrocarbon (methane, ethane) obtained can be used as "hythane" a mixture of hydrogen/natural gas, as a clean alternative fuel in existing transportation sectors.

Scale-up studies demonstrated on tin electrodes show profitable production of formic acid. Interestingly, the capital cost to be invested in the electrochemical reduction of carbon dioxide is less than carbon capture and storage. The products formed from this process are easily separable and its installation is economical when compared to other sophisticated methods like biochemical process. The electrochemical reduction of carbon dioxide can be made energy (grid) independent by combining with solar to perform as a photoelectrochemical reduction in a photovoltaic cell.

Carbon dioxide drives climate change when released to the atmosphere. Alternatively, carbon dioxide could be captured and utilized as carbon source for chemicals. Arne Kätelhön et al. provided a global assessment of the technical climate change mitigation potential of carbon capture and utilization (CCU) in the chemical technology. They developed an engineering-level model of the global chemical industry representing 75% of current greenhouse gas (GHG) emissions. The model allows us to analyse the potential disruptive changes through large-scale carbon dioxide utilization and resulting emission reductions. The study shows that CCU has the technical potential to lead to a carbon-neutral chemical industry and decouple chemical production from fossil resources. This transition, however, would cause increased mass flows and demand for low-carbon electricity.

Chemical production is set to become the single largest driver of global oil consumption by 2030. To reduce oil consumption and resulting GHG emissions, carbon dioxide can be captured from stacks or air and utilized as alternative carbon source for chemicals. Arne Kätelhön et al showed that carbon capture and utilization has the technical potential to decouple chemical production from fossil resources, reducing annual GHG emissions by up to 3.5 Gt CO_2-eq in 2030. Exploiting this potential, however, requires more than 18.1 PWh of low-carbon electricity, corresponding to 55% of the projected global electricity production in 2030. Most large-scale CCU technologies are found to be less efficient in reducing GHG emissions per unit low-carbon electricity when benchmarked to Power-to-X efficiencies reported for other large-scale applications including electromobility (e-mobility) and heat pumps. Once these other demands are satisfied, CCU in the chemical industry could efficiently contribute to climate change mitigation.

Rising temperatures and changing weather patterns have made the effect of climate change more evident, creating an alarm for the need of quick action for its mitigation. CCUS is an effective step in this direction. One of the key reactions in the heart of aqueous CCUS technologies is the reversible hydration of carbon dioxide. The reaction is catalysed biologically by the enzymes of the carbonic anhydrase family and for its industrial realization, catalysts including organometallic compounds and metallic nanoparticles have been widely investigated. These catalysts enhance the reaction rates at moderate conditions of temperature, pressure and pH for suitable applications for carbon dioxide separation and mineralization. However, the use of catalysts is limited by their industrial scale production costs and/or reusability. In this review, I present a materials perspective of different catalysts that have been developed worldwide for carbon dioxide hydration with a specific emphasis on the biomimetic mechanism of the catalytic activity of the catalysts.

The bottom-up model of chemical technology yields future production pathways to produce the 20 large-volume chemicals: acrylonitrile, ammonia, benzene, caprolactam, cumene, ethylene, ethylene glycol, ethylene oxide, methanol, mixed xylenes, phenol, polyethylene, polypropylene, propylene, propylene oxide, p-xylene, styrene, terephthalic acid, toluene and vinyl chloride. Production pathways are represented by more than 160 processes based on engineering-level data. Thereby, flows of energy and materials are determined in detail throughout entire supply (Figure 2.2).

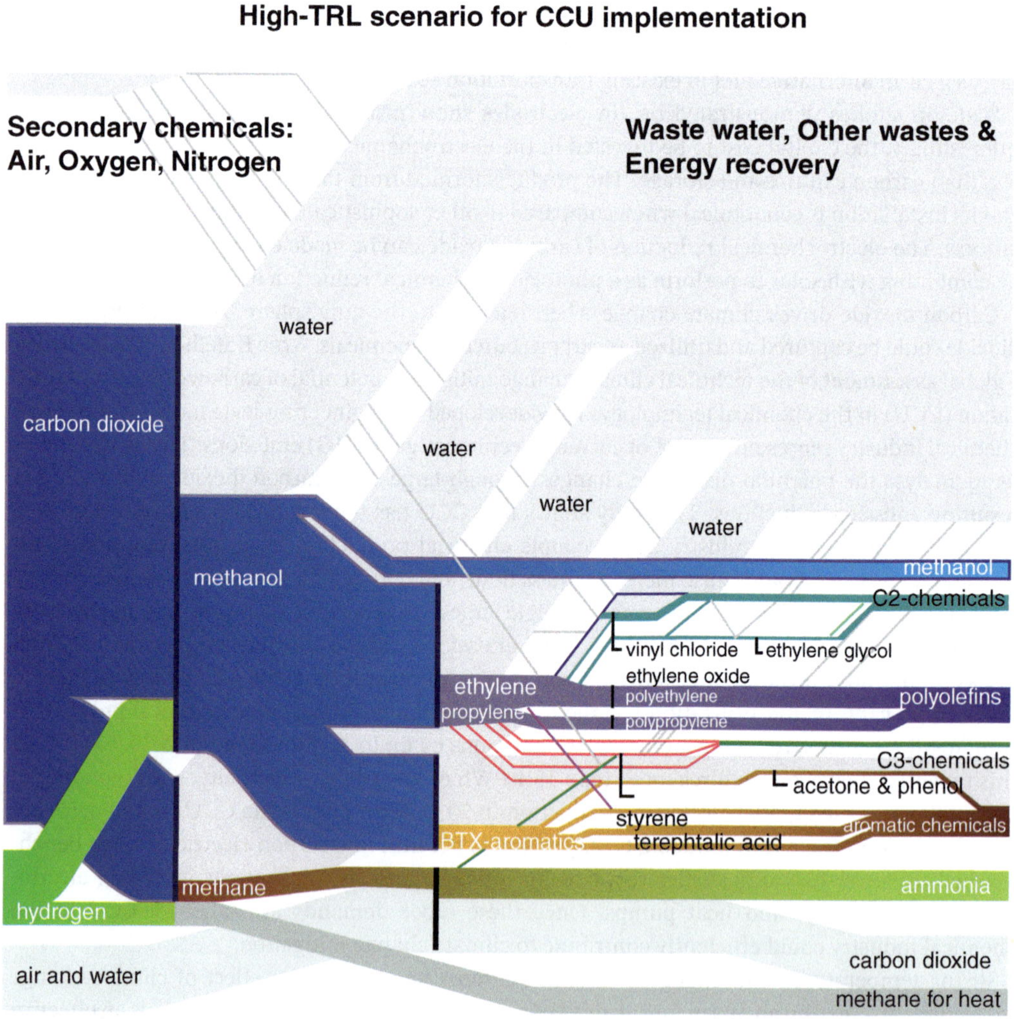

Figure 2.2 Mass flows within the chemical industry in the high-TRL model for CCU implementation. The projected final demand for 20 large-volume chemicals in 2030 is produced.

In conventional chemical technology, the chemical industry converts 1.12 Gt of fossil resources into chemicals and generates about 2.0 Gt CO_2-eq start-to-end emissions using conventional production technologies that are already commercialized today. Due to increased demand, this estimate is higher than current GHG emissions of the chemical industry at about 1.5 Gt CO_2-eq in 2013 and is in line with previous estimates for 2030.

In the model, where the production of the 20 chemicals relies on CO_2-based methanol and methane, feedstock mass flows increase substantially by 287% due to the carbon dioxide used and water coproduced (water technologies) (Figure 2.2): 3.72 Gt carbon dioxide and 0.59 Gt hydrogen from water electrolysis are annually converted into methanol and methane. Methanol is further processed into olefins and BTX, which are the basis of the other large-volume chemicals. Methane substitutes natural gas to produce both heat and ammonia. Carbon dioxide can be captured from two different sources representing the upper and lower bounds for the climate impact of carbon dioxide supply: a highly concentrated industrial point source (~100% carbon dioxide) and ambient air with about 400 ppm carbon dioxide. The integration of carbon dioxide via methane and methanol would require comparatively low research and development efforts and allows use of much of the existing petrochemical infrastructure and technologies.

Technologies are chosen to minimize GHG emissions, enabling the determination of the theoretical maximum climate change mitigation potential by CCU compared with a non-CCU scenario. For each chemical, both conventional and CO_2-based production technologies are considered. For the conventional production pathways, only the best available technologies (BATs) from a climate perspective are included, i.e., the industrialized technology.

2.6 Partial Oxidation of Methane

The alternative route to methanol production is the direct oxidation of methane. The progress in catalyst development with a focus on the metal-organic-framework supported catalysts, in relation to other types of catalysts, has been indicated in several studies. Other catalyst formulations have been receiving attention as well. In these applications (Furimsky et al.), the catalysts supported on CNTs, rGO and AC exhibited superior performance. This route of methanol production may have merits in locations where methane is readily available (e.g., as shale gas).

Cu-exchanged MAZ zeolite-catalysed methane partial oxidation has been proved to have a relatively higher methanol yield, but it needs to be further improved (Wang et al.). Here, the doping of the transition metals Pd, Pt, Fe, Co, Ni, Au and Ag in dinuclear active sites is investigated by density functional theory study. The whole process of $[Cu–O–M]^{2+}$-MAZ-catalysed methane partial oxidation is calculated. The geometric and electronic structural properties of hetero-metallic $[Cu–O–M]^{2+}$ and several activity descriptors for reactivity, such as the $\angle$Cu–O–M angle, μ-O spin density and proton affinity, are discussed. The calculation results prove an efficient process for promoting the catalytic performance of Cu-exchanged MAZ zeolite in methane direct conversion to methanol (MAZ-natural zeolites, Mazzite-Mg named in 1974 after Fiorenzo Mazzi).

2.7 Biorefining

The efficient and clean transformation of renewable raw materials into functional chemicals and fuels has blossomed from a theoretical construct to an interdisciplinary field that is fuelled by fundamental innovations in synthetic chemistry and guided by practical applications of biorefineries.

The requirements of biorefineries for flexibility of feedstocks, relatively mild operational conditions and low environmental impact have resulted in highly innovative methods for defunctionalizing and refunctionalizing biomass feedstocks to value-added chemicals and materials. On the defunctionalisation front, Yang et al. report an organocatalytic method for photochemical C–O bond cleavage of the β-O-4 linkage in lignin scaffolds. This method offers a metal-free strategy to prior reports and is applicable to continuous flow processing (Yang et al.). Cellulose-derived platform chemicals can also be obtained more efficiently from using a redox-switchable biocatalyst for controllable oxidation or reduction of 5-hydroxymethylfurfural (HMF) into high-value derivatives (Xu et al.).

2.8 Sustainable Production to Advance the Circular Economy

2.8.1 Introduction

Sustainable production to advance the circular economy is particularly bold. It recognizes that scientific data must take into account a systems approach to reducing environmental impact through intentional design of chemical products, not just considering how raw materials are sourced and manufactured and the use of industrial and consumer goods, but also how these materials and goods may be reused, recycled, or upcycled. The embracing of life cycle thinking as a goal among the green chemical technology community comes with the backdrop of the realization of limited resources and a climate crisis the likes of which most of us still fail to fully comprehend. We believe these are stepping stones on the pathway to advance closed-loop economies while still serving as models for innovation at a fundamental level within their respective chemical technology subdisciplines.

A significant change is coming to the hydrocarbons industry. The circular economy is in contrast to the traditional linear economy in the twenty-first century. The argument is that circular business models can be as profitable as linear models, allowing consumers to continue enjoying similar products and services. Reducing the resources used and the waste and leakage created conserves resource and helps to reduce environmental pollution (Table 2.1).

Up to 43% of the global conventional polymer production expected by 2040 may disappear due to circular economy (CE) initiatives. In addition, due to the growth in the use of wind turbines and solar panels and in lightweight composite materials in cars and planes, advanced materials waste is just beginning. For example, the EU must soon manage up to 300,000 tpy of used wind turbine blades. As a result, oil and gas companies stand to lose up to half of demand growth – however, they may also have the most to gain. Two drivers of this significant change centre on rising awareness of waste and the transition to cleaner energy sources.

Table 2.1 The development of environmentally friendly processes.

1. Eco-friendly catalysts
2. Hydrogen
3. Alternative feedstocks
4. Alternative technologies
5. Feedstock recycling
6. Functionalization of hydrocarbon
7. Biorefining
8. Circular economy

2019 was a year of successes within the recycled polyester (RPET) industry. The continued war against plastics and the heightened awareness of plastic pollution have driven an unprecedented level of interest in RPET. As such, significant momentum has been building behind collection and recycling commitments. Beyond regulatory commitments, such as the EU's single-use plastics directive, ambitious recycled content sustainability goals have been publicized by many brand owners. Some brand owner targets exceed the EU single-use plastics directive, further highlighting the momentum behind the shift towards achieving a circular economy.

2.8.2 Circular Economy

A circular economy is an economic system aimed at eliminating waste and the continual use of resources. Circular systems employ reuse, sharing, repair, refurbishment, remanufacturing and recycling to create a closed-loop system, minimising the use of resource inputs and the creation of waste, pollution and carbon emissions. The circular economy aims to keep products, equipment and infrastructure in use for longer, thus improving the productivity of these resources. All "waste" should become "food" for another process: either a by-product or recovered resource for another industrial process or as regenerative resources for nature (e.g., compost). This regenerative approach is in contrast to the traditional linear economy (Figure 2.3), which has a "take, make, dispose" model of production.

Proponents of the circular economy suggest that a sustainable world does not mean a drop in the quality of life for consumers and can be achieved without loss of revenue or extra costs for manufacturers. The argument is that circular business models can be as profitable as linear models, allowing consumers to continue enjoying similar products and services.

2.8.2.1 Sustainability

Intuitively, the circular economy would appear to be more sustainable than the current linear economic system. Reducing the resources used and the waste and leakage created, conserves resources and helps to reduce environmental pollution. However, it is argued by some that these assumptions are simplistic; that they disregard the complexity of existing systems and their potential trade-offs. For example, the social dimension of sustainability seems to be only marginally addressed in many publications on the circular economy. There are cases that might require different or additional strategies, like purchasing new, more energy-efficient equipment. Reviewing the literature shows that there are at least eight different relationship types between sustainability and the circular economy.

2.8.2.2 Scope

The circular economy covers many areas. Findings from the literature show that researchers have focused on:

- different areas such as industrial applications with both product-oriented and services, practice and policies;
- better understanding the limitations that the CE currently faces;
- strategic management for details of the circular economy; and
- different outcomes such as potential re-use applications and waste management.

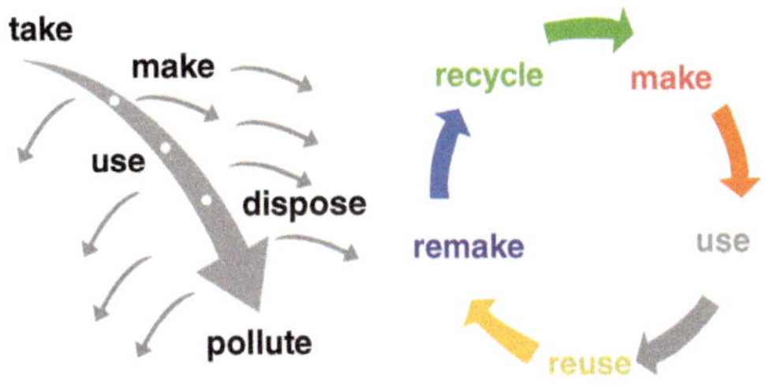

Figure 2.3 Linear versus circular economy.

The circular economy includes products, infrastructure, equipment and services and applies to every industry sector. It includes "technical" resources (metals, minerals, fossil resources) and "biological" resources (food, fibres, timber, etc.). Most schools of thought advocate a shift from fossil fuels to the use of renewable energy and emphasize the role of diversity as a characteristic of resilient and sustainable systems. The circular economy includes discussion of the role of money and finance as part of the wider debate and some of its pioneers have called for a revamp of economic performance measurement tools. One study points out how modularisation could become a cornerstone to enable circular economy and enhance the sustainability of energy infrastructure.

One example of a circular economy model is the implementation of renting models in traditional ownership areas (e.g., electronics, clothes, furniture, transportation). Through renting the same product to several clients, manufacturers can increase revenues per unit, thus decreasing the need to produce more to increase revenues. Recycling initiatives are often described as a circular economy and are likely to be the most widespread models.

2.8.2.3 Background of the Circular Economy

As early as 1966, Kenneth Boulding raised awareness of an "open economy" with unlimited input resources and output sinks, in contrast with a "closed economy", in which resources and sinks are tied and remain as long as possible a part of the economy. Boulding's essay "The Economics of the Coming Spaceship Earth" is often cited as the first expression of the "circular economy", although Boulding does not use that phrase.

The circular economy is grounded in the study of feedback-rich (non-linear) systems, particularly living systems. The contemporary understanding of the circular economy and its practical applications to economic systems evolved incorporating different features and contributions from a variety of concepts sharing the idea of closed loops. Some of the relevant theoretical influences are cradle to cradle, laws of ecology, looped and performance economy (Walter R. Stahel), regenerative design, industrial ecology, biomimicry and blue economy.

The circular economy was further modelled by British environmental economists David W. Pearce and R. Kerry Turner in 1989. In *Economics of Natural Resources and the Environment*, they pointed out that a traditional open-ended economy was developed with no built-in tendency to recycle, which was reflected by treating the environment as a waste reservoir.

In the early 1990s, Tim Jackson began to pull together the scientific basis for this new approach to industrial production in his edited collection *Clean Production Strategies*, including chapters from pre-eminent writers in the field, such as Walter R. Stahel, Bill Rees and Robert Constanza. At the time still called "preventive environmental management", his follow-on book *Material Concerns: Pollution, Profit and Quality of Life* synthesised these findings into a manifesto for change, moving industrial production away from an extractive linear system towards a more circular economy.

2.8.2.3.1 Emergence of the Idea

In their 1976 research report to the European Commission, "The Potential for Substituting Manpower for Energy", Walter Stahel and Genevieve Reday sketched the vision of an economy in loops (or circular economy) and its impact on job creation, economic competitiveness, resource savings and waste prevention. The report was published in 1982 as the book *Jobs for Tomorrow: The Potential for Substituting Manpower for Energy*.

Most frequently described as a framework for thinking, its supporters claim it is a coherent model that has value as part of a response to the end of the era of cheap oil and materials, moreover, contributing to the transition for a low carbon economy. In line with this, a circular economy can contribute to meeting the COP 21 Paris Agreement. The emissions reduction commitments made by 195 countries at the COP 21 Paris Agreement are not sufficient to limit global warming to 1.5°C. To reach the 1.5°C target it is estimated that additional emissions reductions of 15 billion tonnes carbon dioxide per year need to be achieved by 2030. Circle Economy and Ecofys estimated that circular economy strategies may deliver emissions reductions that could halve the gap.

2.8.2.3.2 Moving Away from the Linear Model

Linear "take, make, dispose" industrial processes and the lifestyles dependent on them, use up finite reserves to create products with a finite lifespan, which end up in landfills or in incinerators. The circular approach, by contrast, takes insights from living systems. It considers that our systems should work like organisms, processing nutrients that can be fed back into the cycle – whether biological or technical – hence the "closed loop" or "regenerative" terms usually associated with it. The generic circular economy label can be applied to or claimed by several different schools of thought, but all of them gravitate around the same basic principles.

One prominent thinker on the topic is Walter R. Stahel, an architect, economist and a founding father of industrial sustainability. Credited with having coined the expression "Cradle to Cradle" (in contrast with "Cradle to Grave", illustrating our "Resource to Waste" way of functioning), in the late 1970s, Stahel worked on developing a "closed loop" approach to production processes, co-founding the Product-Life Institute in Geneva. In the UK, Steve D. Parker researched waste as a resource in the UK agricultural sector in 1982, developing novel closed-loop production systems. These systems mimicked and worked with the biological ecosystems they exploited.

2.8.2.3.3 Towards the Circular Economy

In 2013, a report was released entitled *Towards the Circular Economy: Economic and Business Rationale for an Accelerated Transition*. The report, commissioned by the Ellen MacArthur Foundation and developed by McKinsey & Company, was the first of its kind to consider the economic and business opportunity for the transition to a restorative, circular model. Using product case studies and economy-wide analysis, the report details the potential for significant benefits across the EU. It argues that a subset of the EU manufacturing sector could realize net materials cost savings worth up to $630 billion annually up to 2025 – stimulating economic activity in the areas of product development, remanufacturing and refurbishment. *Towards the Circular Economy* also identified the key building blocks in making the transition to a circular economy, namely in skills in circular design and production, new business models, skills in building cascades and reverse cycles and cross-cycle/cross-sector collaboration.

Another report by WRAP and the Green Alliance (called "Employment and the circular economy: job creation in a more resource efficient Britain"), done in 2015 examined different public policy scenarios to 2030. It estimates that, with no policy change, 200,000 new jobs will be created, reducing unemployment by 54,000. A more aggressive policy scenario could create 500,000 new jobs and permanently reduce unemployment by 102,000.

On the other hand, implementing a circular economy in the United States has been presented by Ranta et al. who analysed the institutional drivers and barriers for the circular economy in different regions worldwide, by following the framework developed by W.R. Scott . In the article, different worldwide environment-friendly institutions were selected and two types of manufacturing processes were chosen for the analysis (1) a product-oriented and (2) a waste management.

2.8.3 Circular Business Models

While the initial focus of academic, industry and policy activities was mainly focused on the development of 6R:

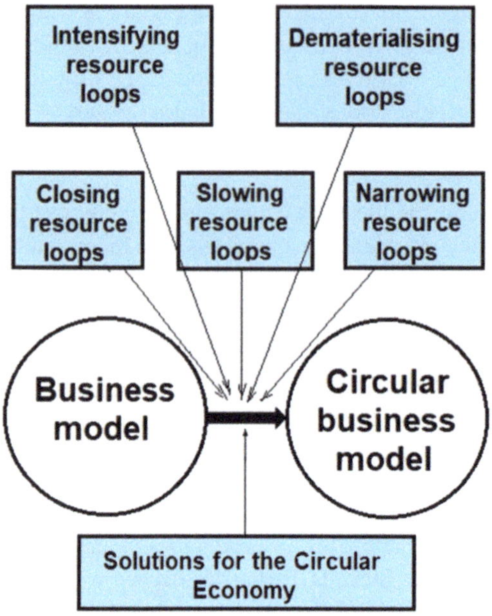

Figure 2.4 Circular business models.

- recycling;
- reduce;
- reprocess;
- remanufacturing;
- reuse; and
- recover technology,

it soon became clear that the technological capabilities increasingly exceed their implementation.

To leverage this technology for the transition towards a circular economy, various stakeholders have to work together. This shifted attention towards business-model innovation as a key leverage for "circular" technology adaption. Rheaply, a platform that aims to scale reuse within and between organizations, is an example of a technology that focuses on asset management and disposition to support organizations transitioning to circular business models.

Circular business models can be defined as business models that are closing, narrowing, slowing, intensifying and dematerializing loops, to minimize the resource inputs into and the waste and emission leakage out of the organizational system. This comprises recycling measures (closing), efficiency improvements (narrowing), use phase extensions (slowing), a more intense use phase (intensifying) and the substitution of products by service and software solutions (dematerializing). These strategies can be achieved through the purposeful design of material recovery processes and related circular supply chains. As illustrated in Figure 2.4 these five approaches to resource loops can also be seen as generic strategies or archetypes of circular business model innovation.

Circular business models can have different emphases and various objectives. For example, extend the life of materials and products, where possible over multiple "use cycles"; use a "waste = food" approach to help recover materials and ensure those biological materials returned to earth are benign, not toxic; retain the embedded energy, water and other process inputs in the product and the material for as long as possible; use systems-thinking approaches in designing solutions; regenerate or at least conserve nature and living systems; push for policies, taxes and market mechanisms that encourage product stewardship, for example "polluter pays" regulations.

2.8.4 Industries Adopting a Circular Economy

EU member states were one of the first groups that started talking about the circular economy, developing the idea to implement and adopt it in the industries of their own states (Table 2.2). Research shows that circular economy can profit around 90 billion Euros and can create 600,000 new jobs by 2030. Countries which contribute the most in circular economy are Germany, UK, France, Austria and Italy.

2.8.4.1 Minimizing Dependence on Fossil Fuels

The results from across chemical literature demonstrate that the design of methods for circularity necessitates multidisciplinary approaches and careful definition of the problem being addressed. One

Table 2.2 Areas and industries adopting a circular economy.

1. Circular economy and plastic solution
2. Textile industry
3. Construction industry
4. Automotive industry
5. Logistics industry
6. Agriculture
7. Furniture industry
8. Minimizing dependence on fossil fuels
9. Minimizing the impact of chemical synthesis and manufacturing
10. Future research needs in developing a circular economy

of the key challenges we face continues to be decreasing our dependence on fossil fuels for chemical and fuel production. This entails both the efficient and clean transformation of renewable (biobased) raw materials into functional chemicals and fuels, as well as efficient carbon dioxide capture and conversion into fuels and chemicals. Several exciting advances in this highlight our progress on both fronts. Wang et al. discuss the development and application of engineered multifunctional nanohybrids of carbon nanomaterials and metal/metal oxide nanoparticles, which exhibit promising multifunctionalities for addressing the critical energy–water–environment (EWE) nexus.

The uptake of reuse within the oil and gas industry is very poor, but the opportunity to reuse is never more evident, or possible, as when the equipment is being decommissioned. Hundreds of thousands of tons of waste are being brought back onshore to be recycled. Unfortunately, what this equates to is equipment, which is perfectly suitable for continued use, being disposed of.

In the next 30–40 years, the oil and gas sector will have to decommission 600 installations in the UK alone. Over the next decade around 840,000 tonnes of materials will have to be recovered at an estimated cost of £25 billion. In 2017 North Sea oil and gas decommissioning became a net drain on the public purse. With UK taxpayers covering 50–70% of the bill, there is an urgent need to discuss the most economic, social and environmentally beneficial decommissioning solutions for the general public.

Organizations such as Zero Waste Scotland have conducted studies to identify areas with reuse potential, allowing equipment to continue life in other industries, or be redeployed for oil and gas.

As is highlighted in Chapter 1, secular changes in demand are forcing the transition from fuels to chemicals. Unfortunately, this shift will place even greater pressure on the nascent bio-based plastics industry, which is already challenged from a cost standpoint. Regardless, the petrochemical industry must find a way to redefine its business models. The circular economy is coming and this industry is still, by and large, linear. It's time for a change. As for bio-based chemical developers, either focus on downstream applications that are bio-advantaged or face a very difficult future. That is true. It seems that the goal posts keep moving for biobased chemicals as the hydrocarbon industry evolves and improves its economics.

2.8.4.2 Minimizing the Impact of Chemical Synthesis and Manufacturing

Another key challenge in developing manufacturing infrastructure for a circular economy remains the minimization of the overall environmental footprint, which must minimize both the environmental impact of the manufacturing processes of chemicals, as well as their potential hazard and persistence of the commodity chemicals produced. While the former has been a cornerstone of research in green chemistry technology from its inception in the early 1990s, it is still a fertile area of research. The latter will be addressed in further detail later.

Sustainability in organic technology, especially in organic synthesis, has been driving innovation for decades. With the amount of waste generated in many synthetic chemistry routes, especially at scale in manufacturing, we are faced with not only an ethical imperative to develop more, sustainable chemical processes and products but also a financial imperative. Metrics to gauge our progress, including process mass intensity (PMI), have been developed that allow all aspects of a process to be compared. For example, conducting a reaction in water may not necessarily be an improvement if several volumes of an organic solvent are needed to extract/purify the product. One could even argue that water can be problematic because its high boiling point makes recycling energy intensive, but we do not know these details without consciously thinking about them. Recent perspectives (Kätelhon et al. and Parker et al.) highlight the need to explicitly include the assessment of sustainability using green chemical technology metrics.

2.8.4.3 Future Research Needs in Developing a Circular Economy

The research publications allow us to identify areas that will require further emphasis to strategically advance the field towards applications that facilitate circularity. Some of these areas include:

- depolymerization and defunctionalisation methods for existing chemicals that can allow circularity, especially for plastics;
- design of circular systems with consideration for human health and ecotoxicity, ideally via rational design of benign commodity chemicals;
- systematic application of life-cycle analysis, or thinking and process metrics in developing new manufacturing/synthetic routes;
- implementation of machine learning and other big-data methods to drive innovation towards new paradigms of circularity.

To allow future contributions to highlight the potential application to circular economies, authors may want to consider explicitly addressing some of following questions in their work. First, consider the source of the materials being used and if it is fossil-fuel-derived, consider if there is a biobased alternative that could provide the same chemical insights. If elements are used that are facing limited supply, consider whether the synthetic routes and processes could help reduce the use of such materials. Second, discuss whether the method described could impact different stages of a chemical's life cycle. For example, is it a step that is potentially reversible based on kinetics and thermodynamics? If it produces by-products, could those be a valuable resource? And finally, consider whether the method or product constitutes a safe product that can be either recycled, reused, or remanufactured. Addressing such questions forces the authors to apply life cycle thinking and to broaden their discussion of the advances presented.

2.9 New Chemical Technologies

When it is time for a major maintenance turnaround or debottlenecking project, plant owners should consider using the latest technologies. This can range from re-rotoring compressors and upgrading turbines to using higher-capacity trays in columns or higher-efficiency heat transfer equipment. The following is a partial list of technology items to consider:

- for new grassroots projects, seek new technologies that provide higher yields and reduced energy consumption, instead of replicating an existing, older, less-efficient design;
- for revamps or new units, use new improved catalysts to increase yield and reduce and recycle, thus reducing energy and GHG intensity;

- for the electrification of compressor drivers on offshore platforms, use offshore wind or onshore green power;
- install once-through steam generators (OTSGs) on open-cycle gas turbines to recover gas turbine exhaust heat. The light weight and small footprint of OTSGs allow for retrofitting;
- use a divided wall column (which uses one distillation column instead of two columns) for reduced capital and operational expenditures;
- install heat pumps on columns separating close boiling temperature components (e.g., ethylene/ethane splitter, propylene/propane splitter);
- use a low-emissions cracking furnace designed to shift more heat into process heating in the firebox and generate less steam in the convection section. This can reduce fuel requirements by 30%. Reduced steam production provides the opportunity to replace condensing steam turbines with efficient motors using renewable power;
- use new monoethylene glycol (MEG) technology, such as a proprietary MEG process (Shell's OMEGA process) that reduces steam usage by 20%;
- utilize new propylene oxide (PO) processes, such as a proprietary PO process (BASF and Dow Chemical Co's hydrogen peroxide to propylene oxide (HPPO) technology) that reduces energy usage by 35% and wastewater by 70% vs. existing PO technologies;
- use new ethylene processes, e.g., methanol-to-olefins, oxidative coupling methane;
- utilize the Allam-Fetvedt cycle for power generation (e.g., an oxy-fired gas turbine using a supercritical carbon dioxide cycle, where carbon dioxide offtake is ready for carbon dioxide pipelines). This creates clean power with no net carbon to atmosphere.

2.9.1 Renewable Power

Renewable power has no carbon dioxide footprint and includes solar photovoltaic systems, as well as concentrated solar, onshore and offshore wind, hydroelectric and geothermal systems. Nuclear power also has no carbon dioxide footprint. One MWh of electricity will power approximately 200 homes on a hot summer day. According to the Center for Climate and Energy Solutions, approximately 29% of global electricity generation was produced via renewables in 2020. In 2021, most of the installed utility-scale power generation in the US from renewable sources was wind (46%), followed by hydro (31%), solar and biomass (7%). Future installed power generation will be predominately solar and onshore/offshore wind. Renewable power can be used in an electrolyser to convert water to green hydrogen and oxygen. Hydrogen is another key decarbonization pathway.

Renewable power is not available at a constant supply rate, so power grids must be upgraded to accept fluctuations in renewable power supply. Solar is not available at night and is reduced on cloudy days. Wind also varies between day and night and from summer to winter. Hydro varies with the seasons. Over a year, the utilization of installed capacity for solar was 20–28%, for wind 35–40% and for hydro 35–45%.

Energy storage is required to stabilize the grid and maximize the use of renewable energy. Batteries can store excess renewable power during the day and then release this power into the grid at night. Significant research in battery energy storage systems is ongoing. Challenges include the amount of power that can be stored (which is limited by battery size and weight) and the amount of the dispatch rate of the power back into the grid when there is sudden demand. Lithium ion batteries are most used in electric vehicles (EVs). Flow batteries are an emerging technology being developed for storing renewable energy. Flow batteries contain a water-based electrolyte liquid that flows between two separate tanks within the battery. When charged, a chemical reaction occurs that enables the energy to be stored and subsequently discharged. In addition to batteries, other methods for storing excess renewable power when grid demand is reduced include the following:

- compressing air in underground storage domes for later use to drive turbines;
- pumping water to elevated storage, to be used later to drive hydropower turbines;
- utilizing molten salts for thermal storage; and
- using electrolysers for green hydrogen generation.

The layout of the wind turbines is critical to avoid secondary air from nearby turbines. Land-based wind turbines presently have a capacity of 3 MW, which will increase to 6 MW as technology improves. With more favourable winds, offshore wind turbines will have a higher capacity that can be scaled up to 17 MW by 2030. An offshore wind turbine can have a rotor diameter of 170 m. Large offshore wind farms are being developed in the North Sea and are planned for installation off the northeast coast of the US (i.e., Massachusetts, New Jersey and New York). One beneficial factor is that offshore oil/gas platform technology can be applied to offshore wind. Integrated oil and gas companies are investing in both solar and wind generation.

Oil companies are moving into the generation and direct sale of renewable electricity to customers. Electrification of transportation vehicles – such as cars, trucks and trains – is a key pathway for reducing emissions resulting from the burning of gasoline and diesel. In the US, the Biden administration has set a goal for half of all new vehicles sold in 2030 to be zero-emissions vehicles, including battery electric, plug-in hybrid electric, or fuel cell electric vehicles. By 2035, General Motors plans to stop making internal combustion cars. The installation of electric charging stations along major highways and at homes and workplaces, will be necessary to achieve these goals – and accomplishing this will constitute a major capital expenditure.

If green power is generated and converted to hydrogen for use in a hydrogen fuel cell electric vehicle, carbon dioxide is reduced by only 42% vs. using green power directly in a battery electric vehicle. Blending 20% hydrogen (vol%) into natural gas as fuel for a combined-cycle plant will reduce carbon dioxide emissions by approximately 7%.

Industrial opportunities for electrification to reduce carbon dioxide emissions include:

- switching condensing turbines to electric drives, considering the impact on steam and fuel balances;
- replacing old and inefficient steam turbines, that cause low-pressure steam venting, with electric drives;
- using electrically driven heat pumps to upgrade unusable low level heat to a higher usable level;
- switching medium-temperature fired heaters to electric power and switching out larger heaters to use green hydrogen as a fuel;
- replacing open-cycle gas turbine drives in LNG plants with large electric drives;
- replacing the gas turbines on offshore platforms with electric motors that use renewable power from offshore wind turbines and land-based renewable power; and
- electrifying ethylene plants, using motor-driven cracked gas and refrigeration compressors.

Further Reading

Abdalla, A.M., Hossain, S., Nisfindy, O.B. et al. (2018). Hydrogen production, storage, transportation and key challenges with applications: a review. *Energy Convers. Manag.* 165: 602–627. doi: 10.1016/j.enconman.2018.03.088.

About Legasea. https://www.legasealtd.com/about.

Allwood, J.M. (2014). Squaring the circular economy. In: *Handbook of Recycling* (eds. E. Worrell and M.A. Reuter), 445–477. Amsterdam: Elsevier.

Alsuhaibani, A.S., Afzal, S. et al. (2020). Impact of the development of catalyst and reaction system of the methanol synthesis stage on the overall profitability of the entire plant: a techno-economic study. *Catal. Today* 343: 191.

Batista, L., Gong, Y., Pereira, S. et al. (2019). Circular supply chains in emerging economies – a comparative study of packaging recovery ecosystems in China and Brazil (PDF). *Int. J. Prod. Res.* 57 (23): 7248–7268. doi: 10.1080/00207543.2018.1558295.

Bazzanella, A.M. and Ausfelder, F. (2017). Low carbon energy and feedstock for European chemical industry, Dechema. Gesellschaft fur Chemische Technik und Biotechnologie, June 2017.

Bjacek, P. (2020). Circular economy: getting in the door of opportunity ahead of a vanishing polymer market. *Hydrocarbon Processing*, January 2020.

Blok, K., Hoogzaad, J., Ramkumar, S. et al. (2017). Implementing circular economy globally makes Paris targets achievable. *Circle Economy, Ecofys*. https://circulareconomy.europa.eu/platform/en/knowledge/implementing-circular-economy-globally-makes-paris-targets-achievable (Retrieved 20 April 2017).

Boulding, K.E. (1966). The economics of the coming spaceship earth (PDF). In: *Environmental Quality in a Growing Economy, Resources for the Future* (ed. H. Jarrett), 3–14. Baltimore, MD: Johns Hopkins University Press. http://arachnid.biosci.utexas.edu/courses/thoc/readings/boulding_spaceshipearth.pdf (Retrieved 6 March 2023).

Buehler, J. (2022). Overview of decarbonization pathways for the oil and gas and petrochemical industries-Part 2. *Hydrocarbon Processing*, June 2022.

China Financial News (2016). Environmental crackdown on spent sulfuric acid from alkylate producers. October 2016.

Cradle to Cradle | The Product-Life Institute (2013). http://product-life.org/en/cradle-to-cradle (Retrieved 22 March 2023).

Dang, S., Yang, H. et al. (2019). Review of research progress on heterogeneous catalysts for methanol synthesis from carbon dioxide hydrogention. *Catal. Today* 330: 61.

De Hemptinne, J.-Ch., Kontogeorgis, G.M., Dohrn, R. et al. (2022). A view on the future of applied thermodynamics. *Ind. Eng. Chem. Res.* 61 (39): 14664–14680.

Fonda, M. (2019). The greening of the downstream industry: a carbon investment case for clean refining. *Hydrocarbon Processing*, July 2019.

Furimsky, E. (2020). CO_2 hydrogenation to methanol and methane over carbon-supported catalysts. *Ind. Eng. Chem. Res.* 59 (35): 15393.

Ganesh, K.N. et al. (2021). Green chemistry: a framework for a sustainable future. *Ind. Eng. Chem. Res.* 60 (25): 8964–8968.

Geissdoerfer, M., Morioka, S.N., de Carvalho, M.M. and Evans, S. (2018). Business models and supply chains for the circular economy. *J. Clean. Prod.* 190: 712721. doi: 10.1016/j.jclepro.2018.04.159. ISSN 0959-6526.

Gentilucci, A., Spagnuolo, L., Hughes, A. et al. (2022). Sustainable steel/chemicals production: removal of impurities for valorization of steel mill gases. *Hydrocarbon Processing*, October 2022.

Hayler, D. et al. (2019). A pharmaceutical industry perspective on sustainable metal catalysis. *Organometallics* 38 (1): 36–46. doi: 10.1021/acs.organomet.8b00566.

Huang, Y., Cheng, Y. and Jinying Zhang, J. (2021). A review of high density solid hydrogen storage materials by pyrolysis for promising mobile applications. *Ind. Eng. Chem. Res.* 60 (7): 2737–2771.

Ince, A.C., Colpan, C.O., Hagen, A. and Serincan, M.F. (2021). Modelling and simulation of power-to-X systems: a review. *Fuel* 304: 121354.

Jackson, T. (1993). *Clean Production Strategies Developing Preventive Environmental Management in the Industrial Economy*. Boca Raton: CRC Press.

Jackson, T. (1996). *Material Concerns — Pollution, Profit and Quality of Life*. Routledge.

Kätelhon, A., Meys, R., Deutz, S. et al. (2019). Climate change mitigation potential of carbon dioxide capture and utilisation in the chemical industry. *Proc. Natl. Acad. Sci. USA* 116 (23): 11187–11194.

Leonzio, G., Zondervan, E. and Foscolo, P.U. (2019). Methanol production by CO_2 hydrogenation: analysis and simulation of reactor performance. *Int. J. Hydrog. Energy* 44: 7915.

Liu, N. (2021). Increasing blue hydrogen production affordability. *Hydrocarbon Processing*, June 2021.

Liu, Z., Zhang, R., Meng, X. et al. (2018). Composite Ionic liquid alkylation technology gives high product yield and selectivity. *Hydrocarbon Processing*, March 2018.

Liu, Z., Zhang, R., Xu, C. and Xia, R. (2006). Ionic liquid alkylation process produces high-quality gasoline. *Oil Gas J* 104.

Majdecki, M. et al. (2020). Highly enantioselective epoxidation of α,β-unsaturated ketones using amide-based *Cinchona* alkaloids as hybrid phase-transfer catalyst. *Org. Lett.* 22 (21): 8687–8691. doi: 10.1021/acs.orglett.0c03272.

McIntee, A. (2020). Five key innovation concepts to impact front engineers in 2020. *Hydrocarbon Processing*, February 2020.

Molchanova, E. (2021). Circular economy as an inclusive economic model. *Econ. Innov.* 23: 3.

North sea oil and gas rig decommissioning & re-use opportunity report. (2015). https://www.zerowastescotland.org.uk//content/north-sea-oil-and-gas-rig-decommissioning-re-use-opportunity-report-0.

Parker, K. et al. (2021). Modeling metal-catalyzed polyethylene depolymerization: $[(Phen)Pd(X)]^+$ (X = H and CH_3) catalyze the decomposition of hexane into a mixture of alkenes via a complex reaction network. *Organometallics* 40 (7): 857–868. doi: 10.1021/acs.organomet.0c00782.

Pearce, D.W. and Turner, R.K. (1989). Economics of natural resources estimating employment effects of the circular economy.

Putten, R. et al. (2021). Manganese-mediated C–C bond formation: alkoxycarbonylation of organoboranes. *Organometallics* 40 (6): 674–681. doi: 10.1021/acs.organomet.0c00781.

Ranta, V., Aarikka-Stenroos, L., Ritala, P. and Mäkinen, S.J. (2018). Exploring institutional drivers and barriers of the circular economy: a cross-regional comparison of China, the US and Europe. *Resour. Conserv. Recycl.* 135: 70–82. doi: 10.1016/j.resconrec.2017.08.017.

Rashid, A., Asif, F.M.A., Krajnik, P. and Nicolescu, C.M. (2013). Resource conservative manufacturing: an essential change in business and technology paradigm for sustainable manufacturing. *J. Clean. Prod.* 57: 166–177. doi: 10.1016/j.jclepro.2013.06.012. ISSN 0959-6526.

Rego De Vasconcelos, B.R. and Lavoie, J.-M. (2019). Recent advances in power-to-X technology for the production of fuels and chemicals. *Front. Chem.* 7. doi: 10.3389/fchem.2019.00392.

Rossen, K. (2019). Greening organic chemistry with process chemistry. *J. Org. Chem.* 84 (8): 4580–4582.

Scott, W.R. (2008). *Institutions and Organization: Ideas and Interest*, 3e. Stanford University: Sage Publications, 50–51.

Siebert, H., Sieggelkow, B. and Dongovianni, E. (2022). Plastics circularity: how to solve the challenge of retaining quality in recaykled polymers. *Hydrocarbon Processing*, April 2022.

Solutions. solve.mit.edu. Retrieved 7 May 2020.

Su, B., Heshmati, A., Geng, Y. and Yu, X. (2012). A review of the circular economy in China: moving from rhetoric to implementation. *J. Clean. Prod.* 42: 215–227. doi: 10.1016/j.jclepro.2012.11.020.

Towards the Circular Economy: an economic and business rationale for an accelerated transition. (2013). Ellen MacArthur Foundation. https://ellenmacarthurfoundation.org/towards-the-circular-economy-vol-1-an-economic-and-business-rationale-for-an (Retrieved 6 March 2023).

U.S. Chemical Safety and Hazard Investigation Board. Factual update: fire and explosions at Philadelphia Energy Solutions refinery hydrofluoric acid alkylation unit. October 2019. https://www.csb.gov/assets/1/6/pes_factual_update_-_final.pdf.

Van Utyvanck, P. (2020). Petrochemicals: was 2019 the year of recycled PET? *Hydrocarbon Processing*, February 2020.

Verma, M., Bhaduri, G.A., Kumar, V.S.P. and Deshpande, P.A. (2021). Biomimetic catalysis of CO_2 hydratation: a materials perspective. *Ind. Eng. Chem. Res.* 60: 4777–4793.

Wang, D. et al. (2021). Next-generation multifunctional carbon–metal nanohybrids for energy and environmental applications. *Environ. Sci. Technol.* 53 (13): 7265–7287. doi: 10.1021/acs.est.9b01453.

Wang, G. et al. (2020). In situ synthesis of Ag–Fe_3O_4 nanoparticles immobilized on pure cellulose microspheres as recyclable and biodegradable catalysts. *ACS Omega* 5 (15): 8839–8846. doi: 10.1021/acsomega.0c00437.

Wang, L., Li, Z., Wang, Z. et al. (2021). Hetero-metallic active sites in Omega (MAZ) zeolite-catalyzed methane partial oxidation: a DFT study. *Ind. Eng. Chem. Res.* 60 (6): 2400–2409.

Weetman, C. (2016). *A Circular Economy Handbook for Business and Supply Chains: Repair, Remake, Redesign, Rethink*. London: Kogan Page.

Wikipedia (2020). Circular economy. https://en.wikipedia.org/wiki/Circular_economy.

Xu, J. et al. (2020). Redox-switchable biocatalyst for controllable oxidation or reduction of 5-hydroxymethylfurfural into high-value derivatives. *ACS Omega* 5 (31): 19625–19632. doi: 10.1021/acsomega.0c02178.

Yang, Y. et al. (2020). Organocatalytic approach to photochemical lignin fragmentation. *Org. Lett.* 22 (20): 8082–8085.

Zhang, S., Wilkinson, L., Ogunde, L. et al. Alkylation technology study final report—South Coast Air Quality Management District (SCAQMD). Norton Engineering, September 2016.

3

Storage Renewable Power-to-Chemicals

CONTENTS

3.1 Introduction

Environmental impact reduction is the major motivation for increasing the use of renewable energies. Since renewable energy sources like solar, wind and marine technologies are intermittent in nature, their increased installation leads to unused surplus power. Therefore, storage for surplus power is required. Since the potential of pumped hydro storage as the only installed large scale storage system is nearly exhausted, a large variety of concepts are currently under investigation to use the surplus power. The proposed concepts span from electrical storage over storage in battery electric vehicles to production of hydrogen. Hydrogen can then be used in several ways including reconversion to power in fuels cells or conversion of hydrogen with carbon dioxide to

Converting Power into Chemicals and Fuels: Power-to-X Technology for a Sustainable Future, First Edition. Martin Bajus.

hydrocarbons. However, surplus power is limited and therefore only storage systems with the highest contribution to the reduction of environmental impacts should be implemented.

In terms of energy generation, wind, solar and marine technologies are the focus of research, development and implementation (Valera-Medina, et al.). Wind energy is one of the most commercialized renewable energy technologies across the world. Wind turbines have become bigger, with taller hub heights and larger rotor diameters. Simultaneously, the dimensions and number of units in offshore facilities keep increasing, with farms, such as Walney Extension (UK), that have reached capacities of 659 MW. Forecasts predict up to an 8-fold increase in wind power generation over the next 20 years, thus emphasizing the high impact that the technology will have by then. However, management of these large resources is critical to improve profitability while supporting the grid with zero-carbon energy in the long term.

Similar to wind energy, solar energy has also gained extensive attention worldwide. Solar energy has become a paramount energy source for the development of renewable energy systems. Solar photovoltaics (PVs), solar thermal electricity and solar heating/cooling are now well-established solar technologies. For example, solar photovoltaic electricity generation increased by about 130 TWh globally in 2019 (from 440 TWh for all renewables), second only to wind in absolute terms, creating 2.7% of the electricity supply.

Finally, another technology that can also be considered is marine energy. However, because of its immature technical development compared to the previous two technologies, only a mere 0.5 GW capacity had been installed by 2018, with 1.7 GW under construction. Nevertheless, better economics and energy management could free the potential of 748 GW power by 2050, creating the way for 160,000 direct jobs.

However, all these technologies are dependent on weather conditions, time, difficult to access locations and seasonal periods. These facts, combined with extreme variability in some regions, can be overcome with the use of proper energy management techniques, such as energy storage, thus enabling the distribution of excess energy when and where it is needed.

There are many ways to store this energy. Concepts go from storage in high power superconductors, managing reasonable energy power (<5 MW) over extremely limited periods (seconds), to pump hydro or chemical storage, capable of storing gigawatts of power for long periods of time (months and years) (Figure 3.1). However, different from mechanical devices, chemical storage presents a unique feature: flexibility. Chemicals can be moved, stored and distributed easily, with many of them having a mature market already in existence for decades. Out of all the zero-carbon options, those commercially assessed are mainly hydrogen, methane, methanol, ammonia and hydrazine as potentials for fuelling the future. Hydrazine presents several issues related to its toxicity and reactivity, making it a difficult fuel that has been relegated only to aerospace applications for auxiliary power during emergency situations.

Renewable energy resources such as solar, wind, biomass and hydropower are considered the main energy supply options to meet the increasing electricity demand (e.g., the global electricity demand is expected to increase from 25,000 TWh in 2020 to around 50,000 TWh in 2050). Today, the share of electricity generation from renewable energy resources is approximately 25% across the world and it is forecast that it will have a share of two-thirds of the electricity generation by 2040. In addition, 100% electricity generation from renewables is considered to be technically and economically achievable in Europe and the USA by 2050.

An important limitation of the use of renewable energy resources is the fluctuating nature, which may lead to temporary surplus or lack of electricity and instability in the electric grid network. In this regard, energy storage technologies play an important role in providing a balance between supply and demand in the grid network at different time scales (seconds and sub-seconds,

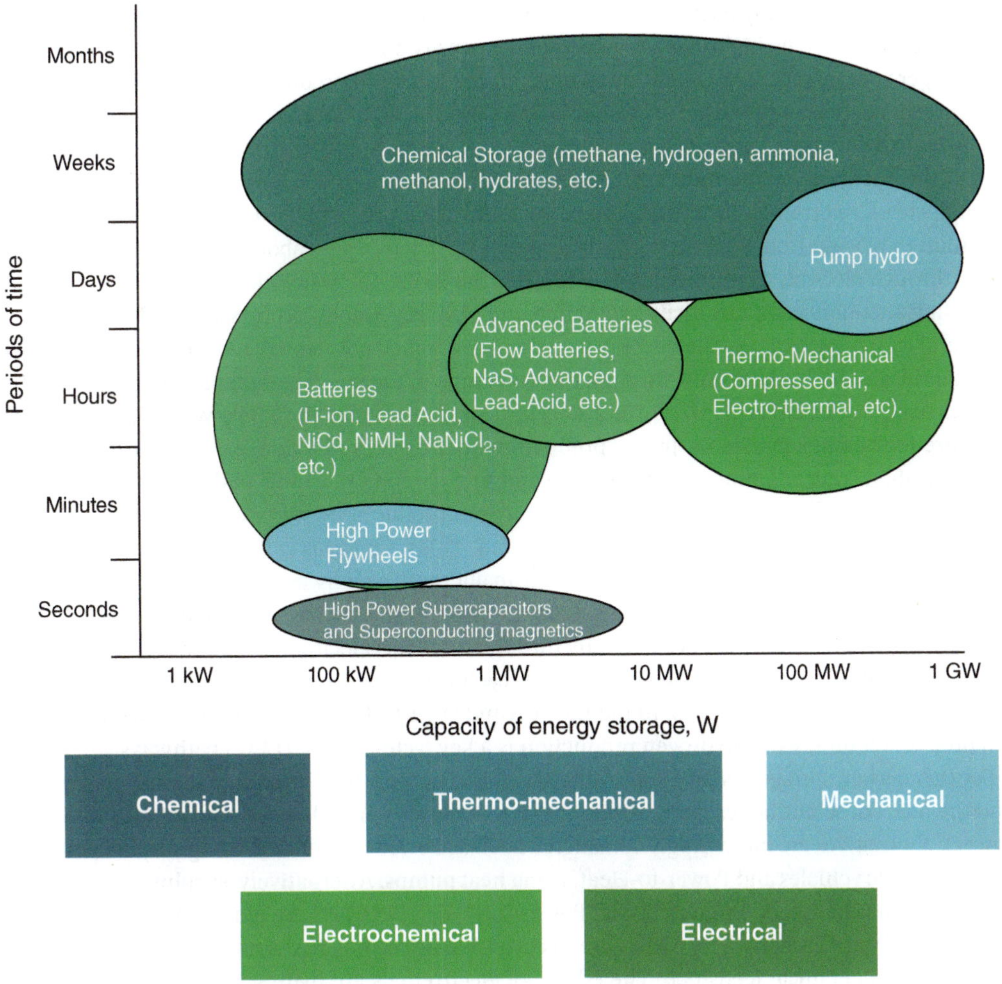

Figure 3.1 Comparison between different energy storage technologies (Valera-Medina and Banares-Alcantara, 2021 / With permission of Elsevier).

hourly, daily and seasonally). Therefore, the capacity of energy storage (e.g., hydrogen, batteries, pumped hydro and compressed air) has increased in recent years (e.g., it increased from 4 GWh in 2017 to 8 GWh in 2018) to ensure a stable grid.

Hydrogen, the most common molecule in the universe, is currently under assessment everywhere. Numerous paths for its production, distribution and consumption present a complex variation and trade-off between costs, emissions, scalability and requirements. Some scenarios foresee the use of hydrogen in up to 50% of end-use demand applications by the middle of the twenty-first century. However, the creation of "Hydrogen Technologies" (Chapter 6) faces several constraints that still require the development of equipment and, overall, infrastructure. In comparison to other molecules, such as ammonia, hydrogen storage is cheaper than the use of ammonia only when salt caves are employed, although this only applies over limited periods of time. Thus, storage costs are 26 and 30 times higher for hydrogen than for ammonia. Moreover, other barriers, such as its high diffusivity, low energy volumetric density, high flammability range, etc., still make it economically unviable for a zero-carbon transition, thus requiring another molecule.

Hydrogen is a promising energy carrier, which is versatile, transportable, clean (if produced from renewables) and has the highest gravimetric energy density among other alternatives, but its energy content per volume is comparatively low. Conventional hydrogen production methods require a significant amount of energy and are at present mainly based on fossil sources, which combined with the lack of hydrogen storage and distribution infrastructure means hydrogen did not played an important role in the energy economy until 2010. More precisely, hydrogen was produced from natural gas through steam reforming (48%), from the petroleum fraction through steam reforming and partial oxidation (30%) and from coal through gasification (18%), while only about 4% was produced from water through electrolysis in the early 2010s. After the 2010s, the water electrolysis process gained increasing attention as the electricity generation cost from renewables decreased significantly.

Several technologies are available for the conversion of renewable energy resources into X options (e.g., liquid, gas, fuels, chemicals) through the utilization of carbon dioxide captured from different sources (e.g., biogas, air, cement, iron and steel industry). In a conventional Power-to-X technology, there are four main subsystems: power production, hydrogen production, carbon dioxide capture and utilization (except ammonia production) and fuel upgrading. In this pathway, X is produced through a chemical plant (reactor) in which hydrogen produced by water electrolysis and carbon dioxide and/or carbon monoxide or nitrogen are fed. This pathway is called water electrolysis followed by a catalytic step (H2-CS, Figure 3.2a). In an alternative pathway, X is produced through the electrochemical reduction pathway (e.g., coelectrolysis) followed by a catalytic step (ERP-CS, Figure 3.2b). X can be also produced directly through the direct electrochemical reduction pathway (D-ERP, Figure 3.2c) in a single electrochemical reactor (e.g., carbon dioxide electrolyser and electrochemical Haber-Bosch reactor). In the ERP-CS and D-ERP, there might not be a separate hydrogen production unit, while hydrogen production is a key factor for the H2-CS pathways.

For this reason, storage systems such as batteries and compressed air energy storage are moving into the focus of scientific research. While these systems also provide power as output (Power-to-Power), alternatives are increasingly considered such as Power-to-Mobility, using surplus power in battery electric vehicles and Power-to-Heat, using heat pumps. Alternatively, surplus power can be used to produce hydrogen by electrolysis (Power-to-Fuel). Hydrogen can either be used directly or converted to hydrocarbons. The hydrocarbons can then be reconverted to electric power, used for mobility, or as chemical feedstock. The educts of electro fuels are hydrogen and carbon dioxide. Therefore, the consecutive chemical processes of synthesis gas (C_1-technology, Chapter 10) are significant in this context.

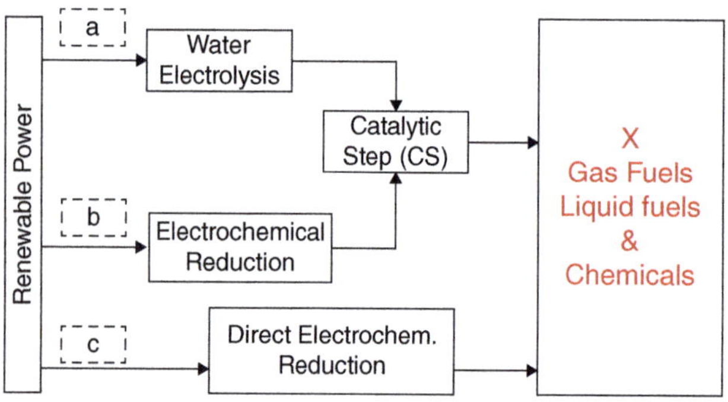

Figure 3.2 Various Power-to-X technologies: a) water electrolyser followed by a catalytic reactor (H_2-CS), b) electrochemical reduction pathway followed by a catalytic step (ERP-CS) and c) direct electrochemical reduction (D-ERP).

A comparative assessment of the proposed energy storage systems is required to support decision makers. However, energy storage systems producing fuels or chemicals cannot be compared directly to energy storage systems producing power due to their different products. Systematic comparisons of energy storage systems are so far only available for energy storage systems with reconversion to power. These energy storage systems are usually assessed by the energy efficiency for surplus Power-to-Power. For the comparison of energy storage systems producing fuels and those producing power, reconversion of the fuel to electrical power is often assumed. However, reconversion to power might not be the most favourable use of the fuel, which could be employed also for transportation or as chemical feedstock.

Power-to-X technology is a number of electricity conversion, energy storage and reconversion pathways that use surplus electric power, typically during periods where fluctuating renewable energy generation exceeds load. Power-to-X conversion technologies allow for the decoupling of power from the electricity sector for use in other sectors (such as chemical technology or transport), possibly using power that has been provided by additional investments in generation. In this context, conversion of renewable electricity to gaseous or liquid fuels (also called e-fuels or electro fuels) and chemicals (e-chemicals or electrochemical) could be a promising alternative.

The X in the terminology can refer to one of the following:

- Power-to-Power;
- Power-to-Mobility;
- Power-to-Heat;
- Power-to-Fuel;
- Power-to-Gas Fuel;
- Power-to-Liquid Fuel;
- Power-to-Gasoline;
- Power-to-Jet-Fuel;
- Power-to-Diesel;
- Power-to-Fischer-Tropsch diesel;
- Power-to-Fischer-Tropsch kerosene;
- Power-to-Hydrogen;
- Power-to-Chemicals;
- Power-to-C_1-chemicals;
- Power-to-Gas (SNG);
- Power-to-Syngas;
- Power-to-Methane;
- Power-to-Methanol;
- Power-to-DME;
- Power-to-OME_n;
- Power-to-Ammonia;
- Power-to-Hydrazine;
- Power-to-Technology;
- Power-to-Gas Demo.

Electric vehicle charging, space heating and cooling and water heating can be shifted in time to match generation, forms of demand response that some term Power-to-Mobility and Power-to-Heat.

3.2 Terminology

Even though there is no doubt that hydrogen has considerable advantages when it is used directly for power generation, another excellent option is the utilization of hydrogen in the production of valuable synthetic options, "X – energy carriers" (that is the final product such as power, heat, fuels and chemicals) through renewable energy technologies. This concept is called Power-to-X. It offers sustainable X products to provide efficient distribution, large-scale energy storage for long periods for fuels or chemicals. It should be noted that in this book X refers to liquid and gaseous fuels as well as chemicals, whereas other options (e.g., hydrogen, heat and power) are out of the scope of the book. Carbon dioxide captured or separated from air, biogas, or flue gas are combined to produce a hydrocarbon based renewable fuel. In addition, nitrogen can be used in this concept to produce synthetic ammonia.

The concept of Power-to-X was first proposed by Koji Hashimoto in 1994 to produce synthetic methane. Prototype and industrial-scale pilot plants for this concept were built in 1995 and 2003, respectively, at the Institute for Materials Research, Tohoku University. Many industrial and demonstration projects have been carried out to produce renewable fuels and chemicals in the last decades. Wulf et al. showed Power-to-X demonstration projects in operation or under planning in 20 different countries in Europe. Here, Germany and France are the leading countries in this technology. They also reported that one of the most feasible applications for Power-to-X is methanation.

Currently, there is no standardized terminology in science or technology for fuels produced by means of Power-to-Fuel processes. As suggested by Ridjan et al., these will be referred to as "(renewable) electro fuels". This term has been used in some recent works. The reason the term "electro fuel" is fitting is that a significant proportion of the energy consumed in the Power-to-Fuel process is electricity from renewable sources.

Three approaches have been mooted to operate the current and near-future fleet of transportation vehicles with significantly less production of greenhouse gases (GHGs):

- renewable fuels;
- electro fuels; and
- electric vehicles recharged using renewably generated electricity.

I use the term "electro fuel" to denote the combining of renewably sourced carbon, hydrogen and renewably sourced electrical energy to generate liquid fuels that could be used in today's vehicles or with powertrains specifically optimized for their use (co-optimization of fuels and engines). Alternately, renewably sourced electricity could directly recharge all-electric powertrains (or the battery of plug-in hybrid vehicles). I do consider indirect fuelling or direct charging, for which the scales of gathering and using energy required for transportation are large enough to affect other aspects of the natural environment. Previous studies have considered land use and water use for making low-carbon fuels from biomass.

I present an estimate of the environmental implications of using of renewably generated electricity for indirect fuelling or direct charging. I use the term "renewable" to denote energy derived from biomass, solar power, wind power and hydropower and I use the term "non-carbogenic" to cover renewable sources plus nuclear power.

The raw material carbon dioxide that is needed in addition to water is recycled and can thus be viewed as renewable, regardless of the source. Use of the term "synfuel" is not optimal, since it is used elsewhere, for example, for synthetic fuels during whose production the synthesis gas originates from coal (CTL), natural gas (GTL) or biomass (BTL). In a study conducted by Agora *Energiewende*, fuels produced from electricity are termed "Stromkraftstoff" ("electricity fuel").

These include hydrogen, methane, Fischer-Tropsch diesel and Fischer-Tropsch kerosene. In this book, the term "(renewable) electro fuels" refers to liquid, non-fossil and non-biological fuels for the transport sector. Thus, when the terms "fuel" or "fuel for the transport sector" are used in this book, they refer to transport fuels used onboard vehicles. The term is widely used in Germany and may have originated there.

Collectively Power-to-X- Technology schemes which use surplus power fall under the heading of flexibility measures and are particularly useful in energy systems with high shares of renewable generation and/or with strong decarbonization targets. Many pathways and technologies are encompassed by the term. In 2016 the German government funded a €30 million first-phase research project into Power-to-X options.

In this context, conversion of renewable electricity to gaseous or liquid fuels, also called electro fuels or e-fuels and electrochemical or e-chemicals, could be a promising alternative. We can consider potential technological pathways for recycling carbon dioxide in the form of e-chemicals and e-fuels by using not only renewable but also nuclear energy (non-carbogenic). USA, China and Germany generate most of their electricity from coal plants. It is expected that the electricity from coal plants will become a "transition state" prior to fully renewable electricity. When the coal power plant is still dominant, some carbon dioxide capture technologies should be promoted. Only France and Slovakia are transitioning from nuclear and Canada and Norway with hydropower. The products of these process pathways are also methane, methanol or hydrocarbons. Hydrocarbons are produced via F-T and methanol-to-gasoline synthesis.

We seek to elucidate the pivotal role of chemical technology in actively pursuing the transition from the fossil energy scenarios to more sustainable energy supply systems. The most relevant issues around primary electricity generation, around electricity distribution and around energy saving strategies are not covered here. The role of nuclear fission, fusion technologies and small modular reactors (Chapter 12) is not discussed in depth. The book intends to clarify some underlying lines of thought and thus refrains from discussing individual examples. Examples can be found in the literature quoted that is selected to illustrate points made in the text and not to cover the issue in any exhaustive way.

This book addresses the basic chemical energy technology with an integrated approach towards understanding and designing catalytic processes for converting small molecules like carbon dioxide, water and nitrogen into energy carriers. It tackles this challenge by cross-linking theory with experiment and in an orthogonal manner with interfacial catalysis. Chemo- electro- and photocatalysis are studied with the aim of generating solar fuels based on understanding the potentials and limitations of these approaches. The photosynthesis reaction suite that evolved in nature is used as inspiration and as a benchmark for the evolution of knowledge.

3.3 Energy Storage Systems

Hydrogen production through water electrolysis is widely discussed as a means of storing renewable electricity in chemical bonds. Hydrogen can be used for transportation in fuel cell vehicles, but it can also react with carbon dioxide to form other fuels. While many concepts have been proposed, detailed comparison of different pathway are still scarce.

Chemical energy is the potential of a chemical substance to undergo a chemical reaction to transform into other substances. Some examples of storage media of chemical energy include batteries, fuels, chemicals and food. Breaking or making of chemical bonds involves energy, which may be either absorbed or evolved from a chemical system.

Energy that can be released or absorbed because of a reaction between a set of chemical substances is equal to the difference between the energy content of the products and the reactants, if the initial and final temperatures are the same. This change in energy can be estimated from the bond energies of the various chemical bonds in the reactants and products. It can also be calculated from $\Delta U^{\circ}_{\text{f reactants}}$, the internal energy of formation of the reactant molecules and $\Delta U^{\circ}_{\text{f products}}$ the internal energy of formation of the product molecules. The internal energy change of a chemical process is equal to the heat exchanged if it is measured under conditions of constant volume and equal initial and final temperature, as in a closed container such as a bomb calorimeter. However, under conditions of constant pressure, as in reactions in vessels open to the atmosphere, the measured heat change is not always equal to the internal energy change, because pressure-volume work also releases or absorbs energy. (The heat change at constant pressure is called the enthalpy change; in this case the enthalpy of reaction, if initial and final temperatures are equal).

Green ammonia, which requires less storage volume than hydrogen and is cheaper to make than other synthetic carbon fuels, seems to be the best candidate as an electro fuel for transportation because the greenhouse gas emissions from an ammonia-driven vehicle could be less than one-third of gasoline or diesel. In the case of heavy freight, because the production sites of the hydrogen carrier (the hydrogen itself or ammonia) will certainly be located far from the consumers, long-distance freight transport with ammonia as fuel is an advantage with the volume density of liquid ammonia more than 2.5 times higher than that of compressed hydrogen at a high pressure. Ammonia can also be a suitable fuel for auxiliary power units because there are fewer issues with storage and the load range is more limited. Reducing the total greenhouse gas emissions from marine transportation is possible using hydrogen and ammonia directly as supplementary fuels or individual fuels. Using ammonia as a dual fuel in marine engines can decrease total greenhouse gas emissions up to 33.5% per ton kilometre, whereas this number increases to 69% if only ammonia is used in the engines. As for heavy-duty trucks, in the case of marine transportations (tankers, super trawlers, or cargo ships), the size of the fuel tanks needed for hydrogen would be too large, too onerous and impossible if the refuelling is not easily achieved. Although more energy-dense than hydrogen, ammonia still occupies significantly more space than diesel for the same amount of propulsion. Nevertheless, ammonia has many advantages to other fossil fuels, for example, to liquefied natural gas (LNG), because it is already produced and transported in large quantities around the world. However and as pointed out through the use of other fuel sources (Table 3.1),

Table 3.1 Energy content and energy density at storage conditions.

	H_2	CH_4	C_2H_5OH	DME	Gasoline	Diesel	HeavyOil	NH_3
Energy content (MJ/kg)	120.0	49.0	29.9	28.0	44.0	42.4	40.2	18.8
Stoichiometric air/fuel	2.4	9.6	14.3	2.3	15.3	14.3		6.05
Energy density (MJ/l)	4.8	9.7	21.3	18.7	32.0	35.2	36.6	11.5
Laminar flame speed (cm/s)[a]	218	35	41	45	34			7
Ignition delay time (ms)[b]	51.4	72.2	6.6	0.76	10	1.12		1383

[a] Experimental measured at initial conditions: $P = 100$ kPa, $T \approx 298$ K and ammonia/air mixtures, $\varphi = 1.0$. The characterization of the laminar burning velocity, (LBV), (also know as laminar flame speed) of ammonia mixtures started in the 1950s.

[b] Under conditions of an homogeneous constant volume reactor at $P = 2$ MPa, $T = 900$ K and $\varphi = 1.0$.

ammonia can be competitive compared with other alternatives but still presents a disadvantage compared with fossil sources, thus requiring more ingenious methods to use all of its properties while raising power and efficiency.

The idea of using ammonia as fuel is nothing new, with the first well-known Belgian use of it in buses during World War II. Several further studies performed during the mid-1960s were focused on evaluating the potential of ammonia as fuel in internal combustion engines. Currently, ammonia-based combustion engine fuelling methods are not ready to be marketed because the use of the substance still poses major problems, not only because of supply and safety issues, but also because of its physical characteristics compared to conventional fuels. Therefore, for many applications, two strategies have been considered for the use of ammonia fuel: supply ammonia either partially in standard engines to limit the carbon footprint or employ it mainly in dedicated engines to reach zero footprint. Another example consists of using ammonia as a reducing agent for NO_x mitigation by direct injection in diesel engines to avoid the use of selective catalytic reactors at the exhaust.

Another useful term is the heat of combustion, which is the energy of the weak double bonds of molecular oxygen released due to a combustion reaction and often applied in the study of fuels. Food is similar to hydrocarbon and carbohydrate fuels and when it is oxidized to carbon dioxide and water, the energy released is analogous to the heat of combustion (though not assessed in the same way as a hydrocarbon fuel). The heat of combustion per mole of carbon dioxide for several fuels or fuel surrogates averages 680 kJ/mol (Table 3.2).

Therefore, making electro fuels from carbon dioxide is challenging in the near term and includes adding significant amounts of clean generating capacity along with energy storage and making the electro fuels energetically efficient.

Chemical potential energy is a form of potential energy related to the structural arrangement of atoms or molecules. This arrangement may be the result of chemical bonds within a molecule or otherwise. Chemical energy of a chemical substance can be transformed to other forms of energy by a chemical reaction. As an example, when a fuel is burned the chemical energy of molecular oxygen is converted to heat and the same is the case with digestion of food metabolized in a biological organism. Green plants transform solar energy to chemical energy (mostly of oxygen) through the process known as photosynthesis and electrical energy can be converted to chemical energy and vice versa through electrochemical reactions.

Table 3.2 Energy required to revert carbon dioxide into a fuels.

Fuel/Enthalpy of Combustion	$\Delta_{comb}H°(kJ/mol_{fuel})$	$\Delta_{comb}H°(kJ/mol_{CO_2})$
Ethanol	1330	665
Methanol	639	639
Methane	802	802
Ethane	1437	719
Octane	5075	634
Dodecane	7520	627
Hexadecane	10700	679

The Gibbs free energy (Chapter 1, Section 1.2.1, Equations. 1.2; 1.3; 1.16) is given by

$$\Delta G^\circ = \Delta H^\circ - T\Delta S^\circ, \tag{3.1}$$

where H is the enthalpy, T is the absolute temperature and S is the entropy.

$$\Delta H^\circ = \Delta U^\circ + p\Delta V, \tag{3.2}$$

where U is the internal energy, p is the pressure and V is the volume.

The Gibbs free energy is the most useful for processes involving a system at constant pressure p and temperature T, because, in addition to subsuming any entropy change due merely to heat, a change in G also excludes the $p\ dV$ work needed to "make space for additional molecules" produced by various processes

The similar term chemical potential is used to indicate the potential of a substance to undergo a change of configuration, be it in the form of a chemical reaction, spatial transport, particle exchange with a reservoir, etc. It is not a form of potential energy itself, but is more closely related to free energy. The confusion in terminology arises from the fact that in other areas of physics not dominated by entropy, all potential energy is available to do useful work and drives the system to spontaneously undergo changes of configuration and thus there is no distinction between "free" and "non-free" potential energy (hence the word "potential"). However, in systems of large entropy such as chemical systems, the total amount of energy present (and conserved by the first law of thermodynamics) of which this chemical potential energy is a part, is separated from the amount of that energy – thermodynamic free energy (which chemical potential is derived from) – which (appears to) drive the system forward spontaneously as its entropy increases.

Power-to-Power storage systems convert electricity to chemical or mechanical energy, which is stored and later reconverted to electric power. Today, large scale Power-to-Power is mostly based on pumped hydro storage (PHS, for example in Slovakia, output 916.4 MW). PHS achieves efficiencies of 65–80% but it is limited by geographical constraints. The other Power-to-Power technology already operating at large scale is compressed air energy storage. However, currently only two plants are in operation worldwide.

Power-to-Mobility storage systems store input electricity in battery electric vehicles (BEV). The most discussed storage units for battery electric vehicles are lithium ion and nickel metal-hydride batteries. This technology seems to be the most promising based on its high efficiency of about 80% and its high cycling capability.

Power-to-Heat storage systems convert power directly to heat. The ratio of the heat obtained to the power input is the characteristic efficiency of these systems, the coefficient of performance (COP).

Power-to-Fuel storage systems produce hydrogen by electrolysis. A suitable technology to absorb the fluctuating surplus power from renewable energies is proton exchange membrane (PEM) electrolysis. For example: 1 MW/h electricity allows the production of 18–22 kg hydrogen. Hydrogen can be used directly as fuel, as chemical feedstock or converted to other fuels.

If hydrogen is used as fuel or chemical feedstock, the fluctuating production has usually to be compensated by storage. Hydrogen storage requires huge storage volumes because of the low volumetric energy density of hydrogen. The construction of additional hydrogen storage capacities can be avoided by directly feeding hydrogen into the already existing natural gas grid (feed-in of hydrogen). However, the fraction of hydrogen in the natural gas grid is limited, for example in Germany to 5–12 vol%.

An alternative option to deal with the low energy content of hydrogen is the conversion to hydrocarbons with higher energy content (Tables 3.1 and 3.2). These hydrocarbons can not only be used as fuels but also as platform chemicals. In addition to hydrogen, the production of hydrocarbons requires a carbon-containing feedstock. A currently discussed C_1-technology for hydrocarbons is carbon dioxide (Pillar C in Chapter 1). Carbon dioxide can be captured from biogas plants, power plants, chemical plants and even air.

The most developed hydrogen conversion process currently is methanation. First demonstration plants are already operating. Hydrogen reacts with carbon dioxide to form methane according to the Sabatier reaction. The reaction is highly exothermic which leads to high conversion losses when the heat is not completely used. The major benefit of this process is that methane can be used in the already existing natural gas infrastructure (feed-in of methane).

Hydrogen can also be converted with carbon dioxide to methanol. Rihko-Struckmann et al. investigated the production of methanol from carbon dioxide based on process simulations and assuming equilibrium conversion. In this case, about 27% of carbon dioxide is converted in the reactor (250°C and 5 MPa). The unreacted carbon dioxide is separated from the raw product gas and recycled to the reactor. Hereby, a total carbon dioxide conversion of 96.8% is achieved. Jadhav et al. report that the equilibrium conversion efficiency can nearly be obtained with a copper catalyst. The required electricity consumption of methanol production (1.33 kW h per kg methanol) is higher than for methanation because additional compressor power is required for the recycle stream. However, methanol synthesis has fewer conversion losses because it is less exothermic. Methanol seems to be a promising storage option because it has a higher volumetric energy density than hydrogen and methane and it can be combusted in gasoline engines. A disadvantage is that currently no large-scale infrastructure exists.

Hydrogen can also be converted to synthesis gas (syngas). Syngas is a mixture of hydrogen and carbon monoxide and, has various applications (Pillar C, Chapter 1) in the chemical industry, e.g., production of methanol. In contrast to methanation and methanol synthesis, this process typically requires hydrogen storage because the reaction is carried out under steady-state conditions today. A dynamic operation is currently not possible due to the high temperatures. In this book I consider the conversion of fuel to three further products:

Fuel-to-Power. Reconversion of fuels to power in fuel cells is discussed. Fuel cells achieve an efficiency of 45–60%.

Fuel-to-Mobility. Fuel-to-mobility has been looked at for the utilization of fuel for vehicles. Hydrogen can be used in a fuel cell vehicle with a consumption of about 0.36 kW/h/km. Methane and methanol can be used in a vehicle with an internal combustion engine. These engines consume about 0.52 kW/h/km and 0.44 kW/h/km, respectively.

Fuel-as-Chemical. Fuel can also be used directly as feedstock for the chemical technologies. The fuel molecules (hydrogen, methanol and carbon monoxide) are all so-called platform chemicals for chemical technologies. In this case, the fuel replaces conventional fossil-based feedstock. The use of fuel as chemical feedstock avoids conversion losses, in contrast to Fuel-to-Power and Fuel-to-Mobility processes. The storage potential of hydrogen, methanol and carbon monoxide (syngas) is limited by the demand of these chemicals. To satisfy the annual global demand of hydrogen and methanol about 3200–5400 TW h and 575–950 TW h are required, respectively. For syngas, no data on the global demand is available. For syngas use in polyurethane production only, a demand of 25–40 TW h would be required.

Various "Power-to-X" technologies are conceivable here. The technologically simplest of these is water electrolysis, which produces hydrogen from water using electricity. It is therefore a central technique to these theoretical energy systems. Water electrolysis is a promising method, particularly when utilizing wind power. Hydrogen is currently primarily produced on an industrial scale via natural gas reforming. The use of surplus electricity from renewable sources to produce hydrogen represents a sustainable process, in contrast to production through natural gas reforming. By converting this back into electrical energy, electrolyser-fuel cell systems can function as a dynamic buffer for load management, with hydrogen itself serving as the link between the energy and transport sectors.

Possible hydrogen-based technologies include Power-to-Fuel (PTF) and Power-to-Chemical (PTC) processes, referring to the production of synthetic fuels and base chemicals from sustainably produced hydrogen, using carbon dioxide as a raw material. PTF technologies thus pave the way for the integration of renewable energies in the transport sector. Investigations carried out by Vandewalle et al. verify this positive effect, affirming the effects that Power-to-Gas technologies have on the electricity market. As with the other Power-to-X technologies, the educts of the Power-to-Gas process are electric current and carbon dioxide. Because the contact points with the energy sector are thus the same, the study's results can be applied to PTF technologies.

Power-to-Fuel is a comparatively new concept which – in contrast to biomass-, coal- and gas-to-liquid (BTL, CTL, GTL) methods – is not usually covered by current specialist literature, studies or review papers concerning alternative fuels and nor does the Federal Government's 2013 Mobility and Fuels Strategy (MFS) make any mention of Power-to-Fuels. Therein, the concept – called Power-to-Liquid (PTL) in this case – can only be found in an appendix containing a schematic overview of energy and fuel pathways.

In summary, both the Energiewende problem of the oil-based transport sector and that of electricity storage can be solved by increased development of PTF processes: surplus energy produced by an increased proportion of renewable energies in electricity generation can be stored in the form of alternative liquid fuels for the transport sector. Additionally, existing value-adding chains can be utilized. There is also the possibility of stabilizing fluctuating electricity supplies by means of chemical storage so that processes can be continuously operated and thus with significantly more efficiency.

The produced fuels can supplement or replace the current, mainly fossil-based, fuels in the transport sector. Hence, PTF technology is a promising option for reducing net greenhouse gas emissions, improving integration of reuses by consumer-side load management, reducing oil dependence and conserving resources, i.e., converting electricity into mobility. In contrast to fossil fuels, the production of PTF is independent of the primary source and thus has great potential for ensuring energy security.

A wide range of propulsion and fuel alternatives for the transport sector are already being discussed, many of which use electric power and contribute to reducing carbon dioxide emissions. Additionally, they offer flexible production options that are economically viable, even if the electricity supply fluctuates. Depending on their intended purpose in the transport sector, the suitability of the various propulsion and fuel combinations differs. According to the German Energy Agency, there is therefore no silver bullet that meets the needs of all means of transport. On the contrary, it is instead necessary to diversify fuel alternatives so that a broad base can be created for the energy and propulsion variants of the transport sector. However, simultaneously offering all possible fuel alternatives with their respective propulsion systems and infrastructures does not offer an economically viable solution. It is therefore necessary to select suitable fuels to keep the number of future standard fuels at a reasonable level. Mollenhauer and Tschöke additionally suggest acceptable upper limits for the number of alternative fuels added to current standard fuels.

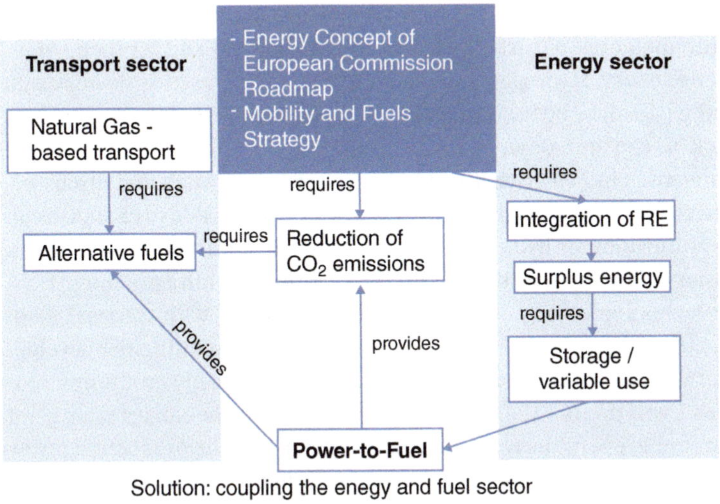

Figure 3.3 Power-to-Fuel, interlinking the energy and transport sectors.

Whether a substance produced using Power-to-X technology qualifies as a future fuel or not depends on two key criteria: on one hand, the substance must be suitable as a fuel and it should be possible to integrate it into existing fuel systems as easily as possible. Also, production of the fuel must be economically viable and therefore competitive with conventional fuels. In this context, its future technology readiness level (TRL, Chapter 1) must also be assessed. In addition, secure supply and adherence to environmental and climate protection requirements must be ensured. For these reasons, the selection of suitable fuels should be undertaken in two steps: first, the fuels should be compared to each other and then selected and second, their production pathways and feasibility should undergo the same process. Selecting the fuels and their production pathways is essential to increasing the market penetration of individual alternative fuels in a targeted manner. Only promising alternatives should be selected and implemented on an industrial scale. This book is concerned with the first of these two steps and offers an outlook for the second.

Figure 3.3 illustrates how the Power-to-Fuel concept interlinks the energy sector (right-hand side) with the transport sector (left-hand side) and simultaneously presents a solution for realizing the Energiewende in the transport sector.

Surplus electric power can be converted to other forms of energy for storage and reconversion. Direct current electrolysis (efficiency 80–85% at best) can be used to produce hydrogen which can, in turn, be converted to methane via methanation. Another possibility is converting the hydrogen, along with carbon dioxide, to methanol. Both these fuels can be stored and used to produce electricity hours to months later. Reconversion technologies include fuel cells. Power-to-Power refers to the round-trip reconversion efficiency. For hydrogen storage, the round-trip efficiency remains limited at 35–50%. Hydrogen produced from electrolysis of water using renewable energy, is a natural technology that allows flexible storage of energy. Electrolysis is expensive and Power-to-Gas processes need substantial full-load hours (say 30%) to be economic. However, while round-trip conversion efficiency of Power-to-Power is lower than with batteries and electrolysis can be expensive, storage of the fuels themselves is quite inexpensive. This means that large amounts of energy can be stored for long periods of time with Power-to-Power, which is ideal for seasonal storage. This could be particularly useful for systems with high renewable penetration, since many areas have significant seasonal variability of solar, wind and hydroelectric generation. Grid-dedicated battery storage is not normally considered a Power-to-X concept.

There is currently a considerable gap as to where bio-based Power-to-Olefins could play a role in the future. Leveraging the future potential of sustainable and low-carbon bio and synthetic fuels, including Power-to-Liquids is necessary. Hydrogen and methane can also be used as downstream fuels, can feed into the natural gas grid, or be used to make synthetic fuel. Alternatively they can be used as a chemical feedstock, as can ammonia.

In this book, the different advances in Power-to-X technologies are investigated and discussed, covering both situations where electricity is used directly to reduce carbon dioxide to chemicals up to technologies relying on the utilization of hydrogen for the reduction of carbon dioxide into a commodity that would be easier to implement into the existing worldwide infrastructure. Even though, in some cases, these technologies may not be entirely competitive with classical fossil fuels, they still represent an pathway towards sustainability and only by promoting the development of these technologies can we ever hope to balance the impact of our society on the environment. In the following sections I will present the recent advances in the technologies mainly for hydrogen production from renewable energy as well as the developments in the production of key chemicals, such as methane, methanol, formaldehyde, formic acid and alkenes via Power-to-X technology.

Among the different methods for sustainably producing hydrogen, water electrolysis (Chapter 6) is the main process investigated since it allows the production of high-purity hydrogen from renewable energy and water. Alkaline electrolysis (AEL), polymer electrolyte membranes (PEM) and solid oxide electrolysis (SOEC) are amongst the three most popular technologies for water electrolysis currently being investigated. AEL has the lowest investment and maintenance costs and is the most mature, being already commercialized at MW scale. However, coupling it with variable renewable energy could be very challenging due to its long start-up time. On the contrary, SOEC is very promising when coupled with renewable energy. Moreover, SOEC systems exhibit a great potential for coupling with exothermic processes as well as for carbon dioxide and water co-electrolysis, but considerable progress is still required to get this technology to the next level. PEM exhibits strategic advantages related to high efficiency and short time response, being the most promising for Power-to-X applications. The competitiveness of electrolysis in comparison to the other hydrogen production technologies should increase in the next few years due to the increase in the production of electrolysers as well as the number of academic and industrial research projects related to this technology. Despite the important advances achieved in the past years, improvements in efficiency as well as cost reduction of electrolysis processes are still required to overcome technical and economic barriers to the successful commercialization of Power-to-X.

Solar photovoltaic systems and wind turbines are unable to supply all the green power required, so there will still be a role for nuclear power after 2035. Development work on the next generation of nuclear reactors is ongoing. For example, NuScale is developing a small modular nuclear reactor, which is a light water reactor that does not need pumps for circulating reactor cooling water. TerraPower is developing its Natrium reactor, which is a sodium-fast reactor combined with molten sodium salt heat storage.

3.4 World Primary Energy Consumption

3.4.1 2019 Briefly

Primary energy consumption rose by 1.3% in 2019 (Figure 3.4), less than half the consumption rise in 2018 (2.8%). Growth was driven by renewables (3.2 EJ) and natural gas (2.8 EJ), which together contributed three quarters of the increase (1 exajoule [EJ] = quintillion joules [1×10^{18}]). All fuels

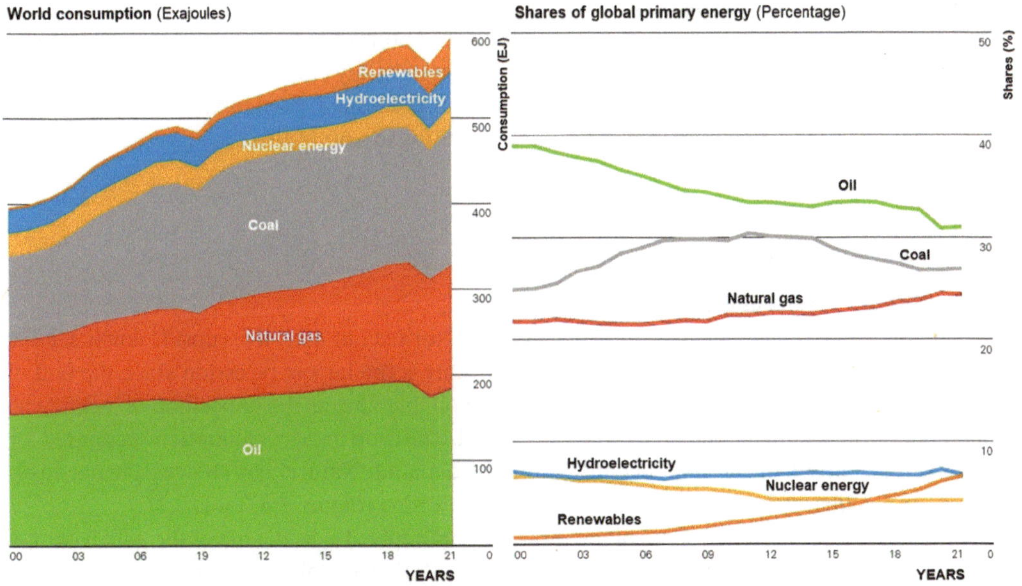

Figure 3.4 World primary energy consumption in 2021 [EJ].

grew at a slower rate than their 10-year averages, apart from nuclear, with coal consumption falling for the fourth time in six years (-0.9 EJ). By region, consumption fell in North America, Europe and CIS and growth was below average in South and Central America. In the other regions, growth was roughly in line with historical averages. China was the biggest individual driver of primary energy growth, accounting for more than three quarters of net global growth. India and Indonesia were the next largest contributors, while the US and Germany posted the largest declines in energy terms. Looking at energy by fuel type, 2019 growth was driven by renewables, followed by natural gas, which together contributed over three quarters of the net increase. The share of both renewables and natural gas in primary energy increased to record highs. Meanwhile, coal consumption declined, with its share in the energy mix falling to its lowest level since 2003. The combination of slower growth in energy demand and a shift in the fuel mix away from coal and towards natural gas and renewables led to a significant slowdown in the increase of carbon emissions.

Oil continues to hold the largest share of the energy mix (33.1%) (Figure 3.4). Coal is the second most used fuel but lost share in 2019 to reach 27.0%, its lowest level since 2003. The share of both natural gas and renewables rose to record highs of 24.2% and 5.0% respectively. Renewables have now overtaken nuclear, which makes up only 4.3% of the energy mix. The share of hydroelectricity has been stable at around 6% for several years.

Growth in energy markets slowed in 2019 in line with weaker economic growth and a partial unwinding of some of the one-off factors that boosted energy demand in 2018. This slowdown was particularly evident in the US, Russia and India, each of which exhibited unusually strong growth in 2018. China was the exception, with its energy consumption accelerating in 2019. As a result, China dominated the expansion in global energy markets, contributing the largest increment to demand for each individual source of energy other than natural gas, where it was only narrowly surpassed by the US. Despite the increase from China, all fuels (other than nuclear) grew at a slower rate than their 10-year averages, with coal consumption declining for the fourth time in six years. Nevertheless, renewables still grew by a record increment and provided the largest contribution (41%) to growth in primary energy, with the level of renewable power generation exceeding nuclear power for the first time. The slowdown in energy demand growth, combined with a shift in the fuel mix away from coal and towards natural gas and renewables, led to a significant slowing in the growth of carbon emissions, although only partially unwinding the unusually strong increase seen in 2018. Energy prices fell on the whole, particularly for coal and gas where growth in production outpaced consumption leading to a build-up of inventories. Oil prices were a little lower.

3.4.2 Energy in 2020

3.4.2.1 Not Just Green but Greening

2020 will forever be remembered as the year COVID-19 turned our daily lives upside down. Global energy demand is estimated to have fallen by 4.5%. This is the largest recession since the end of World War II, driven by an unprecedented collapse in oil demand, as the imposition of lockdowns around the world decimated transport-related demand. The drop in oil consumption accounted for around three-quarters of the total decline in energy demand. Natural gas showed greater resilience, helped primarily by continuing strong growth in China (Figure 3.4, Table 3.3)

History has borne witness to some of the most dramatic episodes in the life of the global energy system: the Suez Canal crisis in 1956, the oil embargo of 1973, soon followed by the Iranian revolution in 1979 and more recently, the Fukushima disaster in 2011. All moments of great turmoil in global energy. But all pale in comparison to the event's of 2020.

Despite the disorder of 2020, renewable energy, led by wind and solar energy, continued to grow prolifically. Remarkably, wind and solar capacity increased by a colossal 238 GW in 2020 – 50% larger than any previous expansion. Likewise, the share of wind and solar generation in the global power mix recorded its largest ever increase. The relative immunity of renewable energy to the events of 2020 is encouraging.

The fall in carbon emissions from energy use was equally striking, with emissions falling by over 6% in 2020 – again, the largest decline since 2012 (Table 3.4). Although unmatched in modern peacetime, the rate of decline in carbon emissions in 2020 is similar to what the world needs to average each year for the next 30 years to be on track to meet the aims of the Paris Agreement.

There are worrying signs that the COVID-induced dip in carbon emissions will be short lived as the world economy recovers and lockdowns are lifted. The challenge is to achieve sustained, comparable year-on-year reductions in emissions without massive disruption to our livelihoods and our everyday lives.

Table 3.3 World consumption of energy in 2019, 2020 and 2021.

	2019		2020		2021	
	Exajoules	**%**	**Exajoules**	**%**	**Exajoules**	**%**
Crude oil	191.89	33.0	173.73	31.2	184.21	30.9
Natural gas	140.54	24.2	137.62	24.7	145.35	24.4
Coal	157.64	27.1	151.42	27.2	160.10	27.0
Nuclear energy	24.93	4.3	23.98	4.3	25.31	4.2
Hydroelectricity	37.69	6.4	38.16	6.8	40.26	6.8
Renewables	28.82	5.0	31.71	5.8	39.91	6.7
Total	**581.51**	**100.0**	**556.63**	**100.0**	**595.15**	**100.0**

Since the Paris Conference of Parties (COP21) in 2015 there has been a huge increase in countries' ambitions to decarbonize. Around 70% of the world's carbon emissions are now covered by net zero targets and pledges. But this rising ambition has yet to be translated into a decisive reduction in emissions. The two big staging posts for that challenge occurred in Glasgow, Scotland (section 3.4.2.3) which was host to COP 26 and in Sharm-el-Sheikh (2022), Egypt was host to COP 27 – arguably the most important UN climate change conference since the Paris meeting. I hope that, in years to come, we all talk about the Glasgow Agreement and Sharm-el-Sheikh Conference with the same importance and reverence as we do the Paris goals.

There is belief that companies with net zero ambitions, coherent plans and near, medium and long-term aims – companies that are committed to "greening" – have a hugely significant part to play in achieving the various COP goals. Yes, the world needs more low carbon companies. But maybe more than anything, it also needs existing energy companies to decarbonize and in so doing use their scale and expertise to help bring about the deep and complex rewiring and replumbing of the global energy system that the world wants and needs to see over the next 30 years. It will take producers and consumers, as well as companies, governments and society, all working together to bring about the necessary change.

3.4.2.2 For Energy, 2020 Was a Year Like No Other

The aim in this analysis is to use recent data to try to shed light on developments, focusing on three key questions.

First, over 2020, we have been bombarded with daily headlines of unprecedented developments and volatility. Standing back from all the noise, what exactly happened that year in the world of energy and how surprising was it?

Second, the global pandemic was the mother of all stress tests. Engineers will tell you that we can learn a lot from how systems behave under extreme pressure. In that spirit, what have we learnt from the response of the global energy system to the COVID-19 crisis?

And finally, this all took place against a backdrop of increasing societal and political demands for an accelerated transition to a net zero energy system. Indeed, Glasgow in Scotland hosted COP26 and Sharm-el-Sheikh hosted COP27 – arguably the two most important UN climate conferences since Paris.

3.4.2.3 Glasgow Climate Pact

On 13 November 2021, the participating 197 countries agreed a new deal, known as the Glasgow Climate Pact, aimed at staving off dangerous climate change. The pact "Reaffirms the Paris Agreement temperature goal of holding the increase in the global average temperature to well below 2°C above pre-industrial levels and pursuing efforts to limit the temperature increase to 1.5°C above pre-industrial levels" and "Recognizes that limiting global warming to 1.5°C requires rapid, deep and sustained reductions in global greenhouse gas emissions, including reducing global carbon dioxide emissions by 45% by 2030 relative to the 2010 level and to net zero around mid-century, as well as deep reductions in other greenhouse gases." However, achieving the target is not ensured, as based on existing pledges the emissions in the year 2030 will be 14% higher than in 2010.

The final agreement explicitly mentions coal, which is the single biggest contributor to climate change. Previous COP agreements have not mentioned coal, oil or gas, or even fossil fuels in general, as a driver, or major cause of climate change, making the Glasgow Climate Pact the first ever climate deal to explicitly plan to reduce uncontrolled use of coal-based power. The wording in the agreement refers to an intention to "phase down" use of unabated coal power, rather than to phase it out. From this wording it implicitly follows that utilizing coal power with "ambition" (net-zero emission), e.g., by neutralizing the resulting carbon dioxide via the CO_2-to-stone process, need not be reduced. However this carbon capture and storage is too expensive for most coal fired power stations. Over 140 countries pledged to reach net-zero emissions. This includes 90% of global GDP. Indonesian President Joko Widodo promised to end and reverse deforestation in Indonesia by 2030. More than 100 countries, including Brazil, pledged to reverse deforestation by 2030.

The final text of the Glasgow Climate Pact include a call for "accelerating efforts towards ... phase-out of inefficient fossil fuel subsidies". Thirty-four countries, along with several banks and financial agencies, pledged to stop international funding for "unabated fossil fuel energy sector by the end of 2022, except in limited and clearly defined circumstances that are consistent with a 1.5°C warming limit and the goals of the Paris Agreement" and increase financing of more sustainable projects, including Canada – the main provider of such finances, France, Germany, Italy and Spain – the biggest financers in European Union.

More than 40 countries pledged to move away from coal. The United States and China reached an agreement about cooperation on measures to stop climate change, including lowering methane emissions, phasing out the use of coal and forest conservation. India promised to draw half of its energy requirement from renewable sources by 2030 and achieve carbon neutrality by 2070.

Governments of 24 developed countries and a group of major car manufacturers such as GM, Ford, Volvo, BYD Auto, Jaguar Land Rover and Mercedes-Benz have committed to "work towards all sales of new cars and vans being zero emission globally by 2040 and by no later than 2035 in leading markets". Major car manufacturing nations like China, the US, Japan, Germany and South Korea, as well as Toyota, Volkswagen, Nissan-Renault-Mitsubishi, Stellantis, Honda and Hyundai had not signed up to the pledge. New pledges for financial help for climate change mitigation and adaptation were announced.

On 9 November 2021, Climate Action Tracker described the results as follows: the global temperature will rise by 2.7°C by the end of the century with current policies. The temperature will rise by 2.4°C if only the pledges for 2030 are implemented, by 2.1°C if the long-term targets are also achieved and by 1.8°C if all the announced targets are fully achieved.

China said it aims to peak carbon dioxide emissions before 2030 and to become carbon neutral by 2060. It was asked to set a clear earlier date as this would have a very large "positive impact" on the Paris Agreement targets. Officials later said the 2030 target was something to "strive to" and not something to be ensured.

Brazil, home to 60% of the Amazon rainforest, promised to halt and reverse deforestation by 2030. (National Confederation of Industry virtual cast from Brasilia on 3 September 2021). Leaders of more than 100 countries with around 85% of the world's forests, including Canada, Russia, the Democratic Republic of the Congo and the United States, agreed to end deforestation by 2030, improving on a similar 2014 agreement that included Brazil, Indonesia, businesses and more financial resources. Signatories of the 2014 agreement, the New York Declaration on Forests, pledged to half deforestation by 2020 and end it by 2030, however in the 2014–2020 period deforestation increased.

Article 6 of the Paris Agreement, which describes rules for an international carbon market (such as for trees in the deforestation agreement) and other forms of international cooperation, is being discussed as it is the last piece of the rulebook remaining to be finalized. Although the parties have agreed in principle to avoid double counting of emission reduction across more than one country's greenhouse gas inventory, exactly how much double counting will actually occur remains unclear. Carrying forward pre-2020 Kyoto carbon credits will be discussed, but is highly unlikely to be agreed. Therefore, Article 6 rules could make a big difference to future emissions.

Climate finance for adaptation and mitigation was one of the principal topics of negotiation. Poor countries want more money for adaptation, whereas donors prefer to finance mitigation as that has a chance of making a profit. Appointed to the role of Climate Finance Adviser was Mark Carney, former Governor of the Bank of England. The Paris agreement included US$100 billion annually for developing countries by 2020. However, wealthy countries failed to live up to that promise, with members of the OECD behind in their commitments and unlikely to reach the agreed amount before 2023. A group of large finance companies committed to net zero portfolios and loan books by 2050. Scotland became the first country to contribute to a loss and damage fund.

South Africa is set to receive $8.5 billion to end its reliance on coal, although details are sparse regarding capping mines, exports and local community support for the workers in the industry. Countries including Chile, Poland, Ukraine, South Korea, Indonesia and Vietnam also agreed to phase out coal in the 2030s for major economies and the 2040s for poorer nations. These nations include some of the world's most intensive users of coal. However they do not include the world's largest users of the fuel, China, India and the United States of America. Japan is to invest $100 million in the transformation of fossil-fired plants into ones based on ammonia and hydrogen fuel.

The US and many other countries agreed to limit methane emissions. More than 80 countries signed up to a global methane pledge, agreeing to cut emissions by 30% by the end of the decade. The US and European leaders say tackling the potent greenhouse gas is crucial to keeping warming limited to 1.5°C (2.7°F). Australia, China, Russia, India and Iran did not sign the deal, but it is hoped more countries will join later. Russia demanded sanction relief on green investment projects for energy companies such as Gazprom. Russia's climate envoy Ruslan Edelgeriyev accused Western countries of hypocrisy for urging Russia "to reduce methane leakages and yet we have Gazprom under sanctions".

Many attendees committed to net-zero carbon emissions, with India and Japan making specific commitments at the conference. India, the third-largest emitter of carbon dioxide by jurisdiction, set the latest target date planning to be net-zero by 2070. Japan is to offer up to $10 billion in additional funding to support decarbonization in Asia. Earlier in October, China – the largest emitter of carbon dioxide by jurisdiction – had committed to net-zero carbon emissions by 2060 and it was believed by the British government that India would issue a similar commitment. However, this was the first time that a date for carbon neutrality had been given as part of India's climate policy. Green hydrogen has emerged as one of the major areas where companies can collaborate to significantly decarbonize industries.

3.4.2.4 Energy in 2020: What Happened and How Surprising Was It

The headline numbers are dramatic: world energy demand is estimated to have fallen by 4.5% and global carbon emissions from energy use by 6.3%.

These falls are huge by historical standards – the largest falls in both energy demand and carbon emissions since 1945. Indeed, the fall of over 2 Gt of carbon dioxide means that carbon emissions last year were back to levels last seen in 2011. It's also striking that the carbon intensity of the energy mix – the average carbon emitted per unit of energy used – fell by 1.8%, also one of the largest ever falls in post-war history.

3.4.2.5 How Should We Think About These Reductions

From a historical perspective, the falls in energy demand and carbon emissions are obviously dramatic. But from a forward-looking perspective, the rate of decline in carbon emissions observed last year is similar to what the world needs to average each and every year for the next 30 years to be on track to meet the Paris climate goals.

Put more solidly, if carbon emissions declined at the same average rate as in 2020 for the next 30 years, global carbon emissions would decline by around 85% by 2050. For those of you familiar with BP's latest Energy Outlook, that is roughly mid-way between the Rapid and Net Zero scenarios, which are broadly consistent with maintaining global temperature rises well below 2°C and below 1.5°C respectively.

Last year's fall in carbon emissions was obviously driven by a huge loss in economic output and activity. A simple calculation comparing the fall in emissions with the decline in world output equates to an implied carbon price of almost $1400 per tonne. Scarily high. The challenge is to reduce emissions without causing massive disruption and damage to everyday lives and livelihoods.

It is interesting to ask how surprising the falls in energy demand and carbon emissions were in 2020. They were the biggest falls seen for 75 years, but they occurred against the backdrop of a global pandemic and the largest economic recession in post-war history. So how surprising were they given everything else that was going on?

The framework uses GDP growth, changes in oil prices (as a proxy for energy prices) and the number of unusually hot and cold days to predict the growth of energy demand at a country level and then aggregates to global energy. The neat thing about the framework is that, despite being very simple, it can explain most of the broad variations in energy demand over the past 20 years or so.

The key feature of 2020's fall in energy demand is that it was surprisingly big. Even after allowing for the collapse in economic activity, the decline in energy demand was close to twice the size of the "predicted" fall: 4.5% compared with a predicted fall of around 2.5%. The source of this unexpected result can be better understood by looking at the size of the falls in the different components of energy demand.

Oil demand is estimated to have fallen by an unprecedented 9.3% (9.1 million barrels/day) in 2020 – far higher than anything seen in history and far higher than the falls in the other demand components. Indeed, the fall in oil demand accounts for around three-quarters of the total decline in energy consumption. It is also the key factor accounting for the near-record fall in the carbon intensity of the energy mix. The fall in oil consumption in 2020 was far bigger than expected based on past relationships. And the extent of that discrepancy was far greater than for any of the other demand components.

The decline in natural gas was pretty much in line with the model prediction and electricity consumption fell by less than predicted. Indeed, for those of you who like to think in statistical terms, the only statistically significant prediction errors were those for total energy demand and oil demand. And the surprise in total energy demand can be entirely explained by the greater-than-expected fall in oil demand.

Of course, for all of us who experienced extended lockdowns in 2020, this is hardly surprising. The lockdowns detracted from oil demand in a completely different way to a normal economic downturn, crushing transport-related demand. Mobility metrics fell across the board. Use of jet fuel and kerosene is estimated to have plunged by 40% (3.2 Mb/d) as aviation across much of the world was grounded

Similarly, gasoline demand fell by around 13% (3.1 Mb/d) as road mobility measures crashed. In contrast, products most closely related to the petrochemicals sector (naphtha, ethane and LPG) were broadly flat, supported in part by increasing demand for personal protective equipment (PPE) and other medical- and hygiene-related supplies.

In comparison, natural gas showed far greater resilience. Natural gas demand is estimated to have fallen by 2.3% (81 billion cubic metres, bcm) in 2020, a broadly similar decline to that seen in

2009 in the aftermath of the financial crisis. Consumption fell in most regions, with the notable exception of China, where gas demand grew by almost 7%. The relative immunity of natural gas was helped by sharp falls in gas prices, which allowed gas generation to gain share in the US power market and hold its own in the EU.

Electricity consumption is estimated to have experienced the smallest fall across the main components of final energy demand, declining by just 0.9% in 2020. The relative resilience of electricity usage was aided by the nature of the lockdowns, with falling power demands in industry and commercial buildings partially offset by increased domestic use by home-based workers and locked-down families.

The relative resilience of overall power generation disguises a more significant shift in the generation mix. Despite the fall in overall power demand, generation from renewables (wind, solar, bioenergy and geothermal energy andbut excluding hydroelectricity) recorded its largest ever increase (358 TWh). This growth was driven by strong increases in both wind (173 TWh) and solar (148 TWh) generation.

Encouragingly, the share of renewables in global energy generation recorded its fastest ever increase. That continues the strong growth seen in recent years. Over the past five years, renewable generation has accounted for around 60% of the growth in global power generation, with wind and solar power more than doubling.

The growth in renewables 2020 came largely at the expense of coal-fired generation, which experienced one of its largest declines on record (405 TWh, 4.4%). In addition to falling power demand and increasing deployment of renewables, coal was also hurt by a loss of competitiveness relative to natural gas, especially in the US and EU.

These trends are exactly what the world needs to see as it transitions to net zero: strong growth in renewable generation crowding out coal. That said, the "more than doubling" in wind and solar generation over the past five years hasn't made even the smallest dent in total coal generation. The level of coal generation in 2020 was essentially unchanged from its level in 2015 as last year's fall simply offset increases from the previous few years. It will take more than just strong growth in renewable energy to remove coal from the global power sector, especially at the pace it needs to happen. There is still a long way to go to squeeze coal out of the power sector.

In the developed world, there needs to be greater progress in energy efficiency. And as we highlighted in the 2020 Energy Outlook, for many emerging market economies to make significant inroads into the role of coal while still ensuring improving energy access, they will likely need to see an expansion in natural gas alongside renewable energy over the next 15–20 years.

Finally, how "surprising" was 2020's fall in carbon emissions? Or put differently, what message should we take from this fall for future carbon trends? Is this the beginning of the much hoped for downward trend in emissions or just a temporary COVID-induced dip?

The two elements combining to produce the near-record decline in carbon emissions – the falls in energy demand and in the carbon intensity of the energy mix – can both be largely traced back to the unprecedented decline in oil demand triggered by the lockdowns. This suggests that as lockdowns around the world eased and economic activity recovers, there is a significant risk that 2020's fall in carbon emissions will be reversed. Indeed, the IEA recently estimated that the level of carbon emissions in December 2020 was already back above pre-crisis levels.

3.4.2.6 What Can We Learn from the COVID-induced Stress Test

Moving to the second question: what can we learn from the behaviour of the energy system in response to the extreme stress test induced by the pandemic?

The focus here is on the supply response: how did different parts of the energy industry react to the sudden, unexpected fall in demand? As already mentioned, it is striking that the relentless expansion of renewable energy was relatively unscathed by the pandemic. Impressive resilience.

The oil and natural gas markets were affected more severely by the events of 2020. Moreover, oil and natural gas markets could become increasingly challenged as the energy transition gathers pace. So, it's interesting to ask whether we can learn anything about their future behaviour from their response to the stresses of last year.

Over the year, global oil production is estimated to have fallen by 6.6 Mb/d – again the largest fall in post-war history. To get a sense of the timing and composition of that supply response it is helpful to split the year into three phases.

Prices responded accordingly. Brent reached a low of below \$20/bbl in April 2020 and oil markets made frontpage news as US WTI prices turned negative for the first time ever. The falls in US tight oil were compounded by falls in conventional supplies.

For us, the main lesson was that OPEC+ was both able and willing to step in and stabilise oil markets. But whether this means it will always be able to do so, depends on the type of shock affecting oil markets. The nature of OPEC's power to shift supply from one period to another means it can offset temporary, short-lived shocks.

Turning to the European gas market, both because it is the largest market in which there is active gas-on-gas competition, and because of the key role it plays as the balancing market for liquefied natural gas (LNG) cargoes. What can we learn from its behaviour in response to the stress caused by the pandemic, with European gas imports falling by over 8.5% in 2020? The gas-on-gas competition in Europe takes the form of pipeline imports – predominantly from Russia – competing against LNG imports – largely from the US as the marginal source of LNG.

3.4.2.7 Progress Since Paris – How Is the World Doing

Finally, question three. We need to put 2020 developments into the broader context of progress made since the Paris COP in 2015. The goals agreed at Paris are widely seen as a watershed in terms of achieving global alignment on ambitions for limiting temperature rises and all countries agreeing to make their contribution to achieving those aims.

So almost eight years on from Paris, how is the world doing? One of the biggest changes since 2015 is the marked increase in focus and ambition on getting to net zero. At the time of the Paris convention, no major country had made a formal commitment to achieve net zero. That accolade went to Sweden, which in June 2017 pledged to reach carbon neutrality by 2045.

Fast forward to today and 10 countries, together with the European Union, have passed net zero targets into law and a further 34 countries have either proposed legislation or outlined formal policy intentions with the same intent. The IEA recently estimated that together these commitments and intentions account for around 70% of global carbon emissions. Although still early days, these are encouraging signs that the collective jolt and huge costs of COVID-19 may have led to renewed determination to prevent an even more damaging global trauma in the form of climate change.

The rising level of ambition is also evident at the corporate level. At last count, the number of companies with stated aims or ambitions to get to net zero had increased more than six-fold since 2019 to more than 3000 companies. This rise in corporate ambition has coincided with growing societal expectations for companies to both increase further their transparency about climate-related risks and demonstrate their strategies and actions are consistent with Paris. One manifestation of those changing societal expectations is the explosion in ESG-related investments.

Until 2020, carbon emissions had continued their unrelenting rise, broadly in line with the policy intentions that had been declared prior to the Paris meeting. Importantly, there was no sign of the decisive shift envisaged by the "less than 2°C" purple scenario. The COVID-induced fall in carbon emissions in 2020 put emissions closer to the 2-degree pathway, but as discussed earlier, there is a good chance that much of that dip proves transitory. Hope and ambition need to be translated into tangible, concrete differences.

Although much of the attention of the Paris Agreement is on the response to climate change, the Agreement stipulates that this response should be in "the context of sustainable development and efforts to eradicate poverty". The UN Sustainable Development Goals for 2030 (UN SDGs), which were adopted around the same time as the Paris COP, provide a natural benchmark for monitoring progress on this aspect of the Paris Agreement. Arguably, the single most important element of the energy system needed to address both aspects of the Paris Agreement – respond to the threat of climate change and support sustainable growth – is the need for rapid growth in renewable energy. I am pleased to say that the progress on renewable energy over the past five or six years has been a perfect example of that tangible, concrete progress I mentioned.

If we start with what happened in 2020, wind and solar energy is where most of the action is. Despite the huge disruptions associated with the global pandemic and the collapse in GDP, wind and solar capacity increased by a colossal 238 GW in 2020 – 50% larger than at any time in history.

3.5 Carbon Dioxide Emissions

Chemical technology is an important component of minimizing carbon dioxide emission. In 1993 Professor Yoichi Kaya from the University of Tokyo proposed the Kaya identity. Four related factors (Equation 3.4) determine carbon dioxide emissions.

$$\mathrm{CO_2} = population \times \left(\frac{GDP}{person}\right) \times \left(\frac{energy}{GDP}\right) \times \left(\frac{\mathrm{CO_2}}{energy}\right) \tag{3.4}$$

where population is the global population and GDP is the global average per capita.

Decarbonization of fuel in the transportation sector, especially in personal vehicles, is believed to be compulsory to minimize greenhouse gases' effect on climate change by reducing the carbon intensity of the energy system (carbon-to-hydrogen, C/H, ratio). The C/H ratio is being reduced from ca. 0.5 (liquid gasoline) to CNG (gas, $C/H = 0.25$), to hydrogen ($C/H = 0$) and electric vehicle (EV, where $C/H = 0$).

Liquid fuel use started around 135 years ago, when Carl Benz successfully managed to produce an internal combustion engine in 1886. Later, there was a tendency to use greener liquid fuel such as ethanol and butanol. Ethanol is produced from biomass using biocatalysts (fermentation). Ethanol is blended with gasoline to produce E15 (15% ethanol and 85% gasoline). ENI, an Italian multinational energy company, proposed Alcohol-20 (A20) which consists of ethanol (5%), methanol (15%) and gasoline (80%).

Carbon emissions from energy use grew by 0.5%, less than half 10-year average growth of 1.1% per year, partially reversing some of the unusually strong increase in 2018 (2.1%). More worrying is the trend for carbon emissions. The slowing of the growth of carbon emissions to 0.5% in 2019 may suggest some grounds for optimism. But this deceleration needs to be seen in the context of the big increase in carbon emissions in 2018 of 2.1%. The hope was that as the one-off factors boosting carbon emissions in 2018 unwound, carbon emissions would fall significantly. That fall did not happen. The average annual growth in carbon emissions over 2018 and 2019 was greater than its 10-year average (Table 3.4).

However, the energy transition is gathering momentum as pressure from society leads policy makers and investors to demand faster and deeper cuts to carbon dioxide emissions from all industrial sectors. Simultaneously, technology is facilitating change, with falling costs of renewable energy resulting in cheaper and greener power and digitalization allowing energy suppliers and consumers to act in a connected, nimble way. Despite this, world demand for energy has increased and will continue to increase, for the foreseeable future. One change in the balance between energy production and consumption will occur as the demand side is poised to make substantial geographical shifts over the coming years.

As the world emerged from the COVID-19 crisis it needed to make decisive changes to move to a more sustainable path. The disruption to our everyday lives caused by many of the world's most polluted cities has improved and air has become clearer. BP estimated that global carbon dioxide emissions fell around by 2 gigatonnes in 2021 (Table 3.4). That has come at considerable cost and as economies restarted and our lives returned to normal there was a risk that these gains will be offset by carbon emissions every other year for the next 25 years. Improvements can be achieved only by a radical shift in all our behaviours. By using resources and energy more efficiently and by implementing the full range of zero and low carbon energies and technologies at our disposal – including renewable energies, electrification, hydrogen, CCUS (carbon capture use and storage), bioenergy and many more. These technologies exist today – the challenge is to use them at pace and scale.

3.5.1 Carbon Footprint

The rise of electric vehicles also may increase the production of carbon dioxide released into the atmosphere because, in different countries, the electricity is generated mostly by coal power plants. Electrical vehicles may produce more carbon dioxide than an internal combustion engine; in this case, development of zeolite adsorbent for post combustion carbon dioxide is recommended. Since 60% of electricity is still generated from the coal, carbon dioxide utilization is paramount. Potentially, porous materials for thermal storage can be applied commercially.

Carbon dioxide production will not be slower with electrical vehicles, as most of the countries with an electrical car program are still generating electricity from coal. The electrical car is a "coal-powered car". United States of America, China, India, Germany, United Kingdom and Netherlands generate most of their electricity from coal plants. Coal to electricity contributed to more than 57% of total electricity produced in China; similarly more than 57% of electricity in India is produced from coal power plants. The percentage of electricity produced from coal power plants is still around 30% in the United States, one of the most developed countries. Therefore, there should be a serious effort to utilize carbon dioxide from power plants. It is expected that the electricity from coal power plants will be the "transition state" prior to fully renewable electricity. Only France and Slovakia are transitioning from nuclear and Norway and Canada with hydropower.

A carbon footprint is the total greenhouse gas emissions caused by an individual, event, organization, service, or product, expressed as a carbon dioxide equivalent. Greenhouse gases, including the carbon-containing gases carbon dioxide and methane, can be emitted through the burning of fossil fuels, land clearance and the production and consumption of food, manufactured goods, materials, wood, roads, buildings, transportation and other services.

In particular, the data compiled in 2018 suggest that global energy demand and carbon emissions from energy use grew at their fastest rate since 2010/11, moving even further away from the accelerated transition envisaged by the Paris climate goals. The carbon footprint grows stronger, wider and deeper. It's significance is indisputable. (Table 3.4). BP's economics team estimate that much of the rise in energy growth in 2018 can be traced back to weather-related effects, as families and businesses increased their demand for cooling and heating in response to an unusually large number of hot and cold days. The acceleration in carbon emissions was the direct result of this increased energy consumption. Even if these weather effects are short-lived, such that the growth in energy demand and carbon emissions slow over the next few years, there seems little doubt that the current pace of progress is inconsistent with the Paris climate goals. The world is on an unsustainable path: the longer carbon emissions continue to rise, the harder and more costly will be the eventual adjustment to net-zero carbon emissions. Yet another year of growing carbon emissions underscores the urgency for the world to change.

In today's world, energy systems should be developed with the following considerations: lower or zero greenhouse gas and air pollution emissions, lower cost, higher efficiency, ability to provide grid-balancing and energy security and reliability. The increase in carbon dioxide emission is a global concern that negatively affects both the environment and human health. The global carbon dioxide emissions were approximately 34,000 Mt in 2018. The primary energy resources that cause carbon dioxide emissions are coal (14,500 Mt), oil (11,377 Mt) and natural gas (6,740 Mt). The sectoral distribution of carbon dioxide emissions is shown in Figure 4.1. In this regard, the electricity and heat sectors contribute most to the carbon dioxide emissions. According to the Paris Agreement, as a response to climate change, global temperature rise should be limited below 2°C and efforts should continue to limit this rise to 1.5°C above pre-industrial levels. Similarly, it is reported by the Intergovernmental Panel on Climate Change (IPCC) that the global carbon dioxide emission is required to be zero after 2050 to limit the global temperature increase to well below 1.5°C above pre-industrial levels. These carbon dioxide emission restrictions enforce governments to transform to the use of clean energy resources and technologies.

3.5.1.1 Climate-driven Warming

One of the most pervasive responses to climate-driven warming is the redistribution of life on Earth, with wide-ranging implications for human well-being, ecosystem function and the climate itself. Climate-driven changes in the distribution of species are a pervasive and accelerating impact of climate change and despite increasing research effort in this rapidly emerging field, much remains unknown or poorly understood. We lack a holistic understanding of patterns and processes at local, regional and global scales, with detailed explorations of range shifts in the southern hemisphere particularly under-represented. Australian waters encompass the world's third largest marine jurisdiction, extending from tropical to sub-Antarctic climate zones and have waters warming at rates twice the global average in the north and two to four times in the south. Here, we report the results of a multi-taxon continent-wide review describing observed

Table 3.4 World carbon dioxide emissions.

Years	CO_2 (million tonnes)
2009	29,745.2
2010	31,085.5
2011	31,973.4
2012	32,273.5
2013	32,795.6
2014	32,804.7
2015	32,787.2
2016	32,936.1
2017	33,275.5
2018	34,007.9
2019	34,169.0
2020	32,284.1
2021	33,884.1

and predicted species redistribution around the Australian coastline and highlight critical gaps in knowledge impeding our understanding of and response to, these considerable changes.

Marine systems are at the forefront of climate-driven environmental change, with the ocean taking up more than 90% of the additional heat trapped in the atmosphere, resulting in rapid warming and an increase in the frequency and intensity of marine heatwaves. Already, some marine species are living under conditions at or near their thermal limits and small increases in ocean temperatures will thus result in a decline in the performance and health of these species. However, these changes are also occurring alongside an acceleration in the extent, intensity and diversity of human uses of marine systems and an increase in the cumulative human impacts on marine ecosystems across most of the ocean. Biological responses are often negatively linked with these concomitant, anthropogenic stressors and already we are observing alterations in marine systems from individual species up to entire ecosystems.

Australia's marine environment encompasses 13.86 million square kilometres, making it the world's third largest marine jurisdiction. A defining characteristic of Australia's approximately 50,000 km long coastline is that it encompasses tropical to subantarctic climate zones and ecosystems and the oceanographic context is unique in that both western (Leeuwin current) and eastern (East Australian Current) Australian coastlines are dominated by poleward flowing boundary currents.

Extensive climate-driven environmental change is now apparent throughout Australian marine systems. The East Australian Current has extended a further 350 km southwards over the last 70 years and both the south-east and south-west regions of Australia are recognized as ocean warming "hotspots", in the top 10% for rates of temperature increase globally. Moreover, the rate of ocean warming around Australia has accelerated in recent decades (Figure 3.5), with the greatest increases in sea surface temperature (SST) occurring off the south-eastern coastline. Increasing sea surface temperatures have been exacerbated by marine heatwaves off south-east Australia in 2015/2016 and 2017/2018, which have lasted up to 251 days and reached a peak intensity of ~3°C above the long-term climatology.

Changes in mean summer and winter sea surface temperature surrounding Australia from 1982 to 2018 based on 5-year means centred on 1984 and 2016 (upper panels) and corresponding time-series for four geographically explicit regions (a–d) of Australia's surrounding ocean. Trends in sea surface temperature data are presented for summer (red lines), autumn (orange lines), winter (blue lines) and spring (green lines) annual averages from each region. Data source: Daily global sea surface reprocessed (level 4) from the Operational SST and Ice Analysis system (OSTIA), downloaded from the Copernicus Marine Environment Monitoring Service (marine.copernicus.eu; product #010_011)

3.5.2 Carbon Emissions in 2020

Carbon emissions from energy use fell of 2.1Gt (by 6.3%) in 2020, to their lowest level since 2011. As with primary energy, this was the largest decline since the end of World War II. From a historical perspective, the falls in energy demand and carbon emissions are obviously dramatic. But from a forward-looking perspective, the rate of decline in carbon emissions observed in 2020 is what the world needs to average each and every year for the next 30 years to be on track to meet the Paris climate goals. Put more concretely, if carbon emissions declined at the same average rate as 2020 for the next 30 years, global carbon emissions would decline by around 85% by 2050.

In 2020 the fall in carbon emissions was obviously driven by a huge loss in economic output and activity. A simple calculation comparing the fall in emissions with the decline in world output equates to an implied carbon price of almost $1400/per tonne. The challenge is to reduce emissions without causing massive disruption and damage to everyday lives and livelihoods.

It's interesting to ask how surprising the falls in energy demand and carbon emissions were in 2020. Yes: they were the biggest falls seen for 75 years, but they occurred against the backdrop of a

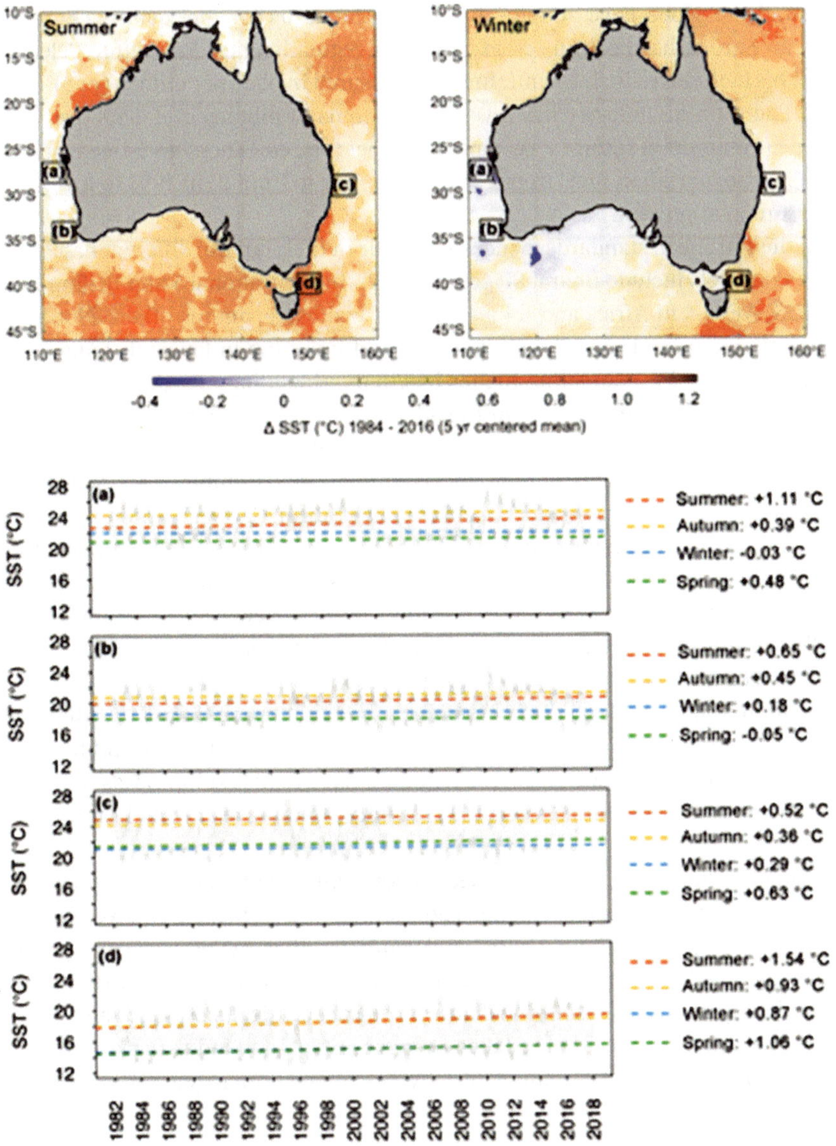

Figure 3.5 The rate of ocean warming around Australia.

global pandemic and the largest economic recession in post-war history. So how surprising were they given everything else that was going on?

3.6 Clean Fuels – the Advancement to Zero Sulfur

The following is a summary of the work that focused on the global increase in policies and regulations to produce low-sulfur and ultra-low-sulfur (ULS) transportation fuels and the subsequent boost in capital expenditures to adhere to new fuel regulations. What are clean fuels? A major aspect of "clean fuels" focuses on mitigating pollutants – carbon monoxide, nitrogen oxide, hydrocarbons and particulate matter – from vehicles' exhausts.

Many refiners around the globe have adopted European standards for fuel quality (Euro 4, Euro 5 and Euro 6), as Europe has been the frontrunner on regulations for low sulfur, "clean" transportation fuels. This includes reducing sulfur limits – measured in parts per million (ppm) – in transportation fuels. The refining industry has invested hundreds of billions of dollars over the past 30 year to reduce the amount of sulfur in gasoline, diesel, jet fuel, etc. These investments have reduced sulfur levels in transportation fuels from 2,000 ppm (prior to Euro 2) in 1993 to less than 10 ppm (Euro 5 and Euro 6) at present.

Most countries use the European standards to measure sulfur in fuels but use different names to describe them. These ULS fuels include Bharat Stage 6 (BS-6) in India, National 5 and Beijing 6 in China, Tier 3 in the US, AFRI 4 in Africa and K4 and K5 in Kazakhstan, among others. Not only are these nations and others around the world, regulating and investing in low-sulfur and ULS fuels production, but they are also mandating higher bioethanol blending rates; an increase in biofuels production and usage; and an increase in the adoption of electric, hybrid-electric, compressed natural gas (CNG), hydrogen-fuelled and/or alternative fuels-powered vehicles.

The push for lower-sulfur transportation fuels has moved into the global shipping sector as well. In 2020, the International Maritime Organization's 2020 Global Sulfur Cap regulation went into effect, which calls for marine vessels to burn low-sulfur marine fuels. Over the past 30 years, the global refining industry has invested hundreds of billions of dollars to reduce sulfur levels in transportation fuels and many new units are expected to begin operations over the next several years. According to OPEC's *World Oil Outlook 2018*, nearly 12 MMbpd of new secondary unit capacity will begin operations by 2024. These capacity additions include more than 6 MMbpd of new desulfurization capacity, more than 3 MMbpd of conversion capacity and more than 1.7 MMbpd of octane-boosting capacity. Most of this new capacity will be in the Asia-Pacific and Middle East regions. Every region has announced new policies and regulations to reduce sulfur levels in transportation fuels, along with emissions-mitigation initiatives and capital investments in new processing capacity.

As environmental concerns emerging from the incessant consumption of fuels have gripped the attention of global leaders, energy producers have been developing cleaner fuel alternatives for use in domestic and commercial settings. The bio-based and synthetic dimethyl ether (DME) petrochemical industry has expanded into numerous applications in this regard, in addition to the compound's conventional deployment as aerosol propellants in pharmaceutical, cosmetics and paint and coatings products. Recently, initiatives undertaken by governments and industry players have resulted in the expansion of research and development activities, aimed primarily at utilizing DME in automobiles and energy production.

3.7 Renewables in 2019

The energy consumption of renewables (which includes biofuels and all traded renewable electricity apart from hydro) continued to grow strongly, contributing its largest increase in energy terms (3.2 EJ) on record (Figure 3.4). This accounted for over 40% of the global growth in primary energy last year, which is larger than any other fuel. As a result, renewables increased its share in the energy mix from 4.5% in 2018 to 5% in 2019. By energy source, wind generation provided the largest contribution to growth (1.4 EJ) followed closely by solar (1.2 EJ). Other sources of renewable electricity (such as biomass and geothermal) grew by 0.3 EJ, while biofuels consumption increased by 0.2 EJ, or 100,000 barrels of oil equivalent per day. China's use of renewables grew by more than any other country, although its increase of 0.8 EJ was below the strong rate of growth seen in 2017 and 2018 (1.2 EJ both years). Solar provided half of China's growth, followed by wind (around 40%). The US (0.3 EJ) and Japan (0.2 EJ) were the next largest individual contributors to growth. Renewables provided the largest increment to power generation, followed by natural gas while coal generation fell. The share of renewables in power generation increased from 9.3% to 10.4%, surpassing nuclear for the first time. Coal's share of generation fell 1.5 percentage points to 36.4% – the lowest in the data set (which started in 1985).

3.8 Hydroelectricity and Nuclear Energy

Hydroelectric consumption rose by 0.8%, below its 10-year average of 1.9% p.a. Growth was led by China (0.6 EJ), Turkey (0.3 EJ) and India (0.2 EJ). The US and Vietnam saw the biggest declines (both -0.2 EJ). Nuclear consumption increased by 3.2%, its fastest growth since 2004 and well above the 10-year average of -0.7% p.a. As in 2018, China recorded the largest increment of any country and, last year, it saw its biggest increase ever (0.5 EJ). Japan also posted notable growth of 0.15 EJ (33%) as it continued to recover from the impact of the Fukushima incident in 2011. Generation of electricity grew by only 1.3% last year, around half of its 10-year average. Growth was weak or negative in most regions, other than in China which increased by 340 TWh (4.7%) accounting for 95% of net global growth (360 TWh).

3.9 Conclusion

Oil continues to hold the largest share of the energy mixture (33.1%). Coal is the second largest fuel but lost share in 2019 to account for 27.0%, its lowest level since 2003. The share of both natural gas and renewables rose to record highs of 24.2% and 5.0% respectively. Renewables have now overtaken nuclear, which makes up only 4.3% of the energy mixture. The share of hydroelectricity has been stable at around 6% for several years. China dominated the expansion in global energy markets. BP's economics team estimate that much of the rise in energy growth in 2018 can be traced back to weather-related effects, as families and businesses increased their demand for cooling and heating in response to an unusually large number of hot and cold days. The acceleration in carbon emissions was the direct result of this increased energy consumption. The average annual growth in carbon emissions over 2018 and 2019 was greater than the 10-year average BP estimated that global carbon dioxide emissions fell around by two gigatonnes in 2021 (Table 3.4)

A variety of different technologies and fuels are likely to be required, including extensive coal-to-gas switching and the widespread deployment of carbon capture, use and storage. All mitigation approaches will need to offset from 100 to 1000 Gt carbon dioxide during the twenty-first century. Mixtures of pentane with monoethanolamine show good solubility and high selectivity for carbon dioxide. Calcium looping uses calcium oxide-based materials as carbon dioxide sorbents. Limestone is a widely available, inexpensive and environmentally benign sorbent.

China has also initiated its Blue-Sky Defense to help curb air pollution. The program includes setting aggressive fuel economy, emissions and quality standards through 2020, retiring many coal-fired power plants and converting them to natural gas; experimenting in utilizing alternatives fuels, such as methanol and CNG; increasing ethanol blending to 10% (known as E10); and increasing the adoption of EVs/HEVs.

Further Reading

Anyanwu, J.-T., Wang, Y. and Yang, R.T. (2020). Amine-grafted silica gels for CO_2 capture including direct air capture. *Ind. Eng. Chem. Res.* 59 (15): 7072–7079. doi: 10.1021/acs.iecr.9b05228.

BP Statistical Review of World Energy. (2019). 68th edition. BP p.l.c., 1 St James's Square, London SW1Y 4PD, UK.

BP Statistical Review of World Energy. (2020). 69th edition. BP p.l.c., 1 St James's Square, London SW1Y 4PD, UK.

BP Statistical Review of World Energy. (2021). 70th edition. BP p.l.c., 1 St James's Square, London SW1Y 4PD, UK.

Buehler, J. (2022). Overview of decarbonization pathways for the oil and gas and petrochemical industries-Part 2. *Hydrocarbon Processing*, June 2022.

Connolly, D., Lund, H. and Mathiesen, B.V. (2016). Smart Energy Europe: the technical and economic impact of one potential 100% renewable energy scenario for the European Union. *Renew. Sustain. Energy Rev.* 60: 1634–1653.

Connolly, D., Mathiesen, B.V. and Ridjan, I. (2014). A comparison between renewable transport fuels that can supplement or replace biofuels in a 100% renewable energy system. *Energy* 73: 110–125.

Ganesh, K.N. et al. (2021). Green chemistry: a framework for a sustainable future. *Ind. Eng. Chem. Res.* 60 (25): 8964–8968.

Gervais, C.R., Champion, C. and Pecl, G.T. (2021 July). Species on the move around the Australian coastline: a continental-scale review of climate-driven species redistribution in marine systems. *Glob. Change Biol.* 27 (14): 3200–3017. doi: 10.1111/gcb.15634.

Halliday, C. and Hatton, T.A. (2020). Net-negative emissions through molten sorbents and bioenergy with carbon capture and storage. *Ind. Eng. Chem. Res.* 59 (52): 22582–22596. doi: 10.1021/acs.iecr.0c04512.

Hasegawa, K. and Shire, T. (2019). Adapting and prospering in the energy transition. *Hydrocarbon Processing*, July 2019.

I.I.E. Agency (2007). Tracking industrial energy efficiency and CO_2 emissions.

Ince, A.C., Colpan, C.O., Hagen, A. and Serincan, M.F. (2021). Modeling and simulation of power-to-X systems: a review. *Fuel* 304: 121354.

IPCC (2018). Global warming of 1.5 °C. Intergovernmental Panel on Climate Change.

Jadhav, S.G., Vaidya, P.D., Bhanage, B.M. and Joshi, J.B. (2014). Catalytic carbon dioxide hydrogenation to methanol: A review of recent studies. *Chem. Eng. Res. Des.* 92: 2557–2567.

Jaraíz-Arroyo, I., Martín-Calvo, A., Gutiérrez-Sevillano, J.J. et al. (2020). OCEAN: an algorithm to predict the separation of biogas using zeolites. *Ind. Eng. Chem. Res.* 59 (15): 7212–7223. doi: 10.1021/acs.iecr.9b06451.

Jun, S.-H. et al. (2020). Stabilized and immobilized carbonic anhydrase on electrospun nanofibers for enzymatic CO_2 conversion and utilization in expedited microalgal growth. *Environ. Sci. Technol.* 54 (2): 1223–1231.

Lee, N. (2019). Business trends: sustainability: clean fuels-the advancement to zero sulfur. *Hydrocarbon Processing*, June 2019.

Mathiesen, B.V., Lund, H., Connolly, D. et al. (2015). Smart energy systems for coherent 100% renewable energy and transportsolutions. *Appl. Energy*, 145: 139–154.

Mollenhauer, K. and Tschoke, H. (2007). Handbuch Dieselmotoren. *VDI-Buch*. Berlin, Heidelberg: Springer-Verlag.

OPEC's *World Oil Outlook 2018*. https://www.opec.org/opec_web/en/publications/340.htm (Retrieved 22 March 2023).

Papalas, T., Antzaras, A.N. and Lemonidou, A.A. (2020). Evaluation of calcium-based sorbents derived from natural ore and industrial wastes for high-temperature CO_2 capture. *Ind. Eng. Chem. Res.* 59 (21): 9926–9938.

Reif, K. (2010). *Konventioneller Antriebsstrang und Hybridantriebe mit Brennstoffzellen und alternativen Kraftstoffen*. Wiesbaden: Vieweg+Teubner Verlag/GWV Fachverlage GmbH Wiesbaden.

Ridjan, I., Mathiesen, B.V. and Connolly, D. (2016). Terminology used for renewable liquid and gaseous fuels based on the conversion of electricity: a review. *J. Clean. Prod.* 112 (Part 5): 3709–3720.

Rihko-Struckmann, L.K., Peschel, A., Hanke-Rauschenbach, R. and Sundmacher, K. (2010). Assessment of Methanol Synthesis Utilizing Exhaust CO_2 for Chemical Storage of Electrical Energy. *Ind. Eng. Chem. Res.* 49: 11073–11078.

Schemme, S., Samsun, R.C., Peters, R. and Solten, D. (2017). Power-to-Fuel as a key to sustainable transport systems – an analysis of diesel fuels produced from CO_2 and renewable electricity. *Fuel* 205: 198–221.

Singh, P. (2019). Petrochemicals: DME market Outlook impacted by clean fuels demand, diesel emissions. *Hydrocarbon Processing*, June 2019.

Sternberg, A. and Bardow, A. (2015). Power-to-what? – environmental assessment of energy storage systems. *Energy Environ. Sci.* 8 (2): 389–400. doi: 10.1039/c4ee03051f.

Sterner, M. and Stadler, I. (2014). *Energiespeicher: Bedarf, Technologien, Integration*. Berlin: Springer Vieweg.

Valera-Medina, A. et al. (2021). Review on ammonia as a potential fuel: from synthesis to economics. *Energy & Fuels* 35 (9): 6964–7029.

Valera-Medina, A. and Banares-Alcantara, R. (2021). *Techno-Economic Challenges of Green Ammonia as an Energy Vector*. Cambridge, MA: Academic Press. doi: 10.1016/c2019-0-01417-3.

Vandewalle, J., Bruninx, K. and D'haeseleer, W. (2015). Effects of large-scale power to gas conversion on power, gas and carbon sectors and their integrations. *Energy Convers. Manage.* 94: 28–29.

Wei, Y., Wang, T., Park, A.-H.A. and Fang, M. (2020). Toward sustainable energy and materials: CO_2 capture using microencapsuled sorbent. *Ind. Eng. Chem. Res.* 59 (21): 9746–9759.

Wikipedia. Chemical Energy. https://en.wikipedia.org/wiki/Chemical_energy.

Wikipedia. Glasgow Climate Pact. https://en.wikipedia.org/wiki/Glasgow_Climate_Pact.

Zheng, J., Chong, Z.R., Qureshi, M.F. and Linga, P. (2020). Carbon dioxide sequestration via gas hydrates: a potential pathway toward decarbonization. *Energy Fuels* 34 (9): 10529–10546.

4

Carbon Capture, Utilization and Storage Technologies

CONTENTS

4.1 Industrial Sources of Carbon Dioxide

A number of industrial sources of carbon dioxide are being considered for carbon dioxide conversion in low-carbon chemical production. Table 4.1 provides an overview of large industrial carbon dioxide sources with typical amounts and carbon dioxide concentrations. From a carbon capture point of view, sources of high concentration appear most attractive, as energy demand and cost of capture depend on the concentration of carbon dioxide.

Distribution by sector of carbon dioxide emissions is shown in Figure 4.1. The electricity and heat sectors contribute most to carbon dioxide emissions. The Paris Agreement, Glasgow Pact and Intergovernmental Panel on Climate Change (IPCC) have reported (see Chapter 3), that the global carbon dioxide emission is required to be zero after 2050 to keep the global temperature increase to below 1.5°C above pre-industrial levels.

Many current chemical processes, especially those via synthesis gas production are emitting substantial amounts of carbon dioxide due to the nature of the respective process and its feedstock. Some of these production routes provide relatively pure carbon dioxide and such sources should be valorised as soon as possible.

Depending on the target product of the low-carbon chemical process, the supply of hydrogen and carbon dioxide requires a specific match of the feed streams. Generally, sources for carbon dioxide as flue gases will be present and the stream needs to be separated, purified and pressurized in order to be used in the subsequent processes. The size of the carbon dioxide stream will define the scale of the hydrogen production. Table 4.2 provides an overview of the typical size of carbon

Converting Power into Chemicals and Fuels: Power-to-X Technology for a Sustainable Future, First Edition. Martin Bajus.

Table 4.1 Large industrial carbon dioxide sources and key parameters.

CO_2 source	Total available amount in EU (Mt)	Amount per unit production	Typical CO_2 concentration in source gas (vol. %)
Power plants	1225	0.7–0.8 t CO_2/MWh	3–4;12–14; for integrated gasification combined cycle
Steel plants	110	1.2–1.5t CO_2/t steel	14–27
Cement plants	114	0.6–1.0t CO_2/t cem.	15–33
Chemical plants, example NH_3	23	1.8 t CO_2/t NH_3	close to 100

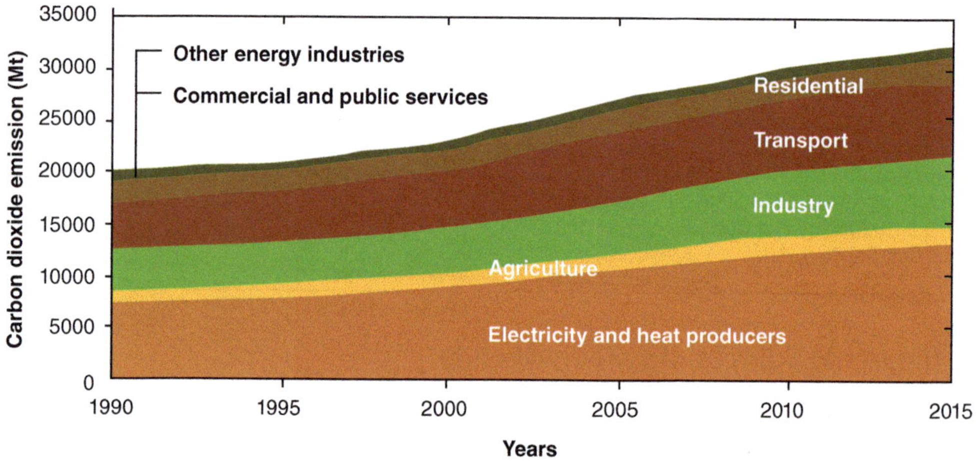

Figure 4.1 Carbon dioxide emissions for different sectors providing close to 100% carbon dioxide include ethylene oxide, ammonia production (the latter often directly utilising the carbon dioxide for subsequent urea production), natural gas cleaning, but also fermentation processes, e.g., bioethanol production. Other industrial sources include steel and cement plants, providing concentrations of carbon dioxide considerably lower than other areas, but often still higher than those in flue gas from fossil fired power plants.

Table 4.2 Typical carbon dioxide sources and their respective stochiometric volume flows for Power-to-Gas and Power-to-Liquid applications.

CO_2 source		Methanation		Methanol production		Fischer-Tropschfuels	
	volume flow m^3/h	H_2-stream m^3/h	CH_4 m^3/h	H_2 – stream m^3/h	CH_3OHt/h	H_2 –stream m^3/h	-CH_2
Biogas plant	500	2.000	500	1.500	0.7	1.500	0.3
Gasification of biomass (+CO)	2.100 (+1.400) CO	8.400 (+4.200) CO	2.100 (+1.400) CO	6.300 (+2.800) CO	2.8 (+1.8) CO	6.300 (+2.800) CO	1.2 (+0.8) CO
Ammonia plant	30,000	120,000	30,000	90,000	39	90,000	17
Coal-fired power station (500 MW, g/kWh)	240,000	960,000	240,000	720,000	314	720,000	137

Standard conditions (298.15 K; 101.300 Pa); for comparison, a typical stack for alkaline electrolysis produces a volume flow of 740 m^3/h H_2

dioxide streams from some potential carbon dioxide sources, the corresponding stoichiometric hydrogen streams for a given process and the corresponding product stream.

4.2 Carbon Capture, Utilization and Storage Technologies

Climate change is likely the key challenge that humanity will face this century. An IPCC special report entitled "Global Warming of 1.5°C" detailed the impacts of global warming of 1.5°C above preindustrial levels (Chapter 3). It also highlighted the importance of limiting the global warming to this value to avoid the most severe effects of climate change and achieving net-zero carbon dioxide emissions by 2050. This is a very demanding scenario, despite the striking development of renewable energy, since more than 80% of global primary energy is still fossil fuel based.

Most decarbonizing scenarios consider a relevant role of carbon capture, utilization and storage (CCUS) technologies. Namely, all mitigation approaches will need to offset from 100 to 1000 Gt carbon dioxide during the twenty-first century. Besides, not implementing CCUS would increase the mitigation costs by 29%–297% to limit the atmospheric carbon dioxide concentration at 450 ppm by the end of the century.

In this context, CCUS will very likely play a relevant role in the next decades. Not surprisingly, carbon capture is one of the fastest growing topics in sustainable chemistry and separations science. To minimize emissions enormous efforts within the scientific community have widely focused on carbon dioxide capture from large stationary sources such as electricity-generating power plants. In recent years, the capture of carbon dioxide from other industrial sources (e.g., cement, steel, refineries, etc.) and also from ambient air, has attracted rapidly increasing attention.

The use of the captured carbon dioxide as a feedstock to produce value-added products beyond a compressed gas for enhanced oil recovery or geological storage has also been an attractive topic for research. Many catalytic applications such as artificial photosynthesis, photocatalysis and the synthesis of chemistry building blocks, fuels and pharmaceutical compounds have been proposed. So, we have an upbeat message on renewables.

But it is important to remember that this pace of progress on renewable energy needs to be matched by the many other aspects of the energy transition: energy efficiency; the growth of new energy vectors, such as hydrogen, to help decarbonize hard-to-abate sectors; and the build-out of carbon capture, use and storage (CCUS). Continued rapid growth in renewable energy is necessary to get to net zero, but it's not sufficient.

4.3 Carbon Dioxide Capture

To address and meet the challenges of global warming caused by an ever-increasing atmospheric carbon dioxide level (Table 4.1), innovative and sustainable solutions must be found. In this regard, the development of carbon capture, utilization and storage, which plays a significant role to meet the goal of the 1.5°C scenario, is extremely important. Current mainstream technologies for carbon dioxide capture from large point sources include carbon dioxide absorption/adsorption, oxy-fuel combustion and chemical looping, all of which are largely reliant on material technologies to allow for their efficient implementation. The most established technique for capturing carbon dioxide is chemisorption using aqueous alkaline solutions such as aqueous carbonate and amine solutions. Although great progress has been achieved in this field, the solvent-based technique still suffers from engineering issues of solvent degradation, a high energy penalty and high equipment cost that limit their deployment to date.

The removal of acid gases (carbon dioxide, hydrogen sulfide) from gaseous streams can be generally classified into two categories: (1) chemical absorption processes and (2) physical absorption processes and there are several such processes which fit into these categories (Table 4.3).

Treatment of natural gas to remove the acid gas constituents (hydrogen sulfphide and carbon dioxide) is most often accomplished by contact of the natural gas with an alkaline solution. The most used treating solutions are aqueous solutions of the ethanolamine or alkali carbonates.

In April 2020, *Ind. Eng. Chem. Res.* presented a special issue on carbon dioxide capture and utilization, compiling more than 40 selected publications submitted by internationally renowned researchers on this multidisciplinary research field, covering novel advances in process optimization, modelling, reactor design and high-performance materials. In the following pages, I will summarize the most relevant features of the main research lines and the findings presented by the contributions to this special issue.

Amine scrubbing is the most developed technology for post-combustion carbon dioxide capture. The typical process flow of this technology is similar to an industrial acid gas removal system using a chemical solvent. Since amine scrubbing relies on thermal swing regeneration, a higher temperature of carbon dioxide absorption reduces energy consumption. This effect of thermal swing is dependent on the temperature of the reboiler, which is limited by the thermal degradation of the solvent. Degradation must be minimized, not only to moderate the cost of the process but also to reduce the emissions of toxic by-products. Degradation behaviour and related toxicity must be well understood before amine absorption is used commercially. Piperazine derivatives show faster carbon dioxide absorption rates and greater resistance to thermal degradation, compared to other second-generation amines, although further investigation is required to assess the corrosion tendency of these advanced amine-based solvents. In addition, less time-consuming and reliable techniques for solvent monitoring are essential to ensure efficient and safe operation. Infrared spectroscopy, combined with principal component regression, allows for a rapid measure of solvent composition, compared to traditional approaches such as chromatography. Amines can also

Table 4.3 Processes for acid gas removal.

Chemical Absorption	**Physical Absorption**
(Chemical Solvent Processes)	(Physical Solvent Processes)
Alkanolamines	
MEA	Selexol
SNPA-DEA (DEA)	Rectisol
UCAP (TEA)	Sulfinol*
Selectamine (MDEA)	
Econamine (DGA)	
ADIP (DIPA)	
Alkaline salt solutions	
Hot potassium carbonate	
Catacarb	
Benfield	
Giammarco-Vetrocoke	
Nonregenerable	
Caustic	

*A combined physical/chemical solvent process

absorb aromatics during gas washing processes. A deep knowledge of carbon dioxide–hydrocarbon mixtures is essential for the design and operation of absorption systems. Binary parameters can be obtained from empirical models that can satisfactorily describe the thermodynamics of this type of amine solutions for wide ranges of temperatures and pressures.

Electrochemically mediated amine regeneration (EMAR) is an alternative method to the conventional thermal regeneration of amines aimed at reducing the energy demand of the carbon dioxide capture process. The EMAR process does not require steam integration, because only electrical energy is used for the regeneration of the solvent. This is a great advantage in the retrofitting of carbon dioxide emitter plants because the capture unit can perform independently of the main plant. A large amount of heat is consumed during the regeneration of aqueous amine solutions, not only to increase the temperature of the solution until the stripping temperature is reached, but also to provide the inevitable energy for water evaporation. A significant reduction in energy demand is possible by using organic solvents instead of water. Mixtures of pentane with monoethanolamine (MEA) show good solubility and high selectivity to carbon dioxide. Moreover, a considerably lower amount of energy must be supplied because this system requires a lower regeneration temperature than the conventional MEA-based process.

Apart from post-combustion applications in large stationary carbon dioxide emitters, chemical solvents are also suitable for capturing the carbon dioxide generated in other sources. Marine transport accounts for 3% of global carbon dioxide emissions. The utilization of amine absorption systems to reduce carbon emissions in ships has recently gained interest. The models of marine diesel and ship energy systems have been developed and validated, showing the potential of solvent-based systems to significantly reduce the carbon footprint at a reasonable cost. The recirculation of exhaust gas greatly reduces energy demand because the heat contained in the recycle is suitable for the reboiler.

Intensive research over the past years has also shown that adsorbent-based capture technologies have the potential to significantly reduce the energy penalty (operating expenditure (OPEX)) and/or equipment costs (capital expenditure (CAPEX)), when compared with state-of-the-art amine-based solvent capture processes. Adsorption-based capture is based on a cyclic process where carbon dioxide molecules interact and bond with the adsorbent material either through weak intermolecular forces (physisorption), or through strong covalent bonding (chemisorption). By altering the thermodynamics of the system, for instance, pressure and temperature, either the adsorption or the desorption step is favoured.

The modification of porous materials via the loading of amines is a well-known technique to obtain solid chemisorbents. These hybrid adsorbents aim to reduce amine corrosivity, evaporation loss and heat-consuming regeneration, while maintaining their selectivity towards carbon dioxide.

Amine functionalization by grafting involves the chemical reaction between a silane and the surface of a porous material. Conventional (dry) and wet grafting of silanes on silica gels have been reported by Yang. Large pores and the wet method were found to enhance carbon dioxide adsorption performance. By grafting linear and branched silanes on to mesoporous silica, it was found that branched compounds yield higher carbon dioxide uptakes. The thermal and oxidative stabilities of the materials were also assessed.

Impregnation is another major functionalization route. In this case, there is no specific reaction with the pore surface, so a higher nitrogen load can be achieved compared to grafting. Large amines or even polymers are used for this purpose. Nevertheless, the adsorbants obtained usually present a poor cycling or stability performance. The most compound most often used for impregnation is polyethylenimine (PEI). The combined effect of PEI and a silane on silica gel was explored, reaching better results than PEI itself.

Furthermore, the Kolle and Sayari have covalently immobilized PEI to porous Al-silica to overcome the disadvantages of the impregnation technique, directing their efforts to investigate the oxidative stability of the materials.

Measuring adsorption kinetics on amine-loaded sorbents is crucial not only for material development purposes but also for reactor design and process development. New methods are currently being explored to measure fast adsorption kinetics in novel micro fluidized bed thermogravimetric analysers. Obtained apparent adsorption kinetics were described by a two-stage model of the linear driving force model.

Non-aminated adsorbents are also used, provided they present a sufficient selectivity towards carbon dioxide. Activated carbons obtained from agricultural wastes are also considered. Date seeds and hazelnut shells were used as raw materials in a laudable effort to achieve adsorbents within the circular economy system.

Microencapsulated carbon sorbents (MECS) are considered promising materials for enhanced carbon dioxide capture owing to their drastically increased gas–liquid contacting area. By reviewing the mass transfer mechanisms of carbon dioxide absorption in terms of materials and reacting systems, Yu et al. established the fundamental principles for designing MECS to lead the selection of solvents and shell materials for enhanced carbon dioxide capture. Current commonly used microfluidic techniques for MECS synthesis are briefly described and other promising methods are discussed as a guide to future studies. Carbon dioxide capture using MECS is substantially affected by the solvents, shell materials and capsule geometry. The effects of shell mass transfer resistance are highlighted there. Detailed process design is fundamental to the scale-up of carbon dioxide capture using MECS and the carbon dioxide capture behaviours in various reactors can be revealed by modelling methods. Finally, the challenges existing in current research are assessed and future directions are given. In recent years, MECS have been proposed to address the process. In the concept of MECS, carbon dioxide solvent is encased in a permeable polymer microfluidic method to generate microcapsules with outer diameters of 100–600 μm and shell thicknesses of 10–50 μm. Owing to their drastically increased specific surface area, MECS are endowed with significant enhancement in carbon dioxide capture kinetics compared with conventional chemisorption systems. Furthermore, MECS transform the liquid solvents physically to solid sorbents combining the merits of liquid solvents in reaction kinetics and solid sorbents in flowing patterns. Therefore, MECS are especially interesting for handling solvents with easy precipitation (e.g., carbonates) or with high viscosity (e.g., ionic liquids), which generally have the advantages of low energy penalty, low degradation, or low corrosion. The improved carbon dioxide capture features of an optimized sorbent such as capacity, kinetics and lifetime might be concealed due to the complex flow field, gaseous composition, or particles movement in real reactors. At present, due to the limited number of experiments, numerical methods have been widely utilized to predict the carbon dioxide capture behaviours of MECS in various reacting systems. Multiscale diffusion-with-reaction based models that range from a single capsule to a 500 MWe coal-fired power plant absorber has been developed. Carbonate MECS showed a competitive carbon dioxide capture performance in a fluidized bed and a packed bed compared with fast-reacting amine solvents. However, it was pointed out that the increase in the gas–liquid contacting area provided by MECS may have been overestimated by about 3–10 times, which results from the underestimation of the surface area available in structured packing. Moreover, the mass transfer resistance from shell membrane (carbonates) that was initially considered not particularly important is significant to the mass-transfer process for MECS such as ionic and aqueous amine solution.

In parallel to the above-mentioned development of adsorbent materials, significant breakthroughs in separation science have been made through the discovery of novel nano porous materials (e.g., metal organic frameworks, MOFs) for low-temperature separations. These breakthroughs have afforded the synthesis of innumerable different nano porous structures. By combining organic ligands and metal-ion nodes in MOFs, over 70,000 different MOFs have been synthesized and even more structures have been proposed computationally. Hence, we can now tailor-make materials with well-defined properties.

However, promising materials must be assessed and modelled at a process level as process integration plays a key role in the energy efficiency and economics of capture processes. The evaluation and screening of adsorbent materials is frequently done by making use of basic material properties metrics (i.e., working capacity, selectivity, ...) or by metrics guided by simplified process models. More-complex process models would provide more insights into the required performance of the materials for cyclic separation processes, but they are much more time-consuming to develop and computationally expensive. Park et al. investigated how a simplified pressure swing adsorption (PSA) model would differ from a detailed model if both were to be used for the screening of MOFs at subambient temperatures for post-combustion carbon dioxide capture. They identified the carbon dioxide swing capacity and the adsorbent performance score as being useful in predicting the ranking of materials for their studied process. Also, to avoid computational time in solving nonequilibrium adsorption-based models, the concept of the Thiele modulus-effectiveness factor from heterogeneous catalysis was adopted to describe mass transfer in an adsorbing nonequilibrium adsorption system. This approximation was proofed to provide a simplified and accurate description of intraparticle mass transfer in adsorptive systems.

In addition, novel advanced materials are normally synthesized in powder form and they are difficult to handle when implemented in the process. For instance, large pressure drops in fixed-bed contactors is a well-known problem when using powder capture adsorbent materials. Those process-derived issues can be addressed by shaping the adsorbents into different structures (e.g., foams, monoliths, polymer beads, etc.). Understanding the optimum structured form of the material for a particular capture process is a key aspect to advance these novel promising technologies and this has been the focus of many ongoing research activities. For instance, a three-dimensional (3D) printing technique based on liquid printing dopes was applied by the group of Lawson to develop 3D-printed MOF monoliths, specifically MOF-74 and HKUST-1, with the potential for possible scaling-up of the monoliths. At the process level, the performance of NaY monoliths was investigated in an intensified vacuum pressure swing adsorption (VPSA) unit to capture carbon dioxide from a high partial pressure stream in a small-scale steam methane-reforming process. Also, spherical MOF beads (CPO-27-Ni) were tested in a moving-bed TSA process and compared to zeolite 13X. Because of the wide variety of adsorbents and working conditions available for carbon dioxide capture, Acuna-Yeomans et al have proposed an algorithm to select the optimal combination by combining molecular simulation and machine learning.

New hybrid materials, shaped by extrusion, suitable to be used in the electric swing adsorption (ESA) process for carbon dioxide capture, were developed. Pellets were produced using activated carbon (AC) and zeolite 13X (with four different compositions). An extensive study was carried out to determine the characterization and adsorption equilibrium of carbon dioxide and nitrogen. The CO_2/N_2 selectivity of the pellets was 62.8, 41.9, 24.6 and 12.2 at 0.15 MPa and 298 K for the 30% AC–7 0% 13X, 50% AC–50% 13X, 70% AC–30% 13X and 100% AC pellets respectively, considering multicomponent adsorption of 20% carbon dioxide of in nitrogen. All pellets demonstrated a temperature increase when an electric current was applied, except for the 30% AC–70% 13X pellets, concluding that these were not suitable for the ESA process because heat regeneration is used and that there is a percolation threshold limit. Binary breakthrough curves developed using 50% AC–50% 13X pellets with electrification have proven that the heating by the Joule effect resulted in significantly faster adsorbent regeneration.

Carbon dioxide capture technologies based on high-temperature solid looping cycles have also undergone great progress in the past decade, showing their potential to capture carbon dioxide at moderate cost and with low energy penalty. Calcium looping uses CaO-based materials as carbon dioxide sorbents. Limestone is a widely available, inexpensive and environmentally benign sorbent. However, this natural material typically shows fast deactivation after multiple carbonation/calcination cycles, because of solid sintering, giving, as a result, residual activities of <10% after

25–30 cycles. The development of high-performance synthetic sorbents is also a matter of great interest for calcium looping applications, especially in those operated with clean fuels (e.g., natural gas, synthesis gas). The production of these upgraded materials is significantly more expensive; therefore, a compromise between durability and cost is required. To reduce costs, several techniques are being developed, such as sol–gel synthesis, precipitation, impregnation, granulation, dry mixing and coating. Particles of well-distributed nanosized calcium carbonate coated with alumina result in stable morphological structures that highly improve the carbon dioxide uptake capacity of the material. Although calcium looping has been mainly developed so far for post-combustion applications, many investigations show using calcium oxide in steam methane reforming or gasification processes is an attractive option for the sustainable conversion of fossil fuels to high-purity hydrogen. The combination of carbon dioxide sorbent and catalyst in all-in-one materials should improve mass and heat transfer in these processes. The development of advanced particle models that accurately describe the kinetics of these combined materials should facilitate the scale-up of this technology once they have been implemented and validated in adequate reactor models. Another interesting possibility is combining calcium looping with thermal storage. The integration with concentrated solar plants allows the energy required for sorbent regeneration to be obtained from a renewable source, giving, as a result, a great reduction in fuel consumption associated with the carbon dioxide capture process.

Papalas et al. focused on the investigation of potential calcium-based, low-cost sorbent materials for post-combustion carbon dioxide capture via carbonate looping (Figure 4.2). Mineral limestone, carbide slag and white mud were evaluated for their carbon dioxide sorption activity, with the latter two being industrial wastes retrieved from chlor-alkali and paper pulp plants, respectively. The mineral magnesite was tested as a structural promoter to improve the cyclic stability of the sorbents. Different methods consisting of hydration, acetic acid treatment and a modified sol–gel auto combustion using glycerol as the combustion agent were applied to modify the structural and morphological characteristics of the mineral- and waste-derived calcium oxide and improve the performance. Preliminary thermogravimetric analysis (TGA) tests proved the superiority of limestone and carbide slag treated with glycerol auto combustion over white mud. The addition of magnesite was considered more beneficial via the acetic acid treatment, since the promoted

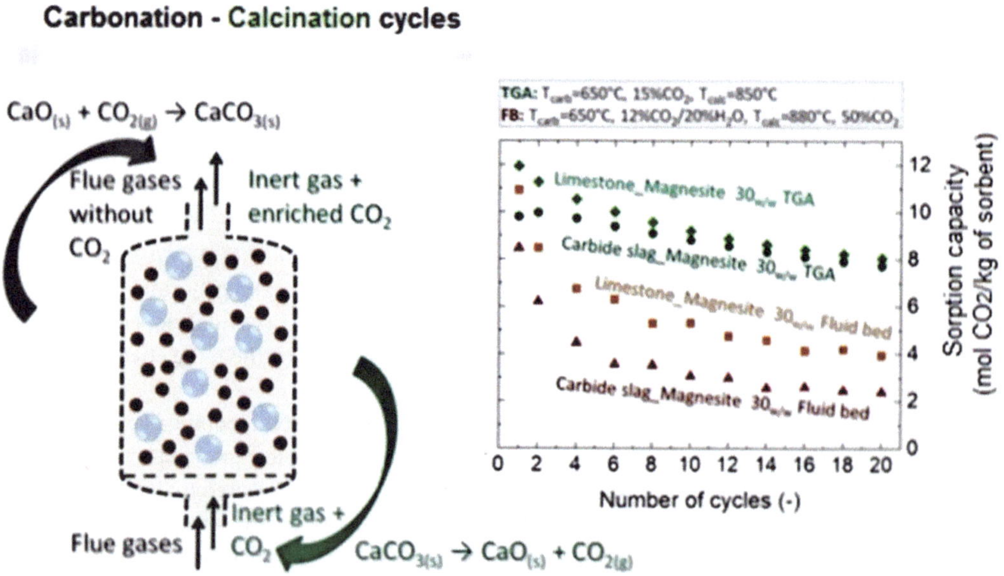

Figure 4.2 Carbon dioxide capture via carbonate looping.

limestone and carbide slag presented adequate stability under 50 sorption/desorption TGA cycles, with a 43.2% and 34.2% deactivation, respectively. The most promising CaO-based sorbents were further evaluated in a bench-scale fluidized-bed reactor, with the promoted materials better withstanding sintering, even under more severe calcination conditions (880°C, 50% carbon dioxide concentration). Magnesite-promoted limestone exhibited an initial carbon dioxide uptake of ~11 mol CO_2/kg of sorbent, which corresponds to ~90% calcium oxide conversion and <35% deactivation after 20 consecutive carbonation cycles, even under harsh calcination conditions. The superiority of the limestone promoted with magnesite was attributed to the enhancement of its pore network due to a reconstruction of its morphology in the presence of steam during the carbonation/calcination cycles.

Chemical looping combustion (CLC) is another promising high-temperature carbon dioxide capture technology, in which fuel is oxidized by a solid oxygen carrier that is reduced. This avoids the dilution of the combustion products with nitrogen from air, thereby generating virtually pure carbon dioxide. Common oxygen materials are transition metals, such as oxides of nickel, copper, manganate, or iron. Most investigations on CLC have been focused on arrangements of interconnected fluidized-bed reactors. Oxygen carriers must show not only high and stable reactivity to ensure complete fuel combustion during operation, but also high mechanical strength to minimize the loss of active material by attrition. The use of suitable supports, such as zirconia or MgO–kaolin, allows the activity of the oxygen carriers to be maintained in the long term, while reducing the particle attrition rate measured in laboratory-scale fluidized-bed reactors. CLC with solid fuels has recently gained great interest. The use of biomass is particularly interesting as the carbon dioxide that is captured can be considered as a negative emission. Detailed simulations of this process allow a deeper knowledge of the optimal operating conditions required, which is essential to facilitate the scale-up of this technology. Although the control of temperature and the replacement of spent materials are much easier in fluidized beds, the generation of power by means of CLC integrated in a combined cycle (i.e., at high pressure) is only feasible using fixed-bed reactors. A suitable reactor configuration, accompanied by appropriate heat management strategies, improves the energy efficiency of the process. Chemical looping technology can also be adapted to produce hydrogen or synthesis gas with inherent carbon dioxide capture. A concurrent moving-bed reactor loaded with an iron–titanium oxygen carrier is able to partially oxidize methane to synthesis gas containing H_2/CO ratios of ~2, while almost complete fuel conversion and negligible carbon deposition are achieved. Carbon monoxide-rich gas can be produced by means of reverse water–gas shift of mixtures of CO_2/H_2 using spinel-structured ternary ferrites. Pure hydrogen can be obtained by means of water splitting CLC of natural gas. An excess of heat can be stored in reduced oxygen carrier particles and their subsequent oxidation generates high-temperature heat suitable for iron ore sintering plants.

Alternative novel technologies for capture applications include membrane-based ones, where carbon dioxide separation is achieved by permeating one or more constituents of the gas stream through a selective membrane that acts as a barrier, while excluding other components of the stream. Membranes benefit from the lack of the regeneration step, a waste-free character and compact size. However, the molecular-level understanding of mass transport phenomena and the development of scalable methodologies for their manufacture are some of the main hurdles to overcome in order to advance next-generation membranes to industrial application. A review study on the current state-of-the-art for carbon dioxide absorption using membrane contactors is provided by Chuah et al. The paper focuses on membrane materials, liquid absorbents, process design and pilot-scale demonstration of membrane contactor processes. Critical challenges, such as membrane fouling and pore wetting, are also discussed. Membrane separation properties are strongly affected by the presence of gases other than carbon dioxide in the gas feed to the membrane system. Single-gas and mixed-gas permeation in DDR-type zeolite membranes (polymer composite material) is also reported. The effect of temperature and carbon dioxide concentration on permeance and selectivity was systematically investigated.

When the process of capturing carbon dioxide from a wide range of emission sources, such as power generation sources, industrial sources and even directly from the air, is coupled with new carbon dioxide conversion routes, carbon dioxide accumulation in the atmosphere could be reduced. If carbon dioxide capture is coupled with permanent storage solutions, even a shift to negative emissions could be possible. For instance, the direct carbon dioxide capture from air (DAC) combined with permanent storage is considered a promising negative emission technology. This technology requires vast contact surfaces with a great affinity and selectivity towards carbon dioxide. Absorption methods generally use a strong alkaline aqueous solvent solution, such as calcium hydroxide, sodium hydroxide, or potassium hydroxide, that reacts with carbon dioxide to yield carbonated salts. The use of membrane electrodialysis for the regeneration of the solvent and carbon dioxide recovery is shown to be less energy-intensive than other methods, but the high cost of these bipolar membranes hinders, for the moment, the scaling up of this technology. Potassium carbonate is also a chemical solvent that can be used in direct air capture. This material typically shows high stability and carbon dioxide absorption capacity, slow corrosiveness, little toxicity and low cost. The regeneration of the saturated solvent demands low temperatures and the carbon dioxide recovered is suitable for subsequent chemical synthesis applications, such as the production of methane by catalytic methanation with hydrogen.

For over a decade, Carbon Engineering (CE) has pioneered a liquid-sorbent-based DAC solution optimized for scale. Using dynamic process modelling software from AspenTech®, engineers at CE have been able to study a highly integrated process, leading to adjustment and modification of the system for improved operability and controllability of this technology on a commercial scale.

Climate change is known to be predominantly caused by the increased concentration of greenhouse gases in the atmosphere, in particular carbon dioxide. To prevent excessive accumulation of carbon dioxide in the atmosphere and the perturbation of natural carbon cycles, carbon capture and sequestration (CCS) is urgently needed. In this section, a brief overview is provided for both biotic and abiotic carbon dioxide sequestration pathways. Special focus is given to sequestration approaches pertaining to clathrate hydrates. Carbon dioxide hydrate, a solid compound made of molecular carbon dioxide enclathrated in crystalline lattices formed by water molecules, is an attractive option for long-term carbon dioxide sequestration due to it having a higher density than seawater, stability below moderate oceanic/permafrost depths and low susceptibility to fluid flow perturbation when formed in sediments. This review compiles and summarizes the research efforts made on carbon dioxide sequestration as hydrates. Various approaches of carbon dioxide sequestration via gas hydrates are discussed, including storage in seawater, sediments under the sea floor, permafrost regions, methane hydrate reservoirs via carbon dioxide–methane exchange and depleted gas fields. The technical feasibility and potential storage capacity of these approaches are analysed. Finally, the key scientific challenges and prospects are identified and highlighted. Issues related to economics, scale-up and relative attractiveness versus non-hydrate methods are touched upon but are not the focus of this section.

More specifically, on the topic of carbon monoxide capture and use, Halliday et al. identify a key challenge for carbon capture and storage (CCS) – the quest for net-negative emissions. They report that bioenergy with carbon capture and storage (BECCS) with molten sorbents offers a unique opportunity to realize net-negative emissions with minimal indirect emissions and low-cost separation of carbon dioxide from other gases. BECCS could remove 300–850 kg of carbon dioxide equivalents from the atmosphere per megawatt-hour of electrical output (kg/MWh_e), making it advantageous to other low carbon technologies and allowing the offset of emissions from hard-to-abate industries. In the search for efficient materials for carbon capture, Jiang et al. report the development of light-responsive metal–organic frameworks (LMOFs), which have tunable structures and performances. While conventional amines cannot achieve controllable adsorption separation, these LMOFs have tunable amine-based active sites, which enhance carbon dioxide capture and control adsorption and separation. Several advances also highlight the improvement of technologies for reduction of carbon dioxide based on electrocatalytic methods, including the use of Ru and Re catalysts (Madsen et al.) and the use of ionic liquids for this application (Cui et al.), as well as exploration of earth-abundant iron catalysts (Iffland et al.). Complementary to electrochemical methods, enzymatic (Jun et al.,) and photocatalytic (Jiang et al.) methods are also shown to have a place in the toolbox for conversion and utilization of carbon dioxide. The capture and conversion of carbon dioxide to bicarbonate is reported by Jun et al. using an environmentally simple system of stabilized carbonic anhydrases loaded onto electro spun polymer nanofibers.

4.4 Developing and Deploying CCUS Technology in the Oil and Gas Industry

Another pathway to decarbonization is through CCUS. This technology plays a key role in decarbonizing hydrogen produced by steam methane reforming (SMR) to produce blue hydrogen. Carbon capture is also key to capturing carbon dioxide in processes, such as natural gas, ethylene oxide, ethanol and hydrogen production, that produce streams of high carbon dioxide concentration. Additionally, CCUS is instrumental in reducing carbon dioxide emissions in the hard-to-decarbonize steel and concrete industries.

The oil and gas industry are leaders in developing and deploying CCUS technology.

In 2020, global anthropogenic (human-made) carbon dioxide emissions totalled 34 gigatons, with only 40 megatons (0.12%) being captured. Oil and gas operations captured 70% of the carbon dioxide emitted. The oil and gas industry has the capability to implement CCUS on a large scale. Depending on the success of implementing the other decarbonization pathways, the International Energy Agency estimates that CCUS may be needed to capture as much as 7.6 gigatons/yr of carbon dioxide to meet 2050 global net-zero goals.

In the US, the 45Q tax credit provides an incentive to install CCUS of \$35/t for carbon dioxide used for enhanced oil recovery (EOR) and at \$50/t for carbon dioxide that goes into storage. These incentives are not large enough for capturing carbon dioxide in diluted carbon dioxide streams. Depending on oil prices, drillers may pay \$20/t–\$30/t for carbon dioxide used in EOR. Carbon dioxide capture cost ranges from \$30/t–\$40/t for streams with high carbon dioxide concentrations and can cost \$60/t–\$90/t for carbon dioxide captured from fired heater flue gas, where carbon dioxide content is only 8%–10%. The cost to capture carbon dioxide in natural gas-fired power generation ranges from \$60/t–\$90/t. Adding carbon capture and storage (CCS) to a combined-cycle power plant derates net power output by 12%–15%. Derating is the result of the heat required to regenerate carbon dioxide absorber solvent (2 Gj/t carbon dioxide) and power for carbon dioxide compression (200 kWh electricity/t carbon dioxide). Note that 2 Gj/t carbon dioxide is roughly equivalent to 1 t steam/1 t carbon dioxide. There are three primary methods of carbon dioxide capture from combustion:

- **Post-combustion:** this method includes burning fuel (methane is used in examples) and capturing carbon dioxide in a fired heater stack. Flue gas contains 8%–9% carbon dioxide, 70% nitrogen, 18% water and 2%–3% oxygen at atmospheric pressure. A gas turbine's exhaust without duct burner firing contains only 3%–5% carbon dioxide. Post combustion capture uses chemical solvents, such as hindered amines. Net carbon dioxide removed from flue gas capture is about 90% of the actual carbon dioxide captured, since absorber solvent regeneration and carbon dioxide compression have a carbon dioxide footprint that reduces the actual carbon dioxide removed from the atmosphere.

- **Oxy-firing system:** this system involves burning methane fuel with oxygen (supplied by an air separation unit) in a fired heater. Flue gas contains carbon dioxide and water but no nitrogen. No carbon dioxide capture unit is required. Compressing flue gas removes water to produce a pure carbon dioxide stream. Leaks in the convection section must be sealed to keep it from sucking in air, which will add nitrogen in the flue gas, leading to the captured carbon dioxide being outside specifications.
- **Pre-combustion:** if hydrogen is burned in a fired heater, the flue gas contains no carbon dioxide, therefore no carbon dioxide capture is required for the heater. The fuel is sent to a reformer, followed by a water gas shift reaction to convert the fuel into hydrogen and carbon dioxide. The carbon dioxide is then captured at the reformer outlet (higher percentage of carbon dioxide and higher pressure makes carbon dioxide easier to capture than from flue gas) by using a physical solvent in an absorber.

  The process, with the carbon dioxide removed, proceeds from the reformer's outlet to a pressure swing adsorption (PSA) unit that recovers pure hydrogen. The PSA tail gas goes to the fuel system. Reforming is an endothermic reaction requiring heat. In an SMR unit, 60% of carbon dioxide is generated by the reformer/water-shift reaction and 40% is from the reformer's firebox, which generates the heat required for the reaction. The carbon dioxide absorber on the reformer's outlet will recover approximately 60% of the produced carbon dioxide. To recover the remaining amount, a second carbon dioxide absorber on the firebox flue gas stream would need to be added to obtain a 90% carbon dioxide capture. An alternative is to fuel the firebox with hydrogen, which produces no carbon dioxide but reduces hydrogen production. Installing a single, large carbon dioxide absorber on the firebox flue gas may be more cost effective. An alternative to the reformer is to use an auto thermal reformer (ATR) to produce hydrogen. An ATR oxidizes (burns) part of the methane feed in the reactor to generate the required reaction heat, thereby eliminating the need for a firebox. The ATR uses oxygen from an air separation unit (ASU) as an oxidant, so there is no nitrogen in the outlet of the reactor. The remainder of the ATR outlet is hydrogen and carbon dioxide, which can be separated like a basic reformer, using a carbon dioxide absorber or a PSA unit. Using partial oxidation of methane with carbon capture is another route to produce hydrogen for fuelling a fired heater.

Carbon dioxide is captured and removed from process streams by either chemical absorption or physical separation. Chemical absorption involves a chemical reaction between the amine solvent and carbon dioxide and requires the regeneration of the solvent – 1 t steam/1 t carbon dioxide, which is equivalent to 2 Gj/t carbon dioxide. Physical separation can include either adsorption, absorption, cryogenic separation, or dehydration and compression. Physical adsorption uses solid surfaces like zeolites, metallic oxides, alumina or activated carbon. After the carbon dioxide is captured by the physical adsorbent, the carbon dioxide is then released by cycling the temperature in temperature swing adsorption (TSA), pressure in a PSA unit, or vacuum in vacuum swing adsorption. Physical absorption utilizes liquid solvents that are regenerated by flashing the solvent to a lower pressure in the regenerator, which does not require steam, thus resulting in less energy being used.

Significant research is being conducted to reduce carbon dioxide capture costs in CCUS. The US Department of Energy's Carbon Negative Shot initiative seeks to reduce the cost of carbon dioxide removed from the atmosphere to less than $100/t by 2030 via either direct air capture or by helping forests, agriculture and energy crops capture and store carbon dioxide. Additional ongoing CCUS research includes:

- absorption with chemical solvents (hindered amines) to lower costs by reducing the energy required in the solvent regenerator for aqueous solvent regeneration;
- absorption with physical solvents, which can be regenerated with vapour flashing and without requiring steam (resulting in lower energy usage);
- utilization of other non-aqueous or water lean solvents to reduce the energy required for regeneration;
- adsorption onto solids, hydrated solid sorbents and metals with thermal or vacuum swing adsorption (solid adsorbents can withstand higher gas temperatures than solvents and can avoid solvent degradation);
- direct air capture, as the low carbon dioxide concentration in air at 420 ppm (0.042%) vs. flue gas at 9% is a challenge (direct air capture requires processing 214 times larger volume of air to remove the same amount of carbon dioxide as flue gas capture, requiring large air contactors, like cooling tower units with air flowing up through packing where it contacts a solvent or solid adsorbent);
- cryogenic separation of carbon dioxide from hydrogen on the outlet of a steam methane reformer; and
- a process for growing algae to absorb carbon dioxide from air (photosynthesis) and then converting algae oil to biofuels.

Large-scale CCUS installations using chemical solvent absorber technology are proven to be effective with 1-MMtpy to 3-MMtpy carbon dioxide capture trains (e.g., Boundary Dam in Saskatchewan, Canada; Petra Nova in Houston, Texas US); QUEST in Alberta, Canada on a hydrogen steam methane reformer (1.1 MMtpy); and LaBarge, Wyoming (US) in a sour gas treating plant (6 MMtpy). Carbon dioxide must be compressed to 10 MPa for pipeline and underground storage. To prevent carbonic acid corrosion in carbon dioxide pipelines, there are tight specifications on oxygen at 50 ppm and on water at 20 ppm. Because of the high capital expenditures of carbon dioxide capture units, pipelines and carbon dioxide storage areas, a "hubs and clusters" concept is necessary to obtain sufficient carbon dioxide volumes for an economically efficient project.

There will be a need for regional carbon dioxide pipelines to collect carbon dioxide and then to either provide it as a feedstock or deliver it to storage. This activity will see competitor companies working together to jointly reduce their carbon dioxide emissions. Several CCS pipeline projects have been announced in the UK and near Rotterdam/North Sea. In the US, discussions are ongoing for a CCS network to be developed in the Houston Ship Channel to capture up to 100 MMtpy of carbon dioxide by 2040 and to store it offshore in the Gulf of Mexico. Additionally, Talos Energy has announced plans to develop carbon dioxide pipelines and storage along the Mississippi River corridor in Louisiana and on the Texas Gulf Coast.

Carbon dioxide storage can include underground saline reservoirs, depleted oil and gas reservoirs, porous rock, subsea or basalt. When carbon dioxide is stored, the operator must measure, monitor and verify the carbon dioxide. No carbon dioxide leakage into the atmosphere or into adjacent underground structures can occur. For any new grassroots facility, a study should be done to identify potential nearby carbon dioxide storage locations. Some uses for carbon dioxide include enhanced oil recovery (EOR) (which uses 0.3 t carbon dioxide/bbl.–0.6 t carbon dioxide/bbl.), along with processes for freeze-drying of food, carbonization of beverages, production of urea to make fertilizer and enhanced photosynthesis in greenhouses. Growing biofuel feedstocks (energy crops) are an effective way to remove carbon dioxide from the atmosphere and to produce low-carbon biofuels. Captured carbon dioxide can be reacted with green hydrogen made from renewable power to produce synthetic fuels with a low carbon dioxide footprint (i.e., Power-to-X or carbon dioxide e-chemistry).

Nature captures carbon dioxide from the atmosphere by photosynthesis, which converts carbon dioxide and water to oxygen and glycose by electron transfer. Research is ongoing to develop electrochemical cells that will replicate and accelerate this electron transfer by using electrolysis to produce synthetic fuels and chemicals. In an electrochemical cell, a catalyst embedded in the cathode will reduce carbon dioxide to carbon monoxide or other products on one side of the cell. On the other side, the anode oxidizes water from the electrolyte to produce oxygen. This is a promising path for using renewable power and captured carbon dioxide to produce chemicals and fuels.

4.5 Sustainable Steel/Chemicals Production: Capturing the Carbon in the Material Value Chain

4.5.1 Valorisation of Steel Mill Gases

Today, most of the global steel production is produced using blast furnace technology. In this process, iron ore is reduced with coke from coal in a blast furnace, followed by the removal of residual-free carbon in the steel by oxygen injection in an oxygen converter. In this process, three major gas streams are co-produced with the target product steel:

- coke oven gas (COG);
- blast furnace gas (BFG); and
- basic oxygen furnace gas (BOFG).

The coke oven gas primarily contains hydrogen and light hydrocarbons, whereas the blast furnace gas is carbon monoxide rich. The largest amount of gas produced in a steel mill is blast furnace gas, which has a 1:1 ratio of carbon monoxide and carbon dioxide in a matrix of ~50% nitrogen. This gas is very abundant in steel mills and is typically combusted in a power plant to produce electricity with a very high carbon dioxide footprint. To create a greener electricity grid, an alternate use for carbon monoxide is required. This is where industrial symbiosis could play an important role. Where companies traditionally optimize their processes and products within their own value chain, industrial symbiosis can take industry to the next level, as it enables the exploration of options to exchange water, waste, energy and material streams in a greater ecosystem by means of system integration and optimization. This not only has economic benefits, but also potentially a positive environmental impact and can therefore optimize the greater industrial system and contribute to energy and feedstock transitions.

Of all the steel mill gases, coke oven gas has the most significant amounts of impurities, such as ammonia, hydrogen cyanide, sulfur species (mainly hydrogen sulfide), naphthalene, light oil and tar. The light oil is mainly characterized by benzene (80%–85%) toluene and xylene (15%–20%). The tar is a highly complex blend that is not easy to process.

Tar and light oil are not the only contaminants that can impact utilization potential. Steel gases like blast furnace gas, basic oxygen furnace gas and coke oven gas also contain a significant number of contaminants, such as ammonia, hydrogen cyanide, hydrogen sulfide, carbonyl sulfide and carbon disulfide, as well as metals such as iron, mercury and arsenic and traces of oxygen. These can affect the operations directly or indirectly and a major objective of a Gentilucci et al. study was to assess that impact as a function of time.

The petrochemical industry has a long history of using carbon monoxide (along with hydrogen) in synthesis gas chemistry. In the past, limited access to oil reserves was a primary driver for the utilization of synthesis gas chemistry, which ultimately led to the development of Fischer-Tropsch chemistry as an alternative route to chemicals and fuels. In periods of high oil prices, synthesis gas chemistry has been explored for converting alternate carbon sources like coal, biomass, natural gas and waste plastics. Therefore, many companies have a long tradition in the development of synthesis gas conversion technologies.

Since 2015, Dow Chemical, ArcelorMittal and Tata Steel have been exploring the use of hydrogen-rich coke oven gas and carbon monoxide-rich blast furnace gas and/or basic oxygen furnace gas as members of an international consortium working on the Carbon2Value project, which explores innovative technology for reducing carbon dioxide emissions across the energy-intensive steel sector by 30–45%. The consortium's goal is to use this technology to separate carbon dioxide streams and to valorise carbon monoxide and potentially carbon dioxide in the future. Based on the knowledge that methanol can be produced commercially from coke oven gas and ethanol from basic oxygen furnace gas and/or blast furnace gas these consortium partners have anticipated that other chemical conversions should be possible.

In parallel, other consortia, such as the Carbon2Chem project in Germany, are also exploring these options. The goal of this project is to avoid carbon dioxide emissions by utilizing carbon monoxide in synthesis gas conversion chemistry, capturing the carbon in the material value chain. This is seen as a more interesting option than post-combustion carbon dioxide utilization, as carbon monoxide is more reactive and requires less energy to be converted.

To utilize steel mill gases in the production of chemicals and materials, several challenges have been identified. First and foremost, the composition of the gas does not meet typical specifications that can be achieved when methane is used to generate synthesis gas via reforming (either partial oxidation, autothermal reforming or steam methane reforming). Carbon monoxide in the blast furnace gas is present in diluted form (with carbon dioxide and nitrogen) and contains many impurities. The steel mill gases are saturated with water and contain dust (sulfided iron oxide) and a variety of metals and sulfur and nitrogen species.

Gentilucci et al. selected amine scrubbing as the technology to achieve the removal of carbon dioxide and specific impurities in a single process. While it is quite common to use the direct reduction of iron to capture carbon dioxide in steel production, to the best of the author's knowledge, amine scrubbing is only employed in one blast furnace gas steel mill and in a few pilot lines. Limited information about these pilot plants is available; however, challenges on the stability of the applied amine, along with details around the quality of the output gas for downstream processing, have not been reported.

For their study, Gentilucci et al. used a commercial amine product (Dow Industrial Solutions' UCARSOL™ AP814 solvent). This is a formulated amine for carbon dioxide and hydrogen sulfide removal that is expected to deliver better performance than the cheaper alternative monoethanolamine. A key area of focus was stability, thus making this an important subject of study.

Amine degradation products can be of particular concern when using commodities like primary and secondary amines, such as monoethanolamine and diethanolamine. Examples of associated, problematic degradants include bicine (bicine is a contaminant in amine systems used for gas sweetening. It is formed by amine degradation in the presence of oxygen, hydrogen sulfide or thiosulfate) and tris-hydroxyethyl ethylenediamine, which can result in corrosion of process equipment when not properly managed. Formulated proprietary amines substantially mitigate these risks (Dow Industrial Solutions' UCARSOL™ GT-10 antifoam; Dow Industrial Solutions' UCARSOL™ amines).

The gas from the steel mill first passes through a set of coalescing filters – two installed in parallel for each feed gas – for high-efficiency removal of sub-micron liquid droplets and solids. Mass flowmeters installed downstream of the filters were used to meter in the gases to the compressor suction.

A process sketch of the acid-gas scrubbing process used by Gentilucci et al. in the pilot plant is shown in Figure 4.3. The mixed gases were compressed in a single-acting, three-stage positive

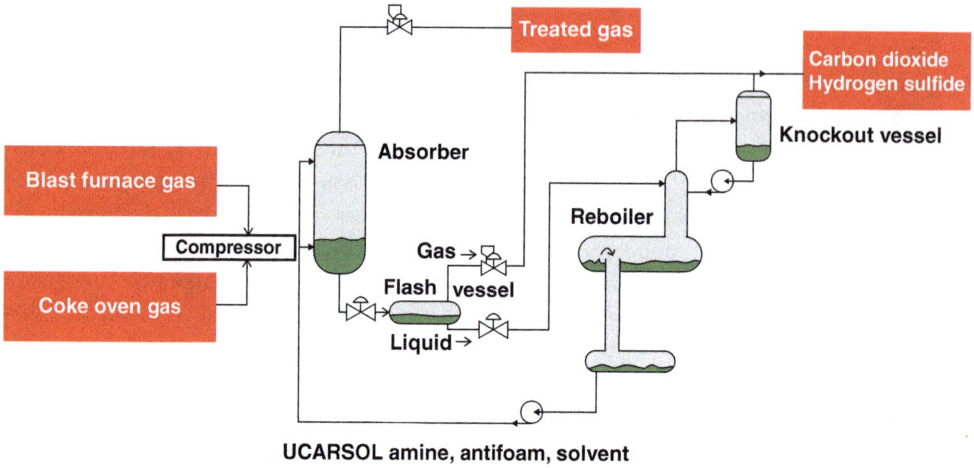

Figure 4.3 Schematic representation of the carbon dioxide removal process.

displacement compressor with interstage cooling and liquid condensate removal. The compressor can deliver gas at a maximum discharge pressure of 3.2 MPa. The compressed gases from the compressor were cooled with chilled ethylene glycol to 40°C in a plate-and-frame heat exchanger. A flowmeter measured the flow of cooled gas before it entered the absorber.

In the absorber, the steel mill gases were contacted with the proprietary solvent to remove the acid gases by chemisorption. The absorber – 2.8 m in height and 5 cm in diameter – had a single packed bed loaded with stainless-steel pall rings for enhanced gas-liquid contacting. The column was fitted with a demister pad above the packed bed for de-entrainment. Pressure drops over the packed bed were monitored using a pressure differential cell to identify potential foaming or fouling in the packed bed. The column was also fitted with a resonating fork for foam detection at the top of the column. This was specified by Dow Industrial Solutions based on successful commercial scale applications. The bottom level in the column was measured using a guided wave radar and controlled via an air-actuated level control valve. The treated gas from the top of the absorber was measured using a flowmeter. The inlet flow measurements, along with the inlet and outlet stream compositional analysis, enabled the determination of the acid gas removal efficiency in the absorber.

The treated gas left the absorption column overhead and was routed to downstream consumers at pressure. The solvent left the absorption column through the bottom nozzle. The solvent is fully saturated (i.e., rich in carbon dioxide) and is called "rich solvent."

The rich solvent left the bottom of the absorber and passed through a particulate filter before being flashed in the horizontal flash vessel. The flash gas was removed from the top of the de-entrainment section of the flash and the pressure was regulated to be 0.45 MPa–0.55 MPa. The rich solvent from the flash vessel was first routed through an activated carbon filter before entering the top of the stripping column. The activated carbon filter removed foam-causing impurities (e.g., dissolved hydrocarbons, degradation products and a variety of surface-active agents) that had entered or formed in the system. Heat tracing was installed on the transfer line from the flash vessel to preheat the solvent to above 90°C before entering the stripper column. Typically, on a commercial unit, preheating the solvent is done by a heat exchanger that recovers heat from the lean solvent after regeneration.

The rich solvent entered the top of the regenerator column and contacted stripped gases and water-amine vapours generated in the reboiler over a single packed bed loaded with stainless-steel pall rings for enhanced gas-liquid contacting. The power input to the electrical reboiler was controlled by maintaining the temperature of the solvent at 120°C. An overflow weir was installed in the vessel to maintain a constant liquid level to submerge the reboiler and ensure adequate liquid residence time in the reboiler. The overflow from the reboiler was routed into a surge vessel that was located directly below the vessel containing the reboiler.

The overhead vapours from the regenerator column were cooled in a plate-and-frame exchanger to 20°C with chilled ethylene glycol. The condensed liquid was recovered from the acid gases in a knockout vessel that had a demister for high-efficiency de-entrainment. The liquid recovered in the knockout vessel was returned to the top of the regenerator column, using a positive displacement pump.

The lean solvent exited the bottom of the surge vessel and was cooled in a plate heat exchanger to 40°C with chilled ethylene glycol. The cooled solvent was pumped to the absorber. The chilling loop system of a chilled water loop, filled with ethylene glycol (30 wt.%), was cooled to 10°C with a chiller unit. The total circulation rate of chilled water was 4 m^3/hr, with a total process flow requirement of 0.4 m^3/hr.

For this small-scale pilot, the heating duty was calculated to be 4.2 GJ/mT of carbon dioxide captured in an idealized configuration. However, the model predicts that this would translate to 2.3 GJ/mT of carbon dioxide captured in a commercial-scale installation. This is well in line with world-class performance for an amine-based carbon dioxide capture unit.

4.5.2 Summary and Outlook

A carbon monoxide/carbon dioxide separation pilot line was developed and operated with co-funding from the Interreg 2 Seas program. The unit was installed in the heart of the ArcelorMittal steel mill in Ghent, Belgium. This enabled the unit to be fed with real industrial waste gases (e.g., blast furnace gas and coke oven gas) directly from the major gas lines in the facility.

In the 2019–2021, the unit was mainly operated on blast furnace gas. It was shown that the inlet filters were effective in removing (acid) condensate and dust from the feed gas. Operation on coke furnace gas demonstrated that the current filter design did effectively remove the tars present in this gas. Consequently, severe process upsets were observed in the run with coke furnace gas, mostly due to foaming and fouling caused by the presence of aromatic tars in the amine solvent. Additional measures to clean coke furnace gas are needed to utilize this gas.

Several thousand hours of operation with blast furnace gas demonstrated that the applied proprietary solvent can capture carbon dioxide from blast furnace gas at greater than 90% efficiency with minimal degradation. More than 95% efficiency was achieved under optimized conditions. In addition, hydrogen sulfide and carbonyl sulfide in the gas were also captured, resulting in a treated gas with hydrogen sulfide and carbonyl sulfide in a low ppm range (i.e., 1 ppm–5 ppm). This enabled the use of the carbon dioxide lean gas for carbon monoxide conversion chemistry, which is typically less tolerant of impurities and requires sub-ppm levels. Although deep removal for chemical conversion is still required, the bulk of the sulfur impurities can be effectively removed using an amine system. A trial with LanzaTech fermentation technology for ethanol production from blast furnace gas was conducted successfully. In a follow-up project, for which the carbon monoxide/carbon dioxide separation pilot line will serve as a feedstock pre-treatment, the project partners will explore the subsequent polishing and catalytic conversion of the carbon monoxide-rich gas.

It was also demonstrated that the relatively poor quality (dust, impurities) of the blast furnace gas had limited effect on the long-term quality of the proprietary solvent (Dow Industrial Solutions' UCARSOL™ AP814 solvent). Consequently, it was anticipated that an excessive make-up of the amine solvent would not be required, which had a positive impact on the operational costs. As expected, while still not critical for the process, the much poorer quality of the coke oven gas showed a higher degradation rate for the solvent. In combination with the observed fouling of the unit when running on this gas, it was decided that amine wash was not a good solution for treatment of coke oven gas and/or blast furnace gas + coke oven gas mixtures without any upstream pre-treatment for tar and oil removal to values of less than 100 ppm.

Operations with blast furnace gas also confirmed that amine carry over to the carbon dioxide was low and that the energy for solvent regeneration was in line with expectations and on par with typical numbers for this technology (i.e., 2.3 GJ/mT–4.2 GJ/mT carbon dioxide captured). This energy use could be significantly improved when heat integration with the exothermic carbon monoxide conversion process was implemented.

The quality of the captured carbon dioxide remains a challenge. As the amine solvent also captured the sulfur species, these will concentrate in the carbon dioxide. Consequently, hydrogen sulfide levels of 50 ppm–250 ppm were observed. To be able to send this carbon dioxide for storage and/or utilization, desulfurization of the carbon dioxide must be considered. A suitable technology for this could be Villadsen's electrochemical desulfurization.

Further Reading

Acuna-Yeomans, E., Calero, S. et al. (2022). Evaluation of ZIF-8 flexible force fields for structural and mechanical properties. *Microporous and Mesoporous Mater.* 348 (9): 112406.

Anyanwu, J.-T., Wang, Y. and Yang, R.T. (2020). Amine-grafted silica gels for CO_2 capture including direct air capture. *Ind. Eng. Chem. Res.* 59 (15): 7072–7079. doi: 10.1021/acs.iecr.9b05228.

Ausfelder et al. (2015). Energiespeicher als Element einer sicheren Energieversorgung. *Chem. Eng. Technol.* 87 (1–2): 17–89.

Bajohr, S., Graf, F. and Götz, M. (2013). Bewertung der Kopplung von PtG-Konzepten mit einer Biomassevergasung. gfw-Gas, 4/2013, 222–226.

Bazzanella, A.M. and Ausfelder, F. (2017). Technology study: low carbon energy and feedstock for the European chemical industry, Dechema Gesellschaft fur chemische Technik und biotechnology e.V., Cefic. 164.

BP Statistic Review of World Energy (2022). 71st edition.

BP Statistical Review of World Energy (2019). 68th edition.

BP Statistical Review of World Energy (2020). 69th edition.

BP Statistical Review of World Energy, (2021), 70th edition.

Buehler, J. (2022). Hydrocarbon processing, overview of decarbonization pathways for the oil and gas and petrochemical industries-Part 2.

Chuah, C.Y., Kim, K., Lee, J. et al. (2020). CO_2 absorption using membrane contactors: recent progress and future perspective. *Ind. Eng. Chem. Res.* 59 (15): 6773–6794. doi: 10.1021/acs.iecr.9b05439.

Cui, Y., He, B., Liu, X. and Sun, J. (2020). Ionic liquids-promoted electrocatalytic reduction of carbon dioxide. *Ind. Eng. Chem. Res.* 59 (46): 20235–20252.

EU ETS data (2015). http://www.eea.europa.eu/data-and-maps/data/data-viewers/emissions-trading-viewer.

Fernández, J.R., Garcia, S. and Sanz-Pérez, E.S. (2020). CO_2 capture and utilization editorial. *Ind. Eng. Chem. Res.* 59 (15): 6767–6772.

Gentilucci, A., Spagnuolo, L., Hughes, A. et al. (2022), Sustainable steel/chemicals production: removal of impurities for valorization of steel mill gases. *Hydrocarbon Processing*, October 2022.

Halliday, C. and Hatton, T.A. (2020). The potential of molten metal oxide sorbents for carbon capture at high temperature: Conceptual design. *Appl. Energy* 280: 116016.

Hasegawa, K. and Shire, T. (2019). Adapting and prospering in the energy transition. *Hydrocarbon Processing*, July 2019.

Iffland, L. et al. (2019). Solvent-controlled CO_2 reduction by a triphos–iron hydride complex. *Organometallics* 38 (2): 289–299.

Ince, A.C., Colpan, C.O., Hagen, A. and Serincan, M.F. (2021). Modeling and simulation of power-to-X systems: a review. *Fuel* 304: 121354.

IPCC (2018). Global warming of 1.5 °C. IPCC.

Jaraíz-Arroyo, I., Martín-Calvo, A., Gutiérrez-Sevillano, J.J. et al. (2020). OCEAN: an algorithm to predict the separation of biogas using Zeolites. *Ind. Eng. Chem. Res.* 59 (15): 7212–7223. doi: 10.1021/acs.iecr.9b06451.

Jiang, Y. et al. (2020). Controllable CO_2 capture in metal–organic frameworks: making targeted active sites respond to light. *Ind. Eng. Chem. Res.* 59 (50): 21894–21900. doi: 10.1021/acs.iecr.0c04126.

Jun, S.-H. et al. (2020). Stabilized and immobilized carbonic anhydrase on electrospun nanofibers for enzymatic CO_2 conversion and utilization in expedited microalgal growth. *Environ. Sci. Technol.* 54 (2): 1223–1231.

Kolle, J.M. and Sayari, A. (2019). Covalently immobilized polyethylenimine for CO_2 adsorption. *Ind. Eng. Chem. Res.* 59 (15): 6944–6950. doi: 10.1021/acs.iecr.9b04669.

Lawson, S., Snarzyk, M., Hanify, D. et al. (2020). Development of 3D-printed polymer-MOF monoliths for CO2 adsorption. *Ind. Eng. Chem. Res.* 59 (15): 7151–7160.

Madsen, N.R. et al. (2020). Evaluation of the electrocatalytic reduction of carbon dioxide using Rhenium and Ruthenium bipyridine catalysts bearing pendant amines in the secondary coordination sphere. *Organometallics* 39 (9): 1480–1490.

OPEC's *World Oil Outlook* (2018).

Papalas, T., Antzaras, A.N. and Lemonidou, A.A. (2020). Evaluation of calcium-based sorbents derived from natural ore and industrial wastes for high-temperature CO_2 capture. *Ind. Eng. Chem. Res.* 59 (21): 9926–9938.

Park, J., Rubiera Landa, H.O., Kawajiri, Y. et al. (2020). How well do approximate models of adsorption-based CO_2 capture processes predict results of detailed process models? *Ind. Eng. Chem. Res.* 59 (15): 7097–7108. doi: 10.1021/acs.iecr.9b05363.

Regufe, M.J., Ferreira, A.F.P., Loureiro, J.M. et al. (2020). Development of hybrid materials with activated carbon and zeolite 13X for CO_2 capture from flue gases by electric swing adsorption. *Ind. Eng. Chem. Res.* 59 (26): 12197–12211.

Yu, W., Wang, T., Park, A.-H.A. and Fang, M. (2020). Toward sustainable energy and materials: CO_2 capture using microencapsuled sorbent. *Ind. Eng. Chem. Res.* 59 (21): 9746–9759.

5

Integrated Refinery Petrochemical Complexes Including Power-to-X Technologies

CONTENTS

5.1 Introduction

Oil refinery-petrochemicals integration is not a new development. The depth and level of integration has varied over the years in different regions of the world depending on site specific drivers and economics. Integrated complexes typically converted 3 to 20 wt % of crude oil intake to petrochemicals and are now pushing this to about 50 wt % utilizing currently available conversion technologies. These hydrocarbon clusters typically start with a mix of light to heavy petroleum diets. The heavier oils and higher target yield of petrochemicals lead to a more complex oil refinery configuration. The petrochemical pillars typically include steam cracker, aromatics complex, C_1 technology and derived chemicals and products. The integrated complexes are driven by refinery conversion units (catalytic processes) as well as opportunities for improving overall economics by upgrading low value refinery streams to high value petrochemicals. Refinery processing objectives can be achieved through a combination of hydrogen addition technologies (like fixed bed, ebulated bed or slurry-based hydrocracking of vacuum gas oil and reside), carbon rejection technologies (like FCC/RFCC, delayed coking, flex fluid coking etc.) and carbon concentration technologies

Converting Power into Chemicals and Fuels: Power-to-X Technology for a Sustainable Future, First Edition. Martin Bajus.

(like deasphalting, visbreaking etc.). Steam cracker plants (Pillar A – see Chapter 1) use a thermal cracking process and are highly flexible and can process a variety of refinery streams starting from light off gases to vacuum gas oils. Aromatics complex (Pillar B – see Chapter 1) is a combination of processing units for converting naphtha and pyrolysis gasoline into benzene, toluene and xylenes. Integrated facilities provide feedstock synergy and flexibility to adapt to future market changes in product quality specifications and product demand shifts.

Oil refinery businesses are looking for product diversification (Pillar D – see Chapter 1) and vertical integration to secure market share, given the uncertainty about demand growth in fuels market. Steam cracking is a dominant technology for production of olefins (A pillar) and will remain so for the foreseeable future. Naphtha is the leading feedstock globally and chemical companies constantly look for advantageous feedstock for building and operating olefin complexes in a competitive and cyclical business environment. Natural gas shale related developments in North America resulted in abundant methane and ethane (Pillar C) feedstocks supply at very attractive pricing based on cheap natural gas. Natural gas price is expected to stay relatively stable at lower levels. This changed the global energy dynamics of natural gas pricing relative to crude. Chemical companies, operating in regions of the world where most of the feedstock is sourced from crude oil, are constantly looking at the options for staying competitive in the changing marketplace. With most of the new refinery capacity being added in those regions of the world, it is a logical option to build integrated refinery petrochemical complexes to stay competitive. The case studies, based on different levels of integration between refinery and petrochemical complex presented in this book, demonstrate improvement in overall economics. The diversification of product (pillar D) portfolio helps in increasing the production of higher value products and minimizing the processing costs and thus positioning integrated businesses to be competitive in a rapidly changing marketplace. Integrated complexes are highly capital intensive and complex to plan and implement. It requires a great degree of understanding of the different refining and petrochemical technologies to evaluate the configuration options, along with the approach towards integration, to meet the desired business objectives over the long-term life cycle of these facilities. The scale of these projects is enormous and some in the industry refer to these as "giga-scale". Most of these large-scale clusters are joint venture ownerships (multiple partners) because of large scale investments (tens of billions of dollars). The success of these joint ventures lies in understanding of all the aspects of these ventures (strengths of each partner, business drivers, synergy, understanding of technologies/operation and facility operating plan, clear and crisp contractual arrangements based on sound understanding of dynamics that impact the projects/plants over the course of their life) early and upfront.

There is a further push to increase the petrochemicals yield to 70–80% level (Saudi Aramco and Chevron Lummus Global signed a technology development agreement to commercialize the crude oil to chemicals technology).

Many chemical companies and technology suppliers have focused their effort on crude and condensate sources to directly feed these to the ethylene plants (pillar A). These developments are targeted to eliminate the needs of separation/processing facilities to reduce the overall investment and energy consumption. Alkene technologies are generally limited to lighter crudes of suitable quality for achieving optimum results. These lighter crudes normally have a heavy, non-volatile tail (typically referred as residue or reside). This residue material needs to be separated from the feed entering the high temperature areas of the steam cracking furnaces as this leads to coke build-up. Most the developments to handle these crudes relate to separating the residue from the feed prior to entering high temperature zones of the steam cracking furnaces. Some plants in Asia and Europe have successfully cracked lighter crudes and condensates. Most of the focus previously (in 1960s and 1970s) was related to the development of technologies that allowed full vaporization of heavier feeds (straight run or hydrotreated/hydrocracked gas oils, vacuum gas oils) at conditions that avoided the problems of heavy fouling in convection sections of the steam cracking furnaces. These technologies have been applied successfully in many industrial applications.

For heavier refinery streams and crude oil cracking, many processes were developed and tested from 1960 to the early 1980s. Most of these processes were based on fluidized bed cracking. Lurgi developed the sand cracker using sand as heat carrier. Ube used inorganic oxide as heat carrier. Kunii process used coke as a heat carrier. Gulf Chemical and Stone & Webster jointly developed a thermal regenerative cracking process using solid heat carriers in the fluid bed. The development of these processes stopped after crude oil prices collapsed in early 1980s and none of these processes achieved commercialization. Steam cracking, to produce basic building pillars – ethylene/propylene/butadiene/butenes/ (pillar A) and aromatics (pillar B), is the most versatile and critical technology that drives the integration approach and effective management of hydrocarbons.

5.2 Synergies Between Refining and Petrochemical Assets

The focus of the closer integration between refining and petrochemical industries is to promote and seize the synergies and existing opportunities between downstream sectors to generate value to the whole crude oil production chain. Table 5.1 presents the main characteristics of the refining and petrochemical industry and the synergies potential.

In recent years the petrochemical industry has been growing at a considerably higher rate than the transportation fuels market. The technological bases of the refining and petrochemical industries are similar, which lead to possibilities of synergies capable of reducing operational costs and adding value to derivatives produced in the refineries.

Figure 5.1 presents a block diagram that shows some integration possibilities between refining processes and the petrochemical industry.

Fuel gas (C_2) is an attractive raw material to the petrochemical industry, as well as materials considered residual to petrochemical industries (butanes, pyrolysis gasoline and heavy aromatics). These can be applied to refiners to produce high quality transportation fuels, which can help the refining industry meet the environmental and quality regulations for derivatives.

The integration potential and the synergy among the processes rely on the refining scheme adopted by the refinery and the consumer market, process units as fluid catalytic cracking (FCC) and catalytic reforming can be optimized to produce petrochemical intermediates. In the case of

Table 5.1 Refining and petrochemical industry characteristics.

Refining industry	Petrochemical industry
Large feedstock flexibility	Raw material from naphtha/NGL
High capacities	Higher operation margins
Self-sufficient in power/steam	High electricity consumption
High hydrogen consumption	High availability of hydrogen
Streams with low added value (unsaturated gases and C_2)	Streams with low value (heavy aromatics, pyrolysis gasoline, C_4s)
Strict regulations (benzene in gasoline, etc.)	Strict specifications (hard separation processes)
Transportation fuels demand that is declining at a global level	High demand products

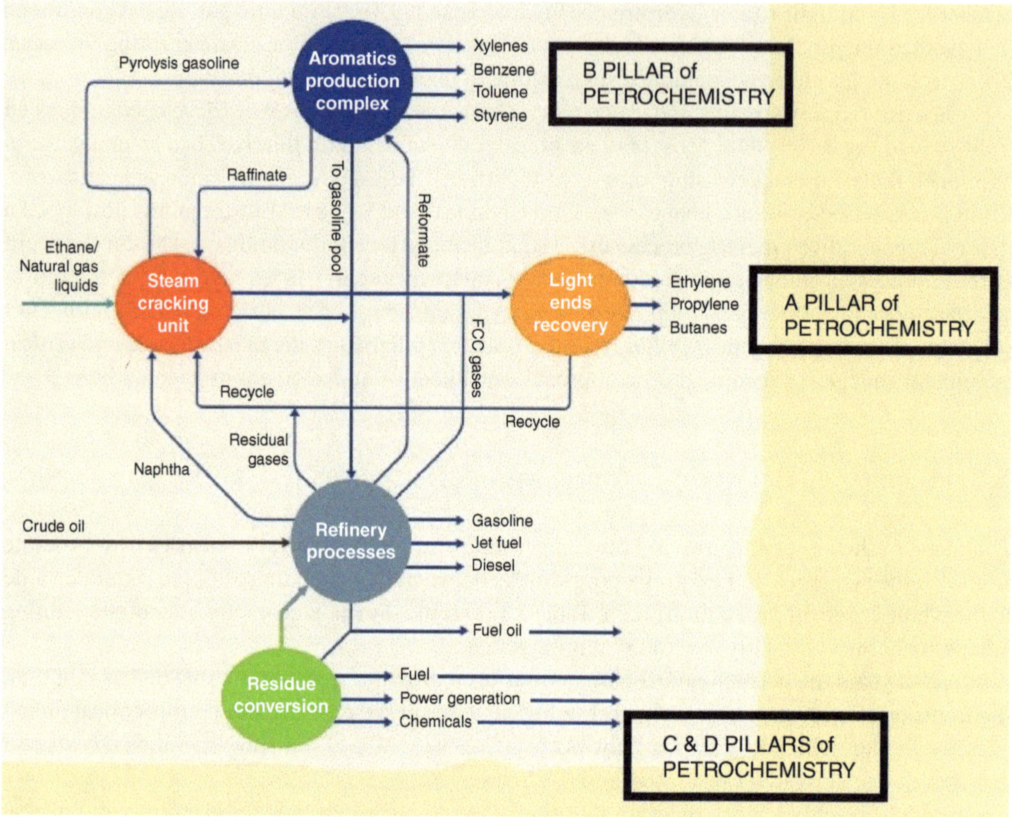

Figure 5.1 Synergies possible between refining and petrochemical processes. A full discussion of the pillars can be found in Bajus, M. (2020). *Petrochemistry: Petrochemical Processing, Hydrocarbon Technology and Green Engineering*. Chichester: Wiley.

FCC, installation of units dedicated to produce petrochemical intermediates, called petrochemical FCC, aims to reduce to the minimum the generation of streams to produce transportation fuels. However, the capital investment is high when the process requires the use of material with excellent metallurgical characteristics.

5.2.1 Reaching Maximum Added Value – Integrated Refining Schemes

Historically, the growth in the refining industry was sustained and focused on transportation fuels. This can explain the profile of the traditional refining schemes. Nowadays, the downstream industry is facing a trend of reduction in the demand for transportation fuels and a growing demand for petrochemicals. This fact is the main driving force to promote the change of focus in the downstream industry.

5.2.1.1 Fluid Catalytic Cracking Alternates

The growing market of petrochemicals has led some refiners to look for a closer integration between refining and petrochemicals assets, aiming to meet market demand, improve revenues and reduce operation costs. To reach this goal, the refiners are implementing integrated refining schemes as presented in Figure 5.2.

As presented in Figure 5.2, the integrated refining scheme relies on flexible refining technologies such as catalytic reforming and fluid catalytic cracking (FCC) that are capable of allowing the

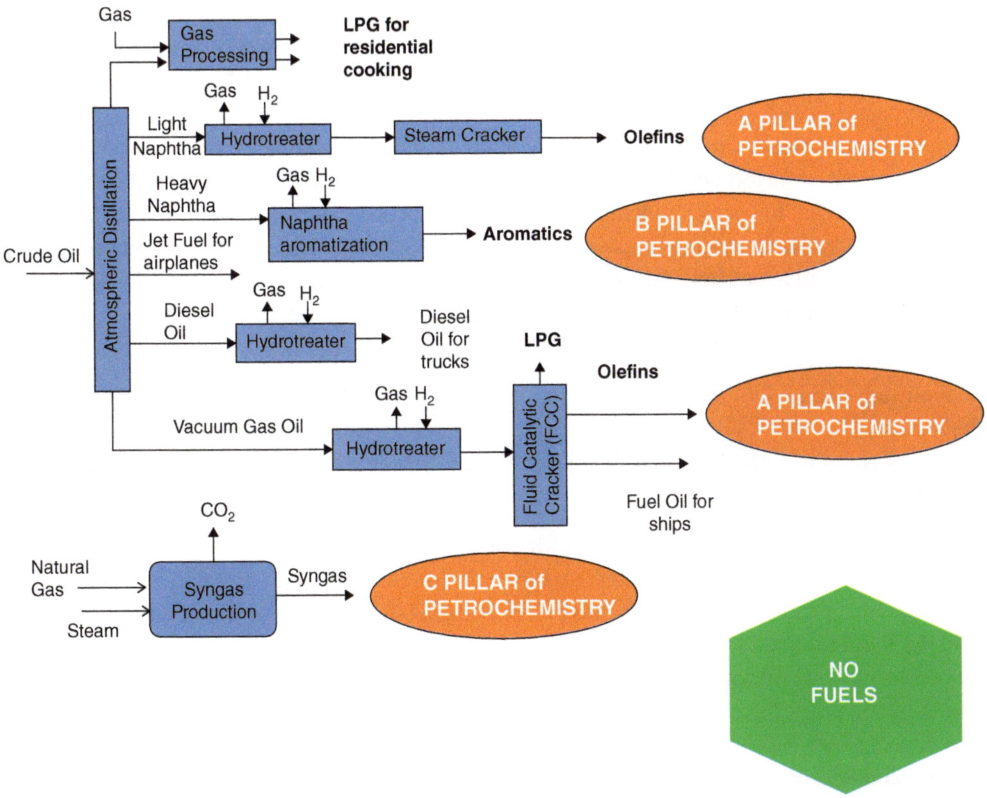

Figure 5.2 Zero gasoline refinery in electric vehicle era.

production of high-quality petrochemicals and transportation fuels, according to market demand. A more integrated refining configuration allows the maximization of petrochemicals, raising the refining margins and ensuring a higher value addition to the processed crude oils. Another fundamental competitive advantage is the operational flexibility reached through integrated refining configurations, allowing the processing of discounted and cheaper crude oils, raising even more the refining margins. The refining configuration presented in Figure 5.2 is considered to be a highly integrated refining system according to the classification proposed in Figure 5.1, one degree below the Crude-to-Chemicals refining asset.

The global demand for hydrocarbons – as petrochemical feedstock, as fuels for transport and for other uses – is expected to increase until at least 2040. These products have an unrivalled energy density and are easy to transport, making them an ideal means to carry and store energy. While alternatives are being developed for some of their current uses (e.g., in passenger cars, where electrification is expected to play a major role), hydrocarbons remain difficult to replace in heavy-duty and marine transport, in aviation and as a feedstock for petrochemical technology.

According to a Wood Mackenzie report in 2017, approximately 7% of global oil demand will be replaced by electric vehicles. Bloomberg Finance also predicted that 7-million barrels of crude oil per day will be replaced by electric vehicles, even though the real growth of electric vehicles depends on numerous factors such as the price of a battery, subsidy from governments and availability of rare-earth and lithium elements. It is also possible to consider a zero-gasoline refinery where a refinery is dedicated to producing olefins and aromatics for petrochemicals, as shown in Figure 5.2. Apart from green reasons, consumerism will also boost the rise of electric vehicles and many countries are opening doors for electric vehicles. However, this policy leaves some work to be done by governments and in chemical technology.

Different analyses predicted different gasoline displacements due to electric vehicle evolution. Total (a French oil company) predicted that 30% of vehicles in 2030 will be personal electric vehicles. Exxon Mobil (an American oil company) recently released a forecast, Energy Outlook 2040, where they said that 40% of light vehicles in 2040 will be electrical vehicles.

To respond to this trend, which was started by Exxon Mobil, national oil companies are adapting their product from crude oil and fuels to petrochemicals or oil to petrochemicals. In 2014, Exxon Mobil built a crude oil to petrochemicals refinery in Singapore to convert sweet crude oil to petrochemicals. In 2018, Saudi Aramco announced grand projects on crude oil to petrochemicals to convert sour crude oil to petrochemicals. Due to the abundant availability of crude oil, SABIC also announced its intent contributed in crude oil and fuels to petrochemicals.

Two impacts of electrical vehicles are (among others):

i) geopolitical energy will shift from Saudi Arabia (22% oil deposit) to Bolivia (70% lithium deposit);
ii) some refineries in Europe will reduce gasoline production and increase olefins and aromatic petrochemicals.

Refiners are continually challenged to maximize economic gain in the face of cost pressures and plant operating constraints. The fluidized catalytic cracking unit (FCCU) plays an integral role in the optimization of the facility. Fluid catalytic cracking catalyst is often one of the largest budget items in the refinery. Therefore, the balance between fluid catalytic cracking unit performance and catalyst costs has a significant impact on refinery profitability.

Fluid catalytic cracking is one of the most important conversion processes used in refineries. This process converts heavy petroleum fractions into lighter, higher-value products, such as gasoline, propylene and others. Fluid catalytic cracking units have the flexibility to process a wide range of feedstocks and at varying severities, to optimize refinery profitability. Maximization of propylene production from fluid catalytic cracking units has become a means for refiners to maintain their competitiveness. This article provides insight into an advanced propylene maximization catalytic cracking (PMCC) technology that integrates process, hardware and catalyst. The technology's propylene yield is higher than any other fluid catalytic cracking.

In a growing and competitive industry, ethylene producers are turning to catalyst suppliers for solutions to innovate. Until recently, a solution to the difficulties inherent in acetylene hydrogenation in ethylene plants has challenged the catalyst industry.

Due to their lower total production cost, "front-end" process flowsheets are now predominantly selected for new projects.

P-xylene production is a highly cyclical business that is subject to changes in the global economy and dependent on feedstock costs, alternative dispositions for xylene capacity added by recently started crude-to-chemicals complexes and the expected impact of the economic recession on commodities demand. Many aromatics complexes face a significant slowdown in the coming months or, possibly, years.

5.2.1.2 Hydrocracking Alternates

Another key refining technology to crude oil to chemicals refineries is the hydrocracking units. Despite the high performance, the fixed bed hydrocracking technologies can be not economically effective to treat crude oils directly due to the possibility of a short operating lifecycle. Technologies that use ebulliated bed reactors and continuum catalyst replacement allow longer cycles and higher conversion rates; among these technologies the most known are the H-Oil and Hyvahl™ technologies developed by Axens Company, the LC-Fining Process by Chevron-Lummus and the Hycon™ process by Shell Global Solutions. These reactors operate at temperatures above of 450°C and pressures up to 25 MPa. Catalysts applied in hydrocracking processes can be amorphous (alumina and silica-alumina) and crystalline (zeolites) and have bifunctional characteristics once the cracking reaction (in the acid sites) and hydrogenation (in the metals sites) occurs.

An improvement in ebulliated bed technologies is the development of slurry phase reactors, which can achieve conversions higher than 95%. In this case, the main available technologies are the HDH™ process (Hydrocracking-Distillation-Hydrotreatment), developed by PDVSA-Intevep, VEBA-Combicracking Process (VCC)™ commercialized by KBR Company, the EST™ process (Eni Slurry Technology) developed by Italian state oil company ENI and the Uniflex™ technology developed by Honeywell UOP.

In the slurry phase hydrocracking units, the catalysts is injected with the feedstock and activated in situ while the reactions are carried out in slurry phase reactors, minimizing the reactivation issue and ensuring higher conversions and operating lifecycle. Figure 5.3 presents a basic process flow diagram for the Uniflex™ slurry hydrocracking technology.

Other commercial technologies for slurry hydrocracking process are the LC-Slurry™ technology developed by Chevron Lummus Company and the Microcat-RC™ process by Exxon Mobil Company.

The steam cracking process has a fundamental role in the petrochemical industry and nowadays most light olefins, light ethylene and propylene are produced through the steam cracking route. Steam cracking consists of a thermal cracking process that can use gas or naphtha to produce olefins (see Bajus, *Petrochemistry*). Naphtha to steam cracking is composed of straight run naphtha from crude oil distillation units, normally to meet the requirements of petrochemical naphtha the stream needs to present high paraffin content (higher than 66 %).

Many great technology developers have dedicated their efforts to improve the steam cracking technologies over the years, especially in relation to steam cracking furnaces. Companies like Stone & Webster, Lummus, KBR, Linde and Technip develop technologies to steam cracking process. One of the most known steam cracking technologies is the SRT™ process (Short Residence Time), developed by Lummus, that applies a reduce residence time to minimize the coking process and ensure higher operational lifecycle.

The cracking reactions occur in the furnace tubes and the main concern and limitation to the operating lifecycle of steam cracking units is the coke formation in the furnace tubes. The reactions are carried out under high temperatures, from 800°C to 900°C, according to the characteristics of

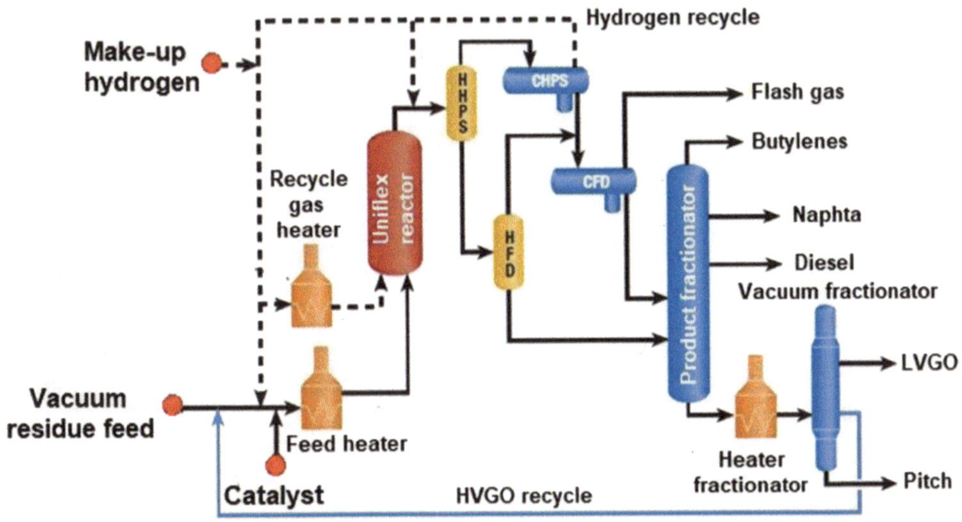

Figure 5.3 Process flow diagram for Uniflex™ slurry phase hydrocracking technology by Honeywell UOP (Honeywell UOP, 2019).

the feed (inlet temperature). For heavier feeds like gas oil, a lower temperature is applied aiming to minimize the coke formation. The combination of high temperatures and low residence time are the main characteristic of the steam cracking process.

Some technology developers are dedicating their efforts to develop commercial Crude-to-Chemicals refineries (Lummus Technology). Other great refining technology developers like UOP, Shell Global Solutions, ExxonMobil, Axens and others are developing Crude-to-Chemicals technologies, reinforcing that this is a trend in the downstream market. Figure 5.4 presents a highly integrated refining configuration, developed by Honeywell UOP, that is capable of converting crude oil to petrochemicals.

As presented in Figure 5.4, the production focus changes to adding of maximum value to the crude oil through the production of high added value petrochemical intermediates or chemicals for general purpose use leading to a minimum production of fuels.

5.2.2 Comparisons and Sensitivities to Product/Utility Pricing

This section introduces the various configurations and how they compare based on investment cost, gross and net margin and internal rate of return (IRR). We evaluate several configurations, highlighting salient differences between them (including economics) and assess their sensitivity to product and feed price fluctuations, cost of utilities, type of crude and a carbon dioxide tax. These options are evaluated industry-wide as the only way to ensure that schemes are compared on a consistent basis and that all interactions and synergies between units are considered.

While the scenarios and investment costs are based on new facilities, the methodology could equally be applied to revamp projects where options to expand refinery operations with petrochemical units are being considered, along with the revamp costs of existing assets.

The base case refinery scheme includes either a residue fluid catalytic cracking unit (RFCCU) or a hydrocracking unit (HCU). The feed to the RFCCU comes from an atmospheric residue desulfurization (ARDS) unit. The hydrocracker schemes include a delayed coking unit (DCU) or residue hydrocracking (RHCU) as a residue conversion unit. The complex includes motor fuel (Euro-5 gasoline and diesel) and jet fuel production. Where required, ethyl tertiary butyl ether (ETBE) is imported as a gasoline pool component. In some cases, fuel-oil components are produced.

The petrochemical facilities produce base petrochemicals such as polyethylene (PE), polypropylene (PP), butadiene, benzene and *para*xylene (*p*-X). The complex includes a hydrogen production unit (HPU) – based on steam reforming – to balance the refinery's hydrogen demand. The complex

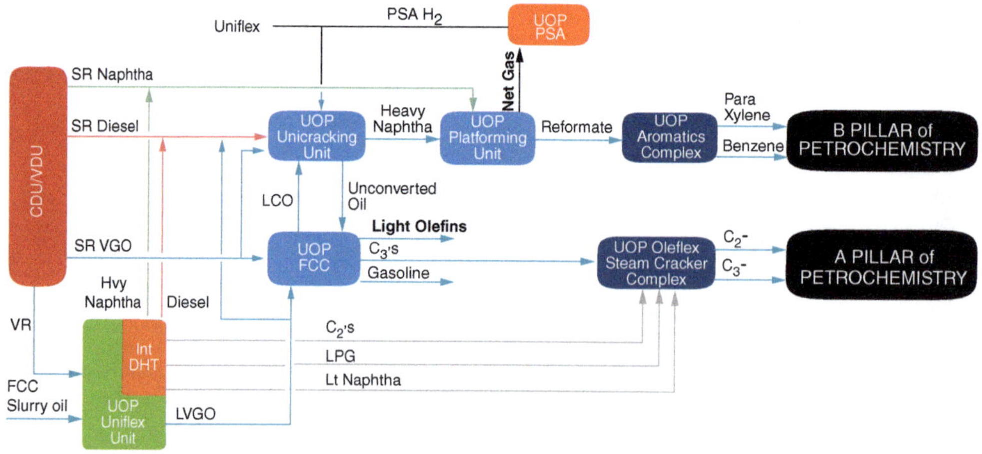

Figure 5.4 Integrated refining configuration based on crude to chemicals concept by Honeywell UOP.

processes 10 MMtpy of Urals crude, priced at $70/bbl. The complex will only import natural gas (for process and utility heater service, as well as for hydrogen production), electric power and raw water.

The economic evaluation depends on the prices of raw materials, products and utilities, as well as the unit prices of steel and the costs of construction, labour, etc. The economic analyses detailed in this article are more representative of the differences between certain schemes, rather than of being correct in the absolute sense.

As a reference case, the obtained internal rate of return results is reported in the first column of Table 5.2. The residue fluid catalytic cracking unit cases (cases A0–A2) typically yield better returns than the hydrocracking unit and naphtha hydrocracking unit cases (cases B0–B4) (see also Figure 5.5). The delayed coking unit and hydrocracking unit case (B0) is penalized by the higher import of ethyl tertiary butyl ether. Adding a steam cracker to the hydrocracking unit (cases B1–B2B) does not close the internal rate of return gap with the fluid catalytic cracking cases.

Table 5.2 Sensitivities: Internal rate of return variation with changes in petrochemical pricing and the costs of fuel gas and power.

Process	Case	Process/processes	$USD	$USD
RFCC	AO	RFCCU	12.4	12.2 (−2%)
	A1	RFCCU + PP unit	16.5	19.1 (+15%)
	A2	RFCCU + PP unit + aromatics complex	18.0	21.9 (+22%)
DCU	BO	DCU + HCU	11.0	8.6 (−22%)
	B1	DCU + HCU + steam cracker	12.4	16.0 (+29%)
	B2A	DCU + NHCU + steam cracker	11.2	16.5 (+48%)
	B2B	DCU + NHCU + steam cracker, gas oil to NHCU	8.3	14.5 (+75%)
	B3	DCU + NHCU + steam cracker + aromatics complex	14.1	20.7 (+47%)
	B4	DCU + NHCU + aromatics complex	15.4	19.9 (+29%)
RHCU	B5	RHCU + NHCU + aromatics complex	17.7	21.8 (+23%)

RFCCU residue fluid catalytic cracking unit, DCU delayed coking unit, HCU hydrocracking unit, NHCU naphtha hydrocracking unit, RHCU residue hydrocracking unit.

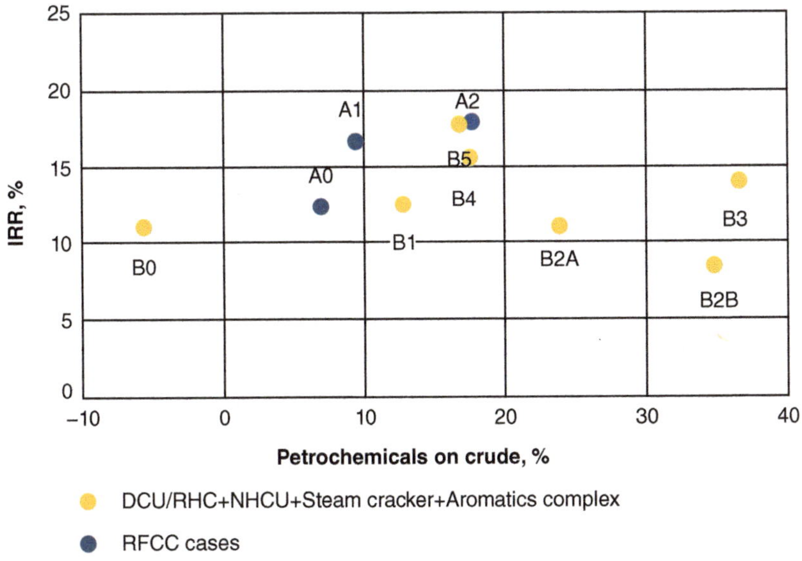

Figure 5.5 Internal rate of return vs. petrochemicals on crude.

Due to the higher capacity of the aromatics complex and the higher naphtha production from the naphtha hydrocracking unit, adding an aromatics unit is more attractive in an naphtha hydrocracking unit, (moving from case B2A to case B3 increases internal rate of return by 2.9%) versus an residue fluid catalytic cracking-based configuration (moving from case A1 to case A2 increases internal rate of return by 1.5%).

The combination of a residue fluid catalytic cracking unit, polypropylene and aromatics complex (case A2) has the highest internal rate of return, followed closely by the residue hydrocracking unit, naphtha hydrocracking unit and aromatics complex (case B5). The economics depend on the pricing scenarios under consideration. Higher petrochemical prices (an additional 20% relative to the other feed/product and utility prices) greatly improve the economics of the schemes having a high petrochemical output (Table 5.2). The internal rate of return is 4%–7% higher as petrochemicals output increases to 20%–40%. With an internal rate of return near 22%, the combination of a residue fluid catalytic cracking unit, polypropylene unit and aromatics complex (case A2) ranks at the top of the list.

If fuel gas costs are halved, the internal rate of return improves by 6% and 33% for all cases, with the high natural gas import cases (B0, B2A and B2B) benefiting the internal rate of return the most (Table 5.2). Relative to the reference case, if electricity prices are halved, internal rate of return improves by 5% and 20% (case B2B).

The effect of electricity and fuel gas prices on economics is most pronounced with the combinations having a high electricity and/or thermal demand. However, the ranking of the cases may change with different assumptions. Whereas case A2 ranks best in the reference case and still looks best in the high petrochemical price scenario, case B5 looks best in a scenario of lower fuel gas prices. In case of low electricity prices, case A2 ranks the highest.

All units have typical fuel gas, steam and electricity consumption levels. No effort has been made to increase electricity consumption by changing steam turbines to electric drives or by considering electric heaters.

5.2.3 Options for Further Increasing the Petrochemical Value Chain

Making specialty chemicals and/or expanding petrochemicals production will further increase petrochemicals output and/or profitability (pillar D – see Chapter 1). Some of the options to be considered include:

- converting butylene and recycling it within the steam cracker;
- sending pyrolysis gas to the aromatic's unit;
- sending hydrotreated pyrolysis gas/raffinate to the steam cracker;
- only producing gasoline blend components;
- incorporating a propane dehydrogenation unit to produce additional propylene; and
- adding metathesis to convert butylene with ethylene to propylene and then to polypropylene.

Forward integration with the following:

- ethylene processed to glycol and its derivatives and/or polyvinyl chloride (pillar D);
- propylene to propylene-oxide and/or its derivatives, which include acrylic acid and acrylonitrile (pillar D);
- propylene and butylene to oxo alcohols (pillar D);
- butylene processed to methyl ethyl ketone (pillar D); and
- butadiene to synthetic rubbers (pillar D).

5.3 Carbon Dioxide Emissions

The carbon dioxide emissions associated with the various configurations are reported in Table 5.3. Carbon dioxide emissions are the total emissions of the complex due to the combustion of natural gas, refinery fuel gas and fluid catalytic cracking coke and exclude the carbon dioxide associated with generating the required electrical power or any other carbon dioxide emissions associated with outside battery limit facilities (such as tank heating, process water demineralization, boiler feed water preparation, flare purging and tracing).

For the fluid catalytic cracking unit cases (A0–A2), the carbon dioxide emissions increase with increased complexity – e.g., increasing from 367 t/hr to 402 t/hr when going from the base case A0 to case A2, primarily due to the contribution of the aromatics complex (Figure 5.6). The residue fluid catalytic unit is the biggest contributor to emissions, representing 45% and 41% in Base Case A0 and Case A2, respectively.

Table 5.3 Specific carbon dioxide emissions for the various cofigurations.

Case	Description	Emission CO_2/t Crude
A0	Residue fluid catalytic cracking unit	0.31
A1	Residue fluid catalytic cracking unit + PP	0.32
A2	Residue fluid catalytic cracking unit + PP + aromatics complex	0.34
B0	Delayed coking unit + hydrocracking unit	0.19
B1	Delayed coking unit + hydrocracking unit + steam cracking	0.38
B2A	Delayed coking unit + naphtha hydrocracking unit[a] + steam cracking	0.56
B2B	Delayed coking unit + naphtha hydrocracking unit[b] + steam cracking	0.79
B3	Delayed coking unit + hydrocracking unit + steam cracking + aromatics complex	0.40
B4	Delayed coking unit + naphtha hydrocracking unit + aromatics complex	0.25
B5	Residue hydrocracking unit + naphtha hydrocracking unit + aromatics complex	0.27

[a]Naphtha hydrocracking unit aimed for maximum naphtha production
[b]Gasoil to the naphtha hydrocracking unit

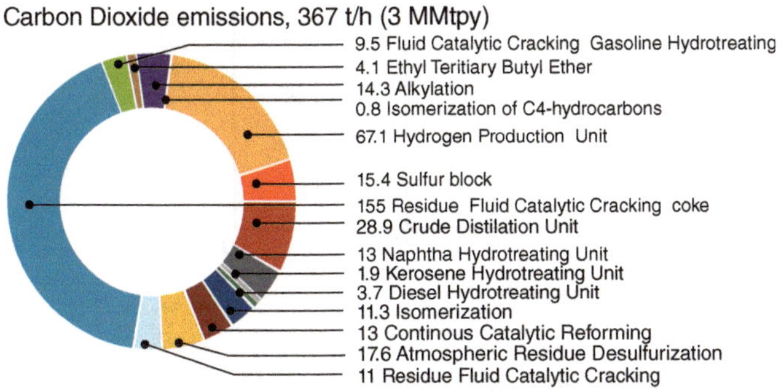

Figure 5.6 Carbon dioxide emissions for case A0: Atmospheric residue desulfurization + residue fluid catalytic cracking unit.

For the hydrocracking unit alternatives, the carbon dioxide emissions gradually increase with increased complexity. In the steam cracking cases (B2A and B2B), carbon dioxide emissions have increased 3–4 times relative to the base case.

In base case B0, the hydrogen production unit and crude distillation/vacuum distillation units are the biggest contributors, accounting for 30% and 24% of total carbon dioxide emissions, respectively. With the addition of the steam cracker and the aromatics complex and converting the hydrocracker to a naphtha-oriented hydrocracker, carbon dioxide emissions increased from 223 t/hr to 472 t/hr (case B3).

In case B3, the steam cracking and aromatics units have become the biggest carbon dioxide emissions contributors, representing 32% and 15%, respectively. In case B4, the aromatics complex is second to the crude distillation/vacuum distillation units in terms of total carbon dioxide emissions. In case B5, the top carbon dioxide emitters are the hydrogen production unit (18%), the crude distillation/vacuum distillation units (16%) and the aromatics complex (15%).

Carbon dioxide emissions increase with increasing petrochemical production (Figure 5.7). The steam cracker hydrocracking unit cases (B1, B2A and B2B) have the highest emissions. The aromatics hydrocracking unit cases (B3, B4 and B5) have the lower emissions at a given petrochemicals output. The fluid catalytic cracking cases (A0–A2) are intermediate.

5.3.1 Effect of a Carbon Dioxide Tax

A carbon tax is a tax levied on the carbon emissions required to produce goods and services. Carbon taxes are intended to make visible the "hidden" social costs of carbon emissions, which are otherwise felt only in indirect ways like more severe weather events. In this way, they are designed to reduce carbon dioxide by increasing prices. This both decreases demand for such goods and services and incentivizes efforts to make them less carbon-intensive. In its simplest form, a carbon tax covers only carbon dioxide emissions; however, they can also cover other greenhouse gases, such as methane or nitrous oxide, by calculating their global warming potential relative to carbon dioxide.

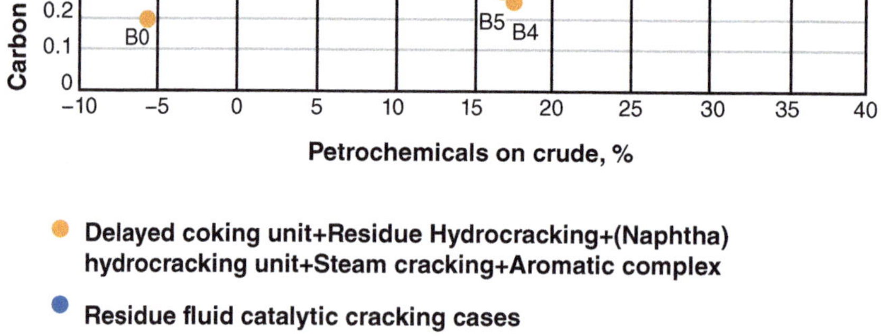

Figure 5.7 Carbon dioxide emissions vs. petrochemical output.

The effect of a carbon dioxide tax on internal rate of return are shown in Figure 5.8. Internal rate of return deteriorates with an increasing carbon tax, but even more so for the high-carbon dioxide cases – e.g., case B2B: delayed coking unit + naphtha hydrocracking unit2 + steam cracking, with emissions of 0.79 t CO_2/t crude oil (Table 5.4). For Case B2B, internal rate of return even becomes

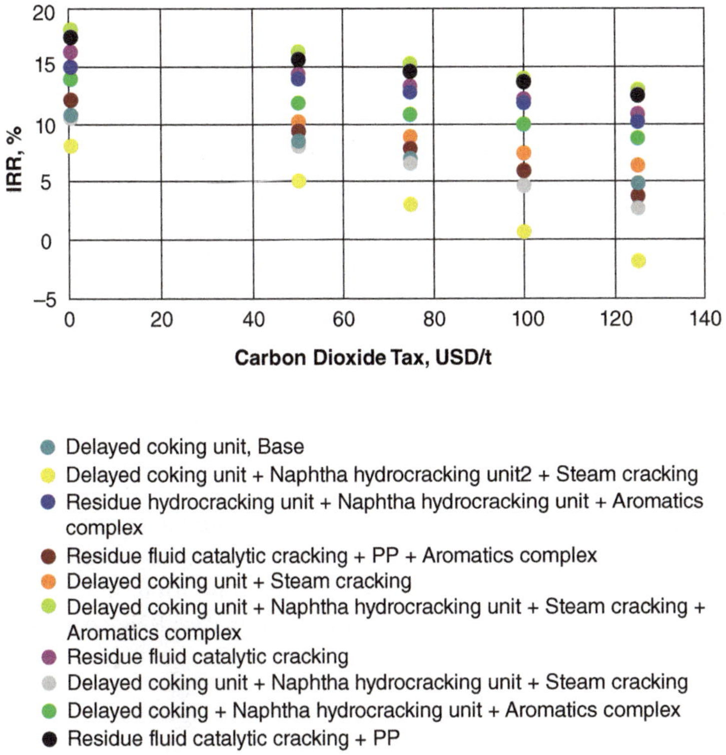

Figure 5.8 Impact of a CO_2 tax on IRR.

Table 5.4 Carbon dioxide emissions for the various configurations with and without carbon dioxide capture.

	Carbon Dioxide Emissions, t/t Crude Oil
Case A2: Residue fluid catalytic cracking + PP + aromatics complex/no capture/	0.34
Pre-combustion capture	0.30
Post-combustion captures	0.11
Case B2B: Delayed coking unit + naphtha hydrocracking unit + steam cracking/ no capture	0.79
Pre-combustion capture	0.68
Post-combustion captures	0.25
Case B5: Residue hydrocracking + naphtha hydrocracking unit + aromatics complex /no capture/	0.27
Pre-combustion capture	0.24
Post-combustion captures	0.09

negative when the carbon dioxide tax reaches \$125/t. For lower-carbon dioxide-intensive configurations (e.g., case B5: residue hydrocracking + naphtha hydrocracking unit + aromatics complex, with carbon dioxide emissions of 0.27 t/t crude oil, or case A2: residue fluid catalytic cracking + PP + aromatics complex, with carbon dioxide emissions of 0.34 t/t crude oil), the internal rate return drops only 5% when the carbon dioxide tax increases from zero to \$125/t carbon dioxide.

The impact of a carbon dioxide tax can be mitigated by reducing carbon dioxide emissions at the source or by opting for carbon capture where carbon dioxide can be reused (e.g., enhanced oil recovery) or stored as part of a carbon dioxide sequestration project. Intrinsic carbon dioxide emissions can be reduced by implementing energy conservation methods, incorporating system electrification (such as replacing steam drivers with electric drivers and using electric heaters instead of gas-fired heaters) and replacing natural gas for bio-based materials or co-processing bio-based materials in the process units.

Two cases of carbon capture are considered: 1) pre-combustion carbon dioxide capture from the process gas in a hydrogen plant (also known as blue hydrogen, which will reduce hydrogen plant carbon dioxide emissions by 50%) and 2) post-combustion carbon dioxide captured from the combined complex flue gases, which reduces total carbon dioxide emissions by approximately 70% in this example (theoretically, higher removal rates are possible). In all cases, carbon dioxide is made available at the complex battery limit at a pressure of 3.5 MPa for further transportation.

Adding carbon dioxide capture comes with increased investment and operational costs, both of which negatively impact economics. Relative to the reference case, the internal rate of return drops from 0.4% (in case A2, the internal rate of return decreases 18% to 17.6% and, in case B5, the internal rate of return decreases from 17.7% to 17.3%) and 0.5% (in case B2B, the internal rate of return decreases from 8.3% to 7.8%), with pre-combustion capture and no carbon dioxide tax enforced (Table 5.5). Post-combustion carbon dioxide capture reduces internal rate of return from 3.5% (in case B5, internal rate of return decreases from 17.7% to 14.2%) and 4.6% (in case B2B, decreases from 8.3% to 3.7%).

Table 5.5 Impact of a carbon dioxide tax on internal rate of return, with and without carbon dioxide capture options-breakeven points are shown for pre-combustion capture (blue) and post-combustion capture (green).

	Internal Rate Return, %				
Carbon dioxide tax, \$/t	0	50	75	100	125
Case A2: Residue fluid catalytic cracking + PP + aromatics complex/no capture/	18.0	16.0	14.9	13.8	12.6
Pre-combustion capture	17.6	15.8	14.9	13.9	12.8
Post-combustion captures	14.4	13.7	13.4	13.0	12.7
Case B2B: Delayed coking unit + naphtha hydrocarbon unit + steam cracking/no capture/	8.3	4.9	2.9	0.7	-21.0
Pre-combustion capture	7.8	4.8	3.1	1.2	1.0
Post-combustion captures	3.7	2.4	1.8	0.1	0.3
Case B5: Residue hydrocracking + naphtha hydrocracking unit + aromatics complex/no capture/	12.7	15.8	14.8	13.7	12.6
Pre-combustion capture	17.3	15.6	14.7	13.8	12.9
Post- combustion capture	14.2	13.5	13.2	12.9	12.6

Internal rate of return decreases as the carbon dioxide tax increases, but less so when carbon dioxide capture is part of the complex. At some point, the internal rate of return with carbon dioxide capture exceeds the internal rate of return without carbon dioxide capture, despite the higher investment and operational costs. For pre-combustion carbon dioxide capture, this happens when the carbon dioxide tax exceeds \$75/t. For post-combustion carbon dioxide capture, this occurs when the carbon dioxide tax approaches \$100/t for a high-carbon dioxide case (e.g., case B2B). For lower-carbon dioxide cases (e.g., cases A2 and B5), the carbon dioxide tax would need to be more than \$125/t.

5.3.2 Crude Oil Effects

The type of crude also influences economics. Both Arab Light and Caspian Pipeline Consortium (CPC) crudes have a higher price. This hurts the simple configurations (e.g., cases A0 and B0), with internal rate of return decreasing 5%–8% (Table 5.6), but much less than in more complex configurations where the internal rate of return decreases only 1%–3%, as they have a higher intrinsic gross margin.

With Caspian Pipeline Consortium crude oil, the residue fluid catalytic cracking case A0 looks poor as the gasoline pool becomes unbalanced. The smaller fluid catalytic cracking unit (and ETBE and alkylation units), relative to the Urals or Arab Light case, now requires the purchase of a significant amount of ETBE as a gasoline pool blending component. The base hydrocracking unit (Case B0) also needs a high ETBE import. This disadvantage disappears when the reformate is sent

Table 5.6 Impact of crude oil type on internal rate of return. The configuration with the highest internal rate of return is in green, followed by the second highest in blue and the lowest in orange.

	Case		Urals	Arab Light	CPC
RFCC	A0	Residue fluid catalytic cracking	12.4	8.9	3.9
	A1	Residue fluid catalytic cracking + PP	16.5	13.5	8.4
	A2	Residue fluid catalytic cracking + PP + aromatics complex	18.0	15.8	15.8
DCU	B0	Delayed coking unit + hydrocracking unit	11.0	5.3	5.8
	B1	Delayed coking unit + hydrocracking unit + steam cracking	12.4	9.6	11.5
	B2A	Delayed coking unit + naphtha hydrocracking unit + steam cracking	11.2	9.0	10.9
	B2B	Delayed coking unit + naphtha hydrocracking unit + steam cracking. Go to naphtha hydrocracking unit	8.3	6.5	7.8
	B3	Delayed coking unit + naphtha hydrocracking unit + steam cracking + aromatics complex	14.1	12.0	14.2
	B4	Delayed coking unit + naphtha hydrocracking unit + aromatics complex	15.4	12.4	16.3
	B5	Residue hydrocracking + naphtha hydrocracking unit + aromatic complex	17.7	15.3	17.3

to an aromatics complex (case A2) or when a steam cracker and/or aromatics complex is included (cases B1–B5). The aromatic cases (A2, B3, B4 and B5) look particularly good when using Caspian Pipeline Consortium crude oil.

As shown, the ranking of technologies changes with crude. In the Caspian Pipeline Consortium case, an advanced delayed coking unit/residue hydrocracking unit scheme (case B4 or B5) is more attractive than an residue fluid catalytic cracking-based scheme (case A2).

5.4 Summary

Nowdays, is still difficult to imagine the global energetic matrix free of fossil transportation fuels, especially in developing economies. Despite this fact, recent forecasts and growing demand by petrochemicals as well as the pressure to minimize the environmental impact produced by fossil fuels creates a positive scenario and acts as main driving force for closer integration between refining and petrochemical assets; in an extreme scenario the zero fuels refineries tend to grow in the middle term, especially in developed economies.

The synergy between refining and petrochemical processes raises the availability of raw material to petrochemical plants and makes the supply of energy to these processes more reliable and at the same time ensures better refining margin to refiners due to the high added value of petrochemical intermediates when compared with transportation fuels. The development of crude to chemicals technologies reinforces the necessity of closer integration of refining and petrochemical assets by the brownfield refineries aiming to face the new market that tends to be focused on petrochemicals against transportation fuels and it's important to note the competitive advantage of the refiners from Middle East that have easy access to light crude oils, which can be easily applied in crude to chemicals refineries. As presented above, crude oil to chemicals refineries is based on deep conversion processes that require high capital spending. This fact can put refiners under pressure with restricted access to capital, again reinforcing the necessity to look for close integration with petrochemical sector aiming to achieve competitiveness (Chapter 1).

At the extreme end of the petrochemical integration trend, there are the zero fuels refineries, as mentioned above and it is still difficult to imagine the downstream market without transportation fuels, but it seems a serious trend and the players of the downstream sector need to consider the focus change in their strategic plans like opportunity and threat, mainly considering the pressure over the transportation fuels due to the decarbonization necessity and new technologies.

Any new refinery complex should integrate petrochemicals manufacturing to produce both transportation fuels and base petrochemical materials. Existing refineries will need to review their options to expand into petrochemical production or integrate with neighbouring petrochemical plants.

Part of this chapter explored several possible grassroots configurations, including a residue fluid catalytic cracking unit, delayed coking unit or residue hydrocracking unit – the latter two combined with a vacuum gasoil hydrocracker. The key petrochemical units included a steam cracker with downstream polyethylene and polypropylene units, as well as an aromatics complex.

While it is theoretically possible to produce large amounts of petrochemicals, it may not always result in better economics, as shown within the confines of this study and regarding market circumstances. A delayed coking unit – combined with a naphtha-oriented hydrocracker that processes middle distillates with the bulk of the gases, LPG and naphtha sent to a steam cracker with downstream polyolefin units – can produce nearly 40% of petrochemicals from crude for a refinery processing Urals crude. However, in this study, a propylene residue fluid catalytic cracking unit-based refinery with an aromatics complex yielding 23% of petrochemicals from crude has better economics.

Fluid catalytic cracking-based schemes typically have better economics than delayed coking unit/residue hydrocracking unit/hydrocracking unit-based schemes. However, this depends on many factors. Other than feed/product and utility pricing, it also depends on the design under consideration. For example, for a light crude, delayed coking unit/residue hydrocracking unit-based configurations may be more attractive.

This comparison is based on commercially available technologies and does not consider technology developments, which continue unabated. In general, the more integrated schemes with a high petrochemical output are more robust to changes in feed/product and utility pricing. Higher petrochemical production is possible by ceasing Euro-5 gasoline production (allowing more naphtha to be sent to the aromatics complex or used as feed for the steam cracker), or by using pyrogas as aromatic feedstock, routing hydrotreated pyrolysis gas and/or raffinate to the steam cracker, changing unit operations (higher residue fluid catalytic cracking unit severity to produce more propylene) or adding other units (e.g., metathesis). These projects can be implemented stepwise, with petrochemical units being built in a second phase. Processing bio-based materials and/or streams from plastics recycling plants presents another opportunity to improve economics and sustainability.

The conclusions drawn in this analysis are based on a particular crude diet, feed and product pricing, as well as on operating and investment costs and could change depending on local circumstances. Feedstock flexibility, along with robustness for feed/product pricing changes, combined with a proper assessment of risks and opportunities associated with each investment, should be part of a proper evaluation.

Any future new refinery complex will likely integrate petrochemicals manufacturing. Existing refineries will need to look at options to expand into petrochemical units or to integrate with neighbouring petrochemical plants. In general, the more integrated schemes – with a high petrochemical output – are more robust to changes in feed/product and utility pricing, as well as to the selection of crude and carbon dioxide tax. Carbon dioxide emissions will increase as petrochemical output increases – the addition of a steam cracker has a large effect on carbon dioxide emissions. A carbon dioxide tax will negatively affect the economics of any refinery/petrochemical complex.

Baars et al. have illustrated how it paid to invest in pre-combustion carbon dioxide capture at the hydrogen plant when carbon dioxide taxes increased to $75/t. Post-combustion carbon capture did not look attractive until the carbon dioxide tax reached $125/t.

These projects can be implemented in phases, with petrochemical units being built in a second phase commensurate with an anticipated reduction in motor fuel demand. Processing bio-based materials and/or streams from plastic recycle plants presents another opportunity to improve economics and sustainability.

The conclusions drawn in this analysis are based on a particular crude diet, feed and product pricing, as well as on operational and investment costs and may change depending on local circumstances. Feedstock flexibility and a robustness for feed/product pricing changes, coupled with a proper assessment of risks and opportunities associated with each investment, should be part of a proper evaluation.

5.5 Power- to-X Technology

Carbon dioxide capture technologies from different carbon dioxide sources (e.g., flue gas, biogas and air) have been discussed in detail in Chapter 4. There are four main capture processes: pre-combustion, post-combustion, oxyfuel combustion process and direct air capture process. After capture, a separation technique (e.g., absorption, adsorption, membrane, or cryogenic) is used to obtain purer carbon dioxide. One of the most important challenges to capture carbon dioxide is the high energy requirement. The membrane process requires the lowest energy among the other processes. However, fouling and lower carbon dioxide purity are the main drawbacks to use the membrane process in the Power-to-X Technologies.

Hydrogen has an essential role in the X production system through the water and the carbon dioxide electrolysers followed by a catalytic reactor (H_2-CS) pathway. Various pathways of the Power-to-X system are illustrated in Figure 5.9. Pure hydrogen can be produced with the use of an electrolyser, which is driven by electricity produced through renewable energy technology. The water electrolyser technology includes three main types that are distinguished with the choice of ion transfer medium: alkaline electrolyser, proton exchange membrane (PEM) electrolyser and solid oxide electrolyser (SOE). The alkaline electrolyser is the most mature technology. This technology is considered the cheapest option for water electrolysis for large plants under the base load as inexpensive materials are used in its manufacturing. PEM electrolyser can respond to the demand quickly. SOE, which operates at high temperatures, can achieve high efficiency and has fast kinetics. In addition, SOE can be operated in a co-electrolysis mode in which water and carbon dioxide reduce simultaneously to carbon monoxide and hydrogen (Co-SOE). Co-SOE is electrochemical device, in which the electrochemical reduction of carbon dioxide and water occur simultaneously in the same unit. However, there are still important challenges with this technology in terms of durability and degradation. Moreover, unlike the PEM electrolyser, coupling it with the intermittent electric input is an important drawback, because SOE operates at a high temperature and thus its regulation or start/stop functioning may present a challenge. The H_2-CS technology has gained increasing interest in recent years as the system produces clean synthetic fuel. In the H_2-CS pathway, after the production of pure hydrogen, either of the two principal ways can be followed according to the system load: hydrogen can be stored in a tank in the case of low fuel

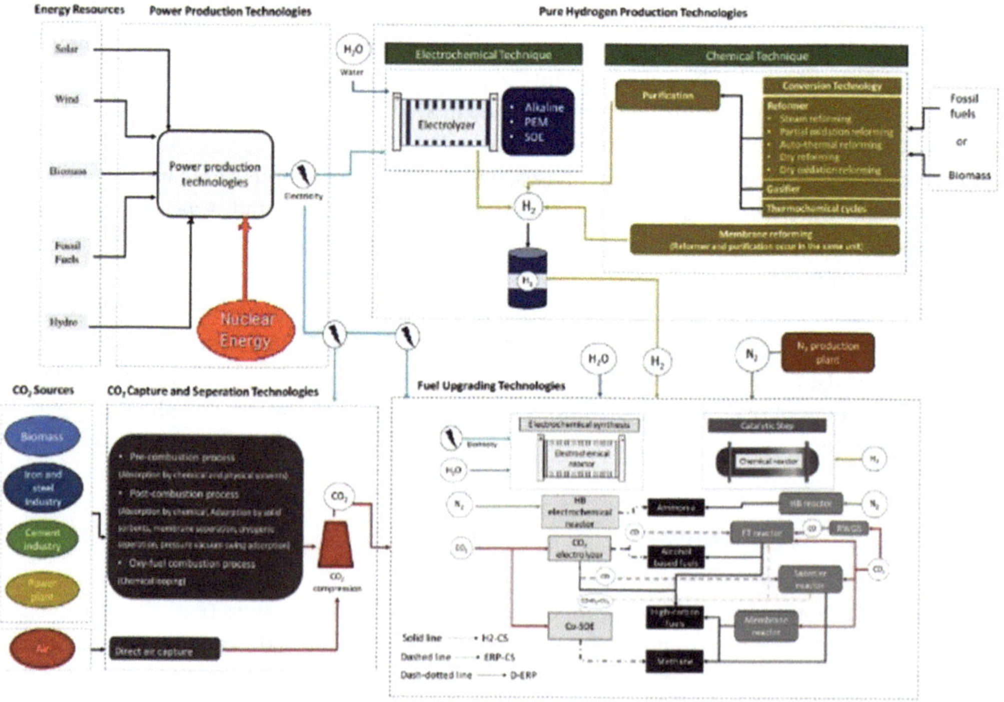

Figure 5.9 Various Power-to-X technologies.

requirement or hydrogen can be sent to chemical reactors (e.g., Sabatier and Fischer-Tropsch reactor), where it reacts with carbon dioxide and/or carbon monoxide to produce hydrocarbon-based fuels and chemicals. These can be further converted to other products. For example, after methanol is synthesized in a chemical reactor, it can be upgraded to other fuels (gasoline, kerosene, diesel and ethers) through several additional chemical processes (additional reactors, separation unit, purification unit, distillation unit, flash unit and absorber) to be used in the transportation and other commercial applications. The chemical processes to produce these liquid fuels will discussed in Chapter 7. The catalysts used in the reactors have another vital role in the type of fuel produced. For example, $CuO/ZnO/Al_2O_3$ is used for methanol synthesis, while a (H-)ZSM-5 catalyst can be used for DME production (methanol dehydration). Alternatively, bifunctional catalysts are available to produce different types of fuels in the same reactor. Hydrogen can also be combined with nitrogen in the chemical Haber-Bosch (HB) reactor to produce ammonia. On the other hand, different catalytic reactor types can be used in the catalytic step for fuel upgrading. For example, fixed-bed, structured, fluidized-bed reactors and micro-reactors can be used for methanation, while a Lurgi tubular reactor can be used for methanol synthesis. The reactors that are used for the catalytic steps are described in Chapter 1. In addition, membrane reactors are a new technology for hydrocarbon-based fuels production. Like chemical FT, HB and Sabatier reactors, hydrogen combines with carbon monoxide and/or carbon dioxide with an appropriate feed gas ratio (H_2:CO_2) in the membrane reactor. Unlike the chemical reactor, the deactivation of the catalyst due to existing water as a by-product can be hindered in the membrane reactor. The reason behind this behaviour of the membrane reactor is that the perm-selective membrane (H-SOD membrane) and sweep gas can remove the water formed in the FT and RWGS reactions. In Chapters 1 and 2, fuels produced through the Power-to-X technology (these fuels are also called e-fuels) were discussed to reveal the potential of e-fuels (e.g., technological maturity, chemical properties and economic aspect). One of the studies revealed that diesel e-fuels have a great capability for improving diesel engine performance. It was also found that oxymethylene dimethyl ethers (OMEs) are promising fuels among the conventional diesel fuel types as they have lower cetane numbers and higher oxygen content as well as lower nitrogen oxide emissions.

The direct electrochemical pathway relies on the electrochemical reduction of water, carbon dioxide and nitrogen, depending on the desired final fuel product. The pathway involved is called electroreduction. The fundamentals of the electrochemistry of electroreduction can be found in detail in the literature. In the electrochemical approach, a system to produce pure hydrogen could not be achieved. Therefore, the management and delivery system of pure hydrogen is not a challenge in this pathway. As this pathway produces fuels directly in a single unit, the electrochemical reduction systems are compact, unitary and simple. On the other hand, there are significant challenges due to the low reaction kinetics and energy efficiency. Most importantly, the performance of electrocatalysts in terms of catalytic activity and stability is low. Early studies showed that the lifetime of electrocatalysts is around 100 h, which is a big technological barrier in terms of commercialization. Therefore, current efforts in the development of electroreduction technology focus on catalytic activity and stability to overcome these obstacles. Several studies were published to discuss the progress of electrocatalysts for the electroreduction of carbon dioxide, water and nitrogen in the last decades. Density functional theory and micro kinetic analysis provide the researchers with a better understanding of the performance of the catalyst, for example, through appropriate materials for the electrochemical synthesis of ammonia. There exist some novel electrochemical devices that use the D-ERP concept to provide the electroreduction of carbon dioxide and nitrogen to produce hydrocarbon-based fuels and ammonia, respectively. For example, low-carbon fuel (e.g., methane, ethylene and methanol) can be produced depending on the catalyst type and operating conditions through a carbon dioxide electrolyser, while ammonia can be produced directly through an electrochemical HB reactor. The electrochemical fuel synthesis in an electrochemical HB reactor or carbon dioxide electrolyser through the D-ERP can be provided by various electrolytes (e.g., solid oxide electrolytes, polymer-type electrolytes and liquid electrolytes). Co-SOE is another electrochemical device in which the electroreduction of carbon dioxide and water occur simultaneously in the same unit, typically with the use of a nickel catalyst, to produce syngas that consists of mainly carbon monoxide and hydrogen, with minor amounts of methane, carbon dioxide and water, depending on the thermodynamic conditions. The amount of methane in the syngas depends highly on the operating conditions (pressure, temperature) of the Co-SOE. On the other hand, Co-SOE is generally used to produce syngas, which can be considered as a pre-step to produce high content methane at the Sabatier chemical reactor. Therefore, Co-SOE can be used in both direct and indirect fuel production. In the case of direct fuel production with Co-SOE, the only fuel that is formed directly is methane. The enhancement of methane yield in Co-SOE is an important engineering task for researchers. In this regard, the studies focus on material development, durability and enhancement of operating strategies. On the other hand, when Co-SOE is used as a pre-step, different fuel types can be produced (e.g., methanol, ethanol and DME as well as methane). Similarly, a carbon dioxide electrolysis system can also be used as a pre-step in producing carbon monoxide with further fuel production through the catalytic step. Here, the electrolysers convert carbon dioxide to carbon monoxide to be used for fuel synthesis in a reactor with high efficiency through different cathodic catalysts (e.g., Ag, Pt, Cu, Au, Zn and Pd). This concept is included in the ERP-CS pathway.

A few recent studies have been published that have addressed the multi-physics model of electrochemical reactors for direct fuel production. The key results show that direct methane production through the Co-SOE depends highly on pressure and temperature. When the operating pressure increases and temperature decreases, the methane yield increases, as expected from thermodynamics. Similarly, high temperature contributes to an increase in the conversion of carbon dioxide to carbon monoxide in carbon dioxide electrochemical reactors with low carbon deposition. The economic assessment of the electrochemical direct fuel production system has been also considered in the literature. For example, key results of the economic analysis have shown that electrochemical direct ammonia synthesis is more economical when compared to the conventional HB chemical reactor system and water electrolysis, which is followed by the chemical HB reactor. The reason behind this finding is that in the electrochemical reduction pathway, the capital cost of subcomponents is not very high. On the other hand, energy efficiency is another important criterion and can be limited by Faradaic efficiency and overpotentials in electrochemical reactors that use ERP. For example, in a study in which 2000 metric ton (mt) of ammonia is produced per day when the Faradaic efficiency is 95% and the operating voltage is 0.6, the energy efficiency of HB-RR is about 62%, which is lower than the energy efficiency of the process that uses H_2-CS. Three pathways (H_2-CS, ERP-CS and D-ERP) as mentioned above can be employed in various pathways of the Power-to-X system. They are illustrated in Figure 5.9. In all of these pathways, power is first produced preferentially from renewable sources (e.g., solar, wind, etc.,) and nuclear sources (Chapter 12). Then, this power is used to produce hydrogen in the H_2-CS pathway. Here, hydrogen can also be produced through other conventional technologies (e.g., reformer, gasifier, membrane reactor and thermochemical cycles); but if electrical power is not used in these technologies, they are out of the scope of the Power-to-X concept. For the ERP-CS and D-ERP, power is used to produce syngas (e.g., through a Co-SOE) and desired fuel (e.g., through carbon dioxide electrolyser and HB-RR), respectively. Other system components and sub-processes (e.g., carbon dioxide capture, fuel upgrading and nitrogen production) can also vary according to the desired output (e.g., methane, methanol, ammonia, etc.) and geographical and economic considerations. In this regard, various Power-to-X system designs have been discussed in the literature.

5.6 The Role of Nuclear Power

Unlike other fuels, hydrogen requires more integration of the fossil, nuclear and renewable energy systems and it will take an integrated approach from all energy sectors to realise the full potential and benefits of hydrogen. A key aspect of the strategy is to enable hydrogen production from a diverse array of low-carbon domestic energy resources, including renewables, nuclear energy and fossil fuels with CCS.

Nuclear power already produces electricity as a major energy carrier with well-known applications. Operating at very high-capacity factors, nuclear energy is well placed to produce zero-carbon hydrogen as an emerging energy carrier with a wide range of applications. The evolution of nuclear energy's role in hydrogen production (Chapter 12) over perhaps two decades is seen to be:

- cold electrolysis of water, using off-peak capacity (needs 50–55 kWh/kg);
- low-temperature steam electrolysis, using heat and electricity from nuclear reactors;
- high-temperature steam electrolysis, using heat and electricity from nuclear reactors; and
- high-temperature thermochemical production using nuclear heat.

In addition, nuclear heat can assist the process which provides most of the world's hydrogen today:

- use of nuclear heat to assist steam reforming of natural gas (methane).

Steam methane reforming (SMR) requires temperatures of over 700°C to combine methane and steam to produce hydrogen and carbon monoxide. A nuclear heat source would reduce natural gas consumption by about 30% (i.e., that portion of feed which would simply be for heat) and eliminate flue gas carbon dioxide emissions. Starting with electrolysis, the efficiency of the whole process (primary heat to hydrogen) moves from about 25% with today's reactors driving electrolysis (33% for reactor x 75% for cell) to 36% with more efficient reactors doing so, to 45% for high-temperature electrolysis of steam, to about 50% or more with direct thermochemical production.

Electrolysis at ambient temperature is being undertaken in at least four US projects at nuclear power plants and is planned for the Kola plant in Russia from 2023. Alkaline and proton exchange membrane (PEM) technology is employed. In August 2021 Nel Hydrogen was contracted to build a 1.25 MW PEM electrolyser at Exelon's Nine Mile Point nuclear power plant to show integrated production, storage and normal use at the plant.

Low-temperature steam electrolysis improves the efficiency of electrolysis at ambient temperatures and utilizes waste heat at up to 200°C from a conventional reactor. The US Department of Energy in October 2020 selected two projects to advance flexible operation of light water reactors with integrated hydrogen production systems to receive cost-shared funding. Two other projects are already under way.

High-temperature steam electrolysis (HTSE, at 550–750°C or more) in solid oxide electrolysis (SOEC) to use both heat and electricity has been demonstrated and shows considerable promise. It is a reverse reaction of the solid oxide fuel cell technology. It requires about one-third less energy than low-temperature electrolysis but has not yet been commercialized due to the poor durability of the ceramic components in a hot hydrogen environment. US research is at Idaho National Laboratory in conjunction with Ceramatec and Fuel Cell Energy in a $12.5 million project part-funded by the US Department of Energy (DoE). Lucid Catalyst shows it producing hydrogen at two-thirds the cost of low-temperature electrolysis across a range of capacity factors. One of the four DOE projects mentioned above will focus on a solid oxide electrolysis cell at high temperature. Idaho National Laboratory will work with Xcel Energy to demonstrate HTSE technology using heat and electricity from one of Xcel Energy's nuclear plants.

The International Atomic Energy Agency (IAEA) has developed the Hydrogen Economic Evaluation Program (HEEP) to assess the economics of large-scale hydrogen production using nuclear energy.

A September 2020 report by Lucid Catalyst said that in addition to deployment of renewables for power generation the only known way to address the "difficult-to-decarbonize" economic sectors is with the large-scale use of hydrogen as a clean energy carrier and as a feedstock for synthetic fuels such as ammonia. But clean hydrogen needs to come from non-fossil sources, at a lower price – under $1/kg – than is possible today (Table 5.7).

The technology readiness level (TRL) is a means to describe the extent of development necessary to reach the state of commercialization. It can range from theoretical principle, at TRL 1, to operational plant, at TRL 9. The TRL was estimated for all hydrogen production candidates. The minimum TRL of any component of the technology was taken, as the limiting factor in development is the component with the lowest TRL. SMR and alkaline electrolysis are the only current commercially available technologies. Electrolysis processes are also more developed than any thermochemical or hybrid process.

Table 5.7 includes a range of cost per unit kg of hydrogen for each method. Thermal-based water splitting methods are expected to be more cost effective because the cost of electricity is significantly higher than the cost of heat (dueto conversion). Similarly, production costs for electrolysis, requiring significant electricity, is highly dependent on electricity prices. As seen in the process flow diagrams, SMR technology incurs extremely high costs due to orders of magnitude more hydrogen produced. SMR is the cheapest technology currently in a $/kg scenario (Table 5.7). Otherwise SOEC and S-I would likely be the cheapest when coupled with a high temperature reactor, as these technologies have previously been chosen for further development. There are situations where hybrid systems may be profitable – when hydrogen, electricity and natural gas prices are high, or severe carbon dioxide costs are included. However, with the volatility of prices, flexible hybrid configurations are not currently valuable enough to overcome capital costs. Further policy will likely need to be implemented (aiming at cleaner energy production) to make hybrid systems profitable. The Department of Energy (DoE), as of late 2019, has issued contracts with three

Table 5.7 Nuclear hybrid energy systems for hydrogen production.

	Alkaline Electrolysis	PEM Electrolysis	Solid Oxide Electrolysis	Steam Methane R eforming	Thermochemical S-I
Technology readiness level	9	6–8	5	9	4
Temperature (°C)	60	60	800	870	910
Pressure (MPa)	0.1	0.1	0.157	0.41	0.385
Efficiency (HHV, %)	30	27	36	79	25
Electricity (MJ)	180	200	146	1.4	75
Heat (MJ)	26	26	30	0	375
Water (kg)	11.5	11.5	83	10.3	9
Natural gas(kg)	0	0	0	2.9	0
CO_2 out (kg)	0	0	0	5–11	0
Production cost	$5.92	$3.56–5.46	$2.24–3.73	$1.54–2.30	$2.18–5.65

nuclear power plants to add electrolysis-based hydrogen production to existing LWR plants, to demonstrate the capabilities and long-term competitiveness of nuclear hydrogen production and improve plant economics.

5.6.1 Small Nuclear Power Reactors

Since the 1950s, nuclear power generation has become established and the size of reactor units has grown from 60 MWe to more than 1600 MWe, with corresponding economies of scale in operation. At the same time there have been many hundreds of smaller power reactors built for naval use (up to 190 MW thermal) and as neutron sources, yielding enormous expertise in the engineering of small power units. The International Atomic Energy Agency (IAEA) defines "small" as under 300 MWe and up to about 700 MWe as "medium" – including many operational units from the twentieth century. Together they have been referred to by the IAEA as small and medium reactors (SMRs). However, "SMR" is used more commonly as an acronym for "small modular reactor", designed for serial construction and collectively to comprise a large nuclear power plant. (The use of diverse prefabricated modules to expedite the construction of a single large reactor is not relevant here.) A subcategory of very small reactors – vSMRs – is proposed for units under about 15 MWe, especially for remote communities. There is strong interest in small and simpler units for generating electricity from nuclear power and for process heat. This interest in small and medium nuclear power reactors is driven both by a desire to reduce the impact of capital costs and to provide power away from large grid systems and the technologies involved are numerous and very diverse.

Small modular reactors (SMRs) are defined as nuclear reactors generally 300 MWe equivalent or less, designed with modular technology using module factory fabrication, pursuing economies of series production and short construction times. This definition, from the World Nuclear Association, is closely based on those from the IAEA and the US Nuclear Energy Institute. Some of the already-operating small reactors mentioned or tabulated below do not fit this definition, but most of those described do fit it. A pressurized water reactor (PWR) may have integral steam generators, in which case the reactor pressure vessel needs to be larger, limiting portability from factory to site. Hence many larger PWRs such as the Rolls-Royce UK SMR have external steam generators (Chapter 12).

This section focuses on advanced designs in the small category, i.e. those now being built for the first time or still on the drawing board and some larger ones which are outside the mainstream categories dealt with in the World Nuclear Association's Advanced Nuclear Power Reactors webpage. Some of the designs described here are not yet actually taking shape, others are operating or under construction. Four main options are being pursued: light water reactors, fast neutron reactors, graphite-moderated high temperature reactors and various kinds of molten salt reactors (MSRs). The first has the lowest technological risk, but the second, a fast-neutron reactor (FNR), can be smaller, simpler and with longer operation before refuelling. Some MSRs are fast spectrum.

5.6.2 Conclusion

With increasing renewable energy sources along with increasing pressure to promote clean, sustainable energy, nuclear hybrid energy systems are a potential way to keep nuclear competitive. With renewable energy, grid power becomes increasingly volatile, leading to fewer base-load power systems, like nuclear, to greenhouse gas emitting coal or natural gas plants. With nuclear hydrogen hybrid technology, an alternative commodity, hydrogen, could be produced to allow for grid load-following. This hydrogen can be used for energy storage and converted back to energy during high electricity prices or demand, or has the potential to be an energy carrier, to be used as fuel for transportation. Hydrogen does not produce carbon dioxide, rendering it a much cleaner fuel than currently used petroleum or natural gas.

The nuclear hybrid energy systems (NHES) candidate technologies (along with the comparison SMR), including temperature, pressure, TRL, efficiency, hydrogen cost and inputs and outputs are summarized in Table 5.7. These technologies require only water, heat and/or electricity, all of which are abundant around a nuclear plant, to produce hydrogen. Alkaline water electrolysis is currently the only widespread, commercially available, large scale clean hydrogen production technology, but other technologies have shown rapid development and promise, along with potentially lower production costs. Additionally, these hydrogen production technologies operate at various temperatures, which allows them to match closely with the current (Generation III) and future (Generation IV) reactor fleet. These nuclear reactors may be loosely subdivided into three categories: low output temperatures (<300°C), medium (<750°C) and high temperatures (<950°C). Within these categories, only low temperature electrolysis is currently available for low temperature reactors, but thermochemical and electrolysis methods could potentially be coupled with high temperature systems. Energy and material process flow diagrams were created for hybrid technologies at all reactor category levels. When coupled with around 2000 MWt power from a nuclear reactor, these technologies can generate around 200–500 tons/day of hydrogen, with no carbon dioxide emissions. Current hybrid technologies are not cost effective and will require either increased electricity prices, carbon dioxide subsidies, or increased hydrogen prices to overcome capital costs. However, new contracts from DoE will provide funding to nuclear power plants to demonstrate hydrogen production capabilities to improve the competitiveness of hybrid technology and improve plant economics. The UK Nuclear Industry Association's Hydrogen Roadmap published in February 2021 showed how the country might achieve 225 TWh (6.8 or 5.7 million tonnes) of low-carbon hydrogen by 2050. It proposed 12–13 GW of nuclear reactors of all types using high-temperature steam electrolysis and thermochemical water-splitting to produce 75 TWh (2.3 or 1.9 Mt) of hydrogen by mid-century.

In Slovakia and Czech Republic, a national hydrogen strategy adopted in April 2021 envisages production from nuclear power by electrolysis.

Russia is planning a new hydrogen industry by 2024. Rosatom produces hydrogen by electrolysis and is planning 1 MW of electrolyser capacity at the Kola nuclear power plant from 2023, then increasing it to 10 MW as a demonstration project for wider adoption.

In the USA, The DoE in October 2020 selected two projects to advance flexible operation of light water reactors with integrated hydrogen production systems to receive cost-shared funding. Two other projects are already under way. One of these four will focus on using a solid oxide electrolysis cell at high temperature. The Idaho National Laboratory will work with Xcel Energy to demonstrate high-temperature steam electrolysis (HTSE) technology using heat and electricity from one of Xcel Energy's nuclear plants.

Further Reading

Andrews, J., Hung, J., Molinier, M. and Lim, Y.T. (2020). Catalytic processes operation during downturns in aromatics complexes. *Hydrocarbon Processing*, September 2020.

Baars, F., Oruganti, S. and Kalia, P. (2021a). Oil refinery/petrochemical integration in a CO_2-constrained world—Part 1, *Hydrocarbon Processing*, July 2021.

Baars, F., Oruganti, S. and Kalia, P. (2021b). Oil refinery/petrochemical integration in a CO_2-constrained world—Part 2. *Hydrocarbon Processing*, August 2021.

Bajus, M. (2020). *Petrochemistry: Petrochemical Processing, Hydrocarbon Technology and Green Engineering*. Chichester: Wiley.

Hydrogen Production and Uses-World Nuclear Association (2021). https://www.world-nuclear.org.

Ince, A.C., Colpan, C.O., Hagen, A. and Serincan, M.F. (2021). Modeling and simulation of power-to-X systems: a review. *Fuel* 304: 121354.

Kapur, S. (2018). Crude oil to chemicals. *Apex PetroConsultant*. https://www.apexpetroconsultants.com/blog/crude-oil-to-chemicals.

Kunugi, D. (1980). Chemical reaction engineering and research and development of gas solid systems. *Chem. Eng. Sci.* 35: 1887–1911.

Lucid Catalyst (2020). https://www.lucidcatalyst.com/hydrogen-report.

Muraza, O. (2021). Highlighting the greener shift in transportation energy and fuel based on novel catalytic materials. *Energy Fuels* 35 (1): 25–44.

Nuclear Industry Association (2021). *Hydrogen Roadmap*. https://www.niauk.org/hydrogen-roadmap.

Pinsky, R., Sabharwallb, P., Jeremy Hartvigsenb, J. and James O'Brien, J. (2020). *Comparative review of hydrogen production technologies for nuclear hybrid energy systems*. https://www.sciencedirect.com/science/article/pii/S014919702030069X.

Shackleford, A., Tan, A. and Eren, B. (2020). Driving FCCU value through fresh and flushing catalyst activity modeling. *Hydrocarbon Processing*, September 2020.

Singh, R., Lai, S., Dharia, D. et al. (2020). Conventional FCC to maximum propylene production. *Hydrocarbon Processing*, September 2020.

Vianna, C. and Adams, D. (2020). Catalyst breakthrough for ethylene production, *Hydrocarbon Processing*, September 2020.

Wagner da Silva, M. (2021). *Petrochemical integration – an essential requirement to the competitiveness in the new downstream industry*. https://www.linkedin.com/pulse/petrochemical-integration-essential-requirement-new-da-silva-mba/?trk=read_related_article-card_title, Published on July 21, 2021.

World Nuclear Association (2021). https://www.world-nuclear.org/information-library/nuclear-fuel-cycle/nuclear-power-reactors/advanced-nuclear-power-reactors.aspx.

6

Power-to-Hydrogen Technology

CONTENTS

Converting Power into Chemicals and Fuels: Power-to-X Technology for a Sustainable Future, First Edition. Martin Bajus.

6.1 Introduction

Hydrogen plays a key role in many industrial applications and is currently seen as one of the most promising energy vectors. Many efforts are being made to produce hydrogen with a zero carbon dioxide footprint via water electrolysis powered by renewable energies. Nevertheless, the use of fossil fuels is essential in the short term. The conventional coal gasification and steam methane reforming processes for hydrogen production are undesirable due to the huge carbon dioxide emissions. A cleaner technology based on natural gas that has received special attention in recent years is methane pyrolysis. The thermal decomposition of methane gives rise to hydrogen and solid carbon and thus, the release of greenhouse gases is prevented. Therefore, methane pyrolysis is a carbon dioxide-free technology that can serve as a bridge from fossil fuels to renewable energies.

The establishment of a sustainable power system to meet the growing energy demand is one of our most critical current challenges. The basis of this energy system is the decrease of greenhouse gas (GHG) emissions to reduce global warming and thus, accomplish the goals set out in the Glasgow Climate Pact (COP26, 2021), Paris Agreement and Kyoto Protocol. Specifically, Germany aims to become carbon neutral by 2050. Since most carbon dioxide emissions derive from the combustion of fossil fuels for power generation, at present much research is concerned with the development of zero-emission fuels. The hydrogen economy concept is based on the use of hydrogen as a potential fuel. Hydrogen is one of the most promising energy carriers and is considered the cleanest fuel because its combustion produces only water. In addition to this, the combustion of hydrogen originates more energy on a per mass basis than conventional fossil fuels. Hydrogen has received special attention for its application in fuel cells and internal combustion engines, enabling the creation of a low-carbon hydrogen economy. The chemical energy of hydrogen can be efficiently converted to electricity and other energy forms without GHG emissions. Hydrogen also plays a major role in the development of new strategies for converting industrial carbon dioxide emissions into important platform chemicals. This is one of the objectives set within the Carbon2Chem project, which aims to convert steel mill gases into base chemicals such as methanol. Thereby, the transformation of industrial emissions into chemicals prevents the release of GHG into the atmosphere.

Although hydrogen is a zero-emission fuel, its suitability also depends on the energy consumed and the cleanliness of the production method. The production of hydrogen via water electrolysis using electricity from renewable energies is the most appropriate technology from an environmental point of view due to its zero-carbon footprint. The global consumption of renewables is predicted to grow in the coming years, but will still remain relatively low for the foreseeable future (Figure 6.1). Drawbacks such as the fluctuations of renewable energies, geographic limitations and problems related to energy storage have to be overcome. For instance, the available renewable energies in Germany, onshore wind and solar energies, could provide electricity up to 2000 hours a year, which is by far too low to deal with the annual power demand. Further electricity may be supplied by offshore wind energy, but the transportation from the coast to inland areas is still a problem. Due to power fluctuations, energy-deficit periods will alternate with excess production seasons. The storage of surplus power could be a solution to the intermittence of renewable energy. However, large-scale energy storage remains a challenge.

Due to the large reserves of fossil fuels and the well-established technologies for their processing, fossil resources will still represent a major part of our energy system in the short term. For this reason, the development of bridge technologies based on fossil fuels with a low environmental impact has been the focus of countless studies in the last years. Among fossil fuels, natural gas is currently the main source of hydrogen (48%) followed by oil (30%) and coal (18%), whereas only

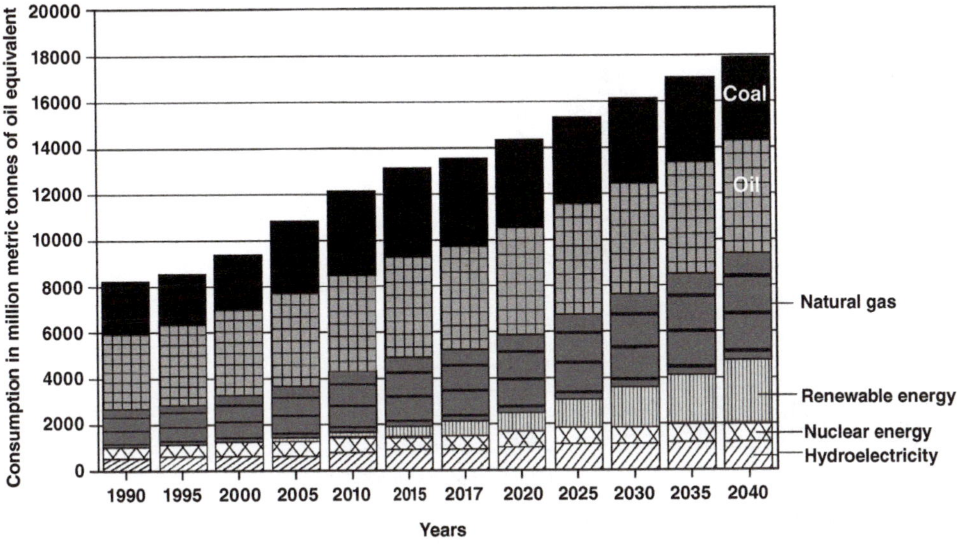

Figure 6.1 World energy consumption by energy source according to the data reported in ref. Sánchez-Bastardo et al., 2021 / American Chemical Society / Licensed under CC BY 4.0.

4% of the global hydrogen production comes from renewable resources. In the next few years, natural gas is expected to gain importance over the use of oil and coal, the share of which within the total energy consumption will suffer a meaningful decline (Figure 6.1). Steam reforming and gasification of fossil fuels are currently the main processes for hydrogen production due to their technological maturity on an industrial scale. Nevertheless, the major drawback is the unavoidable generation of large amounts of GHGs.

6.2 Traditional and Developing Technologies for Hydrogen Production

At present fossil fuels undoubtedly predominate over renewable resources for hydrogen production due to their high availability and cost-effective industrially implemented processes. About 96% of global hydrogen production is obtained by coal gasification, oil/naphtha reforming and steam reforming of methane. However, these processes have a dramatic environmental impact owing to the large amount of carbon dioxide emissions. Coal gasification has the largest carbon dioxide footprint but also the lowest product costs. Because natural gas has a lower carbon content than coal, steam reforming of methane has a much smaller carbon footprint than coal gasification, although the product costs are slightly higher (Figure 6.2). Steam methane reforming has a high energy efficiency (75%), but the need for carbon capture and storage (CCS) systems to obtain high purity hydrogen and decrease the GHG emissions results in a significant energy efficiency drop (60%) (Table 6.1) and may increase the cost of hydrogen production by more than 30%. In addition, the high electricity consumption of the conventional techniques intended for CCS increases the total environmental impact of the process. Dry methane reforming is an additional technology for hydrogen production using natural gas as a feedstock. This process generates less carbon dioxide emissions than steam methane reforming, but it is highly endothermic and is generally intended

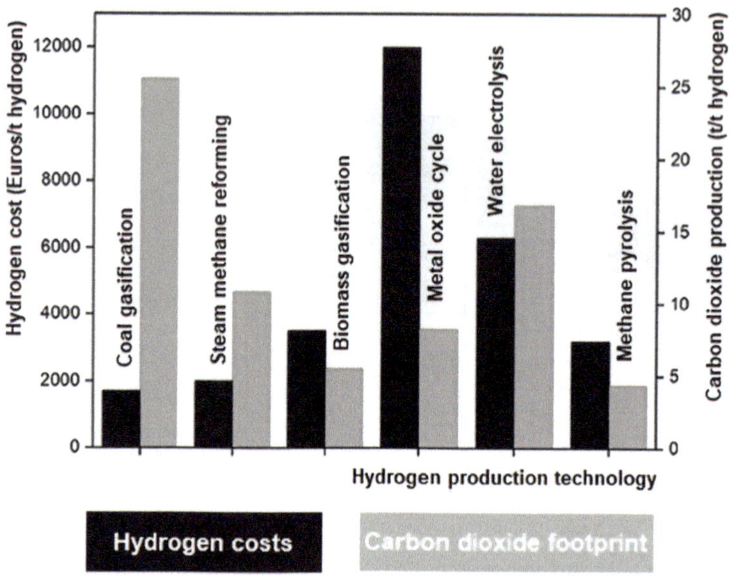

Figure 6.2 Product costs and carbon dioxide footprint of different hydrogen production technologies. Note: the water electrolysis, metal oxide cycle and methane pyrolysis processes produce carbon dioxide-free hydrogen. The carbon dioxide footprint of water electrolysis and the metal oxide cycle corresponds to the production of electricity from the grid. In the case of methane pyrolysis, in addition to the carbon dioxide emissions derived from the generation of electricity, those resulting from the extraction and transportation of natural gas are also taken into account.

Table 6.1 Energy efficiencies of different technologies for hydrogen production.

Technology	Energy efficiency transformation (%)	In energy efficiency CCS (%)
Coal gasification	60	43
Steam methane reforming	75	60
Biomass gasification	35–50	
Thermochemical watersplitting	20–45	
Water electrolysis	50–70	
Methane pyrolysis	58	58

to produce synthesis gas, rather than for pure hydrogen, suitable for the synthesis of higher hydrocarbons and oxygenated derivatives. Biomass gasification is a mature technology that converts biomass into a hydrogen containing gas mixture. Although biomass is considered as a renewable raw material, its limited availability and the complex logistics for transportation to decentralized industries are factors that affect negatively its applicability and the final product costs. Moreover, the production of hydrogen is accompanied by significant amounts of carbon dioxide that force the implementation of further separation and purification steps. For this reason, biomass gasification is usually intended for direct syngas production to generate energy or synthesize fuels rather than for obtaining pure hydrogen.

A great effort is being made to establish industrial strategies for hydrogen production with near-zero carbon dioxide emissions. Thermochemical water splitting and water electrolysis are developing technologies that generate only hydrogen and oxygen. The most common thermochemical water splitting process consists of redox-active metal oxide cycles, which require high temperatures (>1300°C) that can be achieved by large-scale sunlight concentration systems. This technology uses two practically inexhaustible sources, namely water and sun. However, according to the economic evaluation of different hydrogen production processes on a large scale, the metal oxide cycle has the highest product costs and requires a high level of investment (Figure 6.2). The integration of solar energy concentration systems with structures able to split water represents an extensive value and positive impact on the energy and economy. Currently this process is not industrially feasible owing to its relatively poor efficiency and high processing costs. Water electrolysis involves the decomposition of water into oxygen and hydrogen by passing an electric current through the water. Hydrogen production from water via electrolysis is a completely carbon dioxide-free alternative only if the required electricity comes exclusively from renewable resources. If the electricity is not 100% emission-free, water electrolysis can even exceed the carbon footprint of steam methane reforming due to the high energy requirements. The problems related to renewable energies, such as solar and wind, are their variability and unpredictability, which leads to difficulties to match energy supply and demand. The storage of energy during periods with power surplus is often seen as a crucial element of future power systems, but it is unlikely to be a universal solution for all demand–supply imbalance problems. In the specific case of Germany, it seems improbable that a total dependence on renewable energy will be achieved. Thus, due to the high electricity consumption, the viability of water electrolysis depends on the price and carbon footprint of the required electricity. Both water processing technologies are still not economically competitive with the use of fossil fuels. They show the highest product costs, whereas the well-developed technologies, such as steam methane reforming and coal gasification, display the lowest product costs (Figure 6.2). Until the use of renewable energies is sufficiently widespread, processes based on fossil resources seem to be fundamental and thus, cleaner economic technologies need to be developed and implemented industrially.

6.3 Dry Reforming of Methane

Steam reforming of methane has been the major source for production of synthesis gas (Equation 6.1). Steam reforming is a mature technology; however, it suffers from severe limitations like high energy requirements, a high H_2/CO product ratio and poor selectivity for carbon monoxide.

$$CH_4 + CO_2 \rightleftarrows 2CO + 2H_2 \qquad \Delta_r H_{293} = 247\,\text{kJ/mol} \tag{6.1}$$

In dry reforming of methane, carbon dioxide serves both as the oxidant and a carbon source. The reaction is highly endothermic and equilibrium limited (Figure 6.3). Both methane and carbon dioxide are consumed and the carbon content of the synthesis gas product is higher based on the natural gas consumed. The resulting synthesis gas is carbon monoxide rich and has hydrogen to carbon monoxide ratio close to unity, making it suitable for the synthesis of valuable oxygenates such as oxo alcohols.

Nickel-based catalysts and noble metal-supported catalysts (Rh, Ru, Pd, Pt and Ir) were found to have promising catalytic performance in terms of conversion and selectivity. However, according to thermodynamic studies, carbon deposition via carbon dioxide disproportionation reaction and methane cracking under reaction conditions of carbon dioxide reforming process (600–850°C) is

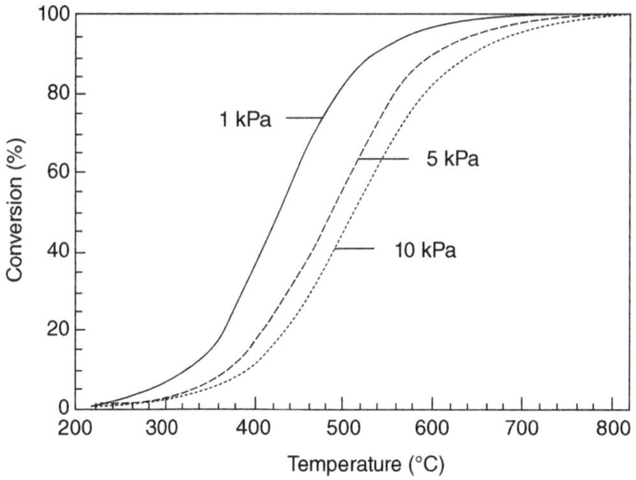

Figure 6.3 Equilibrium conversion for methane reforming with carbon dioxide with the feed ratio of CO_2/CH_4 = 1 at various pressures.

unavoidable on metallic catalysts, in particular nickel. Nevertheless, considering the high cost and limited availability of noble metals, it is more practical to develop improved nickel supported catalysts that exhibit stable operation for a long period of time.

Temperature-programmed surface reaction (TPSR) and temperature-programmed desorption (TPD) studies have shown that the decomposition of methane results in the formation of at least three kinds of surface carbon species on supported nickel catalysts, namely, carbidic, carbonaceous and carbidic clusters with different mobility and thermal stability and the distribution and features of these species depends on the nature of the transition metals and the conditions of methane adsorption. Carbidic carbon is a very active and important intermediate in carbon dioxide reforming with methane and the carbidic cluster species might be the precursor of the surface carbon deposition. CO_2-TPD indicated that carbon dioxide was a weakly adsorption state of carbon dioxide formed.

Mozammel et al. have demonstrated that the addition of rhodium and cobalt to the nickel phase supported on mesoporous alumina (MAl) to form bimetallic and trimetallic Rh-Ni/MAl, Ni-Co/MAl and Ni-Co-Rh/MAl catalysts and alloying of nickel, was found to enhance the catalyst activity and high resistance towards coking. All these catalysts were synthesized using a triblock copolymer P123 templated precipitation of aluminium isopropoxide and metal precursors, followed by the removal of the template by calcination. The high conversions of methane with carbon dioxide into synthesis gas and the nature of carbon formation on these bimetallic catalysts were found to be different from those of the monometallic catalysts and the ratios of hydrogen and carbon monoxide in the synthesis gas were observed to be 0.9–1. The enhanced stability and activity were attributed to the bimetallic phase formation, which was supported by XRD, XPS, TPR, TGA, TEM and N_2-sorption analysis of the fresh and spent catalysts. From these results, a plausible mechanism was proposed to explain the enhanced catalytic activity and stability, which was mainly due to the nature of alloy formation. Nickel was found to form a homogeneous alloy phase with cobalt, while it formed a heterogeneous bimetallic phase with rhodium. Alloying with cobalt moderates the coking in the case of Ni-Co/MAl, while alloying with rhodium increases the carbon gasification by hydrogen spill over effect in the case of Ni-Rh/MAl.

A lot of effort went into the development of an improved nickel-based catalyst with stable and long-term high activity. Currently, SPARG and CALCOR processes are the two commercial processes involving carbon dioxide reforming of methane.

In the sulfur-passivated reforming (SPARG) process, the natural gas is desulfurized and then mixed with steam to undergo pre-reforming. In the tubular reformer, the reforming catalyst is passivated by adding sulfur to the feed to block carbon formation sites. A molar ratio of $H_2/CO = 1.8$ is achieved by controlling the molar ratio CO_2 /hydrocarbon C = 0.9 and the molar ratio CO_2/ hydrocarbon C = 0.54.

Sulfur passivation of nickel catalyst decreases the carbon formation rate more rapidly than unaltered nickel catalysts and thereby reduces the coking rate on nickel surface during the methane steam reforming reaction. Because the ensemble size on the metal surface necessary for carbon formation is larger than that for methane reforming, the effectiveness of the carbon suppression during the SPARG process for carbon dioxide reforming of methane is attributed to the control of the ensemble size on the metal surface. That is, bigger nickel sites favouring carbon formation are poisoned by sulfur while a fraction of smaller nickel sites sufficient for the reforming reaction remain intact.

The CALCOR standard process is a catalytic reforming process of methane or liquefied petroleum gas (LPG) at high temperature and low pressure. It combines steam reforming with carbon dioxide recovery or recycling. The feed is hydrodesulfurized before it is mixed with carbon dioxide. When passing the reformer tubes, which are charged with reforming catalysts, the feed mixture is converted into synthesis gas consisting of carbon monoxide, hydrogen, carbon dioxide, water and trace methane. Then the carbon dioxide is removed and recovered after the synthesis gas is cooled to ambient temperature. Finally, carbon monoxide purification is conducted to remove hydrogen, methane and traces of carbon dioxide. The recovered carbon dioxide is recycled to the reforming process and the tail gas from the purification unit is burned as fuel.

Besides the metal, the catalyst support is also an important parameter for carbon dioxide reforming. It has been found that CeO_2-promoted nickel catalyst on nano-ZrO_2 support shows a good stability for this reaction under identical conditions but otherwise coke formation and catalyst deactivation were rapid.

6.4 Tri-reforming of Methane

Tri-reforming of methane (TRM) combines steam reforming, dry reforming and partial oxidation of methane in a single reactor. Methane or natural gas reforming is one of the processes to produce synthesis gas with desired H_2/CO ratios for clean fuels and chemicals. In general, there are three processes for production of synthesis gas, i.e., steam reforming, carbon dioxide reforming and partial oxidation of methane or natural gas. The novel process called tri-reforming is a synergetic combination of these three reactions in a single reactor for effective production of industrially useful syngas. Both experimental test and computational analysis show that tri-reforming can not only produce synthesis gas (carbon monoxide + hydrogen) with desired H_2/CO ratios (1.5–2.0), but also could eliminate carbon formation, which is usually a serious problem in the carbon dioxide reforming of methane. These two advantages have been demonstrated by tri-reforming of methane in a fixed-bed flow reactor at 850°C with supported nickel catalysts. Over 95% methane conversion and about 80% carbon dioxide conversion can be achieved in tri-reforming over Ni catalysts supported on an oxide substrate. The type and nature of catalysts have a significant impact on carbon dioxide conversion in the presence of water and oxygen in tri-reforming in the temperature range of 700–850°C.

The H_2/CO product ratio usually ranges from 1.5 to 2.5, which makes the TRM process highly versatile, since the synthesis gas quality can be altered by simply changing the feed composition,

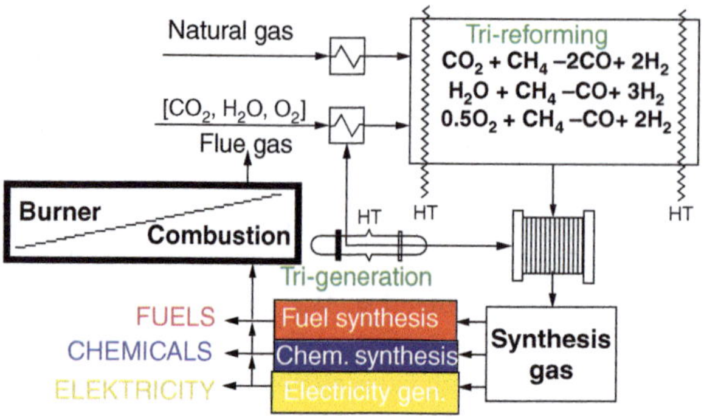

Figure 6.4 Process concept for tri-reforming process.

hence permitting a wide range of applications. A H_2/CO ratio of two is usually desired to produce liquid fuels such as methanol and dimethyl ether (DME). The operational studies of the influence of the gas hourly space velocity (GHSV) and the feed composition were carried out, with the aim of obtaining an H_2/CO ratio suitable for downstream processes, as well as to achieve satisfactory carbon dioxide conversion, which is vital for CCSU (CO_2 capture, storage and utilization) strategies. The catalyst used for these studies was nickel supported on $MgAl_2O_4$ spinels promoted with $CeZrO_2$ (Ce/Zr molar ratio = 4). Reaction tests were performed changing the O_2/CO_2 and H_2O/CO_2 molar ratios in the reactor feed, keeping the GHSV constant at 2.95 mol/g_{cat}/h. Each test was run for 5 h at 750°C and atmospheric pressure. The increase of the O_2/CO_2 molar ratio from 0 to 1.5 in the feed reduced the amount of coke by about 30-fold, while increasing the H_2O/CO_2 molar ratio from 0 to 1.4 led to only an 8-fold decrease of carbon deposition, which showed that increasing the supply of oxygen was more effective for reducing the carbon deposition, compared to increasing the amount of water steam. The feed composition $CH_4/CO_2/H_2O/O_2/N_2 = 3:1.5:1.4:0.25:1$ ($O_2/CO_2 = 0.17$) could be considered the most stable operational condition among those studied in this work, producing syngas with $H_2/CO \sim 1.7$ and significant CH_4 and CO_2 conversions (77 and 60%, respectively).

Figure 6.4 illustrates the proposed process concept for tri-reforming of natural gas using flue gas from coal-fired or natural gas-fired electric power plants and effluent gases from other industrial processes. This process could be applied, in principle, to the production of industrially useful syngas (for synthesis of methanol and dimethyl ether, for Fischer-Tropsch synthesis and for high-temperature fuel cells) by reforming of natural gas using gas mixtures (containing carbon dioxide, water and oxygen) as co-feeds.

The feasibility of tri-reforming reactions has been experimentally demonstrated in independent studies in other industrial academic laboratories. It was proposed that tri-reforming could also be applied for converting low-quality carbon dioxide-rich natural gas into industrially useful syngas. Recently, tri-reforming has been demonstrated at pilot plant scale in Korea.

6.5 Greenfield Technology Option → Low Carbon Emission Routes

Hydrogen is a key element to produce value-added products such as methanol, formic acid, formaldehyde and liquid fuels out of carbon dioxide. Methane can also be used for the same purpose. However, using hydrogen would be preferable from an economical point of view since it allows the direct production of chemical and fuels, as opposed to methane, which allows the production of syngas. Hydrogen is also an important and versatile energy vector with a low heating value (LHV) of 119.9 MJ/kg, which is more than two times higher than the LHV of methane.

Hydrogen is not available in a pure state in the environment and needs to be synthesized. The main pathway currently employed at an industrial scale for hydrogen production is a well-established process called steam methane reforming (SMR) in which natural gas (or other fossil fuel) is reacted at high temperature (>700°C) with water vapour to produce hydrogen in the presence of a metal-based catalyst. This process has a conversion efficiency range of 65–75%.

Partial oxidation (POX) and autothermal reforming (ATR) are also processes used for hydrogen production from fossil fuels. In the POX process, hydrocarbons are reacted with steam and oxygen at varying concentrations to produce hydrogen. Unlike SMR, such technology can be operated without a catalyst. However, very high temperatures are required (>1000°C) to get a hydrocarbon conversion. Also, a lower H_2/CO around two is obtained. This process has been reported to have an efficiency around 50%.

The autothermal reforming (ATR) is a combination of SMR and POX. In this process, the energy released during the POX step is used to cope to the endothermic part of the SMR step. This process is also operated at high temperatures generally in the range of 950 and 1,100°C and has an efficiency around 60–75%. None of these three processes produce pure hydrogen since carbon monoxide is also produced at a H_2/CO ratio varying between two and three. Thus, a second step involving a water-gas shift reaction is required to convert the carbon monoxide into hydrogen.

Hydrogen at a 97% purity can also be produced from coal gasification, where the Koppers-Totzek process is the leading technology. In this process, an entrained flow gasifier is used to convert the carbon into a gas mixture composed of methane, hydrogen and carbon monoxide at temperatures around 1600–1900°C, temperature of which usually enhances the reaction rates.

This section considers three technology options for greenfield blue hydrogen projects: SMR, autothermal reforming and a proprietary gas POX technology. Steam methane reforming is a proven catalytic technology widely applied for grey hydrogen production, uses steam in a multitubular reactor with external firing for indirect heating. More than 48% of hydrogen production is from natural gas, with SMR being the most common production technology. Post-combustion carbon capture can be retrofitted to convert grey hydrogen production to blue and is proven to capture nearly all the carbon dioxide (99%) from low-pressure, post-combustion flue gas.

However, for greenfield blue hydrogen applications, oxygen-based systems, such as ATR and gas POX technology, are more cost-effective than SMR, a conclusion backed by numerous studies and reports. The cost of carbon dioxide makes grey hydrogen via SMR more expensive than blue hydrogen from Shell gas partial oxidation process (SGP technology). The cost advantage of oxygen-based systems over SMR increases with scale because the relative cost of the air separation unit decreases with increasing capacity. Another advantage is that more than 99.9% of the carbon dioxide can be captured using a lower-cost, pre-combustion solvent technology.

Autothermal reforming uses oxygen and steam with direct firing in a refractory-lined reactor with a catalyst bed. It is more cost-effective than SMR, but it requires a substantial feed gas pretreatment investment and the fired heater produces carbon dioxide emissions. ATR can be combined with pre-combustion carbon-capture technology to convert grey hydrogen production to blue.

Gas POX technology is also an oxygen-based system with direct firing in a refractory-lined reactor, but it is a noncatalytic process that does not consume steam and has no direct carbon dioxide emissions. It, too, can be combined with pre-combustion carbon-capture technology for blue hydrogen production. Compared with SMR, gas POX technology saves money by maximizing the carbon-capture efficiency and simplifying the process line-up, both of which offset the cost of oxygen production.

As oxygen-based systems offer clear benefits over SMR, this section considers the advantages of the proprietary gas POX technology over ATR for blue hydrogen production. A key advantage is that the POX reaction does not require steam as a reactant. Instead, high-pressure steam is generated by using waste heat from the reaction, which can satisfy the steam consumption within the blue hydrogen process, as well as satisfy some internal power consumers.

With no need for feed gas pre-treatment, gas POX technology has far fewer process contaminants such as sulfur and can accommodate a large range of natural gas qualities, thereby giving refiners greater feed flexibility to decarbonize refinery fuel gas.

Gas POX technology provides substantial savings compared with ATR – a 22% lower levelized cost of hydrogen. These savings come from a 17% lower CAPEX owing to the potential for a higher operating pressure leading to a smaller hydrogen compressor (single-stage compression), carbon dioxide capture and carbon dioxide compressor units and a 34% lower OPEX (excluding the natural gas feedstock price) from reduced compression duties and more steam generation for internal power. Gas POX technology consumes 6% more natural gas, but this is offset by power generation from the excess steam.

The proprietary blue hydrogen process is an end-to-end line-up that maximizes the integration of the gas POX and solvent technologies. Compared with an ATR unit, modelling shows that a line-up producing 500 tpd of pure H_2 would have: $30 MM/yr lower OPEX, 35% less power import, > 99% CO_2 capture and 10%–25% lower levelized cost of hydrogen.

The choice between a methanator or a pressure-swing adsorption (PSA) unit for the hydrogen purification step depends on the required hydrogen purity. For example, a PSA unit is necessary to achieve the > 99.97% purity required for the hydrogen used in fuel cells. The off gas is predominantly hydrogen, with trace containments such as carbon monoxide, carbon dioxide and nitrogen. In the ATR process, this off gas is typically burned to preheat the natural gas, which produces direct carbon dioxide emissions.

In a methanator, the purity of the final hydrogen is lower (> 98%, depending on the feed gas properties). However, it avoids the direct carbon dioxide emissions from burning the PSA off gas. The main advantage of choosing a methanator is that hydrogen is not lost via the PSA off gas. Consequently, it reduces natural gas consumption for the same hydrogen production. In addition, a methanator is commonly applied in industry, as it satisfies the H_2 purity requirements of most industrial consumers.

The gas POX technology is mature and "low-carbon," which makes it eligible for government funding. It has a long history of development and usage. For example, research into oil gasification was being conducted in the Netherlands as early as 1956.

Today, the proprietary gas POX technology has more than 30 active residue and gas gasification licensees and more than 100 gasifiers using the technology have been built worldwide. For example, the Pearl Gas-to-Liquids (GTL) plant in Qatar has 18 trains, each with an equivalent pure hydrogen production capacity of 500 tpd. Pearl GTL has been operating since 2011. The product is defined as pure hydrogen production – i.e., not including any inerts, methane, carbon dioxide or carbon monoxide, which will also be present, depending on the final purification step.

Since 1997, the Pernis refinery in the Netherlands has been operating a 1-MMtpy carbon-capture program using the technology. The carbon dioxide is used in local greenhouses. The carbon dioxide stream is an essential part of the Pernis CCS project.

No matter how cost-effective the hydrogen production and carbon-capture technologies, without sequestering the carbon dioxide directly or through enhanced oil recovery, the hydrogen remains grey. Many CCUS projects are in operation at various stages throughout the world. For example, since 2015, the Shell Quest facility in Canada has captured and stored more than 5 MMt of carbon dioxide.

Hydrogen will be part of the future energy mix and several mature technologies are available for producing cost-effective, low-carbon blue hydrogen for greenfield applications. SMR is an inefficient method of producing blue hydrogen owing to poor carbon dioxide recovery and scalability; oxygen-based systems offer better value (an independently backed conclusion).

The proprietary blue hydrogen process, which integrates proprietary gas POX and solvent technologies, offers key advantages over ATR, including a 10%–25% lower levelized cost of hydrogen, a 20% lower CAPEX, a 35% lower OPEX (excluding natural gas feedstock price), > 99% carbon dioxide captured and overall process simplicity. The process, which is now available to third-party refiners, is proven at the 500-tpd scale.

6.5.1 Water Electrolysis

Water electrolysis is the best-known electrochemical process for producing hydrogen using renewable electricity and it will play a crucial role in the development of the hydrogen economy and of the Power-X-Technology since it produces high-purity hydrogen suited not only for applications in the metallurgical, fine chemicals and aerospace industries, but also for hydrogen filling stations. This technology allows onsite hydrogen production from renewable energy, contributing to the use of hydrogen as an energy storage medium as well as to the use of renewable hydrogen on the production of chemicals and fuels via Power-to-Technology.

There are three main electrolysis technologies used for hydrogen production classified according to the electrolyte (Figure 6.5): alkaline water electrolysis (AEL), polymer electrolyte membrane electrolysis (PEM) and solid oxide electrolysis (SOEC). AEL is already a mature technology with commercial large-scale systems. PEM systems are also commercially available but only for small scale hydrogen production while SOEC is still at prototype stage. These technologies are compared in Table 6.2. The following sections will briefly discuss the recent advances in each technology as well as their potential for Power-to-X applications.

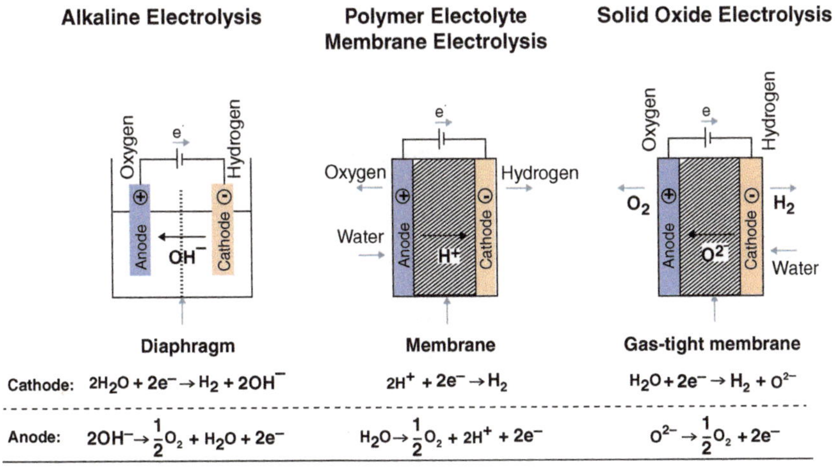

Figure 6.5 Technologies for water electrolysis.

Table 6.2 Comparison between different processes for hydrogen production.

Main Techniques	Cell Voltage	Power Consumption (kWh/m^3 H_2)	t (°C)	P_{max} (MPa)	Efficiency (%)	TRL	Advantages/ Disadvantages
Alkaline	1.8–2.4	3.8–8.2	<100	69	59–79	7–9	**Advantages** Low cost, mature technology, possible application in large plant sizes **Disadvantages** Low current density, low dynamics, corrosive electrolyte
PEM	1.8–2.2	4.4–7.1	<150	40	62–82	7–8	**Advantages** High power density, high pressure, rapid system response, no corrosive substances **Disadvantages** High cost, fast degradation of membranes
SOEC	-	3.7	>500	3	Up to 100	6–7	**Advantages** High efficiency, co-electrolysis of CO_2 and steam, possible integration of waste heat **Disadvantages** Low long term cell stability, not suited for fluctuating systems
Microbial electrolysis	0.2	-	<55	$P_{atm.}$	-	Lab. 2–3	**Advantageas** Use of organic waste as substrate, low energy consumption **Disadvantages** Design of efficient and scalable prototypes
Biomass electro-reforming	<1	<2.4	<100	$P_{atm.}$	-	Lab. 2–3	**Advantages** Flexible feedstock, low energy demand **Disadvantages** Low caloric value of biomass

6.5.1.1 Alkaline Electrolysis

Alkaline electrolysis is the most mature and reliable electrolysis technology and has been used at commercial MW-scale for decades with stacks currently available up to 6 MW (1,400 Nm^3/h) This technology relies on the use of two electrodes immersed in a liquid alkaline electrolyte, usually KOH or NaOH solution and separated by a diaphragm, which avoids the mixing of the product gases (Figure 6.5). Electrons flow from the anode to the cathode, where they are consumed by H^+ ions to form hydrogen. Hydroxide ions (OH^-) are then transferred through the alkaline electrolyte solution from the cathode to the anode, where they are oxidized into oxygen and water. High purity hydrogen (up to 99.9%) can be obtained with this technology. Oxygen is also produced with purity in the range of 99–99.8%, which can be increased to 99.999% using catalytic purification. The electrolyte concentration usually varies between 25 and 30 wt % for working temperatures in the range of 70–100°C and for typical pressures from 0.1 to 3.0 MPa. This type of electrolyser can either operate at atmospheric or high pressure, with high-pressure electrolysers operating up to

69 MPa (Table 6.2). The production of pressurized hydrogen is particularly interesting due to its higher energy efficiency when compared to hydrogen pressurization after production, which is usually required to further use the hydrogen produced or for direct grid injection. Hydrogen needs to be compressed up to 50 MPa for transportation purposes as well as filling stations. Moreover, the gas compression along with the transportation and storage of hydrogen represent a significant cost in the hydrogen production chain. Nevertheless, pressurized electrolysers have lower efficiency and produce lower purity hydrogen when compared to atmospheric pressure electrolysers.

The main advantages of using AEL systems are the readily availability, the durability (55,000–120,000 h) as well as the use of mature stack components. This technology has also the advantage of low capital costs when compared to PEM and SOEC, having investments costs around 800–1.500 €/kW and maintenance costs of 2–3% of the annual investment costs. Some Power-to-Gas plants rely on AEL for hydrogen production. The Audi e-gas plant, the world's biggest Power-to-Gas plant, produces hydrogen with three AEL systems with a total power of 6 MW. Similarly, the BioCatProject, which is a partnership between companies, such as Electrochaea, Audi and Hydrogenics, is currently developing a Power-to-Gas plant producing 1 MW.

Despite being a well-stablished technology, AEL has a few drawbacks, such as low partial load range and limited current density, which have a negative impact on the hydrogen production costs. AEL electrolysers also have a limited dynamic operation, which renders them difficult to adapt to variable renewable energy sources, such as solar and wind power and decrease the system efficiency as well as the gas purity.

Recent studies on AEL electrolysis have mainly focused on the development of new diaphragm and electrode materials to improve the performance of the system. The development of new diaphragm materials has taken into consideration features such as performance, cost and health hazard. Asbestos was initially used as diaphragm. However, it has been replaced by other materials, such as polytetrafluoroethylene (PTFE), reinforced polyether sulfone (PES) membranes, glass reinforced polyphenylene sulfide (PPS) compounds, nickel oxide layer on a mesh with titanium oxide and potassium titanate. The electrode shape, composition and electronic properties have also been studied, aiming to enhance and stabilize the electrode activity. Modifications on the electrode surface, such as addition of slits and holes, have shown to help the dissemination of gas bubbles, which are one of the major causes for extra ohmic losses. Nickel is the most used electrode material due to its high activity and stability in alkaline media and it has been used in commercial systems. The addition of iron has been proved to increase the stability of nickel electrodes by preventing the formation of a nickel hydride phase at the surface of the electrode. Noble metals, such as platinum and ruthenium, are also used as electrocatalysts and are known to improve the activity for the oxygen evolution reaction. In summary, the electrode activity can be improved by using different transition and/or noble metals as well as alloys and physical modifications of the electrode can improve gas removal and decrease the ohmic losses. Further studies are still required to reduce the overpotential of the two half reactions, to improve the electron and ionic transfer as well as to reduce the ohmic losses.

6.5.1.2 Polymer Electrolyte Membrane Electrolysis

In this type of electrolyser, the two half-cells are separated by a proton exchange membrane, often made of Nafion polymer (Figure 6.5). Water is split into oxygen, protons and electrons. Protons are then transferred from the anode to the cathode through the proton exchange membrane and electrons flow from the anode to the cathode via external direct current power source. Protons and electrons then recombine at the cathode to produce hydrogen. The polymer electrolyte membrane has the role of lowering the gas crossover, providing high proton conductivity and allowing high pressure operations as well as Faraday efficiency close to 100%. Since the gas crossover is limited in this type of electrolyser, high purity hydrogen can be produced. Using proton exchange membranes instead of liquid electrolytes allows a quick response to the power input and hence the use of a wide range of power input. The systems are usually operated at temperatures lower than 150°C and with typical pressures from 2 to 5 MPa (Table 6.2). High pressure systems can work up to 40 MPa. This technology has become very promising for hydrogen production due to its compact design, high efficiency and high hydrogen output pressure. It also provides a better coupling with intermittent systems as compared to AEL and SOEC technologies, making it a good candidate for Power-to-X applications. Due to its potential, companies such as Siemens, Proton OnSite, Hydrogenics, AREVA HydrogenGen etc., have invested in development of this technology. A recent joint venture between Enbridge and Hydrogenics has taken this technology one step further in a 2.5 MW Power-to-Gas facility. The Markham Energy Storage facility built in Canada uses a 1.25 MW PEM electrolyser to produce hydrogen that can further be stored, shipped directly to refuelling stations or to industrial and commercial customers. Air Liquide has also recently announced the construction in Canada of a 20 MW PEM electrolyser for zero-carbon hydrogen production using hydrogenic technology. This unit will be the world's largest PEM electrolyser.

The major drawbacks of this technology are related to the high costs of materials and components that need to resist the low pH conditions as well as the high over voltages. These conditions lead to the use of expensive titanium-based components as well as of noble metals-based electrocatalysts, such as platinum and iridium. According to investment costs for these systems vary from 1,400 to 2,100 €/kW while the maintenance costs represent 3–5% of the annual investment costs.

Literature on PEM electrolysis is not abundant. However, the possibility of coupling this technology to intermittent renewable energy has led to an increase of the number of studies on the topic, which have focused on the reduction of the high costs related to this technology and more specifically on the design and synthesis of electrocatalysts. A few studies have also focused on the improvement of materials for current collectors and separator plates. The replacement of noble metals-based electrocatalysts as well as decreasing the noble metal content are approaches currently under investigation. Transition metals-based materials, such as molybdenum sulfide, cobalt sulfide, nickel-molybdenum alloys, iron and cobalt phosphates, have been investigated as promising low-cost alternatives to noble metals electrocatalysts. Addition of low-cost oxides, such as SnO_2, Nb_2O_5 and TiO_2 to noble metals-based electrocatalysts, such as IrO_2 and RuO_2, has also been reported aiming to reduce the noble metal loading while maintaining similar catalytic and electronic conductivity properties. Finally, different synthesis methods have also being investigated to enhance the catalysts performance as well as to reduce the precious metal loading. In summary, in order for this technology to achieve large scale, challenges related to the use of expensive catalysts and to low corrosion resistance and high-cost collector and separator plates still need to be overcome.

6.5.1.3 Solid Oxide Electrolysis

In this technology, the solid oxide electrolysis cell operates at high temperatures (700–1,000°C), reducing the equilibrium cell voltage and thus the electricity demand to lower than 4 kWh/Nm^3 hydrogen (Table 6.2). SOEC electrolysis is in fact the reverse process occurring in a fuel cell, where water (steam) reacts at the cathode with electrons from an external power source, producing hydrogen and oxygen ions which are transported through a gas-tight membrane to the anode, where oxygen ions combine to form oxygen and liberate electrons (Figure 6.5). Hydrogen is produced in this process at high Faraday efficiency around 100%. Another advantage of this technology is the possibility of heat integration with exothermic processes, such as methanol production. Heat released from these processes can be used to heat the steam to the temperature of the SOEC process. SOEC systems also offer the possibility of flexible operation between electrolysis and fuel cells. Since hydrogen produced through this process could later be reconverted in electricity using a fuel cell, this technology could represent an opportunity to store renewable electricity surplus generated by wind or solar power. SOEC also presents the capacity of co-electrolysis of carbon dioxide and steam for syngas production, which can be later converted into value-added products, such as liquid fuels. This capacity of co-electrolysis renders this technology very attractive for Power-to-X applications.

Contrary to AEL and PEM, which are already functioning at a commercial level, SOEC is still at the prototype stage and parameters, such as lifetime, cycling stability and pressurized operation still need to be validated. The major drawback preventing this technology from reaching large scale is the fast material degradation and thus low stability related to the high temperature used and to long-term operation. Studies have proven that this degradation can be limited when the current density is low ($<1\ Acm^{-2}$). Hence, recent studies have focused on improving the stability of materials at high current densities.

Electrode materials must be ionic and electronic conducting, to facilitate electron and mass transport as well as to allow the migration of O^{2-} species. In general, electrode materials consist of mixed oxides with perovskite structure. Ni-YSZ (yttria-stabilized zirconia) and LSM (lanthanum strontium manganite) are the most used materials for cathode and anode, respectively. LSM has a coefficient of thermal expansion close to the one of the electrolytes, stabilizing the electrolysis cell. Moreover, LSM has very low chemical reactivity with YSZ, increasing the lifetime of the material. Double perovskites, fluorites and metals have also been studied. Mixing different oxides at different amounts has shown to improve activity and stability of the materials. Adding Co_3O_4 to LSM-BCZYZ, for example, significantly improve the electrode processes and enhances the current density under a certain voltage. Mixed ion-electron conducting electrodes, such as lanthanum strontium copper ferrite and lanthanum strontium cobalt ferrite have also been used to improve the ionic conductivity property of electrodes

Electrolytes are generally composed of zirconia-based materials doped with CaO, MgO, Sc_2O_3 as well as rare-earth oxides. However, the most used electrolyte is yttria-stabilized zirconia due to its good ionic conductivity and mechanical properties. $La_{1-x}Sr_xGa_{1-y}Mg_yO_{2.85}$ (LSGM) has been identified as a promising electrolyte material due to its high ion conductivity that can be five times higher than that of classical YSZ. However, further investigation is still needed to prove the durability of this material.

The flexibility regarding the operation mode of this technology, the possibility of CO_2 and steam co-electrolysis as well as the higher energy efficiency of SOEC when compared to AEL and PEM have led major companies to invest in the development of SOEC. Sunfire, for example, has recently investigated this possibility by developing a SOEC module in a demonstration project at Salzgitter Flachstahl GmbH aiming to evaluate the potential of the module for energy balancing and load management. The module has a hydrogen production capacity of 40 Nm^3/h and with an input power of 150 kW. The electrical efficiency of the system is higher than 80%. The system can also be reversed into a fuel cell with an output power of 30 kW. Haldor Topsoe, Fuel Cell Energy and Toshiba have also worked on the development of this technology.

The second step of the Power-to-X process is the carbon dioxide hydrogenation into chemicals and fuels either via a Fischer-Tropsch pathway or using methanol as an intermediate (methanol pathway). The carbon dioxide hydrogenation process is more advanced than other processes, such as carbon dioxide electrochemical reduction and it may benefit from the existing infrastructure of the classical Fischer-Tropsch process. The main research on this process focuses on the development of more efficient catalysts capable of withstanding the harsh reaction conditions and able to selectively produce the desired products. Companies, such as Audi, Carbon Recycling International and ZSW have successfully built and operated Power-to-X commercial plants to produce methane and methanol.

Alternatively, carbon dioxide can be directed converted into chemicals and fuels through the carbon dioxide electrochemical reduction coupled with renewable energy. Unlike the carbon dioxide hydrogenation process, this approach is still being developed at a laboratory scale. The main drawbacks that hinder the scale up of this approach are the very low selectivity to the desired products, low efficiency and the high overpotentials required. Considerable technical and catalytic advances have yet to be made to get this technology to a larger scale.

Further development of the Power-to-X technology will mainly require a decrease in the capital costs as well as an improvement of process efficiencies, especially electrolysis, leading to a reduction of the high production costs of fuels and chemicals produced through this technology. Validation of the different steps at large scale is also required. Despite the increase in the number of research projects in the area in partnership with industries, data related to the scale-up of the different steps of the technology is still missing. Finally, the integration of these steps to validate the whole Power-to-X process is mandatory to help the technology reach higher technology readiness levels.

Primary energy consumption growth slowed to 1.3% in 2019, less than half the rate of growth in 2018 (2.8%). The increase in energy consumption was driven by renewables and natural gas, which together contributed three quarters of the expansion. All fuels, apart from nuclear, grew at a slower rate than their 10-year averages. Carbon emissions from energy use grew by 0.5%, less than half 10-year average growth of 1.1% per year, partially reversing some of the unusually strong increase in 2018 (2.1%). More worrying is the trend for carbon emissions. The slowing in the growth of carbon emissions to 0.5% in 2019 may suggest some grounds for optimism. The carbon footprint grows stronger, wider and deeper. It's significance is indisputable.

Amine scrubbing is the most developed technology for post-combustion carbon dioxide capture. The typical flowsheet of this technology is similar to an industrial acid gas removal system using a chemical solvent. The electrochemically mediated amine regeneration is an alternative method to the conventional thermal regeneration of amines aimed at reducing the energy demand of the carbon dioxide capture process. Intensive research over the past years has also shown that adsorbent-based capture technologies have the potential to significantly reduce the energy penalty and/or equipment costs, when compared with state-of-the-art amine-based solvent capture processes. Microencapsulated carbon sorbents are considered to be promising materials for enhanced carbon dioxide capture owing to their drastically increased gas–liquid contacting area. New hybrid materials shaped by extrusion, suitable to be used in the electric swing adsorption process for carbon dioxide capture, were developed. Carbon dioxide capture technologies based on high-temperature solid looping cycles have also undergone great progress in the past decade, showing their potential to capture carbon dioxide at moderate cost and low energy penalty. Chemical looping combustion is another promising high-temperature carbon dioxide capture technology in which fuel is oxidized by a solid oxygen carrier that is reduced. Alternative novel technologies for capture applications include membrane-based concepts, where carbon dioxide separation is achieved by permeating one or more constituents of the gas stream through a selective membrane that acts as a barrier to other components of the stream.

The following is a summary of the presentations, which focused on the global increase in policies and regulations to produce low-sulfur and ultra-low-sulfur transportation fuels. Renewables energy consumption (which includes biofuels and all traded renewable electricity apart from hydro) continued to grow strongly, contributing its largest increase in energy terms on record. Biofuels consumption rose by 6%. Western Europe is pushing for a carbon neutral economy by

2050, which includes programs such as the EU Renewable Energy Directive (RED 2), an increase in the use of biofuels, EVs/HEVs, etc.

The "hydrogen economy" concept, where hydrogen is used as an energy vector, is however not new and was first mentioned in the 1970s after an oil crisis. However, mainly due to the low price of fossil fuels along with the high costs and technical challenges related to the use of renewable resources, fossil fuels remained the primary resource for hydrogen production. During the last few years, interest in the topic again has risen mainly due to climate change issues showing the need to develop other energy scenarios as well as to progress on materials and renewable power generation. However, in light of the actual environmental concerns, the advantages of using hydrogen as an energy vector are still highly dependent on how it is produced. There are different approaches currently under investigation for the sustainable production of hydrogen, water electrolysis being the most promising one. Other approaches to sustainably produce hydrogen are biomass electro-reforming and the use of microorganisms in a bio-electrochemical approach. Table 6.2 summarizes the key operational parameters as well as the main advantages and disadvantages of each technology.

Transitioning from classical electricity production to sustainable models also presents its share of challenges. In opposition to traditional approaches such as nuclear, gas or coal power, renewable electricity production systems such as photovoltaic panels (PV) and wind turbines (WT) are constrained by the alternating flux of electricity produced. While the production is optimal during some peak periods, the supply doesn't always fit with the demand, which makes relying solely on these options a rather risky option, especially if such sustainable options are to be implemented in large cities. Other sources of renewable electricity such as hydroelectricity represent a never-ending flow of current and, in some specific cases, the quantity of green electricity produced exceeds the local demand which could eventually represent opportunities for supplying other locations. However, transportation of electricity over large distances is costly considering the significant losses that are encountered over long distances. In addition, some of the production facilities generating this renewable electricity are already in remote locations on the Earth, which in turn involves significant losses from the start. Hence, green electricity, either flowing or peaking, is constrained by storage, which has become over the years a significant source of concern for industry and governments, unavoidably transferring to academia.

Mechanical storage of electricity has been known for decades and options as simple as using peak energy to pump water in an elevated basin to release it during peak hours have been investigated and even implemented. Such options, although simple, require a suitable landscape and are often criticized for their low efficiency. Batteries are also widely spread across the globe and are omnipresent in our lives going from cell phones to computers and of course, electric cars. Although originating from a historical technology, batteries have been improving over the years both in efficiency, size and price, which is good for the average consumer. It goes without saying that batteries will have their place in the upcoming mixed-energy system to where we are heading, however some of the best technologies actually on the market rely on lithium ions, which in turn still represents a finite element. Recuperation of ions has become a major concern in this field although still far from solving another big challenge, which would be to produce a battery big enough to sustain a full large-scale city.

6.5.2 Methane Pyrolysis

A cleaner technology based on natural gas that has received special attention in recent years is methane pyrolysis. The thermal decomposition of methane gives rise to hydrogen and solid carbon and thus, the release of greenhouse gases is prevented. Therefore, methane pyrolysis is a carbon dioxide-free technology that can serve as a bridge from fossil fuels to renewable energies.

Unlike these fossil fuel-based technologies, the pyrolysis of methane appears to be a potential process in aiding the transition to a sustainable hydrogen economy. Methane pyrolysis involves the thermal decomposition of methane into its components, that is, hydrogen and carbon. Therefore, the greatest benefit is the generation of CO_2-free hydrogen with solid carbon as the only by-product. This is a unique advantage of methane pyrolysis over the conventional steam methane reforming (SMR) and coal gasification processes. Depending on the characteristics and price of the carbon product, its sale may improve the economics of the industrial methane pyrolysis process. Here, the main problem is the absence of markets that can accommodate such large amounts of carbon and hence, new applications become necessary. The use of carbon for soil enrichment and environmental remediation may be the only alternative that would offer a huge market for the carbon resulting from methane pyrolysis. However, further studies are needed to prove the suitability of the pyrolysis carbon for such applications. On the other hand, if carbon is to be stored, the costs associated with solid carbon storage would be lower than the sequestration of carbon dioxide derived from steam methane reforming. Owing to the depletion of natural gas, the production of hydrogen via methane pyrolysis is not a sustainable process in the long term. Nevertheless, it may be a temporary solution and probably the most cost-effective bridging technology over the transition period towards renewable energies. Crucial aspects for the industrialization of the process are the use of suitable catalysts and the implementation of feasible processes on an industrial scale and the finding of realistic applications for the carbon product.

The thermal decomposition of methane is a promising emerging route towards the production of hydrogen with a low carbon footprint. Methane or other lower hydrocarbons are decomposed in a high temperature pyrolysis process generating hydrogen and solid carbon (Equation 6.2).

$$CH_4 \rightleftarrows 2H_2 + C \qquad \Delta H^{\circ}_{298} = 37.4\ \text{kJ}/\text{mol}\,H_2 \tag{6.2}$$

The enthalpy of reaction is significantly lower than for water splitting due to the much lower energy demand for hydrogen production. As for water electrolysis, no direct carbon dioxide emissions are generated in the reaction and the required energy can be supplied by electricity. The methane feedstock could stem from fossil sources, but also biogas, landfill gas or synthetic methane. The type of carbon product depends on the applied technology and can be considered for utilisation, for example, in tyre manufacturing, as a reducing agent in metallurgical processes or for storage. The carbon use generates a commercial benefit for the hydrogen production and replaces other fossil raw material sources in industrial areas other than the chemical industry. Based on today's electricity generation mix, the carbon footprint for hydrogen and carbon production can be reduced by ~50% if carbon use and heat integration within the process is successful. If only indirect emissions are allocated to hydrogen and carbon, an even higher reduction for hydrogen is achieved. Different technical approaches are followed and R&D and pre-commercial projects, for example thermal decomposition by a consortium of BASF, Linde and ThyssenKrupp and plasma pyrolysis by the US-based company Monolith, are ongoing. Other approaches include thermocatalytic decomposition (Muradov) or liquid metal processes (KIT, IASS). Since methane pyrolysis is only partly in a pre-commercial state, it is not taken into account for the scenario calculations in this study.

6.5.2.1 Process Concepts for Industrial Application

Concerning the technical implementation of methane pyrolysis on an industrial scale, different configurations may be possible, such as fluidized-bed reactors, moving-bed reactors, or liquid bubble column reactors (Figure 6.6). Fixed-bed reactors are usually the preferred option on a laboratory scale. They are simple to operate and provide a better understanding of catalyst performance, reaction kinetics and the effect of different operating conditions. However, the main drawback is the filling of the reactor with the carbon product during long-term experiments. This increases the pressure drop and eventually blocks the gas flow.

Fluidized-bed reactors overcome this problem and provide additional advantages that are crucial for the industrial operation. For instance, a continuous addition and withdrawal of catalyst particles is possible in a fluidized-bed reactor. Therefore, the pressure drop does not increase significantly and operation for longer times is possible. Additionally, the vigorous movement of the particles allows the efficient heat and mass transfer between the gas and the solid catalyst. Consequently, the temperature can be successfully controlled and the formation of hot spots is prevented. A potential configuration for the industrial operation with a carbon catalyst using a fluidized-bed reactor is presented in Figure 6.6A. Natural gas is introduced at the bottom of the reactor containing the carbon catalyst particles. The outlet gas, which is composed of unconverted methane and hydrogen, is passed first through a cyclone to remove the possible entrained carbon particles. Afterward, the gaseous product stream flows through a membrane to separate methane and hydrogen. The recovered methane is recirculated and fed back to the reactor together with a fresh natural gas stream. The carbon catalyst and carbon deposits are collected at the reactor bottom, cooled and stored. A small part of the carbon product can be introduced into the reactor after grinding and/or reactivation via a regeneration method. Another possible configuration consists of two parallel reactors (Figure 6.6B). These reactors operate in a cyclic mode by switching the natural gas feed and the regeneration agent stream (air, steam) between the two reactors. Methane

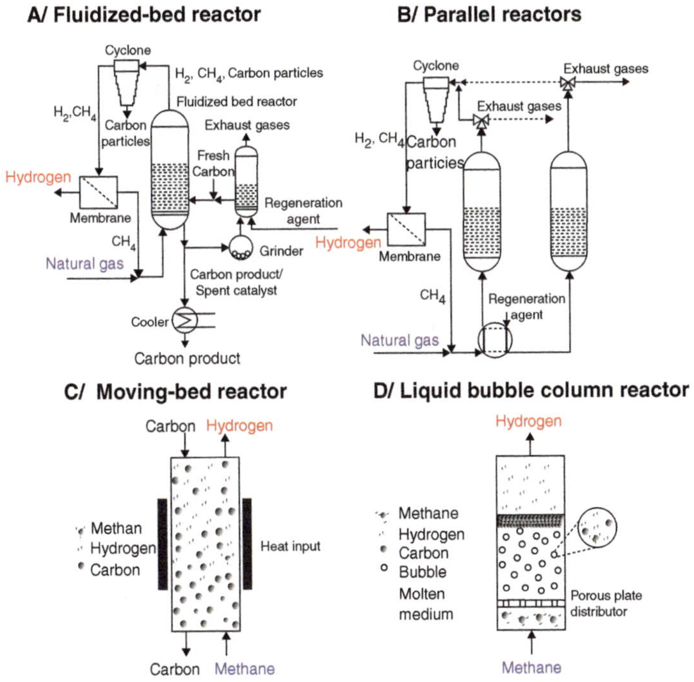

Figure 6.6 Potential reactor configurations for the industrial implementation of methane pyrolysis. (A) Fluidized-bed reactor with a catalyst regeneration unit. (B) Parallel reactors operating in a cyclic reaction-regeneration mode. (C) Moving-bed reactor. (D) Liquid bubble column reactor.

pyrolysis takes place in the first reactor, whereas the catalyst in the second reactor is regenerated. After a certain time, natural gas is fed into the second reactor, where the pyrolysis occurs over a regenerated catalyst. In the same way, the regeneration agent is introduced in the first reactor to recover the activity of the spent catalyst. Therefore, the operation in both reactors alternates between methane pyrolysis and catalyst regeneration. Despite the promising results obtained after catalyst reactivation, the existing regeneration techniques result in undesirable CO_x emissions and should be avoided industrially. For this reason, the resulting carbon product should be intended for further applications or stored until other regeneration methods are available. In any case, obtaining a safe carbon product has to be guaranteed and toxic metal catalysts such as nickel or cobalt must be avoided. Iron and carbon catalysts are therefore the best choice to industrialize the pyrolysis process of methane.

An interesting reactor concept from the industrial point of view is the moving-bed reactor (Figure 6.6C) developed by different partners in a recent BMBF-funded joint project. In a moving-bed reactor natural gas and carbon granules flow in counter current throughout the reactor. Natural gas is fed at the bottom of the reactor and carbon particles are fed at the top. The carbon is electrically heated by a direct current in the reaction zone. This heat is released and transferred to the feed gas. Consequently, the pyrolysis of methane takes place. The carbon originating from the reaction is accumulated on the carbon particles and continuously removed at the reactor bottom. The gaseous product is cooled down at the reactor top upon contact with the cold fresh carbon. As a consequence, the carbon is preheated by the outlet gas before entering the reaction zone and being electrically heated. The moving-bed reactor has important advantages. This kind of reactor provides a very good heat transfer between the gas and the solid catalyst. In addition, there is no back-mixing and the residence time can be controlled for both phases. The counterflow operation allows the energy integration of the reactor and the excellent heat transfer between the gas and the solid particles guarantees a thermally efficient process. Although methane pyrolysis for hydrogen production is still not industrialized, BASF plans to build a large-scale plant by 2030. They have investigated methane pyrolysis since 2010 and from 2013 to 2017 were part of the BMBF joint project based on a moving-bed reactor. Currently BASF is working on a test facility, the results of which will determine if methane pyrolysis can also succeed on an industrial scale.

A reactor design that is gaining importance for methane pyrolysis in recent years is the liquid bubble column reactor (Figure 6.6D). This kind of reactor operates with molten media, such as molten metals (Ti, Pb, Sn, Ga), molten metal alloys (Ni–Bi, Cu–Bi) or molten salts (KBr, NaBr, NaCl, NaF, $MnCl_2$, KCl). Molten metals and salts act as heat transfer fluids and avoid temperature losses or gradients along the reactor. Furthermore, they could also serve as potential catalysts for the reaction. The main advantage of liquid bubble column reactors is the easy separation of the carbon by-product from the liquid medium due to density differences. Moreover, the low volatility and the solubility of carbon in liquids enables its deposition on the top of the molten medium and facilitates its separation and handling. The preferred liquid medium is based on molten metals since their density is very different from that of carbon and thus, the carbon separation is more feasible. However, compared to molten metals, molten salts are less expensive and are being further investigated. The operation of liquid bubble column reactors is based on the formation of bubbles by the contact between the uprising methane gas and the liquid medium. The pyrolysis reaction occurs at the gas–liquid interface between the bubbles and the molten metal or salt. The bubbles ascend through the reactor and their size increases as a consequence of the molar expansion caused by the production of hydrogen and due to bubble coalescence. At the same time the resulting carbon is deposited at the gas–liquid interface. When the bubbles reach the surface of the molten medium, they open and release both hydrogen and carbon. Hydrogen leaves the reactor, whereas carbon is accumulated in layers on top of the molten metal/salt bath. High residence times of the gas in the liquid and the use of a porous plate distributor are key parameters in the design of liquid bubble column reactors. The porous plate homogeneously distributes the gas phase along the reactor and produces small bubbles that increase the gas–liquid interfaces for the reaction and consequently, the conversion of methane is improved. The constant removal of carbon from the liquid medium is possible in a bubble column configuration, so that blockage of the reactor due to carbon agglomeration is avoided. Additionally, the molten metal/salt is neither contaminated nor deactivated with the carbon product. This is an important advantage over solid catalysts. Solid catalysts deactivate and to be reused a regeneration process to burn away the carbon deposits is needed, which generates undesirable carbon dioxide emissions. According to a life cycle assessment to evaluate the environmental impacts of the liquid-metal technology, the pyrolysis of methane could reduce the global warming impact by up to 64% compared to the steam reforming process. Although this technology shows good prospects for the future and may be competitive with steam methane reforming, the industrial scale implementation is still a challenge. One important drawback is the limited stability of the molten media at the required high operating temperatures (>900–1000°C). Also, the corrosion at such high temperatures, especially in steel-based material reactors, limits the applicability of molten metals and salts.

6.5.2.2 Perspectives of the Carbon Coproduct

The quality and sale of the carbon coproduct are aspects that may improve the economic efficiency of the industrial pyrolysis of methane. The characteristics of the carbon depend on the catalyst used and the reaction conditions. The formation of carbon nanotubes and nanofibers usually occurs over metal catalysts. At high operating temperatures the diameter and length of the carbon nanofilaments decrease and their crystallinity and graphitization degree increase. Carbons with different morphologies are obtained in carbon-catalysed reactions. The morphology of carbon deposits over carbon materials depends on the catalyst nature. The formation of carbon blacks takes place over activated carbons, whereas carbon blacks produce amorphous turbostratic structures. The use of carbon nanotubes as a catalyst favours the growth of their walls, leading to the formation of multiwalled carbon nanotubes Carbon products derived from methane pyrolysis, such as carbon black, carbon fibres and carbon nanotubes, have different applications and markets (Table 6.3). Carbon black is mainly composed of carbon (≥97 wt %) in the form of colloidal particles. Approximately 90% of the total carbon black produced is used in rubber applications (as a filler and a strengthening/reinforcing agent in the manufacture of tyres and other rubber and plastic products), 9% as a pigment (to enhance formulations and coatings) and the remaining 1% for various applications. Carbon black is also a valuable product for the metallurgical industry and can

Table 6.3 Global market and price for potential carbon products.

Carbon product	Global market (metric tons)	Expected price ($/ton)
Carbon black	12,000,000 (2014) 16,400,000 (2022)	400–2,000
Carbon fibres	70,000 (2016) 100,000 (2020)	25,000–113,000
Carbon nanotubes	5,000 (2014) 20,000 (2022)	10,000–600,000,000

be used as a reducing agent to produce SiC and as a carbon additive/carburizer in the steel industry. Although the price of carbon black depends on its features, a range can be established depending on the quality and final use. For instance, the price of ASTM-grade carbon black for use in tyres ranges from \$400 to over \$1000 per ton, whereas the price of specialty grade carbon black can exceed \$2000 per ton. The global demand for carbon black, which was nearly 12 million metric tons in 2014, was expected to increase up to 16.4 million metric tons by 2022.

Carbon fibres are polycrystalline, two-dimensional planar hexagonal networks of carbon containing 92%–100% carbon by weight. They have applications in many fields, such as aerospace, automobiles, carbon-reinforced composite materials and textiles. Carbon nanofibers are also used as hydrogen storage materials and have applications in nano electrical devices, nanowires and nano sensors. The global market demand for carbon fibres was 70,000 t in 2016 and was projected to exceed 100,000 t in 2020. Carbon nanotubes (single-walled and multiwalled carbon nanotubes) find applications in polymers, electronics, plastics and energy storage. They are mostly intended to improve the thermal, electrical and mechanical properties of polymers, although they are getting special attention for their applications in lithium-ion batteries and renewable energy storage. The global market demand for carbon nanotubes was slightly over 5,000 tons in 2014 and was projected to surpass 20,000 tons by 2022.

The commercialization of the carbon by-product may positively influence the economics of the global process by reducing the hydrogen costs. This would be possible if the global market could absorb the huge amounts of the carbon product resulting from the pyrolysis process to meet the corresponding hydrogen demand. If the annual global production of hydrogen (approximately 60 million tons in 2017) comes from methane pyrolysis, close to 180 million tons of carbon would be generated. From a different point of view, if only the hydrogen production from steam methane reforming (approximately 48% of the total hydrogen production) is replaced by methane pyrolysis, approximately 86 million tons of carbon would be obtained yearly. Since the current world consumption of all carbon products is only 15–20 million tons per year, in none of the previous cases would the market be able to accommodate such large amounts of carbon. The only realistic application for such quantities of carbon is its use in the area of soil enrichment and environmental remediation, although its suitability has yet to be proven. The use of carbon for the amelioration of soil may accommodate large amounts of carbon and provide a virtually unlimited market for the methane pyrolysis carbon coproduct. The addition of carbonaceous products to soil can significantly improve seed germination, plant growth and crop yields. The application of carbon to soil increases the nutrient and water retention capacity as well as the microbial abundance. Carbon can store nutrients from the soil and serve as a potential slow-release fertilizer. For this reason, the amount of fertilizer applied to soil could be significantly decreased. Furthermore, the addition of carbon to soil may reduce the carbon dioxide in the atmosphere by enhancing its sequestration and by reducing the carbon dioxide emissions from the soil to the environment. Therefore, carbon would act as a long-term carbon dioxide sink. The structure of the amorphous carbons obtained in methane pyrolysis under certain operating conditions is similar to that of the carbons commonly used for soil improvement and thus, they are expected to have a positive effect on plant growth. Nevertheless, further investigations are required to validate the suitability of the carbon coproduct of methane pyrolysis for soil improvement and environmental clean-up. The application in this area requires that the carbon from the decomposition of methane is not poisoned with toxic metals, such as nickel or cobalt. Consequently, the use of iron and carbon catalysts is the only possibility to obtain a carbon product free of harmful compounds with a potential application for soil improvement and environmental remediation. Carbon storage as a back-up energy source for the future is another alternative solution for the carbon by-product. Unlike the carbon dioxide capture and sequestration systems required in the steam methane reforming process, no significant energy consumption is expected from carbon storage. Furthermore, for the same hydrogen production the volume required for the storage of carbon dioxide in liquid form obtained via SMR would be 10 times larger than the volume needed to store the carbon produced by methane pyrolysis.

Methane pyrolysis has a good outlook for future as a zero-emission technology for hydrogen production. Nevertheless, at present there are still many open questions for its industrial implementation. For instance, the role of the impurities of natural gas in the catalytic performance needs to be unravelled. In addition, a suitable experimental setup must be found to industrialize the process and the possible commercialization or storage of the carbon product, as well as its impact on the economics of the process, must be ascertained.

6.5.3 Thermochemical Processes

Splitting of water can be achieved at high-temperatures. However, the 2000°C temperature requirement for direct water splitting process needs to be reduced by catalytic thermochemical cycles. Many materials can serve as the catalyst as long as they match the redox-chemistry and stability requirements at the process temperatures. Thermochemical processes also exist for the splitting of carbon dioxide. In combination, these processes allow the generation of synthesis gas. As with electrolytic hydrogen generation, the question of sustainability or carbon-neutrality rests with the heat generation. Solar heat or waste heat would be preferable.

Thermochemical processes have been developed for solar heat supplies. The concept is well proven but should also be considered for high-temperature industrial waste heat, where new reactor concepts need to be developed. New catalytic thermochemical cycles are expected to reduce temperature requirements. The efficiency and durability of reactant materials for thermochemical cycling need to be improved. Efficient and robust reactor designs compatible with temperatures and heat cycling need to be developed. For solar thermochemical systems, the cost of the concentrating mirror systems needs to be reduced. Demonstration scale process will probably emerge by 2050.

6.5.4 Photocatalytic Processes

Photocatalytic reactions use solar light energy to split water at the surface of a catalyst. A lot of fundamental research activities have been dedicated to various concepts of photocatalytic water splitting. Current materials do not display sufficiently high yields for technical processes and still tend to work best at the ultraviolet part of the solar spectrum. One of the most well-known photocatalysts used is titanium dioxide, but it has its drawbacks, such as the need to add sacrificial agents into solution.

The carbon dioxide photoreduction is a promising way to convert one of the most abundant greenhouse gases to valuable chemicals. The photoreduction in the liquid phase is limited by the low solubility of carbon dioxide in water, but this point is overcome here by using an innovative photoreactor, which allows one to work up to pressures of 2 MPa, improving the overall productivity. The photoreduction was performed in the presence of Na_2SO_3 and using in primis commercial titanium dioxide (P25) and a set of titania catalysts functionalized by surface deposition of either monometallic or bimetallic cocatalysts. The gaseous products were hydrogen and traces of carbon monoxide, while, in the liquid phase, formic acid/format, formaldehyde and methanol were quantitatively detected. The pH was observed to shift the products distribution. A neutral environment led mainly to hydrogen and methanol, while, at pH 14, format was the most abundant compound. The trend for monometallic cocatalysts showed enhanced productivity when using noble metals (i.e., gold and platinum). To limit the cost of the catalytic material, bimetallic cocatalysts were explored, adding titania with Au + Ag or Au + Pt. This may open to the possibility of performing the reaction with a smaller amount of the most expensive metals. Later, we give some conclusions on the cost of the photocatalysts employed here, to support the overall feasibility assessment of the process.

$SrTiO_3$, Bi_2S_3 and $SrTiO_3/Bi_2S_3$ heterojunctions were prepared for efficient photocatalytic hydrogen production. The $SrTiO_3/Bi_2S_3$ heterojunction was prepared by different weight percentages of Bi_2S_3 (1, 3, 5 and 7%) with $SrTiO_3$. $SrTiO_3$ and Bi_2S_3 are exhibited in cubic and orthorhombic crystal structures, respectively. The morphologies of $SrTiO_3$ and Bi_2S_3 have existed as nano cubic and bar-like structures, respectively. Transmission electron microscopy (TEM) and X-ray photoelectron spectroscopy (XPS) results further confirmed the formation of the $SrTiO_3/Bi_2S_3$ heterojunction. The photoluminescence (PL) and electrochemical impedance spectroscopy (EIS) results indicated that the $SrTiO_3/Bi_2S_3$ heterojunction reduces the rate of photogenerated electron–hole pair recombination and electron-transfer resistance than pure $SrTiO_3$ and Bi_2S_3. The formation of the $SrTiO_3/Bi_2S_3$ heterojunction effectively separates the photogenerated charge carriers, which enhances the photocatalytic hydrogen production under UV light. The $SrTiO_3/Bi_2S_3$ heterojunction shows superior photocatalytic hydrogen production compared with bare $SrTiO_3$ and Bi_2S_3 due to the synergistic effect. The photocatalytic hydrogen production of $SrTiO_3/Bi_2S_3$ (5%) gives 7.7 mmol/g within 180 min under UV irradiation.

In recent years, lead halide perovskites have been widely used as appealing candidates in solar energy conversion. However, further development of these materials is restricted by their instability and toxicity. Therefore, searching for efficient lead-free perovskites with excellent photoelectric properties is a key goal. We report a post synthesis visible light irradiation on $Cs_2AgBiBr_6$ to form a defect rich $Cs_2AgBiBr_6$ via inducing surface defects in $Cs_2AgBiBr_6$. The formation of surface defects can promote surface charge separation by tuning the local atomic arrangement and electronic structure and optimizing the surface structure, thereby boosting the photocatalytic efficiency of the defect rich $Cs_2AgBiBr_6$. The photocatalytic hydrogen evolution efficiency is enhanced by about 5.27, 5.51 and 5.48 times compared to the bulk $Cs_2AgBiBr_6$, $Cs_2AgBiBr_6$/Pt and $Cs_2AgBiBr_6/Mo_3S_{13}^{2-}$, respectively. This defect rich $Cs_2AgBiBr_6$ presents strong stability with no performance decrease even after an 80 hour photocatalytic reaction. These findings provide an accessible strategy for preparing highly efficient and stable lead-free perovskites.

6.5.5 Biomass Electro-Reforming

Biomass electro-reforming for hydrogen production has gained increased attention due to its renewable character and to the flexibility of the feedstock. Different feedstocks, such as energy crops and forestry and agricultural residue, can be used. In this process, organic compounds originated from biomass, such as alcohols and sugars, are oxidized in aqueous media coproducing hydrogen in the cathode of the electrolysis cell. The advantages of this process are the use of much lower temperatures (<100°C) then the classical routes (steam reforming and partial oxidation) and the production of high-purity hydrogen. This process is also less energy-intensive than traditional water electrolysis process. Gutierrez-Guerra et al. compared the performance of the electrochemical reforming of ethanol with the classical catalytic reforming of ethanol. They reported that the electrochemical reforming process provided pure hydrogen in a single step while the catalytic process required additional steps. Moreover, the energy consumption and the amount of feedstock material were lower for the electrochemical process. Caravaca et al. investigated the hydrogen production from the electrochemical reforming of a bioethanol/water solution at 80°C in a PEM electrolyser. They showed that the electrochemical reforming of bioethanol led to the production of hydrogen and that the PEM electrolyser could perform for long operation times (6 h).

6.5.6 Microorganisms

Bio-electrochemical systems use microorganisms to catalyse the oxidation-reduction reaction at the cathodes and anodes. These systems are divided into microbial fuel cells (MFCs) and microbial electrolysis cells (MECs). In both processes, the bacteria decompose the organic material at the anode and hydrogen is produced at the cathode. However, additional electricity is required to the MEC system in order to suppress the production of methane and oxygen, which lower the selectivity to hydrogen. Despite the promising aspects of this technology, the design of efficient and scalable prototypes is still pending.

6.5.7 Hydrogen from Other Industrial Processes

Several processes in the chemical and petrochemical industry produce hydrogen in various degrees of purity as side streams. Depending on the site structure, this hydrogen might enter other processes, be captured and sold or just be flared. Refineries generate and use hydrogen on-site to desulfurise their fuels.

Other industrial sectors, like steel, produce significant amounts of hydrogen-heavy off-gases that generally get burned to provide process heat. Some of these processes might add potential for a close cooperation between different sectors for more efficient use of resources.

Hydrogen is mostly used as "captive" hydrogen that is on the same industrial site. Estimates assume 90% of global hydrogen production being used on site and only a relatively small amount being sold and transported to other locations.

6.5.8 Hydrogen Production Cost

Production costs of hydrogen strongly depend on two main factors: i) electricity costs and ii) utilisation rate of the electrolyser (operating time, particularly relevant in case renewable electricity is used and supply is fluctuating). Smolinka et al. have investigated the sensitivity of hydrogen production costs as a function of the utilisation rate and other factor and showed a cost range of 3–9 €/kg hydrogen for different scenarios. The relative CAPEX costs can be high at low utilisation rates, 65% and more and a factor five in hydrogen production costs was observed depending on the utilisation rate, showing a strong need to ensure a continuous operation. The calculation provided for alkaline and PEM electrolysis was performed using data from the IEA *Technology Roadmap: Hydrogen and Fuel Cells.* Cost strongly increases at very low operating range. The comparison shows the cost of hydrogen production for methane steam reforming in comparison with PEM and alkaline electrolysis at full operating hours based on the same cost of energy for the different options, showing the economic gap of nearly a factor of two for alkaline or even higher for PEM electrolysis. Cost reduction is therefore an absolute priority for future innovation.

6.5.9 Electrolysers

A simple calculation shows how the scale of chemical production needs to be matched to a reasonable scale of electrolysers. Typical sizes for methanol production plants (mega methanol, giga methanol plants) produce in the range of 3,000–10,000 tons per day. At a stoichiometric hydrogen

Table 6.4 Carbon footprint of hydrogen production.

Electrolysis		Carbon Footprint (CF) [$tCO_2eq\ tH_2$]	
Efficiency [$kWh/Nm^3\ H_2$]	4.3	CF of required power at 11.2g CO_2eq/kWh,	0.56
Required power [Mwh tH_2]	50.4	Construction and operation of assets, not include in the book	0.213

demand of 189 kg H_2/t methanol, an electrolyser would need to deliver 560–1900 t (or 6.6–22 million Nm^3) hydrogen per day. The electricity demand of such an electrolyser unit would be 28–95 GWh/d and an installed power of 1.3–4.3 GW would be necessary at continuous (8,000 h/a) operation, corresponding to 1–4 large power plant units. This corresponding power would need to be supplied by low-carbon sources such as those that are renewable, to allow a low-carbon chemical production. The electrolyser plant footprint would be at least $1{,}000 \times 700\ m^2$.

Chemical production plants based on hydrogen generated by water electrolysis are therefore likely to be realized in smaller, decentralized installations rather than world-scale type production plants. The fact that electrolysers only leverage moderate benefits from scale effects is advantageous in this respect. However, the relative capital investment of the chemical production plants will increase, so the economy of scale is reduced. The optimum size may differ from local specific infrastructures, availability of low-carbon electricity, carbon dioxide source, industrial symbiosis opportunities with other industry sites nearby, etc.

6.5.10 Carbon Footprint

Hydrogen produced from low-carbon electricity is not carbon neutral. Even for renewable electricity a carbon footprint must be calculated. For instance, the carbon footprint of electricity cradle to gate from a wind power plant encompasses the wind turbine production (e.g., accounting for steel and concrete used in its construction), operation of the wind power plant and transport of the electricity to the industrial site. Ecoinvent data suggests a carbon footprint of 11.2g CO_2eq/kWh for a wind powerplant including infrastructure. For delivering 1 ton of hydrogen at 4.3 kWh/Nm^3, 50.4 MWh electricity are required, corresponding to a footprint of 0.56 tCO_2eq. This is the carbon footprint per ton of hydrogen for low-carbon hydrogen production used in the subsequent sections of this book. A partial allocation of this carbon footprint to oxygen as the by-product of water electrolysis is not considered.

It should be mentioned that other authors performing a complete life cycle analysis on hydrogen from renewable electricity also included in their calculations electrolyser construction and operation with 0.043 kg CO_2eq/kg H_2 and hydrogen compressors and storage tanks with 0.17 kg CO_2eq/kg hydrogen. These relatively low impacts are disregarded in this book for simplicity reasons, as similar considerations would also be necessary for the conventional fossil processes, adding a high level of complexity to the analysis. Table 6.4 summarizes these numbers.

6.6 Advances in Chemical Carriers for Hydrogen

Is hydrogen the fuel of the future? This most-abundant element in the universe has, for many years, played a key role in the international space program and in more down-to-earth applications for refinery upgrading and chemicals production, such as methanol and ammonia. However, production of renewable hydrogen at a competitive cost and the transport thereof, present key issues that are being studied and progressed by almost every developed country in the world.

The benefits are obvious: hydrogen can be made by many and varied processes and burns completely and cleanly, with only water vapour as the by-product of combustion. Japan has progressed its hydrogen economy by developing detailed plans to replace nuclear energy and fossil fuel combustion with clean-burning hydrogen, both at the industrial and residential scale. Europe has, for several years, harnessed wind energy for power production; this technology also can be used for hydrogen production where desired. A project is already underway in Germany to provide hydrogen to an existing refinery for fuels upgrading, using wind energy to power the electrolyser that will produce the renewable hydrogen.

In the United States, California leads the way with initiatives that are not only policy-based but are also state law. Almost 9,000 fuel cell personal vehicles (2021) traverse the California freeways today, with fuel available for a 5-minute fill-up from 42 stations. By 2025, it is anticipated that there will be many more fuel cell electric vehicles (FCEVs) and many more refuelling stations in use, along with hydrogen use in city buses and heavy transport vehicles, supported by state laws driving these developments.

North-eastern states in the United States will likely follow California's lead and provincial energy ministries in Canada's eastern and western regions are already progressing initiatives that are modelled, to a degree, after those in California. While the mid-section of the United States would not be a likely place for hydrogen to take root, a United States Department of Energy (DoE)-sponsored study is already underway in oil- and gas-rich Texas to consider and demonstrate hydrogen as a fuel for data centre backup power and mobility uses. Additionally, the Texas study will consider how the Port of Houston may benefit from material transport using hydrogen, similar to what is already being done in Long Beach, California.

We will introduce liquid organic hydrogen carrier (LOHC) technology and how it has been applied to the world's first global hydrogen supply chain demonstration project.

6.6.1 Demand Drivers

Demand for hydrogen in the United States alone is more than 11 metric MMtpy. At present, the two largest uses for hydrogen are hydro processing in refineries (57%) and chemicals production (38% – ammonia and methanol production combined). Future additional demand will be driven by several applications:

- personal mobility;
- heavy transport of commercial goods;
- mass transport fuels;
- warehouse forklifts and port heavy goods movement;
- electric power backup;
- clean power generation; and
- residential and industrial applications.

Estimates project that United States hydrogen demand will grow to about 17 metric MMtpy by 2030 and to 63 metric MMtpy or more by 2050. Considering mobility applications alone, several automakers are producing personal FCEV cars – Toyota (Mirai), Honda (Clarity) and Hyundai (Nexo). While less than 10,000 FCEVs are on United States roads at present, projections call for up to 150,000 FCEV sales by 2025 and more than 1 MM sales by 2030.

A personal FCEV requires a fill-up of approximately 5 kg of hydrogen. Considering the current proven range of these vehicles and average annual use, the 2030 FCEV hydrogen demand in California alone can be calculated in the range of 250,000 metric ton for passenger cars only. When heavy-duty vehicles for material transport are factored into the mix, demand will be considerably higher.

By 2030, fuel cell-powered forklifts used by large-warehouse companies, such as Walmart and Amazon, could number more than 300,000. Total added hydrogen demand by 2030 could be approximately 3 metric MMt, although this number is expected to increase by around 45 metric MMt by 2050 – a significant volume.

Investment in the hydrogen economy is forecast in the range of $8B–$10B by 2030 and hundreds of thousands of new jobs could be created. While these numbers are not concrete, they impart a sense of scale for the potential of hydrogen in the United States and other forward-thinking countries.

Expected benefits from the integration of hydrogen into the United States energy picture include not only decarbonization, but also a bolstering of the overall economy, preservation and strengthening of the United States energy position and obvious improvements to health, the environment and quality of life. Benefits on a more global basis would be commensurately similar.

6.6.2 Options for Hydrogen Deployment

While the future for hydrogen looks bright, production, transportation, distribution and a host of associated infrastructural issues need to be considered. Assuming hydrogen is available, how to get it from where it is made to where it is most needed (and at a reasonable cost) is of paramount interest.

Historically, hydrogen has been transported either as a compressed gas in tankers or as a cryogenic liquid. Both methods have pros and cons, but transportation choice and cost are ultimately determined by how much hydrogen must be transported and over what distance.

Prior studies undertaken for localized distribution have considered and compared these conventional methodologies with newer, more promising options of transporting hydrogen bound by a chemical carrier, known generically as a liquid organic hydrogen carrier (LOHC).

Hydrogen is liquefied to 20 K (–253°C/–423°F) for transport as a cryogenic liquid. This is an energy-intensive operation that uses around 30% of the contained hydrogen energy to complete the liquefaction process. The cryogenic hydrogen is then transported from the liquefaction point to the point of use by truck. As per United States Department of Transportation (DoT) limitations on truck transport of liquefied hydrogen, approximately 4,000 kg of hydrogen can be transported per truckload. At the delivery site, the liquid can be cryo-pumped to the dispensing point for subsequent vaporization. At present, transportation of liquid hydrogen is available only by truck; therefore, the scale of transportation is limited and further technology development is required for large-scale liquid transportation by ship.

Alternatively, hydrogen at approximately 54 MPa can be transported by tanker from a terminal to a dispensing point where final compression (to around 70 MPa) is accomplished prior to ultimate utilization. United States DoT regulations also limit the amount of hydrogen that can be transported per tanker load to approximately 1,050 kg of hydrogen per truck.

Lastly, the LOHC option may be considered as a viable alternative. LOHC is a suitable technology for large-scale, long-distance hydrogen transportation, or for large-scale or daily to seasonal hydrogen storage. In all of these cases, the distance of transport, the volume of material being transported and the distribution particulars will drive the best choice for the particular hydrogen application at hand.

6.6.3 Advances in Hydrogen Storage/Transport Technology

A trademarked hydrogen process is a hydrogen storage and transportation technology for large-scale and long-distance transportation. Hydrogen is chemically fixed to toluene and converted to methylcyclohexane, according to Equation 6.3. Methylcyclohexane is a convenient carrier for hydrogen, as it is easy to store and transport under ambient temperature and atmospheric pressure. In the process, hydrogen is stored and transported in large-scale quantities at a competitive cost, since cryogenic liquefaction or compression to very high pressures is not required.

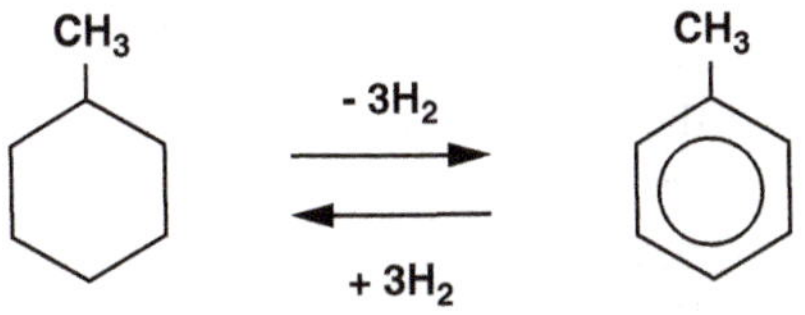

$\Delta H = 205$ kJ/mol (6.3)

The process is based on a simple configuration. Methylcyclohexane and toluene react in fixed-bed, tubular-type reactors in the vapour phase. Figure 6.7 shows simplified process flow diagrams of the hydrogenation and dehydrogenation processes for the overall process. In the hydrogenation process, a toluene feed is vaporized in the vaporizer (1) and mixed with hydrogen, including recycle gas. The mixed feed is superheated to the reaction temperature (2) and then enters the top of the reactor (3), which is a fixed-bed, tubular reactor charged with a semi-conventional hydrogenation catalyst. In the reactor tubes, toluene reacts with hydrogen to produce methylcyclohexane

Hydrogenation of toluene is an exothermic reaction. The generated heat is removed by cooling water to control the reaction temperature and the heat is subsequently recovered as medium-pressure steam, which can be further utilized as needed. The generated steam is clean energy without carbon emissions. The effluent gas from the tubular reactor is cooled, the condensed methylcyclohexane is separated (6) from the recycle gas and the liquid product (methylcyclohexane) is sent to storage tanks. Recycled gas is then returned to the reactor after being mixed with fresh hydrogen feed. Very high product yield is achieved in this step, which contributes to the efficiency of the hydrogen process. Toluene losses within the process are minimal.

Produced methylcyclohexane is then transported to the site of hydrogen use. Hydrogen is recovered from the methylcyclohexane via catalytic dehydrogenation. The dehydrogenation process can be described as follows: the methylcyclohexane feed is vaporized in the vaporizer (11) and superheated in the charge heater (12) before entering the catalyst-packed tubes of the dehydrogenation reactor (13). This reactor is a tubular, quasi-isothermal reactor design similar to that employed in the hydrogenation process. The dehydrogenation catalyst was specially developed in-house for this process. Since dehydrogenation is an endothermic reaction, an external heating source is required. Typically, a hot oil system is applied (14) to supply the necessary reaction heat.

Key to the success of the trademarked hydrogen[a] process was the development of a dehydrogenation catalyst that was not only robust, but also exhibited superior selectivity and stability.

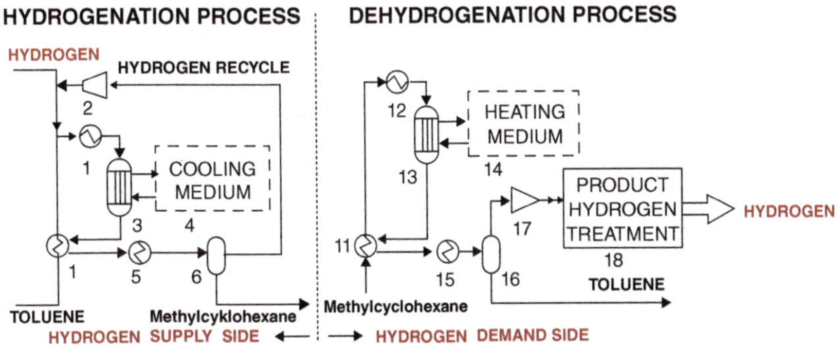

Figure 6.7 Hydrogenation and dehydrogenation processes for the trademarked hydrogen[a] process.

Methylcyclohexane conversion of greater than 95% was eventually achieved at a selectivity of 99.9%, which provides for a very low level of reaction by-products.

Accelerated, long-term testing showed remarkable stability even after 10,000 hr onstream. Commercially, a catalyst life of two years could be reasonably expected. Once a suitable candidate catalyst was developed, the next steps to eventual commercialization included scale-up for manufacturing and pilot-level demo of both the catalyst and the equipment. 10,000 hr of pilot plant operation were successfully completed and the expected performance and life of the catalysts were confirmed. In the pilot-scale test facility, methylcyclohexane and toluene were continuously dehydrogenated and rehydrogenated at a consistent rate of 50 Nm^3/hr of hydrogen.

6.6.4 Global Supply Chain

The companies Mitsubishi, Mitsui and Nippon Yusen established the Advanced Hydrogen Energy Chain Association for Technology Development (AHEAD) and subsequently initiated the world's first global hydrogen supply chain demo project, for which construction was completed in 2020.

AHEAD's key project is aimed at demonstrating the transport of hydrogen via a chemical carrier over a distance of 5,000 km, from the production site in Brunei to Tokyo Harbour. The demo project is funded by the Japanese New Energy and Industrial Technology Development Organization (NEDO).

The production and shipment of methylcyclohexane in ISO containers were started in early 2020. The flow of material is from Brunei to Tokyo Harbour and the return of toluene to Brunei. Hydrogen was produced at a pre-existing process plant site in Brunei through conventional steam reforming of natural gas; however, the trademarked hydrogen process can be applied to hydrogen from any production means.

The planned capacity of the demo system is 210 metric tpy (210,000 kgy), which is approximately the amount required for a single fuelling of around 40,000 FCEVs. In the case of the demo project, the hydrogen is burned in a power-generating gas turbine located within an existing site in Kawasaki, Japan. The two production sites – the hydrogenation site in Brunei and the dehydrogenation site in Kawasaki – operated continuously throughout 2020, fulfilling all of the original goals of the demo program.

6.6.5 Power-to-Gas Demo

In parallel to demonstrating the trademarked hydrogen process as part of a global supply chain, additional programs were undertaken to further demonstrate the robustness of the technology. A small-scale Power-to-Gas demo project was carried out in Yokohama, Japan and funded by NEDO.

Energy from a simulated wind farm was used to power an alkaline electrolyser for hydrogen production. The recovered alkaline hydrogen was then purified and sent to the existing demo plant. These steps demonstrate the ability to harness wind energy, which is then stored as hydrogen and can be transported to any site desired. In the demo project, the recovered hydrogen was purified and then utilized in a lab-scale solid oxide fuel cell (SOFC) for power production. In a real-world utilization scenario, for example, offshore wind energy could be harnessed and converted to transportable hydrogen that could then be transported by LOHC for use in clean power generation at another desired location. In this case, it is further demonstrated that having dehydrogenation adjacent to the SOFC makes possible efficient heat integration between the fuel cell, which generates waste heat and fuels the endothermic dehydrogenation plant.

6.6.5.1 Hydrogen Fuelling Stations

An additional demo project funded by NEDO targeted the use of LOHC to transport hydrogen for use in distributed fuelling stations. These stations could service passenger FCEVs or even larger fuel cells used in heavy-duty trucks or other large transport vehicles. For this study, a small-scale, packaged dehydrogenation facility was developed for testing. In the process flow, methylcyclohexane from storage is dehydrogenated in the compact recovery plant and then recovered hydrogen is purified to meet fuel cell specifications before being compressed to approximately 70 MPa prior to use.

6.6.5.2 Pathway to Commercialization

The technology, having been proven at both the pilot and demo scale, was deemed ready for commercial roll-out. In the Kawasaki demonstration, hydrogen produced remotely was transported via LOHC over 5,000 km and ultimately consumed in a gas turbine used for generating electric power. The demo plant's capacity of 210 metric tpy was simply a capacity of convenience – i.e., large enough to prove the capabilities of the supply chain while limiting the initial investment. Much larger and more ambitious transport programs have already been evaluated; these vary in size according to the specifics of the various applications.

The next step, envisioned for 2025–2026, provides for utilization in mobility and industrial applications, along with gas turbine consumption of clean-burning hydrogen of up to 30% in the fuel. Plans for this second chain provide for up to 1 BNm^3/yr of hydrogen (90,000 metric ton).

By 2030, up to 3.3 BNm^3/yr of hydrogen utilization is planned. This volume is equivalent to 300,000 metric tpy, according to Japan's Hydrogen Strategy outlined in 2017. In addition, Japan announced a target to introduce 3 MMtpy of hydrogen use by 2030, as per the Ministry of Economy, Trade and Industry's "Green growth strategy towards 2050 carbon neutrality," released in December 2020. Significant hydrogen acceleration and volume increase is expected to meet this target.

Future applications consider greatly expanded mobility uses both in the form of private and public transport, accelerated use in materials and goods movement both at ports and in long-haul overland transport, clean industrial consumption and gas turbine fuel with up to 100% hydrogen.

6.6.5.3 Transportation Studies in North America

Production of hydrogen via conventional methodologies, recovery of by-product or through greener approaches, such as water electrolysis, have a marked effect on both the cost and availability of hydrogen. When considering hydrogen as an alternative fuel, it is normal that the first question asked is usually, "What is the delivered cost of hydrogen?" How to best produce hydrogen for the future is a broad topic not covered by this section. The focus is to explore how to best deliver produced hydrogen reliably and at a competitive cost.

Concurrent to the demonstration of LOHC technology, a number of wide-ranging studies have been undertaken to explore competitive options for transportation. Both locally focused and long-haul transport have been considered and reviewed. Truck transport, rail, inland waterway barges, pipelines and ocean-going tankers have been compared in a number of studies.

In 2019, a detailed study was concluded with the government of British Columbia in Canada that considered the efficient transport of hydrogen produced from low-cost hydropower. The trademarked hydrogen technology served as the basis and transport of hydrogen from a coastal location was considered for delivery to Japan, to the city of Vancouver, Canada and to the port of Los Angeles in California, United States.

Another study was completed in 2019 with a US-based industrial partner to produce hydrogen from excess nuclear power in the Midwestern United States. The LOHC was then transported to an industrial port located on the United States Gulf Coast for eventual dehydrogenation and subsequent distribution for both industrial use and/or clean power production. In this case, rail and barge transport options were investigated.

Lastly, a research laboratory of the United States DoE was engaged to complete another detailed transportation study. In this study, the recovery of hydrogen from existing US Gulf Coast sources (cracker by-product and chlor-alkali off gas) was compared with green production of hydrogen using both solar and wind energy, for transport to California for use as mobility fuel and power. Various transport options were considered, including rail and ocean-going, long-range chemical carriers.

With reference to the DoE study, several findings were noted:

- transmission of methylcyclohexane and toluene by large product tankers (115,000 deadweight tonnage, or DWT) is 50% less expensive than transmission by rail at \$0.7/kg vs. \$1.53/kg. It is also important to note that greenhouse gas emissions are reduced by half when utilizing ships for transmission, compared with rail transport;
- use of by-product hydrogen incurs the lowest cost among all of the pathways analysed. Using ships as the transmission mode, the cost (reflected by natural gas substitution only) could be a delivered total cost of below \$2/kg, including transmission cost.

6.6.6 Future Applications

The technology has strength in large-scale transportation by ocean-going, long-range chemical carriers, as well as the potential to provide hydrogen for many applications beyond the obvious uses in mobility, material movement and clean power generation. Hydrogen production by green or traditional methods can be utilized in remote locations for convenience and competitiveness, with shipment to any other location for centralized or decentralized recovery of hydrogen, depending on the desirable end use at hand.

Large-scale hydrogen storage is one method to consider for storage of fluctuating renewable power, such as seasonal fluctuation, at the point when renewable power is widely introduced in the power grid network. The hydrogen also can be used in the gas grid network.

The future for hydrogen, both in the United States and internationally, is more a matter of "when" than "if". Production and transportation costs for hydrogen will come down as usage ramps up based on increasing demand. We believe that it is better to familiarize users with hydrogen sooner rather than later, in anticipation of the transition from blue to green or renewable hydrogen, as consumption rises. For now, the focus must be on adaptation.

The US DoE study results show the possibility of meeting the prescribed DoE target of \$3.10/kg of hydrogen (which dates to 2015) in a large-scale hydrogen supply chain by utilizing recovered off gas hydrogen and transport from point of production to point of use. Blue hydrogen can be made from conventional steam reforming of natural gas for under \$0.50/kg (variable production cost), assuming gas at \$2.50/MMBtu.

Hydrogen consumption in fuel cells for residential power, commercial applications, materials movement, personal mobility, mass transit and power generation will increase rapidly over the course of the next 10 years. Challenges notwithstanding, these will be exciting times for both early adopters and infrastructure providers as we accelerate into a clean new world.

6.7 Ammonia Fuel Cells

Fuel cells have a high potential to generate electricity from electrochemical reactions of hydrogen and oxygen supplied continuously from external sources. Fuel cells show very high electrical efficiency (~60%), which can be increased using combined heat and power (CHP) systems. Although they do not employ combustion processes for the release of energy, chemical reactions, in some cases similar to those previously presented, take place in these devices, thus deserving a special place in the study of ammonia as an energy vector.

Fuel cells are categorized mainly on their type of electrolytes, e.g., polymer electrolyte membrane fuel cell (PEMFC), solid oxide fuel cell (SOFC), molten carbonate fuel cell (MCFC), alkaline fuel cell (AFC), etc. Among the different types of fuel cells, proton-conducting fuel cells (PCFCs), SOFCs and AFCs can be used for power generation from ammonia.

Hydrogen is the most often used fuel for the fuel cells, but it has a low volumetric density of 0.0813 g/l at ambient temperature and pressure (300 K and 0.1 MPa). Thus, hydrogen needs high-pressure storage. The alternative solution can be to store and transport the molecule in the liquid state, which will increase its volumetric energy density as well as transportability. Carbon-based fuels, e.g., methanol and ethanol, are the most common hydrogen-rich alcohols, where hydrogen is chemically bonded with carbon. Because the carbon-based fuels generate carbon dioxide during the synthesis process and in the power conversion, these fuels are not very attractive for fuel cell applications. As previously stated, ammonia is an interesting alternative to the use of carbon-based fuels because of the absence of carbon in its molecules. Ammonia generates nitrogen during the three-phase reaction in fuel cell anodes with an additional loop around the nitrogen cycle. Nitrogen can be easily generated from air to feed the chemical process, usually employing the Haber–Bosch process. After energy consumption, nitrogen is released back to the atmosphere, which makes the overall system completely carbon-free. Another advantage to using ammonia as a fuel for electricity generation using fuel cells is based on its relatively low cost and reduced flammability when compared to other fuels. Any leakage of ammonia is easily detectable from its odors, making it safer than hydrogen.

This section presents an investigation of a storage option of green hydrogen in the form of ammonia based on system simulations considering detailed technical properties. Experimental data from a 1 MW PEM electrolyser is used to develop an electrolyser model and downstream a kinetic reactor for ammonia synthesis. The electricity is supposed to be supplied by a hydropower plant in Norway. An economical evaluation of the plant is performed for optimized process parameters. The costs of hydrogen and ammonia are calculated based on electrolyser investment costs published for deployment between 2020 and 2030 and several electricity cost scenarios. From the results, a broad economic optimum of the operating range of the electrolyser in the entire plant is obtained. This work provides a basis for the future evaluation of complex plants with electrolysers to produce green ammonia as well as strategies for the reduction of costs of green ammonia in the future.

Ammonia can be used directly in SOFCs and direct ammonia fuel cells (DAFCs). Polymer electrolyte membrane fuel cells (PEMFCs) need reforming of the gas into nitrogen and hydrogen before hydrogen is fed into the fuel chamber. As a result of the operation principle of fuel cells, the presence of nitrogen in the fuel stream acts only as a diluent of hydrogen and does not generate any pollutant nor cause significant performance decay, which allows fuel cells to operate with ammonia.

6.7.1 Proton-Conducting Fuel Cells

Proton-conducting fuel cells (PCFC) can be categorized into two different types, i.e., PEMFCs and proton-conducting solid oxide fuel cells (PC-SOFCs). PEMFCs work at low temperature ranges

(360–490 K) using pure hydrogen as fuel. The problem in using ammonia is that it may poison the Pt/C anode catalyst and react with the acidic membrane electrolytes. Therefore, ammonia itself may not be a suitable fuel for PEMFCs but can be used as the source for hydrogen production. Produced hydrogen can be used as PEMFC fuel for power generation. Usually, hydrogen from methane reforming is employed in most of these units.

Concurrently, the overall cost of the production of ammonia could be reduced using alternative hydrogen sources. Ammonia can be produced from nitrogen and HO using Nafion membranes as the proton-conducting electrolyte and Ru/C as the cathode at 360 K. Unfortunately, during the industrial production of ammonia, the used catalyst can be easily poisoned by oxygen and sulfur, which increases the cost of preparation and purification of the feeding gas. Thus, further research is required in this area.

6.7.2 Polymer Electrolyte Membrane Fuel Cells

Ammonia can be used as a hydrogen source for some fuel cells. For example, it can be employed in vehicles. In polymer electrolyte fuel cell (PEMFC) vehicles, the acceptable ammonia concentration in hydrogen is about 0.1 ppmv. Lithium-exchange X-type zeolite (Li-X) can be used for high-purity hydrogen production. Promising results were reported using ammonia fuel in both H^+ and O^{2-} solid electrolyte cells. High-temperature polymer electrolyte fuel cells (HT-PEMFCs) also show promising results using ammonia fuel. Furthermore, low-temperature polymer electrolyte membrane fuel cells (LT-PEMFCs) use the traditional Nafion membrane electrolyte previously mentioned. Nafion can be substituted with a phosphoric-acid-doped polybenzimidazole membrane, which operates at moderate temperature, i.e., in the temperature range of 490–470 K.

Tolerance to impurities normally increases with the temperature, e.g., carbon monoxide tolerance in LT-PEMFC is only a few parts per million volumes (ppmv), a level that increases up to 3% in HT-PEMFCs. HT-PEMFCs are an interesting technology that can employ methanol reforming as a result of this higher tolerance to impurities. The effects of different impurities (i.e., carbon monoxide, carbon dioxide and methanol) have been found to provide satisfactory tolerances for single cells and stacks. However, HT-PEMFCs were not very popular with ammonia as a fuel. Recently, technological developments have led to strong interest in HT-PEMFCs as a result of the possibility of using ammonia as a fuel.

6.7.3 Proton-conducting Solid Oxide Fuel Cells

In proton-conducting solid oxide fuel cells (PC-SOFC), hydrogen moves through the porous anode to the anode–electrolyte interface, where oxidation reactions occur to generate protons. In PC-SOFC, ammonia decomposes to hydrogen and nitrogen in the anode side according to thermodynamic equilibrium at specific temperatures and pressures. Nitrogen acts as an inert diluent. Protons move from the anode to cathode through the dense electrolyte and react with oxygen in the cathode compartment. Water is produced in the cathode side, where it exits the cell as steam.

This is a major advantage of proton-conducting fuel cells because hydrogen is not produced at the anode side and is not diluted by the steam. For example, gadolinium-doped barium cerate, a proton-conducting solid electrolyte, was used in an intermediate-temperature direct ammonia fuel cell. A schematic diagram of the direct ammonia-fed SOFC with a proton-conducting electrolyte is shown in Figure 6.8. Various materials have been found to improve the process. For example, doped-barium cerate or doped-barium zirconate have been found to be highly proton conducting at intermediate temperature ranges with good stability in carbon dioxide and humidified

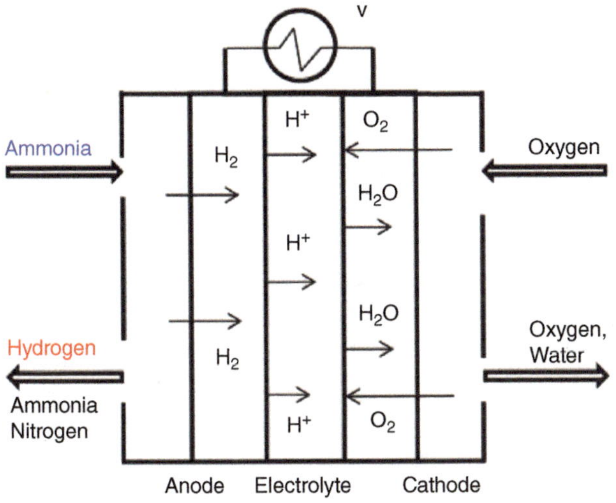

Figure 6.8 Schematic diagram of the ammonia-fed PC-SOFC.

environments. Fuel cells using these electrolytes conduct protons as well as prevent oxygen mobility from the cathode to the anode while stopping the contact with the anode fuel or catalysts, thereby inhibiting NO_x formation.

Doped $BaCeO_3$- and $BaZrO_3$-based PC-SOFCs are good options to avoid the formation of nitric oxide when using ammonia. The anodic reaction is shown in Equation 6.4.

$$2NH_3 \rightarrow 6e^- \; N_2 + 6H^+ \tag{6.4}$$

Thermodynamic analyses on PC-SOFC and oxide-ion-conducting SOFC show that the theoretical performance of the ammonia-fed PC-SOFC is a superior choice for the current generation. However, the high operating temperatures and long start-up time mean is not suitable to use direct ammonia SOFC for power in electrical vehicles, although it might be used as an auxiliary power unit for lorries, coaches, planes, or ships. Oxide-ion-conducting SOFCs operate indirectly, oxidizing fuel by thermally decomposing ammonia to extract hydrogen, which will be electrochemically oxidized in the fuel cell. In these units, the concentration of nitrogen gas will depend upon the thermodynamic equilibrium of the ammonia decomposition reaction at the given temperature and pressure. The reversible potential can also be reduced as a result of the dilution of the hydrogen concentration by inert nitrogen, which will impact this parameter. Furthermore, oxygen from air can move across the cathode layers to the cathode–electrolyte interface, thus reducing the oxygen ion. Then, the oxygen ions move through the electrolyte to react with hydrogen located at the anode, hence producing a cycle to generate electricity continuously until ammonia and oxygen are supplied.

6.7.4 Alkaline Fuel Cells

Alkaline membrane fuel cells (AFCs) transfer hydroxide ions through the electrolyte during the fuel cell operation. The most common type, potassium hydroxide solution electrolyte ammonia fuel cells, were first created in 1968. The working temperature of sodium hydroxide/potassium hydroxide fuel cells using ammonia fuel is from 470 to 720 K. The highest power density of these units, i.e., 40 mW cm^{-2} at 720 K, was observed using a nickel catalyst on both cathode and anode sides. However, the durability of direct ammonia alkaline fuel cells (operating with potassium hydroxide-based electrolytes) is a major issue as a result of the reaction between hydroxide electrolytes and carbon dioxide. Ammonia has also been tested as a fuel for alkaline membrane fuel cells using a nickel anode and manganese dioxide cathode.

Similar to other fuel cells, these ammonia fuel cells also have three main sections: cathode, electrolyte and anode. The cathode section is fed with oxygen, mainly from air, which combines with water and electrons (e^-) to produce hydroxides (OH^-) that move through the electrolyte, as shown in Equation 6.5. In the anode section, the supplied ammonia reacts with hydroxides (OH^-) producing nitrogen, water and electrons. These electrons move through electrical load to provide power, as shown in Equation 6.6. The overall reaction of the ammonia fuel call system is shown in Equation 6.7.

$$O_2 + 2H_2O + 4e^- \rightarrow 4OH^- \tag{6.5}$$

$$2NH_3 + 6OH^- \rightarrow N_2 + 6H_2O + 6e^- \tag{6.6}$$

$$4NH_3 + 3O_2 \rightarrow 2N_2 + 6H_2O \tag{6.7}$$

Interestingly, when operating at room temperature, the power density observed in these units has been very low as a result of the low catalytic activity of the electrode materials. The catalytic activity of Pt/C was found to be better than Pt–Ru/C and Ru/C in alkaline membrane fuel cells.

One of the drawbacks of the alkaline fuel cell is the reaction between alkaline membranes with carbon dioxide to form CO_3^{2-} ions, which reduce the OH^- ion conductivity. Moreover, potassium carbonate or sodium carbonate precipitates can also form because of the reaction with carbon monoxide in air. The crossover of ammonia through the polymeric membrane electrolyte may reduce the open-circuit voltage and efficiency. Furthermore, the oxidation of diffused ammonia in the cathode section may generate toxic nitrogen monoxide. Other materials and electrolytes have also been assessed to improve these effects. For example, direct ammonia borane fuel cells based on a sodium hydroxide electrolyte were first reported by Zhang et al. A study of the use of ammonia borane as a fuel in an alkaline fuel cell showed a power density of 315 mW cm^{-2} at ambient pressure and near ambient temperature, thus showing continuous improvement and promise for the use of these fuel cells employing ammonia as a fuel.

6.7.5 Direct Ammonia Solid Oxide Fuel Cell

Direct ammonia solid oxide fuel cells (DA-SOFCs) using yttrium stabilized zirconia (YSZ) electrolytes and platinum electrodes were first introduced by Wojcik et al., which showed a maximum power density of 50 mW cm^{-2} at 1070 K. When these units are employed, it is possible to avoid the formation of NO_X in DA-SOFC systems, producing only nitrogen gas as the nitrogen-containing product. NO_X is not produced by the decomposition of ammonia at the Ni-YSZ anode. However, there are still many unanswered questions requiring further research to improve the YSZ-based fuel cell performance before full commercialization.

Unfortunately, the ionic conductivity of YSZ decreases considerably and is not suitable as an electrolyte at moderate temperatures. Therefore, different electrolytes have been used to overcome the obstacles and for better overall performance and stability at intermediate temperatures. High-temperature SOFCs show the best performance for DA-SOFC because ammonia can be used directly. The anodic reaction in Equations 6.8 and 6.9 in SOFC using ammonia as a fuel follows the reactions shown in the equations below:

$$2NH_3 + 5O_2 \rightarrow 2NO + 3H_2O + 10e^- \quad (6.8)$$

$$2NH_3 + 3NO \rightarrow 5/2N2 + 3H_2O \quad (6.9)$$

Among these reactions, Equation 6.7 is the rate-limiting step because of the slow diffusion of oxygen through the electrolyte; therefore, a small amount of nitric oxide will be produced at the anode side. This NOx can exceed 60% concentration.

Furthermore, DA-SOFC based on an oxide-ion-conducting $Ce_{0.8}Sm_{0.2}O_{1.9}$ electrolyte can be used at intermediate temperature ranges while showing high conductivity. The difference in power density is very small between SOFC with ammonia or hydrogen as fuels.

6.7.6 Equilibrium Potential and Efficiency of the Ammonia-Fed SOFC

Under open-circuit conditions, the equilibrium potential (voltage) of any fuel cell is related to the partial pressures of the species involved in the electrochemical reaction, which can be written as

$$E = E^{\circ} + \frac{RT}{2F} \ln \left[\frac{P_{H_2} \left(P_{O_2} \right)^{1/2}}{P_{H_2O}} \right] \quad (6.10)$$

where

$$E^{\circ} = -\frac{\Delta G}{nF} \quad (6.11)$$

where n is the number of electrons involved per electrochemical reaction, F is the Faraday constant, G is the change in Gibbs free energy, $E°$ is the voltage at standard pressures and P_{H_2}, P_{H_2O} and P_{O_2} are the partial pressures of hydrogen, steam and oxygen, respectively. P_{H_2O} refers to the partial pressure of H_2O at the anode side of an oxide-ion-conducting SOFC, while the parameters refer to the cathode side for PC-SOFC.

During the operation of SOFC, the maximum electrical work (W) (Equation 6.12) can be calculated as

$$W = q.E \quad (6.12)$$

where q is the electrical charge passing through the outer circuit to give electric power. The efficiency (Equation 6.13) of the ammonia fed SOFC can be calculated as the ratio of the maximum electrical work (W) extracted from the SOFC to the maximum conversion of the chemical energy of the fuel fed to the SOFC

$$\eta(\%) = \frac{W}{-\Delta H^{\circ}} \times 100 \quad (6.13)$$

where $H°$ is the lower heating value of the fed ammonia under standard conditions (320.1 kJ/mol). The fuel utilization is the ratio of consumed fuel (ammonia) and oxidant (oxygen), whereas the oxygen utilization is based on the ratio of the feeding fuel and oxidant.

6.8 Conclusions

The present strict regulations to reduce greenhouse gas emissions are forcing the development of a sustainable low-carbon economy. In this context, hydrogen plays a key role in the transition to clean energy. Hydrogen has received special attention as a zero-emission fuel in fuel cells and internal combustion engines. In addition, hydrogen is essential for converting industrial carbon dioxide emissions into platform chemicals such as methanol, which prevents carbon dioxide from being released into the atmosphere. Most of the current global hydrogen production comes from coal gasification and steam methane reforming. However, both processes are accompanied by large carbon dioxide emissions and must be avoided to comply with restrictions on GHG emissions. Water electrolysis based on renewable energies is the greenest technology for hydrogen production, but total dependence on renewables is not possible, at least, in the near future. Consequently, alternative fossil fuel-based processes with a low carbon footprint become essential. Methane pyrolysis is a suitable technology for converting natural gas into hydrogen without carbon dioxide emissions. Although methane pyrolysis is not a sustainable process due to the depletion of natural gas reserves, it can be an appropriate temporary solution until renewable energies are well established. The decomposition of methane has been extensively studied over different metal (Ni, Co, Fe) and carbon catalysts. From an industrial point of view, only the use of iron and carbon catalysts is viable due to their nontoxicity. This allows a safe storage of the carbon product or its potential application, for example, for soil improvement and environmental remediation. Nevertheless, severe temperatures are required to obtain satisfying hydrogen yields over iron and carbon catalysts. Although the reaction mechanism has been widely investigated, no clear conclusions have been drawn on the elementary reactions and the rate-limiting steps. Therefore, further research is needed to elucidate these issues. This will enable the development of suitable catalysts to optimize the activity under milder operating conditions. Furthermore, the role of natural gas impurities in the production of hydrogen has to be investigated in more detail. Finding an adequate reactor configuration and a thorough understanding of how the sale of the carbon product may affect the economics of the process are additional issues that will determine the industrial implementation. Although the industrialization of methane pyrolysis still has to overcome many challenges, this CO_2-free technology is a promising process that can serve as a bridge in the evolution towards a sustainable hydrogen production based on renewable energies.

Further Reading

Abdalla, A.M., Hossain, S., Nisfindy, O.B. et al. (2018). Hydrogen production, storage, transportation and key challenges with applications: a review. *Energy Convers. Manag.* 165: 602–627. doi: 10.1016/j.enconman.2018.03.088.

Arunkumar, P., Aarthi, U., Rengaraj, S. et al. (2019). Review on solid oxide electrolysis cell: a clean energy strategy for hydrogen generation. *Nanomater. Energy* 8: 1–73. doi: 10.1680/jnaen.18.00009.

Azwar, M.Y., Hussain, M.A. and Abdul-Wahab, A.K. (2014). Development of biohydrogen production by photobiological, fermentation and electrochemical processes: a review. *Renew. Sustain. Energy Rev.* 31: 158–173. doi: 10.1016/j.rser.2013.11.022.

Ball, M. and Weeda, M. (2015). The hydrogen economy – vision or reality? *Int. J. Hydrog. Energy* 40: 7903–7919. doi: 10.1016/j.ijhydene.2015.04.032.

Baykara, S.Z. (2018). Hydrogen: a brief overview on its sources, production and environmental impact. *Int. J. Hydrog. Energy* 43: 10605–10614. doi: 10.1016/j.ijhydene.2018.02.022.

Bazzanella, A.M. and Ausfelder, F. (2017), Technology study: low carbon energy and feedstock for the European chemical industry. Dechema Gesellschaft fur Chemische Technik und Biotechnology e.V., Cefic. 164.

Berkeley, L. (2017). Sunfire SOEC steam electrolysis module for Salzgitter Flachstahl. *Fuel Cells Bull.* 2017: 12. doi: 10.1016/s1464-2859(17)30269-9.

Bessarabov, D. and Millet, P. (2018). The PEM water electrolysis plant. In: *PEM Water Electrolysis* (ed. B. Pollet), 1–31. San Diego, CA: Academic Press.

Bode, A., Agar, K., Buker, K., et al. (2014). Research cooperation develops innovative technology for environmentally sustainable syngas production from carbon dioxide and hydrogen. In Proc. of the World Hydrogen Energy Conference Gwangju. South Korea, June 2014. Curran, New York.

Buttler, A. and Spliethoff, H. (2018). Current status of water electrolysis for energy storage, grid balancing and sector coupling via power-to-gas and power-to-liquids: a review. *Renew. Sustain. Energy Rev.* 82: 2440–2454. doi: 10.1016/j.rser.2017.09.003.

Caravaca, A., De Lucas-Consuegra, A., Calcerrada, A.B. et al. (2013). From biomass to pure hydrogen: electrochemical reforming of bio-ethanol in a PEM electrolyser. *Appl. Catal. B Environ.* 134–135: 302–309. doi: 10.1016/j.apcatb.2013.01.033.

Chourashiya, M.G. and Urakawa, A. (2017). Solution combustion synthesis of highly dispersible and dispersed iridium oxide as an anode catalyst in PEM water electrolysis. *J. Mater. Chem. A* 5: 4774–4778. doi: 10.1039/c6ta11047a.

Conte, F., Villa, A. et al. (2022). Effect of metal cocatalysts and operating conditions on the product distribution and the productivity of CO_2 photoreduction. *Ind. Eng. Chem. Res.* 61 (8): 2963–2972.

Core, R. (2018). Hydrogen for energy storage and transportation at scale. https://www.energy.gov/sites/prod/files/2018/08/f54/fcto-h2-scale-kickoff-2018-7-del-core.pdf (Accessed 23 December 2018).

Coutanceau, C., Baranton, S. and Audichon, T. (2018). Hydrogen production from water electrolysis. In: *Hydrogen Electrochemical Production* (ed. B. Pollet), 17–62. San Diego, CA: Academic Press. doi: 10.1016/b978-0-12-811250-2.00003-0.

Datta, M.K., Kadakia, K., Velikokhatnyi, O.I. et al. (2013). High performance robust F-doped tin oxide-based oxygen evolution electro-catalysts for PEM based water electrolysis. *J. Mater. Chem. A* 1: 4026–4037. doi: 10.1039/c3ta01458d.

Di Giovanni, C., Reyes-Carmona, Á., Coursier, A. et al. (2016). Low-cost nanostructured iron sulfide electrocatalysts for PEM water electrolysis. *ACS Catal.* 6: 2626–2631. doi: 10.1021/acscatal.5b02443.

Dincer, I. and Zamfirescu, C. (2016). Hydrogen production by electrical energy. In: *Sustainable Hydrogen Production* (ed. I. Dincer and C. Zamfirescu), 99–161. Cambridge: Elsevier.

Ferreira-Aparicio, P., Benito, M.J. and Sanz, J.L. (2005). New trends in reforming technologies: from hydrogen industrial plants to multifuel microreformers. *Catal. Rev.* 47: 491–588. doi: 10.1080/01614940500364958.

Ganapathy, M. et al. (2021). Preparation of $SrTiO_3/Bi_2S_3$ heterojunction for efficient photocatalytic hydrogen production. *Energy Fuels* 35 (18): 14995–15004. doi: 10.1021/acs.energyfuels.1c00979.

Götz, M., Lefebvre, J., Mörs, F. et al. (2016). Renewable power-to-gas: a technological and economic review. *Renew. Energy* 85: 1371–1390. doi: 10.1016/j.renene.2015.07.066.

Gutierrez-Guerra, N., Jimenez-Vazquez, M., Serrano-Ruiz, J.C. et al. (2015). Electrochemical reforming vs. catalytic reforming of ethanol: a process energy analysis for hydrogen production. *Chem. Eng. Process. Process Intensif.* 95: 9–16. doi: 10.1016/j.cep.2015.05.008.

Hansen, J.B. (2015). Solid oxide electrolysis—a key enabling technology for sustainable energy scenarios. *Faraday Discuss.* 182: 9–48. doi: 10.1039/c5fd90071a.

He, Z. et al. (2021). Lead-free $Cs_2AgBiBr_6$ Perovskite with enriched surface defects for efficient photocatalytic hydrogen evolution. *Energy Fuels* 35 (18): 15005–15009. doi: 10.1021/acs.energyfuels.1c01482.

Hu, H., Fan, Y. and Liu, H. (2008). Hydrogen production using single-chamber membrane-free microbial electrolysis cells. *Water Res.* 42: 4172–4178. doi: 10.1016/j.watres.2008.06.015.

IEA (2015). IEA Technology Roadmap: Hydrogen and Fuel Cells. https://www.iea.org/reports/technology-roadmap-hydrogen-and-fuel-cells (Accessed 8 March 2023).

Lavoie, J.-M. (2014). Review on dry reforming of methane, a potentially more environmentally friendly approach to the increasing natural gas exploitation. *Front. Chem.* 2: 1–17. doi: 10.3389/fchem.2014.00081.

Li, S., Yan, R., Wu, G. et al. (2013). Composite oxygen electrode LSM-BCZYZ impregnated with Co_3O_4 nanoparticles for steam electrolysis in a proton-conducting solid oxide electrolyzer. *Int. J. Hydrogen Energy* 38: 14943–14951. doi: 10.1016/j.ijhydene.2013.09.082.

Liu, N. (2021). Increasing blue hydrogen production affordability. *Hydrocarbon Processing*. June 2021.

Martinson, C.A., Van Schoor, G., Uren, K.R. and Bessarabov, D. (2014). Characterisation of a PEM electrolyser using the current interrupt method. *Int. J. Hydrogen Energy* 39: 20865–20878. doi: 10.1016/j.ijhydene.2014.09.153.

Moçoteguy, P. and Brisse, A. (2013). A review and comprehensive analysis of degradation mechanisms of solid oxide electrolysis cells. *Int. J. Hydrogen Energy* 38: 15887–15902. doi: 10.1016/j.ijhydene.2013.09.045.

Mozammel, T., Dumbre, D., Hubesch, R. et al. (2020). Carbon dioxide reforming of methane over mesoporous alumina supported Ni(Co), Ni(Rh) bimetallic and Ni(CoRh) trimetallic catalysts: role of nanoalloying in improving the stability and nature of coking. *Energy Fuels* 34 (12): 16433–16444.

Nakamura, J. and Uchijima, T. (1993). Reforming of methane with carbon dioxide on active carbon-supported Co, La catalysts and reactivity of the catalysts. *J. Catal. Soc. Jap. (Shokubai)* 35: 478.

Ogawa, T., Takeuchi, M. and Kajikawa, Y. (2018). Analysis of trends and emerging technologies in water electrolysis research based on a computational method: a comparison with fuel cell research. *Sustainability* 10: 1–24. doi: 10.3390/su10020478.

Ohmori, T., Tachikawa, K., Tsuji, K. and Anzai, K. (2007). Nickel oxide water electrolysis diaphragm fabricated by a novel method. *Int. J. Hydrogen Energy* 32: 5094–5097. doi: 10.1016/j.ijhydene.2007.07.055.

Paladino Lino, A.V., Assaf, E.M. and José Assaf*, J.M. (2020). Adjusting process variables in methane tri-reforming to achieve suitable syngas quality and low coke deposition. *Energy Fuels* 34 (12): 16522–16531.

Rashid, M.M., Al Mesfer, M.K., Naseem, H. and Danish, M. (2015). Hydrogen production by water electrolysis: a review of alkaline water electrolysis, PEM water electrolysis and high temperature water electrolysis. *Int. J. Eng. Adv. Technol.* 4: 80–93.

Rego de Vasconcelos, B. and Lavoie, J.-M. (2019 June 05). Recent advances in power-to-X technology for the production of fuels and chemicals. *Front. Chem.* doi: 10.3389/fchem.2019.00392.

Sánchez-Bastardo, N., Schlögl, R. and Holger Ruland, H. (2021). Methane pyrolysis for zero-emission hydrogen production: a potential bridge technology from fossil fuels to a renewable and sustainable hydrogen economy. *Ind. Eng. Chem. Res.* 60 (32): 11855–11881.

Sapountzi, F.M., Gracia, J.M., Weststrate, C.J.K.J. et al. (2017). Electrocatalysts for the generation of hydrogen, oxygen and synthesis gas. *Prog. Energy Combust. Sci.* 58: 1–35. doi: 10.1016/j.pecs.2016.09.001.

Schneider, R.V., Kurosaki, D. and Oki, M. (2021). Advances in chemical carriers for hydrogen. Is hydrogen the fuel of the future? *Hydrocarbon Processing*. August 2021.

Smolinka. et al. (2011). NOW-Study: stand und Entwicklungspotenzial der Wasserelectrolyse zur Herstellung von Wasserstoff aus regenerative Energien. Fraunhofer ISE.

Song, C. (2006). Global challenges and strategies for control, conversion and utilization of CO2 for sustainable development involving energy, catalysis, adsorption and chemical processing. *Catal. Today* 115: 2–32.

Sousa, J., Waiblinger, W. and Friedrich, K. (2022). Techno-economic study of an electrolysis-based green ammonia production plant. *Ind. Eng. Chem. Res.* 61 (39): 14515–14530.

Spath, P.L. and Mann, M.K. (2004). Life cycle assessment of renewable hydrogen production via wind/electrolysis. Milestone Completion report-NREL/MP-560-35404.

Valera-Medina, A. et al. (2021). Review on ammonia as a potential fuel: from synthesis to economics. *Energy & Fuels* 35 (9): 6964–7029.

Wojcik, A., Middleton, H., Damopoulos, I. and Van Herle, J. (2003). Ammonia as a fuel in solid oxide fuel cells. *J. Power Sources* 118: 342–348. doi: 10.1016/S0378-7753(03)00083-1.

Zangeneh, F.T., Sahebdelfar, S. and Ravanchi, M.R. (2011). Conversion of carbon dioxide to valuable petrochemicals: an approach to clean development mechanism. *J. Nat. Gas Chem.* 20 (3): 219–231.

Zeng, K. and Zhang, D. (2010). Recent progress in alkaline water electrolysis for hydrogen production and applications. *Prog. Energy Combust. Sci.* 36: 307–326. doi: 10.1016/j.pecs.2009.11.002.

Zhang, X.-B., Han, S., Yan, J.-M. et al. (2007). A New fuel cell using aqueous ammonia-borane as the fuel. *J. Power Sources* 168 (1): 167–171. doi: 10.1016/j.jpowsour.2007.03.009.

Zhang, X.-B., Yan, J.-M., Han, S. et al. (2008). A high performance anion exchange membrane-type ammonia borane fuel cell. *J. Power Sources* 182 (2): 515–519. doi: 10.1016/j.jpowsour.2008.04.032.

7

Power-to-Fuels

CONTENTS

Converting Power into Chemicals and Fuels: Power-to-X Technology for a Sustainable Future, First Edition. Martin Bajus.

7.1 Introduction

Definitions of revolutionary fuels for transportation and/or evolution of fuels in personal vehicles are:

- liquid gasoline and oxygenate blending → oil production and Otto-cycle, bioethanol from corn and molasses, biodiesel from vegetable oil. Fuel generation 1.0; Industry 1.0;
- CNG → lowering carbon pollution by adopting compressed natural gas as fuel. Fuel generation 2.0; Industry 2.0;
- hydrogen → using hydrogen as energy carrier in hybrid cars. Fuel generation 3.0; Industry 3.0;
- electricity → solar, tide, hydropower, geothermal, biomass, coal, nuclear energy is used to produce electricity for electric vehicles. Fuel generation 4.0; Industry 4.0.

Definitions of revolutionary fuels for transportation are presented in Table 7.1.

Fuels used for transportation have limitations as highlighted in Table 7.2. In general, different fuel options are also good for energy security, which can be evaluated from affordability, availability, price stability, dependence on imports and dependence of fuel oil. Affordability and availability

Table 7.1 Definition of revolutionary energy and fuels for transportation, learning from design principles in Industry 4.0.

Industry 4.0	Design principles in Industry 4.0	Energy and Fuels 4.0
Cyberphysical system	Interoperability	Infrastructure decentralization (fuel and electric station), Intersection of bioeconomy and circular economy
Digitalization and the Internet of Things	Service orientation	Affordability (price), energy subsidy, energy justice, fuel poverty
Smart factory (modularization)	Decentralization, modularity	Decarbonization, distributed production, economy of scale
Big data	Complexity minimization by using databases	Availability, modelling using artisicial intelligence, design of catalytic materials for energy production

Table 7.2 Limitation of transportation fuels for personal vehicles.

Generation	Energy and Fuels	Limitation	Ways Forward
Fuel 1.0 (liquid)	Gasoline	CO_2 creation	Butanol from biomass
Fuel 2.0 (gas)	Natural gas	People's resistance	Better materials for methane storage
Fuel 3.0 (gas)	Hydrogen	Hydrogen is not primary energy carrier. Limited resource of platinum as electrode material	Hydrogen from electrolysis consumes energy. Only applicable in developed countries with oversupply of energy from nuclear and renewables
Energy&Fuel4.0 (electric)	Electricity	Only in a country with energy oversupply. Limited resource of lithium. Potential disruption in cobalt supply	Improved battery efficiency. Electricity from sun or wind. Battery with earth abundant elements

are important parameters relating to people's perception. The stability of fuel prices is important, as many countries are still dependent on imports and still have a large portion of fuels from petroleum, compared to gas and other alternative fuels. With fluctuations in the price of crude oil from 30 to 100 dollars/bbl. in a short period of time, price stability has become an important concern. Reliance on imports becomes a very important parameter. Hence, energy diversification is essential.

Power-to-Electrochemical technologies and Power-to-Electro fuel processes, thus pave the way for integration of surplus renewable energies into the petrochemical and transport sector. Electrofuel technologies could first be introduced to enhance the carbon of today's fuels derived from biomass and wastes. Approximately 7% of global oil demand will be replaced by electric vehicle (EV) usage even though the real growth of electric vehicles depends on numerous factors such as the price of a battery, subsidy from government and the availability of rare-earth and lithium elements. It is also possible to consider a zero gasoline refinery where a refinery is dedicated to producing olefins, aromatics, synthesis gas and hydrogen for petrochemicals.

The integration of carbon dioxide via methanol and methane would require a comparably low research development effort and allow use of much of the existing petrochemical infrastructure and hydrocarbon technology. The designed chemical models include carbon capture and utilization (CCU) technologies for the direct conversion of carbon dioxide into olefins, BTX aromatics, carbon monoxide and hydrogen, ethylene oxide and styrene. These electrochemical technologies are currently at early research and development stages with TRLs below 7.

After The Glasgow Climate Pact in November 2021, (Chapter 3, Section 3.4.2.3) most countries decided to commit to act on their greenhouse house gases (GHG) emissions to cope with the threats that are represented climate change. From all nations around the world, a small handful decided not to pledge specific reductions or control over their emissions up to 2030. However, reducing greenhouse gases may represent a far more ambitious challenge than it seems because of the significant consumption of energy that is being used. In addition, such a tendency is increasing proportionally with the portion of the Earth's population reaching for the middle class.

Energy is considered one of the most important fundamental requirements for human survival and therefore making this transition represents colossal modifications in our systems. Energy is used to produce electricity, heat and cooling and a large amount of it is used for transportation, ranging from individual cars to planes, trains and boats. As energy cannot be created, we always deal with conversion processes. Many of them involve molecular or solid energy carriers and thus it evident that chemical technology is at the centre of the energy challenge. Chemical technology can control the energetic "cost" of the conversion of energy carriers. After all, our present fossil energy carriers were created by biochemical processes in photosynthesis and later converted to its present high-density form geochemical processes.

The global demand for liquid hydrocarbons – as fuels for transport, as petrochemical feedstock and for other uses – is expected to increase until at least 2040. These products have an unrivalled energy density and are easy to transport, making them an ideal means to carry and store energy. While alternatives are being developed for some of their current uses (e.g., passenger cars, where electrification is expected to play a major role), liquid hydrocarbons remain difficult to replace in heavy-duty and marine transport, in aviation and as feedstock for the petrochemical processing.

This book seeks to elucidate the pivotal role of chemical technology in actively pursuing the transition from the fossil energy scenarios to more sustainable energy systems. It is now recognised by everybody that global warming is a large-scale experiment that mankind started without knowing the boundary conditions and complexity of the interactions in the climate system. The cause of climate change is chemical conversion of carbonaceous materials into thermal and derived forms of energy. The often-used word "sustainability" is not only a term describing the balance between input of energy carriers and output of waste but contains a kinetic term describing the time constant of equilibration between input and output.

When specifically aiming at fuels used for transport, the world demand is gigantic (584 Exajoules, 2019, BP) although alternatives are available and increasingly implemented around the world. In countries where green electricity is abundant the opportunity for implementation of a larger pool of electric vehicles certainly has advantages. Electric vehicles, however, are still constrained by their capacities, limiting their widespread distribution in locations where population is largely distributed over a large area. Other initiatives can include the production of biofuels from renewable carbon sources or biofuels. The latter are usually classified in three different "generations" where the first generally involves the utilization of ethanol (produced from grain or sugar-rich plants) as oxygenate in gasoline (Sections 12.2.8 and 12.2.9) and fatty acid methyl ester (biodiesel) as a partial replacement for diesel (Sections 3.3.3.1 and 12.2.9). The second generation is usually related to the utilization of non-edible sugars (such as cellulose) as a sugar source to replace the contested first-generation feedstocks. Significant efforts have also been dedicated to converting all sorts of residual carbon sources such as waste plastics (Chapter 13), municipal solid waste, residual agricultural and residual forest biomass (Chapter 12) to alkanes through techniques such as pyrolysis, gasification and liquefaction all of which all have their opportunities and challenges. While it might be challenging for one of these options to cope with all the demand in transportation over the next decade, combining different opportunities may lead to good results and in time significantly reduce GHG emissions. Natural gas (Section 4.2) and shale gas (Section 4.5) also represent a finite opportunity for the transportation sector since natural gas could represent a significant source of carbon and although the latter is still fossil based it is generally considered as a more sustainable option than classical oil and could share part of demand in the near future. Natural gas can be used directly in some engines or can be transformed to gasoline, diesel and jet fuel through reforming and Fischer-Tropsch synthesis.

Power-to-X technologies can be used to produce value-added fuels from carbon dioxide through two main approaches: carbon dioxide hydrogenation and carbon dioxide electrochemical reduction. Carbon dioxide hydrogenation process is performed in two steps, the first of which is where hydrogen is produced using renewable energies, such as solar, wind, tide energy and hydro through processes, such as water electrolysis. In the second step, the hydrogen produced is used to convert the carbon dioxide created by industrial processes or power plants. This approach has gained increased attention due to its fast kinetics when compared to other approaches such as electrocatalysis and photocatalysis and to its flexibility. However, this technology still faces some challenges, in particular the sustainable hydrogen production. As presented in the section on water electrolysis (Chapter 3), water electrolysis is the main technology investigated for sustainable hydrogen production. However, improvements in efficiency and cost reduction are still required to overcome technical and economic barriers to its commercialization. Finally, integration of the different steps of the technology (carbon dioxide capture, water electrolysis and carbon dioxide hydrogenation) is also still an issue.

Carbon dioxide electrochemical reduction is a more recent and less advanced approach when compared to carbon dioxide hydrogenation. In this process, carbon dioxide reduction is connected to an oxidation reaction (usually water oxidation), where water is oxidized in oxygen, forming protons and electrons at the anode. The electrons flow to the cathode where they combine with carbon dioxide to form different reduced products, such as methane, alkanes, gasoline, jet fuel and diesel. In this case, renewable electricity is used to convert carbon dioxide directly into fuels, which represents a great potential for renewable energy storage and for lowering the GHG emissions into the atmosphere. In addition to its environmental advantages, the electrochemical reduction of carbon dioxide can be performed at ambient temperature and pressure and can lead to the desired products by adjusting some parameters such as the electrocatalyst, the operating potential and the electrolyte. Furthermore, the systems used with these processes are compact and could easily be scaled up taking the technology close to the demonstration phase. However, considerable challenges such as high overpotentials, the slow kinetics of carbon dioxide electroreduction and poor product selectivity could slow down the use this technology at large scale. In addition, the direct electrochemical reduction of carbon dioxide generates many different products, depending mainly on the reaction medium and on the catalyst. Hence, the development of electrocatalysts, allowing an efficient and selective reduction of the carbon dioxide, is essential for the development of this technology at large scale.

Hydrogen production through water electrolysis is widely discussed as a means of storing renewable electricity in chemical bonds. Hydrogen can be used for transportation in fuel cell vehicles, but it can also be reacted with carbon dioxide to form other fuels.

The purpose of this section is to give a combined technical, environmental and economic evaluation of different alternatives for introducing renewable hydrogen into the transportation sector and to compare light-duty transportation and diesel fuels produced from renewable hydrogen and green carbon dioxide. The evaluation is based on a set of common assumptions and considers both the conversion processes for fuel production and the application in transportation. Beside direct use of hydrogen in a fuel cell, we consider conversion to methane and methanol for use in spark ignition engines, as well as dimethyl ether for use in compression ignition engines. Specifically, the goal is to provide insight into the factors determining the overall performance of the process chain. The list of fuels considered is not intended to be exhaustive but is instead limited to a few promising concepts for which sufficiently detailed information is available on both production processes and engine performance. Hence, the selected fuels are reasonably close to a practical application.

This section highlights Power-to-Fuel processes and their products as a promising solution to the problem faced by the transport sector with regard to the Energiewende. The second part of this section focuses on the products. Determining the educts and products is the first step in process design and the products of Power-to-Fuel processes can be used as alternatives to conventional fuels in the transport sector. The greenhouse gas emissions and oil dependence of the transportation sector can thus be reduced. Additionally, they also offer potential for increasing engine efficiency.

With respect to their suitability as fuels for the transport sector, the Power-to-Fuel products DME, OME_{3-5} and n-alkanes are recommended. The second part of this section focuses on diesel, since it will continue to play the most important role amongst possible fuels for the model region of Germany (Schemme et al.). Thus, the first step in Power-to-Fuel process design is provided: determining the educts and products. Although the products are not different with respect to the educts employed, their production processes do differ considerably.

Transportation currently accounts for more than a quarter of the world end-energy use and causes a similar fraction of the anthropogenic greenhouse gas emissions, with significant future growth expected. This motivates the search for more sustainable alternatives to conventional transportation concepts that are largely based on fossil fuels. Battery electric vehicles charged using renewable electricity are being discussed as a promising alternative for short-distance transportation. However, long-distance transportation is likely to continue to rely on liquid or gaseous fuels in the near-to-midterm future because of the required storage capacities.

Hydrogen is being considered both as a fuel and as an intermediate in the production of a variety of organic fuels. Since it can be produced via water electrolysis, it also provides a means of coupling the electricity and transport sectors. Electrochemical production has thus been suggested as a means for utilizing fluctuating renewable electricity and providing chemical energy storage, as well as providing renewable energy for other industries in both grid connected and isolated configurations. Direct use of hydrogen as a fuel has obvious advantages since it avoids further conversion losses and allows for high tank-to-wheel efficiencies when used in fuel cell vehicles. However, it requires a dedicated hydrogen infrastructure for distribution and storage as well as a fleet of fuel cell vehicles. Therefore, there has been increasing interest in recent years in the conversion of hydrogen to organic fuels that are more compatible with the current distribution infrastructure and vehicle fleet.

To produce organic fuels, hydrogen has to be combined with a carbon source. One option is to use carbon dioxide, which can, at least in principle, either be extracted directly from the atmosphere or captured from point sources. The latter can include fossil power plants, other fossil anthropogenic sources (e.g., chemical processes with an inherent production of carbon dioxide), or biogeneous sources (e.g., carbon dioxide that occurs as a by-product in biogas production).

So far, the discussion of the production of electro fuels (e-fuels) from renewable hydrogen and their application in transportation has mostly either been focused on specific routes (e.g., methanation, methanol production, or ammonia production), on certain technical aspects relevant to different routes (e.g., catalysts), or on region-specific analyses for energy system integration. Existing well-to-wheel (WTW) analyses comparing different fuels, often did not consider hydrogen-based electro fuels in detail. The recent studies of Connolly et al. and Grahn et al. did include hydrogen and derived fuels in their energetic and economic analyses. However, their evaluations do not include a well-to-wheel greenhouse gas (GHG) analysis. On the other hand, Matzen and Demirel and Sternberg and Bardow, did provide a life cycle assessment for different fuels produced from renewable hydrogen, but they did not consider fuel production cost. Parra et al. conducted a thorough techno-economic and environmental study of hydrogen and methane production for energy storage, but they did not consider transport applications or compare to other e-fuels.

Data compiled from such separate studies is not necessarily suitable for direct comparison because they are based on different assumptions and use very different levels of detail. This was recently highlighted by Artz et al. in their comprehensive study of life cycle assessments for carbon dioxide utilization and by Brynolf et al. in their study of production costs estimates for electro fuels. In this context, Albrecht et al. recently advocated the use of a standardized methodology for improving comparability of techno-economic assessments of alternative fuels. This is certainly even more desirable when combining both techno-economic and life cycle analyses to provide a more comprehensive picture that can aid decisions towards practical implementations or guide further applied research. To this end, such an analysis should also contain a discussion of engine performance and the key factors affecting the performance of the fuel production processes, which is lacking in most system level studies.

While many concepts have been proposed, detailed comparisons of different pathways are still scarce. Bongartz et al. presented a technical, environmental and economic comparison of direct hydrogen use in fuel cells and production of methane, methanol and dimethyl ether for use in internal combustion engines for light-duty vehicle applications. The scenario considered uses renewable electricity for water electrolysis and carbon dioxide which is supplied continuously from biogas upgrading. All four fuels enable significant reductions (79–93%) in well-to-wheel greenhouse gas emissions as well as pollutant formation compared to fossil fuels, but they require very cheap hydrogen to be competitive with fossil fuels, confirming intuitive expectations. While direct use of hydrogen has obvious advantages (no conversion losses, high efficiency of fuel cells compared to internal combustion engines) in terms of overall electricity consumption, emissions and fuel cost, its drawbacks compared to the other fuels are the need for an hydrogen infrastructure, the high fuelling pressure and lower driving range. Among the three combustion engine fuels, DME has the lowest fuel cost and electricity consumption per distance driven because of the more efficient use of hydrogen in its production, as well as the highest volumetric energy density, while methane has slightly lower greenhouse gas emissions. Cost and energy demand are dominated by hydrogen supply, meaning that integrated solutions could be more attractive than separate electrolysis and fuel production.

In the following section, we first give some details on the scenario and the assumptions used and the selection of fuel candidates considered and describe the methods used for the evaluation. Next, the performance of processes for converting hydrogen into methane, methanol and DME is discussed and the entire process chains for the different fuels are evaluated according to their fuel cost and well-to-wheel greenhouse gas emissions. Finally, these analyses are combined with an evaluation of the performance of the different fuels from an application perspective including aspects related to engine performance, safety and infrastructure.

In principle, an idealized electro fuel could be made using renewable energy (sunlight, wind, tide, or other non-carbogenic inputs, such as biomass) to collect and revert carbon dioxide back into a fuel. Reverting carbon dioxide back into a transportation fuel would require adding at least the heat of combustion of the fuels, more likely a multiple of that energy to account for the energy required to gather and concentrate carbon dioxide and for energy losses in the reversion. Electrochemical processes exhibit energy efficiencies of <40% for making a simple fuel, formic acid. Replacing just the enthalpy of combustion of the fuels will require at least doubling the total amount of electrical energy generated in the US (Holladay et al.) (27.2 EJ/14.2 EJ) and quintupling the amount of non-carbogenic (renewable + nuclear) electricity generated in 2019 [27.2/ (2.5 EJ + 2.9 EJ)]:

• Production capacity of electrical energy	14.2 EJ
• Renewable	2.5 EJ
• Nuclear energy	2.9 EJ
• Transportation fuels used	28.8 EJ
• Biofuels	-1.6 EJ
• Natural gas	-0.06 EJ

Again, both factors represent the minimum energy that must be added. Holladay et al. expect the actual increase will be at least 100–300% times larger to account for process inefficiencies, such as overpotentials.

Gathering carbon dioxide from its many sources in manufacturing, production of power and production of fuels might be addressed ultimately by direct air capture. Until then, significant costs associated with capturing dilute carbon dioxide could be obviated using carbon dioxide that is emitted in a concentrated stream from fermentation process that produces ethanol (Equation 7.1). Fermentation produces about 1 molecule of carbon dioxide per molecule of ethanol.

$$C_6H_{12}O_6 \rightarrow 2C_2H_5OH + 2CO_2 \tag{7.1}$$

Accordingly, the 15 billion gallons of ethanol produced each year in the United States is accompanied by the production of 1 Tmol/year (44 Mt/year) carbon dioxide. The heat of combustion per mole of carbon dioxide for several fuels or fuel surrogates averages 680 kJ/mol (Chapter 3, Table 3.1); therefore, converting 1 $Tmol_{CO_2}$ × 680 kJ/mol requires an energy supply of 0.7 EJ/year.

Therefore, making electro fuels from carbon dioxide is challenging in the near term and includes at least adding significant amounts of clean generating capacity along with energy storage and making the electro fuels energy efficient.

The educts of electro fuels are hydrogen and carbon dioxide. Therefore, the consecutive chemistry of synthesis gas (mixture of hydrogen and monoxide) is significant in this context. Nitrogen-based fuels such as ammonia, hydrazine and carbazole are also conceivable, in addition to substances composed of hydrogen, carbon and oxygen. However, with regard to safety, implementation strategy and environmental compatibility, these are critically considered and so will therefore not be considered further.

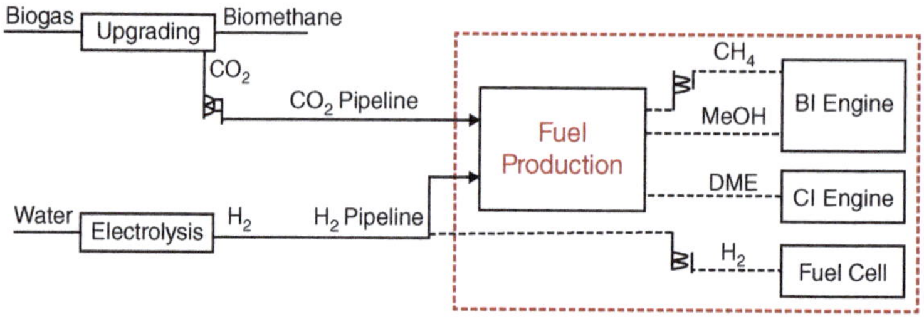

Figure 7.1 Hydrogen production via electrolysis is assumed to operate dynamically using renewable electricity. All other steps operate at steady state and use grid electricity.

Four different options for introducing renewable hydrogen into the transport sector are assessed by comparing state-of-the-art process chains with different fuels for one specific scenario (Figure 7.1).

7.2 Selection of Fuel Candidates

Grube et al. have assumed that hydrogen is produced from renewable electricity via electrolysis, stored in a cavern storage facility and delivered continuously to a centralized large scale fuel production plant (where applicable) or gas stations through a pipeline at a final pressure of 7 MPa. As a carbon source, Bongartz et al. assume the use of carbon dioxide that occurs as a (currently unused) by-product of biogas upgrading. From the fuel production plant, the fuel is transported to gas stations either via pipeline (methane) or via truck (methanol and dimethyl ether). For direct use of hydrogen, the energy demand reported by Grube et al. for pipeline transport to the fuel production plant is taken to represent the distribution to gas stations via pipeline instead. At the gas stations, methane and hydrogen have to be further compressed to their respective fuelling pressures (Table 7.3). All steps except for electrolysis are assumed to operate at steady state using grid electricity.

Table 7.3 Basic hydrogen and methane properties.

Properties of the Fuels	Unit	Hydrogen	Methane
Engine		FCV	SI
Storage (gas)	MPa	70	20
Boiling point	K	20.4	111.5
Density, NTP	kg/m^3	0.0899	0.7188
Evaporation heat	kJ/m^3	450	510
Internal energy, NTP	kJ/m^3	10.760	40.000
Lower heating value	MJ/kg	120	49.8
Volumetric energy density	MJ/l	5.1	7.1
Autoignition temperature	K	850	807
Igniton energy	$J.10^{-6}$	20	300
Flame temperature	K	2400	2190
Flame speed	m/s	2.75	0.37

In a study considering Germany in 2035, Grube et al. estimated that 1.9×10^5 t/a of hydrogen can be produced utilizing renewable electricity that would otherwise stay in northern Germany (which has a high concentration of wind power), because it cannot be utilized, stored, or transported elsewhere at the time of production. Depending on the fuel to be produced, the required amount of carbon dioxide for converting this hydrogen is of the range 1.0×10^6 t/a to 1.4×10^6 t/a, which is roughly 10% of the envisioned availability of carbon dioxide from biogas upgrading in Germany by 2030. However, biogas plants are typically relatively small and carbon dioxide would have to be collected from roughly 50 to 300 individual biogas plants, depending on the exact sizes. Based on this, the maximum transport distance of carbon dioxide from the biogas plants to the fuel production plant is estimated to be on the order of 75 km. This assumed distribution through a pipeline network transporting gaseous carbon dioxide with a final pressure of 2.25 MPa.

The fuels considered in detail in this study are listed in Tables 7.3 and 7.5, along with some basic characteristics. Hydrogen can be used directly in fuel cell vehicles (FCV), in which it is stored at pressures of up to 70 MPa. Work on FCVs is in progress both in terms of research and development efforts as well as practical applications with first models becoming commercially available. The interest in hydrogen as a fuel for spark ignition (SI) engines has decreased in recent years because of the lower tank-to-wheel efficiency compared to FCV. Methane can relatively easily be produced from hydrogen and carbon monoxide and it has already found wide-spread application in SI engines in the form of compressed natural gas (CNG). Methanol, which can also be produced directly from hydrogen and carbon dioxide, can be used as liquid spark ignition fuel or additive to gasoline. Dimethyl ether is a promising compression ignition (CI) fuel and can also be produced from hydrogen and carbon dioxide. It is typically stored as a liquid at conditions like liquefied petroleum gas (LPG).

There are several other options for electro fuels produced from hydrogen and carbon dioxide. These include Fischer-Tropsch gasoline or diesel, the conversion of methanol to gasoline, as well as polyoxymethylene dimethyl ethers (OME or POMDME) or different chain lengths that have recently been discussed as potential diesel fuels or additives. Ethanol is attractive from an engine perspective, but its synthesis from hydrogen and carbon dioxide or carbon monoxide is challenging in terms of conversion and selectivity and research activities have so far been limited to the development of better catalysts.

7.2.1 Fuel Production Processes

Methane, methanol and DME can be produced from hydrogen and carbon dioxide according to the overall reactions (Equations 7.2–7.4):

$$4H_2 + CO_2 \rightleftarrows CH_4 + 2H_2O, \quad \Delta G^{\circ}_{298\,K} = -164.9\,\text{kJ/mol} \tag{7.2}$$

$$3H_2 + CO_2 \rightleftarrows CH_3OH + H_2O \tag{7.3}$$

$$6H_2 + 2CO_2 \rightleftarrows CH_3 - O - CH_3 + 3H_2O \tag{7.4}$$

To assess the performance of the respective fuel production processes, state-of-the-art process concepts are modelled and simulated in AspenPlus®. Hydrogen and carbon dioxide are assumed to enter the processes at their stoichiometric ratio at ambient temperature and their respective distribution pressures neglecting any contaminants. The scale of the process is selected to match the assumed hydrogen availability described in Section 7.2.For the present study, Bongartz et al. consider stand-alone processes that use excess heat from the reactions to power a steam power cycle to cover part (or all) of their electricity demand. Note that when these plants are located close to other plants with a net heat demand, it might be more beneficial to export this heat in the form of steam instead.

Several processes have been developed for producing methane (or substitute natural gas (SNG)) from syngas (i.e., hydrogen and carbon monoxide) obtained by coal or biomass gasification. Recently, similar processes to produce methane from hydrogen and carbon dioxide (often termed "Power-to-Gas") have received attention since methane can be stored and distributed more conveniently than hydrogen. In 2013, the first commercial plant producing methane as a transportation fuel from hydrogen and carbon dioxide started operation in Germany. The process considered by Bongartz et al., is based on the TREMP® process by HaldorTopsøe and the process presented by Schaaf et al.

7.3 Power-to-Methane Technology

Methane is one of the most important energy vectors of our society, being used to produce heat, electricity and value-added chemicals. Methane is mainly obtained through natural gas (shale gas), a fossil fuel source with very low cost (Chapter 4). If produced in a sustainable way, methane, also called substitute natural gas (SNG), has also a great potential for reducing the GHG emissions, since it can be more readily used than renewable hydrogen, for example. Indeed, SNG can be injected directly into the natural gas grid, benefiting existing natural gas facilities. Power-to-Methane technology has shown to be a promising pathway for a sustainable production of methane, using carbon dioxide and renewable energy to produce SNG. Among the different existing routes for Power-to-Methane, catalytic carbon dioxide hydrogenation (methanation) has been extensively investigated and demonstration plants are already in operation in different countries. The carbon dioxide electrochemical reduction route is still at lab-scale validation stage. However, the results obtained over the last few years have highlighted the promising aspects of this process.

The Power-to-Gas process plays a significant role in transforming renewable electric energy into methane via electrolysis and methanation (Jeong et al.). Critical issues in the PTG process include the fluctuating nature of renewable electric energy and the safe operation of the methanation reactor. The high exothermicity of methanation and the variable load caused by renewable energy imperil the stable operation of the reactor. A cooled recycle reactor with a multi-layered catalyst is introduced for methanation. The intensified reactor provides better performance than conventional reactors, but its complexity poses challenging design and control problems. An optimization problem is formulated for the design of the reactor. The catalyst dilution profile and the product recycle ratio are optimized to improve both the reactor efficiency at steady state and the flexibility to load changes while satisfying the major constraint, i.e., safe operation. The objective function considers the reactor performance in terms of the total annual cost and flexibility against a load change. Load variation is a major disturbance affecting the operation of the methanation reactor. A dynamic control strategy is proposed to address load variation. The proposed control system stabilizes the reactor temperature in each layer and properly distributes the reaction load to the layers, enabling the flexible operation of the reactor under a load change.

7.3.1 Carbon Dioxide Electrochemical Reduction

One of the possible ways to sustainably produce fuels (such as methane from carbon dioxide) is by an electrochemical reduction of carbon dioxide. Thermodynamically, carbon dioxide can be electrochemically reduced to methane with a standard potential of +0.17 V vs. RHE. However, since hydrogen evolution reaction (HER) is thermodynamically possible at 0 V (vs. RHE), both reactions will be in competition at all negative potentials. The low carbon dioxide solubility in water (0.03 M, 25°C and 0.1 MPa) is also a challenge. So, studies are still needed to improve carbon dioxide solubility and to develop catalysts able to improve selectivity to methane and to reduce the high potentials required. Alternative approaches to carbon dioxide electrochemical reduction, such as photo irradiation, use of ionic liquid electrolytes and use of biological microorganisms have also been treated elsewhere and are not the focus of this section. There have been experimental works on carbon dioxide electrochemical reduction by heterogeneous catalysts. These performed carbon dioxide reduction tests at 18.5°C in a 0.1 M $KHCO_3$ medium using different metals, such as Cu, Au, Ag, Zn Pd, Cd, Ni, Pt, etc. They proved that methane could be obtained by carbon dioxide electrocatalytic reduction on metal surfaces and that the choice of the metal of the electrode had an impact of the product's selectivity. The only metal investigated that presented a relatively high Faradaic efficiency to methane (33.3%) was copper at −1.44 V vs. normal hydrogen electrode (NHE). Pd, Cd and Ni also produced methane but with Faradaic efficiencies lower than 3%, mainly due to the HER occurring as side reaction. The good performance of copper was attributed to the fact that carbon dioxide is intermediately reduced to carbon monoxide, before being reduced to methane (and other hydrocarbons and alcohols). When other metals, such as Ni and Pt were used, the CO was adsorbed on the surface of the electrodes, preventing its further reduction into hydrocarbons and alcohols.

To the best of my knowledge, the best methane Faradaic efficiency obtained up to date used copper electrodes. The performance of copper nanoparticles supported on glassy carbon (n-Cu/C) was compared with classic copper foils. The electrochemical tests were performed in a two-compartment electrochemical flow cell separated by a Selemion membrane and in a 0.1 M $NaHCO_3$ electrolyte. A Faradaic efficiency for methane of 76% was obtained using n-Cu/C (−1.35 V vs. RHE), which was significantly higher than the Faradaic efficiency of 44% obtained with the copper foil. The good performance of the n-Cu/C catalyst was attributed to the formation of nanoscale aggregates on the n-Cu/C upon polarization, exposing its catalytic sites to methanation. On the other hand, in copper foils highly connected networks of fused particles are formed upon polarization, leading to fewer exposed catalytic sites.

The electrolyte concentration as well as the local pH have also shown to have great influence on the selectivity of the carbon dioxide electrochemical reduction. It has been proved that the methane formation depends on proton concentrations, being favoured when acidic or neutral solutions are used. Thus, the effect of the electrolyte can be directly linked to the local pH. The effect of the concentration of the electrolyte in controlling the selectivity of the carbon dioxide electroreduction was investigated on copper. $KHCO_3$ electrolyte solutions were used with concentrations varying from 0.05 to 0.2 M $KHCO_3$. The highest concentrations of electrolyte favoured the selectivity to methane. Methane Faradaic efficiency of ≈ 70% at −1.43 V (vs. NHE) was obtained when a 0.2 M $KHCO_3$ solution was used due to its higher buffer capacity. It was hypothesized that concentrated $KHCO_3$ solutions had a higher buffer capacity and favoured the electron/proton couple transfer, resulting in a higher methane (and hydrogen) production. On the other hand, methane formation was less favoured in diluted $KHCO_3$ solution with low buffer capacity, due to the low concentration of protons close to electrode surface.

Issues related to the low solubility of carbon dioxide in water often lead to mass transfer limitations for large current densities. To overcome such problems, carbon dioxide reduction has been performed using gas diffusion electrodes (GDE). Electrolysis experiments were carried out at 3 MPa in a stainless-steel autoclave using a 0.5 M $KHCO_3$ aqueous electrolyte and a gas diffusion electrode containing Pt electrocatalysts. The Pt catalyst layer was directed towards the carbon dioxide gas phase while the gas diffusion layer faced the aqueous electrolyte. Methane was produced with a Faradaic efficiency of 34.8% at a −1.92 V (vs. Ag/AgCl). However, when the order of the layers was reversed (Pt catalyst layer facing aqueous electrolyte), negligible amounts of methane were produced, mainly due to the contact between the water and the Pt catalyst, favouring the hydrogen evolution reaction.

Also aiming to improve carbon dioxide solubility, there have been investigations into the electrochemical reduction of carbon dioxide to methane at low temperatures in methanol, since it is a better solvent for carbon dioxide compared to water. Experiments were performed in a H-type cell with a Nafion 117-type ion exchange membrane as diaphragm using Cu electrodes. The authors reported a methane Faradaic efficiency of 58% at −2.0 V (vs. Ag QRE). The authors further investigated the use of different sodium salts ($NaNO_3$, NaH_2PO_4, $NaHCO_3$, NaCl, NaBr, NaI, NaF, Na_2SO_4, NaSCN, $NaClO_4$ and CH_3COONa) to further increase the carbon dioxide solubility in methanol. The electrolysis tests were performed at 243 K using Cu electrodes. The authors reported that for all the sodium salts tested, methane Faradaic efficiencies ≥43.4% were obtained, 70.5% being the highest efficiency depressed the hydrogen evolution reaction that competes with the carbon dioxide reduction, thus increasing the methane formation (Vasconcelos and Lavoie 2019; Kaneco et al., 2002; Kaneco et al. 2006).

7.3.2 Carbon Dioxide Hydrogenation

Power-to-Methane can also be achieved by combining water electrolysis for hydrogen production with methanation process in a two-step process. In the first step, hydrogen is produced using renewable energy, such as solar, wind and tide power and nuclear energy. In the second step, carbon dioxide is hydrogenated in a methanation process to produce methane (Table 7.4, Equation 7.2). Since the substitute natural gas (SNG) produced has high purity, it can be injected directly into the natural gas grid, stored, or used in natural gas facilities allowing this technology to connect the electrical and gas grids in a single system, allowing great flexibility to the balance of the grid.

Methanation can be done either biologically or catalytically. However, the focus of this section will be on catalytic methanation since it is (to the best of my knowledge) the most advanced technology. Catalytic methanation (Equation 7.2) is an exothermic reaction usually performed from 200°C to 550°C and at pressures up to 10 MPa. Due to its exothermicity, temperature control inside the reaction is an issue that can lead to thermodynamic limitations and catalyst deactivation by sintering. Therefore, different reactor types, such as fixed bed, fluidized bed, three-phase and

Table 7.4 Summary of the Power-to-Methane plants current in operation and of the Power-to-Methane projects being developed.

Plant/ Company	Technology	Eff.(%)	CO_2 Conversion	CH_4 Production Rate	Source of Power	Power Input (MW)	Electricity Consumption (kwh/kg CH_4)	H_2 Source
Audi E-GAS/ Audi	CO_2 methanation	54	2.8 kt/y	1 kt/y (max:325 Nm^3/h)	wind	6	13.85	AEL
ZSW 250-kWel	CO_2 methanation	-	-	-	-	-	-	AEL
Store&Go	CO_2 methanation	-	-	-	-	1		AEL
HELMETH	CO_2 methanation	>85	-	1.08–5.42 m^3/h	-	-	-	SOEC

structured reactors have been used to improve the efficiency of the process. An adiabatic fixed-bed reactor is the most studied reactor type and is employed by companies such as Sasol, Linde and Haldor Topsoe, to produce SNG from coal or naphtha. However, catalysts must be resistant against sintering due to the large temperature range they have to support because of the adiabatic mode. Usually, there are at least two fixed bed reactors connected in series for a good control of the reaction temperature, which is done by recirculating the reactor outlet gas stream and by intermediate gas cooling steps. Fluidized bed reactor is the second reactor type most studied where the methanation reactions occur in a fluidized catalyst bed, improving heat and mass transfer during the reaction and thus the control of the process. Nevertheless, abrasion and entrainment of catalyst bed particles are some drawbacks of this type of reactor.

The catalyst has also a significant impact on the efficiency of the methanation process and nickel-based catalysts are commonly used for this purpose due to their high activity and low cost. However, they impose limits to the methanation process, such as operating temperatures that must be kept in the range 200°C and 550°C since potentially highly toxic nickel compounds can be formed at 200°C and catalyst deactivation by sintering and coking can occur above 550°C. So, studies have been carried out on catalyst development, aiming at synthetizing better performing and temperature-resistant catalysts. There have been investigations into the effect of the zeolite structure on the performances of Ni-based catalysts for carbon dioxide methanation. Commercially available USY, BEA, ZSM-5 and MOR zeolites ion-exchanged with Na^+ and Cs^+ and impregnated with 15 wt % Ni were tested at temperatures ranging from 250°C to 450°C and GHSV of 43,000 h^{-1}. The authors concluded that the performance of the catalysts was intimately related to the structure of the zeolites. USY zeolite showed the best catalytic performance with carbon dioxide conversions of about 70% and CH_4 selectivity close to 100%. The good catalytic performance was attributed to its weak interaction with water, which has an inhibitory role in the reaction and to the high dispersion of the Ni particles over the catalyst surface. BEA zeolite led to the highest Ni particles dispersion and thus presented similar catalytic performances to the USY zeolite. Also, the catalytic performance of the BEA zeolite was improved when its hydrophobicity was increased by increasing its Si/Al ratio. Finally, ZSM-5 and MOR zeolites presented the lowest Ni dispersion, leading to lower methane selectivity around 60–70%.

Hydrotalcite materials have also been investigated due to the possibility of receiving high amounts of active phase while keeping high metal dispersion favouring carbon dioxide conversion and preventing the catalyst deactivation. These have compared the performance of Ni-Al/hydrotalcite catalysts with a classical Ni/Al_2O_3 catalyst both doped with 75 wt. % nickel. The Ni-Al/hydrotalcite presented the best catalytic performance with a carbon dioxide conversion of 85% and methane selectivity of about 85% at 275°C, while the Ni/Al_2O_3 commercial catalyst presented a carbon dioxide conversion of around 80% and methane selectivity of about 75% at the same temperature. The difference in the catalytic performances was attributed to a higher nickel dispersion and a higher metal surface area of the Ni-Al/hydrotalcite catalyst. There have been comparisons of the performance of a Ni-Al/hydrotalcite catalyst with a Ni/Al_2O_3 catalyst both doped with 78 wt % of Ni. Ni-Al/hydrotalcite catalyst presented a better catalytic performance with carbon dioxide conversion of 82.5% and methane selectivity of 99.5% at 350°C, which was also attributed to a high nickel dispersion.

The influence of the composition of Lewis basic $(Mg,Al)O_x$ mixed oxide supports on carbon dioxide and carbon monoxide methanation performance of (Ni,Mg,Al)-hydrotalcite-derived catalysts was investigated by Vols et al. The hydrotalcite-derived and related precursors are synthesized by pH-controlled coprecipitation, calcination and subsequent reduction. The catalysts with different supports prove suitability for carbon dioxide and for carbon monoxide methanation but display significant differences in respect to the temperature sensitivity of the reactions. The influence of the systematically varied support composition was studied by kinetic measurements, X-ray photoelectron spectroscopy (XPS) and in situ diffuse reflectance infrared Fourier transform (DRIFTS). Performance differences between carbon dioxide and carbon monoxide methanation were investigated for three different catalyst compositions. Different adsorption patterns in the DRIFTS measurements for these catalysts allow us to assign the increase in performance in the presence of magnesium oxide primarily to the enhancement of carbon dioxide adsorption on the catalyst support. Temperature-dependent DRIFTS measurements display a correlation between the ratio of format to carbonyl species on the catalyst surface and the activity of the catalyst in the case of magnesium-containing catalysts. The result suggests the role of carbon monoxide as a key intermediate for carbon dioxide methanation for magnesium-containing nickel hydrotalcite-derived catalysts. XPS measurements demonstrate that the magnesium content significantly affects the reducibility in the precursor material to generate the active component.

Noble metals have also been extensively investigated due to their good catalytic performance for carbon dioxide methanation at low temperatures and good resistance to carbon formation. There have been studies of the performance of $Rh/\gamma\text{-}Al_2O_3$ catalyst with Rh content varying from 1 to 5 wt % at a temperature range of 50–200°C. They reported that selectivity to methane was about 100% in all conditions investigated. They also reported that the turnover frequency (TOF) of methane formation decreased as the Rh dispersion decreased (increase in Rh content) in a temperature range of 135–165°C. However, the TOF was independent of the Rh dispersion in a higher temperature range (185–200°C). There have been investigations into the effect of mechanically mixing $Rh/\gamma\text{-}Al_2O_3$ with Ni/activated carbon (AC), a catalyst widely used for hydrogenation reactions. The objective was to verify the synergetic effect of the two catalysts in the carbon dioxide methanation reaction. The reactions were performed at 125°C and 0.2 MPa. The authors reported a higher methane production (9.5 $\mu molCH_4/g_{cat}$) when both catalysts were used compared to when only $Rh/\gamma\text{-}Al_2O_3$ (6.8 $\mu molCH_4/g_{cat}$) or Ni/AC (0 $\mu molCH_4/g_{cat}$) was used. Since no new structures were formed when the catalysts were mixed, the synergy was attributed to the cooperation between the two catalysts during the reaction. $Rh/\gamma\text{-}Al_2O_3$ was highly efficient for adsorbing carbon dioxide while Ni/AC was able to adsorb high quantities of hydrogen. These two properties together improved the carbon dioxide conversion and the methane formation. The authors suggested the hydrogen species would migrate from the surface of the Ni/AC towards the $Rh/\gamma\text{-}Al_2O_3$, reacting with the adsorbed carbon dioxide to produce methane. Basic promoters, such as K, Ba and CeO_2 have also been studied to enhance the performance of the catalysts by modifying the surface basicity, the metal-support interaction as well as the metal dispersion. All these parameters would lead to a better activity, stability and resistance to extreme conditions of the catalysts (Vascancelos and Lavoie, Karelovic and Ruiz, Frontera et al., Ghaib and Ben-Fares, Swalus).

Cerium-based bimetallic oxides of the type $3NiO \cdot CeO_2$, $Co_3O_4 \cdot CeO_2$ and $1.5Fe_2O_3 \cdot CeO_2$ were prepared by two sol–gel methods (epoxide addition and Pechini methods) and by the electrospinning technique, aiming at the synthesis of compounds with different morphologies (nanoparticles or nanofibres) and the evaluation as catalysts for the methanation of carbon dioxide (Branco et al.). The results obtained show a catalytic performance that depends on the type of d block element, method and catalyst morphology. Regardless of such parameters, the best results were those obtained over the nickel-based catalysts prepared by the electrospinning technique, which are remarkably more active than a commercial rhodium catalyst used as a reference and tested in the same conditions. Moreover, the cobalt-based catalysts obtained either as nanoparticles (Pechini method) or fibres (electrospinning technique) present also a good activity, comparable to that measured over the reference catalyst. All catalysts were very stable to deactivation during at least 60–70 hours of reaction, which is clearly an advantage that could make them an alternative to be considered for future developments in this area.

Metal organic frameworks (MOFs) are a very attractive catalyst for the conversion of carbon dioxide into methane due to their high surface area, tenable chemical composition, high porosity and well-ordered structures. They are also photo-responsive materials. This review discusses the various strategies and modifications implemented to further ameliorate the thermal, photo and photothermal catalytic performance of MOFs. Initially, three main catalytic approaches, namely thermal catalysis, photocatalysis and photothermal catalysis, are thoroughly discussed to understand the mechanism and the differences between them with their characteristics and limitations. Then, a comprehensive review was carried out on various strategies employed to augment the performance of MOFs for carbon dioxide methanation, such as metal addition and incorporation, MOF templating, surface sensitization, formation of heterojunctions and organic linker modifications via functionalization. Comparisons between MOF-based catalyst and traditional catalyst were carried out to elucidate the beneficial properties of MOFs towards carbon dioxide methanation. The selectivity control for methane production was then extensively reviewed in terms of operating parameters, type of catalyst and reactor. Finally, the mechanism, pathways, intermediates and adsorbed species involved for carbon dioxide methanation are thoroughly discussed with the help of DRIFTS analysis and density functional theory (DFT) calculations. Therefore, metal organic frameworks are highly promising porous crystalline materials for the carbon dioxide methanation reaction and have countless possibilities for further enhancement and development to maximize the production of renewable methane.

As previously stated, there has been a lot of work on the different steps of the Power-to-Methane process, such as electrolysis and carbon dioxide methanation. However, a lot of work still needs to be done on the integration of these steps. Different countries, especially those in Europe, driven by the need to increase the share of renewable energy, started research projects to provide proof of concept to the Power-to-Methane, in operation and others are still being developed. A non-exhaustive list of these plants and projects is provided in Table 7.4.

The largest Power-to-Methane facility in the world is the 6 MW Audi e-gas plant. In this plant, hydrogen is produced by alkaline electrolysers using wind power. The source of the carbon dioxide is the biogas from the EWE Biogas GmbH & Co. KG biomethane plant. The methane production is limited to 1,000 t/y due to availability of renewable energy. However, the maximum methane production capacity of the plant is 325 Nm^3/h. The Audi e-gas plant in Germany presented a Power-to-Methane concept utilizing pure hydrogen from an alkaline electrolyser driven by wind turbines and carbon dioxide from a biogas plant. The facility is capable of producing 1000 tons of synthetic methane per year with a system efficiency of 54%.

In 2012 ZSW launched a Power-to-Methane demonstration plant composed of a 250-kWel alkaline high-pressure electrolyser, a carbon dioxide methanation unit and a process control system. In this plant, two different reactor types (tube bundle and plate reactors) can be operated separately or in combination. Also, a membrane gas processing stage is used to enrich the methane in the product gas and to recycle the hydrogen-rich gas.

The Store&Go project funded by the European Union developed a metallic honeycomb-like carrier-based reactor aiming at overcoming the issues related to removing and reutilizing the heat from the methanation reactor. The proof of concept was done at laboratory scale. The technology was scaled up to a 1MW SNG plant, which started operations in 2018.

The HELMETH project, co-financed by the European Union and by the Fuel Cells and Hydrogen Joint Technology Initiative, aims at determining the conditions for an economic feasibility of the Power-to-Gas process and at demonstrating the technical feasibility with a conversion efficiency >85%. The innovation of this project is the thermal integration of a high temperature electrolysis (SOEC) with methanation. The heat released during the methanation process will be used to vaporize the water that will be fed to the electrolyser, helping to balance the endothermal and exothermal process and allowing a better energy-efficiency storage of the renewable energy.

Despite the key role that this technology might play in the future energy sector, further studies are still needed to validate the whole system as well as to overcome its main drawbacks, such as low efficiency and high costs.

7.4 Power-to-Methanol

Methanol is produced from synthetic gas around the world in plants with a capacity of up to 5,000 t/d (Mega Methanol Technology, Chapter 12). Currently, there are two demonstration plants producing methanol from hydrogen and carbon dioxide, in Japan and Iceland. Two different approaches for methanol production from hydrogen and carbon dioxide have been discussed in the literature. The first approach consists in conducting the reverse water gas shift (RWGS) reaction (Equation 7.5):

$$H_2 + CO_2 \rightleftarrows CO + H_2O \tag{7.5}$$

and removing the water to obtain syngas of the desired composition and then conducting a conventional methanol synthesis. The second approach uses the mixture of hydrogen and carbon dioxide directly to synthesize methanol, employing conventional or modified methanol synthesis catalysts. The direct approach has the advantage that only one reactor is needed and that the selectivity is higher, thus simplifying purification. On the other hand, the once-through reactor conversion is lower than for synthesis from syngas. The process considered in here follows the direct approach and is based on the processes presented by Otto and by Pontzen et al. The basic methanol properties are showed in Table 7.5.

Replacing fossil fuels with renewable energy sources has a fundamental role in creating a sustainable and carbon-free economy. The catalytic hydrogenation of carbon dioxide has great potential to reduce an enormous amount of carbon dioxide and contribute to a green economy by converting carbon dioxide into a variety of useful products. It is very important to develop new and

Table 7.5 Basic methanol properties.

Properties of the Fuel	Unit	Methanol
Engine		SI
Storage, (l)	kPa	101.3
Boiling point	°C	64.6
Density, TNP	$g.cm^{-3}$	0.7915
Octane number, (RM)		112
Octane number, (MM)		91
Cetane number		3
Reid vapour pressure	kPa	32
Evaporation heat	kJ/mol	37.43
Higher heating value	kJ/mol	726.1
Lower heating value	kJ/mol	638.1
Volumetric energy density	MJ/l	15.8
Flash point, closed vessel (TCC)	°C	12
Flash point, closed vessel (TOC)	°C	15.6
Autoignition temperature	°C	470
Flammable limits (in air), lower	(v/v)%	6
Flammable limits (in air), upper	(v/v)%	36.5

highly efficient catalysts for the catalytic hydrogenation of carbon dioxide. Recently, the catalytic hydrogenation of carbon dioxide has attracted an enormous amount of attention, which has been mainly focused on the development of efficient, selective and stable catalysts. Atsbha et al. summarized the current developments and improvements in the catalytic conversion of carbon dioxide by hydrogen used towards the synthesis of carbon monoxide, methanol and hydrocarbons in terms of the catalyst performance, selectivity and stability. The experimental procedures used for the three main pathways for the catalytic hydrogenation of carbon dioxide (carbon dioxide to carbon monoxide via the reversible water gas shift reaction, carbon dioxide to methanol synthesis and carbon dioxide to hydrocarbons via the Fischer-Tropsch reaction) using different catalysts are discussed. Furthermore, the industrial application of carbon dioxide hydrogenation processes including their energy and economic analysis are also discussed.

In another project called methanol fuel from carbon dioxide ($MefCO_2$) renewable methanol is produced by capturing more than 1.5 ton of carbon dioxide per day from flue gas emitted from a coal plant. The $MefCO_2$ pilot plant can be considered as one of the largest plants in the European Union that converts hydrogen with a PEM electrolyser and carbon dioxide from flue gas into methanol. In the project named "From Residual Steel Gases To Methanol" (FReSMe), the reactor used in the $MefCO_2$ project was used.

7.5 Power-to-Dimethyl Ether

Dimethyl ether (DME) has traditionally been produced by dehydration of methanol through the reaction seen in Equation 7.6. To produce dimethyl ether from hydrogen and carbon dioxide, a combination of a methanol plant like the one considered here with a conventional methanol dehydration plant could be used. However, in recent years the so-called one-step process for direct synthesis of dimethyl ether from either syngas or hydrogen and carbon dioxide has gained increasing attention. This approach is promising because only one reactor is needed instead of two and because the thermodynamic synergy between reactions shown in Equation 7.3 and Equation 7.6 enhances conversion. For this reason we consider a process for direct DME synthesis from hydrogen and carbon dioxide similar to the one presented by Otto, with a modified sequence of distillation columns to simplify heat integration.

Most of the data reported in the literature about carbon dioxide hydrogenation into diesel-like fuels are related to the production of dimethyl ether, which is considered a greener alternative to traditional diesel fuels due to its lower NO_x emissions, near-zero smoke, lower carbon particulates and lower engine noise. The methanol pathway is the main route investigated, being used either in a two-step process in which methanol is produced from carbon dioxide hydrogenation in the first step (Equation 7.3) after which methanol is dehydrated in a second step reaction (Equation 7.6), or in a single-step process, in which both steps are done simultaneously in the same reactor.

$$2CH_3OH \rightleftarrows CH_3OCH_3 + H_2O \qquad \Delta H_{298K} = -23.4\ kJ/mol \tag{7.6}$$

Since carbon dioxide conversion could be thermodynamically limited (Chapter 1) at low pressures if methanol is not continuously removed from the reaction medium, its further conversion into DME in a single step process is preferred. So studies for catalyst development for this process have focused on the synthesis of bifunctional catalysts capable of performing both the reaction in Equation 7.3 and the reaction in Equation 7.6 in a single reactor. In general, catalysts used for each reaction separately are combined in a single bifunctional catalyst, which is synthesized by homogenous mixing of both catalysts, by sequential arrangement or even by homogenous mixing and grinding followed by pelletizing. It is widely accepted that a simple homogenous mixing results in better performing catalysts since they maintain their initial properties. Copper and zinc-based catalysts, for example, have been intensively investigated for use in carbon dioxide hydrogenation into methanol (see Section 7.4.). Zinc oxide is responsible for adsorbing carbon dioxide while copper adsorbs hydrogen. The second step (methanol dehydration) is favoured by acid catalysts, such as alumina and zeolites. So, the functions of both catalysts in a bifunctional catalyst prepared by a physical mixture of CuO-ZnO-Al_2O_3-ZrO_2 and HZSM-5 for the carbon dioxide hydrogenation into dimethyl ether have been merged into a single-step process. Reaction temperature and pressure varied from 483 to 543 K and 2 and 5 MPa, respectively. The best catalytic results were obtained at 543 K and 5 MPa with carbon dioxide conversion of about 27% and dimethyl ether selectivity of 15.8%, proving that the experimenters could successfully synthesize dimethyl ether from one-step carbon dioxide hydrogenation. Similarly, there have been investigations into the performance of CuO-TiO_2-ZrO_2/HZSM-5 mixed oxides catalyst for producing dimethyl ether from carbon dioxide hydrogenation in a one-step process. The catalyst with a Ti/Zr ratio of 50/50 presented the best catalytic performance with dimethyl ether selectivity of 47.5% (yield = 7.41%) and carbon dioxide conversion of 15.6%. The results were attributed to the higher reducibility of this catalyst. Despite the advances on dimethyl ether synthesis from carbon dioxide hydrogenation over the last few years, low carbon dioxide conversions and low dimethyl ether selectivity and yield remain a bottleneck in this process.

7.6 Chemical Conversion Efficiency

As a metric for the performance of the conversion processes, we consider the chemical conversion efficiency as the amount of fuel energy that is obtained from a given quantity of hydrogen (Equation 7.7):

$$\eta_{CCE} = \frac{\dot{m}_{Fuel} LHV_{Fuel}}{\dot{m}_{H_2} LHV_{H_2}} \tag{7.7}$$

Where $\dot{m}_{Fuel}$ and $\dot{m}_{H2}$ denote the mass flow rates of produced fuel and consumed hydrogen, respectively.

7.6.1 Exergy

Maximal possible conversion of heat to work, or exergy content of heat (Equation 7.8), depends on the temperature at which heat is available and the temperature level at which the reject heat can be disposed, i.e., the temperature of the surroundings.

Exergy exchanged (q_e) is then:

$$q_e = q \cdot \left(1 - \frac{T_o}{T_{\text{source}}}\right) \tag{7.8}$$

where T_{source} is the temperature of the heat source and T_o is the temperature of the surrounding.

Higher exergy content tends to mean higher energy prices. Exergy in a sense can be understood as a measure of the value of energy. Since high-exergy energy carriers can be used in for more versatile purposes, due to their ability to do more work, they can be postulated to hold more economic value. This can be seen in prices of energy carriers, i.e. high-exergy energy carriers such as electricity tend to be more valuable than low-exergy ones such as various fuels or heat. This has led to substitution of more valuable high-exergy energy carriers with low-exergy energy carriers, when possible. An example is heating systems, where higher investment to heating systems allows using low-exergy energy sources. Thus high-exergy content is being substituted with capital investments.

Exergy of a system is the maximum useful work possible during a process that brings the system into equilibrium with a heat reservoir. The concept of material input per unit of service (MIPS) is quantified in terms of the second law of thermodynamics, allowing the calculation of both resource input and service output in exergy terms. This exergetic material input per unit of service (EMIPS) has been elaborated for transport technology. The service not only considers the total mass to be transported and the total distance, but also the mass per single transport and the delivery time. The applicability of the EMIPS methodology relates specifically to transport system and allows an effective coupling with life cycle assessment. The exergy analysis according to EMIPS allowed the definition of a precise strategy for reducing environmental impacts of transport towards more sustainable transport. Such a strategy requires reducing the weight of vehicles, sustainable styles of driving, reducing the friction of tyres, encouraging the use of electric and hybrid vehicles, improving the walking and cycling environment in cities and by enhancing the role of public transport, especially electric rail.

7.6.2 Exergy Efficiency

Additional energy streams entering or leaving the process can be accounted for by comparing the exergy efficiency of the processes (Equation 7.9):

$$\eta_{exergy} = \frac{\dot{m}_{\text{Fuel}}\, q_{e(\text{Fuel})} + P_{\text{out}}}{\dot{m}_{\text{H}_2}\, q_{e(\text{H}_2)} + \dot{m}_{\text{H}_2}\, q_{e(\text{CO}_2)} + P_{\text{in}}} \tag{7.9}$$

Where q_{ei} are the specific exergies of the produced fuel or utilized hydrogen and carbon dioxide and P_{in} and P_{out} are the net streams of electrical power entering or approximated by the change in Gibbs free energy upon combustion assuming liquid water in the products (ie., based on the higher heating value). This decision was made as in the processes some hydrogen is converted to liquid water as a stoichiometric side product of the reactions in Equations 7.2–7.4. Therefore, calculating fuel exergies based on the lower heating value would underestimate the work that can be extracted from hydrogen and the resulting exergy efficiencies could exceed unity.

7.6.3 Economic and Environmental Evaluation

Several quantities of interest require an evaluation of the entire process chain. We consider fuel cost and WTW GHG emissions (expressed in CO_2 equivalents) for the present model (Bongartz et al.) (Figure 7.1).

The comparison is done both per MJ of fuel used and per km driven with the respective fuels. While the former only captures differences in the fuel production processes, the latter is also influenced by engine efficiency. To this end, driving cycle fuel consumptions for the internal combustion engine fuels are taken from Huss et al. for the "2020+" configuration for compact cars defined therein. This configuration is based on the expected short-term technology development according to Huss et al. and was chosen for the present study since it corresponds best to the scenario considered herein. The corresponding values are 142 MJ/100 km for gasoline, 145 MJ/100 km for methane, 119 MJ/100 km for diesel and 122 MJ/100 km for dimethyl ether. For methanol, the same value is used as for gasoline. This is probably a conservative estimate, since, at least at rated power, slightly increased efficiencies compared to gasoline are usually reported for methanol. For H_2-FCV, the fuel consumption reported for the Toyota Mirai of 91 MJ/100 km is used, which is currently the lowest consumption value of a commercially available car. It is thus assumed to be a likely average fuel consumption for the next FCV generations considering different car classes and different performance characteristics.

For the economic evaluation, fuel production and distribution costs are considered. Changes in vehicle costs are beyond the scope of this evaluation. Fuel production costs include total cost of investment (TCI) of the fuel production plant, operation and maintenance costs (O&M) as well as feedstock (hydrogen and carbon dioxide) and electricity costs. The TCI is estimated based on the costs of the major plant components and according to Turton et al. they are expected to be accurate only within 15% to 70%. However, as will be shown below, the impact of TCI on overall fuel cost is so small that this uncertainty does not impact the results significantly. The specific costs for hydrogen production and transport from Grube et al. are used as a base case that is complemented by a sensitivity analysis. For carbon dioxide supply, the costs for biogas upgrading via pressure swing adsorption reported by Urban et al. are fully attributed to the carbon dioxide. Reiter and Lindorfer calculated a range of 1 to 10€/(t 100 km) for carbon dioxide transport by pipeline. Bongartz et al. use the upper value, since the pipeline network in this study is small and branched. Literature data is used for transport and gas station cost, respectively, of hydrogen, methane, methanol and dimethyl ether.

The evaluation of the fuels in terms of GHG emissions is done following life cycle assessment (LCA) methodology. LCA is a method to analyse the environmental impacts inherent in products, considering the entire life cycle of a product (raw material extraction, production, use, waste treatment, recycling and final disposal). LCA is standardized in DIN EN ISO 14040 and DIN EN ISO 14044. Specifically, Bongartz et al. employ the IPCC 2013 (100 years) method. For the their study, emissions from the construction of plants or infrastructure as well as vehicle manufacturing are not taken into account. It is assumed that only carbon dioxide is emitted during the combustion of the fuel, neglecting non dioxide emissions of other greenhouse gases. As a base case, the greenhouse gases emissions due to the energy demand of carbon dioxide separation from biogas are fully attributed to the car used as a feedstock for the present processes. The data for greenhouse gases emissions associated with hydrogen supply as well as grid electricity in Germany 2035 are taken from Grube et al. A summary of the parameter values and the life cycle inventory data derived from the process simulations can be found in the presentation of Bongartz et al.

7.6.4 Fuel Assessment

The successful introduction of new fuels requires broad consumer and public acceptance, which depends on a variety of factors beyond the economic and environmental performance over the entire process chain. Therefore, the assessment is complemented with selected engine measurement results as well as literature data on a few criteria related to the application of the synthetic fuels. Since new fuels also must compete with conventional fossil fuels, gasoline and diesel fuel are also included as reference cases. Bongartz et al. consider the following main categories:

1) Handling/Safety: safety and ease of use of the fuel in everyday life; compatibility with existing vehicle fleet (either directly or retrofitting);
2) Environmental impact: WTW GHG emissions; impact of the fuel on humans and nature when released unburned (i.e., spilled);
3) Infrastructure: compatibility with existing infrastructure for fuel transport and dispensing, current availability;
4) Efficiency: efficiency of fuel production and engine use;
5) Energy/power density: driving range per tank volume; driving power per mass of powertrain, fuel and tank; and
6) Pollutant formation: carbon monoxide, unburned hydrocarbons, nitrogen oxides (NO_X) and particulate matter.

While some categories can easily be assessed quantitatively, others like compatibility with infrastructure are qualitative by nature. For the latter categories, trends are given as to whether the fuels perform better or worse than the reference fuels gasoline and diesel in the respective category. Quantitative assessment of factors like efficiency or pollutant formation is possible but challenging because the results of simulations and test bench studies are highly dependent on the operating point and engine configuration considered. To this end, average changes at rated power compared to the respective conventional fossil fuels are compiled from literature data and translated into positive or negative trends.

As an additional category, the synthetic ICE fuels are also compared with the conventional fossil fuels with respect to several technical aspects related to their combustion behaviour. However, since these are not applicable to FCV, they are listed separately.

7.6.5 Performance of Fuel Production Processes

In the following, Bongartz et al. first compare the efficiency of the fuel production processes and discuss possible causes for the observed differences, as well as potential for improvements. While derived with the present scenario in mind, these results are expected to be largely independent from the scenario assumptions (with the exception of steady state operation). Next, the results for fuel cost and WTW GHG emissions are presented. Finally, the performance of the fuels from an application perspective is discussed to give an overview of many relevant factors.

Overall conversion of hydrogen and carbon dioxide to the desired products is essentially complete. The mass flows entering and leaving the fuel production processes thus deviate only slightly from the stoichiometry of the reactions shown in Equations 7.2–7.4 because of impurities in the product streams and losses to by-product streams. For methanol and dimethyl ether, about 3% and 1.5%, respectively, of the total mass input are lost to purge streams. The difference between the two corresponds to the lower recycle ratio in the case of dimethyl ether because of the improved equilibrium conversion. In methane production, around 1% of the total methane produced is dissolved in the water streams leaving the flash units. In methanol and dimethyl ether production, this loss is smaller (the water contains methanol corresponding to a 0.2–0.3% product loss) because water is removed via distillation. Therefore, in the latter processes, the loss to the water stream could be reduced further through suitable column design at of expense of capital and energy cost.

Since the aforementioned losses are relatively small, all three processes exhibit chemical conversion efficiencies close to the ideal values determined by the stoichiometry and enthalpy of reactions shown in Equations 7.2–7.4. Accordingly, when producing dimethyl ether, about 10% more fuel energy can be obtained from a given amount of hydrogen than when producing methane. The ideal chemical conversion efficiency corresponds to that of a process with full conversion and perfect selectivity and is determined by enthalpy of reaction in the reactions shown in Equations 7.2–7.4 per unit of hydrogen converted ($\Delta H_{r,298K}$) assuming liquid water in the products (Table 7.6).

The exergy efficiencies of the present processes are relatively high already (Table 7.6), but also show that some further improvements are possible. All three processes require a gross electrical power input on the order of 1–3% of the fuel energy. For methanol, the high reactor pressure of 7 MPa entails a high electricity demand for compressing the incoming carbon dioxide as well as recompressing the internal recycle stream. For dimethyl ether, the refrigeration cycle needed for the cryogenic separation of dimethyl ether from carbon dioxide is the largest power consumer. Thus there is potential for future process improvements by using alternative separation technologies.

In methane production, power generation in the steam power cycle is significant because of the strong exothermicity of the reaction in Equation 7.1 and the absence of energy intensive separation steps. Since gross power consumption is low because of the low-pressure once-through setup, this

Table 7.6 Comparison of performance of different processes.

Metric	CH_4	CH_3OH	DME
$\Delta H_{r,298K}$ (kJ/mol_{H2})	−63.2	−43.6	−42.4
η_{CCE} (%)			
Presented process	82	85	90
Ideal	83	88	92
η_{EXERGY} (%)			
With turbine	87	91	89
Without turbine	81	89	89
Capacity (kt/a)	374	970	707
Investment costs (10^6 €)	60	130	230

leads to a net production of electrical power. External heat input is not required for any of the three processes.

Because of its simple once-through setup, the methane production plant is significantly cheaper than the other two (Table 7.6). Note also that the product mass flow rates of methanol and dimethyl ether are significantly larger than that of methane because of their higher C-to-H ratio and molar mass (recall that the capacity is based on the same hydrogen input). For methanol, the large compressors and heat exchangers that are required because of the high reactor pressure and recycling rate are major cost drivers. For dimethyl ether, around 50% of the investment cost is associated with the refrigeration cycle. The investment cost estimates are in good agreement with the ranges given by Brynolf et al. for large-scale plants.

The two major assumptions of the scenario affecting the analysis of the fuel production processes are steady-state operation and pure reactants neglecting contaminants. If the processes are to be operated dynamically, substantial changes in design may be required and the performances may well differ from the steady-state case. For hydrogen, neglecting contaminants is most likely a good assumption because of the very high purities achieved by common electrolysis processes. For carbon dioxide, the quality of this assumption will depend on the carbon dioxide source and the clean-up processes employed. When using biogas, the biogas would at least need to be desulfurized to prevent contamination of the carbon dioxide with sulfane, which can poison the fuel synthesis catalysts. While the investment cost does depend on the scale of the system and hence the scenario used, we do not expect a change in the ranking of the three processes in this respect. Furthermore, since plant investment causes only a very small fraction of total fuel cost, Bongartz et al. also do not expect this to change the ranking with respect to resulting fuel cost.

7.6.6 Process Chain Evaluation

Based on the assumed hydrogen availability described in Section 7.2.1, the total fuel production is on the order of 2×10^{10} MJ/a, which is around 1% of the total German gasoline and diesel fuel consumption in 2013. This suggests that rather than relying on such "surplus" renewable electricity, designated renewable electricity production is required for supplying a significant fraction of the total fuel consumption within the time frame (and thus penetration of renewables) considered. When using designated renewable electricity for electrolysis, the present analysis would still hold true to a large extent. The only expected changes are hydrogen cost (which is captured in the sensitivity analysis presented below) and possibly slightly different GHG emissions of hydrogen supply resulting from the different evolution of hydrogen generation and resulting changes in hydrogen storage over time.

7.6.7 Fuel Cost

The total fuel cost of the electro fuels as well as the contributions of the different cost factors are summarized in Table 7.7 for the base case of the considered scenario. The cost per MJ fuel energy is very similar (within 2%) between the three ICE fuels, while that of H_2 for direct us is 15% lower. When comparing fuel cost per distance driven, the differences are larger and DME is 17% cheaper than methane and methanol. This is due to its higher tank-to-wheel efficiency since it is used in a CI engine instead of a SI engine. Direct use of hydrogen in an FCV is in turn 47% cheaper per distance driven than the use of methane because of the high efficiency of fuel cells compared to ICEs. However, it should be emphasized that this does not include changes in vehicle cost that will likely be higher for FCVs than for ICE vehicles.

By far the largest share of the total costs results from hydrogen supply (at least 76% for the three ICE fuels). Electricity, investment (based on the assumed lifetime of 15 years and interest rate of 10%) and O&M for the fuel production plant play only a minor role. This agrees with the conclusion of Brynolf et al. for plants with a high-capacity factor (0.95). The contribution of the distribution costs to the total costs is low for the ICE fuels (maximum 8%), but considerable for direct use of hydrogen (28%) because of the expensive infrastructure and extremely high fuelling pressure. If the electricity that is coproduced by the methane synthesis plant is sold at the assumed market price without grid fees, the cost of methane can be reduced by 0.09 €-ct./MJ_{fuel} (i.e., around 1.5%) relative to the total cost listed in Table 7.7. However, this does not change the ranking of the fuels.

Table 7.7 Results of the economic evaluation for the base case of the discussed scenario. Note that "hydrogen supply" includes energy and capital cost of electrolysis as well as hydrogen storage and pipeline transport to the fuel production plant (where applicable).

Electrochemicals/Electro fuels	CH_4	CH_3OH	DME	H_2
Fuel production costs (€$_{-ct}$/MJ_{fuel})				
TCI	0.04	0.09	0.15	
O&M	0.01	0.02	0.04	
Electricity		0.002	0.05	
CO_2 supply	0.55	0.71	0.68	
H_2 supply	4.5	4.4	4.1	3.4
Totals	5.1	5.2	5.0	3.4
Fuel distribution costs (€$_{-ct}$/MJ_{fuel})				
Transport	0.02	0.15	0.12	0.31
Gas station	0.40	0.28	0.28	1.00
Totals	0.43	0.43	0.40	1.31
Total fuel cost				
(€$_{-ct}$/MJ_{fuel})	5.5	5.6	5.4	4.7
(€$_{-ct}$/km)	8.0	8.0	6.7	4.3

Figures (7.2) and (7.3) show sensitivity analyses with respect to hydrogen and carbon dioxide cost as the two main cost drivers. Depending on hydrogen supply cost, direct use of hydrogen in an FCV is 35 to 50% cheaper per distance driven than the three ICE fuels. Among the ICE fuels, dimethyl ether is the cheapest for all but the lowest hydrogen costs and its cost advantage compared to methane and methanol increases with increasing hydrogen cost. This results from a more efficient use of hydrogen (i.e., a higher chemical conversion efficiency) in dimethyl ether production than in methane or methanol production. The sensitivity to the cost of carbon dioxide supply is less pronounced and only the ranking between methanol and methane changes slightly with changing carbon dioxide cost due to the different carbon content.

For comparison, the ranges of net prices of conventional gasoline and diesel in Germany from 2012 to 2017, as well as the fossil fuel selling prices including tax are indicated by the light and dark grey areas in Figures (7.2) and (7.3). For the electro fuels to be competitive with net fossil fuel costs, the lines corresponding to the electro fuels need to be within the light grey areas. For the three ICE fuels, this is the case only if hydrogen is supplied at a cost below 1 to 1.5 €/kg including hydrogen transport (Figure 7.2), which is on the lower end of fossil hydrogen costs. For direct hydrogen use in an FCV, the allowable hydrogen cost for competitiveness with fossil fuels is around 3 €/kg. In contrast, finding cheaper carbon dioxide sources is not sufficient for the ICE fuels to be competitive with net fossil fuel costs (Figure 7.3).

For the base case hydrogen cost of 4.5 €/kg, none of the electro fuels are competitive with gasoline or diesel. However, comparing to fossil fuel selling prices including tax, both hydrogen and dimethyl ether are competitive in terms of fuel cost per distance driven. This gives an indication for the potential of tax subsidies. It should also be mentioned that the base case hydrogen cost of 4.5 €/kg considered here lies in the lower part of the range of typical cost estimates for renewable hydrogen.

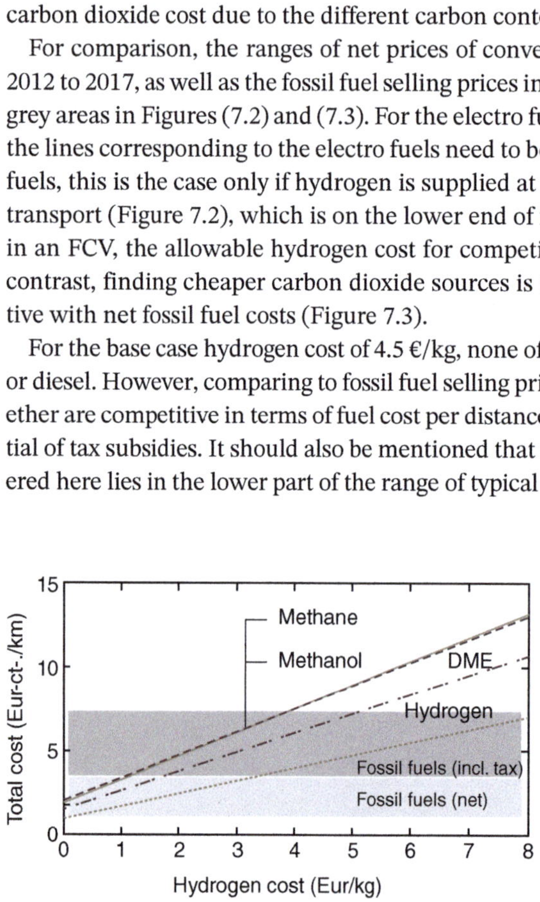

Figure 7.2 Sensitivity analysis for the impact of hydrogen cost (including transport) on total fuel cost. The base case cost for hydrogen from electrolysis using renewable electricity. according to Grube et al. is 4.5 €/kg including transport. The ranges fuels are given for Germany in 2012–2017. Adapted from Grube et al., 2018.

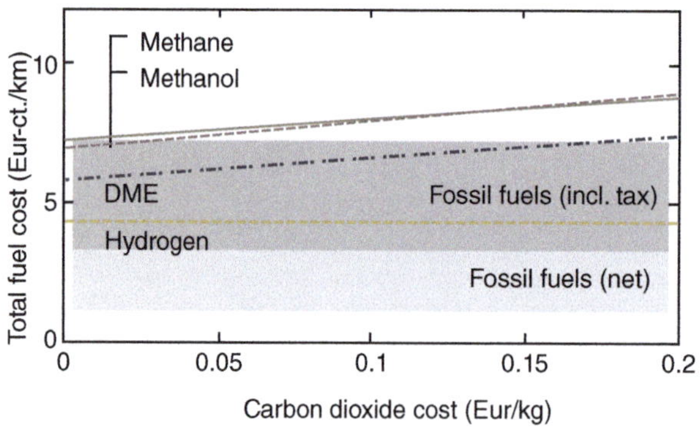

Figure 7.3 Sensitivity analysis for the impact of carbon dioxide cost (using the base case cost of 4.5 €/kg for hydrogen) on total fuel cost. The base case cost for carbon dioxide from biogas upgrading (including transport) is 0.1 €/kg, while typical ranges for carbon dioxide from power plants are 0.04 to 0.1 €/kg. Adapted from Rubin et al., 2012.

7.7 Well-to-Wheel Greenhouse Gas Emissions

Table 7.8 summarizes the well-to-wheel (WTW) greenhouse gases emissions of the electro fuels that were calculated. They can be attributed to the different life cycle phases of hydrogen supply (electrolysis and transport), carbon dioxide supply (separation and transport), fuel synthesis and fuel distribution. Since direct emissions of carbon dioxide during synthesis and carbon dioxide release upon combustion equal the carbon dioxide uptake from biogas upgrading, these values have been omitted from Table 7.8 to enable locating the sources of "true" emissions.

The application of electro fuels leads to reductions of 80–90% in greenhouse gases emissions per fuel energy compared to fossil fuels, the WTW emissions of gasoline and diesel are around 88 g CO_2 eq./MJ fuel, corresponding to 124 g CO_2 eq./km and 105 g CO_2 eq./km, respectively, based on the respective fuel consumptions. This is of a similar order as the findings of previous studies (Bongartz et al.). Under the present assumptions, methane causes the least WTW GHG emissions of the ICE fuels.

This is due to the fact that the methane production plant consumes no electricity and requires less feedstock carbon dioxide (because of the low C/H ratio of methane). The carbon dioxide supply makes up the largest share of the emissions for the ICE fuels, for methanol and dimethyl ether. For dimethyl ether, the high electricity consumption of the fuel production plant due to the refrigeration cycle is an additional drawback. The direct use of hydrogen leads to the lowest GHG emissions since carbon dioxide supply as well as fuel synthesis can be omitted. In this case, fuel distribution is the most significant life cycle phase (79% of the total GHG emissions) because of the high fuelling pressure and hence power demand. When comparing per distance driven, the emissions of dimethyl ether are smaller than those of methanol because of the higher engine efficiency. Also, the advantage of direct hydrogen use increases further (93% reduction compared to gasoline).

The current treatment of the emissions caused by carbon dioxide separation corresponds to the avoided burden method. As an alternative, the emissions caused by biogas production and upgrading can be allocated based on the exergy of the two products biomethane and carbon dioxide. In this case, virtually all emissions of biogas production and upgrading are attributed to the biomethane. Carbon dioxide transport to the fuel synthesis plant still counts towards the electro fuels. This would also be consistent with the notion of carbon dioxide being an (otherwise unused) waste product of biogas upgrading.

Table 7.8 Well-to-wheel GHG emissions caused by the electro fuels. For the organic ICE fuels, "H_2/CO_2 supply" includes H_2/CO_2 transport to the fuel production plant. "Synthesis" denotes emissions due to the electricity consumption of the plant. "Distribution" includes the emissions from fuel transport and compression at the gas station (for methane and hydrogen).

Electro fuel	Methane	Methanol	DME	Hydrogen
Contributions ($gCO_2eq./MJ_{fuel}$)				
H_2 supply	4.2	4.0	3.8	1.9
CO_2 supply	8.4	10.9	10.3	
Synthesis		0.1	2.8	
Distribution	1.5	1.7	1.3	7.0
Total (gCO_2 eq./MJ_{fuel})	14.1	16.7	18.2	8.9
Total (gCO_2 eq./km)	20.4	23.8	22.2	8.1

Bongartz et al. presented the relative performance of the different fuels from an application perspective. In addition to the properties relevant to all fuels, the synthetic ICE fuels also exhibit more desirable performance than the conventional fuels in several combustion-related categories relevant to ICE design.

The increased tank pressure of dimethyl ether (like LPG, e.g., around 1 MPa), methane (20 MPa) and hydrogen (70 MPa) pose a disadvantage regarding both handling (more complex infrastructure and tanks) and safety compared to methanol, gasoline, or diesel, which are stored at atmospheric pressure.

Further differences arise regarding the compatibility with today's vehicle fleet, which in Germany, as in most countries, consists almost exclusively of gasoline and diesel cars. When using methanol or methanol-gasoline blends, the main drawback is the high corrosiveness (especially for rubber parts), which calls for potential changes within the engine. Retrofitting is routinely done for converting gasoline cars to run on methane, but it does require a new tank system as well as modification to the fuel supply and engine calibration. Similarly, dimethyl ether also requires a new tank and a new fuel injection system because of its different physical properties. The FCV fleet is currently negligible and retrofitting of other power trains is not possible.

7.7.1 Environmental Impact

Beyond the environmental impact due to WTW GHG emissions discussed in Section 7.6.6, additional considerations arise when fuel is released unburned because of a leak, a spill during fuelling, or an accident. Toxicity is a concern when fuel comes into contact with humans or animals. While dimethyl ether and especially hydrogen are very benign, toxicity is a major impediment for methanol. Methane and hydrogen are classified as being not hazardous to water, while methanol and dimethyl ether are classified as slightly hazardous and diesel and gasoline as hazardous. For methane, the global warming potential (i.e., the impact of the fuel on global warming when released unburned relative to the impact to an equal mass of carbon dioxide) is an additional concern in case of fuel leakages. While hydrogen is in between carbon dioxide and methane in this respect, the global warming potential of dimethyl ether is below unity.

7.7.2 Infrastructure

The current fuel distribution grid is completely adapted to gasoline and diesel fuel. Methane could build on the natural gas infrastructure for fuel transportation and the existing yet limited CNG infrastructure regarding gas stations. Methanol could use existing infrastructure if blended with gasoline, but if it is to be used as a pure liquid, the distribution and transport equipment have to be adapted. Dimethyl ether requires a fuel infrastructure similar to LPG. An infrastructure for hydrogen is expected to be the most costly due to the high pressures required. However, for the latter, several initiatives are underway for promoting the introduction of such an infrastructure.

The fuel availability (although possibly from different, including non-renewable, sources) at the required quality is an important issue for the successful introduction of a new fuel both from a vehicle manufacturer and a consumer perspective. Obviously, availability is highest for gasoline and diesel. While methane (or at least natural gas) is available in large quantities, it has to compete with use in industry power generation and residential heating. Methanol is an important intermediate in the chemical industry and as such being produced in relatively large quantities. However, fuel standards have so far only been specified for methanol-gasoline blends, not for pure methanol. Fuel standards are available for dimethyl ether and hydrogen. However, these substances are currently produced in lower quantities than methanol.

7.7.3 Efficiency

The well-to-tank efficiency is higher for the fossil fuels (e.g., around 85% for gasoline) than for the electro fuels (55–57% for the ICE fuels and 66% for hydrogen based on the calculations above; defined as the lower heating value of the fuels divided by the electricity inputs for electrolysis, carbon dioxide separation from biogas, hydrogen and carbon dioxide transport, conversion to the desired fuels as well as fuel distribution), although comparison is difficult because completely different starting materials are used.

Quantitative assessment of tank-to-wheel efficiency is complicated since it depends on many factors including the vehicle class and operating conditions considered. To this end, for internal combustion engines, average changes in efficiency at rated power compared to reference gasoline or diesel engines are compiled from literature data. For the reference engines, average values for C-segment vehicles in the worldwide harmonised light vehicle test procedure (WLTP) (high) driving cycle are taken from the Institute for Combustion Engines (VKA) database. Since methanol can also be blended with gasoline, mixtures containing different volume fractions of methanol in gasoline are also considered. Pure methanol or methane result in an average increase in the tank-to-wheel efficiency at rated power of up to four percentage points compared to the reference gasoline engine because of their higher-octane numbers that enable higher compression ratios. When using methanol-gasoline blends, the efficiency increases with increasing share of methanol. For DME, the majority of sources also show a slightly higher efficiency compared to the reference diesel engines. Note that the reference diesel engines in general have significantly higher tank-to-wheel efficiencies (here taken to be 38%) than the reference gasoline engines (here taken to be 31%). Comparing this difference to the average efficiency increases of methane and methanol compared to gasoline, DME is thus the most efficient of the ICE fuels considered here.

For FCVs, the efficiency at rated power can be close to 50%. However, besides exhibiting a different part load behaviour than ICE vehicles, for FCVs an even broader range of values for fuel consumption can be found in the literature. As for ICE vehicles, this is partially due to the fact that the baseline vehicles considered range from compact cars to sport utility vehicles. However, for FCVs the variability is even larger because the technology applied represents diverse statuses of technological maturity. In terms of driving cycle fuel consumption, earlier vehicles that were mostly operated in large demonstration projects achieved values of 120–158 MJ/100 km and driving ranges of 320–380 km. Assessments for near to long-term FCV fuel economies that are mostly based on detailed powertrain simulations come to different conclusions. JRC report fuel consumption for a compact car of as low as 54 MJ/100 km in their comprehensive well-to-tank assessment report. Another extensive fuel economy assessment comparing passenger cars with ICE, hybrid powertrains, battery electric vehicles and FCV concludes with a consumption value of 98 MJ/100 km.

7.7.4 Energy/Power Density

The volumetric energy densities of all electro fuels (Tables 7.3 and 7.5) are lower than those of the fossil fuels. Methanol and dimethyl ether are oxygenated fuels, while methane and in particular hydrogen have very low volumetric energy densities due to their gaseous nature despite the high storage pressures. On the other hand, the mixture heating values of the electro fuels are comparable to those of the conventional fuels, which also leads to comparable specific performances. However, in terms of range, the lower energy densities of the electro fuels cannot be offset by the increased tank-to-wheel efficiency. Vehicles thus have to be equipped with larger tanks to achieve comparable driving ranges as with gasoline or diesel. Based on the fuel consumptions listed above, the required tank volume per range increases from that of gasoline by a factor of around 1.5 for dimethyl ether, 2 for methanol and 4–5 for methane and hydrogen.

For the specific power we consider engine power divided by mass of the engine, transmission, tank system and fuel. Data is taken from Huss et al. for all fuels except for methanol which is estimated analogously. Methane and methanol have a disadvantage compared to gasoline because of the heavy high-pressure tank and increased fuel mass and tank size, respectively. The CI engine fuels exhibit lower values than the SI engine fuels because of the increased engine mass. For dimethyl ether, further drawbacks are the pressurized tank and increased fuel mass. The specific power of FCV is lower still because the fuel cell, electric machine and battery combined are heavier than the ICEs and because of the heavy high-pressure tank.

7.7.5 Pollutant Emissions

In assessing pollutant emissions, a distinction must be made between raw gas emissions and exhaust tailpipe emissions. Reductions in raw gas emissions can translate either into reduced exhaust emissions, or simplified exhaust gas treatment. For H_2-FCV, there are no local pollutant emissions since the only reaction product formed is water.

Analysing trends in literature data on pollutant emissions of internal combustion engines enables only a very rough approximation and further experimental investigations are needed for validation wherever possible. A particular challenge is to find values related to the same base vehicle. For example, the SI engines taken for the present analysis are direct injection engines, as they have the largest market share. However, literature studies refer to various vehicles and technologies, which makes it difficult to compare the results and therefore these values have to be interpreted with care.

However, for methanol, the following statements show reliable tendencies based on literature data: when burning pure methanol, nitrogen oxide (NO_x) emissions can be halved compared to gasoline. The emissions of carbon monoxide and total hydrocarbons (THC) are also lower than those of gasoline, but higher than those of blends containing up to 50% of methanol in gasoline.

To analyse the emission behaviour of methane, test bench measurements were performed on a single cylinder gasoline research engine. The engine speed was set to 1500 revs/min, whereas the indicated mean effective pressure (IMEP) was varied. The results were compared with a conventional gasoline fuel (ROZ98), which was burned at the same compression and air-to-fuel ratio as methane (14.7:1 and $\lambda = 1$). The resulting THC emissions of methane combustion were lower compared to gasoline over the full range of IMEP. The highest reduction of around 30% could be found at medium loads compared to the baseline gasoline fuel and the lowest reduction of around 15% improvement were given at maximum load. On the other hand, the carbon monoxide emissions were higher compared to gasoline over the full range of IMEP. A nearly constant offset of 15–20% higher carbon monoxide emissions for methane was identified. The highest reduction potential was found for the NO_X emissions which were around 30% lower for methane compared to gasoline over the full range of IMEP.

For dimethyl ether, test bench measurements were performed as well on a research diesel engine. Emissions of THC, carbon monoxide and NO_X were nearly identical to those measured with conventional diesel fuel. However, particulate matter (PM) emissions were reduced by more than 95% compared to diesel because of the oxygen content in the fuel. The strong reduction in PM can help to overcome the PM/NO_X trade off in CI engines. This could enable higher exhaust gas recirculation rates and therefore potentially stronger NO_X reductions.

7.8 Gasoline Electrofuels

Gasoline (C_5-C_{11} hydrocarbons) is one of the most important transportation fuels used worldwide. Besides the classical production through petroleum refining, it can be synthesized via Fischer-Tropsch (direct route) or via methanol as intermediate (indirect route). This section will present the recent advances on gasoline production through these two pathways.

Recent work has been done on the production of gasoline via direct carbon dioxide hydrogenation over a multifunctional Na–Fe_3O_4/HZSM-5 catalyst. Hydrogenation tests were performed in a fixed-bed reactor at 320°C and 3 MPa. The catalysts exhibited high selectivity to C_5-C_{11} of 78%, very low selectivity to methane and carbon monoxide (<10%) and high stability over 1,000 hours of time on stream (TOS). The authors concluded that the Fe_3O_4 sites enabled the reverse water gas shift (RWGS) reaction while the zeolite sites favoured the oligomerization/aromatization and isomerization. Moreover, the gasoline fraction was composed mainly of paraffins and aromatics, having a positive impact on the octane number. The authors further reported that the composition of C_5-C_{11} could be tuned by using the zeolites with different topologies. HZSM-5 produced 61% of aromatics in the gasoline fraction while HMCM-22 produced mainly paraffins (≈46%). Despite the high selectivity of the catalysts, carbon dioxide conversion reached only 22% (Vasconcelos and Lavoie, Wei et al.).

The production of C_{5+} isoalkanes, which can be used as gasoline additives, over a Fe-Zn/zeolite core-shell catalysts prepared by cladding method, has been investigated. Different zeolites (HZSM-5, H Beta, HY) were investigated in order to verify the influence of their structure in the carbon dioxide hydrogenation reaction at 340°C and 5 MPa for 2 h. Fe–Zn–Zr@HZSM-5–H Beta double-zeolite shell catalyst with a HZSM-5/H Beta ratio of 4 showed the highest isoalkanes/total hydrocarbons ratio (83.1%). These results were attributed to the synergistic effect of the HZSM-5 that promotes the production i-C_{5+}, with H Beta zeolites that favours the formation of i-C_4. However, carbon dioxide conversion was very low (around 14%). The authors also reported that the cladding method used for the preparation of the catalysts was very promising due to its simplicity and low cost, being suitable for use at large scale (Vasconcelos and Lavoie, Wang et al.).

Work has been done on the performance of a bifunctional catalyst containing partially reducible metal oxides (In_2O_3) and H-form Zeolite Socony Mobil-5 for the conversion of carbon dioxide into liquid fuels using methanol as intermediate. The catalysts showed a high selectivity gasoline-range hydrocarbons (C_{5+}) of 78.6% along with very low methane selectivity of 1% for over 150 h of TOS at 340°C and 3 MPa. The authors reported that the oxygen vacancies on the In_2O_3 activated the carbon dioxide and the hydrogen, producing methanol. Then, a C-C coupling occurs in the zeolite pores, converting methanol to C_{5+} hydrocarbons via a hydrocarbon-pool mechanism. As for all the studies presented up to date on the Power-to-Liquid for gasoline production, carbon dioxide conversion was also very low around 13% (Vasconcelos and Lavoie, Gao et al.).

7.9 Diesel Electrofuels

The educts of diesel electro fuels are hydrogen and carbon dioxide (Schemme et al.). The products of the well-established Biomass-to-Liquid (BTL), Coal-to-Liquid (CTL) and Gas-to-Liquid (GTL) processes will initially serve as models for electro fuels. Synthesis gas is also produced as an intermediate product of these processes. Here, BTL fuels are second-generation biofuels. According to Ajanovic, the products of these processes have much better potential for market penetration than first-generation biofuels. Ajanovic predicts the introduction of second-generation biofuels to the European market between 2020 and 2030. Since the products of the BTL and Power-to-Liquid processes are in principle equivalent, the projection of the market introduction can be transferred to Power-to-Liquid fuels.

In general, any desired gaseous and/or liquid fuel can be produced from synthesis gas. According to Kaltschmitt and Streicher, methanol synthesis, Fischer-Tropsch synthesis and methanation were the most important processes practiced in Central Europe in 2009. Graves et al., consider potential technological pathways for recycling carbon dioxide in the form of fuels by using renewable or nuclear energy. The products of these process pathways are also methane, methanol, or hydrocarbons. Hydrocarbons are produced via Fischer-Tropsch and methanol-to-gasoline synthesis

Wesselak et al. specify four main groups or pathways for BTL fuel synthesis. In addition to methanation, as well as methanol and Fischer-Tropsch synthesis, they include dimethyl ether synthesis. Depending on temperature and the catalyst used, the products of Fischer-Tropsch synthesis are alcohols, alkenes and paraffins with various C-chain lengths. Sterner and Stadler consider dimethyl ether to be part of the consecutive chemistry of methanol. Meanwhile, Bozem et al. do not mention dimethyl ether, but include only Fischer-Tropsch and methanol synthesis. Trippe, in contrast, only lists dimethyl ether and Fischer-Tropsch synthesis. In addition to hydrogen and methane, a recent study by the Agora Energiewende initiative concerning electricity storage for the Energiewende also lists methanol, Fischer-Tropsch diesel and Fischer-Tropsch kerosene as potential chemical energy storage systems. The Federal Government's Mobility and Fuels Strategy lists hydrogen, methane, gasoline, diesel, kerosene and ethanol as electro fuels.

In addition to liquefied natural gas, Fischer-Tropsch and HVO diesel and dimethyl ether, Peters et al. also list oxymethylene ether as potential alternatives to fossil diesel fuels.

Also known as → Dimethoxymethane or methylal, CH_3-O-CH_2-O-CH_3
→ Polydimethylene ether or polyoxymethylene dimethylether;
Abbreviated → OME_1, OME_2, $OMDME_2$, POMDME, $POMDME_n$, or $PODE_n$

A number of previous studies that present groups of synthesis gas-based fuels, in contrast, do not mention OME_n. Connolly et al. analyse various alternative fuels for an energy system based on 100% renewable sources, but also do not to consider OME_n.

However, OME_n are considered promising alternatives to conventional diesel fuel in a significant number of publications; especially with respect to their exhaust gas values. In addition to OME_1 (dimethoxymethane or methylal), Härtl et al. and Ren et al. investigate the potential of 18 further fuel alternatives with high oxygen contents, including nine other ethers, for reducing emissions. However, higher OME_n with $n > 1$ were not included and OME_1 is revealed as the best option for reducing pollutant emissions. Arteconi et al. compare the emissions of diesel blends with OME_1 and ten other ethers. According to these investigations, there is no "perfect ether" that simultaneously improves performance and emissions as a diesel additive and substituent. However, OME_1 is only one of several ethers examined that does not cause nitrogen-oxide emissions to rise. OME_n is composed of H, C and O atoms and can also be produced from synthesis gas. The question of whether OME_n must be produced indirectly following a methanol synthesis or via direct synthesis must be clarified. In addition to oxymethlyene dimethyl ethers (OME_n), oxymethylene diethyl ethers (OMDEEs) may also play a significant role. $OMDEE_n$, in comparison to OME_n, have considerably lower cetane numbers and higher flash points. OME_n are therefore significantly better suited for use as a diesel fuel. For this reason, $OMDEE_n$ are not further considered in this section. Although there is great interest in using OME_n as an additive/substituent for diesel fuel, there are as yet no fundamental and comparative studies that consider OME_n as a pure component. The reason for this is the currently low demand/availability.

With respect to the implementation strategy, it aims to identify substances that can gradually replace the fossil diesel in use today. Mixing must be possible, which is not easy with methane, as it does not have the "drop-in" quality, meaning that dual-fuel engines require two different tanks. Methane synthesis is therefore not considered in greater detail here. In the following, fuels known from the literature and produced from synthesis gas in well-established processes are expanded upon and subdivided into the three following main groups:

1) Hydrocarbons (olefins, paraffins, aromatic compounds);
2) Alcohols (methanol, ethanol...); and
3) Ethers (DME, OME_n)

These main groups represent the final products and therefore include consecutive chemical processes. Thus, dimethyl ether produced from synthesis gas via methanol is part of the ethers group.

Similarly, hydrocarbons produced via dimethyl ether-to-hydrocarbon or methanol-to-hydrocarbon processes belong to the hydrocarbons group. Saturated hydrocarbons are preferable due to their higher cetane number. Therefore, n-alkanes are considered in the following and olefins and aromatics ruled out. The three different substance groups are further diversified in the next part of this section and examined with respect to their suitability for future fuels.

A fuel with parameters such as high energy density, lower viscosity and freezing point is desirable to be used in military aircraft at higher altitudes. In this regard, exotetrahydrodicyclopentadiene, a major constituent of JP-10 fuel, has ideal properties to be used as missile fuel and can be produced from its endoisomeric form known as endotetrahydrodicyclopentadiene. Dicyclopentadiene is used as a feedstock to produce endo- or exotetrahydrodicyclopentadiene either in one or two steps. This brief review (Khan et al.) is a compilation of the research so far, conducted by various research groups and provides knowledge on the dicyclopentadiene transformation to endo- and exotetrahydrodicyclopentadiene over various hydrogenation–isomerization catalysts at various reaction conditions. The role of the catalyst and its deactivation is discussed and the use of a catalytic reactor is also mentioned.

7.10 Electrofuels and/or Electrochemicals

As is the case for fossil fuels and biofuels, the requirements made on electro fuels are different according to their intended purpose. As in the preceding section that aims to generate a basis for discussion, this section will continue to address diesel as an example. After listing all potential substances, the next step is investigating their potential for fulfilling the requirements of diesel engines. The basis for estimating the potential is provided by DIN EN 590 for diesel fuels, on the one hand and by substance-specific investigations, on the other. At best, the alternative fuel can be mixed with the fossil fuel without any further additive and has better fuel properties than fossil diesel. Electro fuels have great potential as energy sources when it comes to future efficiency and emission improvements to diesel engines.

In contrast to fossil diesel, which to some extent contains branched and aromatic molecules, the products of the processes listed consist of unbranched molecules. Unbranched molecules have better self-ignition properties, which lead to a higher cetane number for diesel fuels. During the combustion process, branched molecules can lead to a local oxygen shortage. This can result in incomplete combustion and therefore in coking and soot emissions, as is the case when vegetable oils (FAME) are used in diesel engines. Furthermore, synthetic fuels contain no sulfur, ash, or other residues, making them cleaner and chemically more stable. This is an advantage when used in hybrid vehicles in which the fuel remains in the tank and fuel-line system for longer periods of time. In particular, the absence of sulfur has considerable positive effects on exhaust gas values. Furthermore, they have a narrower boiling range than fossil diesel and can even have a defined vaporization point. If there is a complete switch from fossil to synthetic fuels, engines can be optimized in a more targeted manner. In addition to engine efficiency, oxygen content in fuel may reduce emissions. The better the fuel quality, the better the engines that can be constructed.

According to Gill et al., many studies show that there is still great potential for the optimization of road transport fuels with respect to both emissions and engine efficiency. The initial challenge is to identify innovative sustainable raw materials, processes and technologies. At this point, PTF is a promising alternative to BTL. According to research by Mathiesen et al., electro fuels produced from fluctuating renewable electricity and carbon dioxide could completely replace biofuels – which are affected by limitations – in an advanced phase of the energy system.

According to a prognosis of Peters et al., in 2050 2.5–7.5 Mt electro fuel/a will be produced in Germany, depending on the share of nationwide available carbon dioxide from industry exhaust gases. The hydrogen supply needed for the higher prognosis result of 7.5 Mt electro fuels/a amounts to 3.2 Mt hydrogen/a. This amount of hydrogen can be produced from surplus electricity from wind power, according to the forecasts and energy system design by Robinius. In this consideration, it was assumed that the major part of the surplus electricity from wind power is already utilized to run fuel cell passenger cars, which have a high share of 75% of all passenger cars in Germany in 2050. Worldwide, Peters et al. predict the possible share of fuels in the transport sector produced from renewable hydrogen and carbon dioxide from industry exhaust gases as 7% in 2050.

Derived from the values given in the EU Directive 2015/652, alternative fuels with hydrogen from non-biological renewable energy electrolysis as the raw material source have a significantly lower average life cycle greenhouse gas intensity than fossil-based fuels.

7.10.1 Physico-Chemical Properties

In contrast to pure hydrocarbons, DME and OME_n contain oxygen and no C–C bonds, permitting the assumption that they have a certain potential for soot-free combustion. According to Härtl et al., stoichiometric combustion of OME_1 results in methane in the exhaust gas. The potential for forming methane is less in higher OME_n. Due to the increased oxygen content, the surface tension rises given that the intermolecular van der Waals forces increase. Regarding the Weber number, an excessive surface tension can have negative effects on atomization during fuel injection into the combustion chamber. However, the surface tension values of OME_n are only slightly higher than those of fossil diesel; Lautenschütz et al. state that this higher surface tension does not cause problems during atomization.

In addition to important fuel properties for engine performance, DIN EN 590 also sets out chemical and safety-related requirements. OME_n and n-alkanes have good material compatibility characteristics and low toxicity, while dimethy ether is more chemically aggressive than fossil diesel fuel. It is, however, not carcinogenic, teratogenic, or mutagenic and is also non-toxic.

Mixture formation with fossil diesel is a basic prerequisite for gradual integration in the existing market. At normal pressure, dimethyl ether exhibits poor solubility in diesel, making diesel/DME blends unstable and storage under pressure necessary. Furthermore, diesel/DME blends have a miscibility gap at 0°C. Diesel/OME_n blends do not have such a miscibility gap. Additionally, OME_n can be mixed with diesel in all concentrations. As a mixture of long-chain hydrocarbons, the molecules of FT diesel are very similar to fossil diesel. Mixing is thus not problematic.

7.10.1.1 Density

As Table 7.9 shows, the densities of dimethyl ether and FT diesel are below the minimum value stipulated by DIN EN 590, while the densities of OME_n are above the maximum values. The reason for the higher density of OME_n is the C–O bonds. These are shorter than C–C bonds, leading to a higher density in OME_n compared to n-alkanes with the same chain length. The lower density of the n-alkane cut of diesel relative to fossil diesel is due to the proportion of aromatics and unsaturated hydrocarbons in the latter.

As Table 7.9 shows, the density can be achieved by blending various fuels. Blending FT diesel or dimethyl ether with OME_n therefore seems promising. The difference in the density of the two substances to be blended must not be too great, as separation may otherwise occur. According to Lautenschütz et al., however, this is not critical for OME_n. The density of dimethyl ether is approximately 15% lower than that of fossil diesel. This is acceptable when compared to other alternatives (mainly alcohols and ethers).

7.10.1.2 Tribological Properties

Similar to the density, the viscosity of OME_n and n-alkanes rises with increasing chain lengths, with the proportion of oxygen playing a negligible role. The DIN EN 590 requirement of a viscosity of 2–4.5 mm^2/s is met by the n-alkanes starting from C13. OME_n meets the standard from $n \geq 5$.

Table 7.9 Properties of diesel fuels.

Properties of the fuel	$T_{boiling}$,°C	LHV,MJ/kg	Cetane Number	Density kg/m^3	Viscosity mm^2/s
Fossil diesel	180–370	42.5	50–54	780–840	3.5–3.9
DME	-25	27.60	55	668	<0.1
OME_1 (methylal)	42	22.44	50	867	0.3184
OME_2	105	20.32	63	961	0.64
OME_3	156	19.14	70	1021	1.05
OME_4	201	18.38	90	1059	1.75
OME_5	242	17.86	100	1100	2.63
n-Alkane cut C_{16}-C_{23}	-	42.68	56.5	830	4.13
GTL-FT diesel	90% at 295–335	43.90	79	783	3.47

The viscosity of dimethyl ether is 20 times lower than that of fossil diesel. Using dimethyl ether, therefore, lubricating the plungers in fuel pumps is practically impossible, which leads to wear. Additives improving lubricity can mitigate this. Otherwise, the fuel pumping and injection systems must be adapted. For small amounts of additives, no modification is necessary. The same also applies to OME_1. This is one reason why, according to Härtl et al., the use of higher OME_n is preferable. If, for example 5%, 30% or 50% OME_2 is added to fossil diesel, then, according to Iannuzzi et al., mixture viscosities of 2.83–1.74 mm^2/s (see Table 7.9) are the result, which to a certain extent meet the requirements of DIN EN 590.

The cold-flow behaviour is evaluated by means of the melting point. The melting and boiling points of OME_n and n-alkanes rise with growing chain lengths. Amongst the n-alkanes, n-undecane(C_{11}) has a melting point of 26°C. The melting points of longer n-alkanes are >20°C and therefore these are not suitable for winter diesel. Amongst OME_n, OME_3 fulfils the requirements for winter diesel. Lautenschütz et al. examined an equal weight-distributed mixture of OME_3, OME_4 and OME_5 in the context of their melting behaviours. This mixture had a melting point of 16°C, coming close to fulfilling the standard. The cold-flow behaviour can be improved by the use of suitable additives.

7.10.1.3 Combustion Characteristics

Diesel engines are based on the principle of the self-ignition of fuel. Ignition performance is described by the cetane number. In addition to the density, cetane number and viscosity of diesel fuels required by DIN EN 590, Table 7.10 lists those of methanol, dimethyl ether, OME_2 and n-alkanes. For OME_n, characteristics suitable to evaluating its potential as a fuel are available for n = 1–5.

For analyses concerning n-alkanes, the following assumes diesel that is produced synthetically via Fischer-Tropsch (FT) processes. Fischer-Tropsch diesel is a mixture of various n-alkanes of different lengths. Its product distribution is particularly dependent on the catalyst used, educt composition, process temperature and pressure. For this reason, no defined standard physical and chemical properties can be given for Fischer-Tropsch diesel. Table 7.10 therefore outlines the density, cetane number and viscosity of various Fischer-Tropsch diesels. These Fischer-Tropsch diesels were produced via BTL, CTL and GTL processes. Synthesis gas is an intermediate product of all of these processes, making the products comparable to those of Power-to-Fuel processes. Gill et al. offer an overview of the combustion characteristics and emissions of various Fischer-Tropsch diesels. The properties of the Fischer-Tropsch diesels considered are in the value range given in Table 7.10.

Table 7.10 Overview of fuel evaluations: DME, OME, Fischer-Tropsch diesel.

Properties/Electrofuels	DME	$OME_{3\text{-}5}$	n-Alkanes (FT Diesel)
Comparatively safe	+	+	+
Existing infrastructure usable	–	+	+
Blending with DME, $OME_{3\text{-}5}$, n-alkanes conventional diesel	+	++	++
Engine modification not necessary	–	+	+
Reduction of HC, CO emissions	++	++	+
Reduction of soot emissions	++	++	+
Potential to avoid the soot-NO_x trade-off	+	++	-
Heating value compared to diesel → fuel consumption	–	–	+
Cetane number → optimization potential of engine efficiency	+	++	++

The self-ignition temperatures of fossil diesel and DME are approximately 250°C and 235°C, respectively, which is reflected in cetane numbers of just above 50. With comparatively high self-ignition temperatures of 450°C and 420°C, the respective cetane numbers of methanol is five (cf. Table 7.10). Despite the low cetane number of ethanol (8), investigations were carried out that showed that a diesel-ethanol dual fuel mode can improve the emissions. With an increase in chain length, the cetane number increase for alcohols leads to a cetane number of 25 for n-butanol, for instance. Alcohols therefore do not meet the requirements of DIN EN 590 and are not suitable as pure diesel fuels. However, according to the literature, higher alcohols such as n-butanol can be mixed with a share of up to 20% (vol.) into conventional diesel without phase separation. Due to the oxygen content, soot, NO_x and carbon monoxide emissions can be reduced. However, due to the lower cetane number, a delay in combustion occurs that leads to higher hydrocarbon emissions. Nevertheless, they are well-suited for use as fuel in other internal combustion engines, such as gasoline engines. Their disadvantages are primarily their high toxicity and corrosiveness. Due to their insufficient suitability as diesel fuel, alcohols will not be further considered.

An investigation of dimethyl ether, OME_n and the various Fischer-Tropsch diesels reveals that their cetane numbers are in principle significantly higher than that of fossil diesel. In the same way as for alcohols, for both OME_n and n-alkanes, the cetane number rises with increasing molecule length. Although the cetane number of dimethy ether is only slightly above that required by DIN EN 590, higher thermal efficiencies are achievable than with fossil diesel. The cetane number of OME_1 is below the required value. Accordingly, adding OME_1 to fossil diesel may lead to poorer ignition performance. The cetane numbers of Fischer-Tropsch diesel and higher OME_n ($n > 3$) are significantly higher than that required by DIN EN 590. Therefore, dimethyl ether, OME_n and Fischer-Tropsch diesel are suitable as diesel fuels. According to Liu et al. and Lautenschütz et al., considerable efficiency improvements over fossil diesel are possible.

Due to the high oxygen content of ethers, their heating value is lower. However, the oxygen content of OME_n increases with growing chain lengths, reducing the air demand and thus resulting in an improved heating value of the air-fuel mixture in the combustion chamber. Table 7.10 shows the heating values of various fuels.

Dimethy ether and OME_n have lower heating values than fossil and Fischer-Tropsch diesel. Due to the lower air demand during stoichiometric combustion, however, the heating value of the fuel-air mixture in the combustion chamber is almost the same. Nevertheless, the lower heating value [MJ/l] of the ethers leads to higher fuel consumption per kilometre.

The flash point of Fischer-Tropsch diesel and OME_n at $n > 3$ is above the required minimum value of 55°C. Safe storage can therefore be ensured.

The low boiling point of dimethyl ether leads to rapid vaporization in the combustion chamber and thus to better mixture formation. In diesel engines, the injection pressure can thus be decreased by approximately 50%, to below 50 MPa. Injection pressures of 20–30 MPa are thus possible, while further reductions may impair the functioning of the injection system by falling short of the vaporization pressure of dimethyl ether. Under environmental conditions, dimethyl ether is gaseous and becomes a liquid at 0.5 MPa. The compressibility depends strongly on temperature. Therefore, fuel systems must be pressurized to 1–2 MPa to avoid the risk of vapour bubbles damaging the fuel system. In contrast to melting points and densities, the boiling points of OME_n and n-alkanes are very similar. Although OME_1 is liquid at room temperature, it is highly volatile due to the relatively low boiling temperature of 42°C. In this respect, higher OME_n is more similar to diesel and thus preferable for mixture formation.

The boiling curve of a mixture of OME_n with $n = 2$–6 is lower and flatter than that of fossil diesel. This accelerates vaporization and a more targeted fuel injection is possible, which in turn may lead to increased efficiency.

7.10.1.4 Combustion and Emissions

Due to their different atomic structures, dimethyl ether, OME_n and n-alkanes have different combustion properties. These are reflected in their respective emissions. Generally, a reduction of pollutant emissions is desirable to reduce environmental pollution and simplify exhaust-gas treatment. Simplifying the exhaust-gas treatment leads to an increase in the total efficiency of the vehicle and thereby saves fuel. Ideally, the vehicle should only emit carbon dioxide and water, in addition to any unused combustion air. The pollutant emissions of fossil diesel include carbon monoxide, uncombusted hydrocarbons, soot particles (PM), sulfur oxides (SO_x), nitrogen oxides (NO_x) and carbonyl compounds. There is is usually a trade-off between PM and NO_x emissions for fossil diesel: higher pressures and temperatures reduce PM emissions, but simultaneously increase NO_x emissions.

To reduce the amount of NO_x emitted by the vehicle, an aqueous urea solution can be injected into the exhaust gas. Thus, NO_x formation can be reduced by up to 90% and the total efficiency increased by 3–5%

In diesel engines, dimethyl ether combusts nearly soot-free due to the high oxygen content. Minor amounts of additives reduce PM emissions significantly. A blend of fossil diesel with only 20% dimethyl ether reduces PM emissions by as much as 60%.

Compared to fossil diesel, dimethyl ether burns much more rapidly and silently. This combustion is both faster and more complete. The other pollutant components – NO_x, hydrocarbons and CO – are considerably reduced, according to Gruden. Using dimethyl ether as diesel fuel offers the potential for optimizing the combustion process so that exhaust-gas treatment would no longer be necessary. According to the review conducted by Arcoumanis et al., hydrocarbon emissions are usually reduced. Nevertheless, carbon monoxide emissions are, amongst other influencers, dependent on the engine and can decrease or increase. The difference in the results given in the literature leads to the conclusion that there is still potential for optimization. Oxidation catalysts can represent an additional remedy. DME usually exhibits lower NO_x emissions. According to Arcoumanis et al., as in the case of carbon monoxide emissions, these may be higher or lower than with fossil diesel, depending on the engine. Due to the good vaporization characteristics of dimethyl ether, lower injection pressures are permissible, reducing the risk of NO_x formation. If the injection pressure is not reduced, then the combustion pressures are higher than with conventional diesel. This, however, leads to increased NO_x emissions. In addition to adapting injection pressures, NO_x can also be reduced by means of more intensive exhaust gas recirculation.

According to Ren et al. and Iannuzzi et al., the reduction of PM emissions correlates with the O/C ratio. Investigations conducted by Iannuzzi et al. reveal that this correlation is not linear, but an increased oxygen content leads to a considerable reduction in soot. Accordingly, higher OME_n reduces PM emissions more strongly. According to Ren et al., OME_n permits a reduction in PM emissions without loss of efficiency. Blending OME_n with fossil diesel does not notably change combustion performance.

Overall, the investigations showed a significant reduction in CO, hydrocarbons and PM/soot/smoke emissions. In investigations carried out by Liu et al., carbon monoxide and HC emissions were reduced by 90% by adding only 20–30% $OME_{2,3,4}$ to fossil diesel. No problems occurred in engine operation that would suggest that there was any need to modify the engine. According to the investigations conducted by Iannuzzi et al., soot reduction increases with increasing OME_n chain length and the proportion of OME_n in the fuel. Adding as little as 5% OME_2 to fossil diesel reduces soot emissions by 30%. Increased blending results in reductions of more than 85%. Additionally, pure OME features almost smoke-free combustion.

For NO_x emissions, no uniform trend was discerned. In addition to the combustion temperature, the duration of combustion also influences NO_2 emissions. On the one hand, an increased proportion of oxygen in the fuel increases the combustion temperature, but on the other, it also shortens the combustion duration. A blanket statement on the influence of the oxygen content on NO_x emissions can therefore not be made.

When fossil diesel is combusted, there is usually a trade-off between PM and NO_x emissions due to the opposing influence of the combustion temperature and pressure. This trade-off was not observed in the investigations conducted by Lumpp et al., who noted that PM and NO_x emissions decreased by 50% and 25%, respectively, when 20% $OME_{3,4}$ was added. In contrast, according to Liu et al., the proportion of NO_x increased with rising OME content. This can be explained by increased combustion temperature arising from the high oxygen content. In principle, the PM-NO_x trade-off curve is less pronounced than that of fossil diesel. According to various articles in the literature, NO_x and PM emissions can be considerably reduced simultaneously: only slight alterations to the injection settings are necessary.

Just as in the case of dimethyl ether or OME_n, practical tests with Fischer-Tropsch diesel confirm emission advantages relative to fossil diesel. According to Wesselak et al., the HC and carbon monoxide emissions can be reduced by up to 75% and, according to Gruden, by up to 90%. In addition, PM emissions are reduced by 35–40%. NO_x emissions can be reduced by 5% in passenger cars and by up to 20% in commercial vehicles. By increasing the efficiency and H/C ratio, carbon dioxide emissions are up to 5% lower than with conventional fossil diesel fuel. Hao et al., investigated carbonyl emissions in the exhaust gas. By using FT diesel, a reduction of 27.2–44.6% was achieved.

Investigations by Gill et al. linked the reduction of various pollutant emissions with the substance properties of n-alkanes. In principle, the decreased density, absence of sulfur and aromatics and the high cetane number had positive effects. The reduction of NO_x, PM, hydrocarbons, carbon monoxide and carbon dioxide emissions was confirmed by the majority of reviewed sources, with many of these also revealing currently unused potential to reduce emissions from road transport fuels. All the sources reviewed by Lapuerta et al. pointed to significant reductions in carbon monoxide, hydrocarbons and PM emissions. However, there was no consensus about NO_x emissions. Eight of the reviewed sources stated a reduction, five no change and one an increase. In addition to the review, Lapuerta et al. examined a blend of fossil diesel with 30 vol% FT diesel from the BTL route. The investigation revealed that adequate blending promises potential optimization of the soot-NO_x trade-off. Rimkus et al. investigated the emission values of a blend of fossil diesel with 15% FT diesel from the BTL route. Due to the lower density, the volumetric fuel consumption increased [litre/km]. Carbon dioxide, hydrocarbons and smoke emissions dropped by 1.1%, 3–7%

and 16–18%, respectively, while NO_x emissions were reduced by 9–12%. The decrease in NO_x emissions was attributed to the higher cetane numbers, which reduce ignition delay, leading to lower combustion temperatures. These values were achieved without modifying the engine. If injection is adapted to maximum engine efficiency, smoke emissions decrease by approximately 40%, but NO_x emissions simultaneously increase by approximately 80%. This highlights soot-NO_x trade-off.

7.10.2 Diesel Engine Efficiency

Due to the higher cetane numbers of the electrofuels dimethyl ether, OME_n and higher n-alkanes, higher efficiencies are conceivable. However, these fuels would require engines to be optimized and further developed. Gill et al. compiled various investigations concerning the use of Fischer-Tropsch diesel in diesel engines. According to many of the sources examined by Gill et al., thermal efficiency increases, but the effective performance remains the same. Neither did the sources compiled predict any loss in efficiency resulting from the use of Fischer-Tropsch diesel. The test engines were designed for fossil fuels with low cetane numbers (>51), which implies the potential to increase efficiency. Due to the low density of the n-alkane cut relative to fossil diesel, Wesselak et al. ascertained a rise in fuel consumption by 4%. Rimkus et al. investigated the use of a diesel blend (85% fossil diesel + 15% Fischer-Tropsch diesel): engine efficiency increased by 2%, thus decreasing fuel consumption by 2.6%.

Using OME_n in diesel engines increases both thermal efficiency and fuel consumption, according to Liu et al. Ren et al. determine that the thermal efficiency particularly increases at full load and further potential was also identified. At medium and low loads, the efficiency is slightly lower than that of conventional operation. Liu et al. used OME_n-diesel blends (15% and 25% OME_{n+} fossil diesel). This also resulted in improved thermal efficiency. However, due to the lower energy density, the fuel consumption increased by 2 wt % with 15 wt % OME_n and by 2.8 wt % with 25 wt % OME_n. Further potential for efficiency increases in engines are conceivable for OME_n.

7.10.3 Potential of Diesel Electrofuels

This part provides a compilation of the decisive substance properties outlined above. The first step is an evaluation of societal acceptance. Furthermore, initial assessments concerning the potential of the various electrofuels are given.

Social acceptance is key to any conceivable implementation strategy. Alternative fuels should affect the consumer as little as possible and ideally have advantages over fossil fuels. These advantages can relate to costs or the environment. Costs are reflected in production costs and consumption and environmental compatibility mainly by emissions. For market penetration, the alternative fuel should exhibit the "drop-in" quality demanded by Deutsche Energi-Agentur, i.e., it must be mixable with conventional diesel. It should be possible to increase the proportions in these mixtures gradually, guaranteeing flexibility. According to Dry, Fischer-Tropsch diesel can be integrated into the existing infrastructure. Due to the similar physical properties, this also applies to OME_n. For dimethyl ether, implementation is more difficult, as it requires a different on-board tank and fuel distribution system. This includes an additional tank, for example, which must be pressurized to 0.5–1 MPa so that DME remains liquid and therefore can be transported at adequate energy density. According to Reif, the development of a dimethyl ether infrastructure is unlikely, although dimethyl ether presents an interesting alternative for urban fleet vehicles such as delivery trucks or buses. Centralized vehicle refuelling considerably simplifies market introduction. In summary, it is to be expected that OME_n and Fischer-Tropsch diesel will be more widely accepted by society than dimethyl ether.

Alcohols are unsuitable as complete replacements of diesel fuels. Adding small amounts of up to 20 vol%, however, they can improve the emission values of diesel fuel. Higher contents are suitable for gasoline fuels.

Due to its vaporization affinity and ignition performance, DME is considered ideally suited for diesel engines by Werner and Wachtmeister. Its cetane number is approximately 10% higher than that required by DIN EN 590, indicating further potential for efficiency optimization. Dimethyl ether also has very good exhaust gas characteristics. However, a reduction in NO_x emissions is only guaranteed when injection pressures are adjusted. The reviews by Arcoumanis et al. and Park and Lee, reveal that dimethyl ether is a promising replacement for conventional diesel for street transport vehicles. Both Arcoumanis et al. and Park and Lee regard further research into optimizing engine efficiency and emissions as essential. Only then can a final decision be made on the potential for mass production for the transport sector.

According to existing investigations, OME_n are very good alternatives for use in conventional diesel engines. The cetane number of OME_n are considerably higher than those of fossil diesel and around the same as the n-alkane cut of Fischer-Tropsch diesel. The high oxygen content of OME_n makes for particularly good exhaust gas characteristics, with a similar viscosity and vapour pressure to that of diesel.

According to Härtl et al. OME_1 cannot be used in engines in an unblended form due to poor lubricity, low cetane number and high volatility. Therefore, the use of higher OME_n is recommended since their physical properties are closer to those of conventional diesel. Higher OME_n has higher O/H ratios and therefore greater potential for soot reduction. A cetane number of >90 is recommended, which corresponds to OME_n, with n_4. Considering the figures given in Table 10.6, an OME_n with ideal cold-flow properties and a perfect flash point would be between OME_3 and OME_4. The boiling points of OME_3 and OME_4 are in the lower middle range of fossil diesel and of the n-alkane cuts. OME_3 and OME_4 were the major components of all OME_n mixtures, with $n > 2$.

Although the viscosity of OME_n, with $n > 4$, is closer to what is required by DIN EN 590, OME_n, with $n \geq 5$, may be problematic when it comes to the cold-flow properties of winter diesel. Both the viscosity and cold-flow properties can be improved by means of suitable additives. To find a compromise, the use of $OME_{3\text{-}5}$ is recommended here. Although the cold-flow behaviour is poorer, viscosity is adapted to the requirement. Furthermore, the more the cetane number increases, the higher the proportion of oxygen improves exhaust gas characteristics. Using higher OME_n is not recommended: although the cetane number continues to rise, OME_6 is solid at normal temperature.

Supplementary additives are not required if n-alkanes are to replace fossil diesel, which simplifies market penetration. The viscosity, lubricity, cold-flow properties, flash temperature and boiling temperature are all very close to those values for fossil diesel. Fundamentally, the density is lower due to the lack of aromatics. Negative effects caused by the lower density were not identified in any of the investigations reviewed.

Although oxygenic fuels have better exhaust gas characteristics, the pollutant emissions of n-alkanes are still lower than those of fossil diesel. The cetane number is also much higher, which is indicative of considerable optimization potential for engine efficiency. The dilemma of the NO_x-soot trade-off is less pronounced than in the case of fossil diesel, but persists nonetheless

It was found that the various alternative diesel fuels entail various advantages and disadvantages. The main advantage of dimethyl ether and OME_n is the considerably improved exhaust gas values, including the possibility of simultaneously improving soot and NO_x emissions. The main disadvantage of ethers is their low energy density. For dimethyl ether, additional modifications of infrastructure and vehicles would be necessary. To use $OME_{3\text{-}5}$, in contrast, the existing infrastructure and vehicle fleet can be used without the need for adaptation.

The highest energy density and therefore the lowest volumetric fuel consumption per kilometre of the fuel alternatives examined are to be found amongst the n-alkanes. Their values are on the same level as those of fossil diesel. With regard to physical and chemical properties, n-alkanes are the most similar to conventional diesel. Blending fossil diesel is, however, in principle possible with dimethyl ether, OME_n and n-alkanes. Therefore, the blending of these substances with each other can be expected to be possible. Blends containing higher OME_n and n-alkanes can be used without the need to modify engines, even with high concentrations. In contrast to gasoline, there is no limit to the oxygen content as for diesel fuel. E_{10} gasoline may contain no more than 22 vol% C_{5+} ether, with a maximum oxygen content of 3.7 wt %. According to Burger et al., $OME_{3,4}$ has better blending properties than dimethyl ether, which is mainly due to its higher viscosity and lower vapour pressure. In an implementation strategy, dimethyl ether is preferable for vehicle fleets (e.g., buses, delivery vehicles) than for use in blending, since the infrastructure obstacles are smaller.

7.11 Maturity, TRL, Production and Electrolysis Costs

The conversion of carbon dioxide to formic acid is a strategic topic as formic acid is potentially used as hydrogen storage. However, the research for thermal catalytic conversion of carbon dioxide to formic acid is still at an early stage to find better catalysts (below technology readiness level, TRL-4), mostly still with noble metals such as ruthenium complexes. The utilization of low-cost metals such as iron is expected in the near future.

Energy policies are needed which are at the expanse of research and development funds in the present, for any country's energy security in the future.

Costs calculated by different groups are generally difficult to compare. By varying more than 100 economic parameters, Albrecht et al. identified the electricity price and electrolyser costs as the parameters with by far most significant impact factors on OPEX and CAPEX for fuel production. This is similar for the chain efficiency from power to fuel, which is significantly determined by the electrolysis efficiency. By varying the electricity price of 105 €/MWh by +/− 70%, the net production costs of jet fuels vary from 1.5 €/l_{GE} to 4.2 €/l_{GE}. The assumed electrolysis efficiency was 0.7 (Power-to-Hydrogen). According to the calculations of Albrecht et al., with an electricity price below 33 €/MWh reaching costs below 1.4 €/l_{GE}, the Power-to-Fuel concept is more favourable compared to the BTL concept. In this case, the PEM electrolysis contributes 60% to the CAPEX but only 8% to the total costs. In a comparison of scientific literature, Albrecht et al. recalculated the production costs calculated by Becker et al. and Koenig et al. by using their own assumptions and came to comparable values. Furthermore, Power-to-Fuel seems to be interesting also for small plants since no significant reduction in production costs is observed due to upscaling.

Tremel et al. analysed the synthesis of various liquid and gaseous fuels based on hydrogen. They assumed a hydrogen price of 3 €/kg. Assuming a lower heating value of 8.5 kWh/l_{GE}, the calculated production costs for Fischer-Tropsch diesel, methanol and dimethyl ether are 1.66 €/l_{GE}, 1.49 €/l_{GE} and 1.47 €/l_{GE}, respectively. Thus, they pointed out DME as the fuel with the lowest production costs.

According to calculations performed by Trippe et al., the investment and operating costs of producing hydrocarbons are lower in synthesis pathways, including dimethyl ether as an intermediate product, than in Fischer-Tropsch pathways. However, this investigation examined the production of synthetic gasoline via DME and synthetic diesel via Fischer-Tropsch synthesis.

Buddenberg et al. calculated the production costs for a direct methanol pathway from hydrogen and carbon dioxide as well as for a methanol-to-gasoline pathway. They assumed an electricity price of 25 €/MWh and an electrolysis efficiency of 0.68. Accordingly, when limiting the costs for hydrogen on electricity costs, the costs are 1.22 €/kg_{H2}. The calculated production costs are 0.81 €/l_{GE} for methanol and 0.99 €/l_{GE} for gasoline produced via methanol.

Schmitz et al. calculated the costs for a Methanol-to-$OME_{3\text{-}5}$ process. In the case of a market price of 300 US$/t, methanol has a share of 61.7% of the total production costs of $OME_{3\text{-}5}$ which is then 614.8 US$/t. The share of steam and cooling water is 20% and the share of CAPEX is just 10%. Schmitz et al. point out OME_{3-5} as an attractive fuel component since the production costs of OME_{3-5} is competitive or even lower compared to conventional diesel over a wide range of the methanol price.

Another important aspect in fuel cost calculation is the technical maturity of the various technologies since costs need to be forecasted. To estimate the technical maturity, the technology readiness levels (TRL) are an appropriate method (Chapter 1). The next section gives an overview on the TRL levels for Power-to-Fuel technologies.

The TRL are defined by the United States Department of Defence as well as by the European Commission in the same way (Chapter 1). The scale of TRL is from 1 to 9. TRL 1 as the lowest level means that basic principles of a technology are observed. TRL 9 as the highest level describes a technology as proven in operational environment.

TRL of different Power-to-Fuel technologies including electrolysis and carbon dioxide supply technologies are given in a study published in 2016 by the German Environmental Agency. Depending on the technology, the carbon dioxide extraction from industrial exhaust gas has a TRL of 6–9, whereas the carbon dioxide absorption from air has a TRL of 6. To give an example, the energy required for the chemical absorption-based (MEA) post combustion capture technology for cement as well as black coal power plants has a TRL of 9 and requires 1.2–1.5 MJ electrical energy per kg of carbon dioxide.

According to the this study, the TRL of alkaline electrolysis, polymer-electrolyte membrane electrolysis (PEM) and high temperature electrolysis (SOC) are 9, 8 and 5, respectively. In contrast to that, according to Sterner et al., in 2014, the TRL of PEM and SOC were 7 and 6, respectively. In a study published in 2016, a full market penetration of the PEM is predicted to occur in 2040.

Due to the current relatively low TRL, it is difficult to assume and forecast electrolysis costs. For 2030, Stolten and Scherer predict investment costs of installed capacity of alkaline and PEM electrolysers of 585 €/kW. According to Mergel et al., in 2013, the costs for alkaline electrolysers were 1000 €/kW and for PEM higher than 2000 €/kW. These costs will drop to 500 €/kW and below 600 €/kW, respectively. According to the PlanDelyKaD study, today's research goal is 500 €/kW. For the time after 2030, the study forecasts a significant cost advantage for PEM electrolysis (305 €/kW) compared to alkaline electrolysis (520 €/kW).

As listed in the published study of the German Environmental Agency, the TRL of various Hydrogen-to-Fuel technologies is relatively high: the Fischer-Tropsch process, the methanol direct synthesis via hydrogen and carbon dioxide, dimethyl ether synthesis from methanol and Methanol-to-Gasoline process have TRL of 9. In contrast to that, Perez-Fortes and Tzimas estimate a TRL of 6–7 and Buddenberget al. a TRL of 8 for the methanol process from hydrogen and carbon dioxide. The estimations of Perez-Fortes and Tzimas as well as of Buddenberg et al. are based on the presence of a pilot plant of the company CRI in Iceland that has been operating since 2007. Furthermore, Buddenberg et al. estimate a TRL of 6 for Methanol-to-Gasoline process. To provide the synthesis gas for the Fischer-Tropsch process, a reactor for the reverse water gas shift reaction is needed. This technology has a TRL of 6 and thus the TRL of the whole process drops to 6. For the direct synthesis of OME_{3-5} from methanol and formaldehyde, Burger et al. estimate a TRL of 4 in 2017.

On one hand, the comparison of different productions costs calculated by different groups shows the need of standardized calculations, especially regarding hydrogen production. Sensitivity analyses as done by Albrecht et al. are important. On the other hand, the key role of the hydrogen supply was highlighted. Mostly, Hydrogen-to-Fuel technologies are already available at high TRL. However, there is need for research in the field of electrolysis to achieve the cost targets.

7.11.1 Summary

We started with outlining what challenges the transport sector must face in terms of the Energiewende. The necessity for alternative non-fossil fuels is addressed. The aim is to find an option for reducing oil dependence, pollutant emissions and greenhouse gases, as well as for improving the integration of renewable energy technologies in the transport sector. One solution to this challenge is coupling the energy and transport sectors by developing Power-to-Fuel (PTF) processes. Many of the requisite technologies for these processes are already available today. To evaluate which PTF process is the most promising for a potential implementation strategy, the desired products must first be determined. These products are alternative fuels for the transport sector, whose integration in the existing market should be as efficient as possible and which could replace conventional oil-based fuel. The educts of these alternative fuels for the transport sector should be renewable electricity, water and carbon dioxide, which is why they suitable electrofuels.

The introduction sets out a basis for discussion for the choice of suitable electrofuels. For this purpose, a model region is first determined and the fuel whose replacement would have the greatest positive impact on the Energiewende is highlighted. Germany is chosen as a model region, with diesel as the fuel to be investigated. Subsequently, the requirements that are to be made on diesel fuel are outlined. DIN EN 590 serves as a basis for this, with the values stipulated by this standard first explained and scrutinized. The final step in outlining the parameters for a discussion is the highlighting and listing of potential electrofuels.

Our next part analysed and evaluated the potentials of the electrofuels that were previously selected as conceivable substitutes for conventional diesel. The substances are examined in terms of their suitability as a fuel and whether they can be integrated into the existing system.

The analysis of the physical and chemical properties revealed that none of the proposed electrofuels fulfils the requirements for diesel fuels set out by DIN EN 590. However, DIN EN 590 does not set requirements for fuel properties that are best for diesel-engine operation, but instead represents a compromise between the automotive industry and fuel producers. The latter conventionally produce fuel from mineral oil. The production process thus involves certain limitations in terms of fuel properties. If the fuels are produced from water/hydrogen and carbon dioxide (i.e., independent of oil), then the products of these processes can be adapted to engine requirements.

The investigations reveal that the lower heating value of fuels containing oxygen has no negative impacts on engine efficiency, not even in existing engines. The electrofuels examined have a significantly higher cetane number than fossil diesel, which usually promises potential for improved engine performance. Exhaust gas values were also greatly improved using electrofuels. Since the tests of engine efficiency and exhaust gas values of all sources reviewed were conducted in test engines that had not been optimized for the used fuel, further potentials regarding emission reduction and the optimization of engine efficiency are expected. With regard to a conceivable implementation strategy, the critical aspects are, above all, viscosity as well as boiling, flash and melting temperatures. Suitable additives can mitigate any connected problems. The n-alkanes and OME_n used are usually not pure substances, but instead mixtures of molecules with different chain lengths. Their composition depends on the production process. Analysing the potential of various OME_n as diesel fuels identifies $OME_{3\text{-}5}$ as the most promising mixture.

Oxymethylene ethers are valuable candidate e-fuels and, therefore, represent an important asset in the current energy transition. As such, the interest into predicting their combustion properties has grown significantly over the past decade. A detailed review of the existing kinetic studies of their combustion is presented, highlighting that a major part of the fundamental studies has been dedicated to dimethoxymethane (or OME_1). This molecule has served as a model molecule to elaborate rate rules towards predictive models of longer linear or cyclic oligomers and detailed analysis of the existing kinetic models and their performance permits elaborating recommendations on the reaction classes to include any existing rate constant data. The current state of the art suggests that additional work is needed on precise determination of rate constants for hydrogen atom abstractions and on experimental and theoretical studies of the longer linear, branched and cyclic oligomers.

For an electrofuel to compete with fossil fuels, a very simple implementation strategy is required. The simplest implementation strategy is gradual market penetration by means of blending conventional diesel with electrofuels. Potential blending combinations highlighted here include:

- Fossil diesel + n-alkane cut;
- Fossil diesel + OME_{3-5};
- Fossil diesel + n-alkane cut + OME_{3-5}; and
- n-alkane cut + OME_{3-5}.

Additionally, higher alcohols such as n-butanol can be suitable diesel admixtures. However, due to low cetane numbers, alcohols cannot be used purely as diesel fuels.

According to the investigations reviewed, using the blends above is possible without the need to modify existing diesel engines. Suitable modifications can increase engine efficiency and reduce pollutant emissions. These modifications include injection pressure, angle and timing. The last blend mentioned, namely a suitable n-alkane cut mixed with OME_{3-5}, has the greatest potential for increasing engine efficiency and reducing pollutant emissions. In addition, fossil diesel would no longer be required.

As a basis for discussion, Table 7.10 offers an overview of the most important evaluation points. Section 7.11 gives a brief insight into production costs and technical maturity of production plants and points out the key role of hydrogen supply in the context of Power-to-Fuel.

7.12 Power-to-Liquid Technology

Renewable power can be used in different ways to contribute to the production of liquid fuels. One is Power-to-Fuel, in which renewables reduce the GHG intensity of industry by reducing carbon dioxide emissions from the electricity consumed and green hydrogen is produced to disulfuric the crude base of gasoline and diesel. Renewables can also use in Power-to-Fuel by producing synthetic fuels (e-fuels) such as syn-diesel through various processes.

Sunfire began the production of its low-carbon electrofuels – also called "wonder fuel" – in its pilot plant in Dresden (Germany). This plant is the world's first Power-to-Liquid production plant. According to the company, the Power-to-Liquid technology that synthesises Blue Crude (a synthetic crude oil) reaches system efficiencies of about 70%.

The centrepiece of the three-stage production process is a reversible electrolyser that generates hydrogen with an efficiency 90%. This green hydrogen reacts with carbon dioxide (captured from the air or waste sources) to produce a mixture of hydrocarbon chains, like those found in conventional crude. This crude can then be processed in refineries (or in a refinery-like process) to produce, for example, a synthetic diesel with similar properties to conventional crude-based diesel. This electrodiesel can be plugged into the existing fuel infrastructure. This new electrodiesel was publicly tested in an Audi A6 in Berlin 2015.

The technology has proved its potential at pilot plant scale. But it faces big challenges in form of scalability and access to green electricity.

The transportation sector is one the sectors having the largest GHG emissions in many countries. In Canada, for example, the transportation sector accounted for 25% (173 Mt CO_{2eq}) of the total national emissions, showing the critical need to displace the fossil fuels used in the sector to effectively reduce the GHG emissions of the country. Electric cars represent a promising alternative; however, these technologies are not yet optimized for long distance trips and are still unfit for heavy transportation. Biofuels have also been considered as a promising approach, though their use has raised a lot of questions regarding the amounts that can really be produced to significantly reduce the GES emissions from this sector. The challenge is even bigger when it comes to the aviation sector, since there is not yet a reliable substitute to the fossil-based jet fuel. The Power-to-Liquids technology could represent a huge opportunity to produce a replacement fuel without the issues related to the biomass feedstock and that could reduce the emissions from the transportation sector, especially the aviation sector.

Hydrocarbons from Fischer-Tropsch synthesis and methanol are the two main production pathways to produce alkanes through Power-to-Liquid technologies (Figure 7.4). To fit with the Fischer-Tropsch pathway, carbon dioxide first is converted to carbon monoxide via a reverse water-gas shift reaction (RWGS) and hydrogen is produced through water electrolysis. Hydrogen and carbon monoxide are then used in a classical Fischer-Tropsch synthesis to produce hydrocarbons that can be upgraded to fuels such as jet fuel and diesel. There are several upgrading processes that are already widely employed for upgrading crude oil to jet fuel that could also be applied in the Power-to-Liquids, such as hydrocracking, isomerization and distillation. In the methanol pathway, hydrogen produced from water electrolysis and carbon dioxide (or carbon monoxide) are used in the synthesis of methanol as an intermediate. Methanol can be then converted to fuels through further steps already used at industrial scale, such as DME synthesis, olefin synthesis, oligomerization and hydrotreating.

To the best of my knowledge, there are no reports relating the direct electrochemical reduction of carbon dioxide into long chain hydrocarbons. Most of the carbon dioxide hydrogenation studies focus on the synthesis of short-chain products, such as methane, methanol, formic acid etc., as presented in the previous sections. Nevertheless, recent studies reported the successful production of gasoline, diesel and jet fuel either via Fischer-Tropsch pathway or via methanol pathway (Figure 7.4). The following sections will present a few recent studies on the production of these alkanes via carbon dioxide hydrogenation.

7.12.1 Power-to-Jet Fuel

Since there is no current alternative to fossil jet fuels, the Power-to-Liquids Technology could represent a huge opportunity to produce a replacement aviation fuel capable of effectively reducing the emissions from the aviation sector. The synthetic paraffinic kerosene produced from this process, for example, can be blended up to 50% with jet fuels.

There isn't yet a proof-of-concept of an integrated Power-to-Jet Fuel technology. However, the individual steps have already high technological maturity levels. Many big industrial concerns have been developing this technology. Shell in partnership with other actors created the SOLAR-Jet consortium aiming to demonstrate a carbon-neutral pathway for producing jet fuel using solar

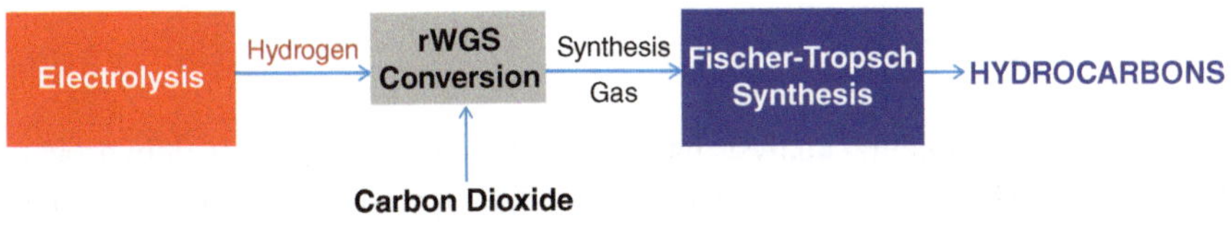

Figure 7.4 Power to FT hydrocarbons process scheme.

energy. From 2011 to 2015 the project explored the solar-thermochemical redox cycles and produced the world's first sample of solar thermochemical kerosene from water and carbon dioxide at laboratory scale. The Sun-to-Liquid project created in 2016 succeeded the first project and aims to design, fabricate and experimentally validate a large scale complete solar fuel production plant.

Carbon Engineering's pilot Air-to-Fuels has successfully produced biocrude from carbon dioxide and water in 2017. In this project, carbon dioxide is captured from the air, purified and thermo-catalytically reacted with hydrogen produced from water electrolysis with renewable energy (solar PV), to produce the biocrude. However, their major challenge is upgrading the biocrude into jet fuel.

Carbon Recycling International's George Olah Renewable Methanol Plant in Svartsengi (Iceland) was completed in 2012 and produces 5 million litres of methanol per year. The plant uses hydro and geothermal energy for producing H_2 from water electrolysis, which is then reacted with carbon dioxide from flue gases to produce methanol. The methanol (vulcanol) produced can then be blended with gasoline for automobiles or used as intermediate in the production of fuels that could potentially be suitable as electro jet fuel.

7.12.2 Power-to-Diesel

Diesel is the main transportation fuel used nowadays for heavy transportation. However, the literature related to carbon dioxide hydrogenation into diesel is very scarce. Recently, Han et al. proposed a new path for direct carbon dioxide conversion into liquid fuels with renewable hydrogen produced via solar water splitting. Carbon dioxide hydrogenation was performed at 300°C and 1 MPa over a new Cu-Fe catalyst, which exhibited excellent catalytic performances with 65% selectivity to C_{5+} liquid hydrocarbons and only 2–3% methane selectivity. The authors reported that the main products of the reaction cover the gasoline (C_5-C_{11}) and diesel range (C_{12}-C_{21}), the product distribution being very similar to the one observed in CO-FT over iron-based catalysts. The performance of the catalyst was attributed to the swift reduction and selective carburization form of the Hagg iron carbide formed, which is the active phase to produce long-chain hydrocarbons in the carbon dioxide hydrogenation process.

One of the most significant studies is the CO_2-to-Diesel process developed by Audi in partnership with Sunfire, generating a carbon-neutral diesel fuel, called e-diesel. The process is performed in three main steps. In the first step, hydrogen is produced from high-temperature water electrolysis using renewable energy. Then, hydrogen reacts with carbon dioxide (from a biogas facility) under high pressure and high temperature, producing long-chain hydrocarbons, called blue crude. In the final step, the blue crude is refined into electrodiesel similarly to the fossil crude oil refining process. The company started the production of e-diesel in 2015 and has produced more than three tons of blue crude up to now.

In 2018, Vazquez et al. presented a demonstration plant on Power-to-Wax, Power-to-Oil and Power-to-Gas within the SOLETAIR project. The plant produced oil and wax of 6.2 kg per day when it was operated for 300 hours. There are several other productions of X through the electrochemical reduction in a single reactor and although this pathway brings important advantages in terms of technological and economical perspectives, the studies are still at laboratory and pilot plant scales.

7.13 Conclusion and Outlook

A technical, economic and environmental analysis has been conducted for direct use of hydrogen in fuel cell vehicles and conversion with carbon dioxide from biogas upgrading to methane, methanol, or dimethyl ether for use in internal combustion engine vehicles. All four pathways enable significant reductions in well-to-wheel greenhouse gas emissions as well as pollutant emissions compared to fossil fuels. However, all of them also require very cheap hydrogen to be competitive with fossil fuels.

When using only curtailed renewable electricity, the predicted fuel production for Germany in 2035 is less than 1% of the German fuel consumption. The carbon footprint of electricity used for electrolysis needs to be extremely low for achieving a green-house gas benefit compared to fossil fuels. This suggests that dedicated renewable electricity is required for producing significant amounts of fuel.

Direct use of hydrogen in a fuel cell vehicle has obvious advantages compared to conversion to the combustion engine fuels in terms of well-to-wheel efficiency, greenhouse gas emissions, pollutant emissions and fuel cost. However, compared to the liquid fuels, the volumetric energy density of hydrogen is very low and driving ranges will be shorter. Additional drawbacks are the need for a dedicated infrastructure, the extremely high fuelling pressure and the increased cost of fuel cell vehicles.

Among the three combustion engine fuels, methane causes slightly lower greenhouse gas emissions because its production needs less feedstock carbon dioxide and no electricity. DME has slightly lower fuel cost and electricity demand per distance driven because its production uses hydrogen more efficiently. Since these differences are rather small, other factors like pollutant formation, range, handling, or infrastructure are important.

Fuel cost and overall electricity consumption are overwhelmingly dominated by hydrogen supply. Therefore, it is a valid approach for using fluctuating renewable electricity to let only electrolysis operate dynamically and all other steps at steady state. However, while systems conducting electrolysis and hydrogen conversion separately cannot be more efficient than electrolysis alone, integrated systems can. Such concepts could therefore be an option for simultaneously improving overall efficiency and fuel cost. Nevertheless, additional questions arise in this case with respect to the optimal design, size and dynamic operation of the fuel production plant, as well as plant location and the logistics. In case of methane production, the net production of electricity in the fuel production plant could open interesting interactions with the electricity grid that need to be explored further.

Regarding other potential fuel candidates, fuels with a low enthalpy of reaction per unit of hydrogen converted are desirable to enable high chemical conversion efficiencies as is the case for dimethyl ether compared to the methane and methanol. Furthermore, the present evaluation underlines the importance of having good engine efficiencies, which tend to be higher for compression ignition engines.

Concerning an implementation strategy, the costs of an alternative fuel are decisive, in addition to its suitability as a fuel. The production costs are a combination of educt and process costs, with process costs consisting of investment and operating costs. As is the case for fuels, the corresponding production pathways should be analysed and evaluated under identical conditions. Only then can comparisons result in recommendations. The choice of initial parameters and conditions is therefore decisive.

Several different production pathways may exist for the same or similar products. For example, dimethyl ether can be produced from the educt's hydrogen and carbon dioxide via direct synthesis or the intermediate product, methanol. Hydrocarbons can be produced from synthesis gas by means of Fischer-Tropsch synthesis, as well as from methanol or dimethyl ether. Different production costs are the result. The product distribution is also different.

The production costs and therefore product costs of the alternative fuels mentioned as today's state of the art, are too high compared to conventional fuel production. They are thus unable to compete with fossil fuels at current oil prices. There has therefore been emphasis on the necessity of developing synthesis concepts for efficient and cost-competitive production pathways. Projections show that global demand for oil will continue to rise in future. The resulting ratio between supply and demand will increase the price of oil and thus make non-fossil fuels more competitive. According to a cost estimate by Connolly et al., electrofuels and biofuels will be cheaper than fossil fuels by 2050.

An effective analysis approach for evaluating fuels in the transport sector is the well-to-wheel method. This not only takes production costs, but the entire active chain, from educt acquisition to production and also including conversion into kinetic energy in the vehicle, into consideration. The focus can be on energy demand, efficiency, or emissions. According to Semelsberger et al., dimethyl ether – excluding methane – has the highest well-to-wheel efficiency of all non fossil fuels in internal combustion engines. In contrast, dimethyl ether and methanol are evaluated more positively than methane by Connolly et al. in terms of their efficiency.

OME_n was not considered in the studies by Connolly et al. or Semelsberger et al. According to a study conducted by Arcoumanis et al., dimethyl ether production has better well-to-wheel efficiency and lower well-to-wheel emissions than CNG or LPG, i.e., methane or propane/butane. According to Lumpp et al., the production of OME_n is cheaper than that of corresponding Fischer-Tropsch hydrocarbons. This highlights the fact that the calculation of production costs and particularly the more extensive well-to-wheel considerations depend heavily on conditions and initial parameters. It becomes apparent that pure literature research does not yield a comparable evaluation of the production pathways. There is therefore a need for a comparative technical and economic assessment of the most promising electrofuels. As this chapter shows, these include alkanes, as well as energy alcohols, mainly for spark ignition engines and ethers for diesel engines. The calculation of production costs should be conducted under the same conditions and initial parameters. To evaluate the influence of conditions and initial parameters, as well as the desired product, sensitivity analyses should be conducted. Thus, points for implementing future process optimization can be recognized and the need for research identified.

Together with the basis for discussion on the choice of suitable non-fossil and non-biological renewable electrofuels, the technical and economic evaluation of PTF processes provided by this chapter results in a scientifically based recommendation for a choice of suitable fuels for future transport systems.

Further Reading

Ajanovic, A. (2013). Renewable fuels – a comparative assessment from economic, energetic and ecological point-of-view up to 2050 in EU-countries. *Renew. Energy* 60: 733–738.

Albrecht, F., König, D., Baucks, N. and Dietrich, R.U. (2017). A standardized methodology for the techno-economic evaluation of alternative fuels – A case study. *Fuel* 194: 511–526.

Arcoumanis, C., Bae, C., Crookes, R. and Kinoshita, E. (2008). The potential of di-methyl ether (DME) as an alternative fuel for compression-ignition engines: a review. *Fuel* 87: 1014–1030.

Arteconi, A., Mazzarini, A. and Di Nicola, G. (2011). Emissions from ethers and organic carbonate fuel additives: a review. *Water Air Soil Pollut.* 221: 405–423.

Artz, J., Müller, D., Thernert, K. et al. (2018). Sustainable conversion of carbon dioxide: an integrated review of catalysis and life cycle assessment. *Chem. Rev.* 118: 434–504.

Atsbha, T.A., Yoon, T., Seongho, P. and Lee, C.-J. (2021 February). A review on the catalytic conversion of CO_2 using H_2 for synthesis of CO, methanol and hydrocarbons. *J. CO2 Util.* 44: 101413.

Bailera, M., Lisbona, P., Romeo, L.M. and Espatolero, S. (2017). Power to gas projects review: lab, pilot and demo plants for storing renewable energy and CO_2. *Renew. Sustain. Energy Rev.* 69: 292–312. doi: 10.1016/j.rser.2016.11.130.

Becker, W.L., Braun, R.J., Penev, M. and Melaina, M. (2012). Production of Fischer-Tropsch liquid fuels from high temperature solid oxide co-electrolysis units. *Energy* 47: 99–115.

BMVBS (2013). The Mobility and Fuels Strategy of the German Government (MFS) – new pathways for energy.

Bongartz, D., Burre, J. and Mitsos, A. (2019a). Production of oxymethylene dimethyl ethers from hydrogen and carbon dioxide – Part I: modeling and analysis for OME_1. *Ind. Eng. Chem. Res.* 58 (12): 4881–4889.

Bongartz, D., Burre, J. and Mitsos, A. (2019b). Production of oxymethylene dimethyl ethers from hydrogen and carbon dioxide – Part II: modelling and analysis for OME_{3-5}. *Ind. Eng. Chem. Res.* 58 (15): 5567–5578.

Bongartz, D., Burre, J. and Mitsos, A. (2020). Corrigendum: "production of oxymethylene dimethyl ethers from hydrogen and carbon dioxide – Part I: modelling and analysis for OME_1 & Part II: modeling and analysis for OME_{3-5}". *Ind. Eng. Chem. Res.* 59 (49): 21615–21616.

Bongartz, D., Doré, L., Eichler, K. et al. (2018). Comparison of light-duty transportation fuels produced from renewable hydrogen and green dioxide. *Appl. Energy* 231: 757–767.

Bozem, K., Nagl, A. and Rennhak, C. (2013). *Energie für nachhaltige Mobilität: Trends undKonzepte.* Wiesbaden, S.L.: Springer Fachmedien Wiesbaden.

Branco, J.B., Ferreira, A.C., Vieira, F. and Martinho, J.F. (2021). Cerium-based bimetallic oxides as catalysts for the preparation method. *Energy Fuels* 35 (8): 6725–6737.

Brynolf, S., Taljegard, M., Grahn, M. and Hansson, J. (2018). Electrofuels for the transport sector: a review of production costs. *Renew. Sust. Energy Rev.* 81: 1887–1905.

Buddenberg, T., Bergins, C. and Schmidt, S. (2016). Power to fuel as a sustainable business model for cross-sectorial energy storage in industry and power plants. In: *5th Conference on Carbon Dioxide as Feedstock for Fuels, Chemistry and Polymers*, Cologne, Germany.

Burger, J. (2017). Large-scale production of OME – state of the art. In: OME InfotagDECHEMA, Frankfurt.

Burger, J., Siegert, M., Ströfer, E. and Hasse, H. (2010). Poly(oxymethylene) dimethyl ethers as components of tailored diesel fuel: properties, synthesis and purification concepts. *Fuel* 89: 3315–3319.

Connolly, D., Mathiesen, B.V. and Ridjan, I. (2014). A comparison between renewable transport fuels that can supplement or replace biofuels in a 100% renewable energy system. *Energy* 73: 110–125.

Deutsche Energi-Agentur (2013). Hintergrundpapier: Energieverbrauch und Energieträger im Straßenverkehr bis 2025.

DoD Assistant Secretary of Defense for Research and Engineering (ASD(R&E)) Tech.

Dry, M.E. (2002). High quality diesel via the Fischer-Tropsch process – a review. *J. Chem. Technol. Biotechnol.* 77: 43–50.

EU (2015). Council Directive 2015/652 – laying down calculation methods and reporting requirements pursuant to Directive 98/70/EC of the European Parliament and of the Council relating to the quality of petrol and diesel fuels. *Off. J. Euro. Union.*

EU (2017). Horizon 2020 – Work Programme 2016–2017, 20. General Annexes: G.Technology readiness level (TRL).

Fenard, Y. and Vanhove, G. (2021). Mini-review on the advances in the kinetic understanding of the combustion of linear and cyclic oxymethylene ethers. *Energy Fuels* 35 (18): 14325–14342. doi: 10.1021/acs.energyfuels.1c01924.

Fernández, J.R., Garcia, S. and Sanz-Pérez, E.S. (2020). CO_2 capture and utilization editorial. *Ind. Eng. Chem. Res.* 59 (15): 6767–6772.

Frontera, P., Macario, A., Ferraro, M. and Antonucci, P. (2017). Supported catalyst for CO2 methanation: a review. *Catalysts* 7: 1–28.

Gao, P., Li, S., Bu, X. et al. (2017). Direct conversion of CO_2 into liquid fuels with high selectivity over a bifunctional catalyst. *Nat. Chem.* 9: 1–6.

Ghaib, K. and Ben-Fares, F.Z. (2018). Power-to-methane: a state-of-the-art review. *Renew. Sustain. Energy Rev.* 81: 433–446. doi: 10.1016/j.rser.2017.08.004.

Gill, S.S., Tsolakis, A., Dearn, K.D. and Rodríguez-Fernández, J. (2011). Combustion characteristics and emissions of Fischer-Tropsch diesel fuels in IC engines. *Prog. Energy Combust. Sci.* 37: 503–523.

Grahn, M., Taljegård, M., Ehnberg, J. and Karlsson, S. (2014). Utilising excess power: the case of electrofuels for transport. In: *Systems Perspective on Renewable Power* (ed. B. Sandén), 128–137. Chalmers University of Technology.

Graves, C., Ebbesen, S.D., Mogensen, M. and Lackner, K.S. (2011). Sustainable hydrocarbon fuels by recycling CO_2 and H_2O with renewable or nuclear energy. *Renew. Sustain. Energy Rev.* 15: 1–23.

Grube, T., Doré, L., Hoffrichter, A. et al. (2018). An option for stranded renewables: electrolytic hydrogen in future energy systems. *Sustain. Appl. Energy* 231: 757–767.

Gruden, D. (2008). *Umweltschutz in der Automobilindustrie: Motor, Kraftstoffe, Recycling.* Wiesbaden: Vieweg +Teubner/GWV Fachverlage GmbH Wiesbaden.

Hao, B., Song, C., Lv, G. et al. (2014). Evaluation of the reduction in carbonyl emissions from a diesel engine using Fischer-Tropsch fuel synthesized from coal. *Fuel* 133: 115–122.

Härtl, M., Seidenspinner, P., Jacob, E. and Wachtmeister, G. (2015). Oxygenate screening on a heavy-duty diesel engine and emission characteristics of highly oxygenate doxymethylene ether fuel. *Fuel* 153: 328–335.

Holladay, J.E., Male, J.L., Rousseau, R. and Weber, R.S. (2020). Synthesizing clean transportation fuels from CO_2 will at least quintuple the demand for non-cabogenic electricity in the United States. *Energy Fuels* 34 (12): 15433–15442.

Huss, A., Mass, H. and Hass, H. (2013). Well-to-Wheels analysis of future automotive fuels and powertrains in the European context. Tank-to-Wheels Report Version 4.0. Tech. Rep. EUR 26027 EN; Joint Research Centre of the European Commission.

Iannuzzi, S.E., Barro, C., Boulouchos, K. and Burger, J. (2016). Combustion behaviour and soot formation/oxidation of oxygenated fuels in a cylindrical constant volume chamber. *Fuel* 167: 49–59.

Jeong, Y., Park, J. and Han, M. (2021). Design and control of a fixed-bed recycle reactor with multicatalyst layers: methanation of carbon dioxide. *Ind. Eng. Chem. Res.* 60 (12): 4650–4667.

Kaltschmitt, M. and Streicher, W. (2009). Regenerative Energien in Österreich: Grundlagen, Systemtechnik, Umweltaspekte, Kostenanalysen, Potenziale, Nutzung. *Praxis.* Wiesbaden: Vieweg+Teubner Verlag/GWV Fachverlage GmbH Wiesbaden.

Kaneco, S., Yabuuchi, M., Katsumata, H. et al. (2002). Electrochemical reduction of CO_2 to methane in methanol at low temperature. *Fuel Chem. Div. Prepr.* 47: 71–72.

Kaneco, S., Katsumata, H., Suzuki, T. and Ohta, K. (2006). Electrochemical reduction of CO_2 to methane at the Cu electrode in methanol with sodium supporting salts and its comparison with other alkaline salts. *Energy Fuels* 20: 409–414.

Karelovic, A. and Ruiz, P. (2012). CO2 hydrogenation at low temperature over Rh/Al_2O_3 catalysts: effect of metal particle size on catalytic performances and reaction mechanism. *Appl. Catal. B Environ.* 113–114: 237–249.

Keen Fan, W. and Tahir, M. (2021). Current trends and approaches to boost the performance of metal organic frameworks for carbon dioxide methanation through photo/thermal hydrogenation: a review. *Ind. Eng. Chem. Res.* 60 (36): 13149–13179.

Khan, A., Ali, S.S., Chodimella, V.P. et al. (2021). Catalytic conversion of dicyclopentadiene into high energy density fuel: a brief review. *Ind. Eng. Chem. Res.* 60 (5): 1977–1988.

Kondratenko, E.V., Mul, G., Baltrusaitis, J. et al. (2013). Status and perspectives of CO_2 conversion into fuels and chemicals by catalytic, photocatalytic and electrocatalytic processes. *Energy Environ. Sci.* 6: 3112–3135. doi: 10.1039/c3ee41272e.

König, D.H., Freiberg, M., Dietrich, R.-U. and Wörner, A. (2015). Techno-economic study of the storage of fluctuating renewable energy in liquid hydrocarbons. *Fuel* 159: 289–297.

Lapuerta, M., Armas, O., Hernández, J.J. and Tsolakis, A. (2010). Potential for reducing emissions in a diesel engine by fuelling with conventional biodiesel and Fischer-Tropsch diesel. *Fuel* 89: 3106–3113.

Lautenschütz, L., Oestreich, D., Seidenspinner, P. et al. (2016). Physico-chemical properties and fuel characteristics of oxymethylene dialkylethers. *Fuel* 173: 129–137.

Liu, H., Wang, Z., Zhang, J. et al. (2015). Study on combustion and emission characteristics of polyoxymethylene dimethyl ethers/diesel blends in light-duty and heavy-duty diesel engines. *Appl. Energy* 185 (2): 1393–1402.

Liu, J., Wang, H., Li, Y. et al. (2016). Effects of diesel/PODE (polyoxymethylene dimethyl ethers) blends on combustion and emission characteristics in a heavy-duty diesel engine. *Fuel* 177: 206–216.

Lumpp, B., Rothe, D., Pastötter, C. et al. (2011). Oxymethylene ethers as diesel fuel additives of the future.

Markterhebung 2007–2008. Tech. Rep.; Fraunhofer-Institut fur Umwelt-, Sicherheits- und Energietechnik (Fraunhofer UMSICHT).

Mathiesen, B.V., Lund, H., Connolly, D. et al. (2015). Smart energy systems for coherent 100% renewable energy and transport solutions. *Appl. Energy* 145: 139–154.

Matzen, M. and Demirel, Y. (2016). Methanol and dimethyl ether from renewable hydrogen and carbon dioxide: alternative fuels production and life-cycle assessment. *J. Clean. Prod.* 139: 1068–1077.

Mergel, J., Carmo, M. and Fritz, D. (2013). Status on technologies for hydrogen production by water electrolysis. In: *Transition to Renewable Energy Systems*, 423–450. Wiley-VCHVerlag GmbH & Co. KGaA.

Muraza, O. (2021). Highlighting the greener shift in transportation energy and fuels based on novel catalytic materials. *Energy Fuels* 35 (1): 25–44.

Otto, A. (2015). Chemical, procedural and economic evaluation of carbon dioxide as feedstock in the chemical industry (Ph.D. thesis). RWTH Aachen University.

Park, S.H. and Lee, C.S. (2013). Combustion performance and emission reduction characteristics of automotive DME engine system. *Prog. Energy Combust. Sci.* 39: 147–168.

Parra, D., Zhang, X., Bauer, C. and Patel, M.K. (2017). An integrated techno-economic and life cycle environmental assessment of power-to-gas systems. *Appl. Energy* 193: 440–454.

Joint Research Centre, Institute for Energy and Transport, Pérez-Fortes, M., Tzimas, E. (2016). *Techno-economic and environmental evaluation of CO_2 utilisation for fuel production: synthesis of methanol and formic acid*, Publications Office.

Peters, R., Grube, T. and Stolten, D. (2016). Sustainable fuels in the transport sector (unpublished). Forschungszentrum Jülich GmbH. Institute of Energy and Climate Research.

Pontzen, F., Liebner, W., Gronemann, V. et al. (2011). CO_2-based methanol and DME-Efficient Technologies for industrial scale production. *Catal. Today* 171: 242–250.

Reif, K. (2010). *Konventioneller Antriebsstrang und Hybridantriebe mitBrennstoffzellen und alternativen Kraftstoffen*. Wiesbaden: Vieweg Teubner Verlag/GWV Fachverlage GmbH Wiesbaden.

Reiter, G. and Lindorfer, J. (2015). Evaluating CO_2 sources for power-to-gas applications – a case study for Austria. *J. CO2 Util.* 10: 40–49.

Ren, Y., Huang, Z., Miao, H. et al. (2008). Combustion and emissions of a DI diesel engine fuelled with diesel-oxygenate blends. *Fuel* 87: 2691–2697.

Ren, Y., Huang, Z.H., Jiang, D.M. et al. (2005). Engine performance and emission characteristics of a compression ignition engine fuelled with diesel/dimethoxymethane blends. *Proc. Inst. Mech. Eng. Part D: J Automob. Eng.* 219: 905–914.

Rimkus, A., Zăglinskis, J., Rapalis, P. and Skačkauskas, P. (2015). Research on the combustion, energy and emission parameters of diesel fuel and a biomass-to-liquid (BTL) fuel blend in a compression-ignition engine. *Energy Convers. Manage.* 106: 1109–1117.

Robinius, M. (2016). *Strom- und Gasmarktdesign zur Versorgung des deutschenStraßenverkehrs mit Wasserstoff*. Jülich: Forschungszentrum Jülich.

Rubin, E., Mantripragada, H., Marks, A. et al. (2012). The outlook for improved carbon capture technology. *Prog. Energy Combust. Sci.* 38: 630,71.

Schaaf, T., Gruning, J., Schuster, M. and Orth, A. (2014). Speicherung von Electrischer Energie im Erdgasnetz-Methanisierung von CO_2-haltigen gasen. *Chem. Ing. Tech.* 86: 476–485.

Schemme, S., Samsung, R.C., Peters, R. and Stolten, D. (2017). Power-to-fuel as a key to sustainable transport systems – an analysys of diesel fuels produced from CO_2 and renewable electricity. *Fuel* 205: 198–221.

Schmidt, P., Weindorf, W., Roth, A., et al. (2016). Power-to-liquids – potentials and perspectives for the future supply of renewable aviation fuel. German Environment Agency.

Schmitz, N., Burger, J., Ströfer, E. and Hasse, H. (2016). From methanol to the oxygenated diesel fuel poly(oxymethylene) dimethyl ether: an assessment of the production costs. *Fuel* 185: 67–72.

Schollenberger, D., Bajohr, S., Gruber, M. et al. (2018). Scale-up of innovative honeycomb reactors for power-to-gas applications—the project store & go. *Chem. Ing. Tech.* 90: 696–702. doi: 10.1002/cite.201700139.

Semelsberger, T.A., Borup, R.L. and Greene, H.L. (2006). Dimethyl ether (DME) as an alternative fuel. *J. Power Sources* 156: 497–511.

Song, Y., Serikawa, K., Imamura, K. et al. (2020). Direct synthesis of C_5-C_{13} iso-paraffins from carbon dioxide over hybrid catalyst in a near-critical n-hexane fluid. *Ind. Eng. Chem. Res.* 59 (26): 11962–11969.

Stan, C. (2015). *Alternative Antriebe für Automobile: Hybridsysteme,Brennstoffzellen, alternative Energieträger*. 4., aktualisierte und erw. Aufl.ed. Berlin: Springer.

Sternberg, A. and Bardow, A. (2015). Power-to-what? – environmental assessment of energy storage systems. *Energy Environ. Sci.* 8 (2): 389–400. doi: 10.1039/c4ee03051f.

Sternberg, A. and Bardow, A. (2016). Life cycle assessment of power-to-gas: syngas vs methane. *ACS Sus. Chem. Eng.* 4: 4156–4165.

Sterner, M. and Stadler, I. (2014). *Energiespeicher: Bedarf, Technologien, Integration*. Berlin: Springer Vieweg.

Stolten, D. and Scherer, V. (2013). *Transition to Renewable Energy Systems*. Weinheim, Germany: Wiley-VCH Verlag GmbH & Co. KGaA, 96.

Swalus, C., Jacquemin, M., Poleunis, C. et al. (2012). CO_2 methanation on Rh/gama-Al_2O_3 catalyst at low temperature: "In situ" supply of hydrogen by Ni/activated carbon catalyst. *Appl. Catal. B Environ.* 125: 41–50.

Tremel, A., Wasserscheid, P., Baldauf, M. and Hammer, T. (2015). Techno-economic analysis for the synthesis of liquid and gaseous fuels based on hydrogen production via elec-trolysis. *Int. J. Hydrog. Energy* 40: 11457–11464.

Trippe, F. (2013) Techno-okonomische Bewertung alternativer Verfahrenskonfiguration zur Herstellung von Biomass-to-Liquid (BtL) Kraftstoffen und Chemicalien. Print on demand ed. KIT Scientific Publishing Technische Informationsbibliotek u. Universitätsbibliotek, Karlsruhe, Hannover.

Trippe, F., Fröhling, M., Schultmann, F. et al. (2013). Comprehensive techno-economic assessment of dimethyl ether (DME) synthesis and Fischer-Tropsch synthesis as alternative process steps within biomass-to-liquid production. *Fuel Process. Technol.* 106: 577–586.

Turton, R., Balie, R., Whiting, W. and Shaeiwitz, J. (2009). *Analysis, Synthesis and Design of Chemical Processes*. Upper Saddle River, NJ: Prentice Hall PTR.

Urban, W., Girod, K. and Lohmann, H., (2008), Technologien und Kosten der Biogasaufbereitung und Einspeisung in das Erdgasnetz. Ergebnise der Markterhebung 2007–2008. Tech. Rep.; Fraunhofer-Institut fur Umwelt-, Sicherheits- und Energietechnik (Fraunhofer UMSICHT).

Vasconcelos, B.R. and Lavoie, J.-M. (2019). Recent advances in power-to-X technology for the production of fuels and chemicals. *Front. Chem.* 7.

Vols, P., Hilbert, S., Stor, B. et al. (2021). Methanation of CO_2 and CO by (Ni,Mg,Al)-hydrotalcite-derived and related catalysts with varied magnesium and aluminum oxide contents. *Ind. Eng. Chem. Res.* 60 (14): 5114–5123.

Wang, X., Yang, G., Zhang, J. et al. (2016). Synthesis of isoalkanes over a core (Fe-Zn-Zr)-shell (zeolite) catalyst by CO_2 hydrogenation. *Chem. Commun.* 52: 7352–7355.

Wei, J., Ge, Q., Yao, R. et al. (2017). Directly converting CO_2 into a gasoline fuel. *Nat. Commun.* 8: 1–8.

Werner, M. and Wachtmeister, G. (2010). Dimethylether – Dieselalternative der Zukunft? *MTZ – Motortechnische Zeitschrift*, 71: 540–542.

Wesselak, V., Schabbach, T., Link, T. and Fischer, J. (2013). Regenerative Energietechnik, 2, erw. u.vollst. neu bearb. Aufl. 2013 ed., Springer, Berlin Heidelberg, Berlin, Heidelberg, s.l.

8

Power-to-Light Alkenes

CONTENTS

8.1 Oxidative Dehydrogenation

8.1.1 Carbon Dioxide as a Soft Oxidant for Catalytic Dehydrogenation

The worldwide development of shale gas is one of the significant energy revolutions of the twenty-first century. Shale gas contains large amounts (typically some 10%) of light alkanes. The dehydrogenation of alkanes to produce alkenes is particularly attractive as these are one of the most important feedstocks for the chemical industry to manufacture a range of important petrochemicals such as polyethylene, ethylene oxide, styrene, acetaldehyde, vinyl acetate, ethylene di(chloride/bromide), ethanol, ethylbenzene and many more. The demand for olefins, especially propylene, is expected to increase rapidly soon.

Steam cracking, fluid catalytic cracking (FCC), catalytic dehydrogenation and oxidative dehydrogenation (ODH) are the primary current methods to obtain olefins from hydrocarbons. Propane dehydrogenation (PDH) is a promising catalytic technology for the conversion of propane into propylene.

Oxidative dehydrogenation of propane via mild oxidation with carbon dioxide is seen as an energy-efficient and environmentally promising approach to convert propane to propylene. Moreover, the concomitant reduction of carbon dioxide to carbon monoxide that occurs during the reaction will lead to syngas production, a key building block for fuels and other chemicals.

Converting Power into Chemicals and Fuels: Power-to-X Technology for a Sustainable Future, First Edition. Martin Bajus.

In addition to acting as a carbon source, carbon dioxide can act as an oxygen source or oxidant in many chemical reactions. It can be used in certain reactions such as the oxidative dehydrogenation reaction and oxidative coupling of methane. These processes should be part of a global effort to recycle carbon dioxide as the carbon source for both chemicals and fuels, a key component of a circular carbon economy.

Carbon dioxide is a thermodynamically stable molecule. Accordingly, most of the chemical reactions involving carbon dioxide are endothermic and hence the application of high-efficiency catalysts is critical in having effective and efficient reactions for the utilization of carbon dioxide. In addition, metal oxide promoters are routinely employed to enhance the propane conversion rate.

Bartholomew reported the activity of carbon dioxide in coke gasification in the order oxygen (105) > water (3) > carbon dioxide (1) > hydrogen (0.003), where the number in the parenthesis shows the relative gasification by oxidizers. This shows that carbon dioxide is a milder oxidant than oxygen. The use of oxygen as oxidant especially with hydrocarbons has two drawbacks: the production of explosive mixtures and the susceptibility to deep oxidation of the feed and/or products and consequently lower product yields.

The oxidative dehydrogenation of propane in the presence of molecular oxygen as an oxidizing agent is favoured in lows temperature reactions. Deep oxide dehydrogenation of propane and propylene to $(CO)_x$ is a major drawback, which often results in loss of propylene selectivity and yield.

Olefins as building blocks of many chemicals such as polyolefins are important feedstocks of petrochemical industries. They may be produced by catalytic (or thermal in the case of lower olefins) dehydrogenation of the corresponding paraffins.

Another example is the oxidative dehydrogenation of lower paraffins, e.g., propane into propylene (Equation 8.1):

$$C_3H_8 + 0.5O_2 \rightleftarrows C_3H_6 + H_2O \tag{8.1}$$

$$\Delta H^\circ_{298} = -117.4\ \text{kJ/mol} \quad \Delta G^\circ 298 = -142.1\ \text{kJ/mol}$$

Essentially, the heat of reaction needed is provided by the formation of water. An alternative is the use of carbon dioxide rather than oxygen (Equation 8.2):

$$C_3H_8 + CO_2 \rightleftarrows C_3H_6 + CO + H_2O \tag{8.2}$$

$$\Delta H^\circ_{298} = 165.6\ \text{kJ/mol} \quad \Delta G^\circ 298 = 115.1\ \text{kJ/mol}$$

In the reaction (Equation 8.2) using K-Cr-Mn on silica catalysts, ethylene, carbon monoxide and hydrogen are produced. This reaction, though rather highly endothermic, has potential merits. It is, in general, difficult to reach high selectivity in oxidative dehydrogenation. In fact, one of the undesired by-products is carbon dioxide. When carbon dioxide is used instead of air or oxygen, carbon dioxide production is suppressed. It is logical to expect that use of carbon dioxide, which is a mild oxidant, may lead to higher selectivity in this type of reaction.

Recent studies of propane dehydrogenation (Equation 8.2) using carbon dioxide have tended to be focused on the development of more active and selective catalysts. ZSM-5 supported chromium oxide catalysts were found to be reasonably efficient catalysts for the oxidative dehydrogenation of propane in the presence of carbon dioxide. Nonetheless, their stability is still a major problem and limits their industrial applications. Moreover, the propylene selectivity of zeolite supported catalysts is not high enough because of the presence of accompanying strong acid sites on and in, the zeolites. Moreover, the Al/Si ratio itself strongly affects the support properties during propane dehydrogenation as it can change both the structures and surface acidity. A strongly acidic site will lead to cracking processes. For natural zeolites, the difficulty in obtaining homogenous materials and the reproducibility of samples obtained from natural ores are the main obstacles in their use as catalytic materials. Acid and ion exchange treatments can be applied to improve the zeolite properties. However, they can also influence the catalytic performance of the zeolite by dealumination and removal of the ion-exchangeable cations. This leads to breaking of Al–O bonds and would mean a loss of crystallinity. Investigations improving zeolites as support for propane dehydrogenation are still in progress. The catalyst support itself plays a decisive role in determining surface metal oxide structures and the surface reducibility by affecting the dispersion of the metal species as the active phase. Irrespective of any metal–support interaction, the surface area of the support obviously affects the metal catalyst dispersion.

The best catalyst performance reached 18% conversion of propane and 20% yield of propylene with over 90% selectivity after 2 hours and about 10% conversion and 15% product yield after 10 hours with still about 93% propylene selectivity. The one-pot synthesis of the calcium oxide-promoted chromium catalyst demonstrated both good stability and good catalytic performance during the dehydrogenation process. The performances of catalysts for propane dehydrogenation with carbon dioxide were examined at a temperature of 550°C in a fixed bed microreactor at atmospheric pressure for a period of 10 hours.

Oxidative dehydrogenation of alkanes using carbon dioxide as a soft oxidant has recently emerged as a potentially attractive alternative to using steam cracking to produce light olefins. To elucidate reaction pathways and their dependence on the operating conditions, carbon dioxide-assisted propane dehydrogenation over a redox-active Cr_2O_3/Al_2O_3 catalyst was examined in a packed bed reactor as a function of temperature, Cr_2O_3/CO_2 feed ratio and residence time. Previous ODH studies have largely focused on carbon dioxide-rich conditions with the aim of preventing coke formation. However, at $t = 600°C$ the present study finds that the use of propane-rich conditions ($1 \leq C_3H_8/CO_2 \leq 2.5$) maximizes propylene production and selectivity while maintaining catalyst stability. It is postulated that the selective Mars van Krevelen dehydrogenation process is optimized at these ratios. Excess carbon dioxide apparently promotes nonselective dehydrogenation and dry reforming pathways that generate additional carbon monoxide, adversely impacting catalyst stability via the Boudouard reaction. This hypothesis is supported by complementary investigations of the reverse water gas shift reaction and thermodynamic analysis. The findings and methodology presented here are likely applicable to related ODH processes with other alkanes and redox-active catalysts.

8.1.2 Carbon Dioxide: Oxidative Coupling of Methane

Because indirect conversion of methane to higher hydrocarbons passes through an energy-intensive reforming reaction, direct routes such as oxidative coupling of methane (OCM) have received attention.

Ethane is the primary product of coupling, which is dehydrogenated to the more desirable product ethylene and then to higher hydrocarbons. The deep oxidation of methane and hydrocarbon products is thermodynamically more favourable and is an important side reaction. Isotope labelling experiments have shown that at low levels of conversion, methane is the main source of carbon dioxide, while at high conversion levels, which are of practical interest, ethylene is the dominant source of carbon dioxide. Therefore, the single-pass C_2^+ yield is limited to around 25% due to this undesired behaviour formulated as the 100% rule.

Oxygen can induce gas-phase radicals leading to deep oxidation of homogeneous reactions, therefore, it is necessary to find an alternative oxidant, such as carbon dioxide, which suppressed the radical induction (Equations. 8.3 and 8.4):

$$2CH_4 + CO_2 \rightleftarrows C_2H_6 + CO + H_2O \qquad \Delta H^{\circ}_{298} = 107.0\ kJ/mol \tag{8.3}$$

$$2CH_4 + 2CO_2 \rightleftarrows C_2H_4 + 2CO + 2H_2O \qquad \Delta H^{\circ}_{298} = 284.0\ kJ/mol \tag{8.4}$$

Equilibrium conversion calculations imply that at 800°C yields of approximately 30% C_2^+ hydrocarbons are possible; however, analysis of many published results shows that the practical yields are much lower, typically less than 10%. This is due to the poor performance of the catalysts presently available.

Unfortunately, over many of the most efficient normal OCM catalysts, the presence of carbon dioxide in the feed decreases the overall rate of catalytic reactions. All component oxidation rate expressions have a negative power of carbon dioxide in the kinetic equations implying some poisoning effect.

Many of oxides and mixed oxides have been tested as potential catalysts in the carbon dioxide OCM reaction. Screening a variety of catalysts prepared by doping magnesium oxide and cerium dioxide-based solids with oxides from alkali (Li_2O), alkaline earth (CaO) and transition metal groups (WO_3 or MnO) showed that 2% Li_2O/MgO catalyst is the most promising with C_2^+ hydrocarbons selectivity and yield at 98.0% and 5.7%, respectively.

A good catalyst should have two functions: redox ability capable of activating carbon dioxide to produce active oxygen species for the conversion of methane and basicity leading to selective formation of C_2 hydrocarbons. Asami et al. suggested two possible reaction mechanisms, depending on the types of metal oxides. One possible mechanism is that carbon dioxide is first adsorbed over the metal oxide and decomposed to form an active oxygen species and carbon monoxide is released. The active species oxidize methane to form ethylene through the methyl radical and ethylene would be formed through the pyrolysis and oxidative dehydrogenation of ethane (Equation. 8.5).

$$CO_2 + * \rightarrow CO + O* \tag{8.5}$$

$$O* + CH_4 \rightarrow CH_3\bullet + HO$$

$$2CH_3\bullet \rightarrow C_2H_6$$

where, * is site on metal oxide catalyst and $CH_3\bullet$ is methyl radical.

The other mechanism would involve the reaction of methane with lattice oxygen in metal oxides to form a methyl radical (Equation 8.6), which, in turn, is converted to C_2 hydrocarbons (Equation 8.7) and the partly reduced oxides are then oxidized by carbon dioxide (Equation 8.8).

$$O(surface) + 2CH_4 \rightarrow 2CH_3\bullet + H_2O + \square \tag{8.6}$$

$$2CH_3\bullet \rightarrow C_2H_6 \tag{8.7}$$

$$\square + CO_2 \rightarrow CO + O\ (surface) \tag{8.8}$$

where, $CH_3\bullet$ is methyl radical and □ is vacancy.

An alternative approach is to combine normal OCM with carbon dioxide-OCM reactions (Equation. 8.6–8.8) using the heat of the former in the latter, i.e., using carbon dioxide as a reactive diluent. The proprietary promoted barium titanate perovskite catalyst developed by NPCRT Company and HRD Corp., for example, has given a stable C_{2+} yield close to 30% for approximately 40% methane conversion, which is in the range suitable for commercialization. However, further development work is necessary to elucidate the effect of pressure, as working at higher pressure is economically more attractive.

A series of gallium doped $MgAl_2O_4$ spinel was prepared via a simple coprecipitation method and characterized by X-ray diffraction (XRD), scanning electron microscopy (SEM), nitrogen adsorption, X-ray photoelectron spectroscopy (XPS) and thermogravimetry (TG) techniques. Their acidity and basicity were measured by ammonia-temperature-programmed desorption (TPD) and carbon dioxide-TPD, respectively. These spinel catalysts exhibit good catalytic performance towards ethane dehydrogenation after gallium-doping either in the absence or in the presence of carbon dioxide. Their dehydrogenation activity is positively associated with the amount of surface acid sites of medium-strong strength, confirming that the coordinatively unsaturated Ga^{3+} cations are the active species in the reaction. The introduction of carbon dioxide shows an obvious promoting effect on the dehydrogenation activity of the catalysts because of their high activity for the reverse water-gas shift reaction, along with an improved catalyst stability because of the coexisting Boudouard reaction. An ethane conversion of 43% and ethylene selectivity of 92% can be achieved and the activity and selectivity of this type of catalysts can be fully recovered by regeneration through calcination in flowing air.

In the search for an abundant and inexpensive alternative feedstock for ethylene production, oxidative coupling of methane has been extensively investigated in the last four decades, primarily using the methane content of natural gas. In this promising process concept, methane will undergo controlled catalytic oxidation to produce ethane and then ethylene. These constitute a net exothermic reaction system in which kinetics is determining the overall methane conversion, yet the selectivity of some reactions such as water gas shift (WGS) reactions is equilibrium limited. The required oxygen reactant can be supplied from air through an inorganic membrane and can be distributed along the catalytic bed to secure a selective methane conversion. A major part of the reactant methane remains unreacted in an OCM reactor, while hydrogen and water as side products and carbon oxides as undesirable products are also produced in significant quantities. Typical the OCM reactor outlet stream is composed of approximately 5% ethylene, 2% ethane and 20–40% methane or carbon dioxide molar fractions depending on the type and level of dilution. Therefore, the mole fraction of the generated ethylene in the OCM reactor outlet gas stream is around 10 times lower than the typically expected mole fraction of ethylene in the outlet of ethane or naphtha crackers. Consequently, the separation cost of the accompanying components, comprising more than 80% of the gaseous species in a demethanizer, amounts to more than 65% of the total specific energy cost invested for processing a ton of ethylene. By increasing the portion of light gases such as nitrogen, hydrogen and carbon monoxide in the OCM reactor outlet or its methane content, the operating cost of cryogenic separation can be dramatically increased due to utilization of a higher methane-to-oxygen ratio in the feed stream. Optimizing the structural and operating characteristics of such energy-intensive separation is necessary, yet not a game-changer. Having considered all of these, an ethylene adsorption unit capable of efficient recovery of such a low ethylene content in the diluted gas stream would be a promising alternative separation concept for this application. The typical performance of a benchmark cryogenic separation as well as ethylene adsorption technique reported for OCM application highlights the feasibility and energy efficiency of this alternative process. Developing a sorbent with high affinity and selectivity towards ethylene is the key requirement for this. To identify the best scenarios for setting up the adsorption unit within an OCM process, the performance of such a sorbent in processing the streams with different gas compositions and operating pressures should be analysed. To address these aspects, a systematic performance analysis of the adsorption separation of ethylene in the OCM process was conducted in this research. This includes a comprehensive experimental study in developing an efficient sorbent and testing its performance in a miniplant-scale adsorber unit.

Sorbents with different characteristics were tested for separation of ethylene from different gas streams, each representing different potential feed streams in the OCM process. Zeolite 13X and its alternative physically and chemically modified forms, the calcined and copper-exchanged forms of zeolite 13X, were tested for ethylene removal from the carbon dioxide-rich and carbon dioxide-free gas streams in the OCM process downstream. Simple physical treatment in the form of calcination of zeolite 13X improved its adsorption capacity to up to 100%. Modifications of zeolite 13X with copper ions increased its adsorption selectivity by reducing its adsorption affinity towards carbon dioxide as indicated by its 25% shorter carbon dioxide breakthrough time. Recording an earlier carbon dioxide breakthrough time for copper-exchanged zeolite 13X was a general observation for the whole range of operating pressures. Using Cu-exchanged zeolite 13X (impregnated with high concentrations of copper) in processing the carbon dioxide-rich feed streams (e.g., with the carbon dioxide content 2.25 times its ethylene content) primarily enhances the adsorption selectivity rather than its capacity such that it has resulted in an ethylene adsorption capacity of 0.46 $mol_{C_2H_4}/kg_S$ with an adsorption selectivity of 0.45 $mol_{C_2H_4} \cdot mol_{CO_2}$. For instance, even in such a low-pressure range ≤ 0.5 MPa, the promising adsorption capacity of 1.4 $mol_{C_2H_4}/kg_S$ and adsorption selectivity of 3.8 $mol_{C_2H_4} \cdot mol_{C_2H_6}$ were observed for the investigated carbon dioxide-free feed compositions representing the OCM reactor outlet gas stream after the carbon dioxide removal section. Therefore, in processing the carbon dioxide-free feed streams, physically treated (calcined) zeolite 13X could provide enough adsorption capacity, while Cu-modified zeolite 13X was preferred to be applied for treating the carbon dioxide-rich streams.

The results of the study indicated that not only in comparison to zeolite 4A and activated carbon zeolite 13X shows superior ethylene adsorption potential, but its absolute value of ethylene adsorption capacity even under low-pressure operation is outstanding. Besides the main performance indicators, such as the stability and capacity of the sorbent, its lower affinity towards carbon dioxide was found to be a crucial factor in improving the energy efficiency of adsorption separation of ethylene in an OCM process, especially in the desorption step when carbon dioxide is used as the sweep gas. This highlights the potential of zeolite 13X to be utilized for efficient ethylene adsorption separation downstream of the industrial-scale OCM process.

Nevertheless, it was demonstrated that, preferably, a sorbent with a lower affinity towards carbon dioxide should be used for adsorption separation of ethylene. It was also demonstrated that the developed adsorption system in the study has an advantage over the cryogenic distillation technique in directly processing the gas streams containing carbon dioxide. When the processed feed gas contains significant portions of light components such as nitrogen or unreacted methane, the advantage of using an adsorption separation would be further highlighted.

Developing an efficient sorbent and adopting an ethylene adsorption separation unit in downstream of an oxidative coupling of methane (OCM) process are the focus of this section. Since the mole fraction of the generated ethylene in the OCM reactor outlet is relatively low, the processing cost of the accompanying components and thereby the separation cost per ton of ethylene using the conventional cryogenic separation are insupportable. Zeolite 13X was modified in this research, demonstrating outstanding ethylene adsorption capacity and selectivity. The conducted adsorption experiments at a miniplant-scale unit enabled monitoring of the adsorption breakthrough times and measuring the sorbents' capacity under different operating pressures in the range of 0.1–0.5 MPa while processing different feed flow and feed compositions, representing the attachment of the adsorption unit to different parts of the OCM process. The modified zeolite 13X showed superior performance than the reference sorbents such as zeolite 4A and activated carbon. Physical treatment of zeolite 13X, by calcining it at 550°C, proved to be efficient in increasing its adsorption capacity. Chemically treating zeolite 13X via copper exchange on the other side increased its ethylene adsorption selectivity in competition to carbon dioxide adsorption. In processing the CO_2-rich feed streams (with the carbon dioxide content 2.25 times its ethylene content), the Cu-exchanged zeolite 13X showed a promising ethylene adsorption capacity of 0.46 $mol_{C_2H_4}/kg_S$

combined with an adsorption selectivity of 0.45 $mol_{C_2H_4} \cdot mol_{CO_2}$. In processing the CO_2-free feed streams (with the C_2H_4 content 2.75 times of its C_2H_6 content), using the calcined zeolite 13X secured the highest adsorption capacity of 1.4 $mol_{C_2H_4}/kg_s$ along with an adsorption selectivity of 3.8 $mol_{C_2H_4} \cdot mol_{C_2H_6}$ under 0.5 MPa adsorption pressure. These indicate the promising potentials of the developed sorbents and the designed adsorption unit for processing the OCM reactor outlet gas stream before and after removal of its carbon dioxide content.

8.1.3 From Carbon Dioxide to Lower Olefins

Ethylene and propylene are large scale petrochemical products and primary building blocks of the pillar A (Chapter 1) opening synthesis pathways; other scale products include polyethylene and polypropylene, styrene (via ethylbenzene), monoethylene glycol (via ethylene oxide) acrylic acid, acrylonitrile, cumene and polyols (via propylene oxide), just to name some of the most important.

Production of ethylene and propylene in Europe is mainly performed by steam cracking with naphtha as predominate feedstock, although LPG is gaining importance in Europe and 25% of feedstock used in the steam crackers in 2015 was LPG. Usually, crackers can use mix of feedstock and very few crackers would use LPG only. Along with ethylene and propylene, steam cracking also yields other products, together referred to as high value chemicals (HVC).

The degree of energy efficiency is measured by the specific energy consumption (SEC) in GJ/t. The SEC in the case of naphtha cracking is the sum of theoretical thermodynamic energy requirement and energy loss. The SEC of naphtha-based steam cracking in Europe is around 16.5 GJ/t HVC. The value for best practice technology (BPT) plants is around 12 GJ/t HVC. The SECs for a typical naphtha steam cracker should be approximately within the range of 14–17 GJ/t HVCs.

Conversion of carbon dioxide to value-added chemicals has been a long-standing objective, direct hydrogenation of carbon dioxide to lower olefins is highly desirable, but is still challenging. In this regard, copper-based catalysts and several newly reported metal-based and metal oxide-based catalysts have shown good performance in methanol synthesis from carbon dioxide hydrogenation. It is well known that the C-C bond formation can take place on zeolite catalyst in methanol conversion, such as hydrocarbons formation on HZSM-5 and lower olefins on SAPO-34. Hydrogenation of carbon dioxide on the composite's catalysts such as Cu-Zn-Cr/zeolite and Cu-ZnO-ZrO_2/zeolite mainly gives alkanes and carbon monoxide. Recently, Zn-CrO_x/MSAPO bifunctional catalyst was used to convert syngas to lower olefins with high selectivity (80% of hydrocarbons) through the gas phase transferring of ketene as intermediate. Later, the Zr-Zn (2:1) binary oxide coupled with SAPO-34 offers around 70% selectivity for lower olefins at about 10% carbon monoxide conversion and methanol or methoxide as intermediates were proposed. Compared with hydrogenation of carbon monoxide to lower olefins by using bifunctional catalysts, the first key difference of carbon dioxide hydrogenation over bifunctional catalyst (methanol synthesis catalyst/zeolite) is the reverse water-gas shift reaction that occurred in the reaction of carbon dioxide hydrogenation, which is kinetically favoured and can result in the producing of amounts of carbon monoxide under reaction conditions of bifunctional catalyst. In addition, the water that formed by carbon dioxide hydrogenation on methanol synthesis catalyst can seriously affect the activity and stability of zeolite for realizing C-C bond formation. Therefore, synthesis of lower olefins with high selectivity from carbon dioxide hydrogenation over bifunctional catalyst is very challenging. It was reported that carbon dioxide hydrogenation on Na modified Fe-based Fischer-Tropsch catalyst and In_2O_3 catalyst combined with H-ZSM-5 exhibited excellent selectivity in gasoline and a CeO_2-Pt@mSiO_2-Co tandem catalyst with two metal-oxide interfaces converted carbon dioxide to C_2-C_4 hydrocarbons with 60% selectivity. However, the goal of achieving high selectivity in lower olefins while suppressing alkane and carbon monoxide productions has not been achieved.

Li et al. developed a ZnO-ZrO_2 mixed metal oxide catalyst that shows considerable methanol selectivity at wide temperatures and a zinc-modified SAPO which can initiate the methanol to olefins reaction. The aim of the proposal is to synthesize the lower olefins by taking the advantages of these two types of catalysis: carbon dioxide hydrogenation to methanol and selective conversion of methanol to lower olefins. Therefore, by considering these dual functional catalysis and chemical engineering factors, Zelong Li with co-workers fabricated a tandem catalyst by dispersing the ZnO-ZrO_2 nanoparticles on the SAPO micro-meter crystals for the carbon dioxide hydrogenation.

Figure 8.1 A (left panel) shows a representative result that the carbon dioxide hydrogenation on the Zn-ZrO/SAPO gives 80% selectivity of $C_2^=$-$C_4^=$ (olefins), 14% C_2-C_4^o (alkanes), 3% methane and 3% C_5+ among all hydrocarbon products, at the carbon dioxide single pass conversion of 12.6% under reaction condition of 2 MPa, 3600 ml/gcat/h and 380°C and carbon monoxide selectivity is suppressed to 47%. Carbon dioxide hydrogenation over the Zn-ZrO mainly produces methanol and carbon monoxide, while neither olefins nor methanol was detected for SAPO alone, indicating that it is the tandem catalyst that efficiently converts carbon dioxide to lower olefins. Figure 8.1 A (right panel) shows that methanol conversion on SAPO or Zn-ZrO/SAPO produces lower olefins with high selectivity (over 90%), affirms that SAPO is the catalyst for methanol to lower olefins conversion. These results lead to the conclusion that tandem reactions take place on the tandem catalyst,

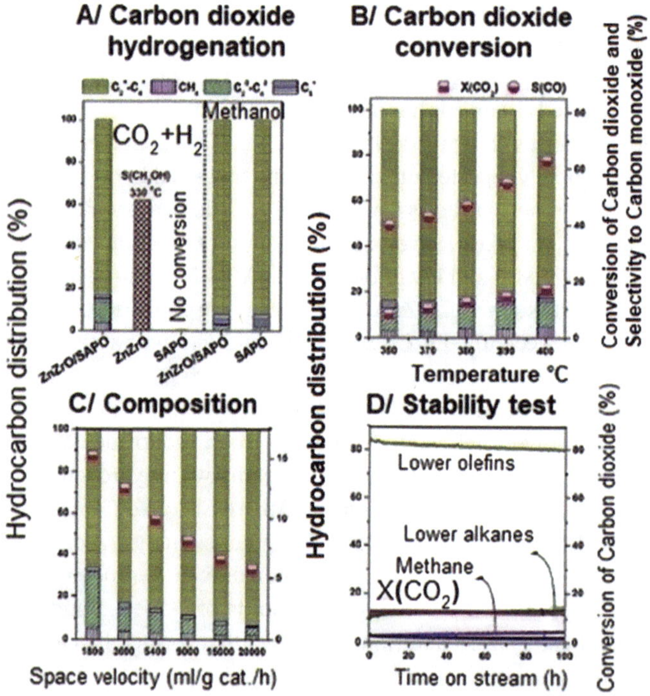

Figure 8.1 Catalytic performance in CO_2 hydrogenation. A. Carbon dioxide hydrogenation on Zn-ZrO/SAPO, Zn-ZrO and SAPO and methanol conversion on Zn-ZrO/SAPO and SAPO. B. Carbon dioxide conversion over Zn-ZrO/SAPO, hydrocarbon distribution and carbon nonoxide selectivity at different reaction temperatures. C. Hydrocarbon distribution and carbon dioxide conversion over Zn-ZrO/SAPO with different space velocity. D. Stability test for Zn-ZrO/SAPO. Reaction condition for Zn-ZrO/SAPO: 380°C, 2 MPa, 3,600 ml/gcat/h; ZnZrO: 330°C, 2 MPa, 3,600 ml/gcat/h (different from 380°C, the reaction temperature for lower olefins production, where the methanol selectivity is about 2%); SAPO: 380°C, 3ml/gcat/h, all catalysts were tested in a tubular fixed bed reactor, with catalyst 0.2 g. (Li et al, 2017, American Chemical Society).

Zn-ZrO/SAPO, where the Zn-ZrO generates methanol (and precursors) from carbon dioxide hydrogenation and the SAPO is responsible for the methanol lower olefins conversion.

With reaction temperatures increasing, carbon dioxide conversion is increased, but the lower olefins selectivity declines while carbon monoxide selectivity grows (Figure 8.1B). This agrees well with the kinetic behaviour of carbon dioxide hydrogenation to methanol on ZnZrO because higher temperature favour the reverse water gas shift reaction on most methanol synthesis catalysts. To achieve a high yield of lower olefins, the reaction conditions were optimized to temperature 380°C and space velocity 3,600 ml/gcat/h, where the single pass carbon dioxide conversion 12.6% and lower olefins selectivity is above 80%. It should be pointed out that the selectivity of lower olefins can be further enhanced from 80% to 93% when the space velocity is increased from 3,600 to 20,000 ml/gcat/h and the selectivity of C_1 and C_5+ are suppressed to marginal (Figure 8.1C). This performance is superior to that of the Fischer-Tropsch process of carbon dioxide hydrogenation in terms of lower olefins selectivity. It is noteworthy that the tandem catalyst shows good stability over 100 hours on stream without obvious deactivation (Figure 8.1D) and it also shows high thermal stability and sulfur resistance property suggesting its potential application in industry.

8.1.4 Low-Carbon Production of Ethylene and Propylene

In contrast to the previously described processes, there is currently no existing ethylene process at an advanced technology readiness level, which directly uses hydrogen and carbon dioxide to produce olefins. Current research efforts such as the project e-Ethylene under the lead of Siemens, funded by the German Ministry of Education and Research, is aimed at a direct electrocatalytic production of ethylene from carbon dioxide and water in a single stage system. While the general proof of concept has been shown, this technology is still at TRL 3–4. Nevertheless, this technology could be a breakthrough, as it omits the need of intermediate products such as methane or methanol as feedstock for olefin synthesis. The production pathway included here is commercially deployed albeit commercial operations are in China and, as yet, no Methanol-to-Olefins (MTO) plant is operating in Europe. The process sequence is depicted in Figure 8.2.

Different processes are licensed, such as the UOP/hydro MTO based on the MTO-100 silicoaluminophosphate synthetic molecular sieve based catalyst and Lurgi's MTP (Methanol-to-Propylene) process based on MTPROP, a proprietary ZSM-5 type of catalyst supplied from Clariant (formerly Sud-Chemie).

The Methanol-to-Olefins reaction is strongly exothermic. To control the heat of reaction and the adiabatic temperature increase, the process follows a two-step dehydration of methanol to dimethyl ether and water, followed by conversion to olefins, as depicted in Equations 8.10–8.12).

$$2CH_3OH \rightleftarrows CH_3OCH_3 + H_2O \tag{8.10}$$

$$CH_3OCH_3 \rightleftarrows C_2H_4 + H_2O \tag{8.11}$$

$$3CH_3OCH_3 \rightleftarrows 2C_3H_6 + 3H_2O \tag{8.12}$$

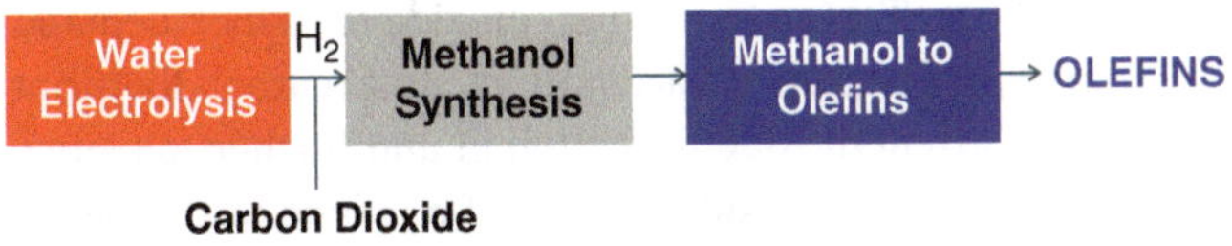

Figure 8.2 Low-carbon process sequence to ethylene and propylene via methanol.

Depending on the catalyst, different target products can be realized. Besides MTO, the Methanol-to-Propylene process, or Methanol-to-Gasoline (MTG) process will be discussed. The two major process concepts are based on a fixed-bed and a fluidized-bed reactor.

The process of olefin synthesis from dimethyl ether was tested on a pilot unit using nitrogen and synthesis gas as a diluent gas (Magomedova et al.) The study was carried out over the catalyst Mg-HZSM-5/Al_2O_3 (150 mL, pellet size of 2.8 x (5–7) mm) and stream temperatures at the fixed-bed reactor inlet were 335°C and 355°C. The effect of diluent gas on the heat exchange in the pilot reactor was demonstrated. The results achieved on the pilot unit were compared with those obtained in the kinetic study on the laboratory unit. The estimation of the effectiveness of porous catalyst Mg-HZSM-5/Al_2O_3 basic on experimental data and theoretical equations was carried out.

Based on fossil feedstock and natural gas specifically, the process sequence would start with steam reforming and methanol production, followed by methanol to olefins. The best practice technology energy consumption of Methanol-to-Olefins plants are at 5 GJ/tHVC (high-value-chemicals). Assuming best practical technology level for plants in Europe seems reasonable, as currently no Methanol-to-Olefins production capacities exists and new plants would have to be built. In addition, the energy consumption for producing methanol as feedstock must be included in the total energy consumption of the process chain, resulting in 17.5 GJ/t HVC, which is higher than for steam cracking. In terms of carbon dioxide emissions, the process sequence of natural gas to olefins via MTO would not be beneficial, as steam cracking emits 0.76 t CO_2/t HVC compared to 2.9 t CO_2/t HVC for the process chain via Methanol-to-Olefins. The latter includes 0.5 t CO_2/t HVC for MTO plus the emissions by production of 2.83 t methanol stoichiometrically required as feed in the Methanol-to-Olefins process.

For the low-carbon route, the Methanol-to-Olefins process would remain unmodified, emission reduction is then predominantly based on providing low-carbon methanol as feedstock.

8.1.4.1 Energy Demand per Unit of Ethylene/Propylene Production via Methanol

Stoichiometrically, 2.28 tons of methanol are required per ton of ethylene or propylene. Energy demand therefore is composed of the energy demand for hydrogen-based methanol production as described in Chapter 10 and energy demand for the MTO process, which is at 5 GJ/t HVC. In total the energy demand is 95.5 GJ/t HVC corresponding to 26.6 MWh/t HVC, based on electricity. This is more than five times the energy demand of naphtha-based process (16.5 GJ/t HVC). Even if the feedstock energy content (42.5 GJ/t HVC) is considered for steam cracking, the SEC of low carbon processes still 40% higher. It must be mentioned that naphtha steam cacking is a net steam exporting process (1.5 GJ/t HVC) and this is accounted for under electricity-based steam generation.

8.1.4.2 Carbon Dioxide Reduction per Unit of Ethylene/Propylene Production

The Methanol-to-Olefins process causes 0.4 t CO_2eq/t HVC compared to 0.80 t CO_2eq/t HVC in the ethane-based and 0.70 t CO_2eq/t HVC naphtha-based processes. The main benefit of the low-carbon process sequence in terms of carbon dioxide emissions would however originate from the low-carbon methanol production, as this process has been shown to be a net carbon dioxide consuming process (negative emissions). The process sequence would therefore benefit from the strong methanol demand of the Methanol-to-Olefins process, which on the other hand causes the very high energy consumption indicated above. Taking the carbon dioxide feedstock demand and footprint for the hydrogen-based methanol production into account, the total carbon footprint of the process chain from water and captured carbon dioxide to methanol and further to olefins amount to −1.89 t, which avoided production of carbon dioxide per ton of ethylene or propylene produced via this pathway.

8.1.4.3 Economics of Low-Carbon Ethylene and Propylene Production

Economic constraints described for methanol (Chapter 10) remain an issue for olefin production, as methanol is the feedstock. For methanol production, costs ranging from 300 to 650 € per ton feedstock will amount to 680 to 1450 € per ton of ethylene or propylene. A recent calculation of production costs for ethylene and propylene from naphtha in Europe, provided by the Joint Research Centre of the European Commission amounted to 816 €/t HVC: an economic gap of least a factor of two must be estimated, making the realization of this pathway very challenging, in particular as new hydrogen-based methanol plants and methanol to olefins plants would be necessary.

8.2 Life Cycle Assessment

As one of the most valuable olefin products of petrochemical industries worldwide, the life cycle assessment (LCA) case study focuses on Carbon Dioxide-to-Ethylene conversion via electrochemical reduction process. The goal of the LCA is:

- projection of net carbon dioxide emissions and other pollutions for the carbon dioxide reduction reaction (CO_2RR) case studies;
- evaluation of the difference in LCA results for both small- and large-scale cases; and
- to highlight the specific environmental issues associated with CO_2RR.

The functional unit to be considered is 1 g ethylene for the small-scale experimental. For both cases, the LCA cradle-to-gate scope starts with the output.In all LCA scenarios, the production, replenishment and treatments of catalysts used for CO_2RR, is not taken into consideration.

8.2.1 Small-Scale Production of Ethylene

The first case study focuses on small-scale set up to produce 1 g of ethylene. Experimental datasets were obtained from two-compartment static setup consisting of high surface area copper oxide-derived copper as working electrode. The reader is referred to Handoko et al. for more details. This setup represents a base-case scenario that can be easily replicated at larger scale, including catalyst preparation and setup. An optimum CO_2RR to ethylene Faradaic efficiency (FE) of 34% was observed around 0.98 V vs. RHE. A more optimistic case was taken from a recent publication based on pyridinium-modified copper as working electrode in a flow electrochemistry setup (Li et al.). This setup allows for CO_2RR to ethylene at 70% FE and –0.82 V using modified copper in aqueous bicarbonate-based electrolyte. This value compares reasonably well with Jhong et al. for the electrochemical reduction of carbon monoxide on copper-mesh electrode in potassium iodide solutions with different concentrations.

The following factors are considered for the LCA cradle-to-gate of the experimental set-up:

- the "cradle" starts from carbon monoxide feedstock sourced from a gas tank and the "gate" ends with 1 g ethylene product;
- the energy requirement for water pump is considered negligible;
- the electrical power required for lab-scale distillation unit is measured as 0.0062 kW h;
- waste treatment is not considered in this setup; and
- the energy needs for electrochemical and distillation can be supplied from

the details of source use, accompanied with associated experimental/process information and output of main and by-products are compiled in Table 8.1. A second investigation was carried out

Table 8.1 Input-output information for small-scale experimental setup.

Resource input	Amount/unit	Process Description
CO_2 input	4.557 g	Sourced from CO_2 gas tank
Water input	5.877 g	At the anode
Energy requirement for electrochemical process	0.0386 kWh	45% voltage efficient x 75% Faradaic efficiency
Output	**Amount (g)**	**Process Description**
Ethylene	1.000	Main product
Hydrogen	0.040	Gaseous by-products
Carbon monoxide	0.377	
Methane	0.006	
Ethanol	0.241	Liquid by-products
Acetic acid	0.067	
Formic acid	0.165	
1-Propanol	0.042	
H_2O	3.276	Wastewater
O_2	5.220	Released to air from anode

for a hypothetical case of production of 1 ton ethylene. It is assumed that the same operating conditions from the small-scale can be employed for large-scale model.

Various technologies employed for carbon monoxide sequestration are introduced and electrochemical reduction of carbon dioxide is selected owing to its technological advantages and simplicity. Background and application of electrochemical reduction of carbon dioxide using homogeneous and heterogeneous catalysts are discussed. In aqueous solution, electrochemical reduction of carbon dioxide is completed by hydrogen evolution reaction. In an attempt to mitigate the hydrogen evolution, deposition of zinc on copper showed enhanced Faradaic efficiency of methane (52%) compared to bare Cu (23%) electrode.

We have presented LCA results for the CO_2RR of 1g ethylene for the small-scale experimental set-up using operating parameters based on the most recent progress in electrocatalysis: a hypothetical case where 1 ton ethylene production was performed with similar setup and considering industry standard methods for carbon dioxide capture and product separation.

8.3 Polymerization Reaction

8.3.1 Carbon Dioxide-Based Polymers

The incorporation of carbon dioxide into polymers, such as polycarbonates, polypyrenes, lactone intermediates and polyurethanes and new solid materials is an active and promising field of research for medium- to long-term fixation of carbon dioxide. The two $C = O$ bonds of the carbon dioxide molecule make it capable of formation of polymers through conversion of one of the $C = O$ π-bonds into σ-bond by reaction with other compounds.

The coupling of carbon dioxide-epoxide (-oxirane) results in cyclic carbonates which can subsequently undergo ring-opening polymerization to form aliphatic polycarbonates. The high strain of the three-membered cyclic ring of these ethers renders them very susceptible to nucleophilic attack by many compounds. Except for Al(III) complexes, only transition and rare earth metal complexes have been reported for the synthesis of carbon dioxide-oxirane copolymers. Zinc-based catalysts are among the first and the most extensively studied catalysts. In the first step of epoxide into the carbonate-metal bond, either re-insertion of carbon dioxide (propagation reaction) or cyclization occurs (Figure 8.3).

Various cyclic carbonates were prepared by cycloaddition of carbon dioxide with different epoxides using imidazolium-based catalysts grafted on a mesoporous silica support through 3-aminopropyltrimethoxysilane and Br–$(CH_2)_n$–Br ($n = 2$ or 4) under efficient and mild conditions. The catalysts were characterized by various physicochemical techniques such as magic angle spinning (MAS) NMR, ^{1}H NMR, Fourier transform infrared spectroscopy (FT-IR), X-ray diffraction (XRD), field-emission scanning microscopy (FESEM), thermogravimetric analysis (TGA) and Brunauer–Emmett–Teller (BET). The reactions were conducted under maximum carbon dioxide pressure of 2.0 MPa at the temperature range of 80–140°C without using any solvent. The influence of different reaction parameters was investigated. The effects of the chain length and different anions on the catalyst activity were also studied. The mechanism of the reaction was investigated theoretically using density functional theory (DFT) calculations in the ONIOM scheme and the influence of hydrogen bonding and the chain length of the linker on the transition state of the rate-determining step was also studied. The selected catalyst was used in subsequent runs without a significant loss of its reactivity.

Polycarbonates are amorphous polymers with excellent properties such as toughness and transparency. The carbon dioxide route avoids the use of highly toxic phosgene as a reactant in conventional methods.

Polycarbonate (PC) is an excellent engineering thermoplastic due to its attractive properties, including high toughness, high glass transition temperature and good transparency. The traditional synthetic process of PC from bisphenol A (BPA) and phosgene is facing environmental pollution problems such as toxic raw materials and large-scale use of dichloromethane. Therefore, the development of green non-phosgene routes for the synthesis of PC is urgently needed. Diphenyl carbonate (DPC) and dimethyl carbonate (DMC) derived from carbon dioxide have attracted great attention as a carbonyl source to replace phosgene. Dimethyl carbonate is not only a less toxic organic synthesis intermediate but also a raw material for the synthesis of diphenyl carbonate. Moreover, phenol, a by-product of the reaction between diphenyl carbonate and bisphenol A, can be removed only under severe conditions, while methanol, as the by-product of the reaction between dimethyl carbonate and bisphenol A, can be removed more easily. Therefore, dimethyl carbonate is considered a more ideal monomer.

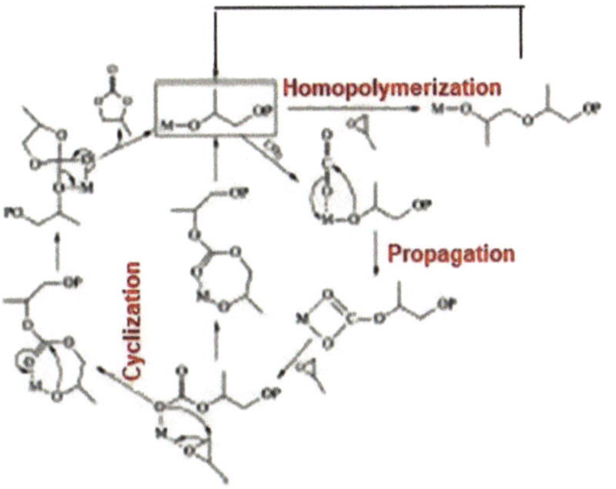

Figure 8.3 Propylene oxide-carbon dioxide copolymerization mechanism and side-reactions.

Metal acetylacetonates were used to catalyse the reaction of dimethyl carbonate and bisphenol A and high conversion of bisphenol A and excellent selectivity of carboxymethylation products were achieved. Among these, zirconium (IV) acetylacetonate ($Zr(acac)_4$) exhibited the best catalytic performance with bisphenol A conversion of 85.2% and carboxymethylation products selectivity up to 99%, which has been the most efficient catalyst so far to the best of my knowledge (Figure 8.4). According to Fourier transform infrared (FT-IR) characterization and density functional theory (DFT) calculations, I found that the most active *cis–cis* interaction species 3 formed between a zircon cation and two oxygen atoms from the CH_3–O moiety of dimethyl carbonate greatly promoted the selective occurrence of the carboxymethylation reaction. It was demonstrated that the catalytic performance of metal acetylacetonate was related to *d*-electrons and interaction mode of the metal and a plausible reaction mechanism was proposed.

The development of bio-based polycarbonates from bio-based epoxide monomers via coupling with carbon dioxide using a sustainable approach is innovative and attractive. Until recently, there were only a few reports of bio-based polycarbonates derived from bio-based epoxide monomers and carbon dioxide. Coates and co-workers developed stereo complexed poly-(limonene carbonate) from limonene oxide and carbon dioxide with a cobalt-based catalyst system, which is pioneering work in the polycarbonates area. Inspired by those works, a type of poly (limonene carbonate) with high molecular weight, good thermal properties and transparency was achieved after treating limonene oxide using appropriate end-capping agents. A group of cobalt catalysts (Salen (CoCl)) for coupling epoxide monomers and carbon dioxide to produce polycarbonates and cyclic carbonates were also investigated. Soybean oil, a low-cost commodity bio-feedstock, has been functionalized as a series of epoxide monomers, including epoxy groups located internally along with fatty acid chains. But such monomers were shown to perform at low activity at copolymerization. Instead, soybean oil-based monomers with epoxy groups locating on the terminals of fatty acid chains have been successfully used for the produce of bio-based polycarbonates (soybean oil-based polycarbonate) via copolymerization with carbon dioxide.

As illustrated in Figure 8.5, the glycerol fraction was obtained from crude glycerol through a simple distillation process, resulting in the leftover residuals rich in fatty acids residues waste (fatty acid fraction), which was designated as crude glycerol-fatty acids waste (CG-FAW). And the refined glycerol can be used to produce epichlorohydrin. An oxirane group from the above epichlorohydrin was then introduced to the terminal of fatty acid chains to produce bio-based epoxide monomers (CG-FAW epoxide monomers); afterwards, crude glycerol-fatty acids waste derived bio-based polycarbonates (CG-FAW bio-based polycarbonates) were obtained via a copolymerization of the aforementioned epoxide monomers coupling with carbon dioxide in the presence of

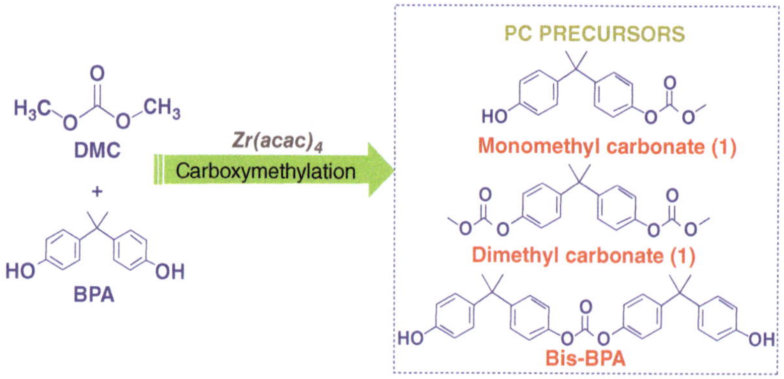

Figure 8.4 The reaction of dimethyl carbonate and bisphenol A.

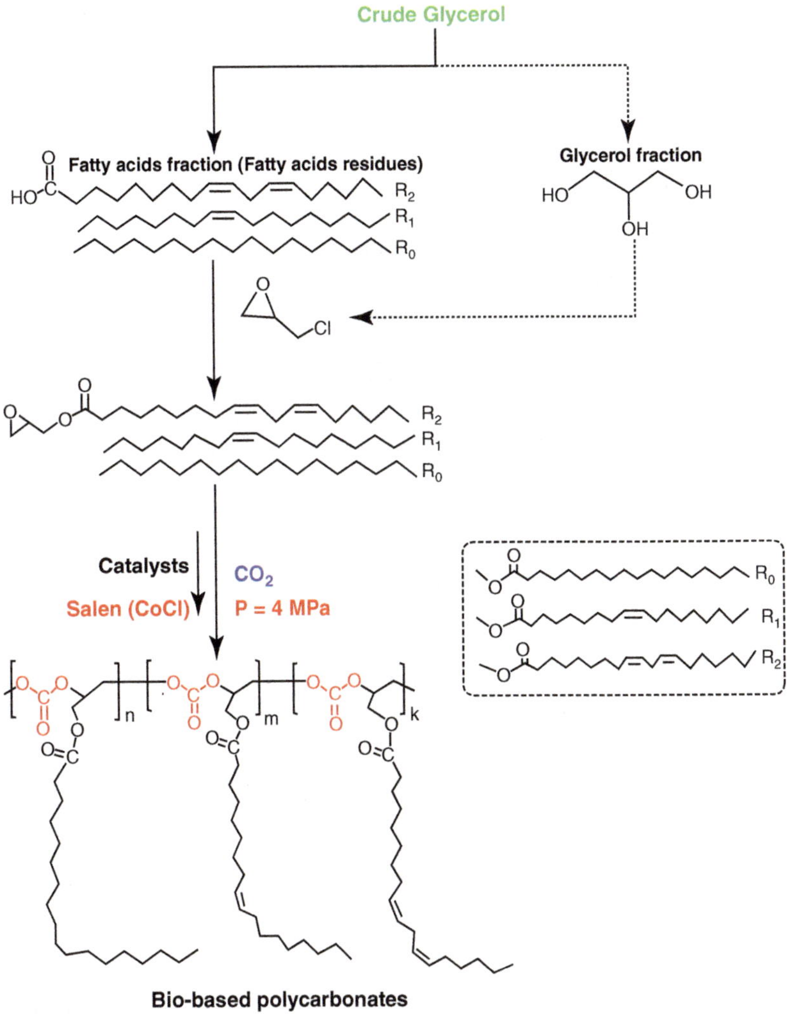

Figure 8.5 The sustainable conversion pathway from crude glycerol fatty acid waste and carbon dioxide to biobased polycarbonates.

catalyst system. In summary, the proposed conversion made full usage of crude glycerol, especially the fatty acid residues waste from crude glycerol refining process and led to a sustainable route to produce bio-based polycarbonates.

Polyurethanes are versatile polymers with a very large market. They can be produced from carbamate precursors. Organic carbamates ($R^1NHCOOR^2$) are also important intermediates for production of fine and commodity chemicals. Commercially, carbamates are synthesized by ammonolysis of chloroformate esters, derived from phosgene and an alcohol. Alternative routes, such as catalytic carbonylation of nitroaromatics and oxidative carbonylation of amines, have also been developed. The reaction of amines, carbon dioxide and alkyl halides are an ecofriendly method to produce carbamates (Equation 8.13):

$$R'NH_2 + CO_2 + n\mathrm{BuBr}(-\mathrm{HBr}) \rightarrow$$
$$R'NH - CO - O - (n\mathrm{Bu}) \rightarrow \text{Polyurethanes} \qquad (8.13)$$

This reaction is catalysed by onium salts, basic catalysts, sterically hindered organic bases, crown ethers or solid cesium carbonate.

Liquid or supercritical carbon dioxide can be used as a medium for certain polymerization reactions. In the presence of carbon dioxide, few polymers, typically amorphous fluoropolymers can be synthesized by a homogeneous solution polymerization. Many insoluble polymers can be synthesized by a heterogeneous chain-growth process, such as precipitation, emulsion, dispersion, or suspension. Due to the solubility of many vinyl monomers and free radical initiators in carbon dioxide and the ability to design appropriate carbon dioxide-soluble polymeric surfactants, dispersion polymerization in carbon dioxide is a common heterogeneous polymerization method. Step growth polymerizations can also be conducted in carbon dioxide with advantages over other processes.

Supercritical carbon dioxide can also be used as a processing solvent for the physical processing of polymeric materials. Impregnation, particle formation, foaming, binding and injection moulding are examples of this category.

8.3.1.1 Perspective and Practical Applications

The field of green chemistry is rapidly developing and the use of green starting materials and the replacement of old technologies with clean catalytic processes are among the most important issues.

In certain processes, where the feed is hazardous or environmentally unacceptable, its replacement with carbon dioxide as a reactant could produce new production paths. A good example is the replacement of the poisonous phosgene with carbon dioxide to produce dimethyl carbonate ($CH_3OCOOCH_3$, DMC). Commercially, it is produced from carbon monoxide with phosgene as the intermediate (Equations 8.14 and 8.15) and in the other two reactions (Equations 8.16 and 8.17) carbon monoxide is used directly. Both carbon monoxide and phosgene are highly poisonous. The replacement of these processes with new routes using carbon dioxide as the feed is shown in Equation 8.18.

$$CO + Cl_2 \rightleftarrows COCl_2 \quad \Delta_r H^\circ = -109\,kJ\,mol^{-1} \tag{8.14}$$

$$COCl_2 + 2CH_3OH \rightarrow CH_3OCOOCH_3 + 2HCl \tag{8.15}$$

EniChem DMC Process (by EniChem-12000 t/y):

$$CO + 0.5O_2 + 2CH_3OH \rightarrow CH_3OCOCH_3 + H_2O \tag{8.16}$$

Ube DMC Process (by Ube Chemical- 3000 t/y):

$$CO + 2CH_3ONO \rightarrow CH_3OCOOCH_3 + 2NO \tag{8.17}$$

New carbon dioxide-based route:

$$CO_2 + 2CH_3OH \rightarrow CH_3OCOOCH_3 + H_2O \tag{8.18}$$

Indala studied 83 papers issued on the laboratory and pilot tests describing new methods for the utilization of carbon dioxide for products of commercial importance. According to the energy required, equilibrium conversion, selectivity, reducibility, catalyst lifetime and price, 20 processes were selected as the candidates for new plants based on carbon dioxide. These processes were simulated with HYSYS software and value-added economic evaluation was performed. From them, 14 most promising cases were selected and incorporated in the current superstructures. That selection includes methanol synthesis, dimethyl ether synthesis, syngas and styrene production, all of which are covered in the study.

The Gulf Petrochemical Industries Company (GPIC), a manufacturer of fertilizers and petrochemicals in Bahrain, has taken a lead in the adoption of CDM by building a carbon dioxide recycling facility in partnership with Italy's Technoment and Japan's Mitsubishi Heavy Industries. The 52 million dollar plant, with a daily production capacity of 450 t, will capture and recycle carbon dioxide from GPIC's methanol plant to be reused as feedstock in the production of urea and methanol project. Construction of the new plant is expected to increase GPIC's production capacity of methanol by 120 t/day and urea by 80 t/day.

Carbon dioxide recovery starts with the feeding of flue gas from the methanol stack at 206°C into the carbon dioxide recovery unit. The recovery units can capture 450 t of carbon dioxide per day and some 90% of the carbon dioxide is in the flue gas. The columns are 4.5 m in diameter and 50 m in length and were fabricated as single pieces. The flue gas is treated with demineralized water through a direct contact quencher. The demineralized water is sent to the packed column through structured packing. The flue gas is then treated with caustic soda to remove the sulfur oxide and is compressed and moved to a two-bed carbon dioxide absorber having a structured packing to absorb the gases. The bottom bed is used for absorbing carbon dioxide and the top bed for treating the exhaust gases. A special solvent called KS-1 is fed from the top of the section to absorb the carbon dioxide. The remaining gases from the flue gas are treated and released into the atmosphere.

The process technology used at the new plant was jointly developed by Mitsubishi Heavy Industries and Kansai Electric Power Company. The process is called the Kansai-Mitsubishi Carbon Dioxide Recovery Process (KMCDR Process) and can recover about 90% of carbon dioxide present in flue gas. A key feature of the technology is that it requires lower energy compared to other technologies. In addition to urea production, carbon dioxide recovery technology can be used for chemical applications, such as production of methanol and dimethyl ether and, in the food and beverage industries, production of carbonated beverages and dry ice.

GPIC will become the first petrochemical plant in the region to use this technology, a process that is expected to reduce its carbon dioxide emissions by more than 100,000 t/y as of 2010. This example, among others, illustrates that the capture and catalytic fixation of carbon dioxide in petrochemical industries are beginning to be realized and implement commercially and further plants will be constructed based on emerging technologies soon in response to the CDM requirements.

The design of plastics to be both biodegradable and sustainably produced from nonpetrochemical sources is another critical circular design area requiring urgent research input. The results by Napper et al. reports a lack of biodegradable plastic formulations used for carrier bags that offered rapid rates of decay compared to conventional plastic bags. For environmental concerns, Grbic et al. provide a novel and simple method to assess the quantities of microplastic fragments in environmental matrices, using hydrophobic iron nanoparticles to extract microplastics from soil, sediments and water magnetically and could be used as a sustainable remediation tool.

Further Reading

An, H., Yang, Z. et al. (2020). Highly efficient and selective synthesis of methyl carbonate-ended polycarbonate precursors from dimethyl carbonate and bisphenol A. *Ind. Eng. Chem. Res.* 59 (31): 13948–13955.

Asami, K., Kusakabe, K., Ashi, N. and Ohtsuka, Y. (1997). Synthesis of ethane and ethylene from methane and carbon dioxide over praseodymium oxide catalysts. *Appl. Catal. A* 156: 43.

Bartholomew, C.H. (1984). Catalyst deactivation. *Chem. Eng.* 91: 96–112.
Cuia, S., Borgemenkea, J., Liua, Z. et al. (2019). Innovative sustainable conversion from CO_2 and biodiesel-based crude glycerol waste to bio-based polycarbonates. *J. CO2 Util.* 34: 198–206.
Dokhaee, Z., Ghiaci, M., Farrokhpour, H. et al. (2020). SBA-15-supported imidazolium ionic liquid through different linkers as a sustainable catalyst for the synthesis of cyclic carbonates: a kinetic study and theoretical DFT calculations. *Ind. Eng. Chem. Res.* 59 (28).
Gao, Y., Jie, X., Wang, C. et al. (2020). One-pot synthesis of Ca oxide-promoted Cr catalysts for the dehydrogenation of propane using CO. *Ind. Eng. Chem. Res.* 59 (28).
Grbic, J. et al. (2019). Magnetic extraction of microplastics from environmental samples. *Environ. Sci. Technol. Lett.* 6 (2): 68–72. doi: 10.1021/acs.estlett.8b00671.
Handoko, A.D., Chan, K.W. and Yeo, B.S. (2017). CH_3 mediated pathway for the electroreduction of CO_2 to ethane and ethanol on thick oxide-derived copper catalysts at low overpotentials. *ACS Energy Lett.* 2: 2103–2109.
Indala, S. (2004). Development and integration of new processes consuming carbon dioxide in multi plant chemical production complexes. Bachelor Thesis. Andhra University, India.
Jhong, H.R.M., Ma, S. and Kenis, P.J.A. (2013). Electrochemical conversion of CO_2 to useful chemicals: current status, remaining challenges and future opportunities. *Curr. Opinion Chem. Eng.* 2: 191–199.
Karakaya, C., Kidder, M., Wolden, J. et al. (2022). Mechanistic interpretations and insights for the oxidative dehydrogenation of propane via CO_2 over Cr_2O_3/Al_2O_3 catalysts. *Ind. Eng. Chem. Res.* 61 (39): 14482–1449.
Li, F., Thevenon, A. et al. (2020). Molecular tuning of CO_2-to-ethylene conversion. *Nature, 577:* 509–513.
Li, Z., Wang, J. et al. (2017). Highly selective conversion of carbon dioxide to lower olefins. *ACS Catal.* 7 (12): 8544–8548.
Magomedova, M., Afokin, M.I., Starozhitskaya, A.V. and Galanova, E.G. (2021). Pilot test of olefin synthesis from dimethyl ether in a synthesis gas atmosphere. *Ind. Eng. Chem. Res.* 60 (12): 4602–4609.
Meerendonk, W.J. (2005). CO2 as a monomer for the phosgene-free synthesis of new polycarbonates: catalyst development, mechanistic investigations and monomer screening, [PhD Dissertation]. Department of Chemical Engineering and Chemistry, Technische Universiteit Eindhoven.
Mokhtarani, B., Repke, J.-U., Son, N.X. et al. (2021). Miniplant-scale demonstration of ethylene adsorption separation in downstream of an oxidative coupling of methane process. *Ind. Eng. Chem. Res.* 60 (31): 11778–11788.
Napper, I.E. et al. (2019). Environmental deterioration of biodegradable, oxo-biodegradable, compostable and conventional plastic carrier bags in the sea, soil and open-air over a 3-year period. *Environ. Sci. Technol.* 53 (9): 4775–4783. doi: 10.1021/acs.est.8b06984.
Qin, Z., Thomas, C.M., Lee, S. and Coates, G.W. (2003). Cobalt-based complexes for the copolymerization of propylene oxide and CO_2: active and selective catalysts for polycarbonate synthesis. *Communications* 42: 5484–5487.
Xie, Q. et al. (2021). Ga-doped $MgAl_2O_4$ spinel as an efficient catalyst for ethane dehydrogenation to ethylene assisted by CO_2. *Ind. Eng. Chem. Res.* 60 (31): 11707–11714.
Young, J.L. and Desimone, J.M. (2000). Frontiers in green chemistry utilizing carbon dioxide for polymer synthesis and applications. *Pure Appl. Chem.* 72: 1357.
Zangeneh, F.T., Sahebdelfar, S. and Ravanchi, M.R. (2011). Conversion of carbon dioxide to valuable petrochemicals: an approach to clean development mechanism. *J. Nat. Gas Chem.* 20: 219–231.

9

Power-to-BTX Aromatics

CONTENTS

9.1 Low-Carbon Production of Aromatics

Aromatics, as the most important platform molecules for polymer industry, are mainly produced from petroleum refinery processes. Benzene, toluene and xylenes (p-xylene) are major compounds used for the production of synthetic fibres, resins, detergent and polymers, including for example polystyrene, polyurethane and polyesters. BTX are produced at a level of 15.7 million tons in Europe with an anticipated growth to 29 million tons in 2050.

In this context, the direct conversion of carbon dioxide to aromatics is particularly attractive, which reduces the demand for fossil resources in aromatics production. Although a lot of progress has been made on the production of basic chemicals, like methanol, through carbon dioxide hydrogenation, the direct conversion of carbon dioxide to value-added aromatics, especially p-xylene, is still a great challenge due to the inert nature of carbon dioxide and high barrier for C-C coupling. The process of carbon dioxide hydrogenation to aromatics can be divided into two types according to different intermediates.

One way is the conversion of carbon dioxide to aromatics via olefin intermediates over a composite catalyst: the Fischer-Tropsch synthesis catalyst combined with HZSM-5 zeolite. Typically, carbon dioxide was hydrogenated to the C_2+ hydrocarbons through the reverse water gas shift (RWGS) reaction and the Fischer-Tropsch synthesis reaction over an iron-based catalyst. Alkali

Converting Power into Chemicals and Fuels: Power-to-X Technology for a Sustainable Future, First Edition. Martin Bajus.

promoters are usually added in a Fe-based catalyst to improve the adsorption of carbon dioxide and selectivity of the hydrocarbon products. Notably, a sodium promoter is used because it facilitates the conversion of carbon dioxide and inhibits the formation of methane. Consequently, the use of a bifunctional catalyst comprising Na-modified iron oxide (Fe_3O_4) and HZSM-5 zeolite for the efficient and direct conversion of carbon dioxide to aromatics is significant.

A low proximity of the two components, a suitable mass ratio of iron/zeolite and a high density of total acidity over the bifunctional catalyst with Na-Fe and hierarchical HZSM-5 zeolite are decisive factors for high selectivity towards aromatics and high catalytic stability. Using the single-step selective conversion of carbon dioxide to aromatics over Na-Fe_3O_4/hierarchical HZSM-5 zeolite catalyst a yield of aromatics as high as 23.5%, with a carbon dioxide conversion of 45.3% (Table 9.1) was obtained. The results indicate that a decrease in the proximity of the two components could facilitate the conversion of carbon dioxide and inhibit product conversion into light hydrocarbons caused by the rapid transfer of intermediate products to zeolite. In addition, the high total acidity of hierarchical HZSM-5 zeolite is conducive to the formation of aromatics. This work offered a highly efficient, bifunctional catalyst to the utilization of carbon dioxide and exhibited a broad industrial-application prospect of the direct conversion of carbon dioxide to aromatics.

This indicated that the further aromatization of olefin intermediates on zeolite requires a high density of total acidity. As Table 9.1 shows, the production of aromatics mainly included methyl

Table 9.1 Hydrogenation of carbon dioxide to aromatics over Na-Fe hierarchical nanocrystalline HZSM-5.

Reaction Conditions	
Temperature, °C	340
Pressure, MPa	3.0
Ratio of H_2/CO_2 in the feed gas	3.3
Mass ratio of iron/zeolite, g/g	1.0
Results	
Conversion of CO_2, %	45.3
Selectivity to CO, %	11.3
Yield of aromatics, %	23.5
Hydrocarbon Distribution %	
CH_4	9.4
$C_2^= - C_4^=$ (olefins)	8.1
$C_2^0 - C_4^0$ (alkanes)	16.5
C_5^+, excluding aromatics	9.8
Aromatics in oil phase, %	88.3
Composition of Aromatic Fraction, %	
Toluene	10.3
Xylenes	26.8
Ethylbenzene	5.2
Trimethybenzenes	14.8
Methyl ethylbenzenes	22.5
Propyl benzenes	–
C_{10}^+– aromatics	22.5

ethylbenzenes and xylenes, followed by trimethylbenzenes, toluene, ethylbenzenes and propyl benzenes. For example, the components of C_{10+} aromatics were numerous, such as diethyl benzene, tetramethyl benzenes, methylpropyl benzenes and so on. As the density of total acidity decreased, the selectivity of long carbon chain aromatics products increased. This could be due to the decrease in acidity which weakens the cracking ability of the HZSM-5 zeolite.

Another process is the production of aromatics from carbon dioxide by integrating carbon dioxide hydrogenation to CH_3OH/DME and CH_3OH/DME to aromatics, i.e., Youming et al. reported an $ZnAlO_x$/HZSM-5 bifunctional catalyst that could directly convert carbon dioxide to aromatics with a selectivity of 73.9% in hydrocarbon products at a carbon dioxide conversion of 9.1%. Li et al. demonstrated direct conversion of carbon dioxide to aromatics by a tandem catalyst combining the Zn-ZrO solid solution and HZSM-5 zeolite and carbon dioxide conversion increased to 14% with aromatic selectivity of 73% in hydrocarbons. However, the yield of aromatics is below 10% due to low carbon dioxide conversion and high carbon monoxide selectivity, which is not economically beneficial in the industry. Therefore, it is still a challenge to obtain excellent catalytic activity and low carbon monoxide selectivity.

9.1.1 Methanol to Aromatics Process

The procedure for methanol to aromatics (MTA) was initiated in the 1970s by Exxon Mobil Company. Later on, in 1985, it was much commercialized in New Zealand. The H-ZSM-5 catalyst was used in MTA reaction and can produce up to 10 carbon chain products. Methanol to hydrocarbons (MTH) production involves the following steps (Equations 9.1 and 9.2).

$$2CH_3OH \underset{+H_2O}{\overset{-H_2O}{\rightleftarrows}} CH_3OCH_3 \xrightarrow{-H_2O} \text{Light olefins} \rightarrow \begin{array}{c} \small n/iso\text{–Paraffins} \\ \small \text{Higher olefins} \\ \textbf{\textit{Aromatics}} \\ \small \text{Naphthene} \end{array} \tag{9.1}$$

A consecutive reaction was carried out between methanol and benzene or toluene. It can be seen that dimethyl ether is first obtained by dehydration of methanol. The equilibrium mixture of methanol, dimethyl ether and water is then converted to light olefins, whereas dimethyl ether acts as an intermediate product responsible for the conversion of light olefins into alkyl aromatics and is paraffins via hydrogen transfer, alkylation, isomerization, other adsorbate, secondary reactions and oligomerization.

The consecutive reaction scheme (Equation 9.2) depicts the hydrocarbon-pool mechanism as initially proposed by Dahl and Kolboe. The consecutive reaction mechanism shows the hydrocarbon-pool mechanism for alkenes' low reactivity in H-SAPO-34. The hydrocarbon pool acted as an adsorbate, considered to have properties like ordinary coke and was considered to have less hydrogen.

It was found that a major part of the propene was formed from arene or some arene derivates. Arstad and Kolboe investigated the stability of species trapped inside the H-SAPO-34 cavities during methanol conversion and methyl benzenes were found to be the main components of the hydrocarbon pool.

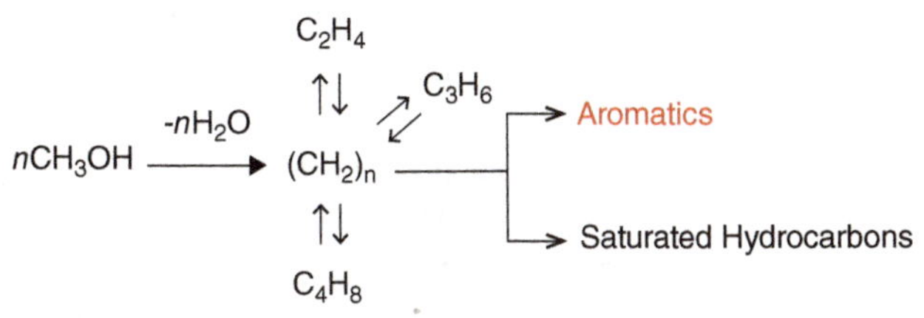

(9.2)

The consecutive reaction scheme (Equation 9.2) depicts the hydrocarbon-pool mechanism as initially proposed by Dahl and Kolboe.

Acid-based catalyst reactions are also involved in the MTA process, as shown in Equation 9.3. ZSM-5 catalyst has optimal acidic sites and pore diameter allowing methanol conversion into BTX. ZSM-5 catalyst was modified using La, Ce, Ga, Mo, Zn and Ag as the promoters, which showed a response for the MTA process.

$$CH_3OH \rightarrow (R - C_6H_5) + H_2O \tag{9.3}$$

The pore size of ZSM-5 was 0.55 nm and is suitable for the transformation of MTG hydrocarbons. The acidity sites are altered by modifying the zeolites and new pores are created. The use and treatment of the H-ZSM-5 (Si/AL = 18) with NaOH gradually diminishes the Al in the catalytic framework to improve catalyst stability; as shown its selectivity was found to be best towards aromatics.

ZSM-5 catalyst is treated by mild hydrochloric acid and sodium hydroxide solution. An increase in the Lewis acid sites after treatment with hydrochloric acid increases the selectivity of p-xylenes as well as improvement in the stability of the catalyst. Ga-HZSM-5 catalyst treated with 0.05 M sodium hydroxide solution gave the aromatics the highest selectivity (60.1%) at 773 K.

MCM-22 catalyst was impregnated with different amounts of phosphorous from 1, 2, 3 and 4 wt % using wet impregnation and treating with diammonium phosphate solution for the MTA reaction. 3 wt % P-H-MCM-22 showed 70% selectivity of aromatics and longer catalytic lifetime (40 h) at 723 K as compared to the other catalyst.

Zirconium, cerium, molybdenum and chromium were impregnated over the Zn-modified HZSM-5 catalyst for the MTA reaction. The higher the methanol conversion, the higher is the aromatics yield, which is achieved through the chromium modified Zn/HZSM-5 catalyst. Chromium helps in enhancing the BTX yield and enhancing the stability using Cr-Zn/HZ. Yttrium and zinc doped ZSM-5 catalyst showed a better yield of olefins and BTX selectivity in the presence of the carbon dioxide environment.

Kinetics and density functional theory (DFT) calculations were also employed to identify first carbon-carbon bond formed during conversion methanol to hydrocarbons in the presence of zeolite catalysts. DFT calculations authenticate the feasibility of the reaction pathways.

9.1.1.1 ZSM-5 Catalyst

ZSM-5 catalyst with varied Si/Al ratios, such as 35, 70, 142, 500 and 1670, is used during the MTO process. The Si/Al ratio significantly reduces the rate of deactivation and improves the catalytic activity. The nano-sized ZSM-5 was treated with 0.1 M sodium hydroxide solution for the MTA reaction. An increase in the catalyst stability and aromatics at 4.7 in an hour.

The MTA reaction was performed in a fixed bed reactor using a nano-ZSM-5 (NZ) catalyst with different Si/Al ratios ranging from 24, 29, 30, 38, 41 and 47. The maximum liquid yield was 29.5% and total aromatics selectivity was 48.5%.

Another important aspect in discovering the overall catalytic performance, such as conversion, selectivity, yield and catalyst stability, is the catalyst's pore size. Nano-ZSM-5 mesoporous structure enhanced the reaction due the reduction in diffusion path length, larger surface area, improving catalytic activity and increasing its durability. Micro-pore HZSM-5 catalyst produce C_6^+ aromatics and paraffins yield, but, on the other hand, lower catalytic stability. The aromatic and alkenes formed within the pores of the catalyst.

Na-ZSM-5 catalyst (Si/Al = 21) used during the MTA reaction and in the present study, two methods were used to change the pore size of the catalyst. The desilication stabilization (DS) method increased the parent zeolite pore size to 0.59 nm from 0.53 nm on treating ammonium chloride with the HZSM-5 catalyst. The desilication stabilization method enhances the production of aromatics from methanol at 748 K. The nano-ZSM-5 catalyst showed maximum selectivity for lighter olefins and the meso-ZSM-5 catalyst represents the maximum yield of alkyl aromatics during MTA reaction. A full conversion of methanol was achieved in the case of nano-ZSM-5, while with meso-ZSM-5, only 70% conversion occurred.

ZSM-5 catalyst was treated with different amounts of sodium hydroxide solutions ranging from 0.2, 0.3, 0.4 and 0.5 mol/l to investigate the effect on the behaviour of acidity and methanol to aromatic transformation. It was found that there was no change in the acid sites and it resulted in 100% methanol conversion. HS200-10 (HZSM-5) catalysts exhibited full conversion and excellent stability due to the lower number of acidic sites than in HS900-10 (micro-ZSM-5). HS900-30 (H_2O/Si = 30 ml/g) showed the highest catalytic stability of 192 hrs and produced 49.11% of aromatics.

The effect of various metal-doped over ZSM-5 catalysts is investigated for conversion of methanol to aromatics. The metal impregnation reduces the number of acidic sites of the catalyst, thereby increasing the catalytic stability, activity and desired product yields. The catalytic performance can be effectively improved in the presence of a second metal.

A bimetallic catalyst using zinc and tin weight percentage doped over the ZSM-5 catalyst at T = 723 K, P = 0.1 MPa and WHSV = 0.8 hours MTA reaction. 1wt % Zn-1wt % Sn doped over the ZSM-5 gives the highest aromatics value up to 64.1% of BTX, while the methanol conversion remains at 100%. Methanol was completely transformed into hydrocarbons, while the aromatics yield was 30%, at 0.8% of iron. An increase in the amount of iron, led to a decrease in the performance of the catalyst and the selectivity of aromatics. Gallium doped ZSM-5 (with ratio from Ga/Al from 0.25, 0.5, 0.75, 1.0) was studied at T = 743 K, P = 100 kPa, WHSV = 4 h for MTA reaction.

A CuO/NH_4-ZSM-5 catalyst with 3, 5, 7 and 9 wt % of copper oxide was studied for the MTA reaction at T = 673 K, P = 100 kPa and 100% conversion of methanol occurred, while aromatics selectivity was 49% using 7 wt % CuO/NH_4-ZSM-5. The gadolinium with 1, 5, 10 wt % was impregnated over HZSM catalyst for the MTA process. All catalysts give 100% methanol conversion. The GdHZ_IE catalyst contains the highest amount of Lewis acid sites due to a higher selectivity for aromatics (35.2%) compared to other catalysts. The formation of polycondensation aromatic enhanced catalyst stability. It was achieved by fabrication with gadolinium over the HZSM-5 catalyst, which reduced the coke formation by 50%. Zinc metal-doped over the ZSM-5 catalyst process in the MTA process at T = 723 K and WHSV = 4.74 h^{-1}. The distribution of zinc oxide species improved the BTX selectivity due to the presence of stronger acidic sites. The maximum selectivity of BTX was 32.5% for the duration of nine hours. After nine hours on stream, the catalyst became deactivated due coke formation, which arises because of the zinc oxide cluster in the channels of ZSM-5, thereby decreasing the formation of the aromatics.

9.1.1.2 Process Variables

The catalyst deactivation rate depends on the process's highest temperature, which varies linearly and increases temperature. NaOH treated the HZSM-5 catalyst with molarity of 0.05 M and 0.20 M for methanol to an aromatic reaction at 643 K. The 0.20 M sodium hydroxide solution showed methanol conversion of 93% and higher yield of hydrocarbons.

The reaction does not affect the catalytic structure, while the lighter olefins selectivity, enhanced to 95% and there was a 100% conversion of methanol. HZSM-5 (Si/Al = 84) catalyst modified by zirconium oxide and phosphoric acid at 723 K to enhance the selectivity of propylene to 45%.

Researchers conducted experiments to find out the effect of pressure during the Methanol-to-Hydrocarbon (MTH) reaction. The lower pressure tends to increase the number of lighter hydrocarbons, while at high pressure, the C_5^+ hydrocarbons are the major products. High pressure helps in the production of aromatics in the methanol to hydrocarbons reaction. The disadvantage of high pressure is that it induces the coke formation over the catalyst surface and in the pores, hence lowering the catalyst stability. Ag-ZSM-5, Zn-ZSM-5 and HZSM-5 catalysts were used during MTA reaction at pressure of 20 kPa. Ag-ZSM-5 catalyst showed the highest aromatic yield (72.5%) due to Ag^+ ions present in the catalyst, while Zn-ZSM-5 and HZSM-5 showed 68% and 48.4%, respectively.

0.5 wt % Zn/nano-HZSM-5 catalyst at 0.1 MPa gives 100% methanol conversion and maximum yield of aromatics up to 51.7%. HSSZ-13 (Si/Al = 13.2), HSSZ-39 (Si/Al = 9), HFER (Si/Al = 10.0) and HBEA (Si/Al = 12.4) catalysts were used for MTA reaction at T = 673 K and P = 1.6 MPa. HBEA catalyst enhanced the aromatics selectivity up to 40% and increased the catalytic lifetime.

The addition of zinc improved the overall catalytic activity due to improved stronger acidic sites, leading to more active sites. The nano-size of the catalyst increases the deactivation period and the diffusion of reactants/products. The use of zinc modified TNU-9 catalysts for the methanol conversion into aromatics was investigated. Zn-TNU-9 catalyst showed the maximum aromatics selectivity and methanol conversion of 34.88% and 99.85% respectively.

The conversion of methanol or dimethyl ether to hydrocarbon is initially slow at low weight hourly space velocity (WHSV), while increasing the WHSV leads to a rapid increase in conversion. Increasing space velocity may deteriorate the product selectivity and also reduce conversion of the reactant.

Zn-Cr/HZSM-5 catalyst was used to co-aromatize methanol and n-hexane at the n-hexane/methanol ratio of 3:7, T = 723 K and WHSV varied from 1–4 hours. With the increase in WHSV, the n-hexane conversion decreased significantly, with an increase in C_3 to C_5 range hydrocarbon yields. At lower WHSV, BTEX's yields (benzene, toluene, ethylbenzene and xylenes) was enhanced, especially xylenes and benzene.

MCM-22 and MCM-36 catalysts were used to react methanol to aromatics at T = 723 K, WHSV = 13 hours. The initial transformation of methanol was approximately 100% with both catalysts. However, the catalyst stability strongly depends on the WHSV. The lifetime of the MCM-36 catalyst was found to be 18 hours and that of MCM-22 was 11.8 hours, at WHSV = 1 hour. The lifetime of MCM-22 and MCM-36 sharply decreased along with an increased WHSV. With MCM-36 at WHSV = 3 hours, methanol's conversion reduces rapidly at the initial reaction time. With the MCM-22 catalyst, the number of C_5^+ (35%) hydrocarbons was higher than the MCM-36 catalyst of 22%. Calcium doped ZSM-5 catalyst was used to produce lighter olefins from methanol at a temperature ranging from 673 to 773 K and WHSV ranging from 4.76, 10 and 14.28 hours. The total olefins selectivity was reduced drastically from 70 to 25%, due to the shorter contact time of methanol (WHSV = 4.76 to 14.28 hours at 773).

9.1.1.3 Kinetic Modelling

The selection of the appropriate reactor and the type of operation to be performed depend upon kinetic expression. The models are divided into two main classes depending on the reaction's kinetics: (a) detained parameter model, which considers individual reaction steps and (b) lumped parameter model, which takes into account complex reaction systems in the concentration of various lumps (Chapter 1). In general, the detailed parameter model takes time for the reaction to complete. For design purposes, lumped parameter models are mainly employed. The kinetic equations require different process variables such as temperatures, pressures and WHSV.

For minimizing the objective function (Equation 9.4) the arithmetic is calculated for different lumps:

$$f = \sum_{i=1}^{n} \left(C_{\text{experimental}} - C_{predicted} \right)^2 \tag{9.4}$$

where f is function, $C_{\text{experimental}}$ stands for experimental and $C_{\text{predicted}}$ stands for predicted.

Chen and Reagan proposed the lumped parameter kinetic model based on such hydrocarbons production as aromatics, paraffin and olefins from oxygenates (methanol + dimethyl ether) using the catalyst ZSM-5 to develop the kinetic model.

Li et al. proposed a simple MTA reaction with an 8-lump parameter model. It is difficult to determine the rate constant obtained in the complex or detailed kinetic model by considering all possible reactions. Recently more reliable solutions can be achieved using a lumped component to develop kinetic modelling of MTO and MTP process. Olefins were the intermediate products of first-order reaction to form paraffin and BTX during the parallel reaction. The 7th reaction (Equation 9.5) represents the formation of coke obtained from aromatics and methanol. As mentioned below, the reaction scheme gives information about the conversion of methanol and the production of olefin and paraffin and various aromatics individually (Table 9.2).

Simulation modelling has been proposed and used to predict the reactor's behaviour for the conversion of methanol into hydrocarbons. The lumped kinetic model proves to be helpful for a large reaction system into model. It helps to analyse the effect of process conditions with respect to an increase in time on stream. The proposed model (Equation 9.5) performs excellently because of the improvements made in kinetic models used to describe the methanol to gasoline, methanol to hydrocarbons, methanol to aromatics and methanol to olefins processes.

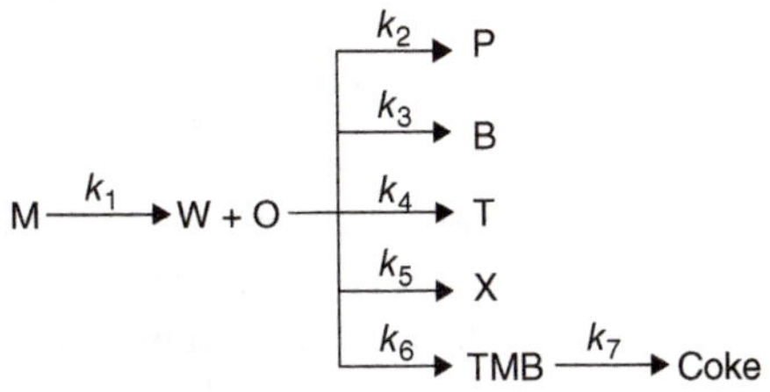

(9.5)

Here M, W, O, P, T, X, TMB, coke and k_i represent methanol, water, olefins, paraffins, benzene, toluene, xylenes, trimethylbenzenes and kinetic constant of the ith individual lump (hours), respectively.

9.1.1.4 Aromatics via Hydrogen-Based Methanol (TRL7)

The production pathway considered here is again based on methanol. The process sequence is depicted in Figure 9.1

Mobil has a methanol-to-aromatics process in which methanol is converted to a range of aromatic compounds using a zeolite catalyst at 370–540°C and 2 to 2.5 MPa. Compared to the methanol to olefins the temperature is lower and higher catalyst acidity is required. Conversion is at 95–100% with aromatics yield of 60–70% of which 80% are BTX, resulting in a total BTX yield of 56%.

Table 9.2 Product spectrum of methanol-to-aromatics over HZSM-5 catalyst.

Product	Wt %	Weighted
Benzene	4.1	0.34
Toluene	25.3	2.51
o-Xylene	9.0	1.01
m-Xylene	22.8	2.55
p-Xylene	10.0	1.12
Total	**71.5**	**7.53**

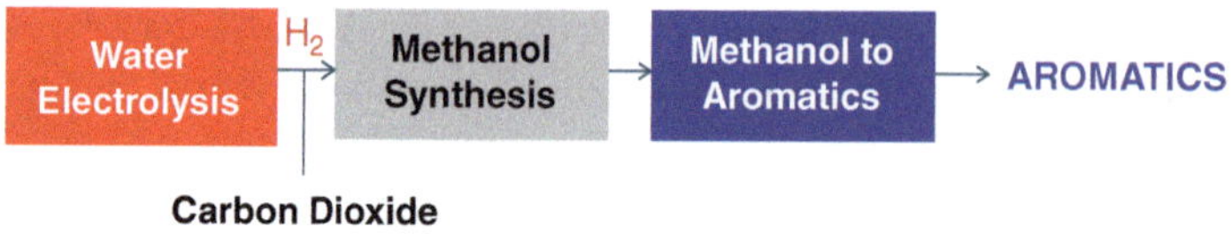

Figure 9.1 Low-carbon production of BTX via hydrogen-based methanol.

Table 9.2 shows a typical product spectrum of methanol to aromatics process, indicating that on average 7.5 methanol molecules are stoichiometrically required per unit of aromatics produced. Considering the total BTX yield of the process, the methanol demand of the process is 4.3 tons of methanol per ton of BTX.

9.1.1.5 Energy Demand per Unit of Low-Carbon BTX Production

Energy demand of the low-carbon production sequence is largely determined by the energy demand of the hydrogen-based methanol production. As 4.3 t methanol/t BTX are required, the energy demand amounts to 171 GJ (or 47.5 TWh) per ton of BTX. The energy demand for the methanol to aromatics process is assumed to be close to the methanol to olefins process, i.e., 5 GJ/t, summing up to 176 GJ/t. This is to be compared to the BTX route from naphtha, which consumes 7 GJ/t.

9.1.1.6 Carbon Dioxide Reduction

The Methanol-to-Aromatics process emissions are higher than those of the naphtha-based process, i.e., 1.13 t carbon dioxide eq/t BTX compared to 0.55 t carbon dioxide eq/t HVC. The major impact comes from the carbon dioxide feedstock demand and footprint of the hydrogen-based methanol. The required 4.3 t methanol per t BTX account for −2.91 t ton carbon dioxide. Taking the production into account, the total carbon footprint of the process chain from water and captured carbon dioxide to methanol and further to BTX via the methanol to aromatics process amounts to −1.7 t, which is the avoided carbon dioxide per ton BTX produced via this pathway.

9.1.1.7 Economics of Low-Carbon BTX Production

As low-carbon methanol is the feedstock for the BTX process described in Section 8.1.1.6, the cost of methanol production and resulting feedstock costs largely determine the cost of BTX production. For methanol production costs from 300 to 650 €/t, feedstock costs will amount to 1300 to 2800 € per ton of BTX, if methanol costs are only based on production costs, i.e., without additional margins. Production costs are accordingly extremely high, a factor of three and higher is to be expected compared to the existing fossil-based technology.

9.2 Production of p-Xylene from 2,5-Dimethylfuran and Ethylene

p-Xylene is a commodity chemical of industrial importance for terephthalic acid production, for which renewable sourcing from naturally abundant lignocellulosic biomass is highly desired. Gulbinski et al. demonstrated that phosphoric acid stabilized on siliceous zeolite supports (e.g., P-BEA, P-SPP) exhibits high selectivity towards p-xylene (>97%) from 2,5-dimethylfuran (DMF) and ethylene. However, the effect of the support and the contribution of heterogeneous versus homogeneous phosphoric acid on the observed catalytic behaviour in the solvated reaction system have not been addressed. Here, we determine the phosphoric acid catalytic activity for 2,5-dimethylfuran conversion and its selectivity to p-xylene when it is supported on a silica support as well as in the absence of a support.

Phosphoric acid catalysis was studied in three different scenarios:

1) phosphoric acid was added in the liquid reaction mixture in the absence of any solid support;
2) phosphoric acid was added in the liquid reaction mixture along with inert silica support including siliceous zeolite (i.e., allowing for phosphoric acid-support assembly to proceed in the reaction mixture); and
3) phosphoric acid was first impregnated on the siliceous zeolite support and then the preassembled supported phosphoric acid catalyst was added to the liquid reaction mixture.

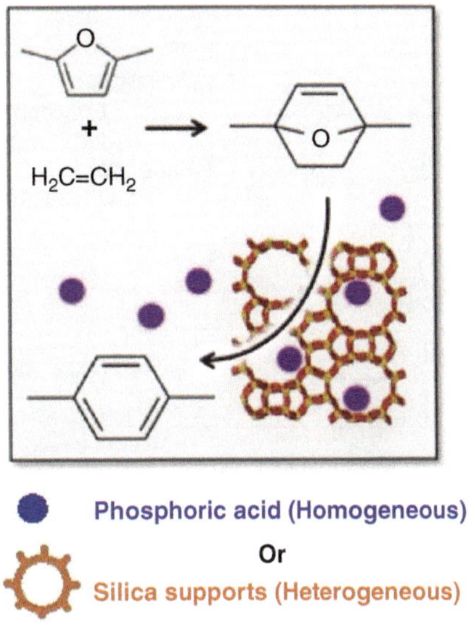

Figure 9.2 Reaction scheme production of p-xylene from 2,5-dimethylfuran and ethylene.

It was found that the reaction rate and selectivity to p-xylene are different in these scenarios, reflecting the effect of the solid support on the catalytic performance of phosphoric acid (Figure 9.2). In scenario 1, a low concentration of phosphoric acid (1.7 mM) in the absence of any solid support exhibited high selectivity to p-xylene (80% selectivity to p-xylene at 60% conversion of DMF), which decreased with increasing acid concentration. The selectivity for p-xylene and activity of phosphoric acid was significantly increased by adding a silica support into the reaction system (scenario 2). This improvement was attributed to phosphoric acid partial association with the surface of the silica support under the reaction conditions (in situ catalyst assembly). Phosphoric acid redeposited on siliceous zeolite supports (e.g., P-BEA, P-SPP) synthesized via impregnation prior to the reaction (scenario 3) catalysed the reaction heterogeneously without noticeable leaching and exhibited the highest activity and selectivity to p-xylene, suggesting an important role of the silica support and the need to ensure that phosphoric acid acts as a heterogeneous catalyst in order to accomplish selective conversion of 2,5-dimethylfuran to p-xylene.

9.3 Carbon Dioxide Dehydrogenation of Ethylbenzene to Styrene

Dehydrogenation of hydrocarbons is an endothermic and equilibrium limited reaction. The conversion of the hydrogen produced to water by oxidation, shifts the reaction to higher yield of the product. For example, dehydrogenation of ethylbenzene (EB) to styrene (Equation 9.6).

$$C_6H_5C_2H_5 \rightleftarrows C_6H_5C_2H_3 + H_2 \quad \Delta H^\circ_{298} = 118\ \text{kJ/mol} \tag{9.6}$$

This reaction can be performed in the presence of carbon dioxide as the oxidant (Equation 9.7).

$$C_6H_5C_2H_5 + CO_2 \rightleftarrows C_6H_5C_2H_3 + CO + H_2 \quad \Delta H^\circ_{298} = 159.2\ \text{kJ/mol} \tag{9.7}$$

Carbon dioxide also acts as the heating medium and diluent. Its advantages over conventional technologies using steam as the diluent include a higher heat capacity compared to steam as heat supplier, higher selectivity to styrene (97% to styrene) and lower cost compared to steam and oxygen. The major disadvantages are endothermicity of reaction, higher operating temperatures and catalyst deactivation. It has been estimated that the energy required for the dehydrogenation of

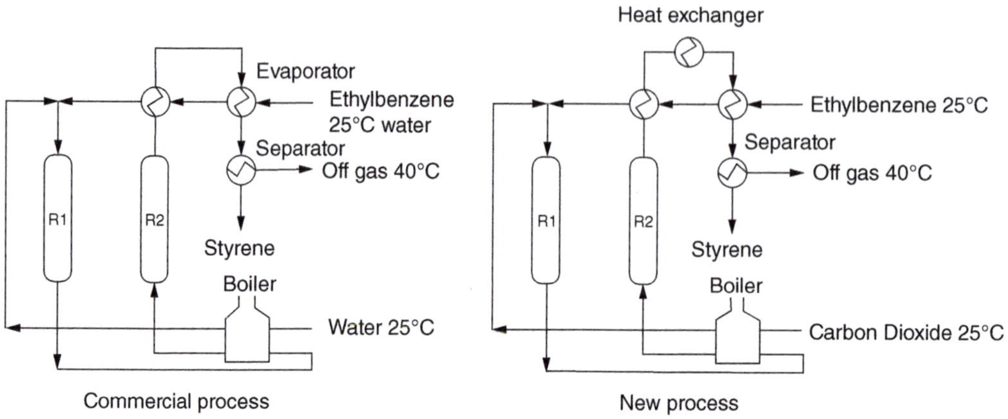

Figure 9.3 Flow diagrams of a commercial process and a new process using carbon dioxide for dehydrogenation of ethylbenzene to styrene.

ethylbenzene to produce styrene using carbon dioxide over Fe_2O_3-Al_2O_3 catalyst and the commercial process using steam is 2.6 GJ/t styrene and 6.3 GJ/t styrene respectively (Figure 9.3), which indicates that the new process using carbon dioxide should be an energy-saving process. The energy required for separation of the mixture of ethylbenzene and styrene presents one of the advantages of using carbon dioxide instead of steam.

The conventional Fe-K catalyst exhibits poor performance in carbon dioxide dehydrogenation of ethylbenzene. However, several catalysts were reported to be effective in dehydrogenation of ethylbenzene in the presence of carbon, such as Fe_2O_3/γ-Al_2O_3, V/MgO, Li/Fe_2O_3/active carbon (AC) and Sb/V_2O_5/AC and iron-doped titanium oxides; most of which are supported catalysts and some are mixed oxide catalysts prepared by an impregnation or deposition method. Carbon dioxide may play a role in the abstractraction of hydrogen from hydrocarbons through the catalytic activation of carbon dioxide to form carbon oxide and oxygen species.

Further Reading

Ali, S. and Zaidi, H.A. (2020). Experimental and kinetic modelling studies of methane transformation to hydrocarbons using zeolite-based catalyst: a review. *Energy Fuels* 34 (11): 13225–13246.

Andrews, J., Hung, J., Molinier, M. and Lim, Y.T. (2020). Catalytic processes operation during downturns in aromatics complexes. *Hydrocarbon Processing*, September 2020.

Arstad, B. and Kolboe, S. (2001a). Methanol-to-hydrocarbons reaction over SAPO-34. Molecules confined in the catalyst cavities at short time on stream. *Catal. Lett.* 71: 209–212.

Arstad, B. and Kolboe, S. (2001b). The reactivity of molecules trapped within the SAPO-34 cavities in the methanol-to-hydrocarbons reaction. *J. Am. Chem. Soc.* 123: 8137–8138.

Chen, N.Y. and Reagan, W.J. (1979). Evidence of autocatalysis in methanol to hydrocarbon reactions over zeolite catalysts. *J. Catal.* 59 (1): 123–129.

Dahl, I.M. and Kolboe, S. (1994). On the reaction mechanism for hydrocarbon formation from methanol over SAPO-34. 1. Isotopic labelling studies of the co-reaction of ethene and methanol. *J. Catal.* 149 (2): 458–464.

Dahl, I.M. and Kolboe, S. (1996). On the reaction mechanism for hydrocarbon formation from methanol over SAPO-34. 2. Isotopic labelling studies of the co-reaction of ethene and methanol. *J. Catal.* 161 (1): 304–309.

Gulbinski, J., Ren, L., Vattipalli, V. et al. (2020). Role of silica support in phosphoric acid catalyzed production of p-xylene from 2,5-dimethylfuran and ethylene. *Ind. Eng. Chem. Res.* 59 (51): 22049–22056.

Li, H., Li, X.G. and Xiao, W.D. (2019). Deactivation kinetics of individual $C_6 - C_9$ aromatics' generation from methanol over Zn and P commodified HZSM-5. *RSC Adv.* 9: 22327–22336.

Li, Z., Qu, Y., Liu, H. and Li, C. (2019). Highly selective conversion of carbon dioxide to aromatics over tandem catalysts. *Joule* 2 (3): 570–583.

Mhatre, S., Warke, V., Molinier, M. and Saple, A. (2020). A new liquid-phase isomerization process for xylene loop debottlenecks and energy saving. *Hydrocarbon Processing*, January 2020.

Wang, Y., Tan, L.T.M., Zhang, P. et al. (2019). Rationally designing bifunctional catalysts as an efficient strategy to boost CO_2 hydrogenation producing value-added aromatics. *ACS Catal.* 9 (2): 895–901.

Wen, Ch., Jiang, J., Chiliu, C. et al. (2020). Single-step selective conversion of carbon dioxide to aromatics over Na-Fe_3O_4/hierarchical HZSM-5 zeolite catalyst. *Energy Fuels* 34: 11282–11289.

Youming, N., Zhiyang, C., Yi, F. et al. (2018). Selective conversion of CO_2 and H_2 into aromatics. *Nat. Commun.* 9 (1): 3457.

Zangeneh, F.T., Sahebdelfar, S. and Ravanchi, M.R. (2011). Conversion of carbon dioxide to valuable petrochemicals: an approach to clean development mechanism. *J. Nat. Gas Chem.* 20: 219–231.

10

Power-to-C_1 Chemicals

CONTENTS

Converting Power into Chemicals and Fuels: Power-to-X Technology for a Sustainable Future, First Edition. Martin Bajus.

10.1 Introduction

After many years of research and development efforts technologies for carbon dioxide utilization are emerging. Carbon dioxide can be considered as a valuable and renewable carbon source. One approach to reduce carbon dioxide emissions could be its capture and recycle via transformation into chemicals using the technologies in C_1 chemistry. Despite great interest, there are difficulties in carbon dioxide separation on the one hand and the thermodynamic stability of the carbon dioxide molecule (Chapter 1) rendering its chemical activity low on the other hand. In carbon dioxide or the dry reforming of methane, carbon dioxide serves both as the oxidant and a carbon source. The reaction is highly endothermic and equilibrium limited. Both methane and carbon dioxide, which are greenhouse gases, are consumed and the carbon content of the synthetic gas product is higher based on the natural gas consumed. The utilization of carbon dioxide does not necessarily involve development of new processes and it is already used in certain processes such as methanol synthesis and methane steam reforming. Methanol and dimethyl ether (DME) synthesis are potentially the most important heterogeneous hydrogenation reactions of carbon dioxide. In other processes, modifications in catalyst and/or processes, or even new catalysts and processes, are necessary. In either case, catalysis plays a crucial role in carbon dioxide conversion and effective catalysts are required for commercial realization of the related processes.

Carbon dioxide has turned out to be a strategic carbon resource for the synthesis of valued chemicals rather than just a greenhouse gas. Utilization of carbon dioxide as a feedstock for the synthesis of chemicals is emerging as a complementary alternative to fossil-derived petrochemicals and carbochemicals. Proposals on carbon capture and utilization (CCU) are receiving increasingly attention worldwide. Lower olefins as basic carbon-based A pillars of petrochemistry (Chapter 1), commonly referring to ethylene, propylene and butylenes, are mainly produced from cracking of hydrocarbon feed stocks like naphtha and dehydrogenation of light alkanes. Synthesis of lower olefins is a practical solution to carbon dioxide storage as olefins can be stored in the form of polymer materials. Although hydrogenation of carbon monoxide to lower olefins, alkenes and aromatics (Pillar B) has been made, progress, as far as I know, towards the hydrogenation of carbon dioxide to lower olefins with high selectivity is rarely reported (Pillar C).

Produce carbon-based products, mostly derived from fossil carbon and they will ultimately need to be decarbonized. Three different routes – carbon capture and storage (CCS route), the use of captured carbon dioxide (CCU route) and a bio route – were qualitatively compared by Gabrielli et al. to produce a generic chemical product and quantitatively compared for the case of methanol production. This section provides a conceptual analysis and a quantitative comparative assessment of three technology chains that enable a carbon neutral chemical industry in a net-zero-CO_2 world. These are based (i) on the use of fossil fuels and current chemical processes and infrastructure coupled with carbon capture and storage; (ii) on the use of captured carbon dioxide as a feedstock together with "green" hydrogen in new power-to-X chemical processes; and (iii) on the use of biomass grown and processed for the specific purpose of making chemicals (BIO route). All routes are feasible and have different pros and cons. Such pros and cons are first evaluated through a qualitative comparison of the three routes for a generic chemical product and are then quantitatively assessed for the specific case of methanol production. In this case, the CCU route results in an electricity consumption 10 to 25 times higher than that of the CCS and BIO routes (excluding the electricity required for heat production), mostly due to the electricity required to produce hydrogen. At the same time, the BIO route requires a land capacity about 40 and 400 times higher than that required by the CCU and CCS routes, respectively. Furthermore, when considering a net-positive-CO_2 emissions world, the carbon dioxide emissions of the CCU route grow about 8 to 10 times faster than that of the CCS and BIO routes. On one hand, they identify key hurdles in all cases. These are (i) the availability, accessibility and acceptance of carbon dioxide storage sites for the CCS route, together with the continued use of fossil fuels; (ii) the very high electricity and energy demand for the CCU route, with the associated strict requirement of very low carbon-intensity of the electricity mix; and (iii) the very high availability of land for biomass growth in the case of the BIO route, with the associated risks of conflict with other uses. On the other hand, they underline that the CCS route offers the possibility of using existing technologies and infrastructures, without the need of a complete reshaping of chemical technologies and of permanently removing carbon dioxide from the atmosphere, hence representing a key element not only in the net-zero-CO_2 emissions world studied here, but also in a net-negative-CO_2 emissions world.

A clear winner was not identified and all routes proved to be feasible and had different pros and cons, with key hurdles being identified for each route. A scenario analysis of CCUS supply chains was specifically performed for Italy and Germany. They minimized an objective function that consisted of remission of an element of carbon tax, economic incentives and revenues subtracted from total supply chain costs, to determine optimal designs for CCUS systems.

Currently, main carbon dioxide utilization research areas include (1) thermochemical conversion and hydrogenation routes, (2) electrochemical and photochemical conversion, (3) carbon dioxide conversion to solid carbonates, (4) biological conversion routes and (5) enhanced hydrocarbon recovery with carbon storage. The development of these technologies to ensure net carbon dioxide emissions is a challenge and there are still many fundamental and technological aspects to be solved. For instance, the development of effective thermal and (photo)electrochemical catalytic reaction pathways, the understanding of the kinetic mechanism for minerals and industrial waste systems for inorganic carbonate formation, or the acceleration of biological carbon dioxide conversion routes, among others.

In routes (1) and (2) above, active research is currently being conducted to develop green solid heterogeneous catalysts that can contain multiple active sites and efficiently convert carbon dioxide to high-value-added products. Zhang et al. contributed to these efforts by reporting a green bifunctional heterogeneous catalyst that can be used for the formation of cyclic carbonates through the carbon dioxide cycloaddition reaction. Similarly, efficient carbon dioxide electrochemical reduction under mild conditions also requires high-performance catalysts. Different catalyst electrodes (Sn/Cu, BiSn/Cu, Bi_2Sn/Cu, Bi_3Sn/Cu, Bi_4Sn/Cu and Bi/Cu) were fabricated by electrodeposition and their activity, stability and selectivity towards carbon dioxide reduction to formic acid was measured. In addition, insights into the proposed mechanism that is catalysed by the electrodes were provided by density functional theory (DFT) calculations. DFT modelling was also the tool used to understand the reaction mechanisms of the direct catalytic conversion of carbon dioxide and methanol to dimethyl carbonate (DMC).

Various cyclic carbonates were prepared by cycloaddition of carbon dioxide with different epoxides using imidazolium-based catalysts grafted on a mesoporous silica support through 3-aminopropyltrimethoxysilane and Br–$(CH_2)_n$–Br ($n = 2$ or 4) under efficient and mild conditions. The catalysts were characterized by various physicochemical techniques such as magic angle spinning (MAS) NMR, ^1H NMR, Fourier transform infrared spectroscopy (FT-IR), X-ray diffraction (XRD), field-emission scanning microscopy (FESEM), thermogravimetric analysis (TGA) and Brunauer–Emmett–Teller (BET). The reactions were conducted under maximum carbon dioxide pressure of 2.0 MPa at the temperature range of 80–140°C without using any solvent. The influence of different reaction parameters was investigated. The effects of the chain length and different anions on the catalyst activity were also studied. The mechanism of the reaction was investigated theoretically using density functional theory (DFT) calculations in the ONIOM scheme and the influence of hydrogen bonding and the chain length of the linker on the transition state of the rate-determining step was also studied. The selected catalyst was used in subsequent runs without a significant loss of its reactivity.

The conversion of carbon dioxide from exhaust gas marine emissions via a hydrogenation route was investigated by Iliuta and Larachi. They explored, by process modelling, the combination of carbon dioxide reduction with hydrogen under microwave-induced plasma conditions and catalytic methanol synthesis in an integrated process. Carbon dioxide hydrogenation to methanol was also investigated by Geng et al., but their focus was the promotional effect associated with bimetallic phosphide catalysts. They showed that the combination of ruthenium and molybdenum in a bimetallic catalyst provides a favourable interaction with carbon dioxide through electronic effects to promote hydrogenation towards methanol. Nickel-based catalysts for methane dry reforming (MDR) are the focus of the work of Li et al. They aimed to suppress the well-known problems of carbon formation and metal sintering, due to the high temperature of the process, by using microwaves as an energy source to drive the MDR reactions under mild conditions (below 220°C).

The carbon dioxide utilization route based on carbon dioxide conversion to solid carbonates involves a complex set of reactions by which carbon dioxide reacts with calcium, magnesium and/or iron oxide-bearing phases to yield the corresponding solid carbonate phase. Both gas–solid and aqueous conversion processes have been extensively studied to perform the carbonation reaction. Direct aqueous mineral carbonation of heat-activated lizardite is the focus of the work by Benhelal et al, where they employed different analytical techniques to characterize carbonation products and to provide insights into the mechanism of formation and the structure of silica-rich by-products. A refining slag feedstock with calcium-containing wastewater was considered by Chen et al., where they proposed a high-gravity carbonation process. Various liquid agents and their effect on calcium ion leaching behaviour was investigated and reaction kinetics were identified by using a surface coverage model.

As an example of biological conversion routes, photoautotrophic microorganisms such as microalgae can convert carbon dioxide to biomass with sunlight via photosynthesis. They simultaneously assimilate carbon dioxide and organic and inorganic pollutants and produce a range of useful intracellular metabolites. They investigated the role of nitrogen availability in the production and release of extracellular matter in three commercially relevant microalgae. Variations of extracellular substances, in relation to microalgae species and cultivation conditions, were reported, with important implications for species selection and optimization of microalgae production, harvesting and dewatering processes.

Regardless of the carbon dioxide utilization route that is followed, to be able to successfully implement CCU, several other factors, such as sustainability of materials and energy processes, economics and permanence of the carbon dioxide in the carbon-neutral or low-carbon materials, must be carefully evaluated, following a standardized approach. The upcycling of carbon dioxide also offers the opportunity for making use of local available renewable energy resources. Ultimately, the development of CCU technologies will also impact the deployment of CCS technologies and all the knowledge that can be gained from research into those areas will certainly help us in achieving our carbon targets by 2050.

10.2 Carbon Dioxide Utilization into Chemical Technology

The increase in atmospheric carbon dioxide content, through utilization of fossil fuels and deforestation activities, resulted in environmental concern due to the global warming effect of this greenhouse gas. This has resulted in worldwide research activities for capture and utilization of carbon dioxide.

According to the Kyoto Protocol, which is an international agreement linked to the United Nations Framework Convention on Climate Change, some industrialized countries and European communities must reduce their greenhouse gas emissions during 2008–2012 to 5% below their emission in 1990. Under the Treaty, countries must meet their targets primarily through national measures. However, the Kyoto Protocol offers them an additional means of meeting their targets by way of three market-based mechanisms. These mechanisms are:

- emissions trading – known as "the carbon market";
- clean development mechanism (CDM); and
- joint implementation (JI).

An Agreement (The Paris Agreement) within the United Nations Framework Convention on Climate Change (UNCCC) was signed in 2016 and deals with greenhouse gas-emissions mitigation, adaptation and finance. The agreement's long-term goals are to keep the increase global average temperature to well below 2°C above the preindustrial level; and to limit the increase to 1.5°C since this would substantially reduce the risk and effects of climate change.

One way to mitigate carbon dioxide emission is its conversion and fixation to value-added products. Carbon dioxide can be considered as an abundant and renewable C_1- building block for chemical synthesis (Figure 10.1). Carbon dioxide has benefits over poisonous C_1 feed stocks such as carbon monoxide and phosgene and research for its chemical conversion is one of most active scientific research fields.

The production of fuels could have the highest influence on mitigation of global carbon dioxide emissions because of the large volumes involved. However, the production of fuels requires considerable energy to be provided, commonly, by fossil fuels. Therefore, the net effect based on well-to-wheel analysis is net production of carbon dioxide unless the required energy is provided from renewable sources such as biomass. Except for the high-purity carbon dioxide-side streams from chemical plants (e.g., in ammonia synthesis), the chemical use of carbon dioxide could imply increasing emissions and still be expensive, due to carbon dioxide recovery costs and high hydrogen price. Carbon dioxide mitigation by means of industrial consumption will remain rather modest and mainly should focus on increasing efficiencies and sustainable energy sources. Probably the most interesting strategies rely on expanding the commercial application of products produced

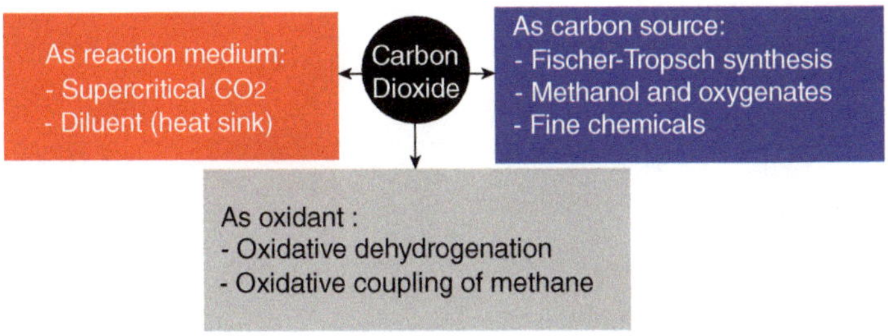

Figure 10.1 Carbon dioxide applications in chemical synthesis.

from carbon dioxide such as urea-based polymers and on developing alternative pathways to produce intermediates such as organic carbonates and end products such as polycarbonates.

Petrochemical technology is an important source of carbon dioxide emissions. However, relatively pure carbon dioxide (over 99%) is available as the by-product of some petrochemical and chemical plants such as hydrogen plants, oxidation processes (for example, from ethylene epoxidation) and ammonia synthesis. This carbon dioxide stream, supplemented by captured carbon dioxide from the plant, can be used for the production of valuable chemicals. Carbon dioxide can be used either as the whole molecule in the reaction, or as the source of carbon or oxygen. Important reactions can be classified as hydrogenation, hydrocarbon synthesis, amine synthesis and hydrolysis. Nearly 110 Mt/y of carbon dioxide is used as the feed for production of limited chemicals such as urea, methanol, polycarbonates, cyclic carbonates and other chemicals (Figure 10.2).

The catalytic conversion of carbon dioxide to value-added chemicals as a potential method for producing carbon-neutral chemicals and thereby reducing carbon dioxide emissions into the environment is reviewed in Figure 10.2. Homogenous and heterogeneous hydrogenation, oxidative dehydrogenation, oxidative coupling of methane and dry reforming of methane are reactions that are being focused on in this book. Potentials and challenges are addressed, examples of our research findings and finally recent technological developments are also provided and discussed.

10.3 Mechanism of Conversion of Carbon Dioxide

The chemistry of carbon dioxide at metal centres, i.e., the coordination of carbon dioxide, its activation and conversion into organic compounds, is a rapidly increasing area of coordination chemistry, organometallic chemistry and of catalysis. Coordination of carbon dioxide to a transition metal is one of the initial steps in the catalytic conversion of carbon dioxide. The electronic structure of carbon dioxide is perturbed by bonding to a transition metal centre. When carbon dioxide coordinates a transition metal, it donates paired electrons to the metal by σ bond resulting in an increase in electron density on the metal. With back donating, excess electrons transfer from the metal to the empty π^* orbital of carbon dioxide. Therefore, the bond between C and O will weaken. The strength and liability of metal-CO_2 bond and the mode of coordination control the reactivity in different carbon dioxide fixation reactions. Carbon dioxide displays several alternate modes of

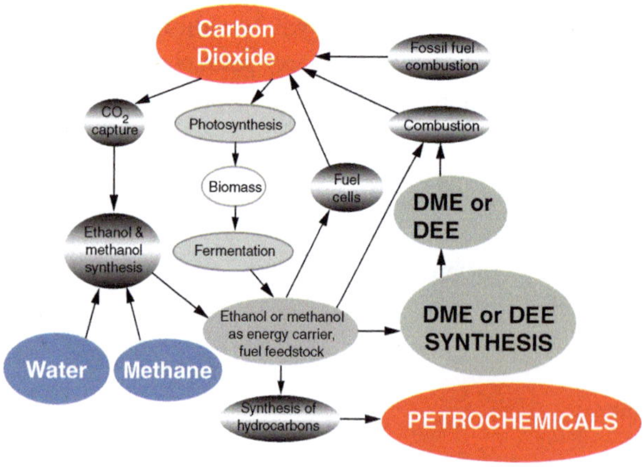

Figure 10.2 Carbon cycle based on methanol and ethanol.

coordination with transitionmetal compounds-via oxygen by donation of the oxygen lone p-electron pair to the vacant orbital of the metal (I), by electron donation from metal to the carbon orbital with formation of the metallo-acid derivative (II), or finally by way of its complex formation via C=O double bond (III):

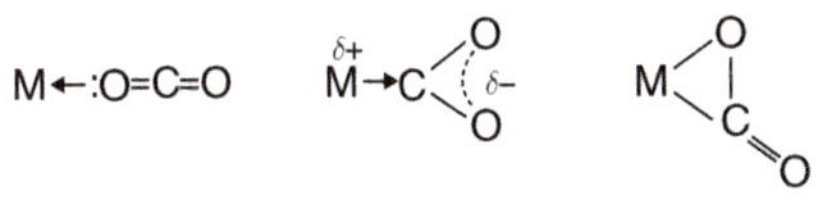

(I) (II) (III)

10.4 Hydrogenation of Carbon Dioxide

Since carbon dioxide is the most oxidized form of carbon, (formal) reduction is the only possible route for its conversion. The chemical reduction of carbon dioxide can be categorized into two groups: heterogeneous reduction and homogeneous reduction. The heterogeneous catalysis is technically more favourable, in relation with the stability, separation, handling and reuse of the catalysts and reactor design. Despite these practical advantages, the range of compounds that have been synthesized from carbon dioxide by heterogeneous catalytic routes is still comparatively narrow and the simultaneous chemical equilibria complicate the equilibrium composition.

10.4.1 Heterogeneous Hydrogenation

Metal-catalysed heterogeneous hydrogenation of carbon dioxide, depending on the conditions and the catalyst, goes directly to methanol or methane. These C_1 chemicals have the potential to become the future energy carriers as well as common feed stocks for petrochemical industries through C_1 chemistry. Methanol and dimethyl ether synthesis are potentially the most important heterogeneous hydrogenation reactions of carbon dioxide. They are versatile multi-source multi-purpose chemicals. The use of carbon dioxide as an alternative feedstock replacing carbon monoxide in methanol production has received attention as an effective way of carbon dioxide utilization. Commercially, methanol has been produced from synthesis gas using natural gas or coal, mainly containing carbon monoxide and hydrogen along with a small amount of carbon dioxide. Dimethyl ether can be produced by dehydration of methanol or directly from syngas over a bifunctional catalyst (Equation 10.1) and carbon monoxide hydrogenation (Equation 10.2) to methanol has a mutual relation through the water gas synthesis (WGS) reaction (Equation 10.3):

$$CO_2 + 3H_2 \rightleftarrows CH_3OH + H_2O \qquad \Delta H^\circ_{298} = -49.16\ kJ/mol \tag{10.1}$$

$$CO + 2H_2 \rightleftarrows CH_3OH \qquad \Delta H^\circ_{298} = -90.77\ kJ/mol \tag{10.2}$$

$$CO + H_2O \rightleftarrows CO_2 + H_2 \qquad \Delta H^\circ_{298} = -41.21\ kJ/mol \tag{10.3}$$

Mechanistic studies have shown that the origin of carbon in the methanol from synthesis gas (mixture of hydrogen and carbon monoxide) is carbon dioxide and not carbon monoxide, with the latter acting as a carbon dioxide source via the WGS reaction. Nevertheless, conventional Cu/ZnO methanol synthesis catalysts exhibit low activity in hydrogenation of both pure carbon monoxide and carbon dioxide to methanol.

Many research efforts are focused on the direct use of carbon dioxide in methanol synthesis. One drawback of hydrogenation of carbon dioxide or carbon dioxide rich feed compared to carbon monoxide is that more hydrogen is used due to the formation of water. The water formed either by methanol synthesis or reverse WGS reaction accelerates the crystallization of copper and zinc oxide in the catalyst, leading to the deactivation of the catalyst by sintering. The addition of a small amount of silica into the catalyst greatly improves the catalyst stability by suppressing the crystallization of copper and zinc oxide.

It is known that enrichment of the synthesis gas with carbon dioxide results in its utilization and increased methanol production. It has been shown that a copper-based catalyst produced by co-precipitation method and promoted by adding a small amount of colloidal silica into the mixed precursor solution of copper and zinc has promising performance. With a loading of 0.6 wt % silicon dioxide to the $Cu/ZnO/Al_2O_3$ reference catalyst, a good activity and stability in conversion of CO_2-rich syngas to methanol was observed.

Alternatively, carbon dioxide can be converted to carbon monoxide via reverse water gas shift reaction and then the produced gas ($CO/CO_2/H_2$) fed to the methanol synthesis reactor after removing the water (Zangeneh et al). This is the basis of the carbon dioxide to methanol via reverse water gas shift reaction (CAMERE) process being developed to form methanol from carbon dioxide (Figure 10.3). Each synthesis reactor has a recycle stream to increase carbon dioxide conversion to carbon monoxide and carbon oxides conversion to methanol, respectively. With the gas feeding of $CO/CO_2/H_2$, the water produced in the methanol reactor is chemically eliminated through a water-gas-shift reaction, increasing carbon oxide conversion to methanol and then decreasing the recycle gas in the methanol reactor. The alignment of two reactors reduced the recycle gas volume, minimizing the purge gas volume. The methanol productivity in the CAMERE process depends on the carbon monoxide concentration in the feed gas of the methanol reactor, which is dependent on the reverse-water-gas-shift reaction conditions, especially the temperature. The yield of methanol increased by 29% as the purge gas volume decreased in the CAMERE process. An active and stable catalyst has the composition of $Cu/ZnO/ZrO_2/Ga_2O_3 = 5/3/1/1$ and showed high activity and stability with the feed of high carbon dioxide concentration. Hydrogen can be produced by catalytic decomposition of methane at 1000–1100°C (Equation 10.4).

$$CH_4 \rightleftarrows C + 2H_2 \tag{10.4}$$

After separation, the hydrogen can be used in the hydrogenation processes, whereas the carbon can be used as a fuel or stored permanently or for future use. The combination of these reactions with methane reforming by carbon dioxide and synthesis gas reactions can lead to low or zero carbon dioxide emission processes for production of chemicals such as fuel methanol.

A promising process for conversion of carbon dioxide to methanol is the CARNOL process (Figure 10.4). It contains processes of methane pyrolysis and methanol synthesis. The feedstock of methane and carbon gasification are applied. Nearly half of the carbon produced in this process is stored and therefore not released to the atmosphere.

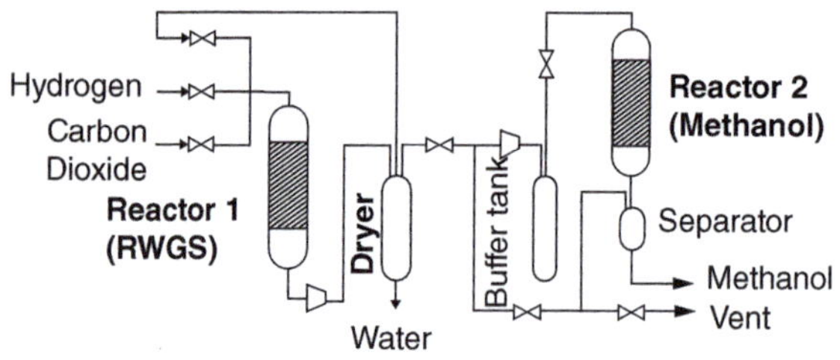

Figure 10.3 CAMERE process diagram for methanol from carbon dioxide.

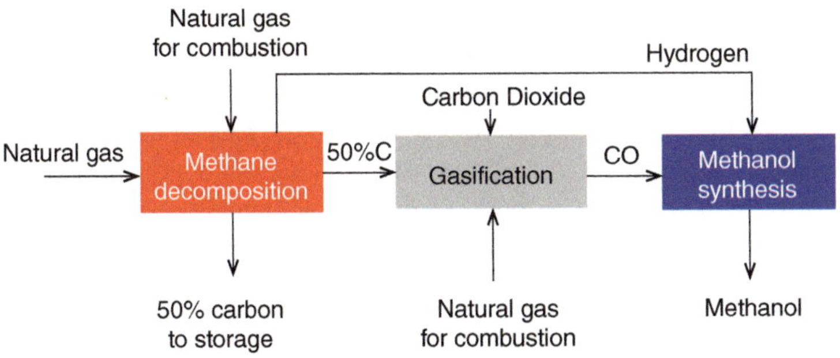

Figure 10.4 The CARNOL process.

Table 10.1 compares the methane efficiency and carbon dioxide emission per mol methanol produced for four processes. From the viewpoint of carbon dioxide emission, the CARNOL process is superior to steam reforming, carbon dioxide reforming or a combination of steam/carbon dioxide reforming of methane. However, the methane efficiency of the CARNOL process is lower than the other processes, since carbon is produced and part of the carbon produced is stored. The separation of the carbon produced is less difficult than for some other processes. Carbon can be sold as a natural commodity.

Metal phosphides have been investigated as promising catalysts for many hydrogenation reactions, including carbon dioxide reduction. There is much work going in to discover active and selective transition metal phosphide catalysts for energy-related reactions and a variety of molybdenum-based and ruthenium-based phosphide catalysts for the hydrogenation of carbon dioxide to methanol in 1,4-dioxane (200°C, 1 MPa carbon dioxide and 3 MPa hydrogen) have been reported. It was determined that from the monometallic catalysts studied (MoP, Mo_3P, RuP and Ru_2P), MoP and Mo_3P displayed higher methanol production rates than RuP or Ru_2P. However, with the addition of Ru to form bimetallic $Ru_xMo_{(2-x)}P$ ($x = 0.8$, 1.0, 1.2), the methanol production rate per carbon monoxide titrated site increased by three-fold, in comparison to MoP. The combination of X-ray photoelectronic spectroscopy (XPS), density functional theory (DFT), carbon dioxide temperature-programmed desorption (TPD) and hydrogenation experiments of reaction intermediates provided evidence that the combination of ruthenium and molybdenum in the bimetallic catalyst provides a favourable interaction with carbon dioxide through electronic effects to promote hydrogenation towards methanol. Lastly, recycling experiments were performed with Ru_1Mo_1P, which showed stable methanol production rates for three consecutive reactions. Overall, this chapter showcases the promotional effect associated with bimetallic phosphide catalysts for carbon dioxide hydrogenation to methanol and provides new directions for catalyst discovery with other metal compositions.

Table 10.1 Comparison of various methanol production processes from methane.

Process	**CH_4 Efficiency (mol CH_3OH/molCH_4)**	**CO_2 Emission (mol CO_2/mol CH_3OH)**
Steam reforming	0.95	1.05
Dry reforming	1.04	0.96
Steam + Dry reforming	1.01	0.99
CARNOL process	0.85	0.68

Gao et al. compared non-Cu_2In nanoalloy counterpart Cu-In/ZrO_2 catalysts with a $Cu_1In_2Zr_4$-O catalyst with the structure of Cu_2In alloy that exhibited an excellent performance for carbon dioxide hydrogenation to methanol. The carbon dioxide conversion (12.8%) and methanol selectivity (72.8%) were outstandingly higher, which indicated that the synergetic effect existed between copper and indium over the $Cu_1In_2Zr_4$-O catalyst. The formation of Cu_2In alloy caused high dispersion of the active species and high surface area of the $Cu_1In_2Zr_4$-O catalyst, enhancing catalyst reduction performance. Strong adsorption of carbon dioxide caused a good conversion property of the $Cu_1In_2Zr_4$-O catalyst. The strong interaction between the indium and copper species formed a nanoalloy and decreased the catalyst reduction temperature, which led to high catalytic performance for carbon dioxide hydrogenation. The Cu_2In alloy rather than metallic copper was the key active site for the methanol formation.

Bifunctional catalysts of Cu/ZnO supported on modified γ-Al_2O_3 were prepared and tested in the direct hydrogenation of carbon dioxide to methanol and dimethyl ether (DME). The γ-Al_2O_3 modification was carried out using a solution of $(NH_4)_2SiF_6$, aiming to incorporate Si atoms in the alumina structure. The metals were added to the support by wet impregnation. The catalysts were characterized by the Brunauer–Emmett–Teller area, chemical composition, ^{29}Si and ^{27}Al magic angle spinning nuclear magnetic resonance, acidity and temperature-programmed reduction. All of the prepared catalysts were able to convert carbon dioxide into methanol and dimethyl ether in the temperature range of 270–290°C and 3.0 and 5.0 MPa of pressure. The yield of dimethyl ether was higher over the catalyst supported on γ-Al_2O_3 containing more incorporated silicon atoms. The results were interpreted in terms of the acid strength of the supports.

Dimethyl ether production has been attracting significant research attention for its broad uses as an important chemical feedstock as well as a promising fuel. Here I report carbon dioxide direct hydrogenation to produce dimethyl ether using a new CIZO–SAPO bifunctional catalytic structure, which consists of Cu–In–Zr–O (CIZO) mixed oxide sites towards methanol synthesis and SAPO-34 zeolite sites for intermediate dehydration to dimethyl ether (Figure 10.5). Compared with CIZO, a significant increase in carbon dioxide conversion was achieved by simply mixing CIZO and SAPO, indicating the existence of synergy within the bifunctional catalyst. The study of mixing ratios and methods further confirmed the synergetic effect being proximity dependent. Mechanistic insight was obtained by conducting in-situ DRIFTS analyses. Variation in the proximity between CIZO and SAPO was discovered to alter the reaction pathways. When CIZO and SAPO were more closely contacted, dimethyl ether could be generated via a shortcut methoxy–dimethyl ether pathway instead of a typical methoxy–methanol–dimethyl ether route, resulting in more efficient dimethyl ether formation. The shortcut pathway was suppressed with an increase in distance between the two components. Therefore, it is proposed that the synergetic effect that leads to boosted dimethyl ether formation in the bifunctional catalyst is determined by the altered

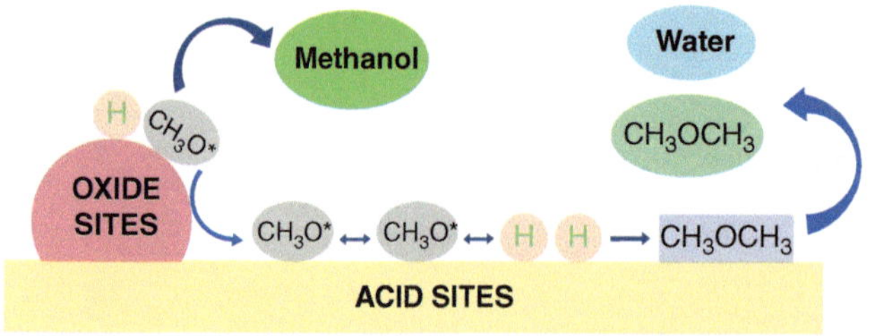

Figure 10.5 Direct hydrogenation of carbon dioxide to produce dimethyl ether using a CIZO- SAPO bifunctional catalyst.

reaction pathway which is controlled by the proximity between CIZO and SAPO active sites. The adjacency of CIZO and SAPO facilitates migration of methoxy intermediates from the former to the latter, so that dimethyl ether would be formed directly without a need of methanol formation. This study unveils the synergetic mechanism within bifunctional catalyst in dimethyl ether synthesis that would provide guidance for new catalyst research.

10.4.2 Homogeneous Hydrogenation

The mild conditions used for homogeneous hydrogenation of carbon dioxide catalysed by transition metal complexes, makes partial hydrogenation of carbon dioxide to formic acid (Equation 10.5) and derivatives feasible:

$$CO_2 + H_2 \rightleftarrows HCOOH_{(l)} \qquad \Delta H^{\circ}_{298} = -31.8\ \text{kJ / mol} \tag{10.5}$$

Further reduction of formic acid under homogeneous conditions is more difficult and very few examples, such as formation of methanol and methane, are known. Many metal complexes have been known to be highly active in carbon dioxide reduction. These are usually hydrides or halides with phosphines as natural ligands and rhodium and ruthenium proved to be the most active metals.

Analogous to synthesis gas chemistry, different CO_2/H_2 ratios are required for different products and varying amounts of water are formed (Chapter 1, Table 1.3), resulting in intermolecular loss of hydrogen which impacts the process economy. Consequently, reactions producing no or little water are economically more attractive.

Electrochemical reduction of carbon dioxide to useful low-carbon fuels can not only mitigate global warming problems but also be used for energy storage. For efficient carbon dioxide electrochemical reduction in mild conditions, high-performance catalysts are required. The electrodeposition of bismuth, tin, or both, on copper sheets to fabricate Sn/Cu, BiSn/Cu, Bi_2Sn/Cu, Bi_3Sn/Cu, Bi_4Sn/Cu and Bi/Cu electrodes was conducted and the electrodes were used as catalysts in carbon dioxide reduction. Experimental results show these electrodes have high activity, stability and selectivity towards carbon dioxide reduction to formic acid. A maximum Faraday efficiency of 90.4% for formic acid production can be achieved when using a Bi/Cu electrode. The mechanism catalysed by such electrodes is also proposed for fundamental understanding and is clearly explained on the basis of density functional theory (DFT) calculations. The results shown in this chapter demonstrate that electrodeposition can be a faster and easier to operate industrial technology for the fabrication of catalyst electrodes, compared with other technologies.

Carbon dioxide is rather inert and its reactions are energetically highly unfavourable. In principle the first problem of inertness can be solved by developing a good catalyst, while the second one is of thermodynamic origin and not solvable by catalysis. However, under certain operating conditions or reaction systems a practical process might be feasible. In conclusion, the following points could be drawn from this discussion.

- The economy of these processes, besides the price of products and reactants, depends on the source of energy and/or hydrogen and the development of new active catalysts.
- In the short term, the best method is to enrich the feed with carbon dioxide which is applicable in certain current processes with minor modifications. However, in the long term, development of new catalysts and processes are required.
- Catalysis plays an important role in activation and chemical utilization of carbon dioxide for its fixation. The current catalytic systems commonly suffer from low activity and/or rapid deactivation.

- Removal of the products through either physical (e.g., by membrane reactor) or chemical approach (conversion to more stable products) would be very desirable.

10.5 Electrochemical Conversion of Carbon Dioxide into Valuable Chemicals

Electrochemical reduction of carbon dioxide is a promising technology that can convert carbon dioxide into useful low-carbon chemicals and fuels. Thermodynamically, carbon dioxide is a highly stable compound (Chapter 1) which necessitates highly active catalysts and energy for conversion into targeted products. Different types of catalysts reported in the literature for activating and reducing carbon dioxide are critically analysed. To start with, metallic electrodes in aqueous solution and nanopores materials are discussed. Copper is unique for the reduction of carbon dioxide as it can favour the breaking of C–O bond in carbon dioxide and hence facilitate hydrocarbon formation. Some products obtained because of reduction of carbon dioxide include methane, ethane, ethylene, hydrocarbons $>C_5$, hydrogen, carbon monoxide, formic acid, methanol, ethanol, propanol, formaldehyde and acetic acid. The feasibility of bringing reported lab bench successes to industrially relevant scale requires careful assessment from a system-based perspective. This section presents a unique life cycle assessment (LCA) case study assessing the potential environmental impacts of both small- and large-scale set ups of CO_2-to-ethylene conversion via electrochemical carbon dioxide reduction reaction (CO_2RR). The LCA parameters are modelled according to the current progress of CO_2RR from a literature study and an industry standard for carbon dioxide capture and product separation. The output of small scale CO_2RR systems are set to 1 g ethylene, while a hypothetical industrial scale is set at 1 ton ethylene production (Chapter 8, Section 8.2.1).

Carbon dioxide, a greenhouse gas, is considered to contribute significantly to climate change and global warming. The carbon footprint grows stronger, wider and deeper and its significance is indisputable. Environmental changes require a reduction of the measure of carbon dioxide in air. The capture, storage and utilization of carbon based on photochemical, biochemical and electrochemical processes are proposed innovative methods to decrease utilization of non-renewables such as coal and oil. The carbon dioxide can be reduced chemically through either a homogeneous or a heterogeneous pathway. In general, photochemical transformation of carbon dioxide involves formation of carrier charges followed by its separation, transportation and finally reduction of carbon dioxide using generated photoelectrons. Photocatalytic reduction of carbon dioxide is a rising area of research. Beginning from the premise of photocatalytic reduction, the investigations into different semiconducting frameworks like oxides, sulfides and phosphides are considered in the research. Biochemical transformation deals with enzymatic conversion of carbon dioxide and electrochemical reduction uses electrical energy for converting carbon dioxide into its reduced form. The enzyme catalytic carbon dioxide change gives an eco-accommodating approach to make carbon-based chemical products. A few favourable circumstances related to enzymatic change incorporate high selectivity, high return, reduced quantity of waste and lower response conditions. However, certain downsides exist, for example, high expense of catalysts and cofactors and longer response times compared with normal strategies. Some products obtained through the reduction of carbon dioxide include methanol, formic acid, carbon monoxide, methane, ethylene and gasoline. In this section, an overview on intrinsically associated methodologies is given and ongoing advancement on the improvement, designing and comprehension of carbon dioxide reduction using photochemical, biochemical and electrochemical is outlined.

Products obtained from carbon dioxide transformation are presented in Figure 10.6.

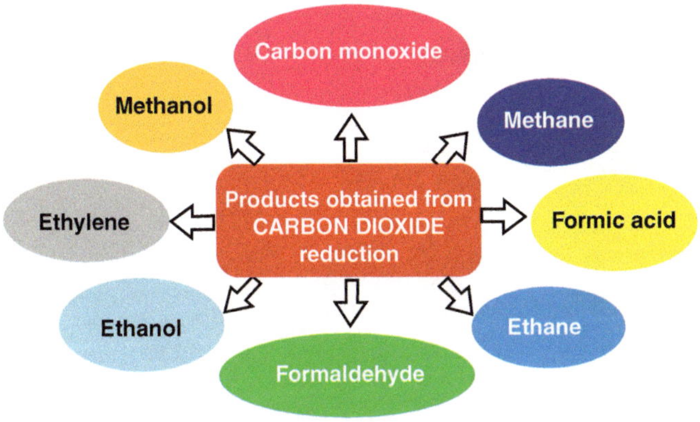

Figure 10.6 List of chemical compounds obtained from chemical conversion of carbon dioxide.

10.5.1 Technologies Available for Carbon Dioxide Reduction

Developing economically feasible technologies to convert carbon dioxide into useful chemicals and fuels can make a significant contribution to addressing the global warming challenge. Also, research on capturing and storing carbon dioxide is equally significant. To mitigate carbon dioxide emission, several treaties have been signed by several nations and organizations. For instance, protocols for NO_x emission as well as SO_x emission and certified emission reductions are in force to attempt to stabilize the situation.

Figure 10.7(a) illustrates the different strategies to control carbon dioxide emission into the atmosphere. Conservation practices include reducing carbon-based fuels or employing reduced carbon sources of energy such as natural gas. One strategy is to capture and store carbon dioxide. Potentially, carbon dioxide can be utilized as an alternative, non-toxic and efficient feedstock for the production of polymers or chemicals that could help with carbon balance.

To convert carbon dioxide into products, an input of energy is required, which can be derived from solar as a sustainable source. Thus, photocatalytic splitting gains interest to convert carbon dioxide into useful fuels.

Various technologies for carbon dioxide conversion to achieve high energy density chemicals are under investigation, including thermochemical, chemical, photochemical, photoelectrochemical, biochemical, biophotoelectrochemical and electrochemical routes. Various techniques in the research and development stage can be grouped as biological processes and non-biological processes as shown in Figure 10.7(b). Some of the emerging technologies which focus on these aspects are:

- biochemical reduction of carbon dioxide;
- chemical sequestration of carbon dioxide;
- thermochemical sequestration of carbon dioxide; and
- photochemical reduction of carbon dioxide

Preferably the development and use of these chemical technologies should use renewable energy sources and be economically viable.

In comparison to heterogeneous photocatalysts for carbon dioxide conversion in the liquid phase, homogeneous catalysts can uniformly supply carbon dioxide to active sites. Homogeneous catalysts are costly metals and necessitate reductants that are sacrificial in nature. Challenges in developing novel reactant materials are reported elsewhere. Carbon dioxide dissolved in the aqueous phase may be electrochemically reduced to form oxygenates, hydrocarbons or carbon monoxide utilizing

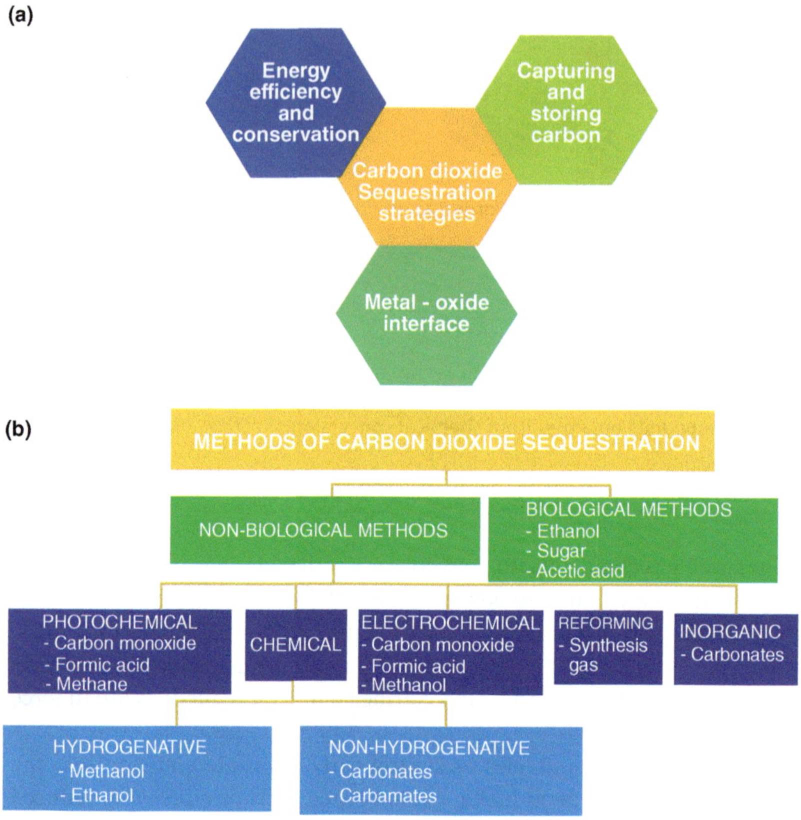

Figure 10.7 Strategies for (a) carbon dioxide management and (b) methods of carbon dioxide sequestration.

both homogeneous as well as heterogeneous catalysts. This approach will be attractive when the source of power is wind or photovoltaic. Another alternative to specifically functionalize carbon dioxide is hydrogenation of carbon dioxide to low molecular weight hydrocarbons or oxygenates by means of modified methanol and Fischer-Tropsch (FT) synthesis. Such procedures have a greater potential to be extended to commercial scale compared with the electrocatalytic or photocatalytic transformations. The challenge in implementing carbon dioxide hydrogenation is the requirement for pure hydrogen at an affordable cost. On the other hand, carbon dioxide can react with methane to yield a blend of carbon monoxide and hydrogen gas (synthesis gas).

A brief introduction to the working principle along with pros and cons of the various techniques follows.

10.6 Electrochemical Technologies

In the electrochemical technologies, gaseous carbon dioxide and protons from aqueous electrolytes are utilized to produce a variety of low-carbon fuels.

By comparing the reactivity of various processes, the reactivity of electrocatalytic reduction (ECR) of carbon dioxide is stated to be higher than bio- and photochemical processes. The

Table 10.2 Summary of electrochemical reduction of carbon dioxide on high hydrogen overvoltage metals.

Electrode	Electrolyte Conditions	Products Formed
Pb	Methanol	HCOOH
High hydrogen overvoltage metals	Aqueous	HCOOH
Pb	K_2CO_3	HCOOH
Sn–Cu	Aqueous	Format and H2, CO, CH_4, C_2H_4
Pb, Zn on Cu	Methanol	Absence of Cu-HCOOH, CO Presence of Cu-hydrocarbons
Pb	Aqueous	Formic acid
Pb in-situ IR studies	Propylene carbonate	Oxalate
Pb, in-continuous	Polymer electrolyte	Format
Sn (pilot plant)	NaOH	HCOOH, Format
Sn, SnOx on Ti	$NaHCO_3$	
Sn	$KHCO_3$, K_2SO_4, KCl, Na_2SO_4, Cs_2SO_4, $NaHCO_3$ and $CsHCO_3$	
Sn	Aqueous	Format

electrochemical reduction of carbon dioxide proceeds by adsorbing and activating carbon dioxide on metal surfaces of electrodes and Table 1.2 shows the basic anodic and cathodic reactions during the reduction of carbon dioxide.

At present, the carbon dioxide catalytic reduction reaction device used in the laboratory is mainly an H-type electrolytic cell as shown in Figure 10.8. This device includes some important parts, such as anode chamber, cathode chamber, three-electrode system and proton exchange membrane. This kind of device not only has the advantages of relatively easy manufacturing, simple operation, adjustable temperature and convenient construction of the three-electrode system, but also makes it easy to optimize the raw materials for carbon dioxide catalysis and study of electrode process dynamics. In the process, carbon dioxide first enters the electrolyte solution through the cathode chamber so that its internal carbon dioxide is always in the saturated state to maintain full contact with the catalyst surface. However, the low solubility of carbon dioxide in aqueous solution would relatively inhibit its effective reduction. To solve this problem, researchers introduced ionic liquids as the electrolyte that have a high solubility of carbon dioxide, which not only increases the electrochemical conversion rate, but also effectively inhibits the hydrogen evolution reaction and improves the Faradaic efficiency of the products.

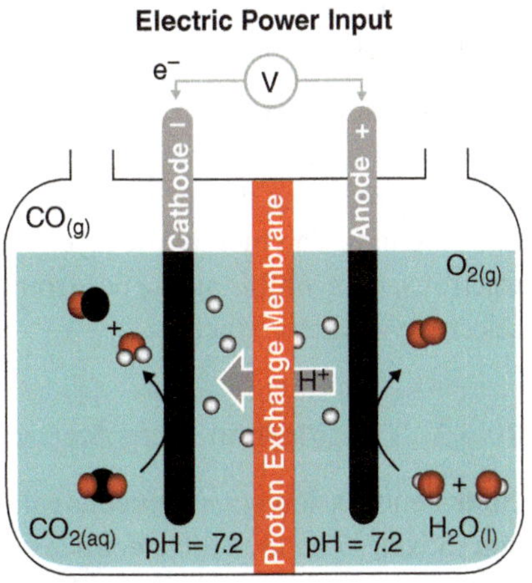

Figure 10.8 ECR electrolytic cell.

According to the type of electrolyte, different carbon dioxide catalytic reduction cells can be designed, such as a liquid electrolyte cell, a solid electrolyte cell and a solid-liquid composite electrolyte cell, respectively. The membrane electrode assembly (MEA) type electrolyte cell contains both a solid electrolyte cell and a solid-liquid composite electrolyte cell. The MEA cell has attracted more and more attention from researchers in recent years and has become one of the research directions of ECR (Chapter 12; Section 12.5).

10.6.1 Roles of Ionic Liquids on Electrochemical Carbon Dioxide Reduction Promotion

The functions of ionic liquids (ILs) in electrochemical carbon dioxide reduction (ECR) process (Figure 10.8) include ionic liquids as carbon dioxide absorbents, electrolytes and electrode modifiers. ILs account for about the same proportion as carbon dioxide absorbents and ECR electrolytes. The reason is that the primary purpose of using ILs as the ECR electrolyte is to have a large solubility of carbon dioxide, which is conductive to the dissolution and absorption of carbon dioxide and enables the system to have a large initial reaction concentration of carbon dioxide. The properties of electrode materials can be further optimized through the modification or preparation of electrode by ILs.

10.6.2 Ionic Liquids as Absorbent for Carbon Dioxide Capture

The factors such as the dissolution and diffusion of carbon dioxide are very important for carbon dioxide conversion and utilization. At present, the most widely used carbon dioxide absorbent in industry is an alkanolamine solution (Chapter 4). However, it is volatile, corrosive and easy to degrade and serious environmental problems are generated in subsequent processing, limiting the use of alkanolamine solution in the ECR system for carbon dioxide conversion. Thus, searching for a stable solvent for carbon dioxide absorption is still desirable. In recent years, extensive studies have shown that ILs process unique features such as high solubility and absorption capacity of carbon dioxide, good thermochemical stability and low saturated vapour pressure.

In the ECR system, most studies focus on imidazolid-based ILs, such as:

[Emim][BF_4] = 1-ethyl-3-methyl-imidazolium tetrafluoroborate;
[Bmim][BF_4] = 1-butyl-3-methyl-imidazolium tetrafluoroborate;
[Bmim][PF_6] = 1-butyl-3-methyl-imidazolium hexafluorophosphate; and
[Bmim][OTf] = 1-butyl-3-methyl-imidazolium triflate.

ILs can increase the concentration of carbon dioxide in the system and improve the efficiency of ECR. Their unique structures can form a complex with carbon dioxide, thus breaking the stable C=O and reducing the formation energy of curved radical anion $CO_2^{\cdot -}$,so that ECR can occur at low potential. The membrane layer formed by cation on the electrode can block H^+ from gaining electrons, thus inhibiting the occurrence of hydrogen evolution reaction (HER) and improving the selectivity of the product.

10.6.3 Classification of the Electrode Material

Homogeneous and heterogeneous catalysts are employed for the electrochemical reduction of carbon dioxide. The metallic electrodes can be categorized into four groups based on the hydrogen overvoltage characteristics for reducing carbon dioxide in an aqueous electrolyte system:

- high hydrogen evolution overvoltage metal;
- medium hydrogen evolution overvoltage metal;

- copper-unique metal; and
- low hydrogen evolution overvoltage metal.

10.6.4 High Hydrogen Evolution Overvoltage Metal

Metals like mercury, lead, tin, indium, thallium, bismuth and cadmium with high hydrogen evolution overvoltage (HEO) characteristics promote the formation of format (Table 10.2).

They have negligible carbon monoxide adsorption and hence are graded as poor catalysts for carbon dioxide reduction. In this category, tin and lead showed promising results yielding formic acid in significant amounts. Formic acid obtained from carbon dioxide through electrochemical reduction may function as a useful fuel as well as an energy-storage medium owing to its technical feasibility and potential scope for commercial applications. Methanol is commonly used as a carbon dioxide absorbent in chemical technologies and the solubility of carbon dioxide in it is high, approximately five times greater than water. Carbon dioxide dissolved in methanol with a lead electrode was reported for its conversion into formic acid. The product was predominant from −1.8 to −2.5 V versus Ag/AgCl. Other products formed were carbon monoxide and methane. The partial current density for carbon dioxide reduction was twenty-two times higher than hydrogen evolution, while the Tafel plot revealed no limitation towards mass transfer in the tested potential range.

10.6.5 Low Hydrogen Evolution Overvoltage Metals

Metals like iron, nickel, titanium, platinum, rhodium, cobalt and iridium exhibit low HEO characteristics and display strong carbon monoxide adsorption. The literature report in Table 10.3 lists the electrochemical reduction of carbon dioxide performed on various metals/metal oxides including Ru, Ru/Cu, Ru/Cd and RuO_2/TiO_2. Due to the low turnover of adsorbed carbon monoxide, they could only evolve hydrogen as the primary product. Some of these metals employed as electrocatalysts for carbon dioxide conversion are discussed in detail later.

10.6.6 Copper Electrodes

Copper is unique in terms of the reduction of carbon dioxide as it can favour the breaking of carbon dioxide and hence facilitate hydrocarbon formation. It has an excellent ability to adsorb and desorb

Table 10.3 Summary of electrochemical reduction of carbon dioxide on low hydrogen overvoltage metals.

Electrode	Electrolyte/Experimental Conditions	Products Formed
RuO_2/TiO_2	$NaHCO_3$	Methanol
Ru, Ru/Cu, Ru/Cd	$NaHCO_3$	Methanol
Ti and hydrogen-storing Ti electrodes	KOH in methanol	Formic acid, CO
Platinum/Nafion on reticulated vitreous carbon (RVC) electrodes	$HClO_4$, CV and CO adsorption experiments	-
Modified platinum mesh electrode	KCl + 10% propylene carbonate	Ethanol and lactic acid
Pt-GDE at high pressures (<5MPa)	$KHCO_3$	Methane, ethanol
Platinum nanoparticles on carbon black	-	Hydrocarbon > C_5
Pt/CNT, Fe/CNT	$KHCO_3$	Isopropanol
Pt/C-TiO_2	0.2 M NaF + 10 mM pyridine	Methanol

metal–carbon bonds in electrochemical carbon dioxide reduction. Comparison of Faradaic efficiency of the products obtained with copper electrode in various electrolytes is listed in Table 10.4. The key products formed during the electrochemical reduction of carbon dioxide are hydrocarbons and alcohols. Research using copper as the base metal has grown as has understanding of carbon dioxide reduction by varying the supporting, lowering temperature, increasing pressure and also the modification by oxidation and deposition.

The possible modes of adsorption of carbon dioxide on metal surface can be either by (a) carbon coordination – with the carbon bonded to the electrode surface, (b) oxygen coordination – with either single oxygen or two oxygen bonded to the electrode surface, (c) mixed coordination – with a carbon and oxygen bonded to the electrode surface. Recently a new mechanistic pathway has been proposed, leading to methane and ethylene with high selectivity during the electrochemical reduction of carbon dioxide on copper electrode.

10.6.7 Other Electrodes for Carbon Dioxide Reduction

Researchers have analysed all possible metals in the periodic table for carbon dioxide reduction and based on the behaviour of reduction, a specific grouping of metals was established. Molecular electrocatalysts including bimetallic copper complexes, rhenium bipyridine complexes, nickel macrocyclic complexes, palladium triphosphine complexes are used. Hara et al. have studied Au, Ag, Zn and In as well as group 8–10 metals comprising Fe, Ni, Rh, Co, Pt and Pd towards carbon dioxide reduction. The final product composition varied with electrode type and the operating conditions (Table 10.5).

Electrochemical reduction of carbon dioxide in aqueous solution is competed by a hydrogen evolution reaction. To overcome this limitation, in our studies, a zinc-based electrode was prepared by constant potential deposition on copper substrate, with high (6 M) concentration (Cu/Zn-H) and low concentration (0.6 M) deposit (Cu/Zn-L). The products from the reduction of carbon dioxide were methane, ethane and hydrogen. Comparison of the effect of electrode potential for the products formed on Cu/Zn-H and Cu/Zn-L revealed the tendency of high hydrogen evolution (~ 59%) and low methane formation (<15%) on low concentration deposit of zinc on copper (Cu/Zn-L). High concentration deposit of zinc on copper (Cu/Zn-H) showed enhanced Faradaic efficiency of methane on Cu/Zn-H (52%) compared to bare copper (23%) electrodes.

Table 10.4 Comparison of Faradaic efficiency of the products for copper electrodes.

Experimental Conditions	Potential (V) versus Ag/AgCl	Faradaic Efficiency (%)			
		CH_4	Ethylene	Ethane	H_2
KCl	-1.9	4.1	NA	38.5	61.1
Cu mesh 3 M KCl (GLS-pH 3)	-2.4	5.1	40.0	0.3	53.2
$KHCO_3$	-2	19.6	NA	8.9	62.5
$KHCO_3$	-1.9	Trace	20	Trace	40
Pulsed electrolysis-$KHCO_3$	-1.8 to -3.2	33.3	25.5	NA	20.5
Methanol	-3	7.5	6	NA	NA
LiCl/methanol at 248K 1.0 MPa	-2	8	12	NA	5

Table 10.5 Product formed at the surface of varies cathodes for electrochemical reduction of carbon dioxide.

Metal	Product
Hg, Pb, Sn, Cd, In	Formic acid
Zn, Au, Ag	CO
Cu	C_2H_4, CH_4, HCOOH, CO and alcohols
Pt and Pd	Adsorbed CO
Ru	Low yield of alcohols, hydrocarbons and formaldehyde
Fe, Mo, Cr, Nb, Ti	Poor performance

Moreover, the efficiency of hydrogen on Cu/Zn (3%) was significantly lower than for the copper electrode (53%). It was proposed that the feasible reaction on zinc electrode is protonation rather than the coupling of carbons to get higher hydrocarbons.

Other than metals, gas diffusion electrodes, semiconductor electrodes, macrocyclic complexes and conducting polymers have also been investigated for carbon dioxide reduction. Among homogeneous and heterogeneous catalysts, the former shows high selectivity and tunability whereas the later show robustness and ease of purification of products. There is an interest to combine these two catalyst surfaces by covalent bonds to address the selectivity issue. Table 10.5 shows the product formed with different metallic structures as the catalyst.

Electrochemical reduction of carbon dioxide to useful low-carbon fuels can not only mitigate global warming problems but also be used for energy storage. For efficient carbon dioxide electrochemical reduction in mild conditions, high-performance catalysts are required. Li et al. used the electrodeposition of bismuth, tin, or both on copper sheets to fabricate Sn/Cu, BiSn/Cu, Bi_2Sn/Cu, Bi_3Sn/Cu, Bi_4Sn/Cu and Bi/Cu electrodes and the electrodes were used as catalysts in carbon dioxide reduction. Experimental results show these electrodes have high activity, stability and selectivity towards carbon dioxide reduction to formic acid. A maximum Faraday efficiency of 90.4% for formic acid production can be achieved when using a Bi/Cu electrode. The mechanism catalysed by such electrodes is also proposed for fundamental understanding and is clearly explained on the basis of DFT calculations. The results in this chapter demonstrate that electrodeposition can be a faster and easier to operate industrial technology for the fabrication of catalyst electrodes, compared with other technologies.

10.7 Power-to-Methanol Technology

Environmental issues related to greenhouse gas emissions are progressively pushing the transition towards a fossil-free energy scenario, in which renewable energies such as solar and wind power will play a key role. However, for this transition to succeed, significant issues related to renewable energy storage must be addressed. Power-to-X Technologies have gained increased attention since they convert renewable electricity to chemicals and fuels that can be more easily stored and transported. Hydrogen production through water electrolysis is a promising approach since it leads to the production of a sustainable fuel that can be used directly in hydrogen fuel cells or to reduce carbon dioxide in chemicals and fuels compatible with the existing infrastructure for production and transportation. Carbon dioxide electrochemical reduction is also an interesting approach, allowing the direct conversion of carbon dioxide into value-added products using renewable electricity.

Batteries allow the conversion of chemical energy to electrical energy, playing on the redox functionalities of ions in solution. However, other opportunities do exist for chemical storage of renewable energy and are now referred as Power-to-X technologies. As the name implies, the concept evolves around converting power (electricity) to chemicals (X), which could be very diverse hence the utilization of the "X." The most well-known technology to produce chemicals out of electricity is water electrolysis producing hydrogen and oxygen. Hydrogen produced through electrolysis has been acclaimed as one of the optimally sustainable fuels since its combustion leads only to the production of water, which involves minimal impact on GHG emissions. Production of hydrogen through electrolysis is still limited in many cases by issues related to cost and storage.

Power-to-X technologies can also involve carbon-based structures, which would make them more compatible to the existing infrastructure both for transportation and for large-scale energy production. One especially abundant carbon-based feedstock suitable for Power-to-X technologies is carbon dioxide, which in turn is also the focus of the worldwide climate change efforts. Using electrocatalytic systems or through the utilization of hydrogen as a reducing agent, carbon dioxide can now be used to produce simple C_1 molecules such as methanol and methane. Reports from the literature show that Power-to-X technologies could eventually lead to the production of gasoline, diesel and even jet fuel (Chapter 7). Hence, these technologies would at the same time allow storage of renewable energy, while reducing carbon dioxide emissions at the source and producing liquid transportation fuels.

Methanol is one of the main commodities produced worldwide, with a price around 470 USD per metric ton (2018, Methanex). It is also an important feedstock in chemical technology for producing olefins, dimethyl ether and liquid fuels, thus becoming an interesting alternative to fossil fuels. Moreover, methanol is a liquid energy carrier, being easier to handle and transport than gases and solid materials.

Methanol is currently produced at commercial scale from fossil fuel-based synthesis gas in a two-step process. In the first step, synthesis gas with a H_2:CO ratio close to 3:1 is produced from steam reforming of natural gas. Then, after adjusting the H_2:CO ratio to around 2:1, the syngas is converted into methanol using copper-based catalysts (Equation 10.2).

One of the simplest ways to sustainably obtain liquid products from carbon dioxide is the production of methanol, mainly electrochemically and through hydrogenation, two pathways that will be detailed in the following sections.

10.7.1 Carbon Dioxide Electrochemical Reduction

As for most of the carbon dioxide electrochemical reactions reported in this book, the development of catalysts able to selectively produce methanol still represents a challenge to be overcome. Copper and copper-based electrodes have proved to be one of the best performing materials for the electrochemical conversion of carbon dioxide into alcohols, including methanol, which was one of the first investigations of the reduction of carbon dioxide to methanol over copper surfaces. The authors showed that an anodized copper foil could successfully reduce carbon dioxide into methanol using a 0.5 M potassium hydrogen carbonate ($KHCO_3$) as solution and −1.9 V versus saturated calomel electrode (SCE) at a rate as high as 10^{-4} mol cm^{-2} h^{-1}. Faradaic efficiency for methanol reached about 240%. Efficiencies higher than 100% were obtained since the theoretical analysis considered six-electron reduction of carbon dioxide to methanol and water and the HER also occurred. Similarly, Le et al. obtained a methanol production rate of 43 μmol cm^{-2} h^{-1} and Faradaic efficiency of 38% in 0.5 M $KHCO_3$ using cuprous oxide thin films at −1.1 V (vs. SCE). The authors showed that this rate of methanol production as well as the Faradaic efficiency were much higher than those obtained with air-oxidized (0.9 μmol cm^{-2} h^{-1}) and anodized (1.5 μmol cm^{-2} h^{-1}) copper electrodes, suggesting that copper (I) species might play a key role in the electrode activity and selectivity to methanol. Malik et al. studied the efficiency of multi wall carbon nanotubes (MWCNTs) impregnated with dioxide cuprous to reduce carbon dioxide into methanol, due to their excellent structural and electrical properties, making them good candidates for electrochemical applications. The authors reported that the MWCNTs acted as active sites for carbon dioxide conversion as well as traps for electrons, increasing the rate of conversion of intermediates into methanol. Also, dioxide cuprous loadings varying from 10 to 50 wt % were investigated. 30% CuO_2-MWCNTs catalyst showed the best catalytic performance, achieving 38% of Faradaic efficiency to methanol at −0.8 V (vs. Ag/AgCl) in a 0.5 M $NaHCO_3$ medium. Dioxide cuprous loadings (higher than 40 wt %) led to an agglomeration of the dioxide cuprous particles and overall larger crystallite sizes, decreasing the surface area of the catalyst and the number of active sites.

The use of copper alloys has also been investigated, since it may enhance the electrochemical carbon dioxide reduction to methanol. Most of the industrial catalysts currently used for methanol production are composed of copper-zinc mixed oxides, highlighting the synergetic effect of the metals to improve methanol production. Watanabe et al. evaluated the capacity of Cu-Ni, Cu-Sn, Cu-Pb, Cu-Zn, Cu-Cd and Cu-Ag alloys to electrocatalytic reduce carbon dioxide into methanol in a 0.05 M $KHCO_3$ aqueous solution. They observed that the product distribution was dependant on the copper alloy. While a copper-nickel alloy produced methanol and formic acid, copper-tin and copper-lead enhanced the production of formic acid and carbon monoxide. A Faradaic efficiency to methanol of 10% was obtained with the Cu-Ni alloy at −0.4 V (vs. Ag/AgCl) while no methanol was formed when pure copper or nickel were used alone even when a wide potential range from −0.5 to −1.5 V was considered. They concluded that this difference in the catalytic performance was related to a mechanism involving the introduction of hydrogen atoms on the alloy surface by nickel sites. Jia et al. also investigated the performance of copper-based alloys. The authors prepared Cu-Au alloys through electrochemical deposition with a nanoporous copper film (NCF) as template. The $Cu_{63.9}Au_{36.1}$/NCF alloy showed a Faradaic efficiency of methanol of 15.9%, which was around 19 times higher than that obtained with pure copper. In fact, the copper-gold alloy favoured not only the carbon dioxide reduction reaction but also the carbon monoxide reduction, improving the overall conversion of carbon dioxide into methanol.

Kuhl et al. further investigated the carbon dioxide electrocatalytic reduction over different metals, such as Au, Ag, Zn, Cu, Ni, Pt and Fe. Experiments were carried out in 0.1 M $KHCO_3$ electrolyte and at different voltages varying from −1.6 to −0.4 V versus reversible hydrogen electrode (RHE). The best results were obtained with gold, presenting almost 100% of Faradaic efficiency to methanol. On the contrary, iron was shown to be inefficient for methanol production under these conditions, presenting 100% of current efficiency to methane production, instead of methanol. The other metals were able to produce both methanol and methane. The authors reported that one of the main parameters that influences the selectivity to methanol or methane is the oxophilicity (capacity to adsorb oxygen) of the catalyst surface. Gold, the metal with the lowest oxophilicity among the metals investigated, produced only methanol whereas iron, the metal with the highest oxophilicity, produced only methane.

Since the hydrogen evolution reaction competes with the carbon dioxide reduction (hence lowering the Faradaic efficiency of the carbon dioxide reduction reaction), Olah and Prakash investigated the possibility of using the HER as an advantage. The authors used a 0.1 M $KHCO_3$ aqueous solution as electrolyte and water was electrolyzed while carbon dioxide was reduced at a gold cathode at −3.2 V (vs. Ag/AgCl), producing syngas with a H_2:CO ratio close to 2:1. They showed that the total Faradaic efficiency for hydrogen and carbon monoxide was close to 100%. The synthesis gas produced could then be converted into methanol using the same process that is currently used at industrial scale (Equation 10.2). The advantages of this process are the high Faradaic efficiencies for syngas production, absence of a purification step since no impurity is present and production of valuable high-purity oxygen at the anode (Olah, Goeppert and Prakash).

Ruthenium and ruthenium dioxide have also shown to be promising materials for the electrocatalytic reduction of carbon dioxide due to their high electrical conductivity, high electrochemical stability, intermediate hydrogen overpotential and capacity to reversibly adsorb hydrogen for carbon dioxide reduction. Bandi reported that electrodes composed of 35% ruthenium and 65% dioxide titanium presented current efficiencies for methanol production as high as 24% when polarized near the equilibrium potential of hydrogen evolution in solutions of 0.2 M Na_2SO_4 saturated with carbon dioxide. The results led the authors to conclude that the first electron transfer is the rate-determining step in electrochemical reduction of carbon dioxide on oxide surfaces at low pH. Finally, the authors suggested that the methanol efficiency could be improved by changing the oxide composition and preparation parameters. Qu et al. investigated the electrochemical carbon dioxide reduction properties of RuO_2/TiO_2 nanoparticles (NPs) and nanotubes (NTs). RuO_2/TiO_2 nanotubes showed a Faradaic efficiency for methanol of 60.5% at −0.8 V (vs. SCE) in 0.5 M $NaHCO_3$, which was about 20% higher than that obtained with RuO_2/TiO_2 NPs. The authors reached similar conclusions to Malik et al., suggesting that the surface structure of the nanotubes composite had an important role for achieving high efficiency and selectivity to the desired products. Table 10.6 summarizes the performance of different catalysts for carbon dioxide electrochemical reduction into methanol reported in the literature.

10.7.2 Direct Carbon Dioxide Hydrogenation into Methanol

The most direct way to produce methanol from carbon dioxide via a Power-to-X process is the catalytic hydrogenation showed in Equation 10.1. The reaction takes place at 250–300°C and 5–10 MPa usually in the presence of $CuO/ZnO/Al_2O_3$ catalysts. In this process, hydrogen is produced by water electrolysis ideally using renewable energy, as presented in the sections on water electrolysis

Table 10.6 Summary of different electrocatalysts used for electrocatalytic reduction of carbon dioxide in methanol.

	Medium (Electrolyte)	E (V)	CH_3OH Production Rate	Faradaic Efficiency (%)
Cu foil	0.5 M $KHCO_3$	−1.9 vs. SCE	10^{-4} mol/cm^{-2} /h	240
Coprous oxide thin films	0.5 M $NaHCO_3$	−1.1 vs. SCE	0.43 × 10^{-4} mol/cm^{-2}/h	38
CuO_2.MWCNT$_s$	0.5 M $NaHCO_3$	−0.8 vs. Ag/AgCl	–	38
Cu-Ni	0.05 M $KHCO_3$	−0.4 vs. Ag/AgCl	–	10
$Cu_{63.0}Au_{36.1}$/NCF	0.5 M $KHCO_3$	−1.1 vs. SCE	–	15.9
Au	0.1 M $KHCO_3$	−0.7 vs. REH	–	~100
Pt RuO_2/TiO_2 NT$_s$	0.5 M $NaHCO_3$	−0.8 vs. SCE	–	60.5
$RuO_2 + TiO_2$	0.5 M H_2SO_4	−0.9 vs. Hg_2SO_4	–	24

(Chapter 1, Section 1.2.4; Chapter 6, Section 6.5.1) and then combined with carbon dioxide waste streams to produce methanol, in a classical Power-to-Methanol process.

Recent process simulations and techno-economic studies have focused on the efficiency of Power-to-Methanol processes and on its comparison with the classical routes (Table 10.7). Rihko-Struckmann et al. investigated the idea of capturing the carbon dioxide from power plants to produce energy storage media, such as hydrogen and methanol. They considered a hydrogen production from water electrolysis and the methanol production from a Power-to-Methanol technology at 220°C and 5 MPa. They achieved a high carbon dioxide conversion of about 97%. However, they concluded that the energetic efficiency of the system using hydrogen as storage medium was higher than the one including methanol. The main advantage of using methanol as chemical storage system would be its simple and cost-efficient storage. Hank et al. evaluated the economic feasibility of a methanol plant with a methanol production capacity of 4–10 kt/y using carbon dioxide and hydrogen as feedstock. The hydrogen used in the process was produced via a PEM water electrolyser with an electricity consumption of 4.76 kWh/m^3 hydrogen. They concluded that the feasibility of the process strongly depended on the costs of electricity and of hydrogen production, on the price of the carbon dioxide (considering the carbon taxes) as well as on the dynamics of the methanol reactor including the necessity of hydrogen storage. They also concluded that locations with high availability and low-cost production of renewable energy will play an important role on the development of Power-to-Methanol technologies in the next few years. Bellotti et al. compared the economic feasibility of three different plant sizes for methanol synthesis from hydrogen, produced by water electrolysis and carbon dioxide, sequestrated from power plants. They concluded that in the case of large plant sizes with a methanol production around 50 kt/y considerable amounts of carbon dioxide (71.6 kt/y) would be sequestrated. However, due to the high cost of the plant, oxygen selling would be mandatory to get economically feasible results. Koytsoumpa et al. showed that a large-scale methanol plant with a methanol production of 50–100 kt/y would require 1.01 MWhth per ton of methanol of thermal energy and 9.74 MWhe/t methanol of electric energy. The techno-economics are highly influenced by the final methanol price, the fuel and chemical market as well as the energy and fuel directives. Kourkoumpas et al. performed a techno-economic evaluation of the Power-to-Methanol concept using carbon dioxide from lignite-fired power plants and hydrogen from water electrolysis. They concluded that the Power-to-Methanol concept is more competitive when large scale plants, low electricity, low carbon dioxide costs and high operating time for both hydrogen and methanol plants are considered.

Table 10.7 Summary of some of the process simulations and techno-economy studies about Power-to-Methanol reported in the literature.

T (°C)	P(MPa)	CO_2 (%) Conversion to CH_3OH	Methanol Production	Electricity Consumption (kWh/kg CH_3OH)	Hydrogen Source	Electricity Consumption (kWh/m^3 H_2)
-	4	90	4–10 kton/y	-	Water electrolysis (PEM)	4.76
240	8	96	97 kg/h	-	Water electrolysis (PEM)	5.2
240	5	96.8	3.03 kmol/h	-	Water electrolysis	-
-	-	-	50–100 kton/y	9.89	Water electrolysis (AEL)	4.4

A few Power-to-Methanol plants are actually already in operation and projects aiming at building pilot and demonstration plants are currently being developed. The European Commission has launched the methanol from carbon dioxide ($MefCO_2$) project aiming to demonstrate the economic feasibility of the Power-to-Methanol technology in a modular intermediate scale and to adapt it to varying plant sizes and gas composition. The technology could later be adapted to work with the existing biomass combustion and gasification system streams, for example, aiming at the production of electric/thermal energy instead of chemical synthesis. Mitsui Chemicals built a pilot plant in 2008 for methanol production from carbon dioxide and hydrogen obtained from water photolysis with a capacity of 100 tons of methanol per year. Their objective is to use the methanol produced as raw material for the synthesis of olefins and aromatics. Carbon Recycling International (CRI) has taken the technology one step further since the company developed an emission-to-liquid technology in which carbon dioxide captured from flue gas of a geothermal power plant and hydrogen produced from water electrolysis using renewable energy are used to produce renewable methanol. The George Olah plant recycles 5.5 thousand tons of carbon dioxide per year, releasing 90% less carbon dioxide than the use of a comparable amount of energy from fossil fuels.

10.7.3 Low-Carbon Methanol Production

The alternative low-carbon pathway to methanol is based on hydrogen, produced by water electrolysis with low-carbon electricity followed by hydrogenation of carbon dioxide as carbon source.

Electrolysis (Equation 10.6):

$$6H_2O + \text{ electrical energy}_{(\text{wind/solar})} \rightleftarrows 6H_2 \text{ (cathode)} + O_2 \text{ (anode)} \tag{10.6}$$

Hydrogenation (Equation 10.7):

$$2CO_2 + 6H_2 \rightleftarrows 2CH_3OH + 2H_2O \qquad \Delta H_{298K,\,5\text{MPa}} = -40.9\ \text{kJ/mol} \tag{10.7}$$

Total (Equation 10.8):

$$2CO_2 + 6H_2 + \text{ electrical energy}_{(\text{wind/solar})} \rightleftarrows 2CH_3OH + 2H_2O + 3O_2 \tag{10.8}$$

Hydrogenation of carbon dioxide is used also in conventional methanol production by adding small amounts of carbon dioxide to adjust the CO/H_2 ratio of the synthesis gas. Synthesis of methanol from carbon monoxide and carbon dioxide are tied through the water gas shift reaction (Equation 10.3).

The anticipated direct synthesis of methanol from carbon dioxide does not use a reverse water gas shift reactor, in which carbon dioxide is reduced to carbon monoxide by hydrogen, to yield classical CO/H_2 synthesis gas for methanol synthesis.

For the hydrogenation of pure carbon dioxide to methanol, catalysts are commercially available and a few pilot plants are in operation, e.g., by Mitsui Chemicals (Japan) and Carbon Recycling International (Iceland) to investigate the feasibility of industrial-scale production. Carbon dioxide hydrogenation to methanol is accompanied by water formation and the recycle gas and crude methanol contain much water compared with the conventional process; that water must be removed by distillation. Figure 10.9 shows a general process scheme, in which carbon dioxide originates from a coal fired power plant. As renewable electricity depicted in the scheme is not continuously available, a temporary storage of hydrogen is foreseen to allow steady-state operation of the methanol reactor. As in the low-carbon production, the electrolysis potentially needs be followed by purification and compression.

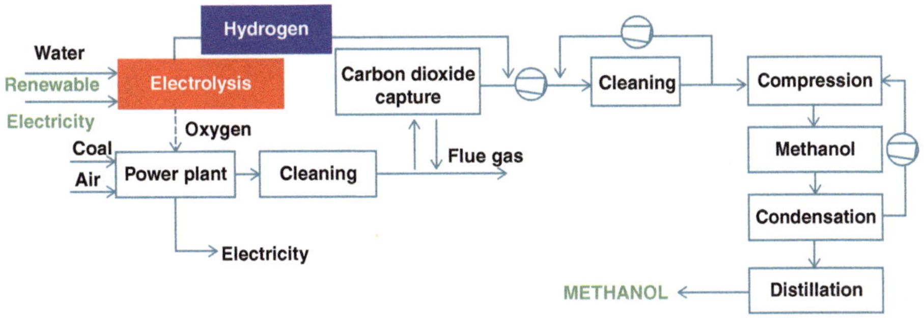

Figure 10.9 Power-to-methanol scheme using renewable electricity.

Alternative concepts to produce conventional synthesis gas are in early development phases and include direct electrochemical reduction of carbon dioxide and electrocatalytic co-reduction of carbon dioxide to carbon monoxide and water to hydrogen. These concepts are investigated in several research institutes on lab-scale, their TRLs are therefore relatively low (TRL 1–3).

In 2011, Carbon Recycling International started operation of the George Olah Renewable Methanol Plant and demonstrated the potential of tapping into Iceland's geothermal energy. The €7.1 million plant (for a capacity of 1,300 metric tons) was designed to produce 4,000 tons of renewable methanol per year. The feed consists of carbon dioxide from geothermal power plant and hydrogen produced by 5 MW water electrolysis fed by a geothermal power plant. All units are operated continuously. The methanol product is mixed into gasoline and substitutes up to 2.5% of Iceland's fuel consumption. Other uses are as feed in biodiesel production or in other methanol-based processes. In comparison to fossil-fuel based methanol, renewable methanol reduces greenhouse gases emissions by 90%.

10.7.4 Energy Demand

Water electrolysis is the main energy intensive step in the Power-to-Methanol production chain. Assuming 4.3 MWh/m^3 hydrogen as energy demand for water electrolysis, 9.52 MWh_{el} (34.3 MJ) are required for producing the 189 kg hydrogen stoichiometrically needed per ton of methanol. Another study quantified 9.2 MWh_{el}/t methanol.

Approximately 1.5 MWh/t (5.4 GJ) are additionally required for utilities (compressor, distillation) resulting in 11.02 MWh/ t (39.7 MJ) methanol in total. The total energy demand of the low-carbon process is substantially higher than for the conventional route (by a factor of three). However, the difference is less pronounced, if the feedstock is included in the energy demand of the fossil process. It must be emphasized that all hydrogen and CO_2-based processes generate the target molecules from the end pipe products water and carbon dioxide, whereas the fossil routes benefit from the energy content of the employed fossil feedstock. If replacing fossil feedstocks is a general goal, the energy content of the target products is part of energy demand of the process. Table 10.8 compares the conventional with the low-carbon methanol route.

10.8 Power-to-Formic Acid Technology

Formic acid is one of the main products that can be produced from carbon dioxide reduction having a market value around 500–900 US$/mt. Formic acid is considered as a C_1 building block and can also be used as a fuel in formic acid fuels cells (DFAFCs). Its production was expected to reach

Table 10.8 Comparison of energy demand and carbon dioxide emissions for methanol production routes.

Per ton of Methanol	Fossil(SMR + Methanol Synthesis)	Low Carbon(Power-to-Methanol)
Energy feedstock (GJ)	25	-
Fuel demand (GJ)	13.9	-
Electricity (GJ)	0.6	34.3
Utilities (GJ)	-	5.4
Steam balance (GJ)	−2	0
Total energy (GJ)	37.5 (12.5 excl. feedstock)	39.7 (41.7 incl. compensation for missing steam export)
Feedstock related CO_2 emission (t)	0.97	−0.79
Process emissions (t)	0.52	0.123
Total emissions (t)	1.49 (1.82 cradle to gate)	−0.67

760 Mt by 2019. Formic acid is also considered as one of the most promising and safest liquid hydrogen carriers with a hydrogen content of 52 g/L, whose release from formic acid is favoured at room temperature ($\Delta G^\circ = -32.8$ kJ/mol). The most common industrial process for formic acid production involves a two-step process, in which methyl formate is synthesized from methanol and carbon monoxide in the first step (Equation 10.9) and then, methyl formate is hydrolysed into formic acid in the second step (Equation 10.10). However, production of formic acid in the second step is not thermodynamically favourable.

$$CH_3OH + CO \rightleftarrows CH_3COOH \qquad \Delta H_r = -29\ \text{kJ/mol} \tag{10.9}$$

$$CH_3COOH + H_2O \rightleftarrows HCO_2H + CH_3OH \qquad \Delta H_r = +16.3\ \text{kJ/mol} \tag{10.10}$$

Formic acid can also be chemically synthesized by several other processes, such as oxidation of biomass, carbon dioxide hydrogenation, carbon dioxide electrochemical reduction, carbon dioxide reduction by biocatalysts and as a by-product of acetic acid production. This section will focus only on carbon dioxide hydrogenation and on the carbon dioxide electrochemical reduction.

10.8.1 Carbon Dioxide Electrochemical Reduction

As showed in Equations 10.9 and 10.10, the equilibrium towards formic acid production is unfavourable in the classical process, leading to the investigation of other possible processes for the direct synthesis of formic acid. The electrochemical reduction of carbon dioxide gained increased attention in the last few decades due to its environmental advantages as well as its potential to directly convert carbon dioxide into formic acid. However, as for all carbon dioxide electrochemical reduction processes, high overpotentials and low product selectivity are the main bottlenecks for the development of this process.

Thus, the development of less energy-intensive electrocatalysts is of importance to lower the overpotentials as well as to increase the selectivity to formic acid. Hori investigated the capability of different metals, such as Pb, Hg, Tl, In, Sn, Au, Ag, Zn, Pd, Ga, Cu, Ni, Fe, Pt, Ti to electrochemically reduce carbon dioxide to formic acid. They found that the product distribution was deeply dependent on the metal electrodes. Pb, Hg, Tl, In, Sn, Cd and Bi showed to be very efficient catalysts for the electrochemical reduction of carbon dioxide in formic acid with Faradaic efficiencies up to 99.5%. However, they required very negative overpotentials around −2.0 V vs. saturated calomel electrode (SCE) to reach good product selectivity. Platinum and palladium showed improved selectivity to formic acid, however, as they also catalysed the hydrogen evolution reaction, a side reaction to carbon dioxide reduction that lowers the Faradaic efficiency towards the formic acid production. Carbon monoxide poisoning is also a problem for platinum electrodes since the carbon monoxide formed during the carbon dioxide reduction covers the platinum surface, reducing the performance of the catalyst. However, Kortlever et al. reported that a palladium-platinum catalytic system reduced carbon dioxide to formic acid using very low overpotentials, starting from −0.05 V vs. reversible hydrogen electrode (RHE) at pH = 6.7. Nevertheless, the main challenge remains to avoid the carbon monoxide poisoning.

Since sulfur poisoning has been reported to be a considerable issue for HER catalysts it could be beneficial for competitive reactions. Thus, Huang et al. studied the influence of sulfur dopants in the performance of copper catalysts, aiming to limit the HER during carbon dioxide electrochemical reduction to formic acid. They showed that a copper-based catalyst, doped with 2.7 wt % of sulfur, exhibited a format current density 46 times higher than the undoped catalyst. The sulfur-doped catalyst showed a Faradaic efficiency of 75% over 12 hours, while its structure was shown to play a role in the selectivity to formic acid. Kumar et al. evaluated the performance of Sn-based catalysts on the carbon dioxide electrochemical reduction and they proved that a nanoporous and high-density grain boundary structure is an important factor to increase the selectivity and the rate of formic acid production from carbon dioxide electroreduction. The dioxide porous tin nanowires synthesized showed a carbon dioxide-to-formic acid efficiency of 56%, being one of the best performing catalysts reported for this process.

Biocatalysts have also been used to produce formic acid from carbon dioxide electrochemical reduction, presenting higher selectivity than conventional metal electrodes. Hwang et al. investigated the use of different methyl bacteria as a whole-cell biocatalysts for the electrochemical conversion of carbon dioxide to format. They found that the *Methylobacterium extorquens* AM1 had a high capability to convert carbon dioxide by suppling electrons, producing 60 mM of format without requiring any additional hydrogen supply. However, the synthesis was relatively slow, taking 80 hours to produce the 60 mM of format with 1.9 g of the catalyst. Similarly, Le et al. investigated the performance of the *Shewanella oneidensis* MR-1 due to it powerful electron transfer system. With an optimization of the *S. oneidensis* MR-1 growth, up to 136.84 mM of format was produced after 72 hours. The reaction rate was 3.8 mM h^{-1} g^{-1}, being almost 10 times faster when compared to the values previously obtained by Hwang et al.

10.8.2 Carbon Dioxide Hydrogenation

The actual industrial process for formic acid production from methanol and carbon monoxide (Equations 10.9 and 10.10) emits around 3,100 kg of carbon dioxide for each ton of formic acid produced. Carbon dioxide hydrogenation to formic acid (Equation 10.11) has gained increased attention since it could reduce the greenhouse gases emissions related to the formic acid production by a 10-fold, especially if coupled with a hydrogen production process using renewable energy, such as electrolysis. Moreover, this process has become a significant milestone to consolidate formic acid as a reversible hydrogen storage carrier. The hydrogenation of carbon dioxide in gas phase is not entropically favoured though since it involves the conversion of two gaseous reactants into liquid products. However, the reaction is favourable in aqueous medium (Equation10.12).

$$CO_{2(g)} + H_{2(g)} \rightleftarrows HCOOH_{(l)} \qquad \Delta G_{298K} = 32.8 \text{ kJ / mol} \tag{10.11}$$

$$CO_{2(ag)} + H_{2(ag)} \rightleftarrows HCOOH_{(aq)} \qquad \Delta G_{298K} = -4 \text{ kJ / mol} \tag{10.12}$$

Hence, the selection of the medium (solvents, water) is of crucial importance since it can allow for the reaction to be more thermodynamically favourable. In addition, basic additives, such as trimethylamine and ammonia are often used to shift the equilibrium towards formic acid production. Noble metals homogeneous catalysts are more commonly used in this process due to their promising catalytic performances. Noble metals, such as ruthenium, rhodium and iridium have shown excellent catalytic results and transition metals, such as nickel, copper and iron, have also been investigated despite their lower activity due to their lower cost compared to that of the noble metals.

Tanaka et al. studied the performance of a Ir(III)-PNP pincer complex in aqueous potassium hydroxide medium for the hydrogenation of carbon dioxide to formic acid. At 120°C and 6 MPa, the catalyst performed well, leading to a high TOF of 73,000 h^{-1} for 48 hours. They also showed that the use of a stronger base (potassium hydroxide instead of potassium phosphate) could increase formic acid yield from 60 to 70%, confirming the strong role of the base in the reaction. Liu et al. developed an efficient iridium complex containing imine-diphosphine ligands for carbon dioxide hydrogenation. The catalyst exhibited a TOF of 450,000 h^{-1} at 140°C for 20 hours in a 5 M potassium hydroxide medium. The catalytic performance of the iridium complex was attributed to the C=N bond of the ligand that acted as an acceptor of hydrogen that further led to the production of formate. Munshi et al. studied the influence of different bases and alcohols on the rate of supercritical carbon dioxide hydrogenation using $RuCl(OAc)(PMe_3)_4$, a ruthenium trimethylphosphine complex. The results revealed that the selection of the appropriate amine and alcohol have a great influence on the rate of the reaction. The use of pentafluoro phenol as alcohol and triethylamine as base at 50°C and 19 MPa for 20 min led to a TOF for formic acid production of 95,000 h^{-1}. The pentafluoro phenol alcohol, for example, could have acted either as a hydrogen donor or as a proton donor, favouring the hydrogenation reaction. To the best of my knowledge, Filonenko et al. developed a ruthenium complex catalyst that has showed the best results for formic acid production. They investigated the performance of a Ru-PNP-pincer catalyst in a batch reactor at 120°C and 2.7 MPa for 1 hour using dimethylformamide (DMF) as solvent and 1,8-diazabicyclo [5.4.0] undec-7-ene (DBU) as base and obtained a TOF for formic acid production of 1,100,000 h^{-1}. In agreement with the finding of Tanaka et al., they stated that strong bases play an important role in this reaction by affecting the rate determining step of the reaction. When a strong base is used, the initial hydrogen recombination is the rate-determining step.

Despite their excellent catalytic performances for hydrogenation of carbon dioxide into formic acid, these homogeneous catalysts are difficult to separate from the products at the end of the reaction and the amount of carbon dioxide hydrogenated per unit of time is still low, hindering their use at large scale. Moreover, these homogeneous catalysts may also promote the reverse reaction, in which the formate produced can be transformed back into carbon dioxide and hydrogen. To cope with the problems related to homogeneous catalysts, studies were recently published on the use of heterogeneous catalysts for the hydrogenation of carbon dioxide.

Umegaki et al. were the first to study the performance of ruthenium nanoparticles (unsupported catalyst) in the hydrogenation of supercritical carbon dioxide into formic acid. The reaction was carried out at 353 K and 13 MPa for 3 hours in a mixture of water and trimethylamine. They observed that the metallic nanoparticles were very active and stable during the test, leading to a TOF of formic acid of 2,117 h^{-1}. The stability of the catalyst was attributed to the fact that the nanoparticles maintained their metallic state after the test.

The influence of ruthenium-based catalysts doped on different supports, such as MgO, Al_2O_3 and activated carbon was studied by Hao et al. Each support was doped with 2 wt % Rh and the hydrogenation tests were performed at 353 K and 13.55 MPa in the presence of ethanol as solvent and triethylamine as base, which were added to extract the formic acid, increasing the reaction rate. The Ru/AC and the Ru/Al_2O_3 catalysts showed good catalytic performances with turnover numbers (TON) of 10 and 91, respectively. The catalytic performance of the catalysts was attributed to the presence of hydroxyl groups on the surface of the catalysts that increased the adsorption of carbon dioxide. The higher yield of formic acid obtained with the Ru/Al_2O_3 catalyst compared to the Ru/AC was in fact due to its greater number of hydroxyl groups. The magnesium oxide support was chosen for its strong basic sites, eliminating the need of using basic additives that need to be neutralized and separated from the product at the end of the reaction. However, no formic acid was produced in the presence of this catalyst, probably due to the absence of hydroxyl groups on its surface as well as the high pH of the solution that led to the formation of inactive ruthenium dioxide species. Similarly, Zhang et al. obtained a good catalytic performance of a Ru-based catalyst doped on an alumina support. They investigated the performance of the Ru-PPh_3/Al_2O_3 catalyst as well as the presence of additives and solvents in the carbon dioxide hydrogenation to formic acid reaction in a batch reactor at 80°C and 12 MPa for 1 hour. The catalyst showed a high turnover frequency of formic acid of about 751 h^{-1} in a mixed solution of ethanol, triethylamine and water and in the presence of potassium dihydrogen phosphate and PPh_3 as additives. They reported that the high TOF obtained was related to the use of PPh_3 as proton donor, potassium dihydrogen phosphate as proton source and trimethylamine as basic additive, increasing the conversion of carbon dioxide.

Su et al. studied the performance of different heterogeneous palladium-based catalysts doped on different supports. In this work, all catalysts were doped with 5 wt % of palladium and the best hydrogenation results were obtained at 20°C for 1 hour in a 1 M ammonium hydrogen carbonate medium. They showed that the carbon dioxide hydrogenation test performed at 2.7 MPa with the Pd/Al_2O_3 catalyst resulted in a TOF of 278 hours while the hydrogenation test, performed at 5.5 MPa with the Pd/AC (activated carbon) nanocatalyst, showed a TOF as high as 1,103 hours. The main reasons for the difference in the performance of both catalysts was the localized higher hydrogen concentrations on the surface of the activated carbon-based catalysts as well as a higher palladium dispersion that favoured the hydrogenation reaction.

10.9 Power-to-Formaldehyde Technology

Formaldehyde is currently used in about 50 industrial processes as a building block to produce products such as paints, cosmetics, resins, polymers, plastics etc. It is mainly produced via the energy and cost-intensive Formox process in which methanol is partially oxidized at 300–400°C (Equation 10.13). Methanol used in the process is produced from synthesis gas, which in turn is produced from steam reforming of natural gas. More than 35% of the world methanol production is used for formaldehyde production. The drawbacks of this process are high temperatures and energy-intensive compression and purification steps, affecting the overall economics of the process.

$$CH_3OH + 1/2O_2 \rightleftarrows HCHO + H_2O \qquad \Delta H_{298K} = -159 \text{ kJ/mol} \tag{10.13}$$

Since formaldehyde is a hydrogen-rich molecule, it has a great potential to be used as a liquid hydrogen carrier, capable of delivering high-purity hydrogen in a hydrogen fuel cell. Moreover, it could be even more interesting than methanol as liquid hydrogen carrier since the formaldehyde reforming process is much less energy intensive than methanol reforming. However, to produce hydrogen with a low carbon footprint, formaldehyde has to be produced more sustainably. There have been recent efforts for developing new processes for sustainably producing formaldehyde, such as carbon dioxide hydrogenation, electrochemical and enzymatic approaches. However, the literature on this subject is relatively sparse, especially when heterogeneous catalysts are considered. The main findings on the formaldehyde production via carbon dioxide reduction will be detailed in the following sections.

10.9.1 Carbon Dioxide Electrochemical Reduction

Very few reports from the open literature are about the electrochemical reduction of carbon dioxide to formaldehyde. To date, Nakata et al. reported the highest Faradaic efficiency to formaldehyde obtained through carbon dioxide electroreduction. The authors investigated the carbon dioxide reduction using boron-doped diamond (BDD) electrodes in various electrolytes, such as methanol and seawater. When seawater was used as electrolyte, a relatively low formaldehyde yield of 36% was obtained due to impurities in the seawater and to the narrow potential window in water. On the other hand, a Faradaic efficiency of 75% to formaldehyde was obtained at −1.5 V (vs. Ag/AgCl). When a glassy carbon electrode was used, a Faradaic efficiency of only 15% was obtained under the same conditions. The performance of the boron-doped diamond electrodes was attributed to the presence of sp^3-bonded carbon, while glassy carbon has large amounts of sp^2-bonded carbon, resulting in low Faradaic efficiencies. Furthermore, boron-doped diamond electrodes presented very good stability over 20 hours, while the surface of the glass carbon electrode dramatically changed after 20 hours of electrolysis, highlighting the potential of the boron-doped diamond electrodes to formaldehyde production via carbon dioxide electrochemical reduction.

An electrochemical carbon dioxide reduction reaction is key to achieving practical renewable energy conversion and storage as well as green chemical production. Sn-based electrocatalysts are believed to be one of the most promising candidates for the electroreduction of carbon dioxide to formate. However, development of novel catalysts operating at low overpotentials with a high selectivity, large current density and high single-pass conversion of carbon dioxide (SPC) simultaneously remains a significant challenge. Here, we report a nanorod@sheet SnO (NRS-SnO) catalysts designed with abundant exposed Sn(II) active sites in the 2D nanosheets for high-performance electroreduction of carbon dioxide to formate. Formate was first reliably detected at an applied potential as low as −0.2 V (vs. RHE) and reached maximum Faradaic efficiency (FE) of approximately 94% at −0.7 V (vs. RHE) with a partial current density (J_{HCOO^-}) of approximately −330 mA/cm^2, giving the highest cathodic CO_2-to-formate energy conversion efficiency (approximately 70%) among known Sn catalysts for CO_2RR to formate. Importantly, a high conversion of carbon dioxide of 39.3% was achieved with the present system, proving for the first time that high SPC can be achieved for GDE based alkaline flow cell.

10.9.2 Carbon Dioxide Hydrogenation

Similar to the electrochemical reduction, there are only few reports in the literature on carbon dioxide hydrogenation to formaldehyde using heterogeneous catalysts. Lee et al. studied the production of formaldehyde via carbon dioxide hydrogenation over a Pt-Cu/SiO_2 catalyst at 423 K and 600 kPa and varying the H_2/CO_2 feed ratio from 3 to 20. The authors reported that the rate of formaldehyde production increased considerably with increase of the H_2/CO ratio, ranging from 0.21×10^{-4} mol/min/gcat at H_2/CO = 3–0.87×10^{-4} mol/min/gcat at H_2/CO = 20. Conversely, the rate of methanol production greatly decreased with an increase of the H_2/CO ratio. These results led the authors to conclude that the relative concentration of surface hydrogen on the catalyst play a major role in the selective production of formaldehyde. For comparative purposes, Cu/SiO_2 catalyst was also tested in the same conditions. However, in this case, when a $H_2/CO = 20$ was used, the formaldehyde production was negligible. The authors concluded that Pt played an important role in the adsorption of hydrogen, which would then migrate to the copper surface, promoting the carbon dioxide hydrogenation into formaldehyde.

Chan et al. recently reported the production of formaldehyde from carbon dioxide hydrogenation in aqueous solution, using formic acid as intermediate of the carbon dioxide hydrogenation process. The performance of the Pt-Cu/γ-Al_2O_3, Pt-Ni/γ-Al_2O_3, Ru-Cu/γ-Al_2O_3 and Ru-Ni/γ-Al_2O_3 catalysts was evaluated at 7 MPa and at a temperature range of 298–363 K. The authors reported that the conversion of formic acid into formaldehyde is possible, despite not been thermodynamically favoured, with Pt-Cu/γ-Al_2O_3 being the catalyst with the highest formic acid conversion and formaldehyde yield ($\approx$ 6 mmol/l/g.cat).

However, all the formic acid converted was not transformed into formaldehyde. The authors suggested that this could be related to the thermal decomposition of formic acid. However, they did not evaluate the formation of other products to confirm this hypothesis.

10.10 Selective Hydrogenation of Carbon Dioxide to Light Olefins

10.10.1 Introduction

Well-defined mesoporous carbon (MPC) was fabricated and used as a support material of iron catalysts to produce light olefins, alcohols and acids from direct CO_2-FTS. The carbon dioxide hydrogenation activity of FeK/MPC was compared with that of the conventional activated carbon and metal oxide-supported iron catalysts.

Increasing shale gas production has helped reduce the delivered costs of natural gas and this has made natural gas attractive for ethylene production in ethane steam cracking units. There were mixtures of steam, carbon monoxide, carbon dioxide, methane, C_2-hydrocarbons (ethane and ethylene), C_3-hydrocarbons (propane and propylene), C_4-hydrocarbons (butane and mixed butylene) and heavier C_5+ hydrocarbons from the outlet of the Fischer-Tropsch reactor.

A high carbon dioxide conversion of 46.7% and C_5-C_9 hydrocarbons selectivity of 48.3% are observed over the hybrid catalyst composed of copper-zinc oxide and ZSM-5 (SiO_2/Al_2O_3 = 23) when the n-hexane fluid at partial pressure of 1.5 MPa is employed. The selectivity for C_5-C_9 hydrocarbons can be further increased by loading palladium promoter on the ZSM-5(23) zeolite. Carbon monoxide and dimethyl ether are dominantly produced at a low carbon dioxide conversion of 28.8% without using the n-hexane fluid. In addition, C_8-C_{13} *iso*-paraffins (jet fuel fractions) has a selectivity of 51.7% under existing the n-hexane fluid, when USY zeolite is used in the hybrid catalyst.

As previously stated, there has been a selective conversion of carbon dioxide to lower olefins through carbon dioxide hydrogenation over a ZnZrO/SAPO tandem catalyst fabricated with a zinc oxide-zircon dioxide solid solution and a zinc-modified SAPO-34 zeolite, which can achieve the lower olefins selectivity as high as 80–90% among hydrocarbon products. This is realized based on the dual functions of the tandem catalyst: hydrogenation of carbon dioxide on the zinc oxide-zircon dioxide solid solution and lower olefins production on the SAPO zeolite. The thermodynamic and kinetic coupling between the tandem reactions enable the highly efficient conversion of carbon dioxide to lower olefins. Furthermore, this catalyst is stable towards the thermal and sulfur treatments, showing the potential industrial application.

Carbon dioxide hydrogenation is one of the promising ways not only to reduce the atmospheric carbon dioxide concentration that is a major cause of global warming, but also to produce to carbon monoxide, methane and liquid hydrocarbons (Figure 10.10). Using a catalytic reactor is also one approach to the mass storage of surplus electricity generated from renewable energy sources. It has been established that some catalytic phases can activate carbon dioxide to able its conversion under hydrogenation conditions. Most recently Furimsky gave detailed accounts of catalyst structures developed for the hydrogenation of carbon dioxide.

Direct hydrogenation of carbon dioxide occurs by the sequential reactions of reverse water gas shift (RWGS) is shown in Equation 10.3) and Fischer-Tropsch synthesis (FTS) in Equation 10.14.

$$nCO + 2nH_2 \rightleftarrows (-CH_{2-})_n + nH_2O\ (\Delta H^{\circ}_{300K} = -166\ kJmol^{-1}) \tag{10.14}$$

In this series of reactions (here after referred to as CO_2-FTS), the equilibrium conversion of RWGS, which is an endothermic and reversible reaction, is limited to approximately 23% for typical FTS reaction conditions of 300°C and 2.5 MPa, while an irreversible exothermic FTS reaction has no thermodynamic limitation. Thus, the overall carbon dioxide conversion might depend on the consumption rate of the carbon monoxide-intermediate via the FTS reaction. I note that another route of carbon dioxide hydrogenation, which includes methanol synthesis and subsequent hydrocarbon formation over the In_2O_3 /HZSM-5 bifunctional catalyst, was recently suggested to improve the long-chain hydrocarbons selectivity.

Iron-based catalysts have been used in the CO_2-FTS process because they have catalytic activities in both RWGS and FTS reactions. The iron-based catalysts are preferred to produce olefin hydrocarbons, but the product distribution depends on the types and combinations of active metals promoters and support materials. The application of various supports allows the modification of the metal dispersion, particle size and metal-support interactions due to the physico-chemical properties of the supports. Metal or non-metal oxide supports such as aluminium oxide, silicon dioxide and titanium dioxide have been used widely in iron-based catalysts with alkali promoters to produce olefin-rich hydrocarbon mixtures. For many researchers, aluminium oxide has been regarded as the best support for iron-based catalysts to obtain a lower proportion of an olefins fraction without sintering of iron particles during the CO_2-FTS reaction due to a strong metal-support interaction. Wang et al. suggested that zirconium dioxide is an effective support for iron catalysts to obtain C_2-C_4 olefins from the CO_2-FTS reaction. However, a strong interaction between the metal and the metal oxide support inhibits the reduction and carburization of an active iron phase during the CO_2-FTS reaction.

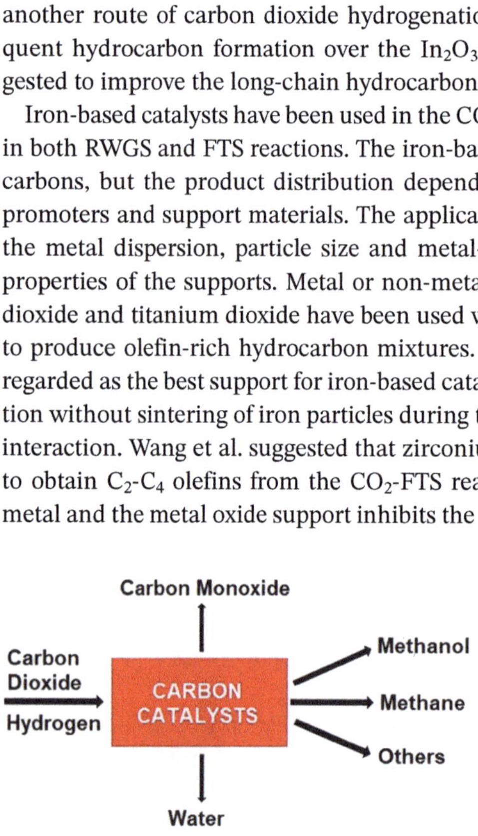

Figure 10.10 Carbon dioxide hydrogenation to methanol and methane over carbon-supported catalysts.

Recently, various types of carbon allotropes such as carbon nanotubes (CNTs), carbon nanofibres (CNFs), activated carbons (ACs), carbon nanotubes (CNTs), and mesoporous carbons (MPCs) have been regarded as promising catalytic supports due to their inert surface properties, thermal stability, weak metal-support interaction and high surface area with a porous structure. However, many studies on the carbon-supported iron catalysts have focused on a conventional FTS reaction rather than the more complicated CO_2-FTS. Most carbon-supported iron catalysts for CO_2–FTS utilize CNTs and nitrogen-doped carbon as a support material. CO_2-FTS with the catalyst using nitrogen-doped carbon supports is beneficial to the carbon dioxide conversion, but it also gave rise to high selectivities of methane and light hydrocarbons due to the unique characteristics of metal-support reactant interactions by an interface charge transfer.

A tunable pore structure of carbon materials is one of the requirements to produce gasoline range hydrocarbons by CO_2-FTS because the pore structure of the support determines the size of active metals and mass transfer properties, both of which can largely affect the product distribution. Activated carbon is one of the most common amorphous carbons, which has microstructures and less graphitic properties. Highly graphitized, ordered mesoporous carbon might also be a promising candidate for the support materials of iron-based catalysts for CO_2-FTS. However, graphitized mesoporous carbon has not been extensively investigated for the application of carbon dioxide hydrogenation, unlike carbon nanotubes (or nitrogen-doped CNTs) and carbon nanofibres (or nitrogen-doped CNFs).

Well-defined mesoporous carbon was fabricated and used as a support material of iron catalysts to produce liquid fuels from direct CO_2-FTS. The carbon dioxide hydrogenation activity of Fe-K/MPC was compared with that of the conventional activated carbon and metal oxide-supported iron catalysts. In addition to basic characterizations of the catalysts prepared, additional physico-chemical properties such as iron phase transition and mass-transfer behaviour were also observed. Our understanding of the carbon-based iron catalyst may shed a light on finding an effective catalyst for obtaining liquid hydrocarbons from CO_2-FTS. Identifying highly efficient catalysts for direct carbon dioxide hydrogenation to liquid hydrocarbons is crucial for carbon dioxide utilization via chemical conversion using renewable resources, the so-called Power-to-Liquids. In the study (Hwang et al.), well-defined mesoporous carbon of 6.9 nm pore size was synthesized as a support material for iron oxycarbide nanoparticles, which serve as active sites for the carbon dioxide hydrogenation reaction, combining reverse water gas shift and Fischer-Tropsch reactions. The unique physicochemical properties of MPC favored the formation of an active carbidic iron phase and the rapid diffusion of produced hydrocarbons which resulted in an enhanced carbon dioxide conversion and long-chain hydrocarbon selectivity. As a result, the MPC supported iron catalyst showed C_5+ hydrocarbon selectivity of 44.5% with a carbon dioxide conversion of 50.6% (300°C, 2.5 MPa, $H_2/CO_2 = 3$).

10.10.2 Carbon Dioxide via FTS to Lower Olefins

Catalytic activity of the catalysts prepared in the study (Fe-K/support) towards CO_2-FTS was evaluated in a fixed-bed reactor with an inner diameter and length of 7.35 mm and 350 mm, respectively. A mixture of 0.3 g of catalyst and 1.5 g of α-Al_2O_3 diluent was loaded at the centre of the reactor for the test. Prior to the CO_2-FTS reaction, the catalyst was reduced with syngas (H_2/CO = 2) at 350°C for 3 hours. After pressurized to 2.5 MPa with helium. The feed stream was switched to a mixed reactant gas of hydrogen (72 vol%)/CO_2 (24 vol%)/Ar (4 vol% as an internal standard). The CO_2-FTS reaction was performed at 300°C and 2.5 MPa with a gas hourly space velocity of 2,000 ml/g_{cat}h.

The catalytic performance of Fe-K/MPC, Fe-K/AC, Fe-K/A12O3, Fe-K/TiO2 and Fe-K/SiO_2 towards the CO_2-FTS reaction was evaluated. Figure 10.11 displays the product yield and the olefin selectivity in C_2-C_4 hydrocarbons. The carbon dioxide conversion was highest on Fe-K/MPC (50.6%), followed by Fe-K/AC (37.5%), Fe-K/Al_2O_3 (36.5%), Fe-K/TiO_2 (30.7%) and Fe-K/SiO_2 (14.1%). Fe-K/MPC showed the highest C_5^+ selectivity (44.5%) as well as the low methane selectivity (15.4%), as summarized in Table 10.9.

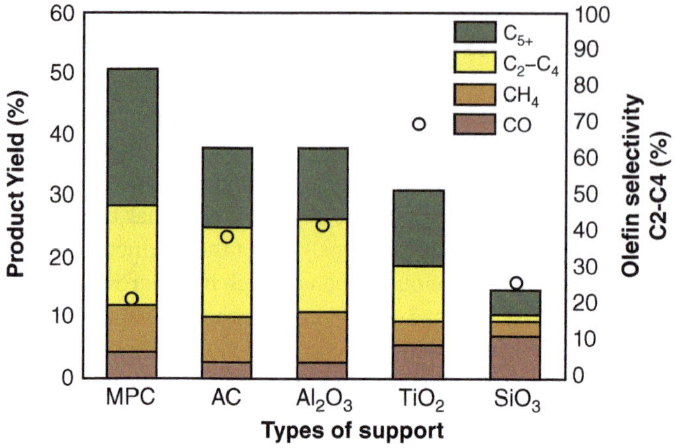

Figure 10.11 Product yield and olefin selectivity in C_2-C_4 hydrocarbon.

Table 10.9 Carbon dioxide conversion (X_{CO_2}) and product selectivity for carbon dioxide hydrogenation reaction (Hwang et al., 2020).

Catalysts	X_{CO_2} (%)	Selectivity					
		CO	C_1	C_2	C_3	C_4	C_{5+}
FeK/APC	50.6	8.2	15.4	10.4	13.1	8.4	44.5
Fe/AC	37.5	6.4	19.8	13.4	16.1	9.6	34.6
FeK/Al_2O_3	36.6	9.6	12.4	10.4	14.0	9.9	43.7
FeK/TiO_2	30.7	17.6	12.6	9.9	11.4	7.3	41.2
FeK/SiO_2	14.1	48.9	17.7	4.2	2.2	0.4	26.6
FeK/MPC[a]	37.2	8.4	18.3	11.4	13.6	8.2	40.2
FeK/MPC[b]	52.4	5.3	23.1	13.0	15.7	9.7	33.2

[a] Reaction condition: 300°C, 2.5 MPa, space velocity = 8000 ml/gcat.h, $H_2/CO_2 = 3$.
[b] Reaction condition: 350°C, 2.5 MPa, space velocity = 2000 ml/gcat.h, $H_2/CO_2 = 3$.

Although Fe-K/AC and Fe-K/Al_2O_3 showed similar carbon dioxide conversion rates, light olefin production was favoured by Fe-K/Al_2O_3. The light olefin selectivity is affected by several factors such as iron particle size (iron carbide) and pore size of catalysts. In the catalyst with large pore structure, the second olefins through hydrogenation, isomerization and chain-growth reaction is suppressed because of enhanced diffusion efficiency in large pores. In their results, the better olefin selectivity in C_2-C_4 of Fe-K/Al_2O_3 was due to the complex interplay of the larger particle size and pore size than that of other catalysts. FeK/SiO_2, which exhibited the lowest carbon dioxide conversion, produced carbon monoxide with the highest selectivity of 48.9%.

Hwang at al., performed CO_2-FTS over the Fe-K/MPC at different reaction conditions, such as higher reaction temperature and space velocity, which were also summarized in Table 10.9. As the reaction temperature increased from 300°C to 350°C, the carbon dioxide conversion slightly increased from 50.6%–52.4%, while the C_5^+ selectivity significantly decreased from 44.5%–33.2% due to accelerated methane formation at 350°C.

Figure 10.12 shows the hydrocarbon distribution of liquid products obtained using Fe-K/MPC catalyst in the range C_{15}–C_{27}. Fe-K/MPC catalyst exhibited similar liquid hydrocarbons distributions

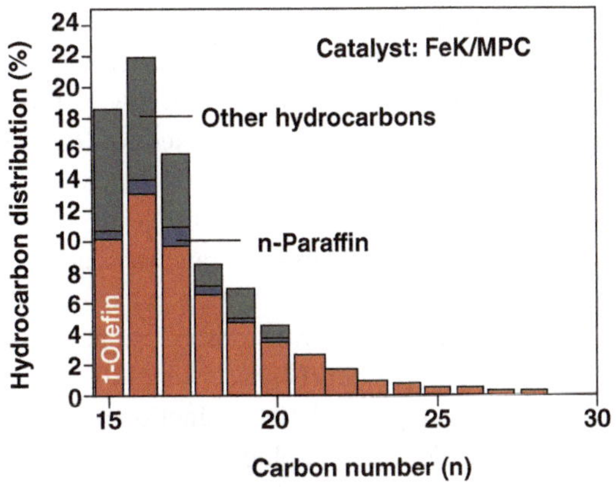

Figure 10.12 Chain growth of Fe-K/MPC catalyst in the range of C_{15} – C_{27}, X_{CO_2} = 0.65 (C_{15} - C_{27}).

showing similar chain growth probability as Fe-K/AC catalyst. This suggests that the difference in carbide phase does not significantly affect the liquid hydrocarbons composition.

The distribution of products and their yields can be influenced by operating conditions. In this case, the type of catalyst has a dominant effect. Gradual conversion of carbon dioxide involves several surface intermediates. Special geometry of the active site is required for maximizing the yield of methanol and for minimizing the reaction runaway to methane. Such a state can be attained over catalysts that contain CuO-ZnO components. Catalysts containing other transition metals and noble metals are less suitable, while the same catalysts exhibit good performance during the hydrogenation of carbon dioxide to methane. In technological applications, catalysts supported on oxidic supports, i.e., aluminium oxide, silicon dioxide, titanium dioxide, etc., have been used. During hydrogenation of carbon dioxide, 1 and 2 mol of water are generated for every mole of methanol and methane, respectively. The stability of catalysts used for hydrogenation of carbon dioxide can be enhanced by replacing oxidic supports with carbon supports. Amorphous carbons, such as activated carbon and carbon nanomaterials, such as carbon nanotubes, carbon nanofibres and graphene-derived solids, have been tested both as catalysts and supports. For methanol synthesis, CuO-ZnO catalysts supported on these carbons have been evaluated. Similar carbon supports have been used for the preparation and testing of methanation catalysts consisting of a wide range of transition and noble metals as active metals. Both monometallic and bimetallic catalysts were evaluated.

10.10.3 Methane via FTS to Lower Olefins

Increasing shale gas production has helped reduce the delivered costs of natural gas ($4.8–5.8 per MMBtu) and this has made natural gas attractive for ethylene production in ethane cracking plants. Recent works found that low olefins could be directly produced through the conversion of syngas by the Fischer-Tropsch (FT) synthesis. Torres Galvis et al. reported that supported iron nanoparticles exhibited a high selectivity for low olefins, which represented over 60 mol % of the total product, with a carbon monoxide conversion rate of 70–80 wt % when the H_2 /CO molar ratio was changed to 1.0 and modifications were made to the reaction conditions such as the pressure and temperature. Recently, Jiao et al. improved the selectivity for low olefins to 80% with a carbon monoxide conversion rate of 15 wt %. However, wide-product-distribution remains a challenge with a theoretical limit of only 58% for C_2–C_4 hydrocarbons. Moreover, a competing water gas shift reaction also occurred with a conversion ratio of over 40% carbon monoxide in the above two kinds of catalysis. Thus, a large ratio of carbon monoxide was converted to carbon dioxide instead of hydrocarbon and this would decrease the total low-olefin production amount.

FT synthesis using metal catalysts such as cobalt, iron and ruthenium has been utilized in the gas-to-liquid (GTL) process to produce liquid hydrocarbon fuels (liquefied petroleum gas, gasoline and diesel) with low sulfur contents from synthesis gas (carbon monoxide and hydrogen) derived from coal, biomass and natural gas.

By using FT synthesis and modifying the reaction conditions and catalysis, they attempted to develop an economical and environmentally friendly process with low olefins as the products. Despite the higher addition value of low olefins, the FT to olefins (FTO) process requires a more complex separation process for mixed low olefins separation, which may increase the capital and operating costs. The competing WGS reaction is also a serious concern, whereby large amounts of carbon monoxide react with water to form carbon dioxide and the production amount of olefins is reduced. Thus, low olefins yield rate was a critical issue that should be improved. A high selectivity to low olefins usually comes with a low conversion rate of carbon monoxide based on the current FTO performance data over a broad range of process conditions (pressure, temperature, H_2/CO molar ratio and gas hourly space velocity). Thus, it requires a feasible way of utilizing the FT synthesis reaction for low-olefin production on a commercial scale combining with high performance of catalysis. Also, methodology for reducing the operating cost should be investigated due to the high energy consumption during the mixed low olefins separation.

Liu et al. have proposed a new commercial-scale process based on FT synthesis reaction for low-olefin production and optimized the process through exergy recuperation and heat integration to minimize the operating costs. Many kinds of FTO catalysts have also been prepared and investigated the effect on the economic performance of the FTO process. Greenhouse gas emission was dramatically reduced by heat integration and recycling by-product carbon dioxide to the combined steam and dry reformer for synthesis gas production. Furthermore, we investigated its feasibility through techno-economic assessment on the FTO process based on two different types of FTO catalysts. Sensitivity analysis of effect of selectivity to low olefins on the economic aspect of the FTO was also conducted.

Iron/cobalt-based catalysts are the main catalysts in the FT reactor used in the GTL to produce long chain hydrocarbons, while the FTO reaction generates short chain hydrocarbons at a higher temperature and lower pressure. Moreover, the FT synthesis in the FTO reaction is a catalytic process that converts synthesis gas at a molar ratio from 1.0 to 2.0, to low olefins and paraffins. The FTO reactor was assumed to be isothermal. A schematic graph of catalyst performance of the FTO catalysis is shown in Figure 10.13. The figure shows that a lower conversion ratio of carbon monoxide results in a high selectivity to low olefins, which means that most unreacted synthesis gas went to the separation part. Recycling of the unreacted syngas improves the conversion rate; however, this requires a larger separation unit and FTO reactor, which would increase the total capital costs. Therefore, trade-offs among the carbon monoxide conversation rate, WGS reaction ratio and selectivity to low olefins should be investigated.

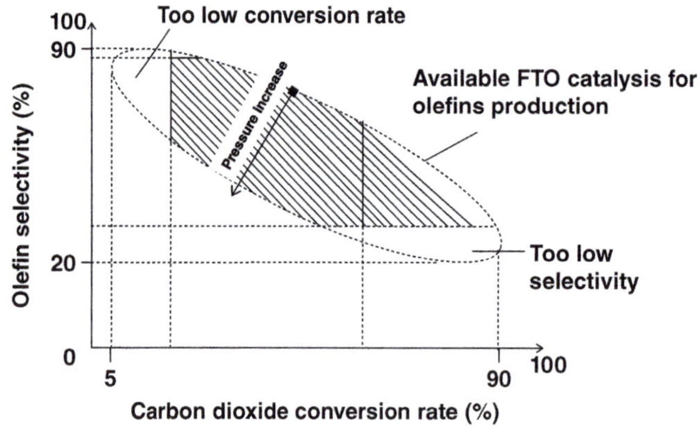

Figure 10.13 FTO catalysis performance plotting map.

Iron-based catalysts are best for FTO with a relative high olefins selectivity and carbon monoxide conversion rate. The catalysis performance data are summarized in Table 10.10 There were mixtures of steam, carbon monoxide, carbon dioxide, methane, hydrocarbons C_2 (ethane and ethylene), hydrocarbons C_3 (propane and propylene), hydrocarbons C_4 (butane and mixed butylene) and hydrocarbons C_5+ (heavier C_5) from the outlet of the FT reactor. The separated condensed water contained the by-product oxygenates, mainly alcohols and organic acids.

10.10.4 Carbon Dioxide via FTS to Liquid *iso*-C_5-C_{13}-Alkanes

Because of environmental issues caused by the excessive consumption of fossil fuels, great effort has been made to develop effective ways to transform carbon dioxide from greenhouse gas into valuable chemicals. This process will not only contribute to manufacture chemicals and hydrocarbon fuels, but also benefit for reducing carbon dioxide emission. A lot of attention has been attracted by hydrogenation of carbon dioxide to produce C_1 chemicals, such as methanol, dimethyl ether, methane and C_2-C_4 hydrocarbons. However, hydrogenation of carbon dioxide to C_5+ hydrocarbons, especially in the range of gasoline and jet fuel, is the least studied and characterized process. Developing an effective catalyst is still the crucial factor for further application of the process. Multifunctional active components should be provided from the desired catalyst for conducting hydrogenation of carbon dioxide to C_5+ hydrocarbons, due to it involving carbon dioxide activation and sequential formation of C-C bonds reactions (Wang et al).

How to maximize the synergistic effect of the different active components is a key challenge of using a hybrid catalyst. Many efforts have been made in controlling the proximity of different active components by changing their integration manner, which includes dual bed, granule-mixing, powder mixing and so on. However, it is not easy to ensure a high synergistic effect by a simple spatial arrangement of the hybrid catalyst in the reaction system, noting that the results of products distribution exhibited significant distinction, even among same integration manner (Zecevic, et al.). It means that other methods should be developed to further improve the performance of the hybrid catalyst. Hydrogenation of carbon dioxide to hydrocarbons is an exothermic reaction and conducted in a fix-bed reactor at a relative low temperature. The rapid extraction of the reaction heat and C_5+ products can strongly promote the productivity of hydrocarbons from carbon dioxide. Therefore, the strategy is to use a solvent in the reaction, which is expected to accelerate the desorption of products from the active sites. Owing to the high solubility and diffusion of substances in supercritical fluids (Darmawan et al.) supercritical fluids are widely used for the extraction and separation in a variety of industrial processes. Previously their studies also showed that hydrocarbons in the LPG fraction and gasoline fraction were directly synthesized from the

Table 10.10 Catalytic results of FTO catalysts.

Catalyst	*T* (°C)	P(MPa)	CO Conv. (%)	CO_2 Selects (%)	CH_4 (%)	C_2-C_4 Olefins (%)	C_2-C_4 Paraffins (%)	C_5^+ (%)
FeMnCuK	330	1	90	27.3	23.4	21.2	7.6	20.4
Fe_xO_y	330	1	39.8	23.4	10.5	25.4	17.7	33.5
Fe/αAl_2O_3 (6%wt Fe)	340	2	77	46	24	35	21	20
Fe/CNT	340	2	88	51	13	52	12	23
$ZnCrO_4$	350	2.5	17	42	2	80	10	8
Co/meso-HZSM-5	240	1	79	2.3	13.4	8.69	12.51	65.4

synthesis gas over a hybrid catalyst when a near-critical n-hexane fluid was employed. These facts inspire us to introduce a near-critical n-hexane fluid into the hybrid catalyst for the hydrogenation of carbon dioxide to liquid hydrocarbons.

In the work, a near-critical n-hexane fluid is employed to directly synthesize C_5-C_{13} *iso*-paraffins over a hybrid catalyst in the hydrogenation of the carbon dioxide reaction. The effect of partial pressure of n-hexane fluid and reaction temperature on the catalytic properties are investigated to demonstrate the role of near-critical n-hexane fluid in the reaction system. The relationship between the pore structure and the acid properties of zeolite and the distribution of hydrocarbons is found by using various zeolites in the hybrid catalysts. A method using a palladium promoter has also been developed to improve the gasoline fractions over the hybrid catalyst. Direct carbon dioxide hydrogenation to hydrocarbons via methanol synthesis is carried out over hybrid catalysts composed of Cu–ZnO and various zeolites by introducing a near-critical n-hexane fluid into the reactor. A high carbon dioxide conversion of 46.7% and C_5-C_9 hydrocarbons selectivity of 48.3% are observed over the hybrid catalyst composed of Cu–ZnO and ZSM-5 ($SiO_2/Al_2O_3 = 23$) when the n-hexane fluid at partial pressure of 1.5 MPa is employed. The selectivity for C_5-C_9 hydrocarbons can be further increased by loading palladium promoter on the ZSM-5(23) zeolite. Carbon monoxide and dimethyl ether are predominantly produced at a low carbon dioxide conversion of 28.8% while without using the n-hexane fluid. In addition, C_8-C_{13} *iso*-paraffins (jet fuel fractions) has a selectivity of 51.7% under existing the n-hexane fluid when USY zeolite is used in the hybrid catalyst.

The catalytic performances of the hybrid catalysts after 6.5 hours of reaction are shown in Figure 10.14. It can be seen from Figure 10.14a that all the hybrid catalysts exhibit similar carbon dioxide conversions, which are 46.7%, 49.0% and 48.3% for the ZSM-5(23), 0.5 wt % Pd ZSM-5(23) and 1.0 wt % Pd/ZSM-5(23), respectively. However, the product distribution is greatly influenced by the loading of palladium. The selectivity for the C_5-C_9 hydrocarbons is promoted from 48.3% to 58.3% and 60.1% by increasing the loading amount of palladium. This trend is consistent with that observed the conversion of synthesis gas to hydrocarbons over the hybrid catalyst containing Pd/ZSM-5. The selectivity for C_5-C_9 hydrocarbons are promoted and the corresponding selectivity for C_3-C_4 hydrocarbons are declined over the palladium added catalysts. A possible reason is that the hydrogenation of the carbonaceous species inside the zeolite channel is enhanced over the palladium modified ZSM-5 due to the hydrogen spillover effect and simultaneously suppressing their cracking. Meanwhile, the loading of palladium on ZSM-5 will also cause an increase of methane formation, it is possible because partially loaded palladium over the external suface of the zeolite enhances the direct methanation of carbon dioxide. In addition, Figure 10.14b shows that the ratios of *iso*-paraffins to n-paraffins is increased by using Pd/ZSM-5, suggesting the isomerization reaction is also promoted by doping palladium.

A highly efficient direct carbon dioxide hydrogenation to liquid fuels ($> C_3^+$) requires the development of effective catalysts with high activity and selectivity for the tandem reactions of RWGS and FTS. Fe-K/MPC exhibited excellent catalytic activity and selectivity towards long-chain hydrocarbons (C_5^+) in direct carbon dioxide hydrogenation. The catalyst performance in CO_2-FTS is highly dependent on the ratio of iron carbide and mesoporous structure of MPC in the catalyst system. The mesoporous structure provides the benefits of fast mass transfer of hydrocarbon molecules, which results in the enhancement of carbon dioxide conversion and C_5^+ hydrocarbon selectivity.

The feasibility of using a newly developed FTO process for direct low-olefin production was investigated based on the performance of existing developed catalysis for low-olefin production.

The hydrogenation of carbon dioxide to hydrocarbons was carried out over hybrid catalysts composed of Cu–ZnO and different zeolites in a near-critical n-hexane fluid. The near-critical

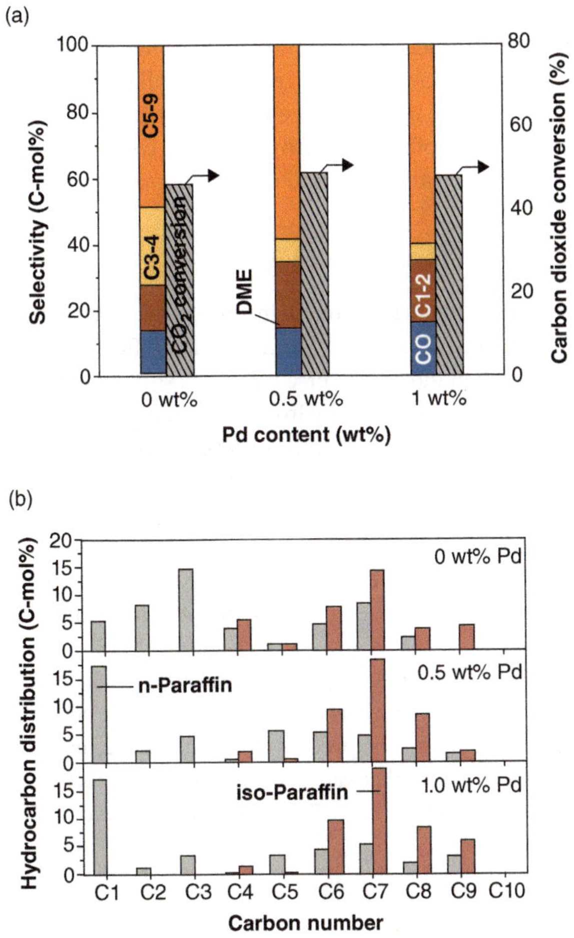

Figure 10.14 (a) Effect of Pd loading amount on the ZSM-(23) zeolite in the hybrid catalyst on the carbon dioxide conversion and the product selectivity and (b) the product distribution of hydrocarbons.

n-hexane fluid played an important role in extraction of hydrocarbon products from active sites in the zeolite. ZSM-5 with abundant acidity contained in the hybrid catalyst was required for selectively producing the C_5-C_9 hydrocarbons (gasoline fraction). In addition, the yield of hydrocarbons in the gasoline fraction was increased by loading palladium promotor on ZSM-5(23). Furthermore, using zeolites with larger micropores such as beta zeolite and USY zeolite in the hybrid catalyst were effective in the selective production of C_8-C_{13} *iso*-paraffins (jet fuel fraction).

The direct hydrogenation of carbon dioxide to lower olefins with outstanding selectivity was realized by constructed tandem catalyst ZnZrO/SAPO, with which carbon dioxide and hydrogen were activated on Zn-ZrO and the C-C bond formation was performed on SAPO. The selectivity of lower olefins can reach to 80% while only 3% methane among hydrocarbon products at a carbon dioxide of 12.6%. Tandem catalysis facilitates the thermodynamics and kinetics coupling through the transferring and migrating of CH_xO intermediate species that not only included methanol, which enable the highly efficient conversion of carbon dioxide to lower olefins. Tandem catalyst usage showed resistance to thermal and sulfur treatments (hydrogen sulfide and sulfur dioxide), suggesting the promising potential application in chemical technology.

Synthetic gas produced from electricity-based hydrogen and carbon dioxide can be used in Fischer-Tropsch synthesis to produce various synthetic fuels. The general process scheme is shown in Figure 10.10. There are several concepts at TRLs of 5–7 to produce synthetic diesel or other hydrocarbons from hydrogen and carbon dioxide. Unlike methanol or ethanol, which are blended into the gasoline to reduce the carbon dioxide footprint of the respective fuels, the corresponding synthetic fuels are drop-in fuels which have almost the same chemical composition as fossil fuels (see Table 10.11) and could replace them completely.

10.10.4.1 Power-to-Liquids

The start-up Sunfire is running the first Power-to-Liquid pilot plant in the world in Dresden, Germany. With the combined operation of a steam-electrolysis reaching an electrical efficiency level of well over 90% (for 10 kW_{el}) under pressure, a carbon dioxide-reverse water gas shift-conversion (CO_2-RWGS) and Fischer-Tropsch-synthesis, the plant can produce hydrocarbons from carbon dioxide, water and renewable energy with an overall efficiency level up to 65% (LHV H_2/kW_{el}). The hydrocarbons can serve the road traffic, shipping, aviation and chemical sector with fuels as gasoline, kerosene, diesel, methanol and methane.

Sunfire highlights the production of a synthetic diesel ("blue crude") that already meets required characteristics and can be used without further adaptations for vehicles. The capacity of the plant for the synthetic diesel is 1 barrel per day. Sunfire claims to save up to 3.14 tons of carbon dioxide for each ton of fuel produced by this approach.

The plant combines a solid oxide fuel (SOFC) and a SOE Cell. The so-called reversible solid oxide cell (SOC) operation is a SOFC and SOE cell in a single device. This combination makes it possible to supply electricity in time of renewable energy lack. Thus, the process can contribute to balancing power for the stabilization of the grid and can enable added value for the supply of electricity in decentralized regions. The first SOC sold operates with a 100kW SOEC power input and SOFC power of output.

10.10.4.2 Energy Demand per Unit of Synthetic Fuel Production

Diesel and kerosene have the same energy density of 43.1 GJ/t. For the reaction scheme described above Sunfire uses 141 kJ energy demand for the water evaporation, 726 kJ for steam electrolysis and 41 kJ for the reverse water gas shift. The exothermic Fischer-Tropsch synthesis releases 147 kJ, which is sufficient to supply the water evaporation if efficient heat recuperation is ensured in process configuration. Taking the 70% overall process efficiency, the energy demand is calculated as 66.3 GJ/t (or 18.4 MWh) in total, of which 62.8 GJ/t (or 17.4MWh) are required for electrolysis and 3.5 GJ/t for the RWGS.

Table 10.11 Characteristics of Sunfire's synthetic diesel from pilot plant.

Characteristics	Diesel (EN 590)	Sunfire Blue Crude
Gravimetric density kg/m^3	820–840	780
LHV MJ/kg	42.5	44.7
Energy density (MJ/L)	34.9–35.7	34.9
Cetane number	>51	65–76
Additional data		Sulphur-free, aromatics content< 1%

10.10.4.3 Carbon Dioxide Reduction per Unit of Synthetic Fuel Production

A life cycle analysis of the Sunfire process clearly shows the essential requirement of using low-carbon electricity for synthetic diesel production. The carbon footprint of synthetic diesel is three times that of fossil diesel, if the current German electricity mix is used. For 100% renewable electricity, emission reductions well-to-wheel of 35% up to 85% have been quantified. Well-to-wheel emissions of conventional diesel amount to 88.6 g CO_2eq/MJ or 3.82 t CO_2eq/t diesel, for kerosene, carbon dioxide emissions are at 71.5 g CO_2eq/MJ or 3.08 t CO_2eq/t. Using a medium reduction potential of 60% compared to the fossil fuel, carbon dioxide reduction is at 2.3 t CO_2eq per ton of synthetic diesel and 1.85 tCO_2eq per ton of synthetic kerosene.

10.10.4.4 Economics

Sunfire has calculated production costs for their synthetic diesel of 1.24 € per litre at electricity costs of 50 €/MWh. A cost range of 1.2 to 1.5 €/L is therefore estimated for both fuels. This is to be compared to 25–60 € per litre for fossil crude exploration, showing production costs higher by a factor of two. Cost parity to fossil fuels is out of scope of the current framework conditions.

10.10.4.5 Comparison of the Hydrogen-Based Low-Carbon Synthesis Routes

Table 10.12 provides an overview of main results obtained for the low-carbon processes investigated in this chapter. Results are presented for the different target products, chemicals and synthetic fuels. For the chemical products, the level of avoided carbon dioxide is very similar. Differences can be seen when avoided carbon dioxide is expressed as function of required electricity or as a function of production costs. Avoided carbon dioxide production per MWh is highest for urea followed by methanol and ammonia, which are on par. Ethylene, propylene and the BTX yield lower efficiency in terms of carbon dioxide avoided per MWh. From the point of view of carbon dioxide and the BTX avoidance costs, the merit order would be the same.

Generally, the energy demand of these low-carbon synthesis routes is very high. In this sense, the low carbon processes are not energy efficient as such, they just have a lower carbon footprint in a defined system. However, part of the high energy demand of the electricity-based low-carbon pathways can be attributed to the fact that the target products are built-up from just water and carbon dioxide, thereby not requiring the high energy content of feedstock that is used in the

Table 10.12 Comparison of hydrogen based low-carbon synthesis routes.

Product	Electricity [MWh]	CO_2 as Feed [t]	Avoided CO_2 [t]	Costs [€]	Avoided CO_2 as kg	
Chemicals	per ton of product				per MWh	per €
Ammonia	12.5	-	1.71	700–800	137	2.1–2.4
Urea	8.1	0.73	2.05	450–500	253	4.1–4.5
Methanol	11.02	1.373	1.53	300–650	139	2.4–5.1
Olefins	26.6	3.2	1.89	670–1900	71	1–2.8
BTX	48.9	5.9	1.7	1300–2800	34	0.6–1.3
Fuels	per ton of product			per litre	per MWh	per €
Diesel	18.4	3.15	2.3[a]	1.2–1.5	125	1.3–1.6
Kerosene	18.4	2.85	1.85[a]	1.2–1.5	100	1–1.2
SNG	26.9	2.7	1.31	2000–3500	49	0.4–0.7

[a]Well-to-wheel

alternative fossil processes. A comparison of the hydrogen-based low-carbon routes with the fossil reference route should therefore be performed based on the total energy demand, i.e., in case of the fossil processes, including the energy content of the feedstock built-in the target product.

Another aspect of the depicted low-carbon routes is potentially significant higher demand for noble metals and rare earth elements, that can be attributed to the

provision of renewable electricity by e.g., wind power and the catalysts required for electrolyser cells. This impact must be considered from a total resource efficiency point of view but has not been investigated further in this study.

10.11 Electrochemical Reduction of Carbon Dioxide to Oxalic Acid

Oxalic acid is an important base chemical that is mainly used for metal treatment, textile treatment, concentration of rare earth elements, bleaching and polymer production with a multibillion-dollar market size. Currently, oxalic acid is predominantly produced from the oxidation of carbohydrates, olefins and carbon monoxide. All three methods require multiple complicated processing steps involving high pressure and/or temperature conditions and acid/base consumption. A more recent approach for producing oxalic acid is based on the electrochemical reduction of carbon dioxide according to the half-cell reaction (Equation10.14):

$$2CO_2 + 2e^- \rightarrow C_2O_4^{2-} \tag{10.14}$$

Note that oxalate formation may involve different reaction steps, including initial electron transfer and radical–radical dimerization of carbon dioxide. On lead or mercury electrodes, oxalic acid is the major product in nonaqueous solvents, but in the presence of water it can further be reduced to higher carboxylic acids like glyoxylic acid (Equation10.15) and glycolic acid (Equation 10.16):

$$C_2O_4^{2-} + 2H_2O + 2e^- \rightarrow C_2HO_3^- + 3OH^- \tag{10.15}$$

$$C_2HO_3^- + 2H_2O + 2e^- \rightarrow C_2H_3O_3^- + 2OH^- \tag{10.16}$$

Carbon monoxide (CO) can also be produced in nonaqueous solvents according to the half-cell reaction (10.17):

$$2CO_2 + 2e^- \rightarrow CO + CO_3^{2-} \tag{10.17}$$

Note that carbon monoxide formation may proceed through several intermediate steps, which are not shown here. Carbon dioxide reduction on the oxalic acid-producing electrodes (i.e., lead or mercury) in aqueous solvents or nonaqueous solvents with a sufficiently high-water concentration shifts the mechanism from oxalate to formate (Equation 10.18):

$$CO_2 + H_2O + 2e^- \rightarrow HCOO^- + OH^- \tag{10.18}$$

In the past decade, the electrochemical reduction of CO_2 has been studied intensively but mostly in aqueous solvents. Data on carbon dioxide reduction in nonaqueous solvents is relatively scarce despite the well-known advantages of these solvents such as high carbon dioxide solubility and suppression of the competing hydrogen evolution reaction.

Boor et al. studied the electrochemical reduction of carbon dioxide to oxalic acid in nonaqueous solvents using a lead catalyst. An H-cell was used to investigate the effects of anolyte, catholyte, supporting electrolyte, temperature, water content and cathode potential on the performance indicators (i.e., Faraday efficiency, current densities and product concentration). The best conditions of these screening experiments were selected to study the carbon dioxide electrolysis performance in a flow cell setup. In addition, they assessed the technical and economic feasibility of oxalic acid production from the electrochemical conversion of carbon dioxide. A process design including carbon dioxide capture, electrochemical conversion and downstream processing of oxalate is presented. The effects of different parameters (i.e., Faraday efficiency, current densities, cell voltage, electricity price, product concentration and electrolyser capital cost) on the net present value and payback time were investigated.

10.11.1 Process Design and Modelling

A schematic of the considered process is shown in Figure 10.15. The process includes carbon dioxide capture, electrochemical carbon dioxide conversion and downstream separation of (by)products, including solvent recycling. Carbon dioxide is captured from a biogas (40 mol % carbon dioxide and 60 mol % methane) stream using propylene carbonate, which is a commercial solvent used in the Fluor Solvent Process. In the classical process, the captured carbon dioxide would be regenerated from the solvent in the desorber. In the integrated process, the carbon dioxide and solvent mixture is sent directly to the carbon dioxide electrolyser (thus eliminating the desorber). In the electrolyser, carbon dioxide is converted to oxalic acid and some by-products, like glycolic acid and glyoxylic acid, which will be neglected in the base case design. The solvent stream containing the electroreduction products are sent to the separation section where the oxalic acid is recovered. The recovery of oxalic acid/oxalate from nonaqueous solutions is not trivial. The selection of the separation method depends on the pH of the solution, which determines the state of the acid. For the separation, it is important to know whether oxalate or oxalic acid is present in the cathode compartment of the electrolyser. Note that the state of the product (dissociated or undissociated) depends on the cell configuration. For example, using an undivided cell with a sacrificial zinc anode will produce zinc oxalate as a product. In the experiments, protons from water oxidation in acidic media (i.e., sulfuric acid) crossed the cation exchange membrane (Nation-117 from Fundtech) and acidified the catholyte (thus producing oxalic acid). To support this hypothesis, Boor et al. extracted the oxalic acid/oxalate from the organic phase (i.e., propylene carbonate) into the aqueous phase by simply mixing the catholyte with water and measuring the pH of the aqueous phase. The measured pH was from 1.4 to 1.7, which corresponds well with the expected pH based on the oxalic acid concentrations. This confirms that in the experiments mostly oxalic acid

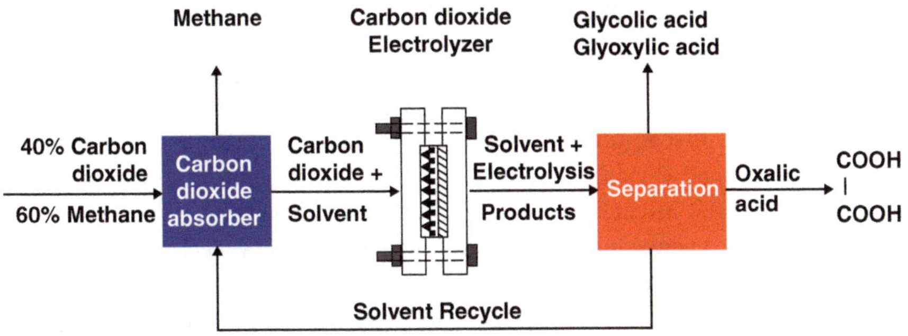

Figure 10.15 Integrated process for carbon dioxide capture, electrochemical conversion and product separation including solvent recycling.

was produced in the cathode compartment. The protonation of oxalate to oxalic acid does not necessarily need to occur on the cathode surface, because this step can equally well happen in the electrolyte.

10.11.2 Carbon Dioxide Absorption in Propylene Carbonate

The absorption of carbon dioxide from biogas with propylene carbonate as a solvent was modelled in Aspen Plus. Boor et al. assumed that the feed with a composition of 40 mol % carbon dioxide and 60 mol % methane enters the absorber at 25°C and 1 MPa. The absorber is designed to process 1 ton/h of biogas with a methane purity of at least 94 mol % to comply with the standards for biomethane injection into the natural gas grid of The Netherlands (6 mol % of carbon dioxide is allowed). This means that roughly 90% of the carbon dioxide should be removed from the biogas. The solvent flow and the number of stages were varied to meet the design specifications. For the property calculations, the Peng–Robinson equation of state was used. The binary interaction parameters of the Peng–Robinson equation of state were fitted to available experimental solubility data of carbon dioxide and methane in propylene carbonate. Note that some methane is absorbed, which will be carried along with the propylene carbonate stream to the cathode compartment of the electrolyser. The mole purity of methane in the product gas was calculated as a function of the solvent to biogas ratio for different numbers of theoretical stages and two pressures (1 MPa and 4 MPa). Operating the column at 4 MPa will significantly reduce the solvent flows, but the feed compression costs and the amount of absorbed methane will increase. In the process design, we selected a pressure of 1 MPa, 10 stages and a solvent to biogas ratio of 30 to meet the design specifications.

Further Reading

Alvarez-Guerra, M., Del Castillo, A. and Irabien, A. (2014). Continuous electrochemical reduction of carbon dioxide into formate using a tin cathode: comparison with lead cathode. *Chem. Eng. Res. Des.* 92 (4): 692–701.

Aresta, M., Dibenedetto, A. and Angelini, A. (2014). Catalysis for the valorisation of exhaust carbon: from CO_2 to chemicals, materials and fuels. Technological use of CO_2. *Chem. Rev.* 114: 1709–1742. doi: 10.1021/cr4002758.

Bandi, A. (1990). Electrochemical reduction of carbon dioxide on conductive metallic oxides. *J. Electrochem. Soc.* 137: 2157. doi: 10.1149/1.2086903.

Bellotti, D., Rivarolo, M., Magistri, L. and Massardo, A.F. (2017). Feasibility study of methanol production plant from hydrogen and captured carbon dioxide. *J. CO_2 Util.* 21: 132–138. doi: 10.1016/j.jcou.2017.07.001.

Benhelal, E., Zahedi, G., Shamsaei, E. and Bahadori, A. (2013). Global strategies and potentials to curb (CO_2) emissions in cement industry. *J. Clean. Prod.* 51: 142–161. doi: 10.1016/j.jclepro.2012.10.049.

Boor, V., Frijns, J.E.B.M., Galent, E.P. et al. (2022). Electrochemical reduction of CO_2 to oxalic acid: experiments, process modeling and economic. *Ind. Eng. Chem. Res.* 61 (40): 14837–14846.

Buttler, A. and Spliethoff, H. (2018). Current status of water electrolysis for energy storage, grid balancing and sector coupling via power-to-gas and power-to-liquids: a review. *Renew. Sustain. Energy Rev.* 82: 2440–2454. doi: 10.1016/j.rser.2017.09.003.

Carvalho, D.F., Almeida, G.C., Monteiro, R.S. and Mota, C.J.A. (2020). Hydrogenation of CO_2 to methanol and dimethyl ether over a bifunctional Cu·ZnO catalyst impregnated on modified γ-alumina. *Energy Fuels* 34 (6): 7269–7274.

Centi, G., Perathoner, S., Winè, G. and Gangeri, M. (2007). Electrocatalytic conversion of CO_2 to long carbon-chain hydrocarbons. *Green Chem.* 9 (6): 671–678.

Chan, F.L., Altinkaya, G., Fung, N. and Tanksale, A. (2018). Low temperature hydrogenation of carbon dioxide into formaldehyde in liquid media. *Catal. Today* 309: 242–247. doi: 10.1016/j.cattod.2017.06.012.

Chaplin, R.P.S. and Wragg, A.A. (2003). Effects of process conditions and electrode material on reaction pathways for carbon dioxide electroreduction with reference to formate formation. *J. Appl. Electrochem.* 33 (12): 1107–1123.

Chen, T.-L., Jiang, W., Shen, A.-L. et al. (2020). CO_2 mineralization and utilization using various calcium-containing wastewater and refining slag via high-gravity carbonation process. *Ind. Eng. Chem. Res.* 59 (15): 7140–7150. doi: 10.1021/acs.iecr.9b05410.

Chen, Y., Li, C.W. and Kanan, M.W. (2012). Aqueous CO_2 reduction at very low overpotential on oxide-derived Au nanoparticles. *J. Am. Chem. Soc.* 134 (49): 19969–19972.

Chi, J. and Yu, H. (2018). Water electrolysis based on renewable energy for hydrogen production. *Chinese J. Catal.* 39: 390–394. doi: 10.1016/S1872-2067(17)62949-8.

Corma, A., Torre, O., De Renz, M. and Villandier, N. (2011). Production of high-quality diesel from biomass waste products. *Angew. Chemie Int. Ed.* 50: 2375–2378.

Cui, Y., He, B., Liu, X. and Sun, J. (2020). Ionic liquids-promoted electrocatalytic reduction of carbon dioxide. *Ind. Eng. Chem. Res.* 59 (46): 20235–20252.

Darmawan, A., Ajiwibowo, A.W., Biddinika, M.K. et al. (2019). Black liquor-based hydrogen and power co-production: combination of supercritical water gasification and syngas chemical looping. *Appl. Energy* 252: 113446.

de Tacconi, N.R., Chanmanee, W., Dennis, B.H. et al. (2011). Electrocatalytic reduction of carbon dioxide using Pt/C-TiO2 nanocomposite cathode. *Electrochem. Solid-State Lett.* 15 (1): B5–B8.

Di Giuliano, A., Gallucci, K. and Foscolo, P.U. (2020). Determination of kinetic and diffusion parameters needed to predict the behavior of CaO-based CO_2 sorbent and sorbent-catalyst materials. *Ind. Eng. Chem. Res.* 59 (15): 6840–6854.

Dokhaee, Z., Ghiaci, M., Farrokhpour, H. et al. (2020). SBA-15-supported imidazolium ionic liquid through different linkers as a sustainable catalyst for the synthesis of cyclic carbonates: a kinetic study and theoretical DFT calculation. *Ind. Eng. Chem. Res.* 59 (28): 12632–12644. doi: 10.1021/acs.iecr.0c01050.

Fernández, J.R., Garcia, S. and Sanz-Pérez, E.S. (2020). CO_2 capture and utilization editorial. *Ind. Eng. Chem. Res.* 59 (15): 6767–6772.

Filonenko, G.A., Van Putten, R., Schulpen, E.N. et al. (2014). Highly efficient reversible hydrogenation of carbon dioxide to formates using a ruthenium PNP-pincer catalyst. *ChemCatChem* 6: 1526–1530. doi: 10.1002/cctc.201402119.

Framework Convention on Climate Change – UNFCCC, 13 August 2010, http://unfccc.int/kyotoprotocol (Accessed August 2010).

Frese, K.W. (1991). Electrochemical reduction of CO_2 at intentionally oxidized copper electrodes. *J. Electrochem. Soc.* 138: 3338–3344.

Furimsky, E. (2020). CO_2 hydrogenation to methanol and methane over carbon supported catalysts. *Ind. Eng. Chem Res.* 59 (35): 15393.

Gabrielli, P., Gazzani, M. and Mazzotti, M. (2020). The role of carbon capture and utilization, carbon capture and storage and biomass to enable a net-zero-CO_2 emissions chemical industry. *Ind. Eng. Chem. Res.* 59 (15): 7033–7045. doi: 10.1021/acs.iecr.9b0657910.1021/acs.iecr.9b05383.

Gangeri, M., Perathoner, S., Caudo, S. et al. (2009). Fe and Pt carbon nanotubes for the electrocatalytic conversion of carbon dioxide to oxygenates. *Catal. Today* 143 (1–2): 57–63.

Gao, J., Song, F., Li, Y. et al. (2020). Cu_2In nanoalloy enhanced performance of Cu/ZrO_2 catalysts for the CO_2 hydrogenation to methanol. *Ind. Eng. Chem. Res.* 59 (27): 12331–12337.

Geng, F., Bonita, Y., Jain, V. et al. (2020). Bimetallic Ru-Mo phosphide catalysts for the hydrogenation of CO_2 to methanol. *Ind. Eng. Chem. Res.* 59 (15): 6931–6943. doi: 10.1021/acs.iecr.9b06937.

Goncalves, M.R., Gomes, A., Condeço, J. et al. (2013). Electrochemical conversion of CO_2 to C_2 hydrocarbons using different ex situ copper electrodeposits. *Electrochim. Acta* 102: 388–392.

Hank, C., Gelpke, S., Schnabl, A. et al. (2018). Economics & carbon dioxide avoidance cost of methanol production based on renewable hydrogen and recycled carbon dioxide – power-to-methanol. *Sustain. Energy Fuels* 2: 1244–1261. doi: 10.1039/C8SE00032H.

Hao, C., Wang, S., Li, M. et al. (2011). Hydrogenation of CO_2 to formic acid on supported ruthenium catalysts. *Catal. Today* 160: 184–190. doi: 10.1016/j.cattod.2010.05.034.

Hara, K., Kudo, A. and Sakata, T. (1995). Electrochemical reduction of carbon dioxide under high pressure on various electrodes in an aqueous electrolyte. *J. Electroanal. Chem.* 391 (1–2): 141–147.

Heyn, R.H. (2003). Carbon dioxide conversion. In: *Encyclopedia of Catalysis*, vol. 2 (ed. I.K. Verbeek), 19. Hoboken: Wiley.

Hori, Y. (2008). Electrochemical CO_2 reduction on metal electrodes. In: *Modern Aspects of Electrochemistry 42* (ed. C.G. Vayenas), 89–189. New York, NY: Springer-Verlag. doi: 10.1007/978-0-387-49489-0.

Huang, M. and Faguy, P.W. (1996). Carbon dioxide reduction on platinum |Nafion®| carbon electrodes. *J. Electroanal. Chem.* 406 (1–2): 219–222.

Huang, Y., Deng, Y., Handoko, A.D. et al. (2018). Rational design of sulfur-doped copper catalysts for the selective electroreduction of carbon dioxide to formate. *ChemSusChem* 11: 320–326. doi: 10.1002/cssc.201701314.

Hwang, H., Yeon, Y.J., Lee, S. et al. (2015). Electro-biocatalytic production of formate from carbon dioxide using an oxygen-stable whole cell biocatalyst. *Bioresour. Technol.* 185: 35–39. doi: 10.1016/j.biortech.2015.02.086.

Hwang, S.-M., Zhang, C. et al. (2020). Mesoporous carbon as an effective support for Fe catalyst for CO_2 hydrogenation to liquid hydrocarbons. *Journal of CO_2 Utilization* 37: 65–73.

Iliuta, I. and Larachi, F. (2020). Enhanced methanol synthesis process via an integrated process involving CO_2 hydrogenation under plasma conditions. *Ind. Eng. Chem. Res.* 59 (15): 6815–6827. doi: 10.1021/acs.iecr.9b04278.

Jia, F., Yu, X. and Zhang, L. (2014). Enhanced selectivity for the electrochemical reduction of CO_2 to alcohols in aqueous solution with nanostructured Cu-Au alloy as catalyst. *J. Power Sources* 252: 85–89. doi: 10.1016/j.jpowsour.2013.12.002.

Jiao, F., Li, J. et al. (2016). Selective conversion of syngas to light olefins. *Science* 351: 1065–1068.

Kaneco, S., Hiei, N.H., Xing, Y. et al. (2002). Electrochemical conversion of carbon dioxide to methane in aqueous NaHCO3 soltion at less than 273 K. *Electrochim. Acta* 48 (1): 51–55.

Kaneco, S., Katsumata, H., Suzuki, T. and Ohta, K. (2006). Electrochemical reduction of Carbon dioxide to ethylene at a copper electrode in methanol using potassium hydroxide and rubidium hydroxide supporting electrolytes. *Electrochim. Acta* 51 (16): 3316–3321.

Kaneko, H., Nozaki, K., Ozawa, T. et al. (1988). U.S. patent no. 4,732,827. U.S. Patent and Trademark Office, Washington, DC.

Keerthiga, G., Viswanathan, B. and Chettz, R. (2015). Electrochemical reduction of CO_2 electrodeposited Cu electrodescrystalline phase sensitivity on selectivity. *Cat. Today* 245: 68–73.

Khoo, H.H., Halim, I. and Handoko, A.D. (2020). LCA of electrochemical reduction of CO_2 to ethylene. *J. CO_2 Util.* 41: 101229.

Köleli, F. and Balun, D. (2004). Reduction of CO_2 under high pressure and high temperature on Pb-granule electrodes in a fixed-bed reactor in aqueous medium. *Appl. Catal. A* 274 (1–2): 237–242.

Kondratenko, E.V., Mul, G., Baltrusaitis, J. et al. (2013). Status and perspectives of CO_2 conversion into fuels and chemicals by catalytic, photocatalytic and electrocatalytic processes. *Energy Environ. Sci.* 6 (11): 3112–3135.

Kortlever, R., Balemans, C., Kwon, Y. and Koper, M.T.M. (2015). Electrochemical CO_2 reduction to formic acid on a Pd-based formic acid oxidation catalyst. *Catal. Today* 244: 58–62. doi: 10.1016/j.cattod.2014.08.001.

Kourkoumpas, D.S., Papadimou, E., Atsonios, K. et al. (2016). Implementation of the Power to Methanol concept by using CO_2 from lignite power plants: techno-economic investigation. *Int. J. Hydrogen Energy* 41: 16674–16687. doi: 10.1016/j.ijhydene.2016.07.100.

Koytsoumpa, E.I., Bergins, C. and Kakaras, E. (2018). The CO_2 economy: review of CO_2 capture and reuse technologies. *J. Supercrit. Fluids* 132: 3–16. doi: 10.1016/j.supflu.2017.07.029.

Kuhl, K.P., Hatsukade, T., Cave, E.R. et al. (2014). Electrocatalytic conversion of carbon dioxide to methane and methanol on transition metal surfaces. *J. Am. Chem. Soc.* 136: 14107–14113. doi: 10.1021/ja505791r.

Kumar, B., Atla, V., Brian, J.P. et al. (2017). Reduced SnO_2 porous nanowires with a high density of grain boundaries as catalysts for efficient electrochemical CO_2-into-HCOOH conversion. *Angew. Chem. Int. Ed.* 56: 3645–3649. doi: 10.1002/anie.201612194.

Le, M., Ren, M., Zhang, Z. et al. (2011). Electrochemical reduction of CO_2 to CH_3OH at copper oxide surfaces. *J. Electrochem. Soc.* 158: 45–49. doi: 10.1149/1.3561636.

Le, Q.A.T., Kim, H.G. and Kim, Y.H. (2018). Electrochemical synthesis of formic acid from CO_2 catalysed by *Shewanella oneidensis* MR-1 whole-cell biocatalyst. *Enzyme Microb. Technol.* 116: 1–5. doi: 10.1016/j.enzmictec.2018.05.005.

Lee, D., Kim, D. and Kim, S.-W. (2001). Selective formation of formaldehyde from carbon dioxide and hydrogen over $PtCu/SiO_2$. *Appl. Organomet. Chem.* 15: 148–150. doi: 10.1002/1099-0739(200102)15:2<148:aid-aoc104>3.0.co;2-n.

Li, C.W. and Kanan, M.W. (2012). CO_2 reduction at low overpotential on Cu electrodes resulting from the reduction of thick Cu_2O films. *J. Am. Chem. Soc.* 134 (17): 7231–7234.

Li, F., Thevenon, A. et al. (2020). Molecular tuning of CO_2-to-ethylene conversion. *Nature* 577: 509–513.

Li, H. and Oloman, C. (2005). The electro-reduction of carbon dioxide in a continuous reactor. *J. Appl. Electrochem.* 35 (10): 955–965.

Li, Q. et al. (2020). Novel Bi, Bi_2Sn, Bi_3Sn and Bi_4Sn Catalysts for efficient Electroreduction of CO_2 to formic acid. *Ind. Eng. Chem Res.* 59 (15): 6806–6814.

Li, W., Wang, H. et al. (2018). A short review of recent advances in CO_2 hydrogenation to hydrocarbons over heterogeneous catalysts. *RSC Adv.* 8: 7651–7669.

Li, X.H., Ma, T. and Chen, C., (2016). Method for producing hydrocarbon. JP5988243.

Liu, C., Xie, J.H., Tian, G.L. et al. (2015). Highly efficient hydrogenation of carbon dioxide to formate catalyzed by iridium(iii) complexes of imine-diphosphine ligands. *Chem. Sci.* 6: 2928–2931. doi: 10.1039/c5sc00248f.

Liu, J., Zhang, A. et al. (2019). Overcoating the surface of Fe-based catalyst with ZnO and nitrogen-doped carbon toward high selectivity of light olefins in CO_2 hydrogenation. *Ind. Eng. Chem. Res.* 58: 4017–4023.

Liu, Y., Kamata, H. et al. (2020). Low-olefin production process based on Fischer-Tropsch synthesis: process synthesis, optimization and techno- economic analysis. *Ind. Eng. Chem. Res.* 59: 8728–8739.

Malik, M.I., Omar, Z., Atieh, M. and Abussaud, B. (2016). Electrochemical reduction of CO_2 to methanol over MWCNTs impregnated with Cu_2O. *Chem. Eng. Sci.* 152: 468–477. doi: 10.1016/j.ces.2016.06.035.

Marcos, M.L., González-Velasco, J., Bolzán, A.E. and Arvia, A.J. (1995). Comparative electrochemical behaviour of CO_2 on Pt and Rh electrodes in acid solution. *J. Electroanal. Chem.* 395 (1–2): 91–98.

Mizuno, T., Kawamoto, M., Kaneco, S. and Ohta, K. (1998). Electrochemical reduction of carbon dioxide at Ti and hydrogen-storing Ti electrodes in KOH–methanol. *Electrochim Acta* 43 (8): 899–907.

Munshi, P., Main, A.D., Linehan, J.C. et al. (2002). Hydrogenation of carbon dioxide catalyzed by ruthenium trimethylphosphine complexes: the accelerating effect of certain alcohols and amines. *J. Am. Chem. Soc.* 124: 7963–7971. doi: 10.1021/ja0167856.

Nakata, K., Ozaki, T., Terashima, C. et al. (2014). High-yield electrochemical production of formaldehyde from CO_2 and seawater. *Angew. Chem. Int. Ed.* 53: 871–874. doi: 10.1002/anie.201308657.

Nanda, M.R., Yuan, Z., Qin, W. et al. (2014). Catalytic conversion of glycerol to oxygenated fuel additive in a continuous flow reactor: process optimization. *Fuel* 128: 113–119.

Nezam, I., Zhou, W., Gusmaäo, G.S. et al. (2021). Direct aromatization of CO_2 via combined CO_2 hydrogenation and zeolite-based acid catalysis. *J. CO_2 Util.* 45: 101405.

Ogura, K. and Endo, N. (1999). Electrochemical reduction of CO_2 with a functional gas-diffusion electrode in aqueous solutions with and without propylene carbonate. *J. Electrochem. Soc.* 146 (10): 3736–3740.

Ohya, S., Kaneco, S., Katsumata, H. et al. (2009). Electrochemical reduction CO2 in methanol with aid of CuO and Cu_2O. *Catal. Today* 148 (3–4): 329–334.

Olah, G.A., Goeppert, A. and Prakash, G.K.S. (2009). Chemical recycling of carbon dioxide to methanol and dimethyl ether: from greenhouse gas to renewable, environmentally carbon neutral fuels and synthetic hydrocarbons. *J. Org. Chem.* 74: 487–498. doi: 10.1021/jo801260f.

Olah, G.A. and Molnár, A. (2003). *Hydrocarbon Chemistry*, chapter 3. Wiley.

Olah, G.A. and Prakash, G.K.S. (2010). *Electrolysis of carbon dioxide in aqueous media to carbon monoxide and hydrogen for production of methanol*. Washington, DC.

The Paris Agreement (2016). An agreement within the United Nation Framework Convention on Climate Change.

Pierpont, A.W., Batista, E.R., Martin, R.L. et al. (2015). Origins of the regioselectivity in the lutetium triflate catalysed ketalization of acetone with glycerol: a DFT study. *ACS Catal.* 5: 1013–1019.

Popić, J.P., Avramov-Ivić, M.L. and Vuković, N.B. (1997). Reduction of carbon dioxide on Ruthenium oxide and modified ruthenium oxide electrodes in 0.5 M $NaHCO_3$. *J. Electroanal. Chem.* 421 (1–2): 105–110.

Qu, J., Zhang, X., Wang, Y. and Xie, C. (2005). Electrochemical reduction of CO_2 on RuO-/TiO nanotubes composite modified Pt electrode. *Electrochim. Acta* 50 (16–17): 3576–3580.

Rego de Vasconcelos, B. and Lavoie, J.-M. (2019 June 05). Recent advances in power-to-X technology to produce fuels and chemicals. *Front. Chem.* doi: 10.3389/fchem.2019.00392.

Rihko-Struckmann, L.K., Peschel, A., Hanke-Rauschenbach, R. and Sundmacher, K. (2010). Assessment of methanol synthesis utilizing exhaust CO_2 for chemical storage of electrical energy. *Ind. Eng. Chem. Res.* 49: 11073–11078. doi: 10.1021/ie100508w.

Ruiz, V.R., Velty, A., Santos, L.L. et al. biomass transformations: valorization of glycerol and glycerol-water mixtures through formation of cyclic acetals. *J. Catal.* 271 (2): 351–357.

Sahebdelfar, S., Tahriri Zangeneh, F. and Takht Ravanchi, M. (2009). Conversion of carbon dioxide to valuable petrochemicals: An approach to clean development mechanism. In: *Proceedings of the 7th Asia Pacific Conference on Sustainable Energy and Environmental Technologies*, Qingdao, China.

Song, Y., Serikawa, K., Imamura, K. et al. (2020). Direct synthesis of C_5-C_{13} *iso*-paraffins from carbon dioxide over hybrid catalyst in a near-critical n-hexane fluid. *Ind. Eng. Chem. Res.* 59 (26): 11962–11969.

Stangeland, K., Kalai, D., Li, H. and Yu, Z. (2017). CO_2 methanation: the effect of catalysts and reaction conditions. *Energy Procedia* 105: 2022–2027. doi: 10.1016/j.egypro.2017.03.577.

Sternberg, A. and Bardow, A. (2015). Power-to-what? —environmental assessment of energy storage systems. *Energy Environ. Sci.* 8: 389–400. doi: 10.1039/C4EE03051F.

Su, J., Yang, L., Lu, M. and Lin, H. (2015). Highly efficient hydrogen storage system based on ammonium bicarbonate/formate redox equilibrium over palladium nanocatalysts. *ChemSusChem* 8: 813–816. doi: 10.1002/cssc.201403251.

Tanaka, R., Yamashita, M. and Nozaki, K. (2009). Catalytic hydrogenation of carbon dioxide using Ir (III)-pincer complexes. *J. Am. Chem. Soc. Commun.* 131: 14168–14169. doi: 10.1021/ja903574e.

Torres Galvis, H.M., Bitter, J.H., Khare, C.B. et al. (2012). Supported iron nanoparticles as catalysts for sustainable production of lower olefins. *Science* 335: 835–838.

Torres Galvis, H.M. and de Jong, K.P. (2013). Catalysts for production of lower olefins from synthesis gas: a review. *ACS Catal.* 3: 2130–2149.

Umegaki, T., Enomoto, Y., and Kojima, Y. (2016). Metallic ruthenium nanoparticles for hydrogenation of supercritical carbon dioxide. *Catal. Sci. Technol.* 6: 409–412. doi: 10.1039/c5cy00994d.

United Nations Climate Change. What is the Kyoto Protocol? http://unfccc.int/kyoto_protocol (Accessed August 2010).

Vázquez, F.V., Koponen, J., Ruuskanen, V. et al. (2018). Power-to-X technology using renewable electricity and carbon dioxide from ambient air: SOLETAIR proof-of-concept and improved process concept. *J. CO_2 Util.* 28: 235–246. doi: 10.1016/J.JCOU.2018.09.026.

Wang, D., Zhou, X., Ji, J. et al. (2015). Modified carbon nanotubes by KMnO4 supported iron Fischer–Tropsch catalyst for the direct conversion of syngas to lower olefins. *J. Mater. Chem. A* 3: 4560–4567.

Wang, J. and You, Z. (2013). Synthesis of lower olefins by hydrogenation of carbon dioxide over supported iron catalysts. *Catal. Today* 215: 186–193.

Wang, Y., Tan, T. et al. (2019). Rationally designing bifunctional catalysts as an efficient strategy to boost CO_2 hydrogenation producing value added aromatics. *ACS Catal.* 9: 895–901.

Watanabe, M., Shibata, M. and Katoh, A. (1991). Design of alloy electrocatalysts for reduction—improved energy efficiency, selectivity and reaction rate for the CO_2 electroreduction on Cu alloy electrodes. *J. Electroanal. Chem. Interfacial Electrochem.* 305: 319–328.

Winter, F., Agarwal, R.A., Hrdlička, J. and Varjani, S. (2019). *CO_2 Separation, Purification and Conversion to Chemicals and Fuels*. Springer Nature Singapore Pte Ltd.

Wu, J., Risalvato, F.G., Ke, F.S. et al. (2012). Electrochemical reduction of carbon dioxide I. Effects of the electrolyte on the selectivity and activity with Sn electrode. *J. Electrochem. Soc.* 159 (7): F353–F359.

Xu, X.D. and Moulijn, J.A. (1996). Mitigation of CO_2 by chemical conversion: plausible chemical reactions and promising products. *Energy Fuels* 10: 305–325.

Yaashikaa, P.R., Kumar, P.S., Varjani, S.J. and Saravanan, A. (2019). A review on photochemical and electrochemical transformation of CO_2 into value-added products. *J. CO_2 Util.* 33: 131–147.

Yano, J., Morita, T., Shimano, K. et al. (2007). Selective ethylene formation by pulse-mode electrochemical reduction of carbon dioxide using copper and copper-oxide electrodes. *J. Solid State Electrochem.* 11 (4): 554–557.

Yao, L., Shen, X., Pan, Z. and Peng, Y. (2020). Unravelling proximity-driven synergetic effect within CIZO–SAPO bifunctional catalyst for CO_2 hydrogenation to DME. *Energy Fuels* 34 (7): 8635–8643.

Zahid, I., Ayoub*, M., Abdullah, B.B. et al. (2020). Production of fuel additive solketal via catalytic conversion of biodiesel-derived glycerol. *Ind. Eng. Chem. Res.* 59 (48): 20961–20978.

Zangeneh, F.T., Sahebdelfar, S. and Ravanchi, M.T. (2011). Conversion of carbon dioxide to valuable petrochemicals: an approach to clean development mechanism. *J. Nat. Gas Chem.* 20: 219–231.

Zecevic, J., Vanbutsele, G., de Jong, K.P. and Martens, J.A. (2015). Nanoscale intimacy in bifunctional catalysts for selective conversion of hydrocarbons. *Nature* 528: 245–248.

Zhang, Y., Su, K., Hong, Z. et al. (2020). Robust cationic calix [4] arene polymer as an efficient catalyst for cycloaddition of epoxides with CO_2. *Ind. Eng. Chem. Res.* 59 (15): 7247–7254. doi: 10.1021/acs.iecr.9b05312.

11

Power-to-Green Chemicals

CONTENTS

Converting Power into Chemicals and Fuels: Power-to-X Technology for a Sustainable Future, First Edition. Martin Bajus.

11.1 Introduction

A completely alternative approach to avoid the use of fossil feedstock as carbon raw materials for the chemical technology is the use of renewable feedstock, i.e., biomass. Biomass is a valuable and limited resource and should be used in a sustainable way. A wide variety of sectors and areas compete for the use of biomass:

- Food and feed supply. Obviously of the highest priority for sugar and starch containing biomass (first generation biomass). As the world population is increasing demand for agricultural land use for food production will increase, making alternative use paths prohibitive.
- Energy. Large amounts of biomass are used for the generation of electricity and heat by combustion of dry biomass including co-firing in power plants. Furthermore, biogas plants and domestic heating contribute to this path; lignocellulosic and waste biomass can be used in these applications, hence competition with food is low.
- Fuel. Large amounts of renewable feedstocks are used for production of biofuels including bioethanol and fatty acid methyl ester by 2020, the European Union aims to have 10% of the transport fuels from renewable sources; the exclusive use of non-food biomass will be a future prerequisite.
- Material and carbon feedstock. This encompasses not only the industrial sector but also the wood (mainly for construction) and the paper industry.

A few factors have to be considered for a sustainable use of biomass. These include environmental aspects, such as soil erosion, water shortage, use of pesticides and eutrophication due to overuse of fertilizers, land availability, indirect land use change, biodiversity etc. It is out of the scope of this study to investigate all these aspects in detail but they are considered with respect to future availability of biomass in Table 11.1. For the large volume basic chemicals, as discussed in Chapter 10, a further consideration concerns the transport of biomass, which should be kept within certain limits to avoid high impact on carbon footprint, cost and logistics. Feedstock should be available within a certain radius of a production site, which is typically below 100 km.

The European chemical industry already uses renewable feedstock, in total about 8 Mt, corresponding to 10% of the total carbon feedstock requirement of 79 million tons. It is anticipated that biomass will play a larger role in the future, however it is limited by the availability of sustainable biomass and the strong competition for use of different biomass use pathways.

As a basis for the scenario work describing a low-carbon chemical technology, it is important to estimate the future availability of biomass and to define reasonable ambitions in terms of the implementation of biomass-based production routes. A few studies have investigated the availability of biomass. As a comprehensive overview, the "Atlas of European Union biomass" developed in the European Union Biomass Futures project has been taken as a baseline. The report identifies different biomass feedstock and provides an inventory of data to quantify and map the technically constrained biomass potentials. For the estimate of potentials, the report takes different sustainability

Table 11.1 Biomass availability in Europe (Mtoe).

Year	Baseline 2005	Reference 2020	Sustainable 2020	Reference 2030	Sustainable 2030
Wastes (Mtoe)	42	36	36	33	33
Agricultural residues	89	106	106	106	106
Rotational crops	9	17	0	20	0
Perennial crops	0	58	52	49	37
Roundwood	57	56	56	56	56
Additional harvestable roundwood	41	38	35	39	36
Primary forestry residues	20	41	19	42	19
Secondary forestry residues	14	15	15	17	17
Tertiary forestry residues	32	45	45	38	38
Landscape care wood	9	15	11	12	11
Total	**314**	**429**	**375**	**411**	**353**

criteria into account that will constrain the future availability of biomass. Two scenarios, "reference" and "sustainable" are provided, the sustainability scenario considering stricter sustainability criteria, e.g., greenhouse gases mitigation requirements including compensation for emissions from indirect land use changes caused by biomass cropping in the European Union. While the report focuses on bioenergy, it is considered as relevant also for the purpose of this study. An overview of the projected biomass availability according to different biomass classes is provided in Table 11.1

Biomass potentials are expected to increase in the reference scenario but will consolidate between 2020 and 2030. In the sustainability scenario the potential will even decline because of constraints on access to land. It is important to note that the contribution of the waste sector is expected to decline.

Within the scope of this study, it is expected that this trend will continue and no additional potentials can be expected beyond 2030, as regulatory constraints will likely increase and pressure from food competition will increase. A biomass potential in Europe of 350 to 400 million-ton oil equivalent is therefore expected. This is also in agreement with a metal study by Zeddies et al., which compares biomass potentials from different studies. A biomass availability in Europe of 15 EJ is projected in 2050, corresponding to 358 Mtoe.

The available biomass will of course have to be distributed to a variety of competing use options, as above. For the scenario work, the availability of biomass is considered and the relative share of biomass use as a function of the ambitions is described.

11.2 Biomethanol Production

11.2.1 Biomethanol Production Process

Production of biomethanol follows the same pathway as coal-based methanol production, i.e. via gasification of the feedstock. A large variety of biomass feedstock can be used for this process and the net yield of wood-based methanol is 1.5 to 2 times higher compared to sugar and starch crops, based on the same heating value. ENI, an Italian multinational energy company, proposed Alcohol-20 (A20) which consists of ethanol (5%), methanol 15% and gasoline (80%).

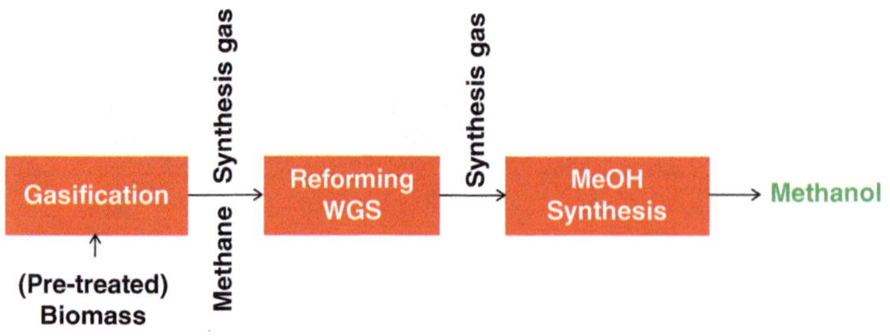

Figure 11.1 Process scheme of methanol production via biomass gasification.

Figure 11.1 depicts a simplified process scheme. The biomass feed needs to be pre-treated. Depending on the biomass type, biomass contains up to 70% water, hence drying of biomass to 15% humidity prior to gasification is required. For gasification, different gasifier technologies are available, varying in gasification temperature and type of used biomass. Using a limited supply of oxygen during the feedstock heating improves the formation of syngas and reduces the amount of carbon dioxide and water. Using oxygen from the electrolysis step of the processes described in Chapter 6 is an option. Following the gasification, tars, dust and inorganic contaminants need to be removed in a gas cleaning step. Unprocessed light hydrocarbons can be further converted in a reformer. A water-gas shift is required to adjust the hydrogen to carbon monoxide ratio to the optimum for methanol synthesis.

A first commercial scale biomethanol plant was announced in 2012 by Värmlands Metanol AB in Hagfors, Sweden with ThyssenKrupp Engineering (former Uhde) as the technology contractor. The conceptual design and feasibility study is based on a pressurised oxygen blown gasifier and predicts 111 MW forest residue as feed to produce 100–110 kt fuel grade methanol, corresponding to a 66–72% efficiency.

To simultaneously produce synthetic natural gas (SNG) and char from biomass, we propose a novel cogeneration process via pyrolysis-coupled hydrothermal gasification. Two typical process configurations were designed and modelled by Aspen Plus. A mathematical model of bio-oil composition involving varieties of typical organic components was established by digging experimental data and then integrating it into the Aspen Plus platform, which can better describe the biomass pyrolysis process. Taking SNG as the main product, this work focused on the effects of pyrolysis temperature, hydrothermal gasification temperature and pressure and feedstock concentration on the composition and yield of synthetic natural gas as well as energy efficiencies. The results show that the pyrolysis temperature significantly affects the yields of synthetic natural gas and char, as well as energy efficiencies. Subsequently, the composition and yield of synthetic natural gas are quite sensitive to the hydrothermal gasification temperature. The hydrothermal gasification pressure has little influence on all process indicators. Finally, the feedstock concentration only has a small effect on methane concentration and synthetic natural gas yield. By application of the flexible operation modules of the cogeneration process, bio-oil can be potentially used as the carrier for seasonal energy storage. This cogeneration process can be regarded as a new approach to upgrade and utilize raw bio-oil.

11.2.2 Energy and Feedstock Demand per Unit of Biomethanol Production

Energy efficiency of the methanol from biomass process is lower compared to the incumbent natural gas process, due to the higher effort in biomass pre-conditioning, the lower hydrogen to carbon ratio of the feedstock and the higher ash and char contents. Process efficiency of the biomass-based process is in the range 50–60%. Azar et al. estimated a conversion plant would have a 66–72% efficiency, which would be on par with the natural gas to methanol efficiency at 64–72%, but this value probably does not include feed preparation steps. As conservative estimate, a 60% efficiency has been assumed. The energy demand of the biomass route can then be calculated as 14.6 GJ/t methanol compared to 12.5 GJ/t methanol from natural gas. For this efficiency, an amount of 2.6 tons dry wood biomass would be required as feedstock per ton of methanol.

11.2.3 Carbon Dioxide Reduction per Unit of Biomethanol Production

Estimates of carbon dioxide emissions from biomethanol production vary widely in the literature, depending on different assumptions. A German study estimated 0.64-ton CO_2eq/t methanol for a production based on short rotation coppice and 0.56 ton CO_2eq/t for forest residues as feedstock, compared to methanol from natural gas at 0.84 ton CO_2eq/t methanol. Taking the higher emission value as a conservative assumption this corresponds to 0.2-ton CO_2eq/t methanol or 24% emission reduction. In addition, the biogenic carbon sequestered in methanol corresponds to 1.37-ton carbon dioxide, yielding an avoidance potential of 1.57-ton CO_2/t methanol.

Analogous to the fossil and hydrogen-based processes discussed in the previous chapters, the carbon dioxide emissions (carbon footprint) of biomass-based processes need to be based on a cradle-to-gate analysis, i.e., must include the feedstock production. In the case of agricultural or forestry biomass, this includes cultivation, harvesting and transport, in the case of residual/waste biomass, the lifecycle starts at the factory gate of the supplying source. Life cycle assessment of biomass tends to be complicated, as many local and seasonal factors affect the result. In addition to the carbon footprint, many other factors such as land use change, biodiversity, water shortage etc., must be considered. In this book, life cycle assessment (LCA) data from different studies and the well-to-tank report of the EU Joint Research Centre have been used. Caution is recommended in trying to generalize these LCA results.

Carbon from biomass sequestered in the target product is generally counted as negative emissions in the production phase of a biomass-based product. These negative carbon dioxide emissions correspond to the positive emissions on total combustion of the target product at the product's end of life. The latter is included in the well-to-wheel data, used for comparing different fuels.

11.2.4 Economics of Biomethanol Production

For basic chemicals, the raw material costs usually represent 60–70% of the production costs. Hence production cost of biomass-derived methanol is largely dominated by the relatively high costs of agricultural and forestry biomass in comparison to fossil feedstocks. The costs of methanol production from biomass have been quantified in many studies and range from 160 €/t methanol up to 940 €/t, very much depending on plant setups and local conditions. Cost estimates at a higher annual production capacity are on the lower end, indicating economy of scale effects. Production costs for biomethanol from waste streams are slightly lower compared to wood, i.e., from €200–500 per ton. Compared to cheapest fossil fuel-based production, biomethanol production costs are at least 1.5 times higher and the capital cost per unit of capacity is at least 3.4 times higher than the capital cost of natural gas-based plants. It was also estimated that biomethanol plants are about 1.8 times more expensive than bioethanol facilities based on the same energy output.

11.3 Bioethanol Production

While ethanol is not among the petrochemical products targeted in this book, the production of bioethanol comprises one of the major renewable feedstock pathways in Europe. Furthermore, bioethanol would be the feedstock for a subsequent synthesis of bioethylene and is a major biofuel contributing to the low-carbon transportation fuel production addressed in this book. The EU is the world's third largest producer of ethanol after Brazil and the US, at 6.4 billion litres in 2015. Ethanol production in Europe has shown strong growth rates in recent years However, a decline has occurred from 2014 (6.9 billion litres) to 2015. The same annual growth rate of 1% as for the petrochemicals has therefore been used for the scenario work. First generation bioethanol is produced from crops such as wheat, corn, sugar cane and sugar beet. In Europe, 37% of the ethanol produced was from corn (maize), followed by wheat (33%) and sugar beets (20%). Due to competition from food use, first generation biofuels are perceived more and more critically and the European Commission has proposed to limit first generation biofuel produced at 7% of energy use in transport. Bioethanol production is at TRL9 for sugar and starch containing crops but is considered at TRL7 for lignocellulosic biomass (second generation biofuels).

11.3.1 Bioethanol Production Process

Biofuels are not new. At 17.5 billion gallons per year ethanol, which is made from corn and mixed with gasoline, is the most-produced alternative transportation fuel in the US Developed in the 1970s and put into wide use by the 1980s, its advantages and cost-related drawbacks are well understood. Today's new green fuels are sourced from agricultural waste and crops that are chemically identical to traditional fossil fuel-based products; however, they provide better performance with a smaller carbon footprint.

Advances in automation technologies make it more feasible to convert previously unusable raw materials (e.g., switchgrass and used cooking grease) into feedstocks that can be processed into high-quality combustible fuels that compete with fossil-based products like jet fuel and automotive gasoline – as well as, increasingly, electric power and natural gas. Meanwhile, leaders in the airline, railroad and trucking sectors have begun powering their fleets with renewable fuels, all of which have created opportunities for savvy downstream operators to stay ahead of the curve.

Carbon intensity (CI) is the amount of greenhouse gas emissions per unit of transportation energy delivered during a fuel's "well-to-wheel" lifecycle – measured in grams of carbon dioxide equivalent per megajoule of energy. Waste-based biofuel feedstocks, such as used cooking oil, have lower carbon intensity scores than those that could be turned into an edible food source (e.g., corn). Other factors affecting carbon intensity include the distance from feedstock source to the processing facility and the energy and greenhouse gas emissions required to process the fuel into a finished product.

Research shows that, depending on assumptions about future policies and societal norms, the decline in the share of hydrocarbons in the global energy system could be dramatic, especially with the corresponding rise in demand for renewable energy as the world increasingly electrifies. If the 2050 net-zero goal is met, then the share of hydrocarbons in primary energy streams could drop from around 85% in 2018 to around 20%–70%, with the share of renewable energy increasing to around 20%–60%.

Government programs such as the European Green Party's Green New Deal and the European Union's Renewable Energy Directive (RED II) that sets the target for renewable consumption at 32% by 2030 are primary reasons why more refiners are exploring these new fuels today. The US government currently offers federal tax credits to biofuel refiners and Canada, California and Oregon have implemented their own low-carbon fuel standards (LCFS), which use a system of incentives and penalties to encourage industries to meet carbon intensity goals in both the manufacture and use of renewable transportation fuels.

Production of ethanol is based on the fermentation of sugar-rich biomass, followed by distillation. The process is depicted in Figure 11.2. The process starts with biomass pre-treatment,

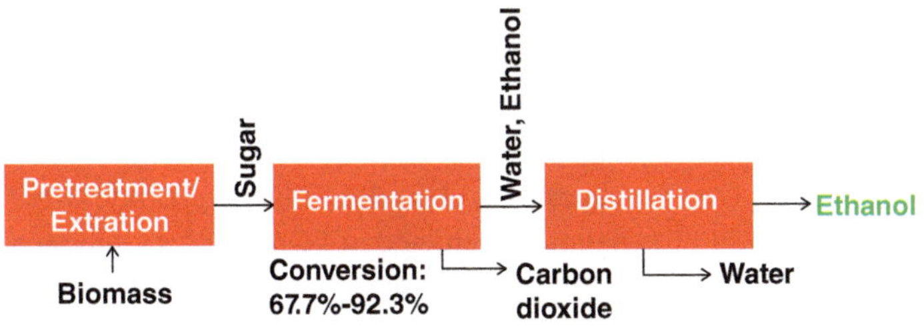

Figure 11.2 Process scheme of bioethanol production.

i.e., extraction of sugar. For sugar beet, the most used sugar crop, sugar is extracted via heat extraction and vapourisation. Starch, from the starch crops wheat, maize or other cereals needs to be hydrolysed into monosaccharides (saccharification). The starch crops are crushed and mashed, then enzymes (e.g., amylases) are added to the mash which dissolve the starch into sugar. Lignocellulose (agricultural and wood residues, wood from forestry, short rotation coppices and lignocellulosic energy crops, such as energy grasses and reeds) require a more complex pre-treatment, due to the different components, i.e., cellulose, hemicellulose and lignin. As common pre-treatment method, steam explosion, is applied, which breaks the structure of the lignocellulosic material through a hydrothermal treatment, using high pressure steam at high temperature for a short time followed by rapid decompression. Alternative methods use supercritical carbon dioxide. The cellulose and the hemicellulose are separated from the lignin and saccharification is induced through enzymatic hydrolysis with cellulases and hemicellulases. The C_6 sugars can be fermented by common yeasts while C_5 sugars need specific microorganisms to get fermented. Lignin is currently usually separated and dried to be used as a fuel for the process or for power generation. The reaction (Equation 11.1) from glucose is as follows:

$$C_6H_{12}O_6 \rightarrow 2\ C_2H_5OH + 2\ CO_2 \tag{11.1}$$

The fermentation yields a diluted aqueous ethanol solution with about 12% ethanol content. At this point the fermentation stops due to the toxic effect of ethanol on the yeasts. Ethanol at 96% is obtained by distillation. To be used as biofuel, 98.7% m/m are required. For this, the remaining azeotropic water is removed by dehydration.

11.3.2 Energy and Feedstock Demand per Unit of Bioethanol Production

For the energy demand, a detailed study has was created for different biomass feedstocks in the well-to-wheel report of the Joint Research Centre (JRC78), including cultivation, feedstock preparation and ethanol production. For sugar beets as feedstock, the energy consumption depends on the process configuration, i.e., the utilisation of the pulp leftover after filtration of the diluted ethanol liquor and the distillation residues ("slops") for electricity cogeneration and biogas production respectively. If credits for these contributions are included, energy consumption well-to-tank is at 0.89 MJ per MJ ethanol or 23.85 GJ/t ethanol respectively; if not, it amounts to 1.42 MJ/MJ ethanol or 38 GJ/t ethanol. For lignocellulosic biomass wood has been used, based on short-rotation forestry of poplar or willow on agricultural land. In this case, 1.78 MJ per MJ ethanol or 47.7 GJ/t ethanol are required. For comparison, the fossil ethanol production via ethylene production from naphtha and dehydrogenation to ethanol results in 21.9 GJ/t ethanol. Stoichiometrically, 51.1% of glucose is converted to bioethanol, meaning that a minimum of 2.12 tons of glucose are needed to produce 1 ton of ethanol. Glucose fermentation is a well-known process and can produce ethanol at 92.3%

yield, resulting in an overall biomass utilisation efficiency of 47.2%. Using wheat straw as feedstock, 0.29 g ethanol per gram of straw have been shown to have a conversion rate of 86% indicating a biomass utilisation efficiency of 29% for lignocellulosic material as feedstock, i.e., 6.75 tons of wheat straw equal 1 ton of ethanol. For wood, 6.05 tons are required. Bioethanol production today is exclusively based on sucrose and starchy biomass. For the scenarios in this book, it has been anticipated that the raw materials for future production of bioethanol will change to lignocellulosic biomass and that the share of lignocellulosic biomass will continuously increase to 25% in 2050.

11.3.3 Carbon Dioxide Reduction per Unit of Bioethanol Production

Carbon dioxide emissions well-to-tank have also been taken from the literature. Emissions range between 16.2 and 38.6 kg CO_2eq/GJ ethanol (0.43-ton and 1.03-ton CO_2 eq/t ethanol) for sugar beet, depending on the process configuration, i.e., if utilisation of the pulp and distillation residues ("slops") for electricity cogeneration and biogas production respectively is considered. In the case of wood, emissions are at 21.2 kg CO_2eq/GJ ethanol or 0.57-ton CO_2eq/t ethanol. The footprint of the fossil route is at 1.06-ton CO_2eq/t ethanol, roughly double that of the carbon dioxide footprint of the most efficient biomass routes. Biogenic carbon sequestered in bioethanol corresponds to 1.91-ton CO_2/t ethanol. As ethanol is included as a fuel in subsequent scenarios, the carbon footprint of bioethanol production has also to be compared to the production footprint of gasoline. The latter is at 12.6 kg CO_2eq/GJ produced gasoline, which is 41% lower than the footprint of bioethanol production. However, this is largely offset by the high gasoline emissions during fuel use.

11.3.4 Carbon Dioxide Reduction for (Partially) Replacing Gasoline with Bioethanol

The European EN228 specification for gasoline allows blending with up to 10% ethanol. Where high ethanol blends (e.g., E85) are used, they can only be used in vehicles specially adapted to use such fuels. For gasoline, carbon dioxide emissions well-to-wheel amount to 87.1 g CO_2/MJ. For ethanol in the beet scenario depicted above, emissions well-to-wheel amount to 17.8 g CO_2/MJ. In total, this corresponds to a reduction of 69.3 g CO_2/MJ of fuel.

11.3.5 Economics of Bioethanol Production

Production costs of ethanol are largely determined by the price of biomass feedstock, which can account for 55–80% of the final price of ethanol. SenterNovem estimated production costs for large plants in Europe at 0.50–0.55 €/l for sugar beet-based processes, 0.55–0.60 €/l for grain-based processes and 0.45–0.55 €/l for processes using residual starch streams. Projections are estimating 0.45–0.50 €/l for sugar beet-based processes, 0.50–0.55 €/l for grain-based processes and 0.40–0.50 €/l for processes using residual starch streams. The International Renewable Energy Agency (IRENA) estimates bioethanol production from lignocellulosic biomass to cost about 750 US$/t or 975 €/t (2012), which would correspond to 0.77 €/l.

Use of liquid fuels started around 137 years ago, in 1886. Later, there was tendency to use greener liquid fuel such as ethanol and butanol. Ethanol is produced from biomass using biocatalysts (fermentation). Ethanol is blended with gasoline to produced E15 (15% ethanol and 85% gasoline). Ethanol blended gasoline in the United States was mandated in 2009 for energy security reason. The most popular blend composition, E10, has been used at least in 20 countries. Most of the time, this blending is limited by the national production of ethanol in specific countries, such as 20 million litres in Ethiopia as compared to the need of 76.2 million litres/year. The popularity of alternative fuels depends on the policy from local government; in the United States of America, ethanol blended gasoline was successfully applied with Biomass Research and Development legislation in 2000. The Bush administration committed 1.7 million USD in 2005 for hydrogen and electrochemistry research. The Obama administration targeted 1 million electric vehicles by 2015, which stimulated electrification in US transportation.

For liquid biofuels, there are at least two blending components for gasoline, ethanol and biogasoline. For ethanol, there are at least two current studies: (i) to convert ethanol to longer chain alcohols (butanol) and aromatics and (ii) biocatalysts to convert lignocellulose waste to ethanol.

11.4 Bioethylene Production

11.4.1 Bioethylene Production Process

Bioethylene production is based on bioethanol as feedstock and hence follows the process steps discussed in Chapter 10. The general process scheme is depicted in Figure 11.3.

Using bioethanol as feedstock, bioethylene is produced via dehydration of the ethanol over an alumina or silica-alumina catalyst at 300°C in a fixed bed or in a fluidized bed reactor. A gas separation is required to remove gaseous by-products and an alkaline scrubbing to remove oxygenates. One ton of bioethylene requires 1.74 t of (hydrated) ethanol. Conversion yields are at 99% with 97% selectivity to ethylene. The reaction is endothermic and requires a theoretical minimum energy of 1.68 gigajoules (GJ) per ton of bioethylene.

The current largest bioethylene producer is Brazil, but the concept has also been implemented in Europe. In 2014, Axens, Total and IFP Energies Nouvelles announced a technology for bioethylene production through dehydration of bioethanol under the technology brand name Atol™, to produce polymer grade bioethylene. The processes can omit the alkaline workup due to the specially developed catalyst showing very high ethylene selectivity. An alternative concept is the gasification of biomass to syngas, followed by methanol production and MTO has to be mentioned, but this process is not favourable in terms of carbon dioxide emissions.

11.4.2 Energy and Feedstock Demand per Unit of Bioethylene Production

For the entire process chain, including production of 1.74 t ethanol followed by ethylene production, the energy demand is 85.5 GJ/t ethylene, which is very high compared to 26–31 GJ/t ethylene (maximum 20–40 GJ/t ethylene) from oil via naphtha (including primary feedstock production).

It is assumed that ethylene production would be based on wood. Feedstock demand in this case is at 10.5 t/t ethylene.

11.4.3 Carbon Dioxide Reduction per Unit of Bioethylene Production

The fossil ethylene production route causes 1.15-ton CO_2eq/t ethylene including feedstock production. For bioethylene, the carbon footprint of the production of bioethanol feedstock (including

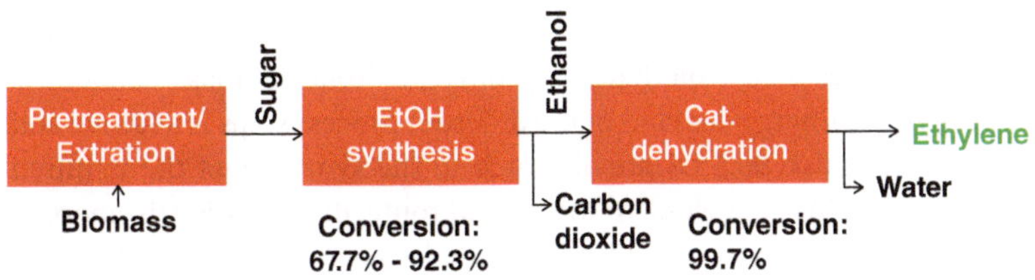

Figure 11.3 Process scheme for bioethylene production.

biomass raw material production) and the emissions caused by the ethanol dehydration process must be considered. In terms of raw material and bioethanol production, the 0.57-ton CO_2eq/t ethanol for wood have been used. As 1.74 t ethanol per ton ethylene are required, the bioethanol production step prior to ethylene synthesis causes around 1-ton CO_2eq/t ethylene. It is estimated that at least 0.2-ton CO_2eq/t is emitted in the ethylene production process, which makes the carbon dioxide emissions from the process relatively close to that of the fossil process. For wood or waste biomass reduced emissions from the biomass route can be expected. The biogenic carbon sequestered into the product, which is counted as negative emission stoichiometrically, amounts to 3.14-ton CO_2 /t ethylene. The amount of avoided carbon dioxide compared to the fossil process is hence at 1.95-ton CO_2/t ethylene.

A case-by-case analysis of the carbon footprint is strongly advised for bioethylene production, as the carbon footprint depends on the type of biomass used and the local production logistics and infrastructure.

11.4.4 Economics of Bioethylene Production

The International Renewable Energy Agency quantifies the production costs of bioethylene in the EU at 3,250 €/t. The biomass feedstock accounts for about 60% of the bioethanol production costs. The bioethanol costs in turn account for about 60–75% of the bioethylene production costs. Bioethanol production from lignocellulosic biomass was estimated to cost about 975 €/t ethanol, assuming mature technical and economic conditions. This leads to a bioethylene production cost of 2,250 to 2,800 €/t ethylene which would be in the range of the current thermochemical production route (pyrolysis) at about 2,500 €/t.

11.5 Biopropylene Production

11.5.1 Biopropylene Production Processes

Biopropylene production can be considered as a further step following bioethylene production. In Brazil, Braskem has announced a production plant for bio-based polypropylene at 30,000 t/a scale. The two-stage process is based on dimerization of ethylene to 1-butene, isomerization to 2-butene and metathesis with ethylene. A new catalyst was reported to allow realization of the process in one stage. As for ethylene, the route via biomass gasification, methanol synthesis and MTO/MTP is also available.

Other routes are available at lower TRL, e.g., the fermentative production of propanol or isopropanol, followed by dehydration, the direct fermentative production of propylene, as investigated by Global Bioenergy or the catalytic conversion of (bio)ethanol to propylene with scandium-loaded indium sesquioxide.

11.5.2 Energy and Feedstock Demand per Unit of Biopropylene Production

The synthesis route via gasification to methanol requires 2.6 t dry biomass/t methanol and the MTO route stoichiometrically requires 2.28 t methanol/t propylene. Total biomass demand is therefore at least 5.9 t per t propylene. Energy demand of the different process steps is additive and amounts to 95.5 GJ/t propylene of which 90.5 GJ are allocated to the synthesis of the required methanol feed and 5 GJ/t HVC for the MTO process. In the second route, dimerization/isomerization with subsequent metathesis, the energy demand is estimated based on the stoichiometric bioethylene demand as no further information on the process specific energy requirement and related emissions is available. The corresponding energy demand would be 130 GJ/t propylene. This can be considered a good approximation, as the energy consumption will be driven by the ethylene synthesis sequence.

11.5.3 Carbon Dioxide Reduction per Unit of Biopropylene Production

As for the process energy, the carbon footprint is composed of the emissions from the different process steps. As a result, process related CO_2 emissions are at 1.86 ton CO_2eq/t propylene, which is 2.5 times that of the fossil route via naphtha steam cracking. Sequestered bio-based carbon amounts to 2.09-ton CO_2eq/t propylene, which slightly overcompensates the process emissions, however, the gain is low. This route is therefore considered as relatively unfavourable in terms of energy and resource efficiency, given the high biomass demand. For the dimerization/isomerization with subsequent metathesis process, related emissions of 1.65-ton CO_2/t propylene would result, which is also well above the fossil route at 0.76 t.

11.6 BTX Production from Biomass

11.6.1 BTX Production Process

The production of BTX from biomass can follow several routes. As the most developed route, gasification of biomass, methanol synthesis and methanol to aromatics (MTA) is used. Figure 11.4 shows the process sequence.

The individual process steps have already been described in Section 11.2.1 (biomass to methanol) and Section 11.6.1 (methanol to aromatics).

An alternative synthesis in four steps of p-xylene can be performed from bioethanol-derived xylene, followed by Ir-catalysed trimerization to 1-hexen, dehydrogenation to hex-2,4-diene followed by Diels-Alder-addition with another ethylene molecule and finally Pt-catalysed dehydrogenation. A yield up to 65% has been reported. In total five ethylene molecules are required to synthesize one molecule of p-xylene.

A seemingly obvious pathway, the selective degradation of lignin thereby releasing the contained aromatic structures, is available only at very low TRL. All attempts based on pyrolysis or hydrogenation have so far failed to result in a technically viable process. All proposed process routes are extremely unselective and yield a large product spectrum.

Another option, fast pyrolysis of lignocellulosic biomass, has been investigated and significant yields of aromatic compounds have been reported. The mixture contains different aromatics with toluene, benzene and xylenes as major components. Anellotech and KiOR are attempting to

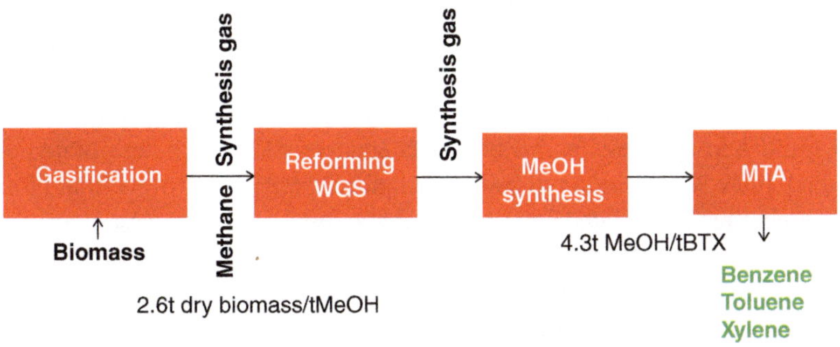

Figure 11.4 Process scheme for BTX production from biomass via gasification.

implement this approach, but KiOR stated they are unable to achieve commercial-scale production. Anellotech's process, developed in collaboration with Johnson Matthey and IFP Energies Nouvelles, is based on the research results of G.W. Huber, University of Wisconsin, Madison, who has reported aromatics yields from wood of up to 20% (C-content).

11.6.2 Energy and Feedstock Demand per Unit of BTX Production from Biomass

Both synthesis routes have been assessed in terms of energy demand and carbon dioxide emissions. As depicted in Figure 11.4, the synthesis route via gasification to methanol requires 2.6 t dry biomass/t methanol and the MTA route requires 4.3 t methanol/t BTX. Total biomass demand is therefore at least 11.2 ton per ton BTX. Energy demand of the different process steps is additive and amounts to 72 GJ/t BTX of which 67 GJ are allocated to the synthesis of the required methanol feed and 5 GJ/ for the MTA process. For the Diels Alder process, 2 t ethylene/t p-xylene are required, yielding an extremely high energy demand of the biomass-based process chain of 174 GJ/t. This is to be compared with the specific energy consumption (SEC) of fossil BTX production, which is around 7 GJ/t HVC.

11.6.3 Carbon Dioxide Emissions per Unit of BTX Production from Biomass

As for the process energy, the carbon footprint is composed of the emissions of the different process steps. As a result, process related carbon dioxide emissions are at 2.21-ton CO_2eq/t BTX for the Diels-Alder route and 2.6-ton CO_2eq/t BTX for the methanol to aromatics route, which is factor of two to three higher than the fossil routes at 0.84-ton CO_2eq/t BTX. Reduction in overall emissions would therefore originate only from the bio-based carbon sequestered in the products, which amounts to 3.3-ton CO_2eq/t BTX, resulting in a total carbon footprint of −0.7-ton CO_2eq/t BTX for the methanol to aromatics route or −1.54-ton CO_2eq/t BTX avoided carbon dioxide compared to the fossil route. Due to very its high energy and feedstock demand, BTX production from biomass based on the two investigated routes is not considered to be beneficial from an energy/resource efficiency point of view. The alternative production routes, selective lignin decomposition and fast pyrolysis, need to be investigated further. This can be regarded as a high priority for a low-carbon chemical production strategy as the hydrogen-based route described in Chapter 6 is also extremely energy demanding.

11.7 Comparison of the Biomass-Based Synthesis Routes

In this chapter, different routes using renewable feedstocks to produce large volume chemicals have been investigated. Table 11.2 provides a summary of the main results obtained.

Results are presented for the different target products. Bioethanol has been included as a fuel. From a biomass use and economic point of view, biomethanol and bioethanol represent the most meaningful use paths of biomass among the investigated routes. The olefins and BTX synthesis based on the suggested multi-step synthesis pathways are characterized by very high biomass demands per ton of product and costs of carbon dioxide avoidance. The BTX synthesis from biomass shows substantially higher process related emissions than the fossil-based processes.

The main issue is the large demand for feedstock for these processes, which is utilised in a relatively inefficient way. A paper by Nova describes the "biomass utilization efficiency" (BUE) as a means of describing the percentage of initial biomass ending up in the product based on the molar mass of the reactant and target bio-based product. The authors conclude

Table 11.2 Comparison of biomass-based synthesis routes.

Product	Energy Demand (GJ)	Biomass as Feed (t)[a]	ΔProcess Related CO_2 emissions (t)[b]	Biogenic Carbon Request CO_2(t)	Cost (€)	Avoided CO_2 (kg)	
Chemicals		Per ton of product				Per t biomass	Per €
Methanol	14.6	2.6	−0.2	−1.37	200–500	600	3.1–7.8
Ethylene	85.5	10.5	+0.05	−3.1	2250–2800	290	1.1–1.3
Propylene	95.5	5.9	+1.1	−2.09	2200–2500	170	0.4–0.45
BTX	72	11.2	+1.76	−3.3	>3000	139	<0.5
Fuels		Per ton of product				per ton	per €
Bioethanol	47.7	6.1	Well-to-wheel emissions: 1.86		975	305	1.9

[a] Dry wood has been as used as a reference.
[b] Compared to the reference fossil process.

a low biomass utilization efficiency of 28.6% for polyethylene, because the ethylene molecule contains no oxygen. Biomass or finally sugar use in such a case encompasses the removal of all oxygen as carbon dioxide from the sugar structure, thereby also eliminating many carbons. This is the case for all large petrochemicals included in this study. The biomass utilization efficiency also tends to decrease with the number of synthesis steps and the biomass based routes described in this study usually encompass several steps. It can be concluded that the production of drop-in petrochemicals from biomass is a rather inefficient way of using biomass, with the biomass utilization efficiency being generally low. This fact is even more relevant considering the limited availability and competition of pathways using sustainably produced non-food biomass or waste biomass.

Much higher efficiencies can be achieved for dedicated biomass-based synthesis routes that maintain the functional units of the feedstock molecules, e.g., oxygen-rich and carbonated molecules such as polylactic acid and succinic acid. Considering the limited availability of biomass as a raw material it is therefore strongly recommended to strengthen the deployment of such routes. The situation might be assessed differently for complex biorefineries, in which different feedstock and production streams would need to be coupled in an intelligent manner.

Converting biomass to chemicals and fuels requires development of highly active and selective catalytic materials for important reactions such as aldol condensation. Aldol condensation is a critical step to convert small molecules such as 5-hydroxymethylfurfural (HMF) and furfuraldehyde (FA) to larger complexes that can be used for fuels or chemicals. Although extensive work has examined amino silica catalytic materials for aldol chemistry, these materials have received limited attention for HMF and FA conversion despite the capability of these materials to catalyse reactions in a cooperative manner. The important structure–function relationships that demonstrate the impact of micropores in these materials have been investigated. Materials are produced using a typical method (REG) and a method that produces limited to no micropores (NMP). The materials are grafted with amino silanes and characterized using standard techniques. It is found that materials with few to no micropores are more active for the aldol reaction and condensation for both FA with acetone and HMF with acetone. Additionally, the materials are tested in recycle experiments that showed the NMP catalyst retained higher activity after four catalytic cycles than the REG catalyst. Overall, the results have important implications for the design of materials for upgrading biomass into chemicals and fuels.

11.8 Biofuels

Different fuel forms will coexist over the next 20 years during the "transition" period prior to a possible pure renewable era, even though numerous analysts considered hydrogen fuel cells as only a current fad. LCA study reported on the methanol and electrical vehicles from well-to-wheel with a note that electrical vehicles have the disadvantage of release of toxic material during manufacturing and maintenance.

The importance of producing "green" gasoline and "green" diesel is that it will lead to energy diversification with lower greenhouse gases. The crucial part of biodiesel development is its technology status. Success in the laboratory phase does not means commercial success of biofuels for national needs. To produce biofuel with national scale production requires close collaboration between the government, chemical technology and course academics on campus and research centres. The development status of a technology is usually classified on basis of technology readiness level (TRL). Results at the laboratory scale are generally still at TRL 3 and TRL 4. Close cooperation with industry is needed to continue to climb to TRL 7. Then the full role of domestic private industry is needed to reach commercial scale in TRL 9. Without investment from industry, research results from research activities at universities and in research institutions will stop at TRL 4. Structured support from government is needed to help biofuel researchers to achieve commercial scale production.

Currently, one of the major priorities for chemical technology is finding sustainable carbon-based energy sources that will lead to a decline in the use of fossil resources and consequent damage to the atmosphere. At present, fossil fuels are preferred as energy sources in most developing countries, owing to their affordable cost and availability. Carbon dioxide emissions are directly associated with human-induced processed like deforestation and combustion of fossil fuels and increased levels of carbon dioxide led to global warming. Many researchers have proposed to eliminate carbon dioxide from the atmosphere and transform it into various energy resources (hydrocarbon fuels). A few alternatives have been suggested to utilize renewable energy resources (solar energy, geothermal or wind energy) for the elimination of greenhouse gases. Among these alternative fuels, biodiesel blends and diesel–ethanol blends have received the most attention due to their high efficiency and biodegradability. However, the biodiesel-blend viscosity is significantly higher than that of diesel, which is a major obstacle in the commercial use of biodiesel. Razzaq et al. suggested the use of binary biodiesel blends to decrease the viscosity and density by adding bioethanol.

Biodiesel is deemed an appealing alternative to petro-diesel due to its sustainability and biodegradability and it has characteristics like those of conventional diesel. The current scenario demands a highly efficient and low-cost fuel that can replace diesel fuel. In this regard, biodiesel seems to be a suitable replacement for diesel fuel owing to its good physicochemical properties (biodegradable, non-toxic and low sulfur content). As well as these advantages, its higher flash point, octane booster ability and renewable nature play essential roles in its commercialization in the automotive industry. Many published studies have shown that biodiesel has a better kinematic viscosity and density as well as a higher cetane rating. A substantial reduction in carbon footprint has been observed during biodiesel production from sustainable sources. Many researchers have stated that 100% biodiesel is very feasible in diesel engines without needing much improvement in the current hardware design. Furthermore, the oxygenated source of biodiesel might give many additional advantages, such as low exhaust emissions (apart from NO_x emissions) and higher lubricity compared to conventional counterparts.

The biodiesel industry has recently grown because it is being used at progressively larger scales within the transportation industry, particularly for street vehicles, in aircraft and in sea transportation, as well as to produce power and for cooking. The worldwide production of biodiesel in 2013 was 110 billion litres. In 2022, the global generation of bioethanol was expected to increase to 168 billion litres and biodiesel generation to 141 billion litres, with the commitment to Renewable Fuel Standard (RFS2) in the United States, together with EU-RED. The renewable energy directive from

2009 (2009/28/EC) supported an overall policy for the production and promotion of energy from renewable sources in the EU. It was predicted that at least 20% of the EU's total energy could be filled with renewables by 2020, mainly to be achieved through the national targets of the member states. All EU countries must also have ensured that at least 10% of their transport fuels came from renewables by 2020. The same directive also specified national renewable energy targets. Even though biodiesel speaks to a renewable, helpful and natural neighbourly alternative to fossil fuels, the cost effectiveness of biodiesel generation needs to be enhanced. This may be achieved by working on certain aspects of the production method, e.g., feedstocks, synthesis itself and the by-products. The process of transesterification was used to synthesize biodiesel from various feedstocks with the help of a solvent (methanol, ethanol) and use of a catalyst (acidic or basic). The produced biodiesel is utilized directly or after mixing with conventional diesel fuel. Biodiesel is usually blended to enhance sustainability, combustion efficiency and lubricity and to increase cetane number. Numerous authors have stated that the emission properties of blended and pure biodiesel lead to a significant reduction of particulate matters and low carbon monoxide and NO_x emissions. Moreover, the high oxygen content present in biodiesel promotes combustion and lowers the calorific value. Despite these advantages, the use of pure biodiesel has some drawbacks due to technical issues related to low volatility, high viscosity and high density. To control the emission of NO_x and particulate matter, fuel additives are introduced which help to reduce knocks and improve engine efficiency, cetane number and lubricity. A comparison of pure biodiesel and their blends is shown in Table 11.3.

During the process of transesterification, crude glycerol was produced at 10% by weight of the total product. It is estimated that the worldwide generation of crude glycerol was more than 41.9 billion litres by the end of 2020. Some impurities like water, inorganic salts, esters, fatty acids and methanol are found in crude glycerol and subsequently treated as the waste stream in the biodiesel industry. Current crude glycerol costs are approximately US $0.04–0.09/lb. The anticipated fast development of biodiesel generation will assist in reducing the glycerol cost.

Hence, unused and conservative paths of utilizing glycerol should be further developed to extend the use of rough glycerol to improve the supportability of biodiesel production industries.

11.8.1 Biodiesel Production

The terms of the Kyoto Protocol on global atmospheric change present basic components of the tool package that is required to minimize the emission of greenhouse gases. Using electricity and raw materials from green resources is one way to reduce carbon dioxide emissions. The chemical

Table 11.3 Physico-chemical properties of diesel versus different mixtures of diesel with biodiesel.

Property	BO (Pure Diesel)	B20 (20% Biodiesel)	B50 (50% Biodiesel)	B100 (Pure Biodiesel)
Cetane number	53.30	54.36	55.95	58.60
Air to fuel ratio	14.66	14.22	13.56	12.49
Density (kg/m^3)	837.90	842.32	848.95	860.00
C (wt.%)	86.35	84.30	81.26	76.31
H (wt.%)	13.65	13.35	12.89	12.15
O (wt.%)	0.00	2.36	5.84	11.54
Viscosity (cSt)	2.64	3.08	3.73	4.82
Higher heating value (MJ/kg)	44.79	43.81	42.35	39.90
Lower heating value (MJ/kg)	41.77	40.86	39.49	37.20

industries have emphasized environmentally sustainable projects and an intensive growth of technologies dependent on ecological materials. At the present time, more and more research on recycled products of natural origin is accessible. This refers to oils that are degradable and a source of new substances for the oleochemical industry. The use of bio-oils as raw materials for industrial growth is important not only from an ecological perspective, it is also of economic concern due to their low market cost and the limited supplies of unrefined oil and natural gas. The blooming of the biofuel sector in the European Union, proposed by the EU's strategic decisions in the field of energy policy, is linked to the decrease in greenhouse gas emissions and to the reduction of their environmental effect. In 2008, the EU Parliament voted on the Environment Plan 3×20 to reduce carbon dioxide emissions by 20% in 2020 relative to 1990. At the same time, the energy share from sustainable resources must rise from 8.5% to 20% and usage of energy must have been decreased by 20%. Moving towards these and other goals, the manufacturing of lubricants and biofuels from renewable resources has increased during the past decade. The process of transesterification of triglyceride with alkyl alcohol is used for production of biodiesel which consists of fatty acid methyl esters. The operation scheme is presented in Figure 11.5.

The result of this reaction is a two-layer product. The top layer is biodiesel and is the main product, the bottom layer contains glycerol, water, soaps, methanol, esters and catalysts. Peptides, proteins, lecithin and phospholipids are also present in minute quantities. Biodiesel synthesis also depends upon the process of esterification of free fatty acids (FFAs) originating mostly from vegetable oils. The types of oils used are classified by the characteristics of crops mainly grown in the area. Rapeseed oil is commonly used for the synthesis of biodiesel in North and Middle Europe, while soya oil is increasingly used in Southern Europe and North America. Many substrates, such as waste oil, animal fats and palm oil, are highly utilized in many regions.

In 2012, per the European Commission, more than 55% of cars were manufactured having diesel engines. In 2013 and 2014, around 50% of newly registered vehicles were equipped with diesel engines and the alternative fuel use rose from 1.4% to 2.1% in the United Kingdom. European Union legislation laid down compulsory emission reduction targets for new vehicles in 2013 and for 2025 there are also plans for a further lowering in CO_2 emissions to 68–78 g/km. Both conditions and regulations make it more likely that conventional oil will be substituted with biofuels. According to the EU Biodiesel Board and Ecofys, the use of 1 kg of biodiesel lowers CO2 emissions by 3 kg (65% to 90% less than with petro-diesel). Also, biodiesel contains a very small quantity of sulfur and its biodegradability is higher than that of petro-diesel, which make it very advantageous with respect to environmental conservation. The boom in the biodiesel industry around the globe has contributed to generate an immense volume of glycerol as a by-product.

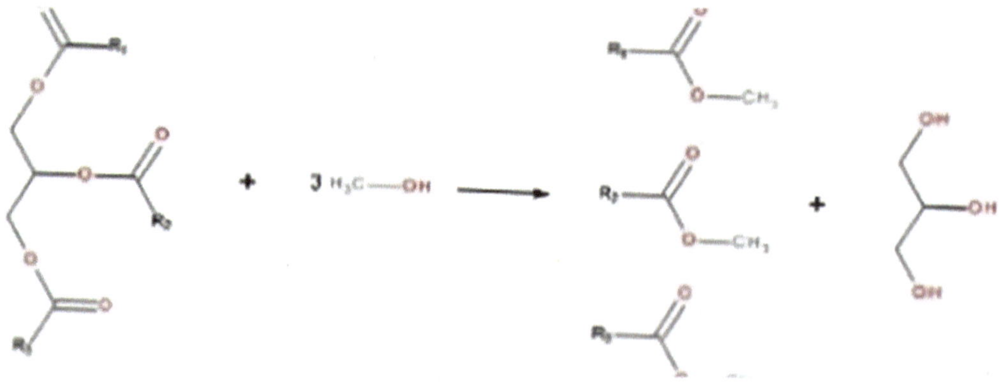

Figure 11.5 Synthesis of biodiesel and glycerol by the transesterification reaction.

The disposal and use of waste glycerol have become a significant concern in relatation to both the ecological and economic responsibility of the biodiesel industry. The production of biodiesel results in a significant quantity of by-products and contaminants, such as wheat meal, sugar, paper, wastewater and diluted glycerol. Natural glycerine and synthetic glycerol are the two ways in which glycerol is manufactured. Natural glycerol is obtained mainly in crude form containing, depending on the production process, water and other residues as impurities. Glycerine is obtained primarily as a by-product via the transesterification process. Usually, crude glycerol produced during this process was sold by the producers of biodiesel.

11.8.2 Purification of Glycerol

The by-product (glycerol) of biodiesel production consists of many impurities that make it unacceptable for use as the feedstock for advanced applications. There are three main forms of purified glycerol, depending on their purity and possible end-use:

a) "Technical classification", utilized as a key element in industrial materials but not applicable for the manufacturing of food or medicine;
b) United States Pharmacopeia (USP) grade, such as vegetable oil or animal fat glycerine, suitable for food and pharmaceutical drugs; and
c) Food Chemicals Codex (FCC) grade, such as glycerine from palm oil supplies, suitable for use in cooking oil.

Glycerol is typically sold 99.5% to 99.7% dry.

Purification is thus necessary to transform impurified glycerol to a functional state that satisfies the purity criteria for current or emerging applications. Various methods such as refining, filtration, ion exchange, adsorption and crystallization are utilized to filter the unrefined glycerol. Each purification procedure addresses specific raw glycerol properties. To attain high purification efficiency, multiple techniques are used together during the purification process of waste glycerol. A combination of the recovery of alcohol via distillation and the elimination of salt through ion-exchange resin and adsorption using activated carbon constitutes a typical purification method. Distillation has been widely used for purification of unrefined glycerol. In general, distillation is done to extract methanol and water at their boiling points during the purification of glycerol. For purifying the crude glycerol most efficiently, distillation is used to remove high salt content and nonglycerol organic matter (MONG). Yong et al. stated that a simple distillation method at a temperature of 120–126 °C yields approximately 141.8 g of glycerol/kg of residual glycerol (14% yield) with an adequate purity of 96.6% glycerol. The other impurities found in purified glycerol are 0.031% ash, 1.01% water vapour, 2.43% non-glycerol organic matter and it has a pH of 3.5.

The authors claimed that for the distillation process, foaming was stopped at the optimal pH < 5. Generally, the process of distillation provides limited wide-scale continuous operation, cheaper material and adaptability to the various grades of crude and final products. However, the refining of raw glycerol is an energy-consuming method due to the high specific heat efficiency of glycerol, which demands a large amount of heat for thermal decomposition and vaporization. Ion-exchange techniques can also be considered for purifying the crude glycerol. Moreover, ion-exchange methods remove the trace impurities, colour and smell by exchanging anions using an ion-exchange resin material. This process utilizes a limited volume of water to remove salts from impure glycerol and therefore does not demand the glycerol vaporization. Isahak et al. used Amberlite 200 and Amberlite IRN-78 resins to refine the crude glycerol and examined the efficiency of the resin. The experiments were conducted by applying a column method, in which the ion-exchange resins were positioned in a vertical column, creating a sheet. High-performance liquid chromatography was used for analysing the glycerol samples. The chromatograms of the distilled samples revealed a single peak of glycerol, showing a very straight baseline, which suited the profile of pure glycerol. These tests showed that impurities like free ions and inorganic salts were eliminated using the ion-exchange resin process.

Additional possibilities for purification are a physical treatment (filtration) and a pure chemical method (protonation of an acid). For chemical processes, the impurified glycerol is treated with an acid to form alkoxide salts of the corresponding free fatty acid. The free fatty acids are extracted by decantation or an extraction process. Ooi et al. stated that purification of crude glycerol with 6% (v/v) sulphuric acid could lower the ash contents via chemically treating at lower pH. Glycerol was recovered from glycerol pitch through a combination of physical and chemical treatments by Hazimah et al. Glycerol with high purity (~99.1–99.8%) was obtained, with fewer impurities (0.11–0.80% (w/w) water, 0.054% (w/w) ash content and 0.56% (w/w) soap). Kongjao et al. stated that a higher ash content, water and non-glycerol organic matter was completely removed by using a combination of physical and chemical treatments with the separation of solvent. Glycerol (~93.34 wt.%) with a high purity level was produced at pH 1 with minimum contaminants (0.00045 wt.% ash and 5.16 wt.% non-glycerol organic matter). Purification of crude glycerol by means of various physical and chemical treatments has the advantages of easy operation, lower operational costs and high performance.

11.8.3 Conversion of Glycerol into Valuable Products

Glycerol is a significant product derived from the oleochemistry industry. In fact, a significant excess volume of glycerol is produced through the processing of biodiesel. The supply of glycerol will increase in the coming years owing to a tremendous rise in the production of biodiesel globally. As the current market for glycerol is small, it cannot absorb the surplus of glycerol produced by the development of biodiesel.

Traditionally, glycerol has been used as an additive in food, tobacco and pharmaceutical industries, which turned it into a valued product. Glycerol is a versatile product and like petrochemical hydrocarbons, has a significant range of valuable chemicals that can be developed through different chemical processes, such as 1,2-propanediol, synthesis gas, glycerol ether, 1,3-dichloropropanol, acetal and ketals, as shown in Figure 11.6. For example, the catalytic oxidation of glycerol

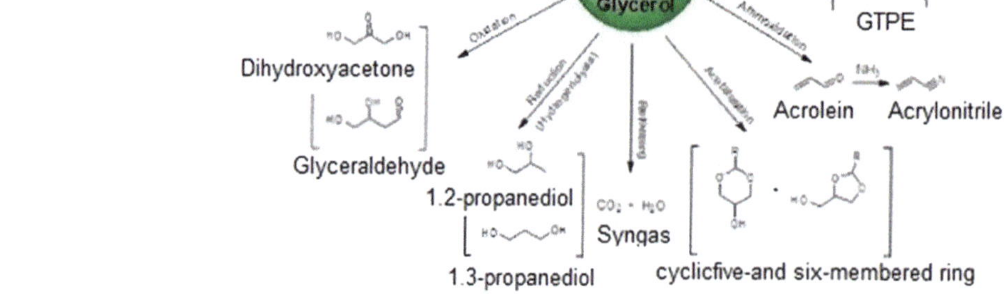

Figure 11.6 Various processes for chemical conversion of glycerol into valuable products.

contributes to the production of different compounds such as dihydroxyacetone and glyceraldehyde. In addition, owing to its outstanding physical and chemical properties, glycerol has gained considerable interest as a pollution-free solvent in synthetic organic chemistry. Its properties, including a high boiling point, minimal vapour strength, fast dissolution and non-hazardous quick handling and easy storage, offer revolutionary ways to reuse glycerol and to utilize it in renewable forms.

1,3-propanediol is a basic organic chemical that is widely used in the manufacturing of cosmetics, polymers, medicines, milk and lubricants. The production of 1,3-propanediol has been reduced and the prices are rising, which implies that glycerol could become a desirable feedstock due to its relatively high abundance and otherwise low demand. Processes used to form 1,3-propanediol include hydrogenolysis, fermentation and dihydroxylation. Glycerol is utilized as a feedstock in the manufacturing of 1,3-propanediol via the fermentation process, which is a two-step enzymatic reaction. In the first step, the catalytic transformation of glycerol to water and 3-hydroxypropionaldehyde (3-HPA) takes place. 3-HPA is broken down into 1,3-propanediol through use of a NADH-dependent enzymes 1,3-propanediol dehydrogenase. 1,3-propanediol can also be produced by glycerol selective hydrogenolysis with a catalyst and hydrogen. In the selective hydroxylation, the glycerol intermediate's hydroxyl group is selectively transformed into a tosyloxyl group, which is then eliminated by catalytic hydrogenolysis. However, in the glycerol dihydroxylation process, glycerol acetalization takes place with C_7H_6O and tosylation of the exposed–OH group of the acetalized glycerol yields 1,3-propanediol.

Glycerol is also transformed into hydrogen via several processes like partial oxidation, reforming, steam reforming and autothermal. Catalytic steam reforming is a highly energy-demanding cycle in which a catalyst is used to combine glycerol and water to create hydrogen. In this process glycerol is converted into hydrogen in three steps: first by dehydrogenation of glycerol, then carbon monoxide is bonded externally to the catalyst, which can undergo desorption, followed by methanation or water–gas shifting. The glycerol dehydration results in acrolein, also known as 2-propenal or acrylic aldehyde, which is the simplest unsaturated aldehyde. To synthesize acrylic acid esters and acrylic acid, acrolein is used an intermediate. It is used in the manufacturing of paper as a slimicide and as a micro-biocide. Acrolein is usually generated from glycerol in the presence of acid via the dehydration process.

The green chemical technology of hydrothermal electrolytic extraction of glycerol through a continuous-flow reactor requires the introduction of metallic catalysts. The glycerol oxidation involves a technological obstacle, whereby a specific catalytic oxidation technique acts on a polyfunctional substance and a basic oxidant. With glycerol, the derived oxygenated products are hydroxyl pyruvic acid, oxalic acid, dihydroxyacetone, tartaric acid, $C_3H_2O_5$ and its intermediates. The most studied metallic catalysts are palladium, gold and platinum, but the major drawback of platinum and palladium is their deactivation during the long reaction duration. To solve this challenge, support materials are embedded on metal catalysts to create a hybrid system. To address the issue of deactivation during longer reaction times, support materials such as activated charcoal and graphite were embedded into the metal catalysts (Pt/C, Pd/C and Pt/graphite) to form a hybrid system. The oxidation selectivity of the secondary –OH group of glycerol was increased significantly by the combination of platinum and other metals, such as bismuth, to yield hydroxy acetone.

The reaction between glycerol and simple C=O compounds forms cyclic products such as 1,3-dioxane (isomeric six) and 1,3-dioxolane (five-membered) as new chemical intermediates. It is reported that 1,3-dioxane and 1,3-dioxolane are not produced in a 50:50 ratio. Research on the variation of the reaction parameters has been encouraged, along with the mole ratio of carbonyl compound to glycerol and temperature, to improve the feasibility of the reaction. However, the acetalization of glycerol did not yield complete conversion without the production of 1,3-dioxane. In fact, the highest selectivity for both 1,3-dioxane and solketal was not reached even at supercritical conditions. At supercritical conditions, the low glycerol conversion and solketal yield might be due to the lack of active acidic sites in acetone. Since ketalization is an exothermic process, temperature is another important factor that affects the equilibrium conversion. The use of catalysts that are highly active at low temperature is another strategy to enhance the economy of the solketal production. The complete formation of isomeric six obtained with aldehydes was induced at higher temperatures and is not considered to be environmentally friendly.

Glycerol esterification was conducted with heterogeneous catalysts: ammonia, resins, multivalent metal powder zeolite, sulfonic ammonia and hetero-polyacids. For the basic catalysis, metal alkoxides ($NaOCH_3$, $KOCH_3$), alkali metal oxides and alkaline catalyst (alkali metal alkoxides and hydroxides combined) are used. These also include other propoxide and butoxide catalysts. The heterogeneous acids used include $H_2SO_4PO(OH)_3$, HCl and RSO_2OH along with others.

The process of glycerol etherification formed oxygenated compounds and low-polymerization products known as polyglycerols, including methyl tert-butyl ether, glycerol tert-butyl ether, 1,2-di-tert-butyl glycerol, 1,2,3-tri-tert-butyl glycerol and 1,3-di-tert-butyl glycerol. The use of heterogeneous catalysts in the etherification process results in better yields. In different industries, glycerol is a desirable by-product since it has a broad variety of uses in manufacturing. Currently, glycerol is used in ~2000 different ways in medicines, skin care products and food industry because it is a non-toxic, nutritious, biodegradable material that can also have major environmental advantages for future applications. Glycerol is mostly used in the pharmaceutical and drug manufacturing industries to increase the viscosity of liquid drugs and to dissolve the drugs. It can also be used for the preservation of tobacco's flavor and to control the moisture content, with a view to eliminate the unpleasant irritating taste. It is also used in beauty care products and hair care products. Moreover, glycerol is the key component of toothpaste to avoid hardening and drying; it is projected that toothpaste takes up to 33.3% of the self-care demand for glycerol. It is applicable as a sweetener, preservative agent and solvent in food and beverage manufacturing. Moreover, glycerol is consumed in the production of coffee and in the extraction of tea, vegetable and ginger materials. It is often used as a softening agent in pasta, cookies, beef, cheese and sweets. There is no concern about the use of glycerol in edible products and beverages, given that it is pre-treated and preferable in the production of food. Glycerol is used as a lubricant and plasticizer in paper processing. Glycerol also used in the garment industries for the shaping, lubrication and softening of wool and fabric.

Additionally, glycerol-based ketals and acetal have been reported as having desirable properties for use as fuel additives (improved cold flow properties and octane number, minimized gum formation and reduced particulate emissions). Indeed, these fuel additives may increase the cleanliness of the various engine parts, ensure increased combustion, decrease fuel gelling and nozzle choking and reduce the corrosion impact on various parts of the engine. These studies indicate that glycerol acetalization is a viable process that could offer substantial economic growth.

11.8.3.1 Solketal Synthesis Process

Glycerol has assorted applications in various fields, particularly within the polymer, food, pharmaceuticals and cosmetics industries. The efficiency mainly depends on the chemical and physical characteristics of glycerol. Glycerol consists of three –OH groups, which allow it to totally dissolve in alcohols and water, although it is insoluble in hydrocarbons. Due to its distinct characteristics, glycerol can be transformed into valuable chemical by various catalytic process such as etherification, reforming, oxidation, carbonylation, hydrogenolysis, acetalization, dehydration and esterification. The major use of inexpensive glycerol is in its conversion into fuel substances. Glycerol ketals and acetals give amazing compounds for biodiesel and diesel blends. Glycerol is used as a feedstock in the synthesis of valuable oxygenated fuel additives.

During acetalization reactions, glycerol reacts with acetone in the presence of acid catalysts to provide 1,3-dioxolane (five-membered cyclic solketal) and 1,3-dioxane (six-membered cyclic). The reaction is reversible and hemiacetal is formed by removing water in the first step. Then, cations are formed by protonation of hemiacetal. It is preferable that the reaction media should be more acidic to promote effective protonation and stabilization of the cationic intermediate. The 1,3-dioxolane (solketal) product formed by this reaction improves cold flow properties, lowers the viscosity of biodiesel, improves combustion processes and boosts octane.

In ketalization reactions, glycerol and acetone react with each other to form solketal as a condensation product (over an acid catalyst). Various catalysts, both homogeneous and heterogeneous, have been used previously for the acetalization of glycerol. Homogeneous catalysts (sulfuric acid, p-toluene sulfonic acid and phosphoric acid) show high activity but have some major drawbacks. The main limitations of this process are catalyst recovery, cost, separation of product, corrosive nature and the environmental concern of effluent disposal. These limitations divert attention towards heterogeneous catalysts (Amberlyst-36, Amberlyst-15, zeolites, mesoporous silica and montmorillonite). These catalysts generally enhance the cold flow properties and octane number and decrease the formation of gum and emissions. Maksimov et al. reported in detail the use of solketal as a flexible and dissolvable suspending material in drugs or medicine industries. Specifically, the aquatic fish toxicity test has shown solketal to be less harmful towards the environment than other fuel additives. The aim of this section is to give an overview of and elaborate on the economical synthesis of solketal by catalytic conversion of glycerol with acetone.

Various processes, preparation of different types of catalysts and various parameters can have an influence of the reaction yield. This section illustrates the foremost challenges during purification and utilization of crude glycerol in the biodiesel industry and highlights some advanced processes which utilized the waste glycerol as a feedstock for solketal production. Moreover, this section critically discusses the latest advances in the conversion of glycerol into oxygenated fuel additives in the presence of various acid catalysts. Furthermore, numerous pathways for the applications of solketal on an industrial scale are elaborated in detail. Finally, the basics of the mechanisms of the reactions between glycerol, acid catalyst and acetone in forming solketal are discussed.

Glycerol consists of three –OH groups, which makes it useful as a fuel because of its lower heating value. Different methods for transforming glycerol into fuel additive components have been studied. Glycerol condensation with cyclic ketals and acetals is among the most promising applications of glycerol for chemical/fuel intermediates. In the ketalization reaction, glycerol and acetone react to produce solketal (2, 2-dimethyl-1,3-dioxolane-4-methanol) with the help of an acid catalyst as a condensation substance. Solketal can be utilized as a fuel additive to minimize particulate pollution and to increase the efficiency of the cold flow properties of liquid transportation fuels. When added to gasoline, it significantly reduces the formation of gum, enhances the stability of oxidation and increases the octane/cetane number.

11.8.3.2 Reaction Mechanism

The world energy requirement is increasing exponentially because of extensive economic growth and population expansion. The use of fossil fuel energy sources to meet the energy demand causes ozone depletion and water and land pollution. Therefore, renewable energy sources like biodiesel are attractive. Biodiesel is produced from various feedstocks using a transesterification process. In the transesterification process, glycerol is produced.

The production of acetals and ketals from glycerol offers a wide range of applications in fuel additives/chemicals. Ketals can be manufactured by the reaction of ketones with glycerol with an acid-catalyst, as shown in Figure 11.7.

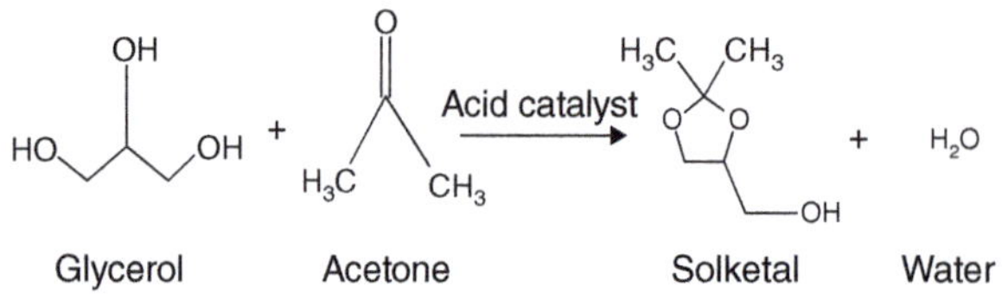

Figure 11.7 Ketalization reaction between glycerol and acetone.

As discussed already, the acidity of acidic catalysts has a strong influence on glycerol's transformation into products and the yield. Therefore, it is very important to study the mechanism of the acid-catalysed reactions. Ketalization of glycerol and acetone forms five-membered ring ketal (solketal) and six-membered ring ketals. The six-membered ring ketal is considered unfavourable due to the presence of one $-CH_3$. The mechanism involves the short lived carbonium ion as an intermediate. Li et al. suggested a mechanism for ketal cation using Lewis's acid catalysts. In this mechanism, Lewis's acid metal active sites play a major part in interacting with and functionalizing the C=O group of acetones. The removal of water from the intermediate forms a five-membered ring, i.e., solketal. The acetone is absorbed on the surface of glycerol and on the catalyst surface. In the first step, a surface reaction takes place in the presence of glycerol and acetone with the catalyst surface, resulting in the formation of hemi-acetal. In the following step, a carbocation is formed on the carbonyl carbon atom by the removal of water. In the last step, a proton is removed to form the solketal.

Nanda et al proposed a mechanism for attack, elimination of water removal and proton migration. A reaction takes place between glycerol and acetone in which the glycerol hydroxyl carbonyl carbon atom to form the intermediate, hemi-acetal. In the next step, water is eliminated, which leads to a carbocation on the carbonyl carbon atom and a proton is captured by the non-ligand electron pair of secondary hydroxyls. These two elimination steps take place simultaneously and result in producing 1,3-dioxolane (solketal). Water is removed from the reaction mixture with the help of entrainers petroleum ether and benzene and a vaporization process under vacuum.

11.8.3.3 Kinetics of Reaction

The general reaction rate of ketalization is described in terms of the Langmuir–Hinshelwood model. The surface reaction step is taken as a rate-determining step. The main reaction steps involved in this reaction are shown below.

a) The surface reaction between glycerol (GF) and acetone (AF), resulting in the formation of hemi-acetal (HF) (Equation 11.2.):

$$\mathrm{GF + AF \rightleftarrows HF + F} \tag{11.2}$$

F refers to free active sites available on the catalyst surface.

b) Hemi-acetal reacts with free available active sites (F) to form absorbed water (WF) (Equation 11.3):

$$\mathrm{HF + F \rightleftarrows WF + IF} \tag{11.3}$$

IF represents the intermediate formed in the reaction.

c) In this step, formation of solketal (SF) takes place (Equation 11.4):

$$\mathrm{IF + GF \rightleftarrows SF + F} \tag{11.4}$$

The generalized rate expression for solketal production using acetalization is expressed below:

$$r = \mathrm{k}\frac{[G][A]-\dfrac{[S][W]}{K_{\mathrm{c}}[G]}}{\{1+K_{\mathrm{w}}[W]\}^{2}} \tag{11.5}$$

In this equation, K_W is the equilibrium constant for the adsorption of water on the surface of the catalyst. The kinetic constant is shown by k and the ketalization equilibrium constant is expressed by K_c.

11.8.3.4 Catalyst Design

Previously, ketalization of glycerol with aldehydes/ketones was done using various mineral acid catalysts, e.g., hydrochloric acid, sulfuric acid, hydrofluoric acid, phosphoric acid, *p*-toluene sulfonic acid, etc. and over homogeneous Lewis's catalysts. Recently, some research has been conducted with the use of heterogeneous corrosive catalysts for acetalization of diverse carbonyl compounds with glycerol:

- ion-exchange resins;
- mesoporous silica hetero-polyacids;
- montmorillonite;
- metal oxide-based catalysts;
- zeolites;
- activated carbons;
- mesoporous silicates with aryl $-SO_3H$ group;
- rare-earth triflates;
- nanoparticles supported on multi-walled carbon nanotubes; or
- ionic liquids.

While examining the impact of the catalyst concentration on the reactant's reactivity, the research centred on the sort of catalyst, porosity, acidity, catalyst stacking and distinctive procedures for the preparation of catalyst and its reusability. Numerous researchers considered the impact of catalytic loading and studies are present on various types of catalyst. Beginning with homogeneous catalysts (solid minerals acids and Lewis's catalysts), which result in erosion and atmospheric issues and present complications in product separation and refinement, numerous considerations were applied with heterogeneous catalysts that sorted out most of the problems associated with the use of homogeneous catalysts. The accessibility, catalyst cost and better activity under mild conditions are also vital aspects along with efficiency and catalyst separation after processing. Ruiz et al. reported the ketalization of glycerol using formaldehyde and three distinctive sorts of corrosive catalysts: zeolites and natural tars-Brønsted corrosive, Lewis's catalysts and a reusable gold complex. A yield of 93% was obtained with a gold chloride catalyst at 80°C after 240 minutes Gold catalysts and solid catalysts are used for biomass transformations, including valorisation of glycerol and glycerol–water mixtures, through formation of cyclic acetals. The reusability of the gold chloride catalyst was studied and it was observed that it can be used in further cycles, whereas the recovery of other catalysts rarely reaches 80%. Gold salts and complexes are superior catalysts compared to other homogeneous and heterogeneous Brønsted and Lewis acids for the synthesis of glycerol acetals, because their better polarity and acidity results in highest yield. Moreover, gold catalysts can be recovered and reused up to seven times. Deutsch et al. reported the acetalization of glycerol using different acid polymer catalysts: Amberlyst-36 and Nafion-H NR50, a clay mineral (montmorillonite) and a zeolite (H-BEA). Amberlyst-36 was used as the catalyst for the reaction due its high activity and the small quantity required in the condensation of glycerol. The solketal yield from glycerol and acetone over Amberlyst-36 was 88% for glycerol–acetone condensation, 77% for glycerol–formaldehyde condensation and 94% for glycerol–benzaldehyde condensation. For glycerol acetalization over heterogeneous catalyst (various zeolites and ion-exchange resins), in the presence zeolite catalysts, the yield of solketal was remarkably high when compared to that obtained with polymer-based cyclopentane because of its lower cost and better activity. With all the considered catalysts, the highest solketal yields (81.4–100%) were obtained from the glycerol condensation with cyclopentanone.

Serafim et al. studied the ketalization of glycerol with butanal for different zeolites with diverse porous structures and acidity. Various zeolites-USY with distinctive Si/Al proportions, ZSM-5 and BEA were utilized as catalyst. By expanding the Si/Al proportion of USY zeolite, the pore volume as well as total surface area was enhanced. The highest glycerol conversion (84%) was obtained with USY 3. Oprescu et al. developed a supported blended oxide solid acid catalyst (SO_4^{2-}/SnO_2) for glycerol ketals of cyclohexane, which is stable at high temperature and water tolerant and a highest yield of 97.5% was obtained. Chavan and partners described preparation of catalyst in their work. Mallesham et al. synthesized SnO^{2-}based solid acid catalysts with addition of WO_3, MoO_3 and SO_4^{2-} to improve the surface properties. The best yields were obtained for glycerol conversion using acetone and furfural: 98% and 99% by SO_4^{2-}/SnO_2.

Fan et al. performed glycerol conversion to acetal up to 95% under optimum conditions over TiO_2–SiO_2 mixed oxides which produced Lewis and Brønsted acidic active sites after adsorption of water. The activity of silica-containing heteropoly acid in acetalization of glycerol using acetone was demonstrated by Ferreira et al., who used a sol–gel method for the preparation of catalysts with tungstophosphoric acid, tungstosilicic acid, molybdophosphoric acid and molybdosilicic acid immobilized in silica. The catalyst lost only 10–13% of its activity after the fourth cycle. The reaction of glycerol and acetone with sulfonic acid-catalyst can be used for three cycles without regeneration. Optimizing the reaction conditions (glycerol: acetone molar ratio of 1:6 at 30°C and 3 wt% of catalyst loading with respect to total reactants weight) yielded 94% glycerol conversion after 120 min over an organic–inorganic hybrid catalyst (tetra propylammonium salt of phosphotungstic acid) because of water, with transformation diminished by only 3%.

Da Silva et al. explained the effect of various catalysts (Amberlyst 15, K10 montmorillonite, zeolite beta and *p*-toluene sulfonic acid) for glycerol condensation using aqueous formaldehyde and acetone. Zeolite beta: Si/Al (1/4) 16:1 revealed the best performance, with glycerol conversion of 95% in a six hour reaction time. Using the sol–gel technique, the reaction of benzaldehyde and glycerol using MoO_3 catalyst supported in SiO2 was reported by Umbarkar et al. Adam et al. used a silica-supported catalyst for acetalization of glycerol and immobilized sulfonic acid with rice husk ash to get silica in a cheap way. In an argon atmosphere, conversion of benzaldehyde was 78% at 120°C in an eight hour reaction time. Rare earth triflate catalysts for the ketalization of glycerol using acetone were examined by Pierpont et al; they were found to give the highest conversion of 100% at room temperature and were water compatible.

Smith et al. used indium (III) triflate (In $(OTf)_2$) catalyst for trans acetalization reaction of glycerol with acyclic acetals in mild conditions to form the cyclic acetals. Various researchers studied the ketalization of glycerol at ambient temperature. They compared the properties of various acid zeolite catalysts and the best performance was shown by HBeta zeolite (average crystalline size 135 nm). After reaction time of one hour with an acetone-to-glycerol ratio of 2:1, the glycerol conversion obtained was 86%. Wang et al. performed the glycerol reaction with benzaldehyde on ionic liquid catalyst and found the yield of acetal was 99.8% at ambient reaction conditions. Silicotungstate-type catalysts also showed better activity at room temperature for glycerol and benzaldehyde reaction when developed and examined by Narkhede et al.

Crotti et al. used organoiridium catalyst for the selectivity of five-membered cyclic ketals. Another catalyst that showed good performance for the condensation reaction of different aldehydes and ketones with glycerol was niobium. After 60 min with a molar ratio of glycerol to acetone 1:4, 74% glycerol conversion was achieved. Glycerol acetalization with niobic catalysts gave the best performances with high conversion of 80%, according to a report by Souza et al.

For selecting catalysts, the activity and separation of catalyst are the major factors, but the operation conditions for industrial-scale catalysts are different. For industrial scale, a catalyst must fulfil the following requirements:

- its production capacity must be large;
- it must have mechanical and thermal resistance;
- it can be used for more than one cycle; and
- it should not cause poisoning or impurities in the reagents.

11.8.3.5 Batch Process

The reaction of ketone with alcohol results in ketals over acid catalyst. Fischer first synthesized solketal by the glycerol reaction with acetone in a batch reactor using hydrogen chloride as a catalyst. 25 years later, Pfähler s and Fischer reported glycerol ketalization by using various catalysts, such as anhydrous sodium sulfate and hydrogen chloride, in the same process. In 1945, Newman and Renoll described a solketal synthesis using a three-neck glass flask equipped with a magnetic stirrer and reflux. Using the same process, another author achieved a higher yield of solketal (87–90%) by using *p*-toluene sulfonic acid (*p*TSA) monohydrate as a catalyst and petroleum ether as a solvent. Distillation at low pressure was then done to separate the product. However, the reaction duration was prolonged to 21–36 h. The work on solketal synthesis persisted without any further study advancements despite the large volume of lowcost glycerol produced from the biodiesel industry.

In 1981–1982, a Spanish patent reported the utilization of a huge volume of glycerol in the synthesis of solketal. The innovators studied the glycerol reaction with acetone in a batch reactor without water entrained at a molar ratio of 1:1.1 under acid catalyst. At equilibrium, water was separated under a pressure of 1.33 kPa. However, the yield of solketal never surpassed 80%, which was a major drawback. To conduct this synthesis, the designed apparatus must operate with lower pressure. Another method was described in which the authors used *p*-toluene sulfonic acid catalyst with an excess of acetone and heated glycerol and the maximum yield of solketal was 56%. The presence of water as by-product was the main reason for the low solketal yield. The formation of water causes thermodynamics and kinetic hindrance for transformation of glycerol into solketal in the ketalization reaction. This reaction was carried out for 33 hours under gentle reflux and then neutralized with sodium acetate and distilled to obtain an 88% yield of solketal.

Garcia et al. described a glycerol reaction with acetone using *p*TSA monohydrate as a catalyst and a molar ratio (A/G) of 3. The mixture was heated under reflux for 16 hours before the reaction. To sustain the amount of the solvent, liquid acetone was introduced in reactor and wet acetone was simultaneously distilled off. After removal of the solvent, the solketal yield was 90%. This reaction is termed as a homogeneous catalysis because of the solubility of catalyst *p*TSA monohydrate in the reactant, which causes difficulty in recovering the catalyst. Moreover, using a homogeneous catalyst leads to many shortcomings like corrosion of the reactor, low catalyst recovery, environmental hazards, high cost and the disposal of residue. Deutsch et al. reported the utilization of the heterogeneous catalyst Amberlyst-36 for ketalization of glycerol in a batch reactor. The experiment was conducted in a 100 ml flask equipped with reflux, catalyst and continual removal of water (by-product). The reaction was conducted at normal temperature and the maximum yield was 88% under the following reaction parameters: 0.1 mol of glycerol, 0.15 mol of solvent (acetone), 17.5 mol of dichloromethane, 0.5 g of catalyst (Amberlyst-36) loading and a duration of 480 min. To achieve high conversion of glycerol requires removal of water, which is obtained as a by-product, during the ketalization reaction to push the reaction towards the product side and increase the yield of solketal. Continuous removal of water breaks the thermodynamic barrier. For this purpose, entrainers have been used to remove water in various processes. Another effective method to increase the yield is feeding in excessive amounts of acetone. Benzene is not an appropriate entrainer for this reaction, as acetone is separated before benzene through distillation. Other entrainers like petroleum ethers and chloroform can be used but these entrainers have some drawbacks like low efficiency due to their higher boiling points compared to acetone. Other catalysts like sodium sulfate and phosphorus pentoxide were also reported and where used as desiccants to separate the water present in the reaction mixture, but the operational cost was high due to the increased consumption of the catalyst. Recently, the use of sieves was introduced, but all these processes are unfavourable on an industrial level.

The overall issues may well be dealt with more successfully by utilizing an overabundance of acetone, which acts as a reactant as well as an entrainer. The surplus acetone can be reused after distillation in the same or a different form. A batch reactor was modified to remove the water into the membrane reactor. The reaction was performed in a three-neck flask (250 ml) connected with a reflux condenser by refluxing a mixture of acetone and glycerol in the presence of the heterogeneous acid catalyst montmorillonite clay K-10 with a zeolite membrane fitted in the central mouth. The membrane permitted saturation of tiny water vapours rather than evaporation. By using a higher molar ratio, A/G 20:1 for 2 hours, the maximum solketal yield was 82%.

Recently, Vicente et al. endeavoured to evacuate water from the mixture continuously by conducting the reaction in a twostep batch reactor. First, glycerol, acetone and a heterogeneous catalyst were mixed under reflux at 70°C in a 100 ml jar and then excessive acetone and by-products were evacuated by vaporization. To sustain the liquid amount in the next step, fresh acetone was added. To find an efficient heterogeneous catalyst for the ketalization, Ferreira et al. performed the reaction over different silica-induced heteropoly acid catalysts (i.e., tungstic-phosphoric acid (P-W), molybdophosphoric acid (P-Mo), molybdic-silicic acid and tungstosilicic acid (Si-W)) in a batch reactor containing a mixer. The major drawback of this reaction is poor miscibility of acetone in glycerol under standard conditions (25°C and 0.1 MPa); only 5 wt.% of glycerol is dissolvable in acetone. Royon et al. suggested the use of supercritical acetone without utilizing any catalyst to produce solketal from glycerol for its better solubility. The reaction was carried out at 235°C and 48×10^5 Pa in a batch reactor for 4 hours at the A/G of >10. The highest yield of solketal was 80% and conversion of glycerol was 28%, with acetone in its supercritical state. The conversion rate and solketal yield were very low at the supercritical condition because of the unavailability of active acid sites in acetone. Since the reaction is exothermic, temperature is an important factor that changes the equilibrium constant. Menezes et al. reported the maximum yield obtained at ambient conditions in a batch reactor (95%), by using 6 M of A/G ratio of 0.5–2 over stannous chloride and C_2H_3N as a solvent. However, the batch reactions described above have specific containments regarding complexity for large-scale solketal synthesis.

11.8.3.6 Continuous Process

As discussed earlier, most processes of solketal production were done in a batch reactor with heterogeneous catalysts, e.g., zeolites, amberlyst, montmorillonite, nation, silica-induced heteropoly acids, etc. However, this process has different drawbacks, e.g., long reaction duration, hence creating difficulty in scale-up and relatively low efficiency. The process of solketal production in a continuous-flow reactor over a heterogeneous catalyst is more favourable, as this process allows mass and heat transfer efficiency and it is easy to scale-up the process as well as decreasing the environmental and economic impacts. Use of the continuous microwave reactor (CMR) to produce solketal gave the highest yield of solketal (84%) in the presence of homogeneous catalyst *p*TSA by using A/G of 13.5 at 132°C, 117 kPa for 1.2 min and at 20 ml/min feeding rate. This method would not be effective for reacting at low temperatures or in the presence of microwave energy.

Clarkson et al. conducted an experiment in a multi-tray reactive distillation column for the synthesis of solketal over catalyst amberlyst DPT-1. In this process, the reaction column was introduced during the reaction to enhance solketal production. As the process is a semi-continuous process, it is difficult to operate at low temperature because of the high viscosity of glycerol.

Monbaliu et al. reported the use of a continuous glass flow reactor of volume 72 ml composed of several glass fluidic modules fitted in series. The major drawbacks of this experimental setup include long reaction duration, difficulty to operate at low temperature, use of homogeneous catalyst and hindrance in separation of the final product, etc. Maksimov et al. described the synthesis of high-octane oxygenated fuel components in a continuous reactor from plant-derived polyols. Recently, Nanda et al. have developed a continuous-flow reactor based on the concept of "Novel Process Windows" with respect to temperature, pressure and/or reactant concentration to enhance the intrinsic kinetics of the reaction for an optimum yield. A homogeneous solution of reactants with solvent was mixed at a specific molar ratio and fed into the reactor through a HPLC pump at a specific flow rate. In each run, the amount of catalyst was determined by the selected weight hourly space velocity (WHSV). The flow reactor can work for a wide range of temperatures and reaction conditions are 25°C, 3.5 MPA, A/G of 4, WHSV of 2 h over catalyst Amberlyst-36 wet, which gives a maximum yield of solketal of 94%. It is clear that a 0.02–0.5 h reaction duration with different catalysts results in the highest yield of solketal (84–94%) in continuous reactor. Therefore, advancement of continuous flow processes on a large scale is promising for the synthesis of solketal.

The acetalization of glycerol with acetone in the presence of various heterogeneous catalysts and production of solketal a valuable green product has been discussed. The research studies described herein mainly focus on innovative and emerging technology for sustainable solketal production, as well as the impact of reaction parameters, removal of water and catalytic activity in both batch and continuous processes. Globally, the potential prospects for the world market should be determined by the availability and demand for glycerol for use in new industries.

Recent advances in catalytic synthesis and characterization have led to remarkable achievements. Despite numerous developments, there are many obstacles to increase the conversion and yield of solketal. In fact, there are tremendous opportunities to synthesize highly active catalyst nanomaterials for glycerol acetalization and its transformation into useful products. The key objective in the future will be to introduce more environmentally sustainable methods and materials for catalytic synthesis. Investigation on the impact of various reaction parameters on catalytic activity has shown that the main obstacle to acetalization is the low equilibrium constant. Consequently, the solution to this problem would be to the shift equilibrium to product side by removing water, which causes hindrance during reaction, from the reaction mixture and by adding an excess of acetone. In a comparison of both continuous and batch reactors, higher product selectivity and shorter reaction times could be achieved by using continuous reactors.Further research on continuous reactors for the acetalization process may be a necessary and large step for its industrialization and commercialization in the future.

11.8.4 Current Issues and Challenges

It has been reported that the production of glycerol around the world is 0.75 million ton per annum, almost completely made from triglycerides; only 12% of global production is synthetic. The demand for glycerine is competitive and its costs are highly dependent on availability. In the immediate future, the availability of impurified glycerol will rely heavily on the biodiesel industry because of the development of biodiesel production facilities where glycerol is extracted as a by-product. During 2010, 12 million metric tons of biodiesel was produced, with 1.2 million metric tons of glycerol. The huge amount of glycerol on the market has led to a steady decline in its cost to around $0.05 per pound. Although the global demand for glycerol is limited in the US, improving living conditions in developed countries presents a major opportunity for innovation in the conventional usage of glycerine. Asia is the world's leading manufacturer and consumer of processed glycerine and accounted for 44% and 35% for global production and usage, respectively, in 2007.

The problems associated with the usage and purification of waste glycerol call for advancement of innovative methods and substitutes to ensure the viability of the biodiesel manufacturing industry. Nowadays, the usage and pre-treatment problems of bioglycerol are being addressed by intensive attempts to find a strategic solution that includes turning waste into useful, cost-effective goods. A typical example of this initiative is the implementation of the glycerol purification method in biodiesel plants. Although "waste" currently has wide implications, one of the key challenges is the purification of crude glycerol and the transfer of crude glycerol to more functional conditions for both existing and new uses. It is mainly for this reason that the bulk of the consumer sector uses processed glycerol as a raw material. As described above, a new solution to the production of biodiesel is required, as the present technology, while capable of generating good-quality biodiesel, still also has a range of drawbacks.

The disadvantages include the use of a base catalyst that causes the production of soap that needs to be discarded, the unnecessary amount of alcohol required to change the balance to more biodiesel generated (more alcohol must be isolated and recycled), the use of a homogeneous catalyst that demands neutralization, which in effect creates a large amount of salt in the waste stream, high operating expenses and a costly separation process. In addition, alcohol and vegetable oil must be moderately anhydrous and have a minimum free fatty acid content since the involvement of water or free fatty acid or both facilitates the forming of soap.

Crude glycerol is obtained from the synthesis of biodiesel by a transesterification process. Novel and sustainable processes or the extraction and utilization of glycerol from the transesterification product are needed. The crude nature of glycerol is one of the key issues limiting its usage, because most industrial processes use refined glycerol, converting it to valuable products like fuel additives. The purification of crude oil is a complex process and has a high cost. As discussed above, ketalization of glycerol is acidic in nature. The higher acidity of acidic catalysts leads to higher yields of solketal. However, the higher acidity of catalysts causes corrosion, which would increase fouling. Nanda et al. studied the leaching of various acidic catalysts including Amberlyst-35 dry, H-Beta zeolite and Amberlyst-36. To understand the process, they measured the acidity of the acids after 24 hours. Mild catalytic activity was observed after a 24 hour deactivation process. The reduction in the activity occurs due to the deficiency of acidic active sites available on the acid particles' surfaces. The leaching of catalyst requires regeneration, increasing the cost of the process. The continuous ketalization process also requires washing of the continuous-flow reactors. Fine particles of disintegrated catalyst cause clogging in the reactor, which can be avoided by washing the reactor occasionally.

The cost of the reactants contributes largely to the production and selling cost of solketal additive. The primary reactant for solketal additive production is the glycerol. The price of refined glycerol is US$500–600/ton, whereas crude glycerol costs around US$40–100/ton. The use of waste glycerol is inexpensive relative to the use of refined glycerol. However, glycerol consists of impurities including water, sodium and potassium salts, esters, fatty acids and alcohols. The usage of waste glycerol for solketal additive production causes the deactivation of catalyst and plugging of the reactor. To study the effect of impurities, de Silva and Mota studied the ketalization behaviour in the presence of impurities. They added 10% water, 15% sodium chloride and 1% methanol, which are commonly found as impurities in impurified glycerol. They studied the influence of these impurities in the presence of HBeta zeolite and Amberlyst-35. Nanda et al. stated that solketal yield decreased significantly because of the presence of the impurities. Therefore, it is necessary to extract the impurities from crude glycerol before using it for solketal production. The development of sustainable purification processes can result in lower costs for the purification of glycerol. Similarly, glycerol purity can also be increased by using sustainable processes to produce biodiesel. Alkali-catalysed transesterification causes soap formation, leading to product contamination. Therefore, sustainable process like supercritical technologies must be implemented for biodiesel production; these supercritical technologies will facilitate the extraction of purified glycerol.

The increase in the synthesis of biodiesel has also contributed to a surplus of crude glycerol, which is a significant bottleneck throughout the supply chain of biodiesel and has produced new problems for its sustainable usage. While there is a broad variety of possible uses of crude glycerol, it is restricted by its purity level, affecting its physical, chemical and biological properties. The conversion of impure glycerol into pure glycerol and potential application of glycerol in value-added products are emerging topics. We illustrate the present scenario for the development of glycerol from the biodiesel sector, the world demand and future emerging markets for glycerol as a commodity chemical.

11.8.5 Future Recommendation

Glycerol can be produced by various methods, including saponification, esterification, hydrolysis and transesterification. It is obtained in large quantities by the transesterification process, where every 10 kg of biodiesel synthesized by means of the transesterification process gives almost 1 kg of crude glycerol. The nature of the feedstock and the catalyst used during the transesterification cycle affect the purity and quality of biodiesel and of glycerol. Several purification methods have been identified for commercial purification of glycerol, with the use of ion-exchange resins or activated carbon having the potential for the development of high-quality glycerol.

Glycerol purification technology creates exciting prospects for the establishment of glycerol as a medium for different applications as well as the conversion of bioglycerol to value-added compounds. It could be used as an additive in various formulations of fuels in addition to its use in biochemical fields and pharmaceuticals. The ketalization reactions take place via an acid-catalysed mechanism; thus, catalysts with high acidity can lead to high conversion of glycerol. Homogeneous and heterogeneous catalysts are used in the ketalization reaction of glycerol and acetone, but the use of heterogeneous catalysts is preferred, as there are a lot of disadvantages with the use of homogeneous catalysts, e.g., catalyst separation, corrosion and concerns about effluent disposal. The limitation in the literature is that only a few researchers used pure or partially filtered glycerol in their experiments; most of them used natural glycerol and a comparison of the amount or percentage of processing costs is not available in the literature. We suggest it would be worthwhile to do brief research on the processing costs and utilization of glycerol ketals on an industrial scale.

11.8.6 Conclusion

A lot of work on the production of fuel additive solketal discusses several applications for solketal in various chemical fields, including polymers, pharmaceuticals and cosmetics, food and fuel industries and addresses some of the major limitations in its industrial synthesis. In fact, we show the potential of a modern revision of the method using unpurified glycerol as feedstock for solketal processing. Moreover, we describe different approaches to transform glycerol into cyclic acetals and ketals, as well as recent progress in achieving high selectivity of the desired products. Most studies were conducted with the use of heterogeneous catalysts to overcome the limitations associated with homogeneous catalysts for the synthesis of solketal from glycerol and acetone. The impact of reaction parameters on the acetalization reaction, product separation and reusability of catalysts is also discussed.

11.9 Higher Alcohols and Ether Biofuels

High molecular weight alcohol and ether fuels with their advanced autoignition propensities and oxygenated molecular structures are promising future fuel candidates for compression-ignition engine application because they can provide improved combustion efficiencies and reduced pollutant emissions. In addition, their production from lignocellulosic biomass as second-generation biofuels offers an improved carbon dioxide balance and avoids the adverse impact of the first-generation biofuel production on the food supply.

The transformation of the transport sector from fossil fuels towards renewable energy sources is one of the most critical challenges in view of climate change and global warming. Despite impressive progress that has been made in the development of battery-electric, hydrogen-fuelled and fuel cell vehicles, liquid fuels storage and distribution infrastructure and mature engine combustion concepts will continue to be an important alternative for providing a massive fraction of transportation energy in upcoming decades, especially in the long-distance and heavy-duty transport sectors, where electrification is difficult. Alcohols and ethers, which can be derived from lignocellulosic biomass via selective transformation, are interesting future transportation fuels, because they offer significant improvements in environmental sustainability.

Many studies have demonstrated the application advantages of short-chain alcohols and ethers, especially ethanol and dimethyl ether in spark-ignition (SI) and compression-ignition (CI) engines, respectively, in terms of thermal efficiency improvement and pollutant emissions reduction. While linear ether fuels generally have good autoignition propensities, indicated by high cetane numbers (CNs) and thus meet requirements of compression-ignition engines, alcohols with up to four carbon atoms show retarded autoignition behaviours at engine-relevant conditions, making them attractive fuel replacements and additives to gasoline fuels with growing chain enhanced reactivities enabling their application in compression-ignition engines. Thus, long-chain alcohols starting from n-pentanol have been primarily regarded as diesel fuel substitutes or blending components. Compared to their short-chain counterparts, longer alcohols and ethers possess higher volumetric energy contents and lower heats of vaporization, which are closer to those of conventional diesel fuels. They have also lower polarity, avoiding phase separation when being blended with petroleum fuels. Consequently, as most of their physical properties lie in the range of diesel fuels, they can be applied in compression-ignition engines without extensive system modifications. Owing to these factors, both high molecular weight alcohols and ethers are very promising alternative fuels for compression-ignition engine application.

Pioneering studies have investigated the compression-ignition engine performance of long-chain alcohol and ether fuels, reporting numerous advantages, such as significantly improved soot/NO_x trade-off with almost soot free combustion and relatively high thermal efficiencies over a wide range of engine loads. While the combustion in compression-ignition engines involves various physical processes, such as spray breakup, fuel evaporation and turbulent flame propagation, it is to a large extent driven by the chemical combustion phenomena of the fuel. Thus, detailed knowledge about these characteristics and a deep understanding of the underlying reaction kinetics are important prerequisites to enhance the engine combustion performance by co-optimizing engines and fuels. Therefore, both the fundamental combustion kinetics and the practical engine application of higher alcohols and ethers have recently attracted increased research attention as evidenced by numerous studies published in the last ten years.

11.9.1 Fuel Production Routes and Sustainability

Biomass-derived oxygenated fuels, including alcohols and ethers, have been considered advantageous due to the possibility of their production from renewable resources. Ethanol as the most used liquid biofuel has been traditionally produced competition with food production. Therefore, recent research has focused on the development of production processes to obtain biofuels from nonedible lignocellulosic biomass. Ethanol production from lignocellulose has been shown to be potentially sustainable and to result in significantly reduced greenhouse gas emissions compared with gasoline.

Short-chain alcohols up to butanol can be obtained through fermentation of carbohydrates after hydrolysis of the lignocellulose. Furthermore, methanol can be synthesized from H_2/CO synthesis gas and advancements in production from synthesis gas have also been made for medium-size alcohols such as butanol. On the other hand, long-chain alcohols larger than C_{12} can be obtained by the reduction of fatty acids. Specific production pathways are also available for other higher isomers.

A promising synthesis route for the tailored production of various bioderived fuel candidates, including the higher alcohol n-octanol, n-octane, as well as the higher ethers including di-n-octyl ether, ethyl octyl ether, 2-butyltetrahydrofuran and 2- are depicted in Figure 11.8. This pathway is based on the initial conversion of the pentose-derived platform molecule furfural with acetone into furfural acetone, which can then undergo hydrogenation to form 4-(2-tetrahydrofuryl)-2-butanol (THFA). Selective deoxygenation of this intermediate can then yield the desired value-added products. For instance, 2-butyltetrahydrofuran (2-BTHF) can be obtained by removing the secondary hydroxy group using adequate catalysts, while n-octanol is subsequently formed through hydrogenolytic ring opening. Also, di-n-octyl ether can be further formed from n-octanol by reversible etherification. On the other hand, di-n-butyl ether (DNBE) can be readily obtained through the dehydration of n-butanol. Obviously, the large-scale success of such promising novel synthesis routes to higher alcohols and ethers inevitably depends upon their energy efficiencies and achievable production costs. For the production route discussed before and shown in Figure 11.8,

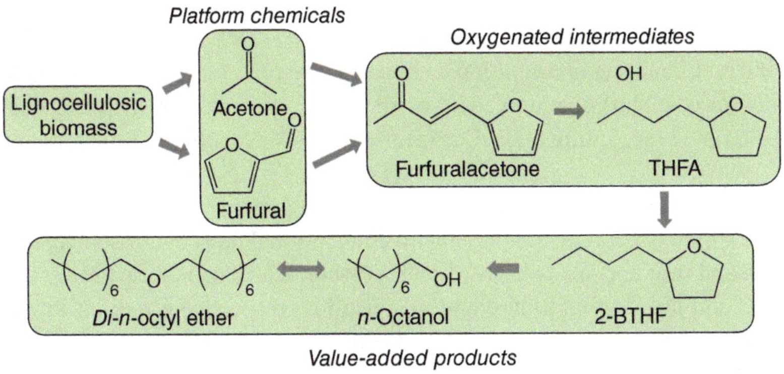

Figure 11.8 Production pathway for the synthesis of higher alcohols and ethers, including 2-BTHF, n-octanol and di-n-octyl ether, from lignocellulose-derived platform chemicals, as proposed by Julis and Leitner and Luska et al.

reaction network flux analysis has shown advantages in terms of costs and energy demand when mixtures of the products, such as n-octanol and 2-BTHF, are produced instead of these corresponding pure components, which is still not competitive with the production of ethanol from fermentation and ongoing work is required for the optimal design of future efficient and sustainable biofuel production pathways.

11.9.2 Lignin

Expanding the gamut of new platform chemicals that can be obtained from biomass. Rio et al. describe the presence of valuable phenolic compounds incorporated into lignin, such as flavonoids, hydroxy stilbenes and hydroxycinnamic amides, which behave as authentic lignin monomers and are potentially available in large amounts from the abundant waste products from processing of agricultural or forest biomass. Nakagawa et al. discuss the utility of erythritol as a C_4 platform in biomass refinery, derived from fermentation of sugars and glycerol.

The direct valorisation of lignin to functional chemicals presents potential new avenues for efficiently converting lignin to chemical products. Dong et al. report a lignin–carbohydrate complex that is a potent antioxidant for scavenging reactive oxygen species in vitro and zebrafish in vivo. Lignin was also used in the manufacture of polyurethane formed by oxidative liquefaction. Many other transformations of biomass to functional chemicals and materials are discussed in these two timely reviews on biomass-derived carbonaceous materials and bacterial cellulose-based composite scaffolds for biomedical applications. On the critical front of transforming our polymer platform to renewable materials, consistent with the circular economy, Zhao et al. discuss how we can narrow the gap for bioplastic use in single-use food packaging, focusing on the most recent development successes in bioplastic materials and highlighting the "gaps" between bioplastics and their conventional counterparts with respect to their properties.

11.9.3 Fuel Properties

Higher alcohols and ethers with more than four carbon atoms exhibit distinct properties, which strongly depend upon the fuel structural characteristics such as carbon chain length. An overview of the most relevant properties for compression-ignition engine applications is provided in Table 11.4 for the C_5–C_9 n-alcohols and the C_8–C_{10} di-n-alkyl ethers. The typical properties of diesel fuel are also shown for reference, since it would be desirable for some of these key fuel properties of the long chain oxygenated components to be in a similar range as conventional diesel fuel for compression-ignition engine application.

A very important fuel property is the (derived) cetane number (CN/DCN), which specifies the autoignition propensity of a fuel. Compression-ignition engines require fuels with sufficiently high cetane numbers, especially for cold-start operation. Therefore, minimum values of 40 (according to the US norm ASTM D975) or 51 (according to the European norm EN 590) are required for current diesel fuels.

An important disadvantage of many small, oxygenated fuel components for application in engines is their relatively low energy content compared with conventional fuels. For instance, the short-chain alcohol ethanol and the short-chain ether, dimethyl ether, which constitute promising fuels for spark-ignition and compression-ignition engine application, respectively, exhibit lower heating values of only about 26.8 MJ/kg, while the values of conventional diesel or gasoline with low oxygen contents lie typically above 40 MJ/kg. In comparison, higher alcohols and ethers are advantageous, as their lower heating values are much closer to those of conventional fuels, owing to their lower oxygen mass fractions in the fuel structure, as seen in Table 11.4.

Table 11.4 Fuel properties of higher n-alcohols di-n-alkyl ethers.

	diesel	n-pentanol	n-hexanol	n-heptanol	n-octanol	n-nonanol	di-n-butyl eth r	di-n-pentyl ether
CN/DCN[a]	54	18.2	23.3	29.5	39.1	46.2	~115	~111
Heat of vaporization,25°C, (kJ/kg)	356	656	598	572	538	532	345	292
Boiling point, (°C)	170–360	138	157	176	194	215	141	187
Vapour pressure, (25°C), mPa	<0.1	6.0	1.2	0.43	0.09	0.03	4.6	1.3
Lower heating value[b] (MJ/kg)	42.8	34.6	35.8	36.7	37.4	38.0	37.4	38.5
Oxygen content, (% mass)	0.8	18.15	15.67	13.78	12.29	11.10	12.3	10.11
Molecular weight, (g/mol)	198	88.2	102.2	116.2	130.2	144.3	130	158.3
Water solubility, (g/l)	Mostly insoluble	22	5.9	1.0	0.54	0.14	0.3	0.5
Density, 25°C, (kg/m^3)	832	811	815	822	822	824	764	775
Dynamic viscosity, (25°C), (mPa.s)	3.3	4.1	4.4	7.1	7.1	11.6	0.66	
Surface tension[c], (mN/m)	29	27	28	29	29	30	24	26

[a] Cetane number/derived cetane number.
[b] Calculated based on atomic composition.
[c] Predicted values.

11.9.4 Concluding Remarks

High molecular weight alcohols and ethers, which can be applied as potential diesel fuel substitutes or additives in compression-ignition engines, have attracted considerable attention in the past ten years. With the comprehensive overview provided here, this chapter aims to deepen the understanding of their fundamental combustion characteristics and practical engine application performance. First, aspects related to the production pathways have been discussed, highlighting the advantage of the missing competition with food supply in case of their synthesis from lignocellulosic biomass. Thereafter, the fuel properties of higher alcohols and ethers have been discussed with respect to carbon chain length and the major literature findings on their engine application characteristics have been summarized. The dependence of important engine combustion properties, such as cetane number, on fuel structure has been highlighted. C_5–C_7 linear alcohols were primarily tested as blending compounds with conventional diesel or biodiesel due to their relatively low autoignition propensity, while n-octanol has been considered as potential pure component fuel. On the other hand, despite its cetane number of above 100, the engine application of pure di-n-butyl ether has also been successfully demonstrated. The excellent atomization behaviour of higher ethers was found to be particularly advantageous. For both higher alcohols and ethers, significantly reduced soot emissions were consistently observed in their application as neat components or blends with other base fuels.

Following these practical application aspects, the fundamental ignition and combustion characteristics of higher alcohols and ethers and their driving kinetics were discussed in detail with respect to the literature findings that have become available in recent years. The dependence of the ignition and combustion behaviours on molecule sizes and types of oxygen functional groups was evaluated based on experimental observations. The available chemical mechanisms were introduced in conjunction with their development processes and the involved assumptions. The significant effect of the alcohol or ether functional group on the C—H bond dissociation energies of the joint impact with the molecule size on the fuel consumption pathways and reactivities were highlighted. Despite the impressive progress made to understand the high- and low-temperature combustion chemistry of C_5–C_8 linear alcohols and ethers, several aspects were demonstrated that deserve further investigation. For instance, the combustion behaviour of these fuels in nonpremixed laboratory-scale configurations remains mostly unexplored, in spite of the crucial importance of the nonpremixed combustion mode in compression-ignition engines. Also, improved knowledge is required for the double-negative temperature coefficient oxidation behaviour observed in jet-stirred reactor experiments and its implications for practical systems. The reliability of chemical mechanisms of higher linear ethers should be further assessed through experimental investigations of the ignition delay times at engine-relevant high pressures and low to intermediate studies on blends with other fuels are necessary to explore the interactions of the components in such mixtures. Finally, the detailed investigation of the kinetics of elementary reactions through high-level quantum chemistry calculations for both higher alcohols and ethers as well as their short-chain counterparts will contribute to an advanced understanding on their reaction kinetics and will allow for the derivation of more accurate rate rules, which can reduce the model prediction uncertainties of chemical mechanisms significantly.

11.10 Biofuels in the World: Biogasoline and Biodiesel

Biofuels production growth (2019 to 2020) averaged 3% (54,000 barrels of oil equivalent per day or boe/d), less than half the 10-year average. Growth was led by Brazil (31,000 boe/d) and Indonesia (32,000 boe/d) but US output declined by 19,000 boe/d. Growth was weighted towards biodiesel, which grew by 34,000 boe/d driven largely by Indonesia. Biodiesel is the dominant fuel in Europe and Asia Pacific (making up 81% and 74% of biofuels respectively in 2019), while ethanol is the main fuel in North America (86% of total) and South and Central America (74%).

Biofuels consumption rose by 6% (100,000 boe/d). As with production, growth was driven mainly by Brazil (42,000 boe/d), most of which ethanol and Indonesia (56,000 boe/d), which was largely biodiesel. At the global level, ethanol made up 63% of biofuels in 2019, but the share of biodiesel has risen continually. For example, biodiesel's share was 23% in 2009 but rose to 37% in2019.

China has also initiated its Blue-Sky Defence to help curb air pollution. The program includes setting aggressive fuel economy, emissions and quality standards through 2020, retiring many coal-fired power plants and converting them to natural gas; experimenting in utilizing alternatives fuels, such as methanol and compressed natural gas; increasing ethanol blending to 10% (known as E10); and increasing the adoption of EVs/HEVs.

Many nations in the Asia-Pacific region are enacting policies to provide their populations with ultra-low sulfur fuels. The move towards clean-fuels production will help mitigate pollution in major cities, as well as adhere to emissions reduction targets set in the Paris agreement.

China is making great strides to mitigate significant smog in major cities. The country has invested in and continues to invest in, capital-intensive refineries to produce ultra-low sulfur fuels, as well as upgrades to existing facilities. This includes the production of China 6 (which is equivalent to Euro 6-standard fuels), as well as removing more than 1 MM high-polluting long-haul trucks by 2020.

Most of the nation's refineries are adapting to the new fuel standard and are investing in new units and/or upgrades to produce compliant fuels. India has approved its National Biofuels Policy (India). This program envisions increasing ethanol blending in gasoline to 10% by 2022 and up to 20% by 2030. The policy will help mitigate fuel imports, which could save India approximately $ 7 billion in oil import costs.

Additional Asia-Pacific clean fuels initiatives include:

- Indonesia. The country has initiated its $25-billion Refinery Development Master Plan, which calls for the construction of greenfield refineries and the upgrade of its existing refining network by 2027.
- Malaysia. The country enacted its B10 biodiesel program, which increased bio-content – primarily palm oil – in diesel from 7% to 10%.

ExxonMobil announced it will make significant investments to its integrated complex in Singapore. These investments include increasing the production of clean fuels. Neste has announced it will invest $1.6 billion in new biofuels production facilities.

In December 2018, Canada announced its Clean Fuel Standard framework. The initiative calls for the mitigation of greenhouse gas emissions of 30% below 2005 levels by 2030. Canada hopes that the plan will encourage the use of biofuels, EVs and alternative fuels, as well as an increase in the investments for clean fuels.

The US has begun to enforce Tier 3 regulations, which lower sulfur content in gasoline from 30 ppm to 10 ppm. Large US refineries – those with a processing capacity of more than 75,000 bpd – had to comply with the regulation in 2017. Refineries with a capacity of less than 75,000 bpd had to comply by 2020. In total, *Hydrocarbon Processing's* Construction Boxscore Database is tracking more than $6 billion in active refining projects to boost the production of ultra-low sulfur fuels in the US. These projects are being executed by companies, such as Andeavor, Chevron, ExxonMobil, Valero, Marathon Petroleum and Meridian Energy.

Mexico has announced an ambitious $11 billion plan to rehabilitate the country's refining network and build a grassroots facility. The rehab plan includes increasing domestic production of high-quality, low-sulfur fuels at all Pemex refineries. Mexico is trying to significantly decrease sulfur levels in domestically produced diesel to adhere to new regulations that call for a reduction in diesel sulfur limits from 500 ppm to 15 ppm. Most capital expenditures will be directed towards a new $8 billion, 340,000-bpd grassroots refinery in the state of Tabasco, enabling the country to boost production for ULS fuels.

With help from China National Petroleum Corp., Brazil's state-owned oil and gas company, Petrobras, plans to finally complete its COMPERJ refining project; the project is in the feasibility study stage. If completed, COMPERJ will produce ultra-low sulfur fuels. The country is also trying to boost the use of biofuels and renewable fuels through its RenovaBio program. The program, which went into effect in 2020, aims to double the use of ethanol to 26 Blpy by 2030, along with an increased use in renewables.

Additional Latin America clean-fuels initiatives include:

- Argentina. YPF is investing $2 billion to add desulfurization units to its La Plata and Mendoza refineries to increase clean fuels production.
- Ecuador. The new units at the Esmeraldas refinery would increase clean fuels production, reduce emissions, etc. If built, the country's 300,000-bpd grassroots refinery would produce ULS fuels.
- Jamaica. Petrojam is investing $1 billion in the Petrojam Refinery Upgrade Project (PRUP). Once completed, the PRUP will increase the refinery's flexibility and conversion to produce high-value, low-sulfur fuels.
- Peru. The country's state-owned petroleum company, Petroperú, is investing $6 billion on the Talara refinery modernization project. The capital-intensive project will increase the refinery's flexibility and conversion to produce fuels that adhere to new sulfur limit requirements in the country.

The Middle East has and continues to build highly complex facilities. These facilities, combined with upgrades and expansions, are designed to minimize fuel oil output and to maximize middle distillate, diesel and gasoline production.

One of the region's main initiatives is the focus on clean fuels production. Most Middle Eastern nations have invested in new units to increase the production of ULS transportation fuels. ULS fuels are becoming mandatory in both developed and developing countries and it is imperative that Middle Eastern refineries produce fuels that adhere to sulfur requirements – especially since surplus fuels will be exported to regions that require low-sulfur fuels (e.g., Asia and Europe).

The region is making capital-intensive investments in secondary unit capacities to increase the production of ULS fuels. According to OPEC, the region will add more than 4.1 MMbpd of secondary unit capacity by 2024. This capacity increase includes 2.8 MMbpd of desulfurization capacity; 700,000 bpd of conversion capacity; and 600,000 bpd of octane-boosting units.

Both Western and Eastern Europe, Russia and the Commonwealth of Independent States (CIS) are investing to increase the production of ULS fuels. However, Western Europe's refining network is much older and more developed, so many Western European countries are focusing on the move towards a carbon-neutral society while Eastern Europe, Russia and the CIS are investing in upgrades and new refining capacity to adhere to higher fuel-quality standards.

Western Europe is pushing for a carbon neutral economy by 2050, which includes programs such as the EU Renewable Energy Directive (RED 2), an increase in the use of biofuels, EVs/HEVs, etc. According to the European Commission, RED 2 has an overall target of 32% use of renewables in the EU total energy mix by 2030, along with a minimum of 14% for renewables usage in the transportation sector.

Most nations in Western Europe have announced major initiatives that call for the increased use of biofuels and alternative fuel technologies and capital investments to increase biofuel production capacity, as well as the outright ban of gasoline and diesel-powered vehicles. Several of these initiatives include::

- Spain – Proposing a ban of gasoline and diesel by 2040. The country's Balearic Islands have announced it will ban the use of diesel-powered vehicles in 2025 and gasoline in 2035.
- Italy – Investing in new biofuels, low-carbon fuels and waste-conversion technologies/plants.
- Germany – The country is allowing cities the right to limit/ban the use of diesel cars to help mitigate emissions. The nation has also announced plans to phase out nuclear power generation for renewables.
- The Netherlands – Implementing zero-emissions vehicles for all public transport by 2025.
- Sweden – Neste and BP will supply low-carbon aviation fuel (e.g., non-palm renewable and sustainable raw materials) to Swedish airlines.
- Finland – Neste is making capital-intensive investments in new biofuels technologies.
- United Kingdom – Announced plans to ban the sales of gasoline and diesel vehicles by 2040. The country will also subsidize the sales of EVs, as well as invest in retrofit public transport, build new charging stations and fund research and development in new low-carbon transportation technologies.
- France – The country has announced plans to ban gasoline-powered vehicles in Paris by 2030, as well as pushing the use of SP95-E10 gasoline (10% ethanol blend).

Russia and the CIS have major ongoing initiatives, as well. Russia is at the tail-end of its $55-billion modernization program to upgrade and modernize its domestic refining network. Since the nation is self-sufficient in fuels production, the program focuses on upgrading and conversion capacity rather than additional crude distillation capacity.

Further Reading

Abate, S., Barbera, K., Giglio, E. et al. (2016). Synthesis, characterization and activity pattern of Ni–Al hydrotalcite catalysts in CO_2 methanation. *Ind. Eng. Chem. Res.* 55: 8299–8308. doi: 10.1021/acs.iecr.6b01581.

Adam, F., Batagarawa, M., Hello, K. and Al-Juaid, S. (2012). One-step synthesis of solid sulfonic acid catalyst and its application in the acetalization of glycerol: crystal structure of cis-5-hydroxy-2-phenyl-1,3-dioxane trimer. *Chem. Pap.* 66 (11): 1048–1058.

Alalwan, H.A., Alminshid, A.H. and Aljaafari, H.A.S. (2019). Promising evolution of biofuel generations. Subject review. *Renew. Energy Focus* 28: 127–139. doi: 10.1016/j.ref.2018.12.006.

An, X., Zuo, Y., Zhang, Q. et al. (2008). Dimethyl ether synthesis from CO_2 hydrogenation on a $CuO\text{-}ZnO\text{-}Al_2O_3\text{-}ZrO_2$/HZSM-5 bifunctional catalyst. *Ind. Eng. Chem. Res.* 47: 6547–6554. doi: 10.1021/ie800777t.

Bacariza, M.C., Maleval, M., Graça, I. et al. (2019). Power-to-methane over Ni/zeolites: influence of the framework type. *Microporous Mesoporous Mater.* 274: 102–112. doi: 10.1016/j.micromeso.2018.07.037.

Bailera, M., Lisbona, P., Romeo, L.M. and Espatolero, S. (2017). Power to Gas projects review: lab, pilot and demo plants for storing renewable energy and CO_2. *Renew. Sustain. Energy Rev.* 69: 292–312. doi: 10.1016/j.rser.2016.11.130.

Bazzanella, A.M. and Ausfelder, F. (2017). Technology study: low carbon energy and feedstock for the European chemical industry. Dechema Gesellschaft fur Chemische Technik und Biotechnology e.V., Cefic.

Cai, L., vom Lehn, F. and Pitsch, H. (2021). Higher alcohol and ether biofuels for compression-ignition engine application: a review with emphasis on combustion kinetics. *Energy Fuels* 35: 1890–1917.

Carugo, M. and Valentine, J. (2022). Accelerating the transition to renewable fuels. Hydrocarbon Processing, October 2022.

Chavan, S.P., Zubaidha, P.K., Dantale, S.W. et al. (1996). Use of solid superacid (sulphated SnO_2) as efficient catalyst in tacile transesterification of ketoesters. *Tetrahedron Lett.* 37 (2): 233–236.

Chen, J.-Y. and Brunelli, N.A. (2021). Investigating the impact of microporosity of aminosilica catalysts in aldol condensation reactions for biomass upgrading of 5-hydroxymethylfurfural and furfuraldehyde to fuels. *Energy Fuels* 35 (18): 14885–14893. doi: 10.1021/acs.energyfuels.1c01934.

Child, M., Bogdanov, D. and Breyer, C. (2018). The role of storage technologies for the transition to a 100% renewable energy system in Europe. *Energy Procedia* 155: 44–60. doi: 10.1016/j.egypro. 2018.11.067.

Clarkson, J.S., Walker, A.J. and Wood, M.A. (2001). Continuous reactor technology for ketal formation: an improved synthesis of solketal. *Org. Process Res. Dev.* 5 (6): 630–635.

Corma, A., Torre, O., De Renz, M. and Villandier, N. (2011). Production of high-quality diesel from biomass waste products. *Angew. Chemie Int. Ed.* 50: 2375–2378. doi: 10.1002/anie.201007508.

Crotti, C., Farnetti, E. and Guidolin, N. (2010). Alternative intermediates for glycerol valorization: iridium-catalyzed formation of acetals and ketals. *Green Chem.* 12 (12).

Das, P. and Tiwari, P. (2018). The effect of slow pyrolysis on the conversion of packaging waste plastics (PE and PP) into fuel. *Waste Manag.* 79: 615–624. doi: 10.1016/j.wasman.2018.08.02.

da Silva, C.X.A., Goncalves, V.L.C. and Mota, C.J.A. (2009). Watertolerant zeolite catalyst for the acetalisation of glycerol. *Green Chem.* 11 (1): 38–41.

da Silva, C.X.A. and Mota, C.J.A. (2011). The influence of impurities on the acid-catalyzed reaction of glycerol with acetone. *Biomass Bioenergy* 35 (8): 3547–3551.

Deutsch, J., Martin, A. and Lieske, H. (2007). Investigations on heterogeneously catalysed condensations of glycerol to cyclic acetals. *J. Catal.* 245.

Dong, H. et al. (2020). Characterization and application of lignin–carbohydrate complexes from lignocellulosic materials as antioxidants for scavenging *in vitro* and *in vivo* reactive oxygen species. *ACS Sustain. Chem. Eng.* 8 (1): 256–266. doi: 10.1021/acssuschemeng.9b05290.

Dulal, H.B., Brodnig, G. and Onoriose, C.G. (2011). Climate change mitigation in the transport sector through urban planning: a review. *Habitat Int.* 35: 494–500. doi: 10.1016/j.habitatint.2011.02.001.

Egbue, O. and Long, S. (2012). Barriers to widespread adoption of electric vehicles: an analysis of consumer attitudes and perceptions. *Energy Policy* 48: 717–729. doi: 10.1016/j.enpol.2012.06.009.

Fan, C.-N., Xu, C.-H., Liu, C.-Q. et al. (2012). Catalytic acetalization of biomass glycerol with acetone over TiO2-SiO2 mixed oxides. *React. Kinet., Mech. Catal.* 107 (1): 189–202.

Ferreira, P., Fonseca, I.M., Ramos, A.M. et al. (2010). Valorisation of glycerol by condensation with acetone over silica-included heteropolyacids. *Appl. Catal., B* 98 (1–2): 94–99.

Fischer, E. (1909). Ueber die verbindungen der zucker mit den alkoholen und ketonen. In: *Untersuchungen Über Kohlenhydrate und Fermente (1884–1908)*, 734–757. Springer.

Fischer, E. and Pfähler, E. (1924). Über Glycerin-aceton und seine Verwendbarkeit zur Reindarstellung von α-Glyceriden; über eine Phosphorsäure-Verbindung des Glykols. In: *Untersuchungen aus Verschiedenen Gebieten* (ed. M. Bergmann), 627–642. Springer.

FortisBC (2018). *Clean Growth Pathway to 2050 Contents.* FortisBC.

Galadima, A. and Muraza, O. (2015). Waste to liquid fuels: potency, progress and challenges. *Int. J. Energy Res.* 39: 1451–1478. doi: 10.1002/er.3360.

Ganesh, K.N. et al. (2021). Green chemistry: a framework for a sustainable future. *Ind. Eng. Chem. Res.* 60 (25): 8964–8968.

Gao, P., Li, S., Bu, X. et al. (2017). Direct conversion of CO_2 into liquid fuels with high selectivity over a bifunctional catalyst. *Nat. Chem.* 9: 1–6. doi: 10.1038/nchem.2794.

Garcia, E., Laca, M., Perez, E. et al. (2008). New class of acetal derived from glycerin as a biodiesel fuel component. *Energy Fuels* 22 (6): 4274–4280.

Ghaib, K. and Ben-Fares, F.Z. (2018). Power-to-methane: a state-of-the-art review. *Renew. Sustain. Energy Rev.* 81: 433–446. doi: 10.1016/j.rser.2017.08.004.

Han, Y., Jeong, Y., Park, H. et al. (2017). Carbon dioxide Fischer-Tropsch synthesis: a new path to carbon-neutral fuels. *Appl. Catal. B Environ.* 202: 605–610. doi: 10.1016/j.apcatb.2016.09.072.

Hara, K., Kudo, A. and Sakata, T. (1995). High efficiency electrochemical reduction of carbon dioxide under high pressure on a gas diffusion electrode containing Pt catalysts. *J. Electrochem. Soc.* 142: 57–59.

Hazimah, A., Ooi, T. and Salmiah, A. (2003). Recovery of glycerol and diglycerol from glycerol pitch. *J. Oil Palm Res.* 15: 1–5.

He, L., Lin, Q., Liu, Y. and Huang, Y. (2014). Unique catalysis of Ni-Al hydrotalcite derived catalyst in CO_2 methanation: cooperative effect between Ni nanoparticles and a basic support. *J. Energy Chem.* 23: 587–592. doi: 10.1016/S2095-4956(14)60144-3.

Hori, Y., Wakebe, H.H.I., Tsukamoto, T. and Koga, O. (1994). Electrocatalytic process of CO selectivity in electrochemical reduction of CO_2 at metal electrodes in aqueous media. *Electrochim. Acta* 39: 1833–1839.

Julis, J. and Leitner, W. (2021). Synthesis of 1-octanol and 1,1-dioctyl ether from biomass-derived platform chemicals. *Angew. Chem., Int. Ed.* 51: 8615–8619.

Kaneco, S., Yabuuchi, M., Katsumata, H. et al. (2002). Electrochemical reduction of CO_2 to methane in methanol at low temperature. *Fuel Chem. Div. Prepr.* 47: 71–72.

Karelovic, A. and Ruiz, P. (2012). CO_2 hydrogenation at low temperature over Rh/-Al_2O_3 catalysts: effect of the metal particle size on catalytic performances and reaction mechanism. *Appl. Catal. B Environ.* 113–114: 237–249. doi: 10.1016/j.apcatb.2011.11.043.

Kondratenko, E.V., Mul, G., Baltrusaitis, J. et al. (2013). Status and perspectives of CO_2 conversion into fuels and chemicals by catalytic, photocatalytic and electrocatalytic processes. *Energy Environ. Sci.* 6: 3112–3135. doi: 10.1039/c3ee41272e.

Kongjao, S., Damronglerd, S. and Hunsom, M. (2010). Purification of crude glycerol derived from waste used-oil methyl ester plant. *Korean J. Chem. Eng.* 27 (3): 944–949.

Lazkano, I., Nøstbakken, L. and Pelli, M. (2017). From fossil fuels to renewables: the role of electricity storage. *Eur. Econ. Rev.* 99: 113–129. doi: 10.1016/j.euroecorev.2017.03.013.

Li, L., Koranyi, T.I., Sels, B.F. and Pescarmona, P.P. (2012). Highly efficient conversion of glycerol to solketal over heterogeneous Lewis's acid catalysts. *Green Chem.* 14 (6): 1611–1619.

Luska, K.L., Julis, J., Stavitski, E. et al. (2014). The selective deoxygenation of biomass substrates. *Chem. Sci.* 5: 4895–4905.

Maksimov, A.L., Nekhaev, A.I., Ramazanov, D.N. et al. (2011). Preparation of high-octane oxygenate fuel components from plant-derived polyols. *Pet. Chem.* 51 (1): 61–69.

Mallesham, B., Sudarsanam, P. and Reddy, B.M. (2014). Eco-friendly synthesis of bio-additive fuels from renewable glycerol using nanocrystalline SnO_2-based solid acids. *Catal. Sci. Technol.* 4 (3): 803–813.

Manthiram, K., Beberwyck, B.J. and Alivisatos, A.P. (2014). Enhanced electrochemical methanation of carbon dioxide with a dispersible nanoscale copper catalyst. *J. Am. Chem. Soc.* 136: 13319–13325. doi: 10.1021/ja5065284.

Menezes, F.D.L., Guimaraes, M.D.O. and da Silva, M.J. (2013). Highly selective SnCl2-catalyzed solketal synthesis at room temperature. *Ind. Eng. Chem. Res.* 52 (47): 16709–16713.

Monbaliu, J.-C.M., Winter, M., Chevalier, B. et al. (2011). Effective production of the biodiesel additive flow process. *Bioresour. Technol.* 102 (19): 9304–9307.

Nakagawa, Y. et al. (2020). Erythritol: another C4 platform chemical in biomass refinery. *ACS Omega* 5 (6): 2520–2530. doi: 10.1021/acsomega.9b04046.

Nanda, M.R., Yuan, Z., Qin, W. et al. (2014). A new continuous-flow process for catalytic conversion of glycerol to oxygenated fuel additive: catalyst screening. *Appl. Energy* 123: 75–81.

Nanda, M.R., Zhang, Y., Yuan, Z. et al. (2016). Catalytic conversion of glycerol for sustainable production of solketal as a fuel additive: a review. *Renew. Sustain. Energy Rev.* 56: 1022–1031.

Narkhede, N. and Patel, A. (2014). Room temperature acetalization of glycerol to cyclic acetals over anchored silicotungstates under solvent free conditions. *RSC Adv.* 4 (37): 19294–19301.

Newman, M.S. and Renoll, M. (1945). Improved preparation of isopropylidene glycerol. *J. Am. Chem. Soc.* 67 (9): 1621−1621.

Ooi, T., Yong, K., Dzulkefly, K. et al. (2001). Crude glycerine recovery from glycerol residue waste from a palm kernel oil methyl ester plant. *J. Oil Palm Res.* 13 (2): 16–22.

Oprescu, E.-E., Stepan, E., Dragomir, R.E. et al. (2013). Synthesis and testing of glycerol ketals as components for diesel fue. *Fuel Process. Technol.* 110: 214–217.

Quak, H., Nesterova, N. and Van Rooijen, T. (2016). Possibilities and barriers for using electric-powered vehicles in city logistics practice. *Transp. Res. Procedia* 12: 157–169. doi: 10.1016/j.trpro.2016.02.055.

Razzaq, L., Farooq, M., Mujtaba, M.A. et al. (2020). Modelling viscosity and density of ethanol-diesel-biodiesel ternary blends for sustainable environment. *Sustainability* 12 (12): 5186.

Rego de Vasconcelos, B. and Lavoie, J.-M. (2018). Is dry reforming the solution to reduce natural gas carbon footprint? *Int. J. Energy Prod. Manag.* 3: 44–56. doi: 10.2495/eq-v3-n1-44-56.

Rego de Vasconcelos, B. and Lavoie, J.-M. (2019 June 05). Recent advances in power-to-X technology for the production of fuels and chemicals. *Front. Chem.* doi: 10.3389/fchem.2019.00392.

Rio, J.C. et al. (2020). Lignin monomers from beyond the canonical monolignol biosynthetic pathway: another brick in the wall. *ACS Sustainable Chem. Eng.* 8 (13): 4997–5012. doi: 10.1021/acssuschemeng.0c01109.

Royon, D. et al. (2011). Ketalization of glycerol to solketal in supercritical acetone. *J. Supercrit. Fluids* 58: 88–92.

Schollenberger, D., Bajohr, S., Gruber, M. et al. (2018). Scale-up of innovative honeycomb reactors for power-to-gas applications—the project store & go. *Chem. Ing. Tech.* 90: 696–702. doi: 10.1002/cite.201700139.

SenterNovem (2006). Biofuels in Europe.

Serafim, H., Fonseca, I.M., Ramos, A.M. et al. (2011). Valorization of glycerol into fuel additives over zeolites as catalysts. *Chem. Eng. J.* 178: 291–296.

Smith, B.M., Kubczyk, T.M. and Graham, A.E. (2012). Metal triflate catalysed acetal exchange reactions of glycerol under solvent-free conditions. *RSC Adv.* 2 (7): 2702–2706.

Souza, T.E., Padula, I.D., Teodoro, M.M.G. et al. (2015). Amphiphilic property of niobium oxyhydroxide for waste glycerol conversion to produce solketal. *Catalysis Today*, 254: 83–89.

Steinmann, W. (2017). Thermo-mechanical concepts for bulk energy storage. *Renew. Sustain. Energy Rev.* 75: 205–219. doi: 10.1016/j.rser.2016.10.065.

Swalus, C., Jacquemin, M., Poleunis, C. et al. (2012). CO_2 methanation on Rh/γ-Al_2O_3 catalyst at low temperature: "*In situ*" supply of hydrogen by Ni/activated carbon catalyst. *Appl. Catal. B Environ.* 125: 41–50. doi: 10.1016/j.apcatb.2012.05.019.

Tervo, E., Agbim, K., Deangelis, F. et al. (2018). An economic analysis of residential photovoltaic systems with lithium-ion battery storage in the United States. *Renew. Sustain. Energy Rev.* 94: 1057–1066. doi: 10.1016/j.rser.2018.06.055.

Thiruvengadam, A., Besch, M., Padmanaban, V. et al. (2018). Natural gas vehicles in heavy-duty transportation—a review. *Energy Policy* 122: 253–259. doi: 10.1016/j.enpol.2018.07.052.

Thorin, E. (2014). Basics of energy. In: *Reference Module in Earth Systems and Environmental Sciences (Editor-in-chief S. Elias)*, 1–2. Elsevier Inc.

Umbarkar, S.B., Kotbagi, T.V., Biradar, A.V. et al. (2009). Acetalization of glycerol using mesoporous MoO3/SiO2 solid acid catalyst. *J. Mol. Catal. A: Chem.* 310 (1–2): 150–158.

Vassileva, I. and Campillo, J. (2017). Adoption barriers for electric vehicles: experiences from early adopters in Sweden. *Energy* 120: 632–641. doi: 10.1016/j.energy.2016.11.119.

Vicente, G. et al. (2010). Acetalisation of bio-glycerol with acetone to produce solketal over sulfonic mesostructured silicas. *Green Chem.* 12 (5).

Wang, B., Shen, Y., Sun, J. et al. (2014). Conversion of platform chemical glycerol to cyclic acetals promoted by acidic ionic liquids. *RSC Adv.* 4 (36): 18917–18923.

Wang, S., Mao, D., Guo, X. et al. (2009). Dimethyl ether synthesis via CO_2 hydrogenation over CuO–TiO_2-ZrO_2/HZSM-5 bifunctional catalysts. *Catal. Commun.* 10: 1367–1370. doi: 10.1016/j.catcom.2009.02.001.

Wang, X., Yang, G., Zhang, J. et al. (2016). Synthesis of isoalkanes over a core (Fe-Zn-Zr)-shell (zeolite) catalyst by CO_2 hydrogenation. *Chem. Commun.* 52: 7352–7355. doi: 10.1039/C6CC01965J.

Wei, J., Yao, R., Wen, Z. et al. (2017). Directly converting CO_2 into a gasoline fuel. *Nat. Commun.* 8: 1–8. doi: 10.1038/ncomms15174.

Yong, K., Ooi, T., Dzulkefly, K. et al. (2001). Refining of crude glycerine recovered from glycerol residue by simple vacuum distillation. *J. Oil Palm Res.* 13: 39–44.

Zahid, I., Ayoub, M., Abdullah, B.B. et al. (2020). Production of fuel additive solketal via catalytic conversion of biodiesel-derived glycerol. *Ind. Eng. Chem. Res.* 59 (48): 20961–20978.

Zeddies, J. and Schonleber, N. (2014). Literaturstudie "Biomasse – Flächen – und Energiepotenziale". December 2014.

Zhao, X. et al. (2020). Narrowing the gap for bioplastic use in food packaging: an update. *Environ. Sci. Technol.* 54 (8): 4712–4732. doi: 10.1021/acs.est.9b03755.

12

Industrial Small Reactors

CONTENTS

12.1 Introduction

A nuclear reactor, formerly known as an atomic pile, is a device used to initiate and control a fission nuclear chain reaction or nuclear fusion reactions. There is strong interest in small and simpler units for generating electricity from nuclear power and for process heat. This interest in small and medium nuclear power reactors is driven both by a desire to reduce the impact of capital costs and to provide power away from large grid systems. The technologies involved are numerous and very diverse.

Since the 1950s, as nuclear power generation has become established, the size of reactor units has grown from 60 MWe to more than 1600 MWe, with corresponding economies of scale in operation. At the same time there have been many hundreds of smaller power reactors built for naval use (up to 190 MW thermal) and as neutron sources, yielding enormous expertise in the engineering of small power units and accumulating over 12,000 reactor years of experience.

The International Atomic Energy Agency (IAEA) defines "small" as under 300 MWe and up to about 700 MWe as "medium" – including many operational units from the 20th century. Together they have been referred to by the International Atomic Energy Agency as small and medium reactors (SMRs). However, "SMR" is used more commonly as an acronym for "small modular reactor", designed for

Converting Power into Chemicals and Fuels: Power-to-X Technology for a Sustainable Future, First Edition. Martin Bajus.

serial construction and collectively to comprise a large nuclear power plant. (Here the use of diverse prefabricated modules to expedite the construction of a single large reactor is not relevant.) A subcategory of very small reactors – vSMRs – is proposed for units under about 15 MWe, especially for remote communities.

The nuclear reactor coupled to hydrogen production has the potential for higher overall energy use efficiency and better utilization of capital equipment. This is because the nuclear reactor can operate at full capacity, but also follow the electricity grid as it varies throughout the day. When electricity prices are low, or negative, the reactor can use a subsystem to convert heat or electricity into hydrogen. This hydrogen may then be converted back to energy during times of high electricity demand using fuel cells or be sold as its own commodity.

In the chemical industry there exist several of routes a nuclear reactor may take for hydrogen production, depending on available reactor outlet temperature. As stated earlier in this book, there are a multitude of ways hydrogen may be produced. The focus of this chapter is on renewable hydrogen production from water that requires only excess heat and electricity that can be produced from the thermal energy of a nuclear reactor. These water-based hydrogen production technologies are also compared to the most common method of hydrogen production available in present-day chemical technology – steam methane reforming, which is used in approximately 48% of current production. This method requires fossil fuels for production and therefore releases greenhouse gases including carbon dioxide.

The temperature of the water is also an important consideration. Nuclear reactors have outlet temperatures that vary from approximately 300°C to 950°C, depending on the design. Generation III reactors primarily consist of light water reactors (LWRs), which are the predominant reactor in the United States. These are considered low temperature reactors and have outlet temperatures estimated at 300°C. Generation IV reactors are pushing on the average outlet temperature. To analysis, the light water reactor and Generation IV reactors have been split into three categories: low (<300°C), medium (<750°C) and high (<950°C, though realistically likely <900°C) outlet temperatures. The outlet temperature can determine what hydrogen production methods are feasible.

12.2 Thermochemical Water Splitting

Thermochemical water splitting is one or more endothermic chemical reactions that require heat to split water. The two contenders considered for this study are the sulfur-iodine (S-I) and calcium-bromide (Ca-Br) cycles. Thermochemical cycles are less developed than electrolysis processes. In these cycles, chemicals are recycled throughout the system, so inputs and outputs consist of water and water products, respectively. Though thermochemical cycles primarily utilize heat, rendering them more cost effective than electrolysis processes, they can face problems such as slow response rates and large size and involve hazardous chemicals that may produce less pure hydrogen.

In the S-I cycle, the sulphuric acid acts as the oxygen carrier while the hydrogen iodide carries hydrogen. Mass balances and optimization studies have been conducted for all reactions to maximize reaction rates and minimize undesired products from alternative reactions occurring. Note that the decomposition reactions are both endothermic, but the Bunsen reaction is exothermic. The limiting reaction is the sulphuric acid reaction, due to its high temperature requirements. This decomposition reaction is why the S-I process may only be coupled to high temperature nuclear reactors if no external heat is desired.

The calcium-bromine cycle was discovered at the University of Tokyo in Japan, which is why it is also referred to as the UT-3 cycle. While considered a promising candidate for nuclear hybrid processes, it is less technologically mature than its competitor, the S-I cycle and therefore has not been further selected in down-selection processes. The minimum temperature required for this process is around 760°C, which potentially could be met with a high temperature reactor and also a potential for medium temperature reactors with external heating. The UT-3 process can be divided into four chemical reactors. A unique advantage of the UT-3 cycle is that it only involves solids and gases, which are easier to separate than the all-liquid reagents of the S-I cycle.

Hybrid water splitting involves a thermochemical cycle in conjunction with electrolysis. The main contenders in the hybrid splitting process are the hybrid sulfur cycle (HyS) developed by Westinghouse and the copper-chlorine cycle. Hybrid cycles have similar disadvantages and advantages as thermochemical cycles alone, mainly: slow response rates, large reactor sizes, utilize hazardous chemicals and are less developed than pure electrolysis, but are more cost effective than electrolysis alone.

The chemical reactions that take place are for sulphuric acid decomposition and sulfur dioxide electrolysis. One potential advantage of the hybrid sulfur cycle exists with the location of hydrogen production. The processes that require heat must be placed relatively close to the nuclear plant to minimize losses, which includes the decomposition step of this process. However, the electrolysis step, which produces the hydrogen, may be placed a safer distance away from the plant as it requires only electricity.

The copper-chlorine cycle, like the Ca-Br cycle, has shown promise but is not as technologically mature as other processes. However, it has a lower minimum temperature requirement, around 500°C, than the other thermochemical processes which would allow it to couple with medium or high temperature as opposed to only high temperature reactors. There are three Cu-Cl cycles proposed: a 3-step, 4-step and 5-step cycle.

12.3 Small Modular Reactors

Small modular reactors are defined as nuclear reactors generally 300 MWe equivalent or less, designed with modular technology using module factory fabrication, pursuing economies of series production and short construction times. This definition, from the World Nuclear Association, is closely based on those from the IAEA and the US Nuclear Energy Institute. Some of the small reactors already operating or tabulated below do not fit this definition, but most of those described do fit it. Pressured water reactor (PWR) types may have integral steam generators, in which case the reactor pressure vessel needs to be larger, limiting portability from factory to site. Hence many larger pressured water reactors such as the Rolls-Royce UK SMR have external steam generators.

Small modular reactors are fissionable reactors that are smaller than conventional nuclear reactors and typically have an electrical power output of less than 300 MW_e or a thermal power output of less than 1000 MW_{th}. They are designed to be manufactured at a plant and transported to a site to be installed. Modular reactors will reduce on-site construction and increase containment efficiency and are claimed to enhance safety. The greater safety should come via the use of passive safety features that operate without human intervention, a concept already implemented in some conventional nuclear reactor types. Small modular reactors also reduce staffing versus conventional nuclear reactors. Small modular reactors are claimed to cross financial and safety barriers that inhibit the construction of conventional reactors.

The term small modular reactor refers to the size, capacity and modular construction only, not to the reactor type and the nuclear process which is applied. Designs range from scaled down versions of existing designs to generation IV designs. Both thermal-neutron reactors and fast-neutron reactors have been proposed, along with molten salt and gas cooled reactor models.

While there are dozens of modular reactor designs and yet unfinished demonstration projects, the floating nuclear power plant Academic Lomonosov, operating in Pevek in Russia's Far East was, as of the end of 2019, the first and only completed working prototype in the world connected to the grid. The plant has two reactors, each with a capacity of 35 MW_e. The concept was based on the design of nuclear icebreakers. The construction of the world's first commercial land-based small modular reactor started in July 2021 with the Chinese power plant Linglong One. The operation of this prototype is due to start by the end of 2026.

One hindrance to commercial use may be licensing since current regulatory regimes are adapted to conventional designs. Small modular reactors differ in terms of staffing, security and deployment time. Licensing time, cost and risks are critical success factors. US government studies that evaluated small modular reactor-associated risks have slowed licensing. One concern with small modular reactors is preventing nuclear proliferation.

Conventional reactors use water as a coolant. Small modular reactors may use water, liquid metal, gas and molten salt as coolants. Some gas-cooled reactor designs drive a gas-powered turbine, rather than boil water. Thermal energy can be used directly, without conversion. Heat can be used in hydrogen production and other commercial operations, such as desalination and the production of petroleum products (extracting oil from tar sands, creating synthetic oil from coal, etc.).

The Integral Molten Salt Reactor (IMSR) is designed for the small modular reactor market. It employs molten salt reactor technology, which is being developed by the Canadian company Terrestrial Energy (Figure 12.1). It is based closely on the denatured molten salt reactor (DMSR), a

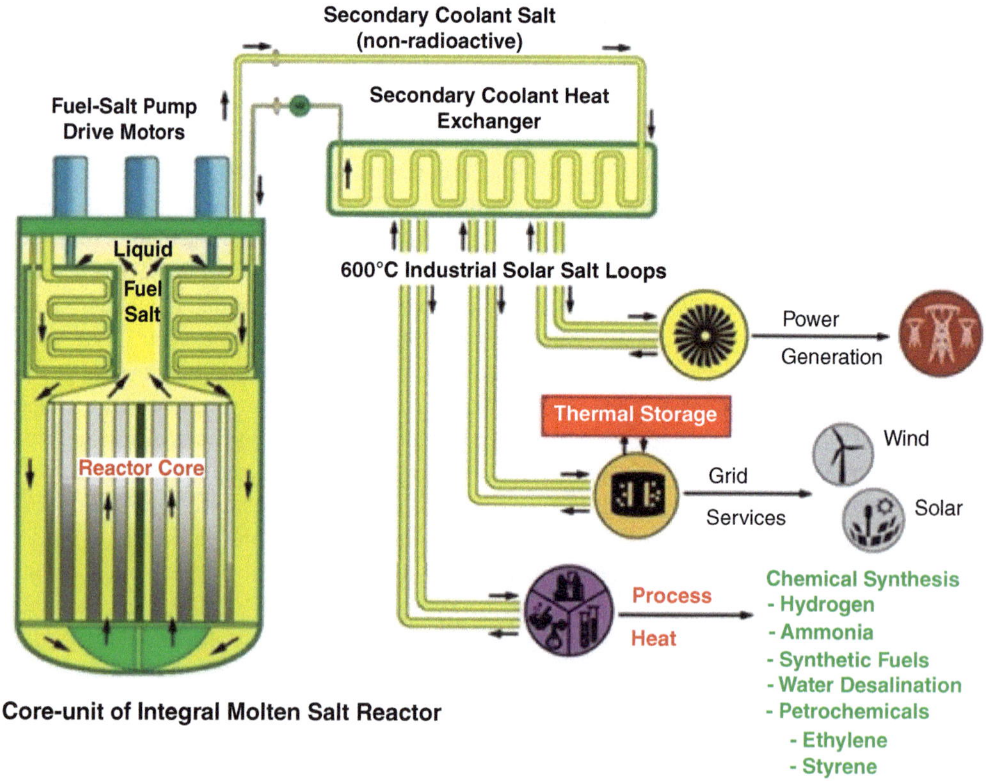

Figure 12.1 Integral molten salt reactor design intended to be used for a variety of heat demand applications, ranging from power generation to cogeneration, or process-heat only.

reactor design from Oak Ridge National Laboratory. It also incorporates elements found in the small, molten, advanced high temperature reactor (SmAHTR), a later design from the same laboratory. The integral molten salt reactor belongs to the denaterated molten salt reactor class of molten salt reactors (MSR) and hence is a "burner" reactor that employs a liquid fuel rather than a conventional solid fuel; this liquid contains the nuclear fuel and also serves as primary coolant.

In 2016, Terrestrial Energy engaged in a pre-licensing design review for the integral molten salt reactor with the Canadian Nuclear Safety Commission and entered the second phase of this process in October 2018 after successfully completing the first stage in late 2017. The company claims it will have its first commercial integral molten salt reactors licensed and operating in the 2020s.

The integral molten salt reactor is so called because it integrates into a compact, sealed and replaceable nuclear reactor unit, called the integral molten salt reactor core-unit. The core-unit comes in a single size designed to deliver 400 megawatts of thermal heat which can be used for multiple applications. If used to generate electricity, then the notional capacity is 190 megawatts electrical. The unit include all the primary components of the nuclear reactor that operate on the liquid molten fluoride salt fuel: moderator, primary heat exchangers, pumps and shutdown rods. The core-unit forms the heart of the integral molten salt reactor system. In the core-unit, the fuel salt is circulated between the graphite core and heat exchangers. The core-unit itself is placed inside a surrounding vessel called the guard vessel. The entire core-unit module can be lifted out for replacement. The guard vessel that surrounds the core-unit acts as a containment vessel. In turn, a shielded silo surrounds the guard vessel.

The integral molten salt reactor belongs to the denatured molten salt reactor class of molten salt reactors. It is designed to have all the safety features associated with the Molten Salt class of reactors including low pressure operation (the reactor and primary coolant is operated near normal atmospheric pressure), the inability to lose primary coolant (the fuel is the coolant), the inability to suffer a meltdown accident (the fuel operates in an already molten state) and the robust chemical binding of the fission products within the primary coolant salt (reduced pathway for accidental release of fission products).

The design uses standard assay low-enriched uranium fuel, with less than 5% U^{235} with a simple converter (also known as a "burner") fuel cycle objective (as do most operating power reactors today). The proposed fuel is in the form of uranium tetrafluoride blended with carrier salts. These salts are also fluorides, such as lithium fluoride, sodium fluoride and/or beryllium fluoride. These carrier salts increase the heat capacity of the fuel and lower the fuel's melting point. The fuel salt blend also acts as the primary coolant for the reactor.

The integral molten salt reactor is a thermal-neutron reactor moderated by vertical graphite tubular elements. The molten salt fuel-coolant mixture flows upward through these tubular elements where it goes critical. After heating up in this moderated core the liquid fuel flows upward through a central common chimney and is then pulled downward by pumps through heat exchanges positioned inside the reactor vessel. The liquid fuel then flows down the outer edge of the reactor core to repeat the cycle. All the primary components, heat exchangers, pumps etc., are positioned inside the reactor vessel. The reactor's integrated architecture avoids the use of external piping for the fuel that could leak or break.

The piping external to the reactor vessel contain two additional salt loops in series: a secondary, nonradioactive coolant salt, followed by another (third) coolant salt. These salt loops act as additional barriers to any radionuclides and improve the system's heat capacity. It also allows easier integration with the heat sink end of the plant; either process heat or power applications using standard industrial grade steam turbine plants are envisioned by Terrestrial Energy.

The integral molten salt reactor core-unit is designed to be completely replaced after a 7-year period of operation. This ensures that a sufficient operational lifetime of materials used in the integral molten salt reactor core can be achieved. During operation, small fresh fuel/salt batches are periodically added to the reactor system. This online refuelling process does not require the mechanical refuelling machinery required for solid fuel reactor systems.

Many of these design features are based on two previous molten salt designs from Oak Ridge National Laboratory – the Oak Ridge National Laboratory denatured molten salt reactor from 1980 and the 2010 solid fuel/liquid salt cooled, small modular advanced high temperature reactor. The denatured molten salt reactor, as carried into the integral molten salt reactor design, proposed to use molten salt fuel and a graphite moderator in a simplified converter design using low enriched uranium, with periodic additions of low enriched uranium fuel. Most previous proposals for molten salt reactors all bred more fuel than needed to operate, so were called breeders. Converter or "burner" reactors like the integral molten salt reactor and denatured molten salt reactor can also utilize plutonium from existing spent fuel as their makeup fuel source. A more recent small, molten, advanced high temperature reactor proposal was for a small, modular, molten salt cooled but solid tristructural isotropic (TRISO) fuelled reactor.

12.4 Nuclear Process Heat for Industry

Nuclear energy is an excellent source of process heat for various industrial applications including desalination, synthetic and unconventional oil production, oil refining, biomass-based ethanol production and in the future, hydrogen production. For most major industrial heat applications, nuclear energy is the only credible non-carbon option.

Light water reactors produce heat at relatively low temperatures in relation to many industrial needs, hence the technology focus has been on high-temperature gas-cooled reactors (HTR) and more recently on molten salt reactors producing heat at over 700°C.

In 2019 there were 79 nuclear reactors used for desalination, district heating, or process heat, with 750 reactor-years of experience in these, mostly in Russia and Ukraine.

The potential application of nuclear heat depends mainly on the temperature required. With reactor output temperatures of up to 700°C there is a wide range of possible applications, at 900°C there are further possibilities and at 950°C an important future application to hydrogen production opens. About 20% of US energy consumption goes into process heat applications, compared with 35–40% into electricity. In this 20%, replacing fossil fuels with nuclear heat promises much in energy security, price stability and reduced regulatory risks and it is the only option if carbon dioxide emissions are to be avoided.

12.4.1 High-temperature Reactors for Process Heat

The most advanced high-temperature gas-cooled reactor (HTR) project is China's HTR-PM, based on its successful HTR-10 prototype. The demonstration unit being built in Shidaowan links twin 250 MWt units with 750°C outlet temperature to a 210 MWe steam turbine.

The Nuclear Cogeneration Industrial Initiative (NC2I), part of the Sustainable Nuclear Energy Technology Platform (SNETP) in the European Union is focused on HTRs producing 550°C steam for a variety of industrial applications. The Gemini+ project launched in September 2017 is an outcome of this, involving the EU and international partners (Japan, South Korea and the USA) and coordinated by the Polish National Centre for Nuclear Research (NCBJ). This project is funded under the Euratom programme and aims to provide a conceptual design of a high temperature nuclear cogeneration system that supplies process steam to industry, a licensing framework for this system and a business plan for a full-scale demonstration.

Poland has 13 large chemical plants which need 6500 MWt of heat at 400–550°C. It is considering the potential role of high temperature gas-cooled reactors for this and very high-temperature gas-cooled reactors (VHTR) for hydrogen production. Deployment of high temperature gas-cooled reactors for industrial heat production was included in the government's July 2016 draft strategy for development. An early target is building a cogeneration high temperature gas-cooled reactor of 200–350 MWt for heat and before this a 10 MWt experimental high temperature gas-cooled reactor at Swierk (Figure 12.1).

In the USA the Department of Energy (DoE) since 2005 has had a programme (with diminishing funding) to develop a next generation nuclear plant (NGNP) which is to be an HTR capable of producing hydrogen, electricity or process heat. Initially this was focused on eventual hydrogen production, which requires very high temperatures with corresponding technological and cost challenges. Other process heat applications would broaden the end-user base for nuclear energy. The prospect of an operating prototype by 2021 receded. Industrial partners in the endeavour, with 11 member companies including reactor vendors Areva and Westinghouse, utility Entergy and potential end-users of electricity and process heat such as Dow Chemical and ConocoPhillips, then set their sights on a more readily achievable 750°C outlet temperature for process heat rather than hydrogen production particularly. The NGNP was to be built at Idaho National Laboratory (INL) with substantial cost share intended from industry.

Early in 2012 the next generation nuclear plant Industry Alliance announced that it had selected the 625 MWt Areva high temperature gas-cooled reactor design, known as Antares, for further development. This would be run with helium as primary coolant at 750°C and a two-loop secondary steam cycle.

More recently, the next generation nuclear plant Industry Alliance has been collaborating with the nuclear cogeneration industrial initiative on the design and regulatory framework for high temperature gas-cooled reactor technology and in September 2017 launched the GEMINI+ project together with an international consortium to demonstrate high temperature nuclear cogeneration with high temperature gas-cooled reactor technology (Figure 12.2). The next generation nuclear plant Industry Alliance has evaluated US sites for a demonstration plant including the Piketon site in Portsmouth, Ohio; Odessa/Midland, Texas; a former Georgia Power coal power plant site; and Idaho Falls. All were found to be suitable, but Piketon was judged best for electricity, process heat and hydrogen.

A 2011 evaluation by INL for a typical high temperature gas-cooled reactor multi-module plant showed that high-temperature process heat and electric power for use by energy-intensive industries can be produced at an energy price that is equivalent to natural gas in the $6-$9/GJ range. Areva's estimates support this.

In October 2015 a DoE assessment said: "The high temperature gas-cooled reactor, with outlet temperature limited to less than 800°C, is suitable for near-term deployment as a demonstration reactor." It noted US government and industry investment in high temperature gas-cooled reactors since the 1960s totalled over $3 billion.

In South Korea the Nuclear Hydrogen Development & Demonstration (NHDD) project is focused on high temperature gas-cooled reactor technology for both hydrogen production and liquid hydrocarbons (Table 12.1). It was launched in 2006 as a key programme supported by the Ministry of Science and Technology (now the Ministry of Science & Information and Communications technology). The Korean Atomic Energy Research Institute earlier submitted a very high temperature reactor design to the Generation IV International Forum. The very high temperature reactor is to produce hydrogen on a large scale, with 300 MWt modules each producing 30,000 tonnes of hydrogen per year. The Korean Atomic Energy Research Institute had expected engineering design to be completed by 2014, construction to start in 2016 and operation

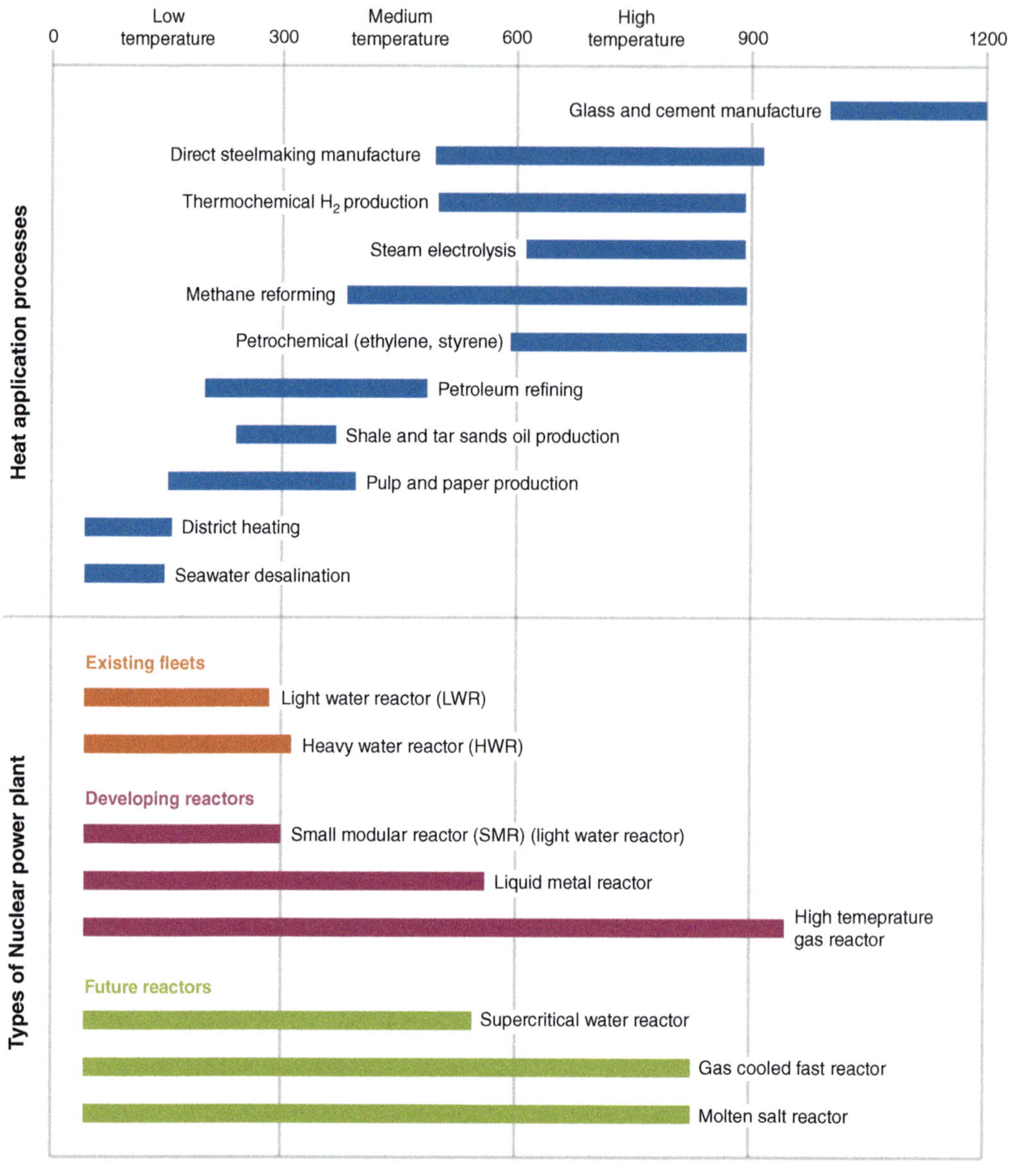

Figure 12.2 Industrial nuclear cogeneration: temperature ranges of heat application processes and types of nuclear power plant.

of a demonstration reactor in 2020, with commercialization about 2025. Research & Development focus is on development of corrosion-resistant high-temperature materials and components and refractory coated-particle fuel.

The Korean Atomic Energy Research Institute also has a research partnership with China's Tsinghua University focused on hydrogen production, based on China's HTR-10 reactor.

In Japan, (Table 12.1) Japan Atomic Energy Agency (JAEA) has confirmed the safety of high temperature gas-cooled reactors and in April 2004 a coolant outlet temperature of 950°C was achieved in its high-temperature engineering test reactor – a world first; and in 2009 it ran at 950°C for 50 days, opening the way for direct thermochemical hydrogen production. In January 2019 it produced hydrogen

Table 12.1 Nuclear process heat application.

Process heat applications

Process temperature	Up to 700°C	Up to 900°C	Up to 950°C
Electricity production	Rankine (steam) cycle	Brayton (direct) cycle	
Utility applications	Desalination	H_2 via steam reforming of methane or high-temperature electrolysis	Thermochemical H_2 production
Oil and chemical industry	Tar/oil sands and heavy oil recovery. syncrude, refinery and petrochemical	Syngas for ammonia and methanol	Thermochemical H_2 production

using the iodine-sulfur process over 150 hours of continuous operation in its integrated process loop of 100 l/h design capacity, operating at 30 l/h. Meanwhile JAEA plans to make hydrogen by steam reforming natural gas, using high-temperature engineering test reactors high temperature.

Based on the high temperature engineering test reactor, JAEA is developing the high temperature reactor-50 nuclear steam supply system to operate at 750°C and produce steam at 538°C and 12.5 MPa. Thermal rating would be up to 200 MWt and it would have the same tri-structural isotropic fuel as the high-temperature engineering test reactor. It involves Toshiba, Fuji Electric and KHI and a lead plant is planned for the 2020s.

Following this and moving to very high temperature reactor, JAEA plans the 600 MWt Gas Turbine High Temperature Reactor (GTHTR300) cogeneration system operating at 850–950°C to enable 300 MWe electricity generation using direct cycle gas turbine, with 370 MWt process heat for hydrogen production (120 t/day), steelmaking (650,000 t/yr) and other industrial applications including district heating and desalination (55,000 m^3/day) using waste heat. The basic design has been completed, involving MHI, Fuji Electric, IHI and NFI and the lead plant is expected in the 2030s. The IS hydrogen system beyond the intermediate heat exchanger would be non-nuclear grade, in line with other hydrogen production facilities.

In 2006 the South African pebble bed modular reactor board formalised the concept of a higher-temperature pebble bed modular reactor process heat plant with reactor output temperature of 950°C. The first plants were envisioned for 2016 and the applications would have been oil sands production and petrochemical industry (process steam), steam methane reforming for hydrogen and eventually thermochemical hydrogen production. Each 600 MWt module would produce about 200 tonnes of hydrogen per day, which is well matched to the scale of current industrial demand for hydrogen. In 2011 the programme was aborted due to lack of funds.

12.4.2 Recovery of Oil from Tar Sands

From about 2003 various proposals have been made to use nuclear power to produce steam for extraction of oil from Alberta's northern tar sands deposits, as well as electricity for the major infrastructure involved.

About 1.05 GJ (28 m^3) of natural gas is typically required to produce a barrel of bitumen by in situ methods (according to the Canadian Energy Resource Institute). This bitumen must then be upgraded to oil. Natural gas is the raw material for hydrogen (by steam reforming) to break down the long-chain hydrocarbons to yield synthetic crude oil. The extraction overall represents a total energy intensity of 1.37 GJ per barrel of oil (each embodying 6 GJ) and in 2014 it accounted for one-third of Alberta's primary energy demand. With production of 2.4 million barrels per day of crude bitumen in 2016 (half mined and half in situ), a great deal of gas is used and the cost exposure is considerable. Production is expected to reach 3.6 million barrels per day in 2026. In fact, Canadian natural gas is inadequate to supply the anticipated expansion in oil sands output and its use has major carbon dioxide implications which are creating public concern – over 80 kg of carbon dioxide per barrel is released in its product.

An intrinsic problem is that the steam supply needs to be semi-portable as tar sand extraction proceeds, so relatively small reactors that could be moved every decade or so may be needed. One problem related to the provision of steam for mining is that a nuclear plant is a long-life fixture and mining of tar sands proceeds across the landscape, giving rise to very long steam transmission lines and consequent loss of efficiency.

One proposal from Energy Alberta Corp. suggested that a single Canada deuterium uranium (CANDU 6) reactor (about 1800 MWt) configured to produce 75% steam and 25% electricity would replace 6 million cubic metres per day of natural gas and support production of 175,000–200,000 barrels per day of oil. It would also save the emission of 3.3 million tonnes of carbon dioxide per year. Other figures from pebble bed modular reactor (PBMR) proponents confirm that each 100 MWt will enable production of 10,000 barrels per day.

The Canadian Energy Research Institute (CERI) published a report in February 2009 saying that employing nuclear energy with (so far untested) carbon capture and storage in tar sands extraction and processing could make oil from that source cleaner than conventional oil with respect to its greenhouse gas and other emissions. The CERI report looked at both very large (1600 MWe) and multiple very small (10 MWe) nuclear reactors.

A more recent proposal is from Canada's Terrestrial Energy, putting forward its integral olten salt reactor (IMSR), which the company aimed to commercialise by 2021. It is considering 300 MWt units, producing simply steam (Figure 12.1).

The main difference between natural gas and nuclear steam generation is that a fuel-intensive process is replaced by a capital-intensive one, with no carbon emissions.

12.4.3 Oil Refining

As well as separating the different components of crude oil by two distillation processes, an oil refinery typically breaks down the residual heavy or long-chain hydrocarbons in the process of hydrocracking. The hydrogen is produced from natural gas and it breaks down the long-chain hydrocarbons to yield synthetic crude oil (about 5 kg is used per barrel). This hydrocracking of heavy crude oil is a major use of hydrogen today. Overall, about 15–20% of the energy value of the crude oil is used in producing refined products.

Hydrogen production is by steam reforming of the natural gas and gives rise to a lot of carbon dioxide. Nuclear power could make steam and electricity and use some of the electricity for high-temperature electrolysis for hydrogen production. (Heavy water and oxygen could be valuable by-products of electrolysis). Oil refineries account for about one quarter of the world's hydrogen demand.

12.4.4 Coal and Its Liquefaction

Here, various processes are used. Indirect liquefaction accounts for most world production, about 260,000 barrels/40,000 m^3 per day. Coal is gasifier to synthesis gas and the product synthesized either to diesel and jet fuel, or to gasoline (Mobil or methanol MTG process). Direct liquefaction processes miss the syngas step and are either hydrogenation or pyrolysis and carbonization.

The Fischer-Tropsch process was originally developed in Germany in the 1920s and provided much of the fuel for Germany during the Second World War. It then became the basis for much oil production in South Africa by Sasol, which now supplies most of that country's diesel fuel. However, it is a significant user of hydrogen, catalysing a reaction with carbon monoxide. The hydrogen is now produced with the carbon monoxide by coal gasification, part of the gas stream undergoing the water shift reaction.

Coal gasification produces both carbon monoxide and hydrogen feed for Fischer-Tropsch and the water shift reaction turns some carbon monoxide into carbon dioxide and more hydrogen, both stages using water. In Fischer-Tropsch itself hydrogen plus carbon monoxide converts to alkane hydrocarbons using a catalyst.

At Erdos, Inner Mongolia, China, Shenhua Coal Liquefaction Corp produces 20,000 barrels/3,000 tonnes of oil per day from nearly 10,000 tonnes of coal from the Shenfu-Dongsheng coalfield, using US technology. The $2.06 billion Erdos Coal-toLiquid (CTL) plant was commissioned in 2010 and in 2013 it produced 866,000 tonnes of oil products. It has total design capacity of 1.08 million tonnes, comprising diesel (621,000 t/yr), naphtha (321,000 t/yr) and LPG (70,000 t/yr). Water supply is a constraint on operation and it requires 7–12 tonnes of fresh water to produce one tonne of product, depending on whether direct liquefaction or indirect (based on FT), along with nine tonnes of carbon dioxide and 4.8 tonnes of contaminated wastewater according to critics. It uses coal hydrogenation at high temperature and pressure with a catalyst.

Shenhua proposed a $7 billion plant in Ningdong, Ningxia, China – the Shenhua CTL plant to produce 80,000 barrels/13,000 m^3/day, using Sasol's indirect liquefaction technology, based on FT. Sasol has withdrawn from the venture.

The hydrogenation (Bergius) process requires hydrogen to react directly with coal at high temperature and pressure. Pyrolysis and carbonization processes give less yield and that usually needs refining. They are not commercial for CTL.

A nuclear source of hydrogen coupled with nuclear process heat would more than double the number of liquid hydrocarbons from the coal and eliminate most carbon dioxide emissions from the process.

Using simply black coal in the Sasol-type plant, 14,600 tonnes produce 25,000 barrels of synfuel "oil" (with 25,000 tonnes of carbon dioxide).

A hybrid system uses nuclear electricity to electrolyse water for the hydrogen. Using some 4,400 tonnes of coal as gasifier using oxygen from the electrolysis to produce carbon monoxide which is fed to the Fischer-Tropsch plant with the hydrogen to produce 25,000 barrels of synfuel "oil". Very little carbon dioxide results and this is recycled to the gasifier.

12.4.5 Biomass-Based Ethanol Production

Commercial-scale ethanol production from cellulose in a biorefinery requires energy input. Two processes can be used: biochemical fermentation, or thermochemical conversion. Either way, the energy input needed to produce ethanol from cellulose materials (such as wood) is equivalent to 25% to 38% of the output, mostly as steam (though the thermochemical process also requires high temperatures).

As well as burning the lignin content of feedstock, some energy can readily be supplied by nuclear plants as cogeneration, on a carbon-free basis. In future, the lignin may be converted to liquid fuels, probably by hydrogenation. In this case, the yield of liquid fuels from biomass can be increased by about 50% per unit of input. Much of the nuclear energy input is as low-temperature steam, essentially a by-product of electricity generation and hence low-cost.

12.4.6 District Heating

Many reactors in cold climates provide hot water for district heating, without a significant penalty in electrical performance. This usually replaces fossil fuel sources, with significant reduction in carbon dioxide emissions. Most of the nuclear district heating has been developed within northern European or former Soviet states as the cold climate and long heating periods create favourable conditions for district heating and cogeneration development.

Apart from new reactors designed specifically for district heat, this service may be from a power reactor's condenser circuit or from tapping into the secondary steam circuit. This is being demonstrated at the Haiyang nuclear power plant in China, to heat the whole city with 300–400 TJ/yr. It is planned for the new CAP1400 plant in Shandong and for plants at Qinshan, Tianwan and Hongyanhe.

If simply tapping the condenser circuit is more widely done, water at around 100°C at low pressure is required.

China has three low-temperature reactor designs specifically for district heating rather than power. China General Nuclear (CGN) has the nuclear heating reactor (NHR-200) (200 MWt) at Daqing City, China National Nuclear Corporation (CNNC) has the deep-pool low-temperature heating reactor (DHR-400) Yanlong (400 MWt) and state power investment corporation (SPIC) has the 200 MWt LandStar-I which delivers hot water at 110°C with convection circulation through a heat exchanger. Ten units are planned in northeaster China, with the first under construction in Jiamusi and Daqing in Heilongjiang province. The two units at Jiamusi will provide steam to a biomass plant in summer.

Russia's low-temperature reactor design specifically for district heating rather than power is the AST-500, proposed for several sites (500 MWt). The 2 x KLT40 floating nuclear power plant at Pevek also produces district heat.

12.5 Microchannel Reduction Cell

Whipple et al. designed a carbon dioxide reduction electrocatalytic cell device with potassium hydrogen carbonate solution as the electrolyte. The device is composed of a cathode plate, carbon dioxide gas diffusion cathode, potassium hydrogen carbonate electrolyte solution layer, oxygen evolution gas diffusion anode and anode plate from top to bottom. Through the design of microchannel technology, its structural advantage is reflected in the fact that the internal electrolyte still has a flow channel even if the distance between cathode and anode is only 1 mm. This technique not only greatly reduces the ohmic polarization loss in the carbon dioxide reduction process, but also maintains a relatively stable pH reaction environment in the cathode area with the flowing electrolyte.

In the experiment, with the use of tin/carbon (Sn/C) as the cathode catalyst at room temperature and a current density of 100 mA/cm^2, the Faradaic efficiency of formic acid reached 89% and the energy efficiency of the electrolytic cell reached 45%. However, the disadvantage is that some products (such as formic acid, ethanol, etc.) of the carbon dioxide reduction at the cathode migrate to the anode by convection and diffusion and be reoxidized, which eventually leads to the reduction of efficiency of the carbon dioxide catalysis.

The cell apparatus is also theoretically applicable to the ionic liquids (IL) electrolytes in the general liquids form, which is determined according to the state of ILs. Although the microchannel space is small and the ionic liquids have a high viscosity, the electrolyte viscosity can be reduced by adding cosolvents such as water and organic solution to meet the operational requirements of the microchannel reactor. For ionic liquids electrolytes in liquid form, the specific reaction effect and the applicability of the reaction device need to be verified in future practical operations.

In practical operation, the reactor (or microchannel reactor) should be designed according to what is need in each specific situation. So far, although the electrochemical carbon monoxide reduction (ECR) technology has not achieved large-scale industrial application, some small reaction devices simulating industrial production have been reported. Liu et al. demonstrated a small continuous reactor at laboratory scale. A pure carbon dioxide or CO_2/N_2 mixture mixed with 0.45 mol/l potassium hydrogen carbonate cathode electrolyte enters the cathode chamber from the cathode end. The anode electrolyte of 1 mol/l potassium hydroxide goes up through the anode chamber and the carbon dioxide reduction reactions are carried out in the reactor.

Electrochemical carbon dioxide reduction to value-added fuels and chemicals pro-vides a "clean" and efficient way to mitigate energy shortages and to lower the global carbon footprint. The unique structures of two-dimensional (2D) nanosheets and their tuneable electronic properties make these nanostructured materials intriguing in catalysis. Various 2D nanosheets are showing promise for carbon dioxide reduction, depending on the preferred reaction product (formic acid, carbon monoxide, methane, methanol, or acetic acid). Sun et al. focused on recent progress that has been achieved in using these 2D materials for electrochemical carbon dioxide reduction technology. We highlight procedures available for tuning catalytic activities of 2D materials and describe the fundamentals and future challenges of carbon dioxide catalysis by 2D nanosheets.

12.6 Conversion of Carbon Dioxide to Graphene

The deployment of graphene has been slow due to its high cost and high carbon footprint. A low carbon footprint, inexpensive synthesis of graphene, consuming rather than emitting carbon dioxide to overcome these obstacles, is introduced. This is accomplished by direct molten carbonate electrolytic splitting of the carbon dioxide to a nano-thin carbon product (carbon nanoplatelets) comprised of 25–125 graphene layers and subsequent electrochemical exfoliation of the nanoplatelets to graphene in a carbonate soluble aqueous solution. The sole products of the carbon dioxide electrolysis are straightforward: high yield carbon nanoplatelets and oxygen. The carbon nanoplatelets provide a thinner starting point than a conventional graphite reactant to facilitate electrochemical exfoliation. The resultant valuable product (graphene) incentivizes utilization of the carbon dioxide.

Graphene has unique properties that are useful for a variety of applications. However, the cost of synthesis and the challenge of isolating the graphene product in its native two-dimensional structure has led to the current high cost of graphene of approximately $1 million per ton.

Liu et al., synthesized graphene from carbon dioxide by:

i) molten carbonate electrolysis to form carbon platelets on a stainless-steel cathode. This is followed by cooling and separation into individual graphene layers by
ii) high yield electrochemical exfoliation in a carbonate dissolving aqueous ammonium sulphate electrolyte.

The electro-synthesized carbon platelets are nano-thin, promoting a higher graphene yield than thicker, conventional graphite exfoliation reactions. Carbon dioxide is the sole reactant used to produce the graphene. This can incentivize utilization of carbon dioxide and unlike alternative products made from carbon dioxide such as carbon monoxide or other fuels such as methane, use of the product does not release this greenhouse gas back into the atmosphere.

Graphene has a high surface area, high thermal and electrical conductivity, strength, elasticity and high charge carrier conductivity that makes it uniquely suitable for energy storage and electronics. Its ability to carry plasmons allows it to strongly interact with light in a non-linear fashion and act both as a transducer and transmitter in optoelectronics. Graphene's 2D honey-comb lattice sp^2 crystal structure possesses extremely high intrinsic charge mobility (250,000 cm^2/Vs), a high specific surface area (2,630 m^2/g), high thermal conductivity (5,000 W/mK), high Young's modulus (1.0 TPa and high optical transmittance (97.7%)). Methods of producing graphene include thermal annealing, unzipping nanotubes, solvothermal and thermal decomposition, ball-milling and chemical exfoliation and chemical vapour deposition (CVD). Chemical vapour deposition is a popular method for producing graphene from a variety of organometallics or other carbon sources using transition metal catalysts. However, conventional chemical vapour deposition can have a massive carbon footprint of over 600 tonnes of carbon dioxide per tonne of nanocarbon produced.

Another form of graphene synthesis is electrochemical exfoliation. As early as 2003, in an exploration of processes detrimental to Li-ion battery anodes, it was noted that electrochemical alkali ion intercalation could lead to peeling off thin layers from graphite and thin graphene-like exfoliated layers were clearly observed. By 2007 the observation of one-atom thick graphene layers by electrochemical exfoliation was observed. Electrochemical exfoliated graphene prepared from graphite is of increasing interest today and is often mechanistically interpreted as an anodic process in which intercalated ions between the graphite layers are oxidized, forming gases which break the weak interlayer Van de Waals bonds and release thin single- or multi-layered graphene sheets into the electrolyte. In 2017, it was observed that compression of graphite flakes prior to exfoliation, such as using graphite powder confined by wax coating, could increase the yield of graphene. Liu et al. use confined carbon nano platelets, rather than graphite, as a reactant to improve graphene yield. These ultra-thin platelets are prepared directly and inexpensively from carbon dioxide as the sole reactant by electrolysis in molten carbonate electrolytes, providing a ready incentive to transform and utilize this greenhouse gas.

Licht et al. recently reported a high yield, high purity, (low) electricity cost constrained, carbon nanotube synthesis carbon dioxide to carbon nanotubes (carbon dioxide to CNTs) (C2CNT). The carbon platelet synthesis introduced in this chapter is an outgrowth of that discovery. Carbon dioxide to carbon nanotube is briefly introduced here: carbon dioxide to carbon nanotube splits carbon dioxide by electrolysis in molten carbonates. Isotopic 13C tracking was used to follow the consumption of carbon dioxide as it is dissolved in molten carbonate and is split by electrolysis to form the building blocks of carbon nanotubes. Carbon dioxide dissolution in molten lithium carbonate is exothermic and rapid, which along with heat generated by the electrolysis provides thermal balance during carbon deposition on the carbon dioxide to the carbon nanotube cathode. Transition metal nucleated electrolysis in lithium carbonate forms carbon nanotubes, oxygen and dissolved lithium oxide:

$$\text{Electrolysis: } Li_2CO_3 \rightarrow C_{CNT} + O_2 + Li_2O \tag{12.1}$$

The lithium carbonate is effective either when pure or with added oxides, added sodium, calcium, or barium carbonates, or added boron, sulfuric, phosphorus, or nitrogen dopants. Carbon dioxide added to the electrolyte dissolves and chemically reacts with lithium oxide to renew and reform lithium carbonate:

$$\text{Chemical dissolution: } CO_2 + Li_2O \rightarrow Li_2CO_3 \tag{12.2}$$

Transition metals, such as nickel or chromium, nucleate carbon nanotube formation and can be variously added to the electrolyte, added to the cathode, or added by leach from the anode. The net reaction of Equations (12.1) and (12.2) is carbon dioxide split by electrolysis to carbon nanotubes and oxygen (Equation 12.3):

$$CO_2 \rightarrow C_{CNT} + O_2 \tag{12.3}$$

Rather than carbon nanotubes, carbon nanoplatelets are formed by molten carbonate electrolysis when other transition metals except zinc are excluded and facilitated by increasing the electrolysis current in a stepwise manner prior to the constant current electrolysis:

$$\text{Electrolysis: } Li_2CO_3 \rightarrow C_{platelets} + O_2 + Li_2O \tag{12.4}$$

The carbon platelets are converted to graphene by exfoliation:

$$\text{Exfoliation (DC voltage): } C_{platelets} \rightarrow C_{graphene} \tag{12.5}$$

The net reaction of Equations (12.4), (12.5) and (12.6) is carbon dioxide split by electrolysis into graphene and oxygen:

$$CO_2 \rightarrow C_{graphene} + O_2 \tag{12.6}$$

Wang et al. discuss the development and application of engineered multifunctional nanohybrids of carbon nanomaterials and metal/metal oxide nanoparticles that exhibit promising multifunctionalities for addressing the critical energy–water–environment (EWE) nexus.

12.7 The Ammonia Synthesis Reactor-Development of Small-scale Plants

Ammonia synthesis is one of the most energy-consuming processes, accounting for nearly 3% of the world energy consumption. Thus, every attempt to decrease energy demand has enormous impacts on resources, savings and globally on the environment. Different strategies are in place for this purpose through the study of novel cycles and materials. New catalysts are needed to further decrease the ammonia synthesis loop pressure, which is now one of the key issues for the development of small-scale plants, which would be better integrated with renewable-based raw materials (e.g., biomass) and energy sources (e.g., water electrolysis supported by renewable electricity). Furthermore, key process parameters, such as the synthesis loop pressure and the gap between the hydrogen synthesis section and the utilization in the loop, become paramount in the new methods of production. Therefore, high-pressure electrolysers should be selected to decrease the compression duty. In addition, the improvement of the thermal management within the loop and overall in the process would allow for considerable savings, especially when thermal energy utilization is envisaged (i.e., high-pressure steam production to support the compression duties). However, as part of a green agenda, green ammonia (a product of renewable processes) will increase its influence in the production of the chemical worldwide.

The ammonia synthesis reactor is the core of the ammonia synthesis loop. Different reactor space velocities are designed, ranging from 12,000 h at 15.0 MPa to 35,000 h at 80.0 MPa, with a temperature range of 620 and 800 K. Accordingly, the typical conversion-per-pass ranges from 12 to 16.5 vol% ammonia, rising to 19–19.5% in advanced processes. The adiabatic temperature raises these ranges to 447 and 505 K (up to 560°C for the high conversion processes). As mentioned, the heat of reaction should be removed, e.g., through interbed cooling and energy recovery. In the most optimized configurations, the energy recovery can be up to 93%.

Direct heat exchange is accomplished in the converter with cooling pipes inside the catalyst bed or with a shell and tube heat-exchanger reactor. Figure 12.3a describes a reactor with axial feed split across the catalyst bed and heat recovery at the outlet, while the option of a cooling system inside the catalyst bed and between consecutive catalyst layers is reported in Figure 12.3b, with a schematic of the expected temperature profile along the reactor.

The cold feed may be also used to quench the catalyst bed, using feed distribution nozzles or between two adjacent adiabatic beds. Otherwise, the catalyst bed can be operated as adiabatic, with the heat removed between the layers.

Specific materials are needed for the converter, to withstand the high pressure in the presence of hydrogen and nitrogen (i.e., embrittlement and nitridation). Alloy steels can be used below 673 K, whereas stainless-steel, Incoloy and Inconel materials, should be used for higher temperatures. To reduce investment costs, a cold-wall arrangement is often employed, as in Figure 12.1, with the hot catalyst bed contained in the internal cartridge, surrounded by the flow of the preheated feed (~573 K). In this way, the pressure difference across the cartridge wall is only a few bars, with the external fluid already being pressurized. The significant pressure difference between the external environment and the internal high-pressure medium is located across the external vessel wall, which is subject to a much lower temperature, thus needing a less expensive metal for its manufacture.

Formerly, the reactors were operated with axial flow and the design of the first radial flow reactor was a key innovation to decrease the pressure drop across the catalyst bed, thus allowing for a lower particle size of the catalyst, with higher effectiveness and practically no diffusional limitations. Mixed axial/radial configurations were instead proposed by Ammonia Casale, with an annular catalyst bed.

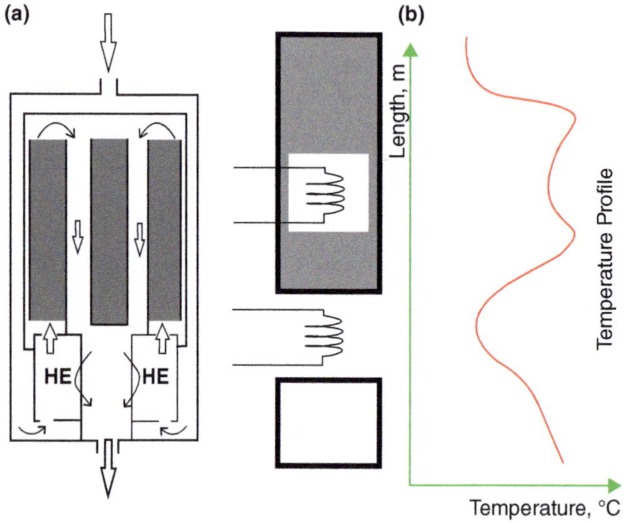

Figure 12.3 (a) Cold-wall axial flow converters with an internal heat exchanger (HE) and (b) qualitative description of intra- and interbed cooling, with the temperature profile (red) along the reactor length. Coils represent cooling heat exchangers and grey boxes represent the catalyst bed. Cooling can be achieved with water to produce high-pressure steam.

Further Reading

Advances in small modular reactor technology developments 2020 Edition. A Supplement to: IAEA Advanced Reactors Information System (ARIS). aris.iaea.org.

Anthony, D.W. (2021). The potential for small modular nuclear reactors to support the hydrogen economy (Unpublished Master's project). University of Calgary, Calgary, AB. https://prism.ucalgary.ca/bitstream/handle/1880/113936/capstone_Anthony_2021.pdf.

Licht, S., Douglas, A., Ren, J. et al. (2016). Carbon Nanotubes Produced from Ambient Carbon Dioxide for Environmentally Sustainable Lithium-Ion and Sodium-Ion Battery Anodes. *ACS Cent. Sci.* 2: 162–168.

Liu, X., Wang, X., Licht, G. and Licht, S. (2020). Transformation of the greenhouse gas carbon dioxide to graphene. *J. CO_2 Util.* 36: 288–294.

Pinsky, R., Sabharwallb, P., Jeremy Hartvigsenb, J. and James O'Brien, J. (2020). Comparative review of hydrogen production technologies for nuclear hybrid energy systems. *Prog. Nuc. Eng.* 123: 103317.

Small modular reactors: challenges and opportunities. (2021), Nuclear Technology Development and Economics, OECD, NEA No 7560, 2021.

Sun, Z., Ma, T., Tao, H. et al. (2017). Fundamentals and challenges of electrochemical CO_2 reduction using two-dimensional materials. *Chem* 3 (4): 560–587.

Valera-Medina, A. et al. (2021). Review on ammonia as a potential fuel: from synthesis to economics. *Energy Fuels* 35 (9): 6964–7029.

Wang, D. et al. (2019). Next-generation multifunctional carbon-metal nanohybrids for energy and environmental applications. *Environ. Sci. Technol.* 53: 7265–7287. doi: 10.1021/acs.est.9b01453.

Whipple, D.T., Finke, E.P. and Kenis, P.J. (2010). Microfluidic reactor for the electrochemical reduction of carbon dioxide: the effect of pH. *Electrochem Solid-State Lett.* 13: B109–B111.

13

Recycling of Waste Plastics → Plastics Circularity

CONTENTS

Converting Power into Chemicals and Fuels: Power-to-X Technology for a Sustainable Future, First Edition. Martin Bajus.

13.1 Introduction

The increasing demand for plastics for widespread applications has ultimately resulted in the accumulation of substantial plastic waste, which remains a concern due to limited efforts, inadequacy and environmental problems of conventional technologies for handling waste plastics remediation. The enhanced production of raw materials for polymer syntheses has a dual impact on our ecosystem by causing rapid depletion of non-renewable petroleum resources and waste generation. To address this situation, researchers have adopted advanced thermochemical recycling processes to produce intermediate products of the petrochemical industries including monomers, fuels and other value-added products. Such practices can potentially serve the purpose of a circular economy. This chapter aims to cover recent highlights in the field of waste plastics pyrolysis including critical observations from the past to provide precise understanding. Consequently, the reactivities and product distributions for plastic feeds, pyrolysis reactors, roles of catalysts and effects of operating parameters on reactivity and selectivity have been covered. Coprocessing of plastic waste with radioactive materials, biomass and heavy petroleum residue is also discussed. Furthermore, an overview on kinetics and mechanistic aspects of plastic pyrolysis is presented with a discussion on relevant analytical techniques. The applications of pyrolysis oil as a fuel or fuel additive are covered in a separate section. Lastly, comparisons of existing chemical recycling technologies, summaries of commercial operations and future projections are provided.

The contribution plastics have made in improving the living standards of our society are remarkable. The use of plastics as cost effective, lightweight and resistant alternatives in numerous applications have transformed human lives, of which the replacement of conventionally used hazardous metals and wildlife-derived materials is worth mentioning. A rapid growth of the world population has direct impact on rising demands of commodity plastics. The substantial development in this sector is further due to innovations in healthcare, electronics, construction, the automotive industry and packaging materials. Nevertheless, the worldwide contribution of plastic waste to overall municipal solid waste is about 10%–13%, mainly composed of polyethylene (PE) and polypropylene (PP) (together 75% of total plastic waste) polymers. As per the recent global plastic production report, the production of plastics was 368 million metric tonnes (MMT) in 2019, with approximately a 20% increase in last five years. As per the estimate, around 181MMT of plastic waste was produced globally in 2015. Alarmingly, the United Nations anticipates that the production of over 400 MMT of plastic waste since 2020 will be doubled in the next 15 years. The World Economic Forum predicts that by 2050 the plastics industry will have a 20% share of global petroleum consumption with 15% contribution to annual carbon emissions.

Considering the fact that almost 98% of synthetic plastics are produced from the chemicals derived from non-renewable resources such as crude oil, natural gas and coal, their intensive usage has an additional cost of non-systematic disposal and little recycling, which has a detrimental effect on lands, rivers, oceans, wildlife and humans. The major issue with synthetic plastics is their very low degradability and the percentage of plastic waste ending up in landfills is still very high. The processing of combinations of plastics with other materials is even more challenging. The garbage mixed with plastics interferes in conventional waste processing facilities and causes problems in landfill operations. Plastic waste leaches out toxic chemicals, like bisphenol A and phthalates, which can cause neurological disorders, cardiovascular disease, cancer, birth defects and hormonal disorders. Moreover, the plastic waste in marine systems can host other poisonous and polluting materials.

Mechanical recycling is used as an alternative to landfills to reduce plastics disposal; however, the percentage of recycling is still very low. Recycling of used plastics is an unattractive approach because of the constraints over water contamination and inadequate separation prior to recycling.

Furthermore, plastics must be separated according to their grades, colour and transparency. The demand of transparent plastics with greater flexibility causes lower market values for recycled dyed plastics. The process of mechanical recycling results in partial degradation of polymeric material, which requires optimization for mechanical properties and often fails to endure the strict regulations of marketing high value products. Incineration is an efficient way to reduce the waste volume and generate energy for heat or power consumption by direct burning of wastes in the presence of excess air at high temperatures. The incinerators reduce the quantity of original waste by 80%–85%. Thus, incineration could be preferred in countries like Japan due to the limited availability of land for landfill applications. However, the process contributes to air pollution and global warming and liberates hazardous chemicals such as dioxin and furans. To address these critical issues, sustainable alternatives to conventional mechanical recycling, landfills and incineration practices are being sought.

Increasing awareness of reducing the exploitation of fossil fuels prompted the development of energy alternatives including solar energy, wind power and geothermal and hydropower technologies. However, biofuels are considered direct substitutes for gaseous and liquid fuels with similar calorific values and have compatibility with existing automotive engines. Plastics are essentially composed of carbon and hydrogen as the repeated monomeric units, which can be either degraded into fuel-like hydrocarbons, monomers, or useful intermediates for chemical industries. The process is known as thermochemical recycling, which is broadly categorized into pyrolysis, gasification, depolymerization and upcycling. Thermochemical recycling provides a sustainable alternative to conventional technologies of waste plastics treatment, which also aids a circular economy. Such processes operate under flexible conditions, with low impacts on the environment and avoidance of intense sorting practices.

Pyrolysis is a reliable and conservational approach to serve its dual purpose by minimizing waste plastics and supplementing increasing energy demands from petroleum industries. The process operates at high temperature under anaerobic conditions. The major products of the pyrolysis process are oil, gas and solid residue. Pyrolysis is a flexible technique used to process a mixture of plastic feed and with an ability to process marginally contaminated plastic feed, which can also be optimized for high yields of certain products for desired applications. Thermal cracking of waste plastics produces liquid oil for boilers, furnaces, generators, agricultural pumps and more without requiring additional treatment. Remarkably, the combustion of gaseous by-products can satisfy the total energy requirements of the pyrolysis plant. Commercially, using good quality plastic feed, the pyrolysis process can be optimized to produce 80%–85 wt % of liquid oil or even more at a temperature around 500°C. Polymers such as polystyrene (PS), poly(methyl methacrylate) (PMMA) and polyethylene terephthalate (PET) have different chemical reactivities due to the presence of more labile chemical bonds and, therefore, can be depolymerized to obtain monomers or chemically upcycled by subsequent repolymerization in the case of polyethylene terephthalate to obtain virgin polymers. Another process, the gasification of waste plastics, produces a stream of hydrogen, carbon monoxide, carbon dioxide, methane and other gases at elevated temperatures. Gasification has even greater flexibility to handle different compositions of plastics or organic waste/biomass mixed feedstocks. This process extends the scope of waste plastics from energy to useful chemicals, as the produced syngas can be processed further to yield ammonia, methanol, fuel and more. Plasma pyrolysis is an advanced technology, based on severe cracking of all types of plastics including biomedical and hazardous wastes by heated plasma in a secure and sustainable manner. A recent Energy, Economy and Environmental index (EEE index) study suggested pyrolysis and liquefaction of plastic waste is the second-best practice after the exercise of integrating it into road pavements and concrete structures.

Businesses and governments around the world have united behind the vision of a circular economy for plastics. Realizing this vision calls for industry-scale innovation to ensure that plastics are reusable, recyclable, or compostable.

Among the many obstacles to overcome is the need to maintain high polymer quality over multiple recycling loops, which is critical to the viability of plastics recycling and to increasing the use of recycling. Demand from brand owners for high-quality recycled polymers remains high. This is in combination with their willingness to reduce plastic waste and use it as future raw material. Therefore, enormous efforts are going into technology, process and product development to achieve these key targets.

13.2 Mechanism Aspects of Waste Plastic Pyrolysis

The clarification of the reaction mechanism and the kinetics of waste plastics pyrolysis offers advantages in terms of understanding structural characteristics, effects of various parameters and product distribution. Recent advances in ascertaining losses of volatiles during polymer degradation and subsequent characteristic analysis involved coupling of two or more conventional technologies. These advanced techniques provide complete information on volatiles, including unstable products and even intermediate species, which is otherwise unfeasible. In this regard, thermogravimetric analysis (TGA) coupled with Fourier transform infrared (FT-IR) offers potential advantages for providing non-destructive and real-time analysis data of volatiles in complex mixtures. A combined study of weight loss and kinetic parameters obtained from thermogravimetric analysis with simultaneous detection of functional groups in volatile species by Fourier transform-infrared provides significant understanding of the mechanism and kinetics of the reaction. Unfortunately, TG-FTIR is not effective in identifying specific volatile substances. Meanwhile, quantitative pyrolysis coupled with more sophisticated characterization techniques such as gas chromatography-mass spectrometry (GC/MS) reveal chemical structural attributes of complex feeds in addition to unique features of volatile species. Similarly, TGA coupled with GC/MS has also been used to identify the chemical formula/structure of volatile products. Undoubtedly, these advanced techniques can provide rapid solutions to process design and optimization, catalyst evaluation, choice of feed and composition and understanding synergetic effects during copyrolysis. During TGA analysis, the weight loss patterns at different heating rates determine the degradation kinetics of plastic pyrolysis.

The degradation of plastics typically consists of:

i) initiation: homolysis of C–C, C–H, or C–X bonds to generate free radicals;
ii) propagation: β-scission, inter- and intramolecular radical migration, radical addition and hydrogen transfer; and
iii) termination: radical–radical coupling/ recombination.

Initial cracking of polyolefins yields long-chain hydrocarbons and subsequent gas-phase degradation and an increase in temperature provides lighter hydrocarbons. Monomer fractions formed this way, including ethylene and propylene, undergo cycloaddition–dehydrogenation to form benzene and its derivatives. Furthermore, even higher temperatures (>750°C) or longer residence times favour the formation of gaseous products such as hydrogen, methane and other light hydrocarbons with solid char and polyaromatic hydrocarbons (PAHs).

In this chapter we have adopted an "all inclusive" approach to impart holistic knowledge by extracting the essence of fundamentals of the pyrolysis process and recent developments. Such a comprehensive version is aimed at assisting researchers in performing quality research in this field. The process optimizations in waste plastics pyrolysis, including pre-treatment of feed, selective depolymerization and decomposition mechanisms of different plastics, are discussed.

A detailed overview of reactor designs for plastics pyrolysis is concluded with a real-time assessment of reactor types in terms of scope and selectivity. The impacts of variable process parameters such as feed type, temperature, pressure, reaction medium and catalysts on pyrolysis are well described. Pyrolysis of plastics mixed with radioactive materials is specifically covered alongside discussions on synergetic effects and challenges of copyrolyzing plastics with biomass and heavy petroleum fractions. Furthermore, real-time applications of pyrolysis products as fuels or fuel blends and kinetics of pyrolysis are examined. More importantly, comparisons of existing chemical technologies, details of technology providers for pyrolysis and future projections of major companies across the world are summarized.

Based on the characteristics of streams, plastic waste is broadly categorized into post-industrial and post-consumer. Post-industrial plastic waste is generally clean, meaning it is of consistent quality with a defined composition; however, consumer plastic waste is dirty and contaminated with other materials including organic waste, biomass, metals and more. Although synthetic plastics have very low degradability under normal conditions, chemical reactivities, for example, aerobic oxidation, partial hydrolysis of certain polymers and thermal and photochemical decomposition of polymer resins, essentially depend on their compositions. For these reasons, post-consumer waste is not ideal for mechanical recycling and requires thorough cleaning, separation and sorting. On the contrary, pyrolysis of plastics allows such plastic waste to be valorised into valuable oligomers, monomers and other molecules with a broader range of feed type. The pyrolysis process tolerates considerably higher levels of contaminants in the feed and, therefore, requires fewer pre-sorting steps.

The structure and functional groups in individual resins (Figure 13.1) define their reactivity and product distribution. However, the processing of plastic waste is generally reported in terms of proximate analysis, which is usually carried out by analysing the fixed carbon, volatile matter, moisture content and ash content of feed. The high volatile fraction and low ash content of plastic waste are responsible for high yields of liquid oil during pyrolysis. More specifically, with the presence of saturated carbon chains in polymer skeletons and as a more hydrogen-rich feed, most of

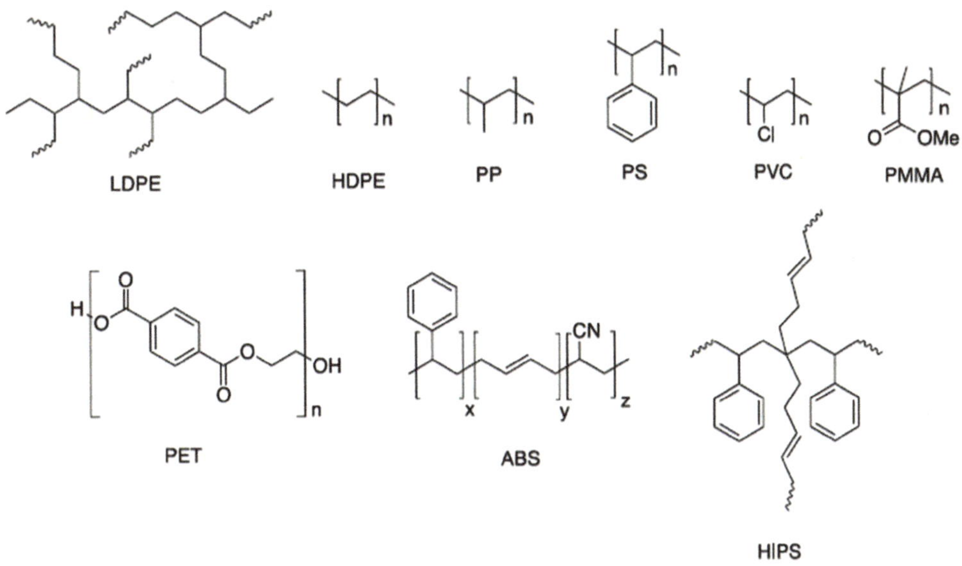

Figure 13.1 Chemical structures of synthetic polymers.

these polymers pyrolyze with a low char/carbon formation compared to petroleum crude, which generally contains polyaromatics and sulfur/metal species. The variability in proximate analysis data for the same polymer might be attributed to adoption of different analytical methods and combustion parameters, degree of deformation/contamination, influences of polymer grades and additives in decomposition mechanism and feed pre-treatment. Product distribution of liquid streams during pyrolysis of different plastic feeds depends on the structure and chemical composition of polymeric materials. The presence of labile/functional groups in polymeric chains or at polymeric skeletons favors depolymerization to produce monomers, as in the cases of PET, PS and PMMA. However, the polymers such as PP and PE provide oligomeric or high-molecular weight products under typical pyrolysis conditions. The factors affecting the reactivity and selectivity in pyrolysis of such polymers are discussed in the following sections.

13.2.1 Polyethylene and Polypropylene

The mechanisms of thermal and catalytic degradation of polyolefins are shown in Figure 13.2. Thermal pyrolysis proceeds via C–C bond homolysis and subsequent free-radical generation, whereas the presence of a catalyst favours C–H activation. The presence of acidic/basic sites in a catalyst promotes the formation of ionic intermediate species. It is also believed that the imperfections in polymeric chains are primarily responsible for degradation. Due to different activation energy barriers, inter/intramolecular hydrogen transfer and β-scission pathways show prominence in the influence of the reaction temperature. The higher energy barrier for β-scission is accountable for increased monomer yields at elevated temperatures.

The macroscopic mechanism of thermal degradation of polyolefins was explained by Murata and co-workers, wherein random and chain-end scissions of C–C bonds take place simultaneously. While the former results in generating high molecular weight hydrocarbons, the latter causes the formation of volatile products. Chain-end scission occurs at gas–liquid interfaces of liquid reactants and gaseous products. Consequently, the reaction pressure determines the degradation behaviour of polyolefins.

The absence of any functional groups or heteroatoms in the polymeric structure make it ideal for the recovery of fuel and petrochemical intermediates with up to 95% overall yield of liquid products. The optimized temperature for the highest liquid yield was 550°C, whereas at 600°C more

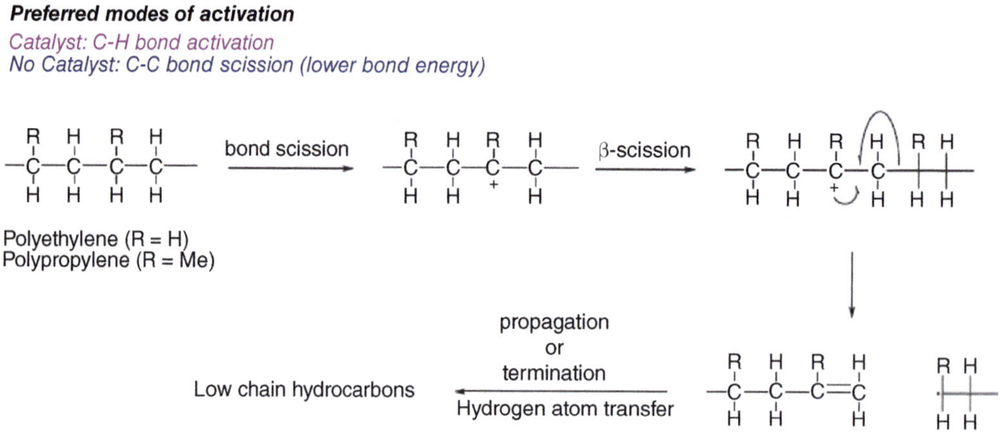

Figure 13.2 Mechanism of pyrolysis of polyolefins: LDPE, HDPE and PP.

cracking resulted in gaseous product formation. The formation of char was highest at 450°C. The major products in pyrolysis oil were alkanes, alkenes (1-alkenes) and aromatics, indicating fuel quality. The gas chromatographic analysis of the gaseous fraction revealed the presence of hydrogen, carbon monoxide, carbon dioxide and C_1–C_5 hydrocarbons. There have been studies of pyrolysis of LDPE at temperatures ranging from 203 to 400°C. The liquid products were obtained in maximum 78.4% yields, followed by 17.2% of solid residue and 4.4% organic gases. The hydrocarbons were mainly aliphatic in the range between C_{10}–C_{27}. High-temperature pyrolysis of LDPE resulted in improved selectivity towards light hydrocarbons.

In summary, the presence of tertiary carbons due to branching and weaker intermolecular forces offers low-temperature degradation of polymeric structures in LDPE. Nonetheless, pyrolysis oil from LDPE degradation shows good resemblance to conventional fuels. The presence of waxes is still observed during thermal pyrolysis in some cases besides the high abundance of diesel-grade hydrocarbons. However, catalytic cracking can improve the selectivity towards gasoline by secondary cracking of long-chain hydrocarbons or jet fuel by dehydrogenation/cyclization/aromatization of cycloalkanes.

Polypropylene is the most widely produced commodity plastic, followed by HDPE and LDPE, all of which contribute much to plastic waste. The demand of PP is increasing day-by-day andresearchers are highly concerned about pyrolysis of polypropylene. The presence of a tertiary carbon in the structure due to the presence of a methyl group results in different reactivity of polypropylene compared to other polyethylene.

Kim and co-workers performed activator-assisted pyrolysis of polypropylene in a continuous two-stage process, consisting of auger and fluidized bed reactors. Basically, the auger reactor elevated the vibrational states of molecules being fed to the fluidized bed reactor. The influences of temperatures in the auger reactor and bubbling zones, fluidizing medium and residence time in the fluidized bed reactor were studied. The reaction conditions were tuned to achieve high yields of gases (81 wt %; mainly of hydrogen and methane), olefins (52 wt %; ethene, propene, 1,3-butadiene and butenes), or pyrolysis oil (33 wt %; containing 92 wt % aromatics). The activation energies of activated molecules in the auger reactor were approximately 17 kJ/mol lower than those of unactivated molecules.

In another approach, pyrolysis of polypropylene was conducted under a helium–oxygen atmosphere in a fixed bed reactor at 250°C for 30 min. The activation energy was considerably lower than that for the anaerobic medium and oil and wax were formed in 76 and 6.0 wt % yields, respectively. The presence of oxygenated hydrocarbons (C=O, C–OH) was also observed.

Nevertheless, the recovery of a liquid product could be maximized at temperatures below 500°C under thermal conditions. A further increase in reaction temperature resulted in enhanced selectivity towards gaseous products, which could otherwise be achieved in the presence of acidic and high surface area catalysts even at less than 400°C. Notably, thermal pyrolysis at less than 350°C resulted in partial degradation of the polymer and/or showed a high presence of waxes in products.

13.2.2 Polyethylene Terephthalate

Pyrolysis of polyethylene terephthalate was studied by several researchers, wherein the gaseous products formed were higher than the liquid fractions. The product analysis further revealed the formation of benzoic acid as a major liquid product. The corrosive properties of benzoic acid and a tendency to sublime causing clogged piping and heat exchangers, could potentially damage the reactor. The presence of benzoic acid and derivatives contributes to the total acid number (TAN)

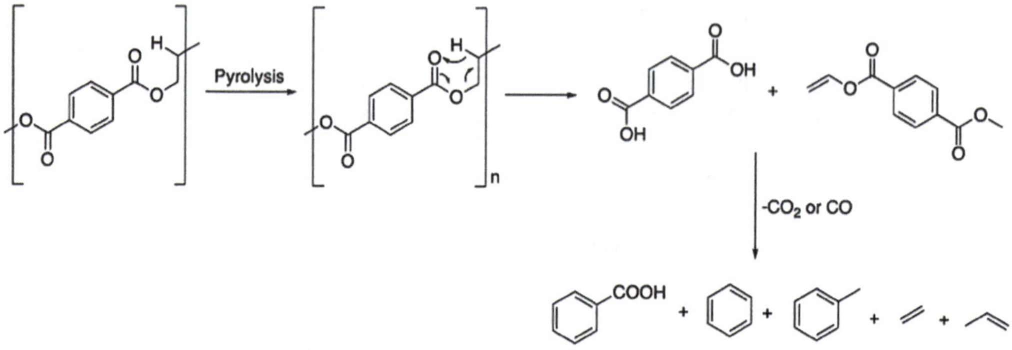

Figure 13.3 The mechanism of thermal degradation of polyethylene terephthalate.

of fuel and therefore, they are not desirable components. However, a catalyst can promote decarboxylation, which effectively reduced the total acid number and increased the yields of monoaromatics. The mechanism of thermal degradation of PET is shown in Figure 13.3.

A lower bond dissociation energy of the C—O bond and the presence of hydrogen at the α-position of carbon facilitate the formation of phthalic acid and its vinyl ester derivatives. Decarboxylation/decarbonylation of carboxyl groups provide benzoic acid, benzene and toluene as major liquid products.

Pyrolysis of PET at less than 500°C provide high yields of liquid products; however, this fraction consists of significant amounts of oligomers and/or carboxylates including benzoic acid. The presence of oxygenates in the oil fraction results in a high TAN number and is thus it is unsuitable for fuel applications. Although decarboxylation of carboxylic acid derivatives and cracking of oligomers can be promoted at higher temperatures or under catalytic conditions, the formation of gaseous products (50%–77%) is considerably higher than liquid oil (23%–39%). Consequently, thermal decomposition of PET is less energy efficient than polyolefins. Nevertheless, chemical upcycling of PET to monomers or value-added chemicals by reacting with alcohol/amine might be an attractive solution.

13.2.3 Polyvinyl Chloride

The mechanism of thermal degradation of polyvinyl chloride is shown in Figure 13.4. The presence of thermally labile segments and defects in polymeric chains, random allylic chlorine atoms or tertiary chlorine atoms, partial oxidation of PVC resulting in the formation of ketoallylic chlorine atoms, internal double bonds and head-to-head structures were proposed as the causes for the high instability of PVC.

Pyrolysis of PVC is not a preferred route due to very low liquid yields (12%–13%), even after minimizing the formation of chlorinated hydrocarbons through two-stage pyrolysis. Furthermore, the recovery of harmful gases including hydrochloric acid and further processing of chlorinated compounds requires additional set-ups. Nevertheless, the exploitation of carbon dioxide as a

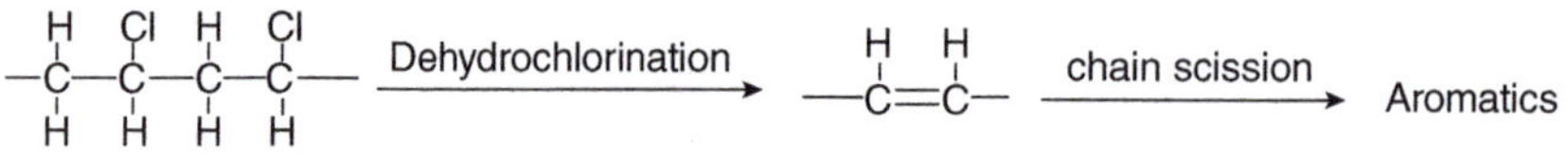

Figure 13.4 The mechanism of pyrolysis of polyvinyl chloride.

reaction medium in the presence of metal oxide/carbonate catalysts offers better control of reactivity and selectivity by suppressing the formation of chlorinated hydrocarbons and PAHs.

Pyrolysis of PVC was performed in a carbon dioxide medium for in situ mitigation and/or reduction of harmful chemical species. The study was carried out in a TG analyser and a tubular quartz reactor, which suggested that the formation of benzene derivatives and PAHs could be controlled at temperatures above 480°C. The formation of carbon monoxide at higher temperatures (>600°C) was observed, suggesting a reaction between carbon dioxide and intermediate species from PVC pyrolysis. In the presence of calcium carbonate, the formation of carbon monoxide was enhanced up to 12 times, whereas the selectivity towards benzene derivatives and polyaromatic hydrocarbons was reduced. This study claimed the possible utilization of carbon dioxide for more useful energy applications by controlling the reactivity during PVC pyrolysis and the amount of air pollutants. Similarly, PVC was pyrolyzed in a carbon dioxide medium at different temperatures using a Fe_3O_4 (magnetite) catalyst. The GC-TOF analysis suggested enhanced thermal decomposition of intermediate hydrocarbons leading to a higher production of syngas.

13.2.4 Polystyrene

Nisar and co-workers pyrolyzed polystyrene waste in the presence of a 2 wt % cuprous monoxide catalyst at 400°C, which provided the highest yield of liquid (87.34%) with low amounts of gases (10.14%) and tar (2.52%). The product distribution was also studied using GC-MS and FTIR. In another effort, the yield of the liquid fraction was increased up to 90.11% in the presence of a natural volcanic tuff catalyst. The GC-MS and FTIR analyses showed the presence of up to 55.62% styrene monomer in the liquid fraction.

The mechanisms of thermal- and acid- and base-catalysed pyrolysis of polystyrene are shown in Figure 13.5. The mechanistic investigations and experimental studies suggested that base catalysed pyrolysis is more selective to the formation of styrene, whereas relatively higher yields of benzene, toluene, xylene and ethylbenzene are observed in the presence of acid catalysts. Thermal homolysis of C—C bonds results in the formation of polymeric radicals; as a result, the bond energies of C—C and C—H bonds adjacent to radical centres reduces by three times the actual bond energies, i.e., 346.9 and 409.6 kJ/mol, respectively. Controlled unzipping of polymeric chains produces a styrene monomer, whereas other processes such as H abstraction, β-decomposition and radical—radical couplings provide benzene, toluene and other alkylated aromatics. Base-catalysed abstraction of protons from polymeric chains results in a straightforward formation of styrene by spontaneous movements of bonding electrons. On the other hand, the protonation of aromatic rings in acid catalysis at β-carbon provides stable 3° carbocation, which undergoes β-scission to provide styrene. However, the reaction is less selective, as the spontaneous cleavage of less stable 2° carbocation at the β-position is feasible under high-temperature conditions.

Pyrolysis of polystyrene can potentially provide the highest liquid yields (up to 99 wt %) under optimized conditions, i.e., less than 500°C. From the product distribution point of view, styrene is obtained as a major product under thermal or catalytic conditions followed by ethylbenzene and BTX. Although a high recovery of styrene could be achieved in the presence of a base catalyst, the subsequent addition of stabilizers and low-temperature storage are required. In this regard, some efforts have been made to improve the selectivity towards BTX and ethylbenzene.

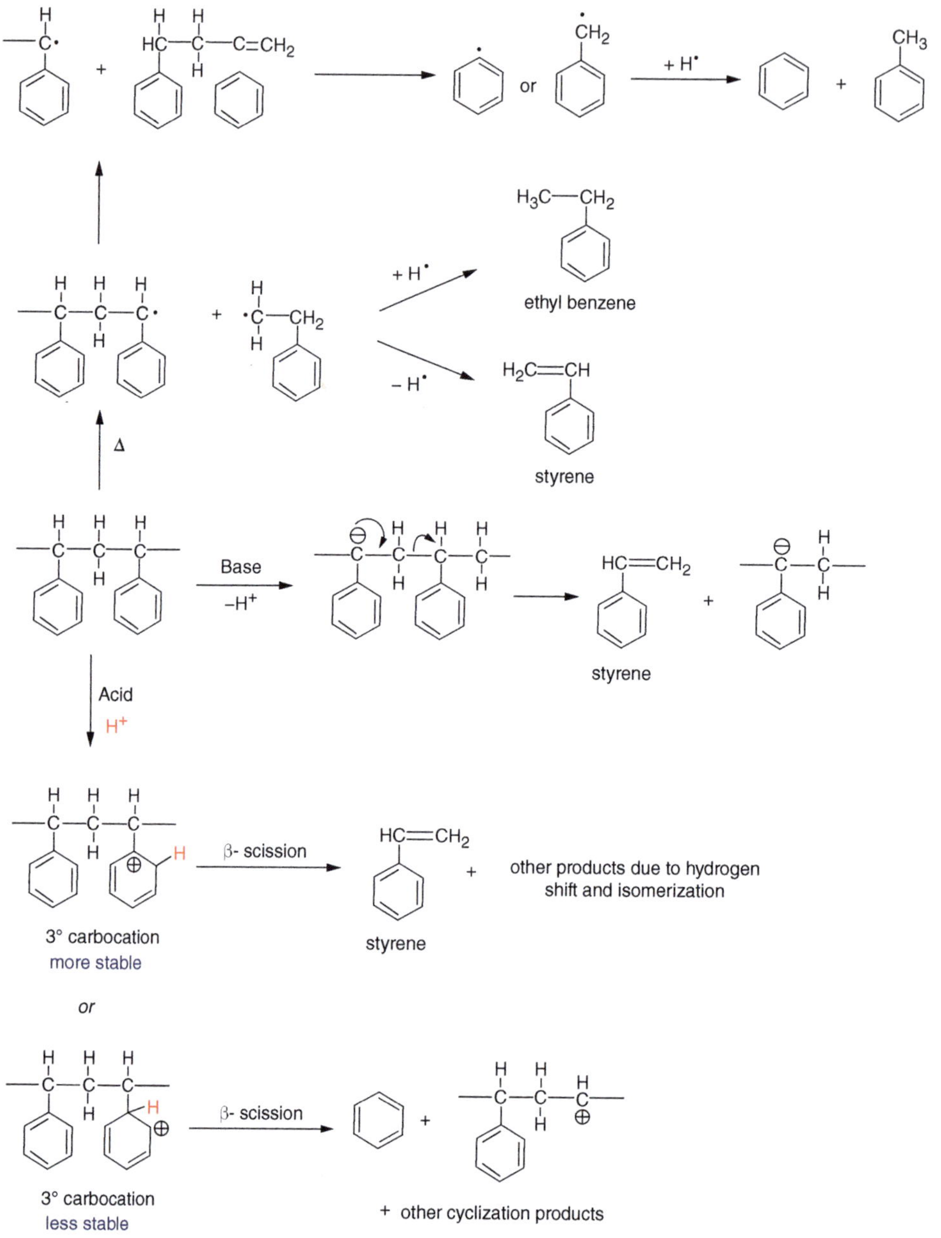

Figure 13.5 Mechanistic pathways of thermal- and acid- or base-catalysed pyrolysis of polystyrene.

13.2.5 Poly (Methyl Methacrylate)

Poly (methyl methacrylate) (PMMA) has high tensile and flexural strength, transparency and good UV resistance. PMMA is used as an alternative to polycarbonate, which does not contain bisphenol A units like polycarbonates.

The mechanism of thermal degradation of PMMA is shown in Figure 13.6. Noticeably, the predominance of chain-end scission results in high monomer selectivity. Selective depolymerization of PMMA by chain-end scission provides 98% selectivity towards methyl methacrylate (MMA) at

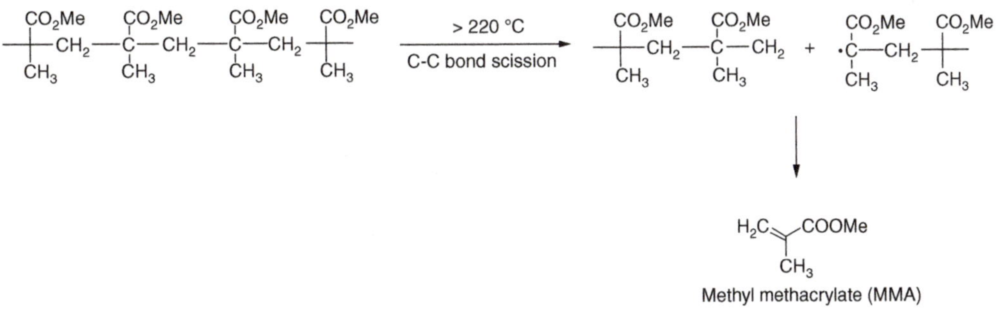

Figure 13.6 The mechanism of pyrolysis of poly (methyl methacrylate).

less than 450°C; however, an increase in reaction temperature results in enhanced recovery of gaseous products.

The presence of a carboxylate functional group at alternate carbons in the polymeric chain facilitates depolymerization of PMMA by selective C–C bond scission, which produces methyl methacrylate as the major product.

13.3 Kinetics

Kinetic models based on the data obtained from analytical techniques are used to determine the activation energies (E_a) of polymer degradation. Thus, E_a for LDPE and HDPE range between 163 and 303 kJ/mol and 206 and 445 kJ/mol, respectively. Also, E_a was reported to be 83–310 kJ/mol for PS and 136–267 kJ/mol for PVC.

Although pyrolysis of common polymers follows first-order reaction kinetics, the degradation behaviour of PVC is distinct and thus more studied. Pyrolysis of PVC typically occurs in three stages:

i) dehydrochlorination in the range of 200–340°C with a mass loss of about 40;
ii) reforming reactions including cracking, isomerization, cross-linking and aromatization at 360–530°C for another 20% mass reduction; and
iii) thermal cracking of larger molecules into low-molecular weight compounds at 530–770°C to reduce mass by another 8%.

Also, increases in pyrolysis peak temperatures were observed at higher heating rates, which also resulted in a reduction of solid residue. For commercial applications, kinetic modelling of mixed plastic waste is a critical factor in establishing the reactor design and operations. However, understanding cross kinetic degradation behaviours as well as having a knowledge of kinetics for individual components is quite challenging. Interestingly, a delayed degradation is observed for polymer mixtures compared to pure polymer samples during TGA studies. The synergistic effect in mixed blends depends on the miscibility of individual polymers, which also alters the degradation mechanisms. In another case, the thermal decomposition of PVC was facilitated in the presence of PS or PE. Accordingly, the kinetic analysis data for divergent reaction mechanisms, nonidentical reaction rates and different temperature dependencies can be utilized in designing unique methods for a controlled and selective pyrolysis of mixed plastic waste by effectively tuning or avoiding the interactions between different polymers/intermediate species.

13.4 Catalysts

The purpose of using a catalyst in a reaction is to provide an alternative pathway requiring low activation energy. The chemical interactions in the formation of intermediate species or transition states are responsible for this phenomenon. The catalyst is regenerated in a reaction and, therefore, can theoretically be used countless times. However, some secondary reactions in a reactor result in the deactivation of the catalyst. Heterogeneous catalysts are usually preferred due to easy catalyst separation and low or no contamination in the product stream but at the expense of lower activity than homogeneous catalysts. The reaction temperature is significantly reduced in pyrolysis of plastic waste, which is an endothermic process. The use of a cost-effective catalyst can save energy and reduce the operating cost. Moreover, the catalytic reactions generally operate through different mechanisms, which can significantly affect the product distribution. In pyrolysis, higher cracking of intermediate oligomers results in the formation of low-molecular weight hydrocarbons and gaseous products with properties like conventional fuel. Heterogeneous catalysts used in pyrolysis of plastic waste are conventional acid–base salts and oxides (ZnO, MgO, $CaCO_3$, $CuCO_3$, SiO_2, Al_2O_3), microporous natural zeolites, mesoporous and mixed synthetic zeolites, amorphous silica–alumina, MCM-41, kaolin, CaC_2 and metal supported on carbon and simple oxides.

13.4.1 Zeolites

Zeolites are the most widely used catalysts in refining industries, which seems true in the case of catalytic pyrolysis as well. A zeolite framework is typically an aluminosilicate made of silica and alumina linked by oxygen atoms at the tetrahedral sites. The ratio between silica and alumina is responsible for its acidity and, therefore, determines the catalytic activity as well as the product distribution. An increase in alumina content results in higher acidity of zeolite, which causes more cracking of polymeric chains to produce lighter hydrocarbons and gases including olefins and naphthene. On the contrary, the use of zeolite with more silica content provides long-chain hydrocarbons, having more resemblance to diesel, base oil and waxes. The catalytic performance of HUSY in pyrolysis of PP after thermal pre-treatment at different temperatures was studied using photoionization time-of-flight mass spectrometry (Py-PITOFMS). In this case, the catalytic activities of HUSY pre-treated at 200 and 800°C were lower than those treated at 350–650°C; however, the yields of alkenes, dienes and aromatics were higher.

Olazar and co-workers studied the effects of zeolite catalysts on selectivity in HDPE pyrolysis. The shape selectivity and higher acid strength of a HZSM-5 zeolite disfavoured secondary reactions including hydrogen transfer and radical coupling, which resulted in a subsequent increase in light olefins and monoaromatics with low coke formation. To the contrary, the hydrogen transfer reactions were favoured in the case of HY and Hβ zeolites, resulting in higher paraffinic products. The catalysts were regenerated by combustion of coke with air below 550°C. To understand the secondary reactions of volatile products, thermal and catalytic pyrolysis of LDPE were performed in a two-stage reactor at 300–600°C with residence times of 17.7–53.0 s. Consequently, high volatilization temperatures and/or residence times were required to obtain lighter hydrocarbons in thermal pyrolysis. The yields of liquid products were higher in the presence of a commercial HY zeolite catalyst at low temperature.

13.4.2 Fluid Catalytic Cracking Catalysts

A fluid catalytic cracking catalyst is made of an acidic silica–alumina matrix and a binder (mostly clay) and zeolite Y crystals. Thermal stability and high product selectivity of zeolite Y prove it to be an ideal catalyst for cracking heavy oil fractions from refineries into gasoline and liquid petroleum gases fractions. The quality of the fluid catalytic cracking catalyst is crucial to produce high-quality hydrocarbons; therefore, the catalyst with a somewhat diminished activity is often rejected from fluid catalytic cracking units and is called a "spent fluid catalytic cracking catalyst". Fortunately, this spent fluid catalytic cracking catalyst is good enough to crack the polymer chains into their oligomers with the properties of waxes or even liquid fuel.

Olazar and co-workers used a commercial fluid catalytic cracking catalyst for catalytic pyrolysis of HDPE in a conical spouted bed reactor under continuous operation. The optimizations of catalyst steaming and pyrolysis temperature were studied to maximize the diesel-like fraction. Pyrolysis at 500°C provided mostly gaseous products (52 wt %) with a light liquid fraction (35 wt %) and the lowest number of diesel-grade hydrocarbons (13 wt %). However, at a pyrolysis temperature of 475°C the mildly steamed catalyst showed higher selectivity towards the diesel-like fraction (69 wt %) with reduced yields of the liquid fraction (22 wt %).

Amorphous silica–alumina catalysts are high surface area catalysts having Lewis acid sites due to the presence of alumina, Brønsted acid sites due to protonated oxygen sites and electron accepting low-valent aluminium sites. The mole ratio of silica and alumina contents affects the acidity of catalyst, which is opposite to zeolite catalysts. The major difference in such phenomenon is due to the increase in Brønsted acid sites of silica–alumina catalysts, which is not prevalent in a zeolite structure. Therefore, a high SiO_2/Al_2O_3 ratio responds to high acidity of the catalyst. The activity of a silica–alumina catalyst is very high due to the morphosis (high surface area) and high Brønsted acidity; however, the same properties are accountable for low thermal and chemical stabilities of the catalyst.

Park and co-workers studied the catalytic pyrolysis of PS over natural clinoptilolite zeolite (HNZ), HZSM-5 and silica–alumina catalysts at 400°C. The activity of the HNZ catalyst was comparable to HZSM-5 in terms of C_5–C_{12} selectivity, having styrene as a major product. However, mesoporous silica–alumina showed higher selectivity towards ethylbenzene over styrene. An increased selectivity towards styrene formation at higher degradation temperature during HNZ-catalysed pyrolysis was correlated to lower yields of ethylbenzene, propyl benzene and styrene dimers.

In the petroleum industry, clays were explored as solid acid catalysts prior to the finding of zeolites. As a cost-effective alternative, the modifications of clays by acid treatment, selective Si/Al leaching, or pillaring are continuously practiced. The use of pillared clays in pyrolysis is more attractive due to the formation of two-dimensional networks in a more uniform structure than in parent clays. The exposure of aluminium sites is responsible for Lewis acidity, whereas protonated oxygen of Si–O sites provide Brønsted acidity to clay. The process of calcination after the chemical treatment of clay provides additional stability, which can be critical to high-temperature reactivity during pyrolysis. An interesting comparison between natural clays and ultrastable (US)-Y zeolite was made by Manos et al., wherein saponite, montmorillonite and their modified pillared clay catalysts performed better for pyrolysis of PE at high temperature and provided higher yields of liquid products with low coke formation.

In a different approach, the catalytic performance of phosphoric acid-activated carbon in pyrolysis of waste polyethylene was studied, wherein a high phosphorus content and increased residence time enhanced the recovery of liquid products. Further, Brønsted acidity of a phosphoric acid moiety promoted hydrogen transfer and P=O sites promoted dehydrogenation reactions resulting in improved yields of aromatics.

Among these catalysts, the fluid catalytic cracking catalyst is most suitable for producing pyrolysis oil from waste plastics feed. Moreover, the use of "spent FCC catalyst" alone or in combination with a fresh catalyst or other catalysts is an economically viable option.

13.5 Parameters Affecting Pyrolysis

The influences of operating parameters such as temperature, residence time, pressure, reaction medium and copyrolysis on productivity and selectivity are imperative and therefore require primary attention for the success of any pyrolysis reaction.

13.5.1 Type of Plastic Feed

The influences of molecular characteristics such as bond energies, polymeric structure and crystallinity are most prominent on pyrolysis behaviours of different plastics. The effect of plastic waste types on the quality and quantity of pyrolysis oil was studied by Miandad et al. who performed the reaction on individual polymers (PE, PP, PS and PET) at 450°C for 75 min in a pilot-scale batch reactor. Interestingly, high-molecular weight PE provided waxy products, whereas others were converted into pyrolysis oil.

Further, mixing PS and PP with PE (2:1:1 ratio) provided liquid products due to supplementary activation of long polyethylene chains by polystyrene derived free radicals. The presence of aromatics as a major component was observed in all cases, wherein polystyrene provided over 95% aromatics including styrene, toluene and ethylbenzene. Similarly, polypropylene produced propyl benzene as the major component along with ethylbenzene, benzene, alpha methyl styrene, xylene, naphthalene, biphenyl and more. The formation of differently substituted benzene derivatives was observed in the case of mixed plastics due to secondary reactions. The comparisons of fuel properties and typical oil yields are provided in Table 13.1.

The calorific values and densities of polyolefins/PS resembled conventional fuel. The research octane number and pour point values of HDPE-, PP- and PS-derived oil were obtained under acceptable limits for gasoline and diesel, respectively. However, viscosities of pyrolysis

Table 13.1 Impact of plastic type on fuel properties and yields of pyrolysis oil.

Physical Properties	LDPE	HDPE	PP	PS	PET	PVC	Diesel	Gasoline
Caloric value (MJ/kg)	39.5	40.5	40.8	43.0	28.2	21.2	43.0	c. 42.5
API gravity	47.75	27.48	33.03	n.r.[a]	n.r.	38.98	38.0	55.0
Viscosity (mm^2/s)	5.56[b]	5.08[c]	4.09[c]	1.4[d]	n.r.	6.36[e]	1.9–4.1	1.17
Density@ 15°C (g/cm^3)	0.78	0.89	0.86	0.85	0.9	0.84	0.807	0.78
Ash (wt %)	0.02	0.00	0.00	0.006	n.r.	n.r.	0.01	–
Octane number MON (min)	n.r.	85.3	87.6	n.r.	n.r.	n.r.	–	81–85
Octane number (RON) (min)	n.r.	95.3	97.8	90–98	n.r.	n.r.	–	91–95
Pour point (°C)	n.r.	−5	−9	−67	n.r.	n.r.	6	–
Flash point (°C)	41	48	30	26.1	n.r.	40.0	52	42
Aniline point (°C)	n.r.	45	40	n.r.	n.r.	n.r.	77.5	71
Diesel index	n.r.	31.05	34.35	n.r.	n.r.	n.r.	40	–
Typical yields, (%)	75–93	75–91	70–92	90–97	23–39	12–13	–	–

[a] n.r. = not reported; [b] At 25°C; [c] At 40°C; [d] At 50°C; [e] At 30°C

oil derived from different plastic feeds, except for PS, were out of range. Also, flash points for PS- and PET-derived oil were lower than conventional fuel. The liquid fractions obtained from polyethylene terephthalate and polyvinyl chloride pyrolysis were not suitable for fuel applications. Although PS-derived oil satisfied almost all major requirements, the aromatic content was much higher than the acceptable limit (35 vol%). Nevertheless, relatively higher viscosities of pyrolysis oil restrict its application as a fuel or high blend. Typical liquid yields were in the order of PS > PP ≃ LDPE > HDPE > PET > PVC. Generally, pyrolysis at 450–500°C provided high yields of liquid products in cases of polypropylene, polyethylene and polystyrene. However, the concentration of oxygenated species in PET-derived oil is reduced under higher temperature (>600°C) or catalytic deoxygenated conditions. In contrast, two-stage pyrolysis of polyvinyl chloride feed restricts the formation of chlorinated hydrocarbons and has excellent hydrochloric acid recovery.

13.5.2 Temperature and Residence Time

Being one of the most significant parameters in plastic waste pyrolysis, temperature plays a decisive role in controlled cracking of polymeric chains. The degradation of plastics mainly consists of initiation, propagation and termination steps. Thermal cracking of polymers typically proceeds with homolysis of C–C, C–H, or C–X bonds once the supplied energy becomes higher than their bond enthalpies. The formation of free radicals is followed by inter- and intramolecular hydrogen/alkyl group transfer, elimination and radical–radical coupling processes to yield high-molecular weight products. A further increase in temperature promotes additional cracking to provide lighter hydrocarbons and aromatics. In the case of polyolefins, an increase in residence time results in coke formation along with gaseous products such as methane and hydrogen. Temperature around 800°C is considered ideal for the gasification process, which provides hydrogen, methane, carbon monoxide, carbon dioxide and light hydrocarbons.

A thermogravimetry analyser is used to measure thermal degradation behaviour of the plastics feed, which produces TG and DTG curves. The operating temperature is strongly dependent on the chemical composition and reactivity of the plastics feed. Similarly, pyrolysis at lower temperatures (350–500°C) results in the formation of liquid products, whereas the yields of gases and/or solid residue are increased at high temperatures (>500°C). Thermographic analysis of zeolite US-Y-catalysed degradation of HDPE was performed by Lin and co-workers. The study concluded that deposition of coke resulted in deactivation of catalyst. Hu and co-workers derived a relationship between the reactivity and compositions of waste plastics feed. The decomposition of mixtures mixed plastics from tyres occurred at different temperatures in a delayed pyrolysis process. Further studies showed the synergy between polystyrene and waste tyres, in contrast to the negative effects in cases of waste tyre mixed with LDPE, HDPE, or PP.

The product distribution in plastic pyrolysis is greatly influenced by the average time the reactant spends under the operating temperature and pressure conditions (also known as residence time). A longer residence time often provides low molecular weight hydrocarbons and gaseous products. There is a possibility of partial or incomplete pyrolysis at moderate temperature, which adversely affects the amount and quality of the liquid product. However, the effect is less prevalent for continuous processes being performed at high temperature. The optimization of residence time for different feeds can improve the quality of desired products. The influence of the temperature and the residence time on the product distribution in thermal pyrolysis of HDPE was performed by Mastral et al. Pyrolysis was performed in a fluidized bed reactor at 650–850°C and a residence time of 0.64–2.57 s. The yield of the liquid containing aliphatic hydrocarbons ($\leq C_{30}$) was highest at 640°C (79.7 wt %), whereas at 780°C only 9.6 wt % of oil containing high aromatic hydrocarbons with high yields of gaseous products (86.4 wt %) was obtained. Similarly, the longer residence times provided enhanced yields of gaseous products. Miskolczi studied the effect of residence time on pyrolysis of HIPS-Br and ABS-Br. In this experiment, a tube reactor combined with a distillation column was operated at 360–440°C. Oxygenated/nitrogenous compounds were identified in light and/or middle distillates along with tribromide antimony but no organ bromine compounds. The process was advantageous over previously reported fluidized/fixed bed pyrolysis in terms of lower concentrations of organ bromine compounds. In principle, individual polymers pyrolyze at different temperatures to provide maximum liquid products. For instance, the ideal decomposition temperature for PP is reportedly 378–456°C, whereas LDPE and HDPE degrade at 437–486°C and 452–489°C, respectively. A further increase in reaction temperatures results in enhanced yields of gaseous products.

13.5.3 Pressure

Pyrolysis in a closed reactor can be performed at high pressure, which can have a significant impact on product distribution, particularly at lower temperatures. Thus, the formation of gaseous products and lighter hydrocarbons was evidently high at elevated pressure. However, the impact was less apparent at higher temperature (above 450°C). On the contrary, the reaction in a continuous batch reactor at sub atmospheric pressure, i.e., under vacuum, helped in effective removal of product vapours. Vacuum pyrolysis at high temperature for a short duration improved the yields of liquid products and suppressed the formation of gases and char. Notably, a high reaction pressure increased the formation of lower molecular weight hydrocarbons and decreased the rate of double bond formation.

Murata and co-workers studied the effect of pressure on HDPE pyrolysis in a continuously stirred tank reactor. The analysis of products revealed that the effect of pressure on the yields of gaseous products was high at lower temperature, i.e., 410°C, with a change in pressure from 0.1 to 0.8 MPa increasing the yield from 6 to 13 wt.%, which was less significant at 440°C. Similarly, the effect of pressure on residence time was less prevalent at high temperature. The lighter hydrocarbons were produced in higher amounts at elevated pressure but with lower yields of olefins.

13.6 Type of Reactors

Pyrolysis of waste plastics has been studied based on different reactor designs, which significantly affect the reactivity and selectivity in the reaction. Notably, the reactors with continuous flow operation are considered better than batch reactors in providing steady state data at a constant temperature, pressure and feed concentration, particularly for investigating the mechanism of pyrolysis and/or stability. The advantages and limitations associated with each reactor are covered in this section. The influence of reaction medium on product distribution is discussed.

13.6.1 Rotary Kiln Reactor

A rotary kiln is a rotating heating tube with good mixing properties. In this case, the transfer of heat is good, but the heating rate is comparatively low as the heat is typically supplied by externally located burners. The use of heat exchangers with the feed is another option to increase the rate of heating. The reactor has a high flexibility over the use of heterogeneous material and residence time. A rotary kiln is cheap and operates at a low maintenance cost. Therefore this reactor is globally used for thermal pyrolysis under industrial operations, including the Gibros PEC process in Germany, Edith process in France, Toshiba Saporro process in Japan, Renewlogy in Canada and PHJK and VadXX in the USA. The reactor must be designed for better control of temperature and heat transfer to avoid the accumulation of sticky polymeric substances at the inner wall.

Yazdani et al. performed thermal pyrolysis of waste tyres in a rotating kiln reactor under a nitrogen atmosphere at a temperature range of 400–1050°C. The maximum yield of pyrolytic oil (44 wt %) was obtained at 550°C, whereas more gases were produced at higher temperature with low char formation. The distillation of pyrolytic oil provided light naphtha (14%), middle distillate (36%) and heavy naphtha (4%), with a major fraction of vacuum gas oil-grade oil (46%, boiling point >350°C). The continuous pyrolysis of a high-scale and complex waste stream, particularly organic waste mixed plastics, make it a first choice for waste management programs. However, the liquid products from thermal pyrolysis in a rotary kiln are generally high-molecular weight, more suitable for waxes, a fuel for boiler and power generation and more.

13.6.2 Screw Feed (Auger) Reactor

Screw feed reactors are also considered modified rotary kilns. The feed is transported by a screw, or an auger positioned coaxially in the fixed vessel (Figure 13.7). The waste plastics feed hopper unit is simultaneously fed with an inert gas to ensure pyrolysis under an oxygen-free environment. The inert gas generates some pressure to assist the subsequent withdrawal of product vapours through the reactor.

The screw provides better control over the handling of different complex plastic mixtures, mechanical mixing and residence time of miscellaneous feeds. Although an external heating system is sufficient for the reactors with small diameter tubes, solid heat carriers such as ceramic and steel pallets are supplied with the waste plastics to internally heat the feed. The solid heat carriers are separated in a sieving pot seperate from the char collector. Twin-screw reactors consist of two intermeshing screws to endorse constant trembling of the feed. The effective mixing of heat carriers and feedstock ensure higher efficiency, productivity and process versatility in terms of feed quality. The effective mechanical mixing of feed prevents choking of these screw feed reactors.

Al-Salem and co-workers performed slow pyrolysis of reclaimed plastic waste from a landfill after pre-treatment in a continuous auger pyrolizer at 500°C. The selectivity to wax products was

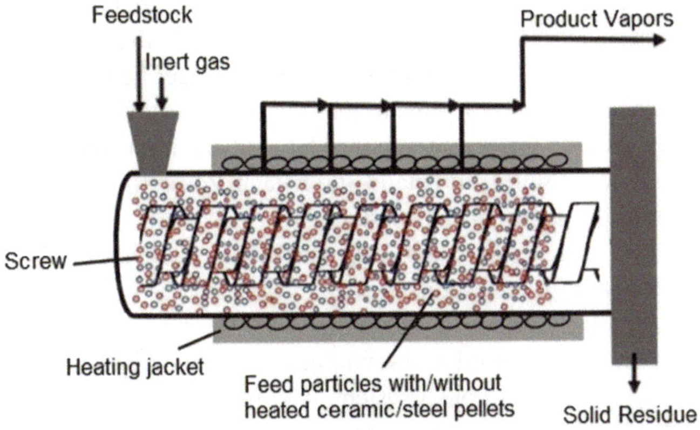

Figure 13.7 Design of a screw (auger) reactor.

very high, i.e., 93.2 wt % with low amount of pyrolysis oil (5.5 wt %) and gases (1.3 wt %). In such cases, the potential of fast pyrolysis in producing oils rather than wax as a major fraction can also be explored.

13.6.3 Fluid Catalytic Cracking Reactor

Fluid catalytic cracking is an important process of petroleum refineries, which converts high-molecular weight and high boiling hydrocarbon residues such as vacuum gas oil (VGO) from the distillation unit into lightweight hydrocarbons including gasoline and olefins as major products. Waste plastics or their waxes are pyrolyzed in a solvent system including vacuum gas oil. The process has the advantage of easy processing of feed in a liquid form, removal of metallic and inorganic contaminants during dissolution and production of high-quality products. However, safety measures are generally high, with an expensive reactor and limited availability of VGO. The process is therefore more suitable for existing refineries.

The effect of the catalyst-to-oil ratio on productivity and selectivity was examined for HDPE/HDPE wax mixed VGO by Rodríguez et al. in separate studies. For instance, the catalytic cracking of HDPE pyrolysis waxes was performed in a CREC riser simulator reactor under industrial fluid catalytic cracking conditions at 500–560°C with a high catalyst/oil ratio (5:1) for 6 s. The formation of naphtha with RON up to 104.8 was observed in 23.8–31.1 wt % yields along with liquid petroleum gases fractions (13.7–18.1 wt %) in the presence of three different commercial fluid catalytic cracking catalysts. A comparison between the cracking of HDPE waxes and vacuum gas oil suggested a lower reactivity in the case of HDPE at a low catalyst-to-feed ratio, whereas the conversion was similar or even higher at a higher ratio. The product analysis revealed comparatively higher yields of light liquid hydrocarbons with low coke formation for HDPE waxes. In another study, the synergetic behaviour of an alkylcarbenium intermediate offered higher yields of naphtha and light olefins.

Furthermore, lower contents of iso-paraffins and aromatics were observed in pyrolysis oil. The practice of copyrolyzing polyolefin waste or its pyrolysis oil with vacuum gas oil in existing refineries is quite attractive, as the process does not require new infrastructure. Furthermore, the synergic cracking of hydrogen-rich polymers with vacuum gas oil reduces the number of aromatics with improved selectivity towards light hydrocarbons and olefins.

13.6.4 Stirred-Tank Reactor

Stirred-tank reactors (STR) are the most common reactors being used for diverse chemical processes. Plastic pyrolysis in these reactors provides flexible operations with respect to the residence time. However, the reactor experiences formation of heat gradients due to improper heat transfer during pyrolysis, which further results in secondary reactions to increase the yields of solid residue. Other disadvantages of this reactor include the requirement of large installations and high maintenance cost. The reactor has been used by Smuda in Poland and Hitachi Zosen in Japan and is currently in use for catalytic pyrolysis at Plastic Energy in London, UK.

Kim and Kim pyrolyzed solid PS and PP in a stirred batch reactor, wherein pyrolysis of polystyrene occurred from 370 to 410°C by heating the reactor at rates of 0.5–2°C/min to provide high yields of styrene and its dimer. A relatively higher temperature (400–460°C) was required for polypropylene due to higher activation energy. Similar observations were made by Kim and co-workers during pyrolysis of solid polystyrene in an open system at 370–400°C under a heating rate of 11.3–12.3°C/min. The selectivity to styrene and 1,3-diphenylbutene was attributed to suppressed scission of volatile diphenyl radicals at short residence times. However, secondary reactions of these radicals are dominant in sealed reactors. In a continuous system, melting and thermal dechlorination of polyvinyl chloride mixed plastic waste in a screw feeder was followed by catalytic pyrolysis in a stirred tank reactor using a Ferestructured clay (Fe-RC) catalyst. A dechlorination efficiency of over 90% was achieved in a screw feeder with a retention time of more than 35 min, whereas the yield of pyrolytic oil was significantly high (83.73 wt %) at 450°C. The product composition showed the presence of kerosene (C_9–C_{12})- and diesel fuel (C_{13}–C_{19})-grade hydrocarbons. The process showed high operational stability and good consistency with product yields at steady state.

13.6.5 Plasma Pyrolysis Reactor

Plasma pyrolysis is typically performed using heated plasma at high temperature in the presence of limited amounts of oxygen. Rapid heating of waste plastics in a highly reactive plasma zone having energy radiation, electrons, ions and excited molecules cracks the high molecular weight molecules into gaseous products including hydrogen, carbon monoxide, carbon dioxide, methane, acetylene and more.

Plasma reactors are not economical due to the high use of electricity and low durability of electrodes with a plasma torch. The major advantage of plasma pyrolysis is safe disposal of hazardous chemical mixed plastic waste at high temperatures; thus, it is most suitable for the remediation of waste streams from hospitals and pharmaceutical industries and, e-waste such as semiconductors, batteries and circuits. Nema and Ganeshprasad reported the safe disposal of medical waste to produce hydrogen, carbon monoxide and hydrocarbon gases at 1200°C using a plasma arc. The hot gases were quenched to avoid recombination of gaseous molecules to inhibit the formation of hazardous dioxins and furans. In another experiment, plasma pyrolysis of plastic waste was performed to produce syngas at extremely high temperature under inert conditions. The formation of small amounts of long-chain hydrocarbons was also observed.

The significance of two-stage pyrolysis of HDPE by coupling a low-temperature (250°C) plasma catalytic steam reformer with a pyrolysis reactor was explored by Aminu et al. and Ratnasari et al. Here, the catalytic steam reforming of in situ-generated pyrolysis gases produced hydrogen gas by utilizing dielectric barrier discharge. The hydrogen yield improved with increasing plasma power even in the absence of a catalyst. Furthermore, an alumina-supported nickel catalyst provided the highest hydrogen yield (1.5 mmol/g) over iron, cobalt and copper catalysts under dry conditions, which increased to 4.56 mmol/g after introducing steam to the system. The plasma gasification of plastic waste and municipal solid waste was studied by Mazzoni and Janajreh in an integrated plasma gasification combined (IPGC) cycle using two RGIBBS (plasma) reactors operating at 2500°C and 1250–1315°C, respectively. Different compositions of air with oxygen or steam were used to analyse the pyrolysis performance, wherein an improvement was observed with an increasing oxygen ratio. The treatment of mixed plastic waste and municipal solid waste in a 3:7 ratio showed maximum efficiency of 38% under a pure oxygen medium.

Wang et al. performed pyrolysis of waste circuit boards consisting of furans, dioxins and other hazardous metals at 1200°C. The decomposition of organic materials completed in few milliseconds, showing the absence of secondary pollutants. The concentration of nitrogen oxide gas remained within national safety limits, whereas the recovery of carbon monoxide was high. In this experiment, a significant reduction in the amount of e-waste was observed and metal like copper, silver, platinum and gold were recovered after cooling the molten slag. In a similar approach, plasma pyrolysis of printed circuit boards containing 10%–25% copper, 2%–5% each of iron and aluminium and other metals was performed at a 1400–1600°C reactor temperature. The plasma reactor was coupled with acid leaching to recover the metallic portions. The leaching of copper, nickel and cobalt metals was as high as 90% of total concentration, whereas it was 40% in the case of iron and aluminium.

13.6.6 Batch Reactor

A batch reactor is usually a sealable vessel, typically with a mechanical stirrer/agitator, designed to operate at high temperature and pressure. The disadvantages of a batch reactor include product variability, high labour costs, low productivity and installation requirements for large set-ups. The high vapour pressure generation in the reactor due to low boiling points of cracked products and gaseous by-products is critical. Moreover, the presence of acid vapours including hydrochloric acid at high pressure for some duration can be dangerous, particularly for polyvinyl chloride and polyvinyl dichloride mixed waste plastics feed. Pyrolysis in such reactors is not energy efficient, as it requires occasional cooling to condense the liquid products. However, a high reaction conversion can be obtained when the reactions can be performed for an extended time. In short, such reactors are suitable for small-scale operations, particularly for lab-scale pyrolysis of waste plastics where the reactions are optimized for a type of feed and other operating parameters. Pyrolysis in the presence of a catalyst increased the efficiency even at low temperature but often resulted in reduced catalytic activity due to high coke formation at the surface.

Anene et al, studied the thermal and catalytic pyrolysis of virgin LDPE, PP and LDPE/PP mixtures in a batch reactor at 460°C. The addition of PP in LDPE resulted in slower decomposition compared to pure LDPE. The degradation temperatures were reduced in the presence of a CAT-2 catalyst. Thermal pyrolysis of plastic feeds produced hydrocarbons in gasoline (C_7–C_{12}) and diesel ranges (C_{13}–C_{20}), whereas enhanced cracking at lower temperatures was observed in the presence of a catalyst with a narrow product distribution (C_7–C_{12}).

The potential of recycled HDPE plastic in slow pyrolysis was investigated by Gutiérrez and co-workers in a batch autoclave reactor. Reactions were performed at 430–490°C for 15–60 min to maximize the recovery of pyrolysis oil, having streams appropriate for refineries. Reaction yields showed higher dependence on temperature than reaction time, which was 85%–90% at 430°C and 15–37.5 min. Interestingly, the product profile of the reaction performed at 430°C and 15 min was similar to vacuum gas oil. The recovery of pyrolysis oil at 430°C and 60 min or at 460°C and 15–60 min was less than 80%, which was suitable for hydroprocessing with light cycle oil (LCO). Furthermore, the naphtha and middle distillates of pyrolysis oil from these reactions were appropriate for blending with commercial diesel and gasoline after reducing olefins in a hydrotreating unit.

13.6.7 Fixed Bed Reactor

A fixed bed reactor consists of a fixed bed, wherein a catalyst is packed in pelletized form. The use of solid particles of plastic feedstock in the reactor causes trouble during the feeding process, which also has limited access to the surface areas of the catalyst pellets. Therefore, fixed bed reactors are designed to have dual reactor sections. The lower part of the reactor is fed with waste plastics, which partially pyrolyzes the feed under pyrolytic conditions. The pyrolysis vapours pass through the catalyst bed at high temperature to further crack the high weight oligomers to low weight hydrocarbons. However, direct pyrolysis of polyvinyl chloride and polyvinyl dichloride mixed plastic is not suitable, where liberated hydrochloric acid can deactivate the catalyst or even damage the bed. A more appropriate application is its use as a secondary reactor, which is integrated with a pyrolyzed unit. Unfortunately, these two step processes are less economical, providing the quality of cracking products analogous to one-step pyrolysis. Sharuddin and co-workers performed pyrolysis of different plastic waste in a fixed bed reactor at 500°C with a heating rate of 20°C/min for 30 min. The use of nitrogen as the carrier gas (200 ml/min) to remove product vapours helped in improving the yields of liquid products by minimizing the secondary reactions of vapours, which were obtained in 29.0–44.3 wt % yields. The addition of polystyrene in waste plastics improved the quality of the liquid feed. Alternatively, thermal pyrolysis of polyvinyl chloride mixed feed was performed in a horizontal fixed bed reactor at 700°C (heating rate 10°C/min), wherein slow pyrolysis resulted in improved liquid yields. Nitrogen gas was utilized as a carrier gas (400 ml/min). The hydrocarbon fraction of pyrolytic oil contained a high number of aliphatic hydrocarbons (26.31% paraffins and 46.02% olefins) and a low number of aromatics (6.36%).

The use of a fixed bed reactor offers the advantage of two-step processing of plastic feed, which can potentially improve the quality of the products. However, deactivation of the catalyst by unique products/ionic species and fine carbon particles produced during thermolysis of complex/contaminated plastic feed limits its use as a versatile pyrolysis approach.

13.6.8 Fluidized Bed Reactor

The fluidized bed reactor is designed to pyrolyze the waste plastics feed in a fluidized state. In catalytic pyrolysis, the catalyst remains as a free-flowing powder at the distributer plates in a stationary phase before operation. An inert gas, typically nitrogen, is used as a medium for fluidization. Unlike a fixed bed reactor, the surface area of the catalyst is mostly accessible to the fluidized feed and the heat transfer inside the reactor is very good. Notably, the feed can be charged as per the requirements based on the conversion of different feedstocks. This feature addresses the limited productivity of a noncontinuous process in a batch reactor and incomplete pyrolysis in continuous feed reactors causing the formation of solid residue or high-molecular weight liquid products. Pyrolysis is complete even at a short residence time and a uniform spectrum of products is formed. Kaminsky and Franck observed that the depolymerization of poly methyl methacrylate in a fluidized bed reactor was much higher than the yields in melting vessels or heated screw feeders, i.e., 50–80 wt %.

del Remedio Hernandez and co-workers performed flash pyrolysis of HDPE in a fluidized bed reactor under different thermal and catalytic conditions using HZSM-5 or HUSY at 500–800°C. The liquid fraction under thermal conditions was composed of C_{10}–C_{40} linear paraffins with very low aromatics; however, the formation of aromatics and branched alkanes and alkenes was favored under catalytic pyrolysis. Moreover, a HUSY catalyst produced a higher amount of aromatics and branched alkanes with narrow product distribution. In this reactor, the rate of catalyst deactivation is low due to low coke formation and therefore, the catalyst is reused several times without replacement. Consequently, the fluidized bed reactors are most suitable for large-scale operation in terms of economy, ease of operation and improved product quality.

13.6.9 Conical Spouted Bed Reactor

A conical spouted bed reactor (CSBR) is an alternative to a fluidized bed reactor; the vigorous cyclic particle movement facilitates the handling feed particles of larger sizes and different textures. The advantages of high heat and mass transfer, low residence time and improved product quality remain the same as a fluidized bed reactor. Also, the design of conical spouted bed reactor prevents agglomeration of sticky solids at the surface unlike a conventional fluidized bed reactor, thus allowing simpler operation. Niksiar and colleagues studied the kinetics of polyethylene terephthalate pyrolysis in a spouted bed reactor, wherein the activation energies of feed particles of sizes 0.1–1.0 and 1.0–3.0 mm were determined as 276.8 and 264.3 kJ/mol, respectively. The study concluded that heat and mass transfer limitations in spouted bed reactors could be minimized. In another study, pyrolysis of HDPE was performed in a conical spouted bed reactor at 500–700°C in a continuous mode. The performance of a conventional fluidized bed reactor was compared with batch and fluidized bed reactors in terms of product yields and compositions, wherein high yields of waxes and fuels with low aromatics were achieved at low temperature and short residence time with better heat and mass transfer. The formation of gaseous products and gasoline-grade hydrocarbons was considerably higher at 700°C.

13.6.10 Microwave Reactor

The use of microwaves for pyrolysis of waste feed is a new technology, which typically requires premixing of a microwave heat-absorbent material such as particulate carbon and silicon carbide with plastic waste. The major advantages of this process are rapid heating, increased productivity at shorter residence times and lower production costs. Furthermore, a microwave reactor is compatible with various contaminants including organic and even metallic (such as aluminium) waste. The dielectric property of plastic feed influences the efficiency of microwave heating, which often requires a microwave absorber to improve energy absorption. Therefore, the process optimization is challenging and still under development for large-scale operations.

Russell et al. reported microwave-assisted pyrolysis of HDPE using catalytic amounts of activated carbon. High cracking at different operating temperatures (400–600°C) provided a lighter liquid product with a narrower range of chain lengths, which was mainly composed of petrol- and diesel-grade hydrocarbons. In another approach, pyrolysis of a polystyrene and polypropylene mixture was performed in the presence of 10 wt.% carbon and activated carbon catalysts, wherein microwave pyrolysis at 900 W produced the highest oil yield of 84.30 wt % in 10 min, with a styrene recovery of 67.58%. Similarly, pyrolysis of LDPE was studied in a microwave assisted two-stage pyrolysis–catalysis system using nickel oxide and HY zeolite catalysts. Thermal pyrolysis at 500°C followed by catalytic cracking at 450°C yielded 56.53 wt % liquid product with 93.80% gasoline-grade hydrocarbon. The degree of aromatization and isomerization was 80.4% for a high HY zeolite-to-LDPE ratio (1:5), which was optimized for oil yield and quality at a 1:10 ratio. The presence of nickel oxide in cocatalysts slightly decreased the oil yield by 5.3–8.5 wt % but increased the alkene and aromatic contents due to improved hydrogen abstraction. However, the costs of operation and maintenance of these reactors are major challenges for their application in pilot-scale operations.

In another effort, the influences of microwave power and graphite susceptor quantity on the heating rate, conversion efficiency and heat losses were investigated during polypropylene pyrolysis. The heating rate was proportional to microwave power and disproportional to susceptor quantity. Pyrolysis oil with a maximum heating value of 44 MJ/kg was produced under these conditions containing C_8–C_{12} as the major component.

Also, solid char with a rich carbon content (96 wt %) showed a high specific surface area of 195 m^2/g. Selectivity towards cyclic and acyclic products was effectively tuned with different power conditions.

Fan and co-workers performed pyrolysis of LLDPE in a continuously stirred batch microwave reactor under similar conditions and the performances showed higher selectivity towards gaseous products including methane as a major fraction. Catalytic upgrading of pyrolysis vapour was studied using a HZSM-5 catalyst, wherein C_7–C_{10} hydrocarbons with major portions of monoaromatics (72%) were selectively formed in a continuously stirred reactor. The major advantage of microwave pyrolysis in processing metal-rich plastic waste, particularly waste electrical and electronic equipment (WEEE), was explored by Wang and co-workers. In this case, pyrolysis of waste printed circuit boards (WPCBs) provided average yields of 15.7 wt % oil, 5.7 wt % gas and 78.6 wt % solid residue. The presence of phenol and derivatives was observed, whereas metals were predominantly found in solid residue. The mechanical processing was applied to reclaim metals (around 95% or higher) from solid residue. In a distinct approach, the performances of conventional and microwave pyrolysis were compared during pyrolysis of waste printed circuit boards coupled with acid or thiourea treatment to recover copper and gold, respectively. Overall weight loss was higher under microwave conditions and copper recovery rates in a two-stage process for conventional and microwave pyrolysis were 75% and 96%, respectively. Similarly, the recovery rates of gold were 69% and 80%, respectively, under conventional and microwave conditions.

Microwave pyrolysis is effective for treating plastic waste from WEEE or aluminium mixed plastics with good compatibility, high liquid yields and considerable reduction in the amounts of solid residue.

13.6.11 Pyrolysis in Supercritical Water

Supercritical fluids are substances at high temperature and pressure, above their critical point, where the medium has the properties of a liquid to dissolve, as well as a gas to diffuse through any solid. Most common supercritical fluids are water and carbon dioxide, which behave very differently with small alterations in temperature and pressure. The replacement of organic solvents with supercritical fluids is promising for many laboratory and industrial processes. For example, water behaves like a supercritical fluid at 373°C and 22 MPa. This condition itself is enough to keep the waste plastics feed in liquid or molten form. The energy provided by supercritical water is sufficient to cleave the carbon–carbon bonds in a few seconds. The formation of char is generally low due to the participation of water molecules in a termination process, which may also result in the formation of oxygenated species. The process is costly and requires the use of a closed semi batch reactor. Therefore, high-scale productivity and formation of a high-quality product are less feasible. Sato et al. investigated pyrolysis of polyvinyl chloride in supercritical water which produced various organic compounds in liquid and gas phases. Interestingly, the selective dehydrochlorination averted the formation of organochlorine compounds. The sequential removal of hydrochloric acid, cracking of resulting conjugate double bonds in the polymer chain and cycloaddition– aromatization result in the formation of aromatics. A similar observation was made by Takeshita et al. when pyrolysis was performed in subcritical and supercritical regions from 250 to 350°C.

The design and operational requirements of pyrolysis reactors at laboratory- or pilot-scales differ from the commercial scale. For instance, batch and fixed bed reactors are used to investigate the effect of various parameters at small scale due to cost effectiveness and ease of operation. Similarly, recent interest in using auger reactors during these studies corresponds to low carrier gas requirements and high energy efficiency. However, the fast pyrolysis of plastic wastes requires continuous operation at controlled temperatures, high heating rates and efficient heat transfers to the feed. In this regard, fluidized bed reactors, rotary kilns and vacuum pyrolyzers are promising for safeguarding operations with good productivity. Similarly, a geometrically modified CSBR is superior to a conventional fluidized bed reactor due to enhanced heat and mass transfer between different phases and reduced particle agglomeration at high gas velocities. These reactors are designed to operate in batch and continuous modes; however, the continuous operation has several advantages in terms of effective heating, better phase interactions, shorter residence times and constant removal of carbon black.

Although microwave and plasma pyrolysis reactors have great potential to treat WEEE and medical plastic waste in effective and safe manners, their large-scale operations still require a significant amount of research. Also, the application of existing FCC reactors in treating plastic mixed VGO is an interesting approach. However, the process is limited to polyolefin feed and that too in a limited proportion.

13.7 Applications of Pyrolysis Products

The concept of utilizing plastic waste to produce energy, fuel and value-added chemicals is fascinating and offers a promise to deal with environmental and health issues. The technology offers processing of almost all kinds of waste plastics in a convenient and person-power-saving manner. Typically, the occurrence of low or no sulfur content in pyrolysis oil offers major advantages over conventional fuel. The process is meant as a sustainable alternative to conventional plastic waste treatments; however, the incompatibility of polymeric materials such as polyvinyl chloride, waste electrical and electronic equipment plastics and contaminated polymeric waste in conventional pyrolysis needs attention. The product stream, particularly in the case of mixed plastic waste, is complex and thus requires further treatment in the case of pyrolysis oil. Further, the gaseous products consist of harmful gases such as hydrochloric acid and carbon monoxide. To address these serious limitations of pyrolysis technology, the capital expenditures (CAPEX) of a plant operation stands high to meet the emissions standards and product worth.

Feeds enriched with hydrocarbons such as LDPE, HDPE, PP and PS have higher energy density and require simple processing. Nonetheless, thermal pyrolysis of plastic waste is an energy-consuming process, which requires a significant amount of energy to crack the polymeric structure. However, the exploitation of efficient heterogeneous catalysts and/or synergistic pyrolysis of plastic waste with oxygen/nitrogen-rich biomass renders great opportunities.

13.7.1 Pyrolysis Gases → Hydrogen and Methane

The gaseous by-products such as hydrogen, carbon monoxide and methane, light hydrocarbons (mainly C_2-C_3 olefins) are generally produced in a pyrolysis reactor, which can be used to heat the reactor or for other energy applications. The major by-product of polyvinyl chloride mixed plastic waste is hydrochloric acid, which is either neutralized or selectively trapped using an adsorbent. Alternatively, the high-temperature gasification of plastic waste is practiced with an aim to produce syngas and/or LPG-grade (C_3–C_4) hydrocarbons.

13.7.2 Pyrolysis Oil → Aromatics and Diesel Fuels

The liquid oil obtained after pyrolysis is primarily composed of aliphatic, monoaromatic and polyaromatic hydrocarbons. However, the presence of some impurities such as chlorine, moisture, acids and solid residues deteriorates the quality of pyrolysis oil and, therefore, delimits their commercial application. The post-treatments, including homogenization to remove solid char particles, total acid number (TAN) reduction by extracting or neutralizing the acidic compounds and distillation of crude pyrolysis oil, can improve the quality of liquid products. The fate of different distillates can be decided as per their physical and specific properties in fuel applications. Also, pyrolysis at low temperature under thermal conditions results in high-molecular weight fractions suitable for wax production, heating applications in boilers and power generation in generators and agricultural pumps. Also, pyrolysis of PS to produce styrene monomers and other useful chemicals including BTX and ethyl benzene is noteworthy. The presence of high amounts of aromatics such as BTX, styrene and ethylbenzene, is suitable for their recovery as precursors in petrochemical industries. Similarly, selective depolymerization of PMMAto its monomer can potentially respond to the present global need for waste management. Characteristic properties of pyrolysis oil obtained from PE, PP, PS and mixed feeds were compared by Quesada and Calero et al. The liquid oil from PE film was more viscous and waxier compared to commercial diesel, whereas PP-derived oil contained high olefin content with a semi waxy appearance. On the contrary, the PS pyrolysis oil consisted of value-added chemicals such as styrene, BTX and ethyl benzene. Furthermore, intermediate properties were observed for mixed polymers.

Sachuthananthan et al. studied the fuel properties of pyrolysis oils obtained from PE, PP and PS and their binary/tertiary mixtures and their blends with diesel fuel including engine and emission behaviours. The amount of PE in feed contributed to a high viscosity and waxy appearance of oil, whereas high fluidity was observed due to aromatics derived from PS pyrolysis. PP pyrolysis oil consisted of olefins with semi waxy appearances. Except for the cetane number and pour point values, other fuel properties of diesel blends were within acceptable limits. Interestingly, the power output of an engine for 5% pyrolysis oil blends was higher than commercial diesel, but the torque output was lower, particularly at high speeds.

Also, NO_X and carbon dioxide emissions increased for blended fuel, but carbon monoxide emissions were lower. In another study, pyrolysis oil from pyrolysis of mixed plastic waste including PE, PP, PS and PET was blended with diesel (10–50 vol%) to determine the engine performance and characteristics. The authors concluded that these blends are suitable for existing diesel engines, although at the cost of a minor efficiency loss and slightly lower specific fuel consumption (SFC) at a higher pyrolysis oil-to-diesel ratio. However, in-cylinder pressure was high due to greater heat generation and delayed ignition for diesel blends. The presence of oxygenated species resulted in reduced emissions, but carbon monoxide emissions increased at higher loads.

The inclusion of certain additives to diesel offers improved thermal efficiency with minimized exhaust pollution. Consequently, metal oxide nanoparticles enhance the ratio value of the surface area to volume, which facilitates the combustion efficiency of any potential fuel including biodiesel. Such additives donate oxygen to minimize carbon deposits besides reducing carbon monoxide, NO_X and hydrocarbon emissions. Following which, the combustion behaviours of pyrolysis oil were improved by blending it with magnesium oxide nanoparticles.The brake thermal efficiency (BTE) of pyrolysis oil with 75 ppm magnesium oxide increased by 2.5% compared to pure plastic oil. The emissions of carbon monoxide, hydrocarbon and smoke were lower; however, the NO_X emissions increased. Bharathy and co-workers analysed power output and emission behaviours of pyrolysis oil having dispersed titanium oxide nanoparticles (25–100 ppm) in a mono cylinder compression ignition engine. In this case, brake thermal efficiency increased by 2.1% for the pyrolysis oil compared to neat pyrolysis oil, with a considerable drop in carbon monoxide, hydrocarbon and smoke emissions. In a similar approach, nano fuels containing aluminium oxide nanoparticles in pyrolysis oil were combusted in a light duty compression-ignition engine at different concentrations of nanoparticles (100–200 ppm). At 200 ppm nanoparticle concentration, lower in-cylinder pressure, delay period and combustion duration were observed compared with using neat pyrolysis oil. Thermal efficiency was comparable and gaseous emissions were reduced, to values even lower than diesel.

Verma et al. studied the performance of pyrolysis oil obtained from waste packaging as a blend for diesel engine applications, wherein a 40% blend of pyrolysis oil with diesel possessed comparable carbon monoxide, NO_x and carbon dioxide emission properties with similar power generation and compressibility. Gala et al. performed thermal pyrolysis of molten industrial plastic waste (IPW) and postconsumer plastic film waste (PCPW) in a continuous flow stirred tank reactor at 80 kg/h scale. The C_7–C_{18} paraffins were obtained in 66.8–72.2 wt %, whereas olefins and aromatics were in the range of 14–23 wt %. Distillates in a boiling range of 180–380°C (about 50 vol%) possessed similar calorific values to diesel with advantages of ultralow-level chlorine, low TAN, good lubricating properties and low sulfur (4–19 ppm) contents. The physical properties, i.e., density, kinematic viscosity, flash point and colder filter plugging point, did not matching diesel specifications. However, blending of the diesel fraction with commercial diesel in a 1:1 v/v ratio satisfied all 21 parameters required for automotive applications.

Meanwhile, the influence of organic additives, particularly oxygen-containing molecules on improving certain fuel properties of pyrolysis oil blends including combustion efficiency and emission behaviours was explored. In this regard, the performance of petrol blended with 10% and 20% pyrolysis oil was analysed by Kumar and Puli in a multicylinder petrol engine in the presence/absence of 5% methanol. For all blends, improved brake specific energy consumption (BSEC) and brake thermal efficiency were reported with reduced carbon monoxide and hydrocarbon emissions. However, the NO_x emissions were lower than the blend without methanol. Bridjesh and co-workers improved the combustion performance and reduced NO_x and hydrocarbon emissions of pyrolysis oil using composite additives, i.e., soy lecithin and 2-ethylhexyl nitrate. Similarly, Gadwal and co-workers blended pyrolysis oil with diesel and ethanol to study the influence of injection timing (IT) and injection pressure (IP) on the performance of a modified common rail direct injection (CRDI) engine. Consequently, IT of 10° bTDC and IP of 90 MPa maximized BTE with lower gaseous emissions for blended samples. In another approach, the effects of engine load, pyrolysis and ethanol blend ratio in diesel and injection pressure on the engine performance and emission properties were studied with the Taguchi methods (statistical methods). Hence the blend of 10 vol% each of injection pressure under full load conditions (Mangesh et al. and Padmanabhan et al.).

Furthermore, attempts were made for successive or simultaneous upcycling of pyrolysis oil. Mangesh and co-workers examined combustion and emission properties of hydro processed polypropylene derived pyrolysis oil as blends in diesel engines. Interestingly, 20% blends of pyrolysis oil matched with European diesel fuel standards (EN590). In another study, the performance of catalytically reformed polyethylene derived pyrolysis oil with natural zeolite (NZ) catalyst was analysed in a four cylinder and four-stroke diesel engine. In this case, diesel blended pyrolysis oil showed higher brake thermal efficiency (BTE) with lower brake specific fuel consumption (BSFC). Furthermore, NO_x emission was lower, but carbon monoxide, hydrocarbon and smoke emissions were higher compared to commercial diesel at higher loads.

The performance and emission behaviours of different blends of pyrolysis oil (10–30 vol%) were compared with commercial diesel in a single cylinder DI four-stroke diesel engine. Brake thermal efficiency was marginally higher, whereas BSFC was lower for a 20 vol% oil blend as compared to commercial diesel. In addition, NO_x and hydrocarbon emissions were lower with increased fuel exergy. However, reduced exegetic efficiency was observed, which improved at higher loads.

In a different approach, fuel properties of pyrolysis oils obtained from a conventional pyrolysis and microwave reactor were compared in a CI engine. BTE values of electrical pyrolysis oil and microwave pyrolysis oil were comparable or lower than commercial diesel. NO_x emissions were significantly lower than conventional diesel, but hydrocarbon emissions and filter smoke numbers were high. Notably, the exergy efficiency of microwave pyrolysis oil was relatively higher than that of conventional pyrolysis oil.

Lately, some researchers evaluated the performances of distilled gasoline/diesel fractions from pyrolysis oil in conventional engines. Dobó and co-workers extracted gasoline fractions of pyrolysis oil from HDPE (37.3%), LDPE (40.8%), PP (42.1%) and PS (70.5%) in a spark-ignition engine. The engine performed well with these fractions, where fuel consumption for PS-derived gasoline reduced by 9.1%–9.4% compared to the commercial gasoline (RON = 95), but NO_x emissions increased by 82%–147%. The carbon monoxide emissions were 91%–96% lower for PS-derived gasoline but higher for HDPE, LDPE and PP fractions. Also, PET pyrolysis failed to produce a gasoline fraction and the gasoline fraction from polyurethane (PU) was inappropriate for real-time applications. Recently, the same group obtained gasoline-like fuels in 473–512 g/kg waste yield after distilling pyrolysis oil obtained from mixed plastic waste (PE, PP and PS). During analysis, low fuel consumption was observed as compared with commercial gasoline (RON=95) but with NO_x emissions. However, the olefin content and distillation properties showed deviations from EN228 standards. In another study (Kunwar et al.), the distillates of oil obtained from pyrolysis of postconsumer polypropylene and polyethylene plastics were compared with commercial diesel and petrol. The blends of 5% and 20% pyrolysis oil distillates with ultralow sulfur diesel (ULSD) exhibited superior low-temperature performance compared to neat ULSD with a cold filter plugging point of (CFPP) less than−50°C and pour point of less than−74°C. The oxidation stability and kinematic viscosity of these blends were well within the ASTM standard limits.

It is noteworthy that certain fuel properties of pyrolysis oil including calorific value, density and octane/cetane number were comparable to petroleum gasoline and diesel. However, major differences were observed for viscosity, API gravity and chemical compositions. Therefore, the use of pyrolysis oil is mainly limited to blending applications. Nevertheless, the quality of pyrolysis oil and its applications are strongly influenced by the chemical structures and reactivities of different polymers present in waste feed; particularly the presence of heteroatoms (O, Cl, N and more) in polymeric chains as well as additives limits their use even as fuel blends. Although the use metal oxide nanoparticles or oxygenated compounds as additives can improve the properties to some extent, such practices are not attractive due to economic and/or environmental concerns. Also, successive processing of pyrolysis oil to produce high-end chemicals is not very cost effective. However, the use of multifunctional catalysts or a combination thereof in simultaneous upcycling of undesired products can be a solution. In addition, the processing of pyrolysis oil in existing refineries and petrochemical plants such as steam reformer, catalytic cracker, hydrocracker, delayed coking and fluid catalytic cracking units can maximize the value addition. Recent interest has been in adopting integrated processes for catalytic or thermal upgradation of pyrolysis oil to obtain high-end chemicals including gasoline-, diesel-, LPG-grade hydrocarbons, as well as value-added chemicals.

13.7.3 Pyrolysis Char → Nanotubes

Recent applications of utilizing by-products to produce carbon materials are worthwhile. Veksha, Yin and co-workers used plastic packaging waste with 11.8 and 27.5 wt % PET content (samples were denoted as PET-12 and PET-18) to produce multiwalled carbon nanotubes (MWCNTs) from non-condensable gases along with pyrolysis oil. Pyrolysis of waste plastics was realized using Fe_2O_3/ZSM-5 catalysts at 400°C to induce decarboxylation of terephthalic and benzoic acids to untraceable levels, by 4.6 and 9.4 wt % for PET-12 and PET-28, respectively. Different fractions of pyrolysis oil possessed properties like gasoline, diesel and heavy oil/wax fractions. MWCNTs were prepared from non-condensable gases over Ni/$CaCO_3$ at 700°C in 2.4 and 1.5 wt % yields from PET-12 and PET-28, respectively. The performance of synthesized MWCNTs in an oxygen evolution reaction was remarkably superior to commercial MWCNTs and Pt-based electrodes. In a distinctive approach, one-step microwave-assisted pyrolysis was performed in the presence of an iron-based catalyst to convert commercial HDPE into MWCNTs with a high hydrogen yield, i.e., over 97%, of the theoretical mass of hydrogen. Nevertheless, the process optimization for producing high end gaseous and carbon by-products, i.e., carbon nanotubes along with liquid hydrocarbons, can potentially contribute to the financial success of pyrolysis technology.

Further Reading

Aishwarya, K.N. and Sindhu, N. (2016). Microwave assisted pyrolysis of plastic waste. *Proc. Technol.* 25: 990–997.

Al-Salem, S.M., Yang, Y., Wang, J. and Leeke, G.A. (2020). Pyro-oil and wax recovery from reclaimed plastic waste in a continuous auger pyrolysis reactor. *Energies* 13 (8): 2040.

Aminu, I., Nahil, M.A. and Williams, P.T. (2020). Hydrogen from waste plastics by two-stage pyrolysis/low-temperature plasma catalytic processing. *Energy Fuels* 34 (9): 11679–11689.

Anene, A.F., Fredriksen, S.B., Sætre, K.A. and Tokheim, L.A. (2018). Experimental study of thermal and catalytic pyrolysis of plastic waste components. *Sustainability* 10 (11): 3979.

Bharathy, S., Gnanasikamani, B. and Lawrence, K.R. (2019). Investigation on the use of plastic pyrolysis oil as alternate fuel in a direct injection diesel engine with titanium oxide nanoadditive. *Environ. Sci. Pollut. Res.* 26 (10): 10319–10332.

Bridjesh, P., Periyasamy, P. and Geetha, N.K. (2018). Influence of chemical composite additive on combustion and emission characteristics of a diesel engine using waste plastic oil as fuel and modified piston bowl. *Orient. J. Chem.* 34 (6): 2806–2813.

Das, A.K., Hansdah, D., Mohapatra, A.K. and Panda, A.K. (2020). Energy, exergy and emission analysis on a DI single cylinder diesel engine using pyrolytic waste plastic oil diesel blend. *J. Energy Inst.* 93 (4): 1624–1633.

del Remedio Hernandez, M., Garcia, A.N. and Marcilla, A. (2007). Catalytic flash pyrolysis of HDPE in a fluidized bed reactor for recovery of fuel-like hydrocarbons. *J. Anal. Appl. Pyrolysis* 78 (2): 272–281.

Dobó, Z., Jakab, Z., Nagy, G. et al. (2019). Transportation fuel from plastic wastes: production, purification and SI engine tests. *Energy* 189: 116353.

Dobó, Z., Kecsmár, G., Nagy, G. et al. (2021). Characterization of gasoline-like transportation fuels obtained by distillation of pyrolysis oils from plastic waste mixtures. *Energy Fuels* 35 (3): 2347–2356.

Elordi, G., Olazar, M., Lopez, G. et al. (2011). Product yields and compositions in the continuous pyrolysis of high-density polyethylene in a conical spouted bed reactor. *Ind. Eng. Chem. Res.* 50 (11): 6650–6659.

Fan, L., Liu, L., Xiao, Z. et al. (2021). Comparative study of continuous-stirred and batch microwave pyrolysis of linear low-density polyethylene in the presence/absence of HZSM-5. *Energy* 228: 120612.

Filip, M.R., Pop, A., Perhaiţa, I. et al. (2013). Investigation of thermal and catalytic degradation of polystyrene waste into styrene monomer over natural volcanic tuff and Florisil catalysts. *Cent. Eur. J. Chem.* 11 (5): 725–735.

Gadwal, S.B., Banapurmath, N.R., Kamoji, M.A. et al. (2019). Performance and emission characteristic studies on CRDI diesel engine fuelled with plastic pyrolysis oil blended with ethanol and diesel. *Int. J. Sust. Eng.* 12 (4): 262–271.

Gala, A., Guerrero, M., Guirao, B. et al. (2020). Characterization and distillation of pyrolysis liquids coming from polyolefins segregated of MSW for their use as automotive diesel fuel. *Energy Fuels* 34 (5): 5969–5982.

Hu, Q., Tang, Z., Yao, D. et al. (2020). Thermal behaviour, kinetics and gas evolution characteristics for the co-pyrolysis of real-world plastic and tyre wastes. *J. Cleaner Prod.* 260: 121102.

Huang, Y.-F., Pan, M.-W. and Lo, S.-L. (2020). Hydrometallurgical metal recovery from waste printed circuit boards pretreated by microwave pyrolysis. *Resour. Conserv. Recycl.* 163: 105090.

Hussain, A.Z., Santhoshkumar, A. and Ramanathan, A. (2020). Assessment of pyrolysis waste engine oil as an alternative fuel source for diesel engine. *J. Therm. Anal. Calorim.* 141: 2277–2293.

Jie, X., Li, W., Slocombe, D. et al. (2020). Microwave-initiated catalytic deconstruction of plastic waste into hydrogen and high-value carbons. *Nat. Catal.* 3 (11): 902–912.

Kaimal, V.K., Vijayabalan, P., Balachandar, M. et al. (2020). Effect of using plastic nanofuel as a fuel in a light duty diesel engine. *Heat Transfer* 49 (2): 726–742.

Kaminsky, W. and Franck, J. (1991). Monomer recovery by pyrolysis of poly (methyl methacrylate) (PMMA). *J. Anal. Appl. Pyrolysis* 19: 311–318.

Kim, S.S. and Kim, S. (2004). Pyrolysis characteristics of polystyrene and polypropylene in a stirred batch reactor. *Chem. Eng. J.* 98 (1–2): 53–60.

Kim, Y.S., Hwang, G.C., Bae, S.Y. et al. (1999). Pyrolysis of polystyrene in a batch-type stirred vessel. *Korean J. Chem. Eng.* 16 (2): 161–165.

Kumar, K.V. and Puli, R.K. (2019). An effect of alcohol-plastic oil petrol blends on SI engine performance and exhaust emissions. *J. Mech. Sci. Technol.* 32 (4): 1849–1855.

Kumar, K.V. and Puli, R.K. Effect of plastic oil-methanol blends operated on petrol engine performance and exhaust emissions. *Australian J. Mech. Eng.* 19: 1–7.

Kunwar, B., Chandrasekaran, S.R., Moser, B.R. et al. (2017). Catalytic thermal cracking of postconsumer waste plastics to fuels. 2. Pilot-scale thermochemical conversion. *Energy Fuels* 31 (3): 2705–2715.

Lee, S.Y., Yoon, J.H., Kim, J.R. and Park, D.W. (2001). Catalytic degradation of polystyrene over natural clinoptilolite zeolite. *Polym. Degrad. Stab.* 74 (2): 297–305.

Lin, Y.H., Sharratt, P.N., Garforth, A.A. and Dwyer, J. (1997). Deactivation of US-Y zeolite by pyrolysis of high-density polyethylene. *Thermochim. Acta* 294 (1): 45–50.

Mangesh, V.L., Padmanabhan, S., Tamizhdurai, P. et al. (2020). Combustion and emission analysis of hydrogenated waste polypropylene pyrolysis oil blended with diesel. *J. Hazard. Mater.* 386: 121453.

Manos, G., Yusof, I.Y., Papayannakos, N. and Gangas, N.H. (2001). Catalytic cracking of polyethylene over clay catalysts. Comparison with an ultrastable Y zeolite. *Ind. Eng. Chem. Res.* 40 (10): 2220–2225.

Mastral, F.J., Esperanza, E., Garcıa, P. and Juste, M. (2002). Pyrolysis of high-density polyethylene in a fluidised bed reactor. Influence of the temperature and residence time. *J. Anal. Appl. Pyrolysis* 63 (1): 1–15.

Mazzoni, L. and Janajreh, I. (2017). Plasma gasification of municipal solid waste with variable content of plastic solid waste for enhanced energy recovery. *Int. J. Hydrogen Energy* 42 (30): 19446–19457.

Miandad, R., Barakat, M.A., Aburiazaiza, A.S. et al. (2017). Effect of plastic waste types on pyrolysis liquid oil. *Int. Biodeterior. Biodegrad.* 119: 239–252.

Miskolczi, N. (2006). Kinetic model of the chemical and catalytic recycling of waste polyethylene into fuels. In: *Feedstock Recycling and Pyrolysis of Waste Plastics: Converting Waste Plastics into Diesel and Other Fuels* (eds. J. Scheirs and W. Kaminsky), 225–247. John Wiley & Sons.

Murata, K., Sato, K. and Sakata, Y. (2004). Effect of pressure on thermal degradation of polyethylene. *J. Anal. Appl. Pyrolysis* 71 (2): 569–58.

Nema, S.K. and Ganeshprasad, K.S. (2002). Plasma pyrolysis of medical waste. *Curr. Sci.* 83: 271–278.

Niksiar, A., Faramarzi, A.H. and Sohrabi, M. (2015). Kinetic study of polyethylene terephthalate (PET) pyrolysis in a spouted bed reactor. *J. Anal. Appl. Pyrolysis* 113: 419–425.

Nisar, J., Ali, G., Shah, A. et al. (2020). Pyrolysis of polystyrene waste for recovery of combustible hydrocarbons using copper oxide as catalyst. *Waste Manage. Res.* 38 (11): 1269–1277.

Olazar, M., Lopez, G., Amutio, M. et al. (2009). Influence of FCC catalyst steaming on HDPE pyrolysis product distribution. *J. Anal. Appl. Pyrolysis* 85 (1–2): 359–365.

Padmanabhan, S., Prabhakaran, S. and Balan, K.N. (2020). Analysis of influencing parameters on CI engine performance fuelled by waste plastic oil and ethanol blends using taguchi method. *Test Eng. Manage.* 82: 7069–7077.

Palos, R., Gutiérrez, A., Vela, F.J. et al. (2019). Assessing the potential of the recycled plastic slow pyrolysis for the production of streams attractive for refineries. *J. Anal. Appl. Pyrolysis* 142: 104668.

Park, K.B., Jeong, Y.S. and Kim, J.S. (2019). Activator-assisted pyrolysis of polypropylene. *Appl. Energy* 253: 113558.

Quesada, L., Calero, M., Martín-Lara, M.A. et al. (2020). Production of an alternative fuel by pyrolysis of plastic wastes mixtures. *Energy Fuels* 34 (2): 1781–1790.

Ratnasari, D.K., Nahil, M.A. and Williams, P.T. (2017). Catalytic pyrolysis of waste plastics using staged catalysis for production of gasoline range hydrocarbon oils. *J. Anal. Appl. Pyrolysis* 124: 631–637.

Rex, P., Masilamani, I.P. and Miranda, L.R. (2020). Microwave pyrolysis of polystyrene and polypropylene mixtures using different activated carbon from biomass. *J. Energy Inst.* 93 (5): 1819–1832.

Rodríguez, E., Gutiérrez, A., Palos, R. et al. (2019). Fuel production by cracking of polyolefins pyrolysis waxes under fluid catalytic cracking (FCC) operating conditions. *Waste Manage.* 93: 162–172.

Rodríguez, E., Gutiérrez, A., Palos, R. et al. (2020). Co-cracking of high-density polyethylene (HDPE) and vacuum gasoil (VGO) under refinery conditions. *Chem. Eng. J.* 382: 122602.

Rodríguez, E., Palos, R., Gutiérrez, A. et al. (2020). Towards waste refinery: co-feeding HDPE pyrolysis waxes with VGO into the catalytic cracking unit. *Energy Convers. Waste Manage.* 207: 112554.

Rodríguez, E., Palos, R., Gutiérrez, A. et al. (2019). Effect of the FCC equilibrium catalyst properties and of the cracking temperature on the production of fuel from HDPE pyrolysis waxes. *Energy Fuels* 33 (6): 5191–5199.

Russell, A.D., Antreou, E.I., Lam, S.S. et al. (2012). Microwave-assisted pyrolysis of HDPE using an activated carbon bed. *RSC Adv.* 2 (17): 6756–6760.

Sachuthananthan, B., Krupakaran, R.L. and Balaji, G. (2021). Exploration on the behaviour pattern of a DI diesel engine using magnesium oxide nano additive with plastic pyrolysis oil as alternate fuel. *Int. J. Ambient Energy* 42: 701–712.

Sachuthananthan, B., Reddy, D.R., Mahesh, C. and Dineshwar, B. (2018). Production of diesel like fuel from municipal solid waste plastics for using in CI engine to study the combustion, performance and emission characteristics. *Int. J. Pure Appl. Math.* 119: 85–98.

Sato, Y., Kato, K., Takeshita, Y. et al. (1998). Decomposition of ployvinylchloride using supercritical water. *Jpn. J. Appl. Phys.* 37 (11R): 6270.

Sharuddin, S.D., Abnisa, F., Daud, W.M. and Aroua, M.K. (2017). Energy recovery from pyrolysis of plastic waste: study on non-recycled plastics (NRP) data as the real measure of plastic waste. *Energy Convers. Manage.* 148: 925–934.

Siebert, H., Sieggelkow, B. and Dongovianni, E. (2022). Plastics circularity: how to solve the challenge of retaining quality in recaykled polymers. Hydrocarbon Processing, April 2022.

Singh, R.K., Ruj, B., Sadhukhan, A.K. et al. (2020). Waste plastic to pyrolytic oil and its utilization in CI engine: performance analysis and combustion characteristics. *Fuel* 262: 116539.

Soni, V.K., Singh, G., Vijayan, B.K. et al. (2021). Thermochemical recycling of waste plastics by pyrolysis: a review. *Energy Fuels* 35: 12763–12808.

Sun, J., Wang, W., Liu, Z. and Ma, C. (2011). Recycling of waste printed circuit boards by microwave-induced pyrolysis and featured mechanical processing. *Ind. Eng. Chem. Res.* 50 (20): 11763–11769.

Syamsiro, M., Saptoadi, H., Kismurtono, M. et al. (2018). Utilization of waste polyethylene pyrolysis oil as partial substitute for diesel fuel in a DI diesel engine. *Int. J. Smart Grid Clean Energy* 8 (1): 38–47.

Takeshita, Y., Kato, K., Takahashi, K. et al. (2004). Basic study on treatment of waste polyvinyl chloride plastics by hydrothermal decomposition in subcritical and supercritical regions. *J. Supercrit. Fluids* 31: 185–193.

Veksha, A., Yin, K., Moo, J.G. et al. (2020). Processing of flexible plastic packaging waste into pyrolysis oil and multi-walled carbon nanotubes for electrocatalytic oxygen reduction. *J. Hazard. Mater.* 387: 121256.

Verma, A., Raghuvansi, A., Quraishi, M.A. et al. (2018). Engine fuel production from waste plastic pyrolysis (WPO) and performance evaluation in a CI engine with diesel blend. *J. Mater. Environ. Sci.* 6: 1712–1721.

Wang, S.B., Cheng, C.M., Lan, W. et al. (2013). Experimental study of thermal plasma processing waste circuit boards. *Adv. Mater. Res.* 652: 1553–1561.

Yazdani, E., Hashemabadi, S.H. and Taghizadeh, A. (2019). Study of waste tire pyrolysis in a rotary kiln reactor in a wide range of pyrolysis temperature. *Waste Manage.* 85: 195–201.

Index

d

i

k

l

m

r

s

Printed in the USA/Agawam, MA
September 16, 2026

910661.191